Integrated management of diseases and insect pests of tree fruit

It is widely recognised that agriculture is a significant contributor to global warming and climate change. Agriculture needs to reduce its environmental impact and adapt to current climate change whilst still feeding a growing population, i.e. become more 'climate-smart'. Burleigh Dodds Science Publishing is playing its part in achieving this by bringing together key research on making the production of the world's most important crops and livestock products more sustainable.

Based on extensive research, our publications specifically target the challenge of climate-smart agriculture. In this way we are using 'smart publishing' to help achieve climate-smart agriculture.

Burleigh Dodds Science Publishing is an independent and innovative publisher delivering high quality customer-focused agricultural science content in both print and online formats for the academic and research communities. Our aim is to build a foundation of knowledge on which researchers can build to meet the challenge of climate-smart agriculture.

For more information about Burleigh Dodds Science Publishing simply call us on +44 (0) 1223 839365, email info@bdspublishing.com or alternatively please visit our website at www.bdspublishing.com.

Related titles:

Achieving sustainable cultivation of apples
Print (ISBN 978-1-78676-032-6) Online (ISBN 978-1-78676-034-0; 978-1-78676-035-7)

Achieving sustainable cultivation of temperate zone tree fruits and berries Volume 1: Physiology, genetics and cultivation
Print (ISBN 978-1-78676-208-5) Online (ISBN 978-1-78676-210-8; 978-1-78676-211-5)

Achieving sustainable cultivation of temperate zone tree fruits and berries Volume 2: Case studies
Print (ISBN 978-1-78676-212-2) Online (ISBN 978-1-78676-214-6; 978-1-78676-215-3)

Integrated management of insect pests: Current and future developments
Print (ISBN 978-1-78676-260-3) Online (ISBN 978-1-78676-262-7; 978-1-78676-263-4)

Chapters are available individually from our online bookshop:
https://shop.bdspublishing.com

climate•SMART•publishing
IN AGRICULTURAL SCIENCE

BURLEIGH DODDS SERIES IN AGRICULTURAL SCIENCE
NUMBER 68

Integrated management of diseases and insect pests of tree fruit

Edited by Professor Xiangming Xu and
Dr Michelle Fountain, NIAB EMR, UK

burleigh dodds
SCIENCE PUBLISHING

Published by Burleigh Dodds Science Publishing Limited
82 High Street, Sawston, Cambridge CB22 3HJ, UK
www.bdspublishing.com

Burleigh Dodds Science Publishing, 1518 Walnut Street, Suite 900, Philadelphia, PA 19102-3406, USA

First published 2019 by Burleigh Dodds Science Publishing Limited

Library of Congress Control Number: 2019945949

British Library Cataloguing in Publication Data
A catalogue record for this book is available from the British Library

ISBN 978-1-78676-256-6 (Print)
ISBN 978-1-78676-259-7 (PDF)
ISBN 978-1-78676-258-0 (ePub)
ISSN 2059-6936 (print)
ISSN 2059-6944 (online)

DOI 10.19103/AS.2019.0046

Typeset by Deanta Global Publishing Services, Chennai, India
Printed and bound by CPI Group (UK) Ltd, Croydon, CR0 4YY

Contents

Series list

Title	Series number
Achieving sustainable cultivation of maize - Vol 1 From improved varieties to local applications Edited by: Dr Dave Watson, CGIAR Maize Research Program Manager, CIMMYT, Mexico	001
Achieving sustainable cultivation of maize - Vol 2 Cultivation techniques, pest and disease control Edited by: Dr Dave Watson, CGIAR Maize Research Program Manager, CIMMYT, Mexico	002
Achieving sustainable cultivation of rice - Vol 1 Breeding for higher yield and quality Edited by: Prof. Takuji Sasaki, Tokyo University of Agriculture, Japan	003
Achieving sustainable cultivation of rice - Vol 2 Cultivation, pest and disease management Edited by: Prof. Takuji Sasaki, Tokyo University of Agriculture, Japan	004
Achieving sustainable cultivation of wheat - Vol 1 Breeding, quality traits, pests and diseases Edited by: Prof. Peter Langridge, The University of Adelaide, Australia	005
Achieving sustainable cultivation of wheat - Vol 2 Cultivation techniques Edited by: Prof. Peter Langridge, The University of Adelaide, Australia	006
Achieving sustainable cultivation of tomatoes Edited by: Dr Autar Mattoo, USDA-ARS, USA & Prof. Avtar Handa, Purdue University, USA	007
Achieving sustainable production of milk - Vol 1 Milk composition, genetics and breeding Edited by: Dr Nico van Belzen, International Dairy Federation (IDF), Belgium	008
Achieving sustainable production of milk - Vol 2 Safety, quality and sustainability Edited by: Dr Nico van Belzen, International Dairy Federation (IDF), Belgium	009
Achieving sustainable production of milk - Vol 3 Dairy herd management and welfare Edited by: Prof. John Webster, University of Bristol, UK	010
Ensuring safety and quality in the production of beef - Vol 1 Safety Edited by: Prof. Gary Acuff, Texas A&M University, USA & Prof. James Dickson, Iowa State University, USA	011
Ensuring safety and quality in the production of beef - Vol 2 Quality Edited by: Prof. Michael Dikeman, Kansas State University, USA	012
Achieving sustainable production of poultry meat - Vol 1 Safety, quality and sustainability Edited by: Prof. Steven C. Ricke, University of Arkansas, USA	013
Achieving sustainable production of poultry meat - Vol 2 Breeding and nutrition Edited by: Prof. Todd Applegate, University of Georgia, USA	014

Introduction

This volume reviews advances in understanding key diseases and insect pests of tree fruit. The chapters summarise current research on what causes key fungal diseases (apple scab, powdery mildew, apple replant diseases, apple canker and brown rot), and viral diseases (apple mosaic virus and plum pox). The volume discusses integrated fruit disease management techniques such as surveillance, breeding disease-resistant varieties, improved fungicide application including the use of biocontrol agents. Other chapters review the ecology of major insect pests (aphids, tortricid moths, mites and fruit flies). Chapters also address a wider range of tree fruit pest and ways of improving integrated pest management (IPM) techniques for tree fruit, from monitoring and forecasting to agronomic practices and methods of biological control.

Part 1 Fruit diseases

The first part of the book focuses on fruit diseases, starting with fungal diseases. Chapter 1 deals with apple scab (*Venturia inaequalis*) which is the most important disease on apple (*Malus* × *domestica*) worldwide. This fungus can infect both leaves and fruit. Although there are major resistance genes identified against *V. inaequalis*, nearly all commercial dessert cultivars are susceptible, thus requiring stringent control measures to minimise crop losses. The chapter introduces apple scab, explains the epidemiology of the disease, briefly describes management strategies.

Moving onto another fungal disease of fruit, Chapter 2 looks at powdery mildew. Tree fruits such as apple, cherry, apricot and peach, are amongst the most susceptible crops to powdery mildews. The fungus mostly affects vegetative tissues but can also infect the reproductive organs (flowers and fruits) of the tree. In the absence of appropriate management, multiple cycles of infections may significantly reduce tree vigour, fruit set and yield. The chapter summarises current knowledge on the biology, epidemiology, host resistance and control of powdery mildew species of apple (*Podosphaera leucotricha*) and cherry (*P. clandestina*).

The subject of Chapter 3 is apple replant disease (ARD) which significantly affects the worldwide apple industry. There are many factors that may contribute to ARD, including several plant pathogens; as a result, ARD is difficult to manage in the absence of broad-spectrum soil fumigants. This chapter summarises the main causes of ARD and the main methods to control ARD with specific reference to China (since ARD management for other regions has been reviewed recently), based on recent developments in knowledge of the physical and chemical properties of soil deterioration, allelopathic effects (phenolic acids), and microbial community structure in apple orchards. The chapter discusses methods to manage ARD, including cultural practices such as soil disinfection, resistance breeding and biological control.

Chapter 4 considers the challenge presented by the apple canker disease in Europe, with a particular focus on European apple canker caused by *Neonectria ditissima*. The chapter discusses *N. ditissima* in the context of the tree production cycle, introducing the aetiology and epidemiology of the pathogen, reviewing the key control measures (covering cultural, chemical and biological control), and identifying a number of future research areas (including detection and diagnostics, and the use of endophytes). The chapter describes changes in the plant protection products available and in the methods of tree production and varietal susceptibility.

Chapter 5 considers the case of apple valsa canker in Asia. Apple valsa canker, caused by *Valsa Mali*, is mainly distributed in China, Japan, South Korea and North Korea. Apple valsa canker mainly damages the cortex of apple branches, causing the cortex to rot. The chapter describes the distribution and damage associated with valsa apple canker before going on to introduce the the pathogen in terms of its biology, epidemiology, infection process and pathogenesis. The chapter concludes with the recent research progresses in China on the management of apple valsa canker.

The final chapter on fungal diseases of fruit, Chapter 6, deals with brown rot. *Monilinia* spp. cause economically important diseases of pome and stone fruits. The chapter focuses on the four main *Monilinia/Monilia* species i.e. *Monilinia fructicola, M. laxa*, *M. fructigena*, and *M. polystroma*. The chapter concentrates on the general and specific features of the four brown rot fungi in relation to their impact on fruit hosts and to the integrated disease management. The chapter covers brown rot-related yield loss and its impact on fruit crops, the characteristics of causal organisms, and the disease's major fruit hosts and host resistance. The chapter goes on to describe brown rot symptoms on specific plant parts, methods for rapid and reliable identification and detection, epidemiology, disease prediction and disease management options.

Moving on from fungal to viral diseases of fruit, Chapter 7 looks at apple mosaic virus. Apple mosaic virus (ApMV) has been detected in apple as well as in a number of other plant taxa including other members of the Rosaceae family. The chapter provides a comprehensive review of how the virus is spread and describes the symptoms displayed by affected host plants. The chapter goes on to discuss the geographical distribution of the virus, its epidemiology and the economic impact of infection on the fruit and nut industry. The chapter reviews ways to obtain virus-free material with meristem culture in combination with thermo- or cryo-therapy used to eradicate ApMV from propagation material. The chapter concludes with a discussion of the most reliable molecular detection methods for ApMV.

The section's final chapter, Chapter 8, considers a further viral disease of fruit: plum pox. Sharka disease, caused by *Plum pox virus* (PPV), is the most important virus causing significant yield losses in plums, with reduced fruit quality and premature fruit drop. The chapter summarises status of this disease around the world, with emphasis on strains, detection methods, orchard management and breeding for resistance. The chapter examines the challenge of PPV control for sustainable cultivation of plums and looks at the genetic and molecular basis of PPV resistance in *Prunus*.

Part 2 Integrated fruit disease management

The second part of the book examines the challenge of integrated management of diseases on fruit crops. Chapter 9 starts with a review of disease monitoring and forecasting in integrated fruit disease management. Regular inspection of fruit crops for diseases is a vital part of integrated disease management. Combining this information with that disease forecast models (usually based on weather information only) can play a significant role in producing perfect fruit with minimal use of fungicides. The chapter assesses fruit disease monitoring, disease warning systems and decision making; it then uses examples of specific diseases in apple and pear to illustrate the role of both orchard monitoring and disease warning systems in integrated disease management. The chapter covers apple scab, apple powdery mildew, European apple canker, sooty blotch, fly speck, fire blight and storage rots. The examples given demonstrate current 'best practice' in orchard monitoring and how disease warning systems can be integrated into integrated disease management programmes. The chapter also discusses the use of future technologies such as smart spore traps, drones and orchard scanners.

Moving on from disease monitoring to preventative breeding strategies, Chapter 10 considers the challenge of breeding fruit varieties with durable resistance to pests and diseases. Resistance breeding can make a lasting contribution to pest and disease control in an integrated management strategy provided that the resistance is durable. The chapter discusses strategies for achieving the breeding objective of durable resistance based on genes that target pathogen fitness, gene pyramiding and non-host resistance. The chapter also examines the application of a range of breeding tools.

The subject of Chapter 11 is on how plant propagation methods may improve disease control in fruit crops. The movement of plant propagation material across countries and continents allows for the exchange of high-quality planting material providing producers with superior germplasm, and enhancing farm sustainability and profits. However, this practice has hidden dangers, including the potential of moving infectious agents to new areas. The chapter takes a systems-based approach on the detection, elimination and safeguarding against systemic pathogens that affect clonally propagated fruit crops. The chapter comprehensively discusses the steps taken to improve propagation material and presents a number of case studies including blackberry yellow vein disease and raspberry crumbly fruit and decline. The chapter discusses new technologies that allow for the rapid and accurate characterization of plant pathogens, including new tools such as high throughput sequencing which has proved invaluable in assessing the health status of high value plants and for the rapid identification and characterisation of unknown viruses.

Moving from plant propagation to a more direct method of disease control, Chapter 12 looks at improving fungicide use in integrated fruit disease management. With the rapid emergence of resistance to the limited number of fungicide classes in many pathogenic fungi, improving stewardship of extant fungicides is a priority. The chapter describes the evolution of fungicide resistance and examines ways of monitoring the development of resistance to fungicides. The chapter focuses on the importance of strain identification in monitoring resistance, and examines the adoption of informed decision-support tools to improve fungicide efficacy.

The subject of Chapter 13 is the use of biocontrol agents and biostimulants in fruit tree disease management. Increasing demand for residue-free fruits, changes in regulation and a decreasing number of available synthetic fungicides favour growth of the market for alternative products. These products include naturally-occurring beneficial microorganisms which are used to control plant diseases. The chapter assesses the current status of biocontrol techniques and their practical application. This chapter reviews main biocontrol mechanisms, including competition, induced resistance, hyperparasitism and antibiosis. The chapter reviews the use of biocontrol agents for control of major fruit tree diseases such as fireblight, post-harvest fruit rots, brown rot, apple scab, European canker and pear brown spot. Commercial biocontrol products are registered mainly for control of fireblight in pome fruit and of post-harvest fruit rots in pome fruit and stone fruit.

The final chapter of the section, Chapter 14, looks at new techniques for managing post-harvest diseases of fruit. Chemical fungicides are traditionally used to preserve the quality of fruit and vegetables during storage or transportation. However, global concern about pesticide use and the reduced efficacy of chemicals due to pathogen-resistant strains have gradually led producers to evaluate safer alternatives for controlling postharvest diseases. Several means, such as natural compounds of animal and plant origin, organic and inorganic salts, antagonistic microorganisms, and physical means, are among the approaches that have been recently evaluated to manage fruit rots. The chapter reviews these new alternative means, and the latest solutions for controlling postharvest diseases in order to provide effective yet eco-friendly solutions for disease management. The chapter deals with physical agents (UV light treatment, heat treatment, electrolyzed water), chemical agents (organic/inorganic salts, chitosan) and biological agents (biocontrol, botanicals).

Part 3 Insect pests of fruit

The third part of the volume is dedicated to insect pests of tree fruit. Chapter 15 looks at the threat posed by aphids (Homoptera, Sternorrhyncha, Aphididae), which currently include 5575 valid species. Aphids are characterized by some peculiar morphological and biological characteristics that make these insects unique. Although often under-reported, it is estimated that aphids cause very high economic losses each year worldwide. The chapter describes what we know about the biology of citrus aphids, apple and pear aphids and stone fruit aphids and describes their management.

Chapter 16 moves on to consider the major pests of tree fruit, tortricid moths. Members of this group are commonly the most destructive insect pests of deciduous tree fruits throughout the world. The chapter summarises recent research on tortricid biology and management. After an introduction to tortricid systematics and general biology, the chapter examines key species, distribution and dispersal mechanisms of tortricids. The chapter describes insecticide use in organic tree fruit production, as well as techniques of physical crop protection and biological control, including mating disruption and precision pest management. Finally, the chapter examines area-wide Integrated Pest Management (IPM) and postharvest management techniques.

Phytophagous (plant feeding) mites are considered in Chapter 17. Mites are a perennial pest of tree fruit crops worldwide. Foliar feeding by mites can lead to decreased photosynthetic activity, reducing plant vigor and yield. For some mite pests, fruit deformation or russeting can occur. Two main families of mites are found on tree fruits: tetranychids (spider mites) and eriophyids (rust, bud, and blister mites). The chapter reviews cultural and biological control tactics for controlling spider mites, which include choosing tolerant varieties and weed management, releasing predators into orchards to increase their numbers, and/or using management practices that conserve existing predator populations. The chapter considers the thresholds to determine when a spray application should be made and argues for biological control as the cornerstone of integrated mite management.

The final chapter on insect pests, Chapter 18, looks at the problems posed by fruit flies. The vinegar fly *Drosophila suzukii* Matsumura (Diptera: Drosophilidae) commonly known as Spotted Winged Drosophila (SWD) is a highly polyphagous invasive pest, native to Asia, which causes significant damage to a wide variety of berry and stone fruit crops. The chapter reviews the significant progress that has since been made in understanding the biology of the fly, providing key information required for the development of sustainable and integrated control methods. The chapter assesses the fly's impact on crops and describes its biology and ecology, host range and host susceptibility, detection methods, and management and control strategies.

Part 4 Integrated management of fruit insect pests

The volume's final section focuses on integrated pest management (IPM) of insect pests of tree fruit. Chapter 19 deals with cultural control and agronomic practices to prevent or manage fruit insect pests. Perennial systems, such as temperate tree fruit, present interesting challenges and opportunities for the practice of cultural pest management tactics. The chapter outlines key cultural tactics for managing arthropod pests in temperate tree fruit, including modification of trees, tree architecture, orchard floor management, cultivation practices, mowing and the cultivation of orchard cover crops. The chapter includes a detailed case study on integrating rotational hog grazing and apple production.

Moving from control to monitoring, Chapter 20 focuses on improving monitoring and forecasting in integrated management of fruit insect pests. The efficient and accurate monitoring of insects and mites is a cornerstone of IPM. Pheromone baited traps are probably the best known, most convenient, and specific method used for monitoring in orchards. However, mating disruption strategies have complicated the interpretation of monitoring via pheromone traps, leading to the development of alternative (bisexual/food based) baits. The chapter describes both manual and automated monitoring systems (sensor and camera-based), models for forecasting the phenological dynamics and spatial distribution of insects and the challenge of understanding the relationships between monitoring data, absolute pest densities and economic threshold values.

The subject of Chapter 21 is biological control in integrated management of fruit insect pests, with a particular focus on the use of semiochemicals. Over the past 25 years, semiochemical-based technologies have emerged as essential components of pest management programmes in deciduous fruit production systems. Pheromone-based mating disruption is by far the most widely adopted behavioral control, with nearly 800 000 hectares of fruit crops treated worldwide. The chapter reviews both the progress that researchers and practitioners have made, along with the hurdles encountered as they conceptualized, developed and implemented behavioral controls. The chapter focuses on the uses of behaviour-modifying chemicals for direct control of insect pests of deciduous tree fruits and berry crops. The chapter pays particular attention to the development and future prospects of mating disruption. The chapter covers pheromone-mediated mating disruption, mass deployment of attractant-baited traps, and the application of attract-and-kill formulations/devices. The chapter concludes with a detailed case study on codling moth (*Cydia pomonella*) mating disruption (CMMD) covering pheromone Identification, early research and adoption, demonstration and areawide programmes, companion insecticides and dispenser development.

The final chapter of the volume, Chapter 22, looks at optimising insecticide use in integrated management of fruit insect pests. The integrated pest management approach was conceived to reduce grower reliance on pesticides by combining and optimising pesticide use with biological, physical and agronomical control tools. Natural enemies are an essential component of the agroecosystem and play a valuable role in controlling insect pests in fruit orchards. Growers need to understand how to protect them from harmful agronomic practices and how these natural allies could be integrated with other control tools. The chapter describes old and new monitoring tools, robust monitoring protocols available for estimating abundance of pest and beneficial insects, use of available phenology models to predict crucial future events important for pest management (e.g. the use of precise timing for surveillance) and pest control. The chapter includes three detailed case studies of common fruit pests as examples of how optimisation of insecticide applications could be implemented within a framework of an integrated management of fruit insect pests.

Part 1

Fruit diseases

Chapter 1

Epidemiology and management of apple scab

Tom Passey and Xiangming Xu, NIAB EMR, UK

1 Introduction

Apple scab, caused by *Venturia inaequalis*, also known as black spot in Australasia, is the most important disease of apples worldwide, especially so in temperate regions with wet and mild climates. This pathogen rarely kills its host, but can significantly reduce fruit yields and quality. This is especially true because of the market demand for blemish-free fruit. Mycelial growth on the fruit surface leads to an olive-brown coloured lesion, which generally forms a brown 'scabby' area (Fig. 1). Severe scabbing of the fruit can lead to cracking of the fruit skin and a potential route for secondary infection from other pathogens (often resulting in post-harvest fruit rot). Infection of young fruit can also lead to a reduction in fruit size and misshapen fruit.

Leaves are also infected (Fig. 2) and, though less frequent, infection of shoots is possible (leading to wood scab). A severe epidemic mid-season can lead to defoliation, potentially reducing wood growth and restricting fruit bud formation for the following year. The wealth of knowledge on the biology, epidemiology and management of apple scab has been captured in a published scab monograph (MacHardy, 1996). Therefore, only details of publications since 1996 are usually explicitly given.

V. inaequalis attacks only *Malus* spp. although *formae speciales* of *V. inaequalis* are known on other species of the Rosaceae family such as *Pyracantha* spp. and loquat (*Eriobotrya japonica)* (Le Cam et al., 2002; Gladieux

http://dx.doi.org/10.19103/AS.2019.0046.01

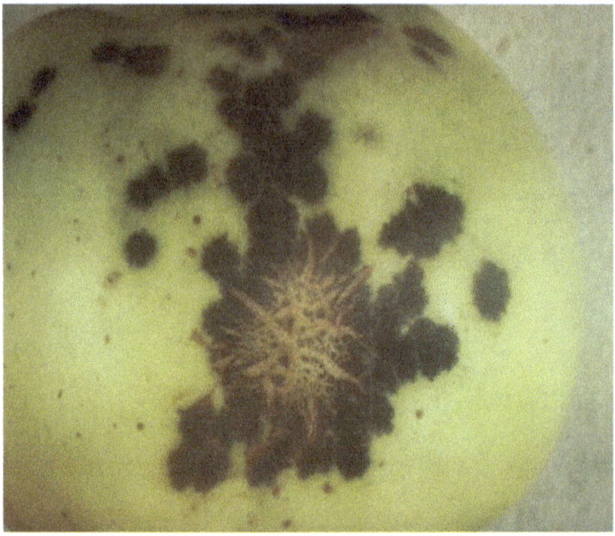

Figure 1 Symptoms of apple scab on apple fruit by *Venturia inaequalis*.

Figure 2 Symptoms of apple scab on leaves by *Venturia inaequalis*.

et al., 2010). The pathogen is likely to have originated in Central Asia (Gladieux et al., 2008) alongside *Malus* spp. (Harris et al., 2002). *V. inaequalis* is a haploid organism, and conventional cytology suggests that it has seven chromosomes. One linkage map of *V. inaequalis* (Xu et al., 2009) has 11 linkage groups spanning 1106 cM in length, while another has 15 linkage groups spanning

972 cM (Broggini et al., 2011). Construction of a bacterial artificial chromosome library estimated the genome size to be 102 Mb (Broggini et al., 2007). Whole genome sequencing has estimated the genome size to be between 40 Mb and 72 Mb (Deng et al., 2017; Passey et al., 2018). It is estimated that 36 Mb of the genome is non-repeated DNA (Bowen et al., 2011), making assembly more complicated. The *de novo* assembly and annotation of the *V. inaequalis* transcriptome have been published (Thakur et al., 2013).

2 Apple scab epidemiology

Figure 3 shows a simplified diagram of the annual life cycle of *V. inaequalis* on apple. Annual epidemics of apple scab start with primary infections of leaves in the spring by overwintered inoculum in leaf litter, buds and wood scab on twigs. These primary infections lead to the first wave of visible lesions, with the length of incubation largely determined by ambient temperature. Conidia produced from these lesions are then rain-splash dispersed and can lead to further infections of leaves, fruit and shoots, namely secondary infection.

This cycle of secondary infections continues throughout the summer, until the leaves fall in the late autumn. The frequency of secondary infection and scab severity are largely determined by cultivar susceptibility, tissue susceptibility and weather conditions.

In most apple-growing regions, *V. inaequalis* survives in winter mainly as pseudothecia on dead, scabbed leaves on the ground (MacHardy, 1996). This pathogen is a heterothallic fungus and hence requires strains of opposite mating types to mate and produce ascospores. Predicting the dynamics of ascospore

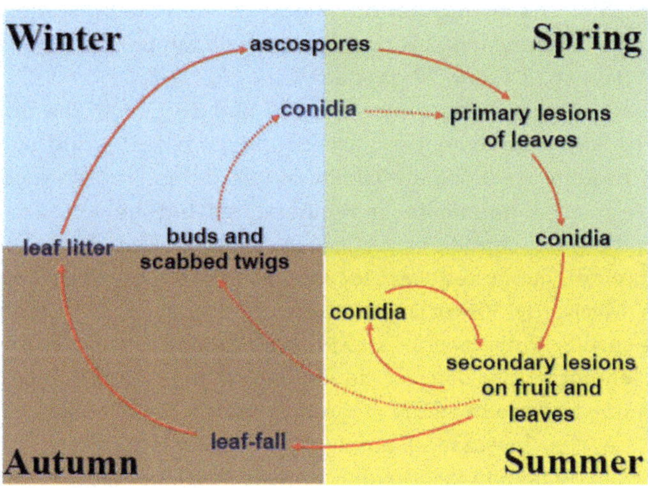

Figure 3 A simplified life cycle of the apple scab fungus *Venturia inaequalis*.

maturation and release has received much attention as these predictions can help growers to determine the timing of primary infections (Schwabe et al., 1989; Stensvand et al., 2005; MacHardy, 1996; Roubal and Nicot, 2016). Pseudothecium formation occurs under a wide range of temperatures with the light enhancing pseudothecial production. The pseudothecial maturation is primarily determined by temperature (i.e. thermal time). Rain is necessary for the growth of mycelium into the leaf lamina and for initiation of the ascigerous stage; however, continuous wetness can delay ascospore maturation. Mature ascospores are discharged during or following rain events and ascospores are more likely discharged in daylight (MacHardy, 1996). Thus discharging (the majority) of ascospores following a rainfall event during the night may be delayed until dawn. The actual level of ascospore inoculum in the spring (i.e. PAD – potential ascospore dose) can also be estimated with scab severity the previous season, winter conditions (temperature, rainfall and snow cover) and the extent of leaf litter degradation over the winter (MacHardy, 1996).

V. inaequalis can also survive in winter as conidia and thus these overwintered conidia can initiate primary infections. Conidia are shown to overwinter on the inside of bud tissues, leading to infections of young leaves in the spring (Holb et al., 2004; Becker et al., 1992). Conidia may have contributed 20–50% of the primary inoculum in one specific orchard under UK conditions (Passey et al., 2017). Thus, it may be prudent not to ignore conidia as a source of primary inoculum in scab management.

Primary infection by ascospores and conidia leads to olive-green/grey lesions on leaves. Hyphae grow between the cuticle and epidermal cell wall resulting in the development of stromata. Conidiophores and, thus, conidia form and rupture the cuticle. *V. inaequalis* is able to sporulate at temperatures from 5°C to 25°C. The latency period length (i.e. the time from infection until the occurrence of sporulating lesions) depends on temperature. For example, it is about 8 days at 18°C and 17 days at 9°C.

Ascospores can germinate between 0°C and 30°C, with the optimum at 15–25°C. The key requirement for ascospore germination and subsequent infection of host tissues is the availability of free or high-moisture conditions. Conidia are able to germinate at temperatures between 10°C and 30°C under high-moisture conditions, usually ≥ 95% RH. The exact duration of high-moisture conditions required for infection differs between conidia and ascospores (MacHardy, 1996). Under a given temperature, the longer the wet period, the severer the infection is. Germinating/infecting spores can survive for various lengths of dry periods, depending on their developmental stage, the temperature and the humidity in the dry periods. Much research has been directed to develop forecasting/warning systems for apple scab in order to assist growers in better timing intervention measures (MacHardy, 1996; Xu et al., 1995; Aćimović and Rosenberger, 2018). Most of these forecasting models are based on the simple lookup table provided by Mills and La Plante (1954); this table gives the duration of surface wetness under a range of temperatures required for scab (light, moderate or severe) infection.

3 Host resistance

In total, 18 gene-for-gene relationships have been identified in the *Malus-V. inaequalis* pathosystem: 17 were described by Bus et al. (2011) and subsequently an additional one (*Rvi18*) added (Soriano et al., 2014). Research in identifying and using major R genes against *V. inaequalis* in commercial breeding has focused on genes from wild *Malus* species rather than domesticated apple. The only exception is *Rvi1* (formerly *Vg*) from 'Golden Delicious', but this gene can be overcome by ca. 87% of the scab population in Europe and is described as "an exception to the premise that narrow spectrum R genes should be excluded from the nomenclature" (Bus et al., 2011). A number of these R genes have now been cloned and characterised, including *Rvi1* (Cova et al., 2015), *Rvi6* (widely referred to as *Vf*) (Malnoy et al., 2008) and *Rvi15* (Schouten et al., 2014). Markers for R genes have been identified for their potential use in breeding programmes (e.g. Padmarasu et al., 2014; Soriano et al., 2014).

There is a range of susceptibility to *V. inaequalis* across domesticated apple cultivars, suggesting that seemingly susceptible cultivars contain potential resistance factors (Koch et al., 2000; Barbara et al., 2008; Papp et al., 2016). Partial resistance against *V. inaequalis* in cultivars, not conditioned by a (known) R gene, might come from multiple loci (including those old 'defeated' R genes) that have a quantitative effect. A number of quantitative trait loci (QTLs) associated with scab resistance have been identified in a cross between 'Discovery' and 'Fiesta' (Liebhard et al., 2003), as well as in a cross between 'Discovery' and the apple hybrid 'TN10-8' (Calenge et al., 2004). Partial resistance from QTLs may be more durable than that of the qualitative resistance offered by an R gene (Bastiaanse et al., 2016), but could still be subject to erosion, although this erosion might be slow (Caffier et al., 2016). Locally adapted, susceptible cultivars may possess effective resistance against *V. inaequalis* strains from other regions (Xu et al., 2008).

The only major R gene that has been incorporated into commercial apple cultivars is the *Rvi6* gene. Although the *Rvi6* resistance was durable for around 50 years, breakdown was first seen in Europe in the last decade of the twentieth century (Parisi et al., 1993; Roberts and Crute, 1994). Since then, breakdown in the *Rvi6* resistance has been reported in many European countries, predominantly in organic orchards; no confirmed breakdown has yet been reported in North American commercial orchards (Gessler and Pertot, 2012). Only a few cultivars with the *Rvi6* are grown commercially (e.g. 'Topaz', a relatively popular cultivar in Germany) as named varieties.

The breakdown of R-gene resistance, as seen with *Rvi6*, shows that the use of a single R gene to confer resistance against *V. inaequalis* is not likely to be a durable solution. Pyramiding scab resistance genes, i.e. incorporating a number of resistance genes within a single cultivar, has been proposed as a way of breeding cultivars with durable scab resistance (Gessler and Pertot, 2012). The erosion of a single broad-spectrum resistance factor in a genotype could allow a rapid epidemic resulting from the virulent pathogen race, whereas pyramiding these resistance factors would require an isolate to have all the necessary virulence factors to overcome the differing resistance factors. Pyramiding of three QTLs for resistance against apple scab increased the efficiency of resistance compared to any of the individual independent QTLs (Laloi et al., 2017). However, it should be noted that these three specific QTLs (Laloi et al., 2017) affected *V. inaequalis* development at different stages. It is, therefore, expected that pyramiding these QTLs should benefit from resistance targeted at different stages of pathogen development.

In addition to inherent genetic resistance against *V. inaequalis*, apple tissues also possess age-related resistance/susceptibility (MacHardy, 1996); this age-related resistance is often called 'ontogenic resistance'. Young leaves are much more susceptible to *V. inaequalis* than old leaves; this ontogenic resistance is mainly manifested in the reduced lesion density and lengthened incubation period on old leaves. Mycelia on old leaves grow so slowly that they may not lead to visual symptoms at the time of leaf-fall, even though some of these symptomless colonies have already produced conidia when viewed microscopically (Li and Xu, 2002). A considerable amount of symptomless infections may therefore be expected on old leaves in autumn, which may have considerable impact on the amount of primary inoculum in the following season. These symptomless colonies may be one of the reasons responsible for the difficulties in estimating PAD accurately. Similarly, apple fruits are highly susceptible in the early stages of development, but become increasingly resistant as they mature (Xu and Robinson, 2005).

4 Managing scab

Although current scab management still relies on frequent applications of synthetic fungicides, there is a consensus now that we urgently need to develop and implement an integrated approach with far less reliance on synthetic fungicides to manage crop diseases. Crop management practices are applied to reduce disease pressure, including reducing inoculum levels, reducing the amount of susceptible tissues, and altering microclimates. Whenever possible, alternative products (particularly microbial products, e.g. biocontrol products) should be used instead of synthetic fungicides. Frequent disease monitoring in conjunction with disease prediction/forecasting is necessary to better time intervention measures.

4.1 Fungicides

For very susceptible cultivars, the most effective control is still through a programme of chemical sprays. It is not unusual for 12–18 spray applications to be applied to an orchard every year. It was Mills who in the 1920s began research on the importance of timing of sprays against scab (MacHardy, 1996). More recently, scab forecasting models have been used to optimise fungicide sprays as well as selection of the type of fungicides (protectant, curative or eradicant) (Berrie and Xu, 2003; Aćimović and Rosenberger, 2018). Multi-year orchard trials showed that integrating scab and powdery mildew forecasting models in managing apple diseases resulted in similar or better control than a routine programme, but with reduced fungicide input (5% to 45%) and costs, even in seasons exceptionally favourable for these diseases (Berrie and Xu, 2003).

Fungicides account for around two-thirds of the total pesticides used in orchard crops in the UK. There is ever-increasing legislation around the use of pesticides and a decreasing number of products available. Demethylase inhibitors, or DMIs, are a group of fungicides including piperazines, pyridines, pyrimidines, imidazoles, triazoles and triazolinthiones. They target the protein 14-α demethylase, which is part of the ergosterol biosynthesis pathway. Its inhibition by DMIs renders isolates unable to make ergosterol, badly damaging the cell membrane (Joseph-Horne and Hollomon, 1997). The DMI family has a strong pre-symptom (curative) activity, preventing early infections from developing into lesions and significantly reducing conidium production (Schwabe et al., 1984; O'Leary and Sutton, 1986). Furthermore, both within-season and post-harvest applications of DMIs lead to reduced production of pseudothecia and therefore ascospores (MacHardy, 1996).

Over-reliance on a number of fungicides for scab control has led to emergence and subsequent spread of resistance against fungicides. For instance, fungicide resistance in *V. inaequalis* has been reported for dodine (Köller et al., 1999; Chapman et al., 2011) and DMIs (Errampalli, 2004; Xu et al., 2010). Some of the resistance mechanisms that are thought to be involved include mutations in the CYP51A1 gene, overexpression of the gene, and transporters that work in the efflux of the antifungal compounds (Jobin and Carisse, 2007; Pfeufer and Ngugi, 2012). The resistance against DMIs in *V. inaequalis* is likely to be under polygenic control (Jobin and Carisse, 2007; Schnabel and Jones, 2001; Xu et al., 2010) as field isolates showed a spectrum of sensitivity levels to myclobutanil (a DMI fungicide) rather than just a sensitive and an insensitive phenotype. Tebuconazole (another DMI fungicide) as a substitute for myclobutanil to control scab may also face an uncertain future, as it has been shown to be losing efficacy in a number of countries (Jobin and Carisse, 2007; Chapman et al., 2011; Pfeufer and Ngugi, 2012).

Nevertheless, most of these products, against which resistance in *V. inaequalis* has been reported, are still recommended for use and will provide acceptable control of scab for commercial production in orchards where fungal resistance against these fungicides has not been observed. Furthermore, some of these products are also effective against other diseases, for example myclobutanil against apple powdery mildew, another important apple disease. The Fungicide Resistance Action Committee (FRAC) recommends mixing the use of fungicides with differing modes of action to improve disease control. Mixed use of fungicides combines the characteristics of the components in the mixture, minimises crop losses due to fungicide resistance, and reduces resistance build-up against individual fungicides.

4.2 Alternative non-fungicide products

Alternatives are required to the common commercially used fungicides, due to the ever-decreasing number of available fungicides because of stringent legislation and fungal resistance to fungicides. Furthermore, alternative products are essential for scab control in organic orchards.

The use of biocontrol agents (BCAs) against apple scab has been widely investigated in the past couple of decades. For example, an isolate of *Cladosporium cladosporioides* was selected as the most effective fungal coloniser, from a range of those found, at reducing conidial production (Köhl et al., 2009) and then shown to reduce the severity of apple scab on leaf and fruit in orchard conditions (Köhl et al., 2015).

A plant extract product, 1% populin from black poplar (*Populus nigra*), can slow down *in vitro* conidial germination and reduce scab on leaves and fruits when applied to run-off (Bálint et al., 2014). Plant retardant Prohexadione-Ca can reduce scab incidence (Bazzi et al., 2003), most likely by upregulating pathogenesis-related proteins and inducing host resistance (Bini et al., 2008). Several alternative products, including potassium bicarbonate, and products based on extracts from *Quillaja saponaria*, orange peel, grapefruit seeds, *Yucca schidigerra*, and *Camellia oleifera* and *Chenopodium quinoa* seeds, can have good protectant efficacy against *V. inaequalis* when applied at 1% concentration (Laurent, 2011). Copper use, even with low doses, is still necessary to control scab on susceptible cultivars in organic orchards.

Although these alternative methods show promise in lab and field trials, they are rarely adopted in conventional commercial apple production, primarily because of variable and often low levels of control efficacy achieved. Like fungicides, there is the chance that pathogens may develop resistance to these products. Although BCAs are thought to be more durable than chemical control, the limited available information suggests this assumption is not always justified (Bardin et al., 2015). These alternative products are often compared to fungicides for disease control; this framework of comparison is not justified for BCAs, ignoring BCAs as living organisms and hence that any biocontrol effect is expected to be gradual. These alternative products should be viewed as components of an integrated pest management system rather than straight replacement of conventional fungicides.

4.3 Orchard sanitation

One of the key scab management objectives is to reduce the level of overwintering inoculum in the form of ascospores on leaf litter on orchard floors. Early effective control of scab epidemics in the spring lays the foundation for effective scab management for the entire season.

A practice that is commonly used in commercial orchards is the use of a 5% urea spray, with application recommended between harvest and leaf fall, following a paper in *Nature* (Burchill et al., 1965) that showed urea treatment of leaves in autumn is effective in inhibiting the sexual reproduction of *V. inaequalis* on leaf litter. This reduction may be due to the inhibition of pseudothecium formation by excessive nitrogen, increased bacterial population antagonistic against *V. inaequalis,* and leaves becoming softer, making them more palatable to earthworms for faster degradation. The use of urea has been shown to reduce ascospore number and/or scab incidence to varying degrees in a number of other studies both as an individual treatment and as part of a multifaceted approach (Sutton et al., 2000; Vincent et al., 2004; Holb et al., 2006).

Shredding all leaf litter with a flail mower in autumn reduced the risk of scab by up to 90%, while shredding as little as 65% of the litter could reduce the risk by 50% (Sutton et al., 2000). Sweeping up leaves with a lawn sweeper and ploughing them into the row reduced the incidence of fruit scab by 50%–80% depending on seasons (Gomez et al., 2007). In seasons with conducive conditions for scab development, the effects of removing primary inoculum are less pronounced.

In addition to ascospores on leaf litter as primary inoculum, overwintered conidia in buds and twigs can also be an important source of primary inoculum. However, there are no effective measures to reduce the level of overwintering conidia in buds. Wood scab can also be observed on some cultivars (particularly in organic orchards) and may produce considerable amount of conidia in the spring. The green stems of potted Cox's Orange Pippin trees were susceptible to infection by *V. inaequalis* from June to August and some isolates showed the conspicuous ability to attack the wood (Cook, 1974). These wood scab lesions may be able to produce conidia throughout the growing season and provide a constant risk of infection. Thus, pruning off these shoots with wood scab lesions can contribute to the reduction of primary inoculum next spring (Holb, 2005), particularly in heavily diseased organic orchards.

Appropriate pruning increases air circulation in the tree canopy and therefore affects tree canopy microclimates, particularly reducing humidity and the length of leaf wetness (Carisse and Dewdney, 2002), contributing to reduced scab development. Tree architecture (Simon et al., 2006) and both winter (Holb, 2005) and summer (Cooley and Autio, 2011) pruning also allow better spray deposition for more effective chemical control. Pruning is carried out predominantly for improved fruit quality with a positive impact on disease control as an additional benefit. However, it should be noted that pruning can also lead to an increase in some diseases such as fire blight (*Erwinia amylovora* – an important disease in North America) and apple canker (*Neonectria ditissima*) (Cooley and Autio, 2011).

4.4 Cultivar mixtures

Reduction of disease development through cultivar mixtures has been shown in a range of crops (Mundt, 2002), predominantly in foliar diseases of cereals, for example, rust diseases (caused by *Puccinia spp.*) of wheat (Cox et al., 2004) and powdery mildew (*Blumeria graminis* f. sp. *hordei*) of barley (Tratwal and Bocianowski, 2018). Disease development in mixtures is reduced through inoculum dilution, barrier effect and induced resistance by non-pathogenic inoculum.

The potential for mixed apple cultivar orchards to reduce scab development was first assessed by simulation (Blaise and Gessler, 1994). A field trial with mixed susceptible and R-gene-carrying resistant cultivars demonstrated that scab lesion density is lower in mixture than in monoculture (Bousset et al., 1997). Similar results were also obtained in later studies (Didelot et al., 2007; Parisi et al., 2013), showing that the magnitude of mixture effects depends on season (climatic conditions), mixture structure and orchard management practices. In general, the higher the disease pressure, the lower the mixture control efficacy is.

Although often significant, the reductions in scab observed in mixed cultivar orchards are usually insufficient for disease control in commercial production. However, they could be of benefit if implemented as part of an integrated disease management programme alongside other strategies, such as sanitation, use of BCAs, and backed up by the use of fungicides in high-risk periods. Mixture could be more appropriate for cider apple production as disease control is not as stringent as for dessert apple. The management costs of a mixed cultivar orchard are likely to be higher than that of monoculture due to differences in the timing of bud-break, flowering and fruit development between cultivars, leading to complications in pest and disease control, crop husbandry (e.g. irrigation, fertilisation) and harvesting. Furthermore, dessert cultivars are sold as named varieties and therefore fruit needs to be separated into individual cultivars on harvesting.

A major concern in adopting mixtures in commercial agriculture is the emergence and subsequent rapid build-up of a scab 'super race' combining virulence factors that overcome most or all of the resistance genes in the host cultivars present in the mixture, rendering the mixture ineffective as a means of managing scab. A simulation study suggested that a super race is not likely to occur within the life time (ca. 20 years) of a perennial tree plantation of cultivar mixtures where annual fungal population crashes occur during overwintering (Xu, 2012). However, the formation of super races is real as artificial crossing showed that it is possible for ascospore progenies to be produced that combine virulence factors to overcome resistance in different cultivars (Barbara et al., 2008).

Recent studies on fungal population structures of *V. inaequalis* over time show that a super race is unlikely to occur in the length of a commercial orchard (Xu et al., 2013; Passey et al., 2016). One likely explanation as to why a super race of *V. inaequalis* is unlikely to occur and become dominant in an orchard is because there is insufficient sexual mating between isolates from different cultivars. Mating among strains of opposite types is more likely to occur and succeed for lesions on the same leaf than on different leaves, resulting in more mating among strains with similar genetic (in terms of virulence factors) background.

5 Future trends

Given the polycyclic nature of apple scab, it is critically important to predict and reduce the level of primary inoculum. Currently, prediction of potential primary inoculum is primarily based on severity of visible scab lesions in autumn and winter conditions. As there are old leaves harbouring micro-scab colonies that are not visible to the naked eye, further research is needed to elucidate the role of these symptomless colonies in overwintering of *V. inaequalis* and in production of ascospores.

Recent research on population changes of *V. inaequalis* in a cultivar mixture orchard suggests that the risk of a super fungal race emerging in the mixture is much less than we anticipated based on the assumption of random mating. One plausible explanation is the lack of mating among strains on different leaves compared to those on the same leaf. We need to study the dynamics of the mating process in terms of successful production of ascospores in relation to time and physical location of lesions. This could have a significant impact on the adoption of mixtures in commercial production.

There is an urgent need to develop alternative products, which are not necessarily a straight replacement of fungicides. In particular, we need to explore the biocontrol options and develop multiple biocontrol products, targeting specific stages of the pathogen life cycle integrated with forecasting models. Availability of multiple products with different epidemiological modes of actions will reduce the risk of over-reliance on individual products. Similarly, long-term large-scale demonstration is necessary to implement an integrated pest management approach and demonstrate the feasibility and benefit of such an approach in commercial horticulture.

6 Where to look for further information

The monograph by MacHardy (1996) is the best resource for apple scab research up to 1996, whereas a brief introduction to the disease can be found in the paper by Bowen et al. (2011) and the website hosted by the American Phytopathological Society on apple scab (https://www.apsnet.org/edcenter/disandpath/fungalasco/pdlessons/Pages/AppleScab.aspx). The following publications are a good starting point for research synopsis on more specific topics on apple scab: (1) host resistance against *V. inaequalis* - Bus et al. (2011), (2) epidemiology of *Venturia* diseases - González-Domínguez et al. (2017), and (3) fungicide resistance management - Beckerman et al. (2015). There are several chapters in this book that also deal with specific areas relevant to apple scab: use of disease forecasting models for practical management (Chapter 9), breeding for resistance (Chapter 10), use of fungicides (Chapter 12) and biocontrol (Chapter 13).

Apple scab is the most important disease on apple worldwide and thus studied in most research institutes/centres globally where apple is grown. Leading research organisations on apple scab include: (1) New York State Agricultural Experiment Station, USA; (2) NIAB EMR (East Malling Research), UK; (3) Plant & Food Research, New Zealand; (4) INRA, Université d'Angers, France; and (5) Research Institute of Horticulture, Skierniewice, Poland. Although scab related research can be presented in many horticultural and plant pathological conferences, the conference with most contributions on apple scab is the workshop organised by the IOBC-WPRS Pome Fruit Disease Subgroup in the Working Group of Integrated Protection of Fruit Crops (https://www.iobc-wprs.org/expert_groups/02_wg_fruit_crops.html).

7 References

Aćimović, S. G. and Rosenberger, D. 2018. An introduction to the RIMpro apple scab prediction model. *Scaffolds Fruit J.* 27, 1-5.

Bálint, J., Nagy, S., Thiesz, R., Nyárádi, I.-I. and Balog, A. 2014. Using plant extracts to reduce asexual reproduction of apple scab (*Venturia inaequalis*). *Turk. J. Agric. For.* 38, 91-8. doi:10.3906/tar-1302-17.

Barbara, D. J. J., Roberts, A. L. L. and Xu, X.-M. 2008. Virulence characteristics of apple scab (*Venturia inaequalis*) isolates from monoculture and mixed orchards. *Plant Pathol.* 57(3), 552–61. doi:10.1111/j.1365-3059.2007.01781.x.

Bardin, M., Ajouz, S., Comby, M., Lopez-Ferber, M., Graillot, B., Siegwart, M. and Nicot, P. C. 2015. Is the efficacy of biological control against plant diseases likely to be more durable than that of chemical pesticides? *Front. Plant Sci.* 6, 566. doi:10.3389/fpls.2015.00566.

Bastiaanse, H., Bassett, H. C. M., Kirk, C., Gardiner, S. E., Deng, C., Groenworld, R., Chagné, D. and Bus, V. G. 2016. Scab resistance in 'Geneva' apple is conditioned by a resistance gene cluster with complex genetic control. *Mol. Plant Pathol.* 17(2), 159–72. doi:10.1111/mpp.12269.

Bazzi, C., Messina, C., Tortoreto, L., Stefani, E., Bini, F., Brunelli, A., Andreotti, C., Sabatini, E., Spinelli, F., Costa, G., Hauptmann, S., Stammler, G., Doerr, S., Marr, J. and Rademacher, W. 2003. Control of pathogen incidence in pome fruits and other horticultural crop plants with prohexadione-Ca. *Eur. J. Hort. Sci.* 68, 108–14.

Becker, C. M., Burr, T. J. and Smith, C. A. 1992. Overwintering of conidia of *Venturia inaequalis* in apple buds in New York orchards. *Plant Dis.* 76(2), 121–6. doi:10.1094/PD-76-0121.

Beckerman, J. L., Sundin, G. W. and Rosenberger, D. A. 2015. Do some IPM concepts contribute to the development of fungicide resistance? Lessons learned from the apple scab pathosystem in the United States. *Pest Manag. Sci.* 71, 331–42.

Berrie, A. M. M. and Xu, X.-M. 2003. Managing apple scab (*Venturia inaequalis*) and powdery mildew (*Podosphaera leucotricha*) using Adem (TM). *Int. J. Pest Manag.* 49(3), 243–9. doi:10.1080/0967087031000101089.

Bini, F., Ragaini, A. and Bazzi, C. 2008. Resistance responses induced by the plant growth retardant prohexadione-Ca in apple against scab infections. *Ann. Appl. Biol.* 152(1), 19–27. doi:10.1111/j.1744-7348.2007.00188.x.

Blaise, P. and Gessler, C. 1994. Cultivar mixtures in apple orchards as a mean to control apple scab? *Nor. J. Agric. Sci.* 17, 105–12.

Bousset, L., Blaise, P., Kellerhals, M. and Gessler, C. 1997. Mixtures of apple cultivars in orchards: effect on the scab epidemics. *IOBC WPRS Bull.* 20, 42–8.

Bowen, J. K., Mearich, C. H., Bus, V. G. M., Beresford, R. M., Plummer, K. M. and Templeton, M. D. 2011. *Venturia inaequalis*: the causal agent of apple scab. *Mol. Plant Pathol.* 12(2), 105–22. doi:10.1111/j.1364-3703.2010.00656.x.

Broggini, G. A. L., Le Cam, B., Parisi, L., Wu, C., Zhang, H. B., Gessler, C. and Patocchi, A. 2007. Construction of a contig of BAC clones spanning the region of the apple scab avirulence gene AvrVg. *Fungal Genet. Biol.* 44(1), 44–51. doi:10.1016/j.fgb.2006.07.001.

Broggini, G. A. L., Bus, V. G. M., Parravicini, G., Kumar, S., Groenwold, R. and Gessler, C. 2011. Genetic mapping of 14 avirulence genes in an EU-B04 × 1639 progeny of *Venturia inaequalis*. *Fungal Genet. Biol.* 48(2), 166–76. doi:10.1016/j.fgb.2010.09.001.

Burchill, R. T., Hutto, K. E., Crosse, J. E. and Garrett, C. M. E. 1965. Inhibition of the perfect stage of *Venturia inaequalis* (Cooke) Wint., by urea. *Nature* 205(4970), 520–1. doi:10.1038/205520b0.

Bus, V. G. M., Rikkerink, E. H. A., Caffier, V., Durel, C. E. and Plummer, K. M. 2011. Revision of the nomenclature of the differential host-pathogen interactions of *Venturia inaequalis* and *Malus*. *Annu. Rev. Phytopathol.* 49, 391–413. doi:10.1146/annurev-phyto-072910-095339.

Caffier, V., Le Cam, B., Al Rifaï, M., Bellanger, M. N., Comby, M., Denancé, C., Didelot, F., Expert, P., Kerdraon, T., Lemarquand, A., Ravon, E. and Durel, C. E. 2016. Slow erosion of a quantitative apple resistance to *Venturia inaequalis* based on an isolate-specific Quantitative Trait Locus. *Infect. Genet. Evol.* 44, 541–8. doi:10.1016/j.meegid.2016.07.016.

Calenge, F., Faure, A., Goerre, M., Gebhardt, C., vande Weg, W. E., Parisi, L. and Durel, C. E. 2004. Quantitative trait loci (QTL) analysis reveals both broad-spectrum and isolate-specific QTL for scab resistance in an apple progeny challenged with eight isolates of *Venturia inaequalis*. *Phytopathology* 94(4), 370–9. doi:10.1094/PHYTO.2004.94.4.370.

Carisse, O. and Dewdney, M. 2002. A review of non-fungicidal approaches for the control of apple scab. *Phytoprotection* 83(1), 1–29. doi:10.7202/706226ar.

Chapman, K. S., Sundin, G. W. and Beckerman, J. L. 2011. Identification of resistance to multiple fungicides in field populations of *Venturia inaequalis*. *Plant Dis.* 95(8), 921–6. doi:10.1094/PDIS-12-10-0899.

Cook, R. T. A. 1974. Pustules on wood as sources of inoculum in apple scab and their response to chemical treatments. *Ann. Appl. Biol.* 77(1), 1–9. doi:10.1111/j.1744-7348.1974.tb01381.x.

Cooley, D. R. and Autio, W. R. 2011. Summer pruning of apple: impacts on disease management. *Adv. Hortic. Sci.* 25, 199–204.

Cova, V., Lasserre-Zuber, P., Piazza, S., Cestaro, A., Velasco, R., Durel, C. E. and Malnoy, M. 2015. High-resolution genetic and physical map of the *Rvi1* (*Vg*) apple scab resistance locus. *Molecular Breeding* 35, 16. doi:10.1007/s11032-015-0245-1.

Cox, C. M., Garrett, K. A., Bowden, R. L., Fritz, A. K., Dendy, S. P. and Heer, W. F. 2004. Cultivar mixtures for the simultaneous management of multiple diseases: tan spot and leaf rust of wheat. *Phytopathology* 94(9), 961–9. doi:10.1094/PHYTO.2004.94.9.961.

Deng, C. H., Plummer, K. M., Jones, D. A. B., Mesarich, C. H., Shiller, J., Taranto, A. P., Robinson, A. J., Kastner, P., Hall, N. E., Templeton, M. D. and Bowen, J. K. 2017. Comparative analysis of the predicted secretomes of Rosaceae scab pathogens *Venturia inaequalis* and *V. pirina* reveals expanded effector families and putative determinants of host range. *BMC Genomics* 18(1), 339. doi:10.1186/s12864-017-3699-1.

Didelot, F., Brun, L. and Parisi, L. 2007. Effects of cultivar mixtures on scab control in apple orchards. *Plant Pathol.* 56(6), 1014–22. doi:10.1111/j.1365-3059.2007.01695.x.

Errampalli, D. 2004. Distribution of myclobutanil fungicide sensitivities among populations of *Venturia inaequalis*, the causal agent of apple scab, in Ontario. *Acta Hortic.* 638, 157–62. doi:10.17660/ActaHortic.2004.638.20.

Gessler, C. and Pertot, I. 2012. Vf scab resistance of *Malus*. *Trees Struct. Funct.* 26(1), 95–108. doi:10.1007/s00468-011-0618-y.

Gladieux, P., Zhang, X. G., Afoufa-Bastien, D., Valdebenito Sanhueza, R. M., Sbaghi, M. and Le Cam, B. 2008. On the origin and spread of the scab disease of apple: out of central Asia. *PLoS ONE* 3(1), e1455. doi:10.1371/journal.pone.0001455.

Gladieux, P., Caffier, V., Devaux, M. and Le Cam, B. 2010. Host-specific differentiation among populations of *Venturia inaequalis* causing scab on apple, pyracantha and loquat. *Fungal Genet. Biol.* 47(6), 511–21. doi:10.1016/j.fgb.2009.12.007.

Gomez, C., Brun, L., Chauffour, D. and Le Vallée, D. D. L. 2007. Effect of leaf litter management on scab development in an organic apple orchard. *Agric. Ecosyst. Environ.* 118(1–4), 249–55. doi:10.1016/j.agee.2006.05.025.

González-Domínguez, E., Armengol, J. and Rossi, V. 2017. Biology and epidemiology of *Venturia* species affecting fruit crops: a review. *Front. Plant Sci.* 8, 1496. doi:10.3389/fpls.2017.01496.

Harris, S. A., Robinson, J. P. and Juniper, B. E. 2002. Genetic clues to the origin of the apple. *Trends Genet.* 18(8), 426–30. doi:10.1016/S0168-9525(02)02689-6.

Holb, I. J. 2005. Effect of pruning on apple scab in organic apple production. *Plant Dis.* 89(6), 611–8. doi:10.1094/PD-89-0611.

Holb, I. J., Heijne, B. and Jeger, M. J. 2004. Overwintering of conidia of *Venturia inaequalis* and the contribution to early epidemics of apple scab. *Plant Dis.* 88(7), 751–7. doi:10.1094/PDIS.2004.88.7.751.

Holb, I. J., Heijne, B. and Jeger, M. J. 2006. Effects of integrated control measures on earthworms, leaf litter and *Venturia inaequalis* infection in two European apple orchards. *Agric. Ecosyst. Environ.* 114(2–4), 287–95. doi:10.1016/j.agee.2005.11.021.

Jobin, T. and Carisse, O. 2007. Incidence of myclobutanil- and kresoxim-methyl-insensitive isolates of *Venturia inaequalis* in Quebec orchards. *Plant Dis.* 91(10), 1351–8. doi:10.1094/PDIS-91-10-1351.

Joseph-Horne, T. and Hollomon, D. W. 1997. Molecular mechanisms of azole resistance in fungi. *FEMS Microbiol. Lett.* 149(2), 141–9. doi:10.1111/j.1574-6968.1997.tb10321.x.

Koch, T., Kellerhals, M. and Gessler, C. 2000. Virulence pattern of *Venturia inaequalis* field isolates and corresponding differential resistance in *Malus x domestica*. *J. Phytopathol.* 148, 357–64.

Köhl, J. J., Molhoek, W. W. M. L., Groenenboom-de Haas, B. B. H. and Goossen-van de Geijn, H. H. M. 2009. Selection and orchard testing of antagonists suppressing conidial production by the apple scab pathogen *Venturia inaequalis*. *Eur. J. Plant Pathol.* 123(4), 401–14. doi:10.1007/s10658-008-9377-z.

Köhl, J., Scheer, C., Holb, I. J., Masny, S. and Molhoek, W. 2015. Toward an integrated use of biological control by *Cladosporium cladosporioides* H39 in apple scab (*Venturia inaequalis*) management. *Plant Dis.* 99(4), 535–43. doi:10.1094/PDIS-08-14-0836-RE.

Köller, W., Wilcox, W. F. and Jones, A. L. 1999. Quantification, persistence, and status of dodine resistance in New York and Michigan orchard populations of *Venturia inaequalis*. *Plant Dis.* 83(1), 66–70. doi:10.1094/PDIS.1999.83.1.66.

Laloi, G., Vergne, E., Durel, C. E., Le Cam, B. and Caffier, V. 2017. Efficiency of pyramiding of three quantitative resistance loci to apple scab. *Plant Pathol.* 66(3), 412–22. doi:10.1111/ppa.12581.

Laurent, J. 2011. Innovative strategies for the control of apple scab (*Venturia inaequalis* [Cke.] Wint.) in organic apple production. University of Liege, Liege, p. 188.

Le Cam, B., Parisi, L. and Arene, L. 2002. Evidence of two formae speciales in *Venturia inaequalis*, responsible for apple and Pyracantha scab. *Phytopathology* 92(3), 314–20. doi:10.1094/PHYTO.2002.92.3.314.

Li, B. and Xu, X.-M. 2002. Infection and development of apple scab (*Venturia inaequalis*) on old leaves. *J. Phytopathol. Z.* 150(11–12), 687–91. doi:10.1046/j.1439-0434.2002.00824.x.

Liebhard, R., Koller, B., Patocchi, A., Kellerhals, M., Pfammatter, W., Jermini, M. and Gessler, C. 2003. Mapping quantitative field resistance against apple scab in a "Fiesta" x "Discovery" progeny. *Phytopathology* 93(4), 493-501. doi:10.1094/PHYTO.2003.93.4.493.

MacHardy, W. E. 1996. *Apple Scab: Biology, Epidemiology, and Management*. American Phytopathological Society, St. Paul, MN.

Malnoy, M., Xu, M., Borejsza-Wysocka, E., Korban, S. S. and Aldwinckle, H. S. 2008. Two receptor-like genes, *Vfa1* and *Vfa2*, confer resistance to the fungal pathogen *Venturia inaequalis* inciting apple scab disease. *Mol. Plant Microbe Interact.* 21(4), 448–58. doi:10.1094/MPMI-21-4-0448.

Mills, W. D. and La Plante, A. A. 1954. Diseases and insects in the orchard. Cornell Univ. Ext. Bull. 711.

Mundt, C. C. 2002. Use of multiline cultivars and cultivar mixtures for disease management. *Annu. Rev. Phytopathol.* 40, 381–410. doi:10.1146/annurev.phyto.40.011402.113723.

O'Leary, A. L. and Sutton, T. B. 1986. Effects of postinfection applications of ergosterol biosynthesis-inhibiting fungicides on lesion formation and pseudothecial development of *Venturia inaequalis*. *Phytopathology* 76(1), 119–24. doi:10.1094/Phyto-76-119.

Padmarasu, S., Sargent, D. J., Jaensch, M., Kellerhals, M., Tartarini, S., Velasco, R., Troggio, M. and Patocchi, A. 2014. Fine-mapping of the apple scab resistance locus *Rvi12* (*Vb*) derived from 'Hansen's baccata #2'. *Mol. Breed.* 34(4), 2119–29. doi:10.1007/s11032-014-0167-3.

Papp, D., Király, I. and Tóth, M. 2016. Suitability of old apple varieties in organic farming, based on their resistance against apple scab and powdery mildew. *Org. Agric.* 6(3), 183–9. doi:10.1007/s13165-015-0126-2.

Parisi, L., Lespinasse, Y., Guillaumes, J. and Kruger, J. 1993. A new race of *Venturia inaequalis*s virulent to apples with resistance due to the *Vf* gene. *Phytopathology* 83(5), 533–7. doi:10.1094/Phyto-83-533.

Parisi, L., Gros, C., Combe, F., Parveaud, C.-E., Gomez, C. and Brun, L. 2013. Impact of a cultivar mixture on scab, powdery mildew and rosy aphid in an organic apple orchard. *Crop Prot.* 43, 207–12. doi:10.1016/j.cropro.2012.09.014.

Passey, T. A. J., Shaw, M. W. W. and Xu, X.-M. 2016. Differentiation in populations of the apple scab fungus *Venturia inaequalis* on cultivars in a mixed orchard remain over time. *Plant Pathol.* 65(7), 1133–41. doi:10.1111/ppa.12492.

Passey, T. A. J., Robinson, J. D., Shaw, M. W. and Xu, X. -M. 2017. The relative importance of conidia and ascospores as primary inoculum of *Venturia inaequalis* in a southeast England orchard. *Plant Pathol.* 66(9), 1445–51. doi:10.1111/ppa.12686.

Passey, T. A. J., Armitage, A. D. and Xu, X.-M. 2018. Annotated draft genome sequence of the apple scab pathogen *Venturia inaequalis*. *Microbiol. Resour. Announc.* 7, e01062-18.

Pfeufer, E. E. and Ngugi, H. K. 2012. Orchard factors associated with resistance and cross resistance to sterol demethylation inhibitor fungicides in populations of *Venturia inaequalis* from Pennsylvania. *Phytopathology* 102(3), 272–82. doi:10.1094/PHYTO-04-11-0117.

Roberts, A. L. and Crute, I. R. 1994. Apple scab resistance from *Malus floribunda* 821 (Vf) is rendered ineffective by isolates of *Venturia inaequalis* from *Malus floribunda*. *Nor. J. Agric. Sci.* 17, 403–6.

Roubal, C. and Nicot, P. C. 2016. Apple scab: numerical optimization of a new thermal time scale and application for modelling ascospore release in southern France. *Plant Pathol.* 65(1), 79–91. doi:10.1111/ppa.12398.

Schnabel, G. and Jones, A. L. 2001. The 14 alpha-demethylase (*CYP51A1*) gene is overexpressed in *Venturia inaequalis* strains resistant to myclobutanil. *Phytopathology* 91(1), 102–10. doi:10.1094/PHYTO.2001.91.1.102.

Schouten, H. J., Brinkhuis, J., van der Burgh, A., Schaart, J. G., Groenwold, R., Broggini, G. A. L. and Gessler, C. 2014. Cloning and functional characterization of the *Rvi15* (*Vr2*) gene for apple scab resistance. *Tree Genet. Genomes* 10(2), 251–60. doi:10.1007/s11295-013-0678-9.

Schwabe, W. F. S., Jones, A. L. and Jonker, J. P. 1984. Greenhouse evaluation of the curative and preventive action of sterol-inhibiting fungicides against apple scab. *Phytopathology* 73, 1347.

Schwabe, W. F. S., Jones, A. L. and van Blerk, E. 1989. Relation of degree-day accumulations to maturation of ascospores of *Venturia inaequalis* in South Africa. *Phytophylactica* 21, 13–6.

Simon, S., Lauri, P. E., Brun, L., Defrance, H. and Sauphanor, B. 2006. Does manipulation of fruit-tree architecture affect the development of pests and pathogens? A case study in an organic apple orchard. *J. Hortic. Sci. Biotechnol.* 81(4), 765–73. doi:10.1080/14620316.2006.11512135.

Soriano, J. M., Madduri, M., Schaart, J. G., van der Burgh, A., van Kaauwen, M. P. W., Tomic, L., Groenwold, R., Velasco, R., van de Weg, E. and Schouten, H. J. 2014. Fine mapping of the gene *Rvi18* (*V25*) for broad-spectrum resistance to apple scab, and development of a linked SSR marker suitable for marker-assisted breeding. *Mol. Breed.* 34(4), 2021–32. doi:10.1007/s11032-014-0159-3.

Stensvand, A., Eikemo, H., Gadoury, D. M. and Seem, R. C. 2005. Use of a rainfall frequency threshold to adjust a degree-day model of ascospore maturity of *Venturia inaequalis*. *Plant Dis.* 89(2), 198–202. doi:10.1094/PD-89-0198.

Sutton, D. K., MacHardy, W. E. and Lord, W. G. 2000. Effects of shredding or treating apple leaf litter with urea on ascospore dose of *Venturia inaequalis* and disease buildup. *Plant Dis.* 84(12), 1319–26. doi:10.1094/PDIS.2000.84.12.1319.

Thakur, K., Chawla, V., Bhatti, S., Swarnkar, M. K., Kaur, J., Shankar, R. and Jha, G. 2013. *De novo* transcriptome sequencing and analysis for *Venturia inaequalis*, the devastating apple scab pathogen. *PLoS ONE* 8(1), e53937. doi:10.1371/journal.pone.0053937.

Tratwal, A. and Bocianowski, J. 2018. Cultivar mixtures as part of integrated protection of spring barley. *J. Plant Dis. Prot.* 125(1), 41–50. doi:10.1007/s41348-017-0139-z.

Vincent, C., Rancourt, B. and Carisse, O. 2004. Apple leaf shredding as a non-chemical tool to manage apple scab and spotted tentiform leafminer. *Agric. Ecosyst. Environ.* 104(3), 595–604. doi:10.1016/j.agee.2004.01.027.

Xu, X.-M. 2012. Super-races are not likely to dominate a fungal population within a life time of a perennial crop plantation of cultivar mixtures: a simulation study. *BMC Ecol.* 12(1). doi:10.1186/1472-6785-12-16.

Xu, X.-M. and Robinson, J. 2005. Modelling the effects of wetness duration and fruit maturity on infection of apple fruits of Cox's Orange Pippin and two clones of Gala by *Venturia inaequalis*. *Plant Pathol.* 54(3), 347–56. doi:10.1111/j.1365-3059.2005.01177.x.

Xu, X.-M., Butt, D. J. J. and Van Santen, G. 1995. A dynamic model simulating infection of apple leaves by *Venturia inaequalis*. *Plant Pathol.* 44(5), 865–76. doi:10.1111/j.1365-3059.1995.tb02746.x.

Xu, X.-M., Yang, J., Thakur, V., Roberts, A. and Barbara, D. J. D. J. 2008. Population variation of apple scab (*Venturia inaequalis*) isolates from Asia and Europe. *Plant Dis.* 92(2), 247–52. doi:10.1094/PDIS-92-2-0247.

Xu, X.-M., Roberts, T., Barbara, D., Harvey, N. G. N. G., Gao, L. and Sargent, D. J. D. J. 2009. A genetic linkage map of *Venturia inaequalis*, the causal agent of apple scab. *BMC Res. Notes* 2, 163. doi:10.1186/1756-0500-2-163.

Xu, X.-M., Gao, L.-Q. and Yang, J.-R. 2010. Are insensitivities of *Venturia inaequalis* to myclobutanil and fenbuconazole correlated? *Crop Prot.* 29(2), 183–9. doi:10.1016/j.cropro.2009.07.002.

Xu, X.-M., Harvey, N., Roberts, A. and Barbara, D. 2013. Population variation of apple scab (*Venturia inaequalis*) within mixed orchards in the UK. *Eur. J. Plant Pathol.* 135(1), 97–104. doi:10.1007/s10658-012-0068-4.

Chapter 2

Powdery mildew: biology, epidemiology, and management of *Podosphaera* spp. of tree fruit

A. Amiri and L. Gañán, Washington State University, USA

1 Introduction

The obligate biotrophic pathogens of the *Erysiphaceae* group, commonly known as powdery mildews, affect more than 10 000 types of angiosperm plants (Amano, 1986; Braun and Cook, 2012; Glawe, 2008; Liberato et al., 2006; Takamatsu et al., 2010). These species have been researched and investigated for more than 250 years and continue to provide fascinating and perplex speciation and knowledge (Glawe, 2008). Major economically important tree fruit, such as apple (*Malus × domestica*), pear (*Pyrus* spp.) and stone fruit (cherry, apricot, peach: *Prunus* spp.), are amongst the most susceptible crops to mildews. The fungus affects vegetative tissues mostly but can also infect the reproductive organs (flowers and fruits) of the tree. In the absence of appropriate management, severe polycyclic infections may significantly reduce fruit set and yield. The disease may also have a chronic effect on tree vigour, resulting in detrimental impact on both the longevity and profitability of an orchard. The species that causes powdery mildew of cherry is the most challenging among tree fruit because severe damage can be caused to the fruit pre- and post-harvest. Mildew species infecting pome and stone fruit are host-specific, although recent studies have reported them on other closely related but also on unrelated hosts (Garibaldi et al., 2005; Jankovics et al., 2011; Liang et al., 2012).

http://dx.doi.org/10.19103/AS.2019.0046.02

Management of powdery mildew of tree fruit can vary between growing regions and the cultivar planted and is best achieved via integrated pest management approaches including cultural practices, host resistance and chemical control. Breeding efforts over the last 50 years have succeeded in identifying resistance sources and selecting some cultivars with sustainable levels of resistance to the powdery mildew fungi (Alston, 1977; Brown, 1959; Dunemann et al., 2007; Monet, 1983; Watkin and Brown, 1956). However, several resistant cultivars have been replaced by new cultivars with different susceptibilities to the powdery mildew fungi. One consequence of cultivar shift is the appearance of new physiological races of the pathogen that may overcome resistance present in specific cultivars (Caffier and Laurens, 2005; Krieghoff, 1995; Lesemann et al., 2004; Urbanietz and Dunemann, 2005). Complete protection of tree fruit against mildew fungi is not achievable by means of host resistance solely as several of the current popular cultivars are susceptible to the disease. Therefore, in many growing regions mildew management still relies on chemical control to avoid yield reduction as high as 50% (Yoder, 2000). Besides the multi-site fungicide sulphur, single-site fungicides from three chemical groups, including the demethylation inhibitors (DMI), succinate dehydrogenase inhibitors (SDHI) and quinone outside inhibitors (QoI), are used to control mildews in pome and stone fruit conventional orchards (Amiri, 2017; Grove and Probst, 2017; Koch et al., 2015; Shoelberg and Haag, 1994; Yoder, 2000). In organically managed orchards, mildew management can be challenging, especially with susceptible cultivars. Although some generally regarded as safe (GRAS) products have shown acceptable levels of control, future efforts should focus on developing more efficient methods of control. This chapter will summarize current knowledge on the biology, epidemiology, host resistance and control of powdery mildew species of apple (*Podosphaera leucotricha*) and cherry (*Podosphaera clandestina*).

2 Hosts and speciation

The *Podosphaera* spp. belong to the *Cystotheceae*, one of the five *Erysiphaceae* tribes, and are known to cause powdery mildew on 250 woody plants among which 90% are from the *Rosaceae* (Takamatsu et al., 2010). The powdery mildew fungus *P. leucotricha* (Ellis & Everh.) E.S. Salmon has long been believed to be apple-specific, although its ability to cause mildew-like symptoms has been reported on six other crops so far (Table 1). The host range of the neighbouring species *P. clandestina* (Wallr.:Fr.) Lév. and *P. tridactyla* (Wallr.) de Barr. seems to be narrower, and specific to the *Prunus* genus, based on the number of hosts reported. Genetic diversity of most *Podosphaera* populations in tree fruits species appears to be low as indicated by results using the amplified fragment length polymorphism (AFLP) technique (Urbanietz and Dunemann, 2005;

Table 1 Hosts reported to be susceptible to the *Podosphaera* spp.

Mildew species	Hosts infected	References
Podosphaera leucotricha	*Malus × domestica* (Apple) *Malus prunifolia* (crab apple) *Pyrus communis* (European pear)	Kwon et al. (2010)
	Pyrus calleryana (Callery pear)	Vajna and Kiss (2008)
	Prunus persica (Peach)	Jankovics et al. (2011)
	Prunus africana (African cherry)	Mwanza et al. (2001)
	Photinia serrulata (Oriental photinia)	Liang et al. (2012)
	Photinia fraseri (Fraser's *Photinia*)	Garibaldi et al. (2005)
Podosphaera clandestina	*Prunus avium* (Cherry) *Cydonia oblonga* (Quince) *Crataegus* spp. (Hawthorn)	Ahmad et al. (1998), Vagelas et al. (2011), Xu and Robinson (2000)
Podosphaera tridactyla	*Prunus avium* (Cherry)	Choi et al. (2015)
	Prunus domestica (Greengage, plums)	Arzanlou and Torbati (2017)
Podosphaera paracurvispora	*Pyrus ussuriensis* (Ussurian or Manchurian pear)	Tang et al. (2017)
Podosphaera pannosa	*Prunus persica* (Peach)	Ogawa et al. (1995)
	Prunus cerasus (European cherry)	Hubert et al. (2012)
	Rosa sp. (Rose)	Dominguez-Serrano et al. (2016)
	Eucalyptus spp. (Eucalyptus)	Bovolini et al. (2018)

Al-Sadi et al., 2012). In contrast, the species *P. tridactyla* is more diverse and genetically distinct from the three other *Podosphaera* species and exclusively pathogenic to the *Prunus,* which may suggest a co-speciation (Cunnington et al., 2005; Takamatsu et al., 2010). Although *P. clandestina, P. leucotricha* and *P. paracurvispora* clustered together in one clade based on the internal transcribed spacer (ITS) region sequences, the latter two were more closely related, suggesting they are the most ancestral species (Takamatsu et al., 2010). The topic of a co-speciation or evolutionary relationship between the *Podosphaera* and the *Rosaceae* is still debated. However, most recent studies tend towards a close evolutionary relationship supported by the evidence that divergence of the *Rosaceae* may have occurred earlier than that of the *Podosphaera* (DeVore and Pigg, 2007; Takamatsu et al., 2010).

3 Life cycle and epidemiology

To a certain extent, *P. leucotricha* and *P. clandestina* have similar disease cycles as they both cause spring and early summer epidemics, from green tip in early season until the end of shoot growth (Fig. 1). One notable distinction is that

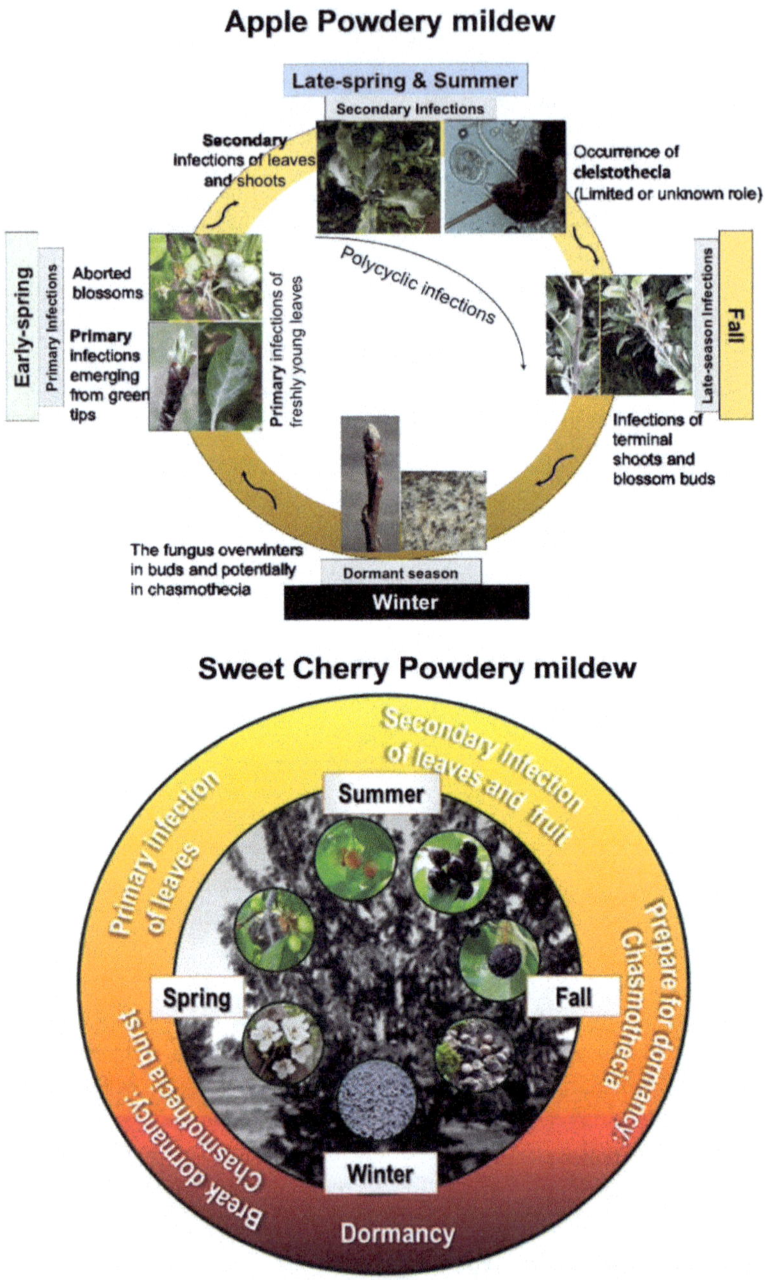

Figure 1 The disease cycles of apple powdery mildew (courtesy of Achour Amiri, WSU) and sweet cherry powdery mildew (courtesy Claudia Probst and Gary Grove, WSU).

P. clandestina most often causes fruit infections, whereas *P. leucotricha* rarely causes fruit russeting on susceptible cultivars under high disease pressure. For *P. leucotricha*, overwintering of mycelia in dormant buds is believed to be the main survival pathway in apples (Hickey and Yoder, 2014). Interestingly, this mode of survival has not been reported on pear trees. Infected buds have been reported to open after healthy ones (Hickey and Yoder, 2014) and conidia released from the former infect new leaves which have emerged earlier from healthy buds, therefore, causing higher infection risks. The fungus may survive as cleistothecia, also called chasmothecia, on stems (Figs. 1 and 2c) or leaves; however, released ascospores have reduced viability as reported in other mildew species (Gadoury and Pearson, 1988; Grove and Boal, 1991). Apple powdery mildew (APM) affects vegetative tissues and under high disease pressure, the fungus can girdle and abort the blossoms (Figs. 1 and 2a,b) and cause fruit russeting, especially in highly susceptible cultivars. If not managed adequately, polycyclic APM infections occurring throughout the growing season can have a chronic effect on tree vigour resulting in detrimental impact on both the longevity and profitability of apple orchards.

Cherry powdery mildew (CPM) caused by *P. clandestina* overwinters mainly as cleistothecia in decomposed leaves or in bark crevices (Fig. 1) in eastern Washington State (Grove and Boal, 1991), whereas dormant buds are the primary inoculum source of *P. clandestina* in hawthorn (Khairi and Preece, 1978). Visible CPM symptoms from primary infections are usually observed 4–6 weeks post-bud break and serve as a source of secondary inoculum for infections that occur later in the season. In contrast to APM, CPM infects green and mature fruit and, therefore, can result in fruit rejection for the fresh market.

Figure 2 Symptoms caused by *Podosphaera leucotricha* on apple: (a) secondary infections on leaves, (b) reproductive organs (note yellow arrows) and (c) cleistothecia (yellow arrows) formed on apple cultivar 'Granny Smith'.

At the end of the summer, nearly all terminal shoots of the cherry trees may become infected by CPM (Grove and Boal, 1991).

Temperature and relative humidity (RH) are the main environmental factors that impact the development of tree fruit mildews the most. While *P. leucotricha* and *P. clandestina* have almost similar temperature requirements, free water has been reported to inhibit spore germination. The optimal temperature for spore release is between 15°C and 20°C and germination is optimal at 20°C for *P. leucotricha* and at 25°C for *P. clandestina* (Grove and Boal, 1991; Grove, 1998; Hickey and Yoder, 2014; Ogawa et al., 1995; Xu and Butt, 1998; Xu and Robinson, 2000). The incubation period of *Podosphaera* spp. can vary from 3 to 12 days depending on the temperature, with an average of 8 days under favourable conditions (Xu, 1996; Xu and Robinson, 2000). Temperatures below 10°C and above 30°C significantly reduce all metabolic activities of both species. Germination and growth of *P. leucotricha* is not greatly impacted by the ambient humidity as germination on apple leaves can occur at RH between 40% and 100% (Butt, 1978; Molnar, 1971; Roosje, 1961; Xu and Butt, 1998). However, conidial dispersal of *P. leucotricha* is negatively affected by higher RH (Sutton and Jones, 1979) and germination is completely inhibited in free water (Butt, 1978). Ascospores of *P. clandestina* are released in free water and germination promoted by high RH and is optimal at lower vapour pressure deficit (Grove and Boal, 1991).

Dispersal of *P. leucotricha* and *P. clandestina* in apple and cherry orchards is mainly diurnal with concentrations of conidia peaking about midday, whereas a second type of dispersal caused by rain has been reported to occur at lower frequencies (Grove, 1998; Sutton and Jones, 1979). The same authors found a negative correlation between the dispersal of both *Podosphaera* species and the RH or leaf wetness. The extensive research conducted on these two major mildew species resulted in the development of useful forecast models for powdery mildew. In Washington State and the U.S. Pacific Northwest, a web-based decision support system for CPM (Jones et al., 2010) has been developed and being used by growers in the region (https://decisionaid.systems/). Another model named Adem™ (Apple Disease, East Malling) was developed based on weather indices for APM and other diseases (Xu and Butt, 1996; Xu, 1999). This model was translated into a PC-based system to forecast risks of several apple pathogens (Berrie and Xu, 2003) but its current implementation in U.K. commercial apple orchards is limited as the disease control in the United Kingdom is more focused on apple scab.

4 Cultivar susceptibility and host-pathogen interactions

Commercial apple cultivars display variable levels of susceptibility ranging from high susceptibility to resistance to *P. leucotricha* (Fig. 3a), whereas no cultivar is known to be immune. Different levels of susceptibility to *P.*

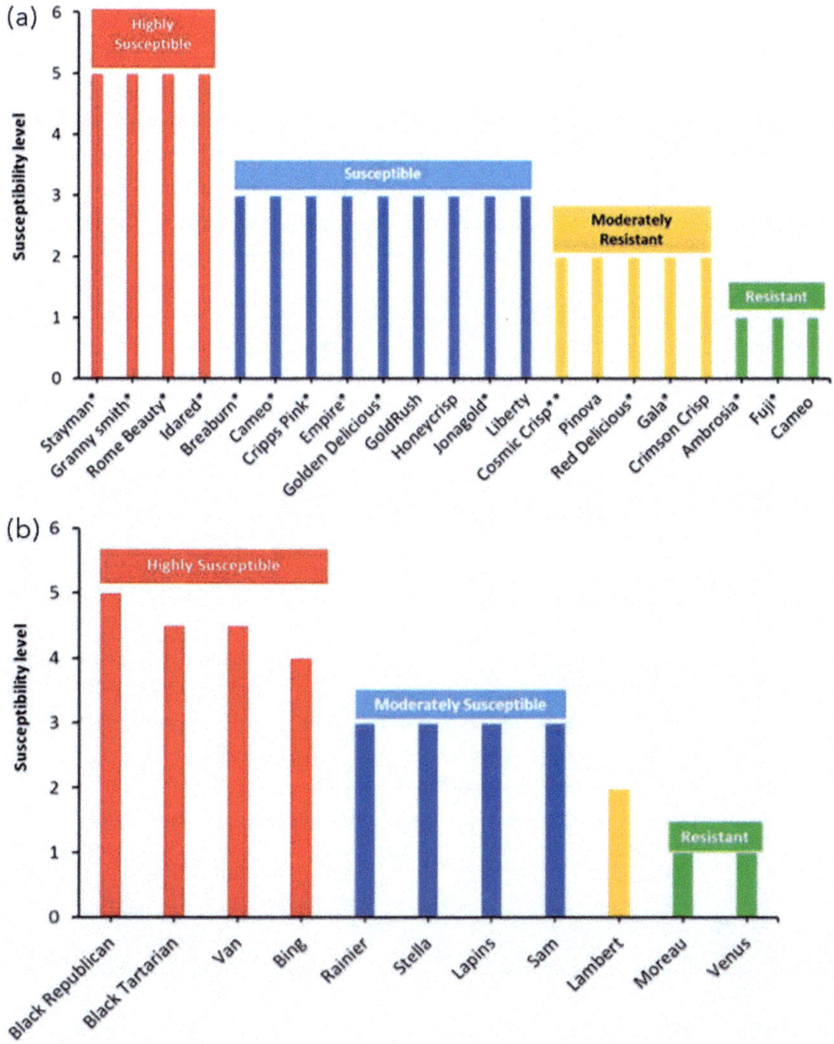

Figure 3 Relative susceptibility of major apple (a) and cherry (b) commercial cultivars to *Podosphaera* spp. Asterisk next to the apple cultivar indicates it is also susceptible to scab. For the new apple cultivar 'WA 38' (Cosmic Crisp^{TM}), the susceptibility level is assumed based on early observation in Washington State. Susceptibility of cherry cultivars was adapted from Olmstead et al. (2000).

clandestina have also been reported in cherry cultivars (Fig. 3b). On pear, *P. leucotricha* can be a concern on 'd'Anjou' which makes more than 90% of the total acreage in central Washington. Other pear cultivars such as 'Louis Bonne' have also been reported to be susceptible to *P. leucotricha* (Hickey and Yoder, 2014).

In general, sources of polygenic resistance to *Podosphaera* spp. used in tree fruit breeding programmes originate from domesticated species, while monogenic resistances are generally derived from wild related species (Table 2). Most of the breeding programmes have been largely reliant on monogenic sources as they offer significantly higher levels of resistance once introgressed into cultivars. However, because fruit quality is the main determinant for the successful introduction of a cultivar, most developed progenies with desired

Table 2 Exploited genetic breeding sources for *Podosphaera* resistance in tree fruits

Mildew species	Hosts infected	Gene	References
Podosphaera leucotricha	*Malus × robusta* MAL 59/9	*Pl-1*	Knight and Alston (1968)
	Malus × zumi MAL 68/5	*Pl-2*	Knight and Alston (1968)
	'White Angel' (ornamental crab apple)	*Pl-w*	Gallot et al. (1985), Batlle and Alston (1996)
	'D12' seedling (from open-pollinated crab apple)	*Pl-d*	Visser and Verhaegh (1977)
	Mildew immune seedling (MIS; from open-pollinated 'Starking Delicious')	*Pl-m*	Dayton (1977), Bus et al. (2010)
	Malus baccata 'Jackii' accession 419 (Siberian crab apple)	*Plbj*	Korban and Dayton (1983), Dunemann and Schuster (2009)
	Malus hupehensis (Tea crab apple)	*MhNPR1*[a]	Chen et al. (2012)
	Malus × domestica clone 40 (from open-pollinated 'Goldparmäne')	Polygenic	Krüger (1994)
	Clone U211 (from open-pollinated 'Primula')	Polygenic	Pitera and Bogdanowicz (1992), Stankiewicz-Kosyl et al. (2005)
	Malus × domestica cv. 'Discovery'	Polygenic	Calenge and Durel (2006)
	Malus × domestica cv. 'TN10-8'	Polygenic	Calenge and Durel (2006)
Podosphaera clandestina	PMR-1 selection[b] (unknown parentage, derived from a group of open-pollinated seedlings)	*Pmr1*	Olmstead and Lang (2002)
P. pannosa	*Prunus ferganensis*	*Vr and Sf*[c]	Dabov (1983)
	Green leaf peach rootstock 'Pamirskij 5'	*Vr2*	Pascal et al. (2010, 2017)
	Prunus dulcis (Almond 'Texas')	*Vr3*	Donoso et al. (2016)
	Prunus davidiana (native peach)	Polygenic	Foulonge et al. (2003a)

[a] Homologous non-expressor of pathogenesis-related gene 1 (*NPR1*); increased disease resistance when overexpressed in transgenic 'Fuji' apple plants (Chen et al., 2012).
[b] Found in the heterozygous allelic state (i.e. *pmr1PMR1*) in the PMR-1 selection (Olmstead and Lang, 2002).
[c] *Vr* is epistatic to *Sf*.

resistance need to be improved to satisfy market and consumer requirements (Kellerhals et al., 2012).

Molecular markers have been used to assist in the selection of progenies carrying major genes of resistance to *P. leucotricha*. However, unreliability in marker-assisted selection (MAS) has been reported for some loci, suggesting that multiple genes may be involved (Dunemann et al., 2007; Calenge and Durel, 2006; Wan and Fazio, 2011). Current applications and challenges of MAS in tree fruit were reviewed by Ru et al. (2015). Growers should be encouraged to establish a mixed cultivar blocks to reduce inoculum pressure of *Podosphaera* spp. in order to minimize adaptation of pathogen populations to a single cultivar extensively planted in a wide area (Holb, 2017). However, planting susceptible cultivars next to somehow resistant cultivars may require additional fungicide sprays and sanitation to avoid high disease pressure and limit the emergence of novel physiological races of the pathogen.

Polygenic resistance appears to be easily transmitted to apple progenies (Krüger, 1994; Calenge and Durel, 2006; Stankiewicz-Kosyl et al., 2005), but can be strongly influenced by environmental conditions and disease pressure (Jeger et al., 1986; Lalancette and Hickey, 1985). Several QTLs have been consistently detected across different populations of tree fruit species when evaluated under multiple environments, with some of them being stable over years (Foulongne et al., 2003b; Calenge and Durel, 2006; Stankiewicz-Kosyl et al., 2005). A better understanding of the defence mechanisms expressed by polygenic resistant genotypes will improve yield and reduce fungicide inputs.

The genetic determinism of partial resistance in tree fruits might also depend on the physiological age of plant tissues, that is ontogenic or age-related resistance. Immature or younger tissues of apple and sweet cherry are more susceptible to infections by *Podosphaera* spp. (Ellis et al., 1981; Lalancette and Hickey, 1985; Hjeltnes, 1988; Olmstead et al., 2000, 2001; Urbanietz and Dunemann, 2005). Interestingly, a recent study reported a case of reverse ontogenic resistance where sweet cherry fruits became increasingly susceptible to *P. clandestina* infection as they matured (Pandey, 2017).

5 Conventional and organic control

The inclusion of durable host resistance, as discussed in the previous section, is the first step towards a more sustainable management of powdery mildews in tree fruit. However, the absence of total immunity and the risk of emergence of physiological races that can overcome specific host resistance, requires, for most of the time, reliance on intergraded pest management tactics to control powdery mildew.

5.1 Cultural control

Although it can be labour intensive, pruning and discarding infected shoots showing first symptoms of powdery mildew in early spring or during the vegetative growth period will help reduce the inoculum load. Research has shown that pruning can significantly reduce APM incidence on both shoots and fruit especially when combined with fungicide applications (Hickey and Yoder, 2014; Holb and Kunz, 2016). Moreover, because *P. clandestina* overwinters as cleistothecia in decomposed leaves or in bark crevices, practices such as leaf shredding and removal will considerably reduce the load of primary inoculum available for the following season. Removal of root suckers in cherry orchards will limit the sources of inoculum that will, potentially, cause further infections on leaves and fruit. Management of irrigation will limit the spread of the fungus and create unfavourable conditions for spore release and germination of the fungus (Grove and Probst, 2017). Moreover, rootstocks and training systems have been documented to affect mildew development in cherry orchards and require close attention (Calabro et al., 2009). Removal of other potential susceptible hosts such as rose bushes near commercial tree fruit orchards is another recommended practice to lower inoculum loads.

5.2 Chemical control

Management of mildews begins at tight cluster when buds are breaking and continuous every 10–14 days until the end of vegetative growth. In wet apple growing regions, management of APM is usually combined with scab management as most registered fungicides are effective against both pathogens. However, reduced fungicide input in dry regions, such as in central Washington State where scab is not a major concern, can lead to increased APM infection risks. Given that mildew is a concern on cherry fruit, sprays for CPM management are recommended from shuck (sepal) fall to harvest. Besides the multi-site carbonates and sulphur, fungicides from three main chemical groups are registered to control mildews of tree fruit and other crops. Thus, fungicides from the demethylation inhibitors (DMI), quinone inhibitors (QoI) and succinate dehydrogenase inhibitors (SDHI) are part of mildew spray programmes (Table 3). In addition to the fungicides listed in Table 3, cyprodinil (FRAC group 9) is pre-mixed with difenoconazole (DMI, FRAC 3) and used in both apple and cherry orchards for mildew control.

Besides the efficacy of a given spray programme, management strategies have to take into consideration two important factors: (1) fungicide resistance management and (2) non-target organisms. Rotation and tank mixing of fungicides with different modes of action is highly recommended to lower the risk of selecting resistant fungal populations. In central Washington State, recommendations include using the DMI fungicides instead of SDHI fungicides

Table 3 Fungicides registered to control powdery mildews in conventional orchards

Chemical group	FRAC group	Fungicide	Resistance risk	Registered for APM[a]	Registered for CPM[a]
Demethylation inhibitors (DMI)	3	Triflumizole/myclobutanil/fena rimol/flutriafol	Medium	+	+
		Metconazole	Medium	−	+
Succinate dehydrogenase inhibitors (SDHI)	7	Penthiopyrad	Medium to high	+	+
		Benzovindiflupyr	Medium to high	+	−
Quinone outside inhibitors (QoI)	11	Azoxystrobin	High	−	+
		Pyraclostrobin	High	−	+
		Trifloxystrobin	High	+	+
		Kresoxim-methyl	High	+	−
DMI + QoI	3 + 11	Tebuconazole + trifloxystrobin	Medium to high	+	+
		Azoxystrobin + difenoconazole	Medium to high	−	+
SDHI + QoI	7 + 11	Fluxapyroxad + pyraclostrobin	Medium to high	+	+
		Boscalid + pyraclostrobin	Medium to high	+	+
		Fluopyram + trifloxystrobin	Medium to high	+	+
Polyoxins	19	Polyoxin D zinc salt	Medium	+	+
Multi-sites	M2	Sulphur	Low	+	+
	NC[b]	Potassium bicarbonate	Low, unknown	+	+
	NC	Calcium polysulphide	Low, unknown	+	+

[a] Fungicide registered for apple/pear powdery mildew (APM) or cherry powdery mildew (CPM). Registered fungicides are shown as solo or pre-mixed with fungicides from other chemical groups. List of registered fungicides and trade names may vary between countries. Fungicides registered for CPM are also effective against mildews species affecting other stone fruit crops.
[b] Non-classified (no FRAC group) fungicides.

in early season to limit selection of resistant populations in other fungal pathogens such as grey mould (*Botrytis cinerea*) and other pre-harvest and storage rots against which Group 7 fungicides have a better activity than the DMIs. The mixtures of boscalid + pyraclostrobin and fluxapyroxad + pyraclostrobin are widely used in the last few weeks prior to harvest to control storage rot pathogens. Their usage in spring and mid-summer should be avoided to limit resistance development in storage pathogens. Another relatively low-risk fungicide, and a good candidate to rotate with fungicides from Group 3, is Polyoxin D (OSO™ 5%SC) which showed a good efficacy against APM in apple (Amiri, 2018). Ultimately, it is important to consider that the nature and the efficacy of a given spray programme may vary between regions and environmental conditions.

In organically managed orchards, managing powdery mildew can be very challenging especially when susceptible cultivars are planted. In central

Table 4 Products registered to manage powdery mildew of pome and stone fruit in organic orchards in the U.S. Pacific Northwest

Type	Active agent/ingredient	Example for trade names[a]
Biological	*Streptomyces lydicus* WYEC 108	Actinovate® AG
	Bacillus amyloliquefaciens strain D4747	DoubleNickel™ LC
	Bacillus subtilis strain QST713	Serenade®
	Bacillus pumilus strain QST 2808	Sonata®
Inorganic and others	Sulphur	Microthiol Dispress®
	Calcium polysulphide	Lime Sulfur Ultra
	Potassium mono/bicarbonate	Kaligreen, Po-Carb/Bicarb

[a] Trade names may vary between countries.

Washington State for example, the susceptible apple cultivars 'Granny Smith', 'Cripps Pink' and 'Honeycrisp' make for nearly 50% of the total organic acreage, whereas the susceptible cherry cultivars 'Bing' and 'Rainier' are predominant. Therefore, enhanced disease management programmes should be implemented. Besides appropriate sanitation practices, preventive applications and rotations of different bio-pesticides from tight cluster to the end of growth of new shoots, in early to mid-summer, should help keep powdery mildew under the economic threshold. Sulphur is widely used for APM and CPM control in organic apple and cherry orchards. In addition to potential phytotoxicity, especially at temperatures above 28°C, sulphur may not effectively control the disease under high disease pressure conditions. A list of other bio-pesticides registered to control powdery mildew in organic orchards in the U.S. Pacific Northwest region is shown in Table 4. However, their efficacy here and under different environmental conditions in different regions requires future assessments.

6 Future trends and conclusion

Conducting research on biotrophic pathogens such as *Podosphaera* spp. will continue to be a challenge for plant pathologists and breeders. New apple cultivars that meet consumers' demands but are more susceptible to APM have been recently selected. These are expected to replace the previously predominant and resistant cultivars such as 'Golden Delicious' and 'Red Delicious', and will cause additional challenges for APM management. The organic tree fruit market which is expanding quickly will put additional pressure on breeders to develop more resistant cultivars and on pathologists to develop more sustainable management approaches. In cherry and pear, current breeding programmes are aimed at developing new rootstocks and scion cultivars, which may pose additional challenges for managing the mildews and other diseases. Additional knowledge about the biology of the *Podosphaera* spp. is needed to identify and examine the distribution of physiological races, mating-type idiomorphs in natural populations. The role of sexual reproduction in genetic diversity of *Podosphaera* spp. in tree fruit crops is underexplored and warrants further studies to enhance mildews management strategies.

7 Where to look for further information

Compendium of Apple and Pear Diseases and Pests (2nd Edition) and *Compendium of Stone Fruit Diseases* published by the American Phytopathological Society are useful additional sources of information on powdery mildews of tree fruit. The Tree Fruit portal (http://treefruit.wsu.edu/) and Decision Aid System (DAS, https://decisionaid.systems/) of the Washington State University contain a valuable amount of updated information, newsletters on mildews and other tree fruit diseases. The American Phytopathological Society website (https://www.apsnet.org/) includes well-summarized fact sheets on the biology and management of tree fruit mildews. FruitBreedomics (http://www.fruitbreedomics.com/) and RosBREED (https://www.rosbreed.org/home) contain information about breeding and pathology topics on major *Rosaceae* species, that is advanced plant material with traits of interest introgressed from different sources.

8 References

Ahmad, N., Sarbhoy, A. K. and Kamal. 1998. A new variety and two new species of powdery mildews from India. *Mycological Research* 102(1), 30–2. doi:10.1017/S0953756297004425.

Al-Sadi, A. M., Al-Raisi, I. J., Al-Azri, M., Al-Hasani, H., Al-Shukaili, M. S., Al-Shuraiqi, S. M., Al-Fahdi, K. O. and Deadman, M. L. 2012. Population structure and management of *Podosphaera pannosa* associated with peach powdery mildew in Oman. *Journal of Phytopathology* 160(11–12), 647–54. doi:10.1111/j.1439-0434.2012.01955.x.

Alston, F. H. 1977. Practical aspects of breeding for mildew (*Podosphaera leucotricha*) resistance in apples. In: *Proceedings of Eucarpia Meeting on Tree Fruit Breeding, Wageningen, the Netherlands*. Institute for Horticultural Plant Breeding, pp. 4–17.

Amano, K. 1986. *Host Range of Geographical Distribution of the Powdery Mildew Fungi*. Japan Scientific Society, Tokyo, Japan.

Amiri, A. 2017. Powdery mildew management in conventional and organic apple orchards. *Washington State University Fruit Matters Newsletter*. Available at: http://treefruit.wsu.edu/article/powdery-mildew-management-in-conventional-and-organic-apple-orchards/.

Amiri, A. 2018. Efficacy of new and experimental products for control of powdery mildew on apple. *Washington State University Fruit Matters Newsletter*. Available at: http://treefruit.wsu.edu/article/efficacy-of-new-and-experimental-products-for-control-of-powdery-mildew-on-apple/.

Arzanlou, M. and Torbati, M. 2017. Powdery mildew on greengage (*Prunus domestica* subsp. *italica* var. *claudiana*) detected in Iran. *Australasian Plant Disease Notes* 12(1), 7.

Batlle, I. and Alston, F. H. 1996. Genes determining leucine aminopeptidase and mildew resistance from the ornamental apple, 'White Angel'. *Theoretical and Applied Genetics* 93, 179–82.

Berrie, A. M. and Xu, X. M. 2003. Managing apple scab (*Venturia inaequalis*) and powdery mildew (*Podosphaera leucotricha*) using Adem™. *International Journal of Pest Management* 49(3), 243–9. doi:10.1080/0967087031000101089.

Bovolini, M. P., Lazarotto, M., Gonzatto, M. P., Borges Junior, N. and Lanna Filho, R. 2018. Susceptibility of clonal seedlings of *Eucalyptus* spp., to powdery mildew disease. *Floresta e Ambiente* 25(3), e20170597. doi:10.1590/2179-8087.059717.

Braun, U. and Cook, R. T. A. 2012. *Taxonomic Manual of the Erysiphales (Powdery Mildews)*. CBS Biodiversity Series 11, pp. 1–707.

Brown, A. G. 1959. The inheritance of mildew resistance in progenies of the cultivated apple. *Euphytica* 8, 81–8.

Bus, V. G. M., Bassett, H. C. M., Bowatte, D., Chagné, D., Ranatunga, C. A., Ulluwishewa, D., Wiedow, C. and Gardiner, S. E. 2010. Genome mapping of an apple scab, a powdery mildew and a woolly apple aphid resistance gene from open-pollinated mildew immune selection. *Tree Genetics and Genomes* 6(3), 477–87. doi:10.1007/s11295-009-0265-2.

Butt, D. J. 1978. Epidemiology of powdery mildews. In: Spencer, D. M. (Ed.), *The Powdery Mildews*. Academic Press, London, pp. 51–81.

Caffier, V. and Laurens, F. 2005. Breakdown of Pl2, a major gene of resistance to apple powdery mildew, in a French experimental orchard. *Plant Pathology* 54(2), 116–24. doi:10.1111/j.1365-3059.2005.01147.x.

Calabro, J. M., Spotts, R. A. and Grove, G. G. 2009. Effect of training system, rootstock, and cultivar on sweet cherry powdery mildew foliar infections. *HortScience* 44(2), 481–2. doi:10.21273/HORTSCI.44.2.481.

Calenge, F. and Durel, C.-E. 2006. Both stable and unstable QTLs for resistance to powdery mildew are detected in apple after four years of field assessments. *Molecular Breeding* 17(4), 329–39. doi:10.1007/s11032-006-9004-7.

Chen, X. K., Zhang, J. Y., Zhang, Z., Du, X. L., Du, B. B. and Qu, S. C. 2012. Overexpressing MhNPR1 in transgenic Fuji apples enhances resistance to apple powdery mildew. *Molecular Biology Reports* 39(8), 8083–9. doi:10.1007/s11033-012-1655-3.

Choi, I. Y., Nam, E. Y., Cho, S. E., Park, J. H. and Shin, H. D. 2015. First report of powdery mildew caused by *Podosphaera tridactyla* on sweet cherry in Korea. *Plant Disease* 99(11), 1648. doi:10.1094/PDIS-01-15-0118-PDN.

Cunnington, J. H., Lawrie, A. C. and Pascoe, I. G. 2005. Genetic variation within *Podosphaera tridactyla* reveals a paraphyletic species complex with biological specialization towards specific *Prunus* subgenera. *Mycological Research* 109(3), 357–62. doi:10.1017/S0953756204002072.

Dabov, S. 1983. Inheritance of peach resistance to powdery mildew. III. Leaf resistance in F1 of JH Hale × nectarine Ferganensis 2. *Genetics and Plant Breeding*. 16, 146–50.

Dayton, D. F. 1977. Genetic immunity to apple mildew incited by *Podosphaera leucotricha*. *HortScience* 12, 225–6.

DeVore, M. L. and Pigg, K. B. 2007. A brief review of the fossil history of the family *Rosaceae* with a focus on the Eocene Okanogan highlands of eastern Washington State, USA, and British Columbia, Canada. *Plant Systematics and Evolution* 266(1-2), 45–57. doi:10.1007/s00606-007-0540-3.

Dominguez-Serrano, D., Garcia-Velasco, R., Mora-Herrara, M. E., Salgado-Siclan, M. L. and Gonzalez-Diaz, J. G. 2016. The rose powdery mildew (*Podosphaera pannosa*). *Agrociencia* 50(7), 901–17.

Donoso, J. M., Picañol, R., Serra, O., Howad, W., Alegre, S., Arús, P. and Eduardo, I. 2016. Exploring almond genetic variability useful for peach improvement: mapping major genes and QTLs in two interspecific almond × peach populations. *Molecular Breeding* 36(2), 16. doi:10.1007/s11032-016-0441-7.

Dunemann, F. and Schuster, M. 2009. Genetic characterization and mapping of the major powdery mildew resistance gene Plbj from *Malus baccata jackii*. In: Socias i Company, R., Espiau, M. T. and Alonso, J. M. (Eds), *Proceedings of the XIIth Eucarpia Symposium on Fruit Breeding and Genetics*. Acta Horticulturae 814, 791–98. doi:10.17660/ActaHortic.2009.814.134.

Dunemann, F., Peil, A., Urbanietz, A. and Garcia-Libreros, T. 2007. Mapping of the apple powdery mildew resistance gene *Pl1* and its genetic association with an NBS-LRR candidate resistance gene. *Plant Breeding* 126(5), 476–81. doi:10.1111/j.1439-0523.2007.01415.x.

Ellis, M. A., Ferree, D. C. and Spring, D. E. 1981. Photosynthesis, transpiration, and carbohydrate content of apple leaves infected by *Podosphaera leucotricha*. *Phytopathology* 71(4), 392–5. doi:10.1094/Phyto-71-392.

Foulongne, M., Pascal, T., Pfeiffer, F. and Kervella, J. 2003a. QTLs for powdery mildew resistance in peach ×*Prunus davidiana* crosses: consistency across generations and environments. *Molecular Breeding* 12(1), 33–50. doi:10.1023/A:1025417507358.

Foulongne, M., Pascal, T., Arús, P. and Kervella, J. 2003b. The potential of *Prunus davidiana* for introgression into peach. *Prunus persica* (L.) Batsch] assessed by comparative mapping. *Theoretical and Applied Genetics* 107(2), 227–38. doi:10.1007/s00122-003-1238-8.

Gadoury, D. M. and Pearson, R. C. 1988. Initiation, development, dispersal, and survival of cleistothecia of *Uncinula necator* in New York vineyards. *Phytopathology* 78(10), 1413–21.

Gallott, J. C., Lamb, R. C. and Aldwinckle, H. S. 1985. Resistance to powdery mildew from some small-fruited *Malus* cultivars. *HortScience* 20, 1085–7.

Garibaldi, A., Garibaldi, G. and Gullino, M. L. 2005. First report of powdery mildew caused by *Podosphaera leucotricha* on *Photinia* × *fraserii* in Italy. *Plant Disease* 89(12), 1362.

Glawe, D. A. 2008. The powdery mildew: a review of the world's most familiar (yet poorly known) plant pathogens. *Annual Review of Phytopathology* 46(1), 27–51. doi:10.1146/annurev.phyto.46.081407.104740.

Grove, G. G. 1998. Meteorological factors affecting airborne conidia concentrations and latent period of *Podosphaera clandestina* on sweet cherry. *Plant Disease* 82(7), 741–6. doi:10.1094/PDIS.1998.82.7.741.

Grove, G. G. and Boal, J. R. 1991. Overwinter survival of *Podosphaera clandestina* in eastern Washington. *Phytopathology* 81(4), 385–91. doi:10.1094/Phyto-81-385.

Grove, G. G. and Probst, C. 2017. IPM note, cherry powdery mildew. *Washington State University Tree Fruit*. Available at: http://treefruit.wsu.edu/crop-protection/disease-management/cherry-powdery-mildew/.

Hickey, K. D. and Yoder, K. S. 2014. Powdery mildew. In: Sutton, T. B., Alswinckle, H. S., Agnello, A. M. and Walgenbach, J. F. (Eds), *Compendium of Apple and Pear Diseases and Pests* (2nd edn.). APS Press, Minneapolis, MN, pp. 12–4.

Hjeltnes, S. H. 1988. Mildew susceptibility in pear progenies. *Acta Horticulturae* 224(224), 223–8. doi:10.17660/ActaHortic.1988.224.27.

Holb, I. J. 2017. Categorization of apple cultivars based on seasonal powdery mildew disease progression in two disease management systems over 12 years. *Trees* 31(6), 1905–17. doi:10.1007/s00468-017-1595-6.

Holb, I. J. and Kunz, S. 2016. Integrated control of apple scab and powdery mildew in an organic apple orchard by combining potassium carbonates with wettable sulfur, pruning, and cultivar susceptibility. *Plant Disease* 100(9), 1894–905. doi:10.1094/PDIS-12-15-1416-RE.

Hubert, J., Fourrier, C., Payen, C., Fournie, J. L. and Loos, R. 2012. First report of powdery mildew caused by *Podosphaera pannosa* on *Prunus cerasus* in France. *Plant Disease* 96(9), 1375–6. doi:10.1094/PDIS-02-12-0218-PDN.

Jankovics, T., Dolovac, N., Bulajić, A., Krstić, B., Pascal, T., Bardin, M., Nicot, P. C. and Kiss, L. 2011. Peach rusty spot is caused by the apple powdery mildew fungus, *Podosphaera leucotricha*. *Plant Disease* 95(6), 719–24. doi:10.1094/PDIS-10-10-0711.

Jeger, M. J., Butt, D. J. and Swait, A. A. J. 1986. Components of resistance of apple to powdery mildew (*Podosphaera leucotricha*). *Plant Pathology* 35(4), 477–90. doi:10.1111/j.1365-3059.1986.tb02045.x.

Jones, V. P., Brunner, J. F., Grove, G. G., Petit, B., Tangren, G. V. and Jones, W. E. 2010. A wed-based decision support system to enhance IPM programs in Washington tree fruit. *Pest Management Science* 66(6), 587–95.

Kellerhals, M., Szalatnay, D., Hunziker, K., Duffy, B., Nybom, H., Ahmadi-Afzadi, M., Höfer, M., Richter, K. and Lateur, M. 2012. European pome fruit genetic resources evaluated for disease resistance. *Trees* 26(1), 179–89. doi:10.1007/s00468-011-0660-9.

Khairi, S. M. and Preece, T. F. 1978. Hawthorn powdery mildew: overwintering mycelium in buds and the effect of clipping hedges on disease epidemiology. *Transactions of the British Mycological Society* 71(3), 399–404. doi:10.1016/S0007-1536(78)80066-7.

Knight, R. L. and Alston, F. H. 1968. Sources of field immunity to mildew (*Podosphaera leucotricha*) in apple. *Canadian Journal of Genetics and Cytology* 10(2), 294–8. doi:10.1139/g68-041.

Koch, A., Felsenstein, F. and Stammler, G. 2015. No evidence of QoI resistance in apple powdery mildew (*Podosphaera leucotricha*). *Journal of Phytopathology* 163(3), 178–84. doi:10.1111/jph.12302.

Korban, S. S. and Dayton, D. F. 1983. Evaluation of Malus germplasm for resistance to powdery mildew. *HortScience* 18, 219–20.

Krieghoff, O. 1995. Entwicklung einer in-vitro-Selektionsmethode auf Resistenz von malus-Genotypen Gegenüber *Podosphaera leucotricha* und in vitro-Differenzierung von Virulenzunterschieden Des Erregers. PhD thesis. Humboldt-University, Berlin, Germany.

Krüger, J. 1994. Observations on different mildew sources used in apple breeding at Ahrensburg. In: Schmidt, H. and Kellerhals, M. (Eds), *Progress in Temperate Fruit Breeding*. Springer, Dordrecht, pp. 7–12.

Kwon, J. H., Park, C. S. and Kim, J. 2010. First report of powdery mildew on *Malus prunifolia* caused by *Podosphaera leucotricha* in Korea. *The Plant Pathology Journal* 26(4), 428. doi:10.5423/PPJ.2010.26.4.428.

Lalancette Jr., N. and Hickey, K. D. 1985. Apple powdery mildew disease progress on sections of shoot growth: an analysis of leaf maturation and fungicide effects. *Phytopathology* 75(2), 130–4. doi:10.1094/Phyto-75-130.

Lesemann, S., Urbanietz, A. and Dunemann, F. 2004. Determining population variation of apple powdery mildew at the molecular level. In: *XI Eucarpia Symposium on Fruit Breeding and Genetics*. Acta Horticulturae 663, 199–204.

Liang, C., Xing, H. H., Cho, S. E. and Shin, H. D. 2012. First report of powdery mildew caused by *Podosphaera leucotricha* on *Photinia serrulata* in China. *Plant Disease* 96(11), 1695. doi:10.1094/PDIS-05-12-0473-PDN.

Liberato, J. R., Barreto, R. W., Niinomi, S. and Takamatsu, S. 2006. Queirozia turbinate (*Phyllactinieae, Erysiphaceae*): a powdery mildew with dematiaceous anamorph. *Mycological Research* 100, 567–74.

Molnar, L. 1971. The effect of environmental factors on the germination of conidia of *Podosphaera leucotricha* (Ell. Et Ev.) Salm. *Ochrana Rostlin (Praha)* 7, 203–10.

Monet, R. 1983. *Le pécher. Génétique et physiologie. INRA Actualités Scientifiques et Agronomiques*. Masson, Paris.

Mwanza, E. J. M., Wiathaka, S. K. and Simons, S. A. 2001. First report of powdery mildew caused by *Podosphaera leucotricha* on *Prunus africana* in Kenya. *Plant Disease* 85(12), 1285. doi:10.1094/PDIS.2001.85.12.1285C.

Ogawa, J. M., Zehr, E. I., Bird, G. W., Ritchie, D. F., Uriu, K. and Uyemoto, J. K. 1995. Powdery mildew. In: Ogawa, J. M., Zehr, E. I., Bird, G. W., Ritchie, D. F., Uriu, K. and Uyemoto, J. K. (Eds), *Compendium of Stone Fruit Diseases*. APS Press, St. Paul, MN, pp. 12–4.

Olmstead, J. W. and Lang, G. A. 2002. *Pmr1*, a gene for resistance to powdery mildew in sweet cherry. *HortScience* 37(7), 1098–9. doi:10.21273/HORTSCI.37.7.1098.

Olmstead, J. W., Lang, G. A. and Grove, G. G. 2000. A leaf disk assay for screening sweet cherry genotypes for susceptibility to powdery mildew. *HortScience* 35(2), 274–7. doi:10.21273/HORTSCI.35.2.274.

Olmstead, J. W., Lang, G. A. and Grove, G. G. 2001. Inheritance of powdery mildew resistance in sweet cherry. *HortScience* 36(2), 337–40. doi:10.21273/HORTSCI.36.2.337.

Pandey, B. 2017. Investigation of *Erysiphaceous* Fungi, *Podosphaera clandestina* and *Erysiphe necator*, as hosts of fungal viruses and effect of inoculum concentration and fruit development stage on fruit infection of sweet cherry by *Podosphaera prunicola*. Ph.D Thesis. Washington State University.

Pascal, T., Pfeiffer, F. and Kervella, J. 2010. Powdery mildew resistance in the peach cultivar Pamirskij 5 is genetically linked with the *Gr* gene for leaf color. *HortScience* 45(1), 150–2. doi:10.21273/HORTSCI.45.1.150.

Pascal, T., Aberlenc, R., Confolent, C., Hoerter, M., Lecerf, E., Tuéro, C. and Lambert, P. 2017. Mapping of new resistance (*Vr2, Rm1*) and ornamental (*Di2, pl*) Mendelian trait loci in peach. *Euphytica* 213(6), 132. doi:10.1007/s10681-017-1921-5.

Pitera, E. and Bogdanowicz, J. 1992. Evaluation of scab-resistant apple progenies for powdery mildew tolerance. In: *Fruit Breeding and Genetics*. Acta Horticulturae 317, 81–8.

Roosje, G. S. 1961. The importance of short time intervals between sprays against apple powdery mildew. *Proceedings of British Insecticide and Fungicide Conferences Volume 1*, pp. 185–96.

Ru, S., Main, D., Evans, K. and Peace, C. 2015. Current applications, challenges, and perspectives of marker-assisted seedling selection in *Rosaceae* tree fruit breeding. *Tree Genetics and Genomes* 11(1), 8. doi:10.1007/s11295-015-0834-5.

Shoelberg, P. L. and Haag, P. 1994. Control of apple powdery mildew (*Podosphaera leucotricha*) in British Columbia by demethylation-inhibiting fungicides. *Canadian Journal Plant Disease Survey* 74(1), 5–11.

Stankiewicz-Kosyl, M., Pitera, E. and Gawronski, S. W. 2005. Mapping QTL involved in powdery mildew resistance of the apple clone U 211. *Plant Breeding* 124(1), 63-6. doi:10.1111/j.1439-0523.2004.01049.x.

Sutton, T. B. and Jones, A. L. 1979. Analysis of factors affecting dispersal of *Podosphaera leucotricha* conidia. *Phytopathology* 69(4), 380-3. doi:10.1094/Phyto-69-380.

Tang, S. R., Jiang, W. T., Qiu, P. L., fu, X. Y., Li, Y., Wang, L. and Liu, S. 2017. *Podosphaera paracurvispora* (*Erysiphaceae*, Ascomycota), a new powdery mildew species on *Pyrus* from China. *Mycoscience* 58(2), 116-20. doi:10.1016/j.myc.2016.11.004.

Takamatsu, S., Niinomi, S., Harada, M. and Havrylenko, M. 2010. Molecular phylogenetic analyses reveal a close evolutionary relationship between *Podosphaera* (*Erysiphales: Erysiphaceae*) and its rosaceous hosts. *Persoonia: Molecular Phylogeny and Evolution of Fungi* 24, 38-48.

Urbanietz, A. and Dunemann, F. 2005. Isolation, identification and molecular characterization of physiological races of apple powdery mildew (*Podosphaera leucotricha*). *Plant Pathology* 54(2), 125-33. doi:10.1111/j.1365-3059.2005.01156.x.

Vagelas, I., Kalorizou, H., Papachatzis, A. and Gravanis, F. 2011. First report of powdery mildew of Cydonia oblonga caused by *Podosphaera clandestina* and *Setoerysiphe kashmeriensis* in Greece. *XXVIII International Symposium on Plant Protection*. Lisbon, Portugal.

Vajna, L. and Kiss, L. 2008. First report of powdery mildew on *Pyrus calleryana* caused by *Podosphaera leucotricha. Plant Disease* 92(1), 176. doi:10.1094/PDIS-92-1-0176B.

Visser, T. and Verhaegh, J. J. 1977. Review of tree fruit breeding carried out at the Institute for Horticultural Plant Breeding at Wageningen from 1951-1976. *Proceedings of the Eucarpia Meeting of Tree Fruit Breeding*, Wageningen, 1976, pp. 113-32.

Wan, Y. and Fazio, G. 2011. Confirmation by QTL mapping of the *Malus robusta* ('robusta 5') derived powdery mildew resistance gene *Pl1. Acta Horticulturae* 903(903), 95-9. doi:10.17660/ActaHortic.2011.903.8] Pl1.

Watkin, W. and Brown, A. G. 1956. Genetic response to selection in cultivated plants: gene frequencies in varieties of *Prunus persica*. *Proceedings of the Royal Society of London. Series B - Biological Sciences* 145(920), 337-47. doi:10.1098/rspb.1956.0046.

Xu, X. M. 1996. The effects of constant and fluctuating temperatures on the length of the incubation period of apple powdery mildew (*Podosphaera leucotricha*). *Plant Pathology* 45(5), 924-32. doi:10.1111/j.1365-3059.1996.tb02903.x.

Xu, X. M. 1999. Modelling and forecasting epidemics of apple powdery mildew (*Podosphaera leucotricha*). *Plant Pathology* 48(4), 462-71. doi:10.1046/j.1365-3059.1999.00371.x.

Xu, X. M. and Butt, D. J. 1996. ADEM™ a pc-based multiple disease warning system for use in the cultivation of apples. *Acta Horticulturae* 416, 293-6.

Xu, X. M. and Butt, D. J. 1998. Effect of temperature and atmospheric moisture on the early growth of powdery mildew (*Podosphaera leucotricha*) colonies. *European Journal of Plant Pathology* 104(2), 133-40. doi:10.1023/A:1008626206164.

Xu, X. M. and Robinson, J. D. 2000. Effects of temperature on the incubation and latent periods of hawthorn powdery mildew (*Podosphaera clandestina*). *Plant Pathology* 49(6), 791-7. doi:10.1046/j.1365-3059.2000.00520.x.

Yoder, K. S. 2000. Effect of powdery mildew on apple yield and economic benefits of its management in Virginia. *Plant Disease* 84(11), 1171-6. doi:10.1094/PDIS.2000.84.11.1171.

Chapter 3

Apple replant disease: causes and management

Zhiquan Mao and Yanfang Wang, Shandong Agricultural University, China

1 Introduction

Apple is an important fruit crop globally, however, its quality and productivity significantly decline when young apple trees are replanted in the same field where the species was grown previously (Tewoldemedhin et al., 2011b); this phenomenon is commonly described as replant disease, replant problem or replant disorder (Mazzola and Manici, 2012; Utkhede, 2006). When the old orchard is replanted, the apple replant disease (ARD) issue cannot often be avoided due to limited virgin apple land (Yin et al., 2016c). ARD may result in reduced growth of young apple trees, restricted root development, yield reduction, predisposition to diseases and pests, and even tree death, eventually shortening the productive life of replant orchards (Rumberger et al., 2007; Wang et al., 2014; Yin et al., 2016c). China is the largest apple producer in the world (Chen et al., 2010; Sun et al., 2017). There were 2.3 million ha under apple cultivation in China in 2015, producing about 43.8 million tons annually, with 'Fuji' accounting for 65% of apple production (Chen et al., 2015). Due to limited land resources, ARD is likely to seriously hinder the sustainable development of the apple industry. Understanding the biology underpinning the ARD complex and developing appropriate management strategies are thus of paramount importance.

http://dx.doi.org/10.19103/AS.2019.0046.03

2 Causes of ARD

ARD has been reported in literature for more than 200 years from the fruit-growing regions worldwide (Börner, 1959; Benizri et al., 2005; Spath et al., 2015), but its causes had not been clearly defined until recently. Studies found that the causes of ARD were complex, an outcome of interactions among plants, pathogens, soil microbial community and soil physio-chemical characteristics (Bai et al., 2009; Mazzola, 1998; Mazzola and Manici, 2012; Rumberger et al., 2007; Tewoldemedhin et al., 2011b; Yin et al., 2016c). Main causes for the occurrence of ARD include soil structure and nutritional imbalances (Benizri et al., 2005), toxins (Gao et al., 2009b; Yin et al., 2016c; Zhao et al., 2009), bacteria, fungi, oomycetes and nematodes (Utkhede and Smith, 2000). More and more findings supported that ARD can be attributed to many microorganisms (Spath et al., 2015; Utkhede and Smith, 2000) (bacteria, fungi and nematode), especially fungal and oomycete pathogens.

2.1 Deterioration of soil physical and chemical properties

Previous studies found that after long-term continuous cropping, soil structure is severely damaged, showing increased bulk soil density and reduction in soil aeration pores. Disproportional increased input of nitrogen and phosphorus relative to potassium in production has led to nutrient imbalances in the soils (Lv et al., 2006; Wu et al., 2002). Fruit trees are perennial with deep and wide roots, and long-term plantation of apple crops is likely to result in accumulation or lack of certain nutritional elements in the soils. Therefore, most of the early studies on ARD focused on the deficiencies in soil nutrients; however, results from these early studies showed that manipulating fertilizer input failed to alleviate ARD (Neilsen and Yorston, 1991; Parchomchuk et al., 1994; Wilson et al., 2004). For example, application of monoammonium phosphate (MAP) increases growth of replant apple trees, but these studies were often carried out in conjunction with soil disinfection or other measures (Brown and Koutoulis, 2008; Neilsen and Yorston, 1991). In one greenhouse trial, the organic matter amended to ARD soil provided no significant benefit compared with apple trees transplanted to ARD soil (Wilson et al., 2004). A field trial was planted on an area which had been under apple cultivation for 15 years. When ammonium phosphate was applied to this soil surface or to the tree planting hole, new trees had a near twofold increase in the extension and radial growth, while organic matter incorporated into ARD soil surface was ineffective. Also, the leaf nutrient content was the same in different treatments, which proves that the tree growth has nothing to do with nutrition (Wilson et al., 2004).

Over application of nitrogen and phosphorus can easily lead to soil acidification (Liang et al., 2013). Although decreasing pH in alkaline soils can improve utilization of trace elements, this decrease in acidic soils may lead to deficiencies in zinc, calcium, magnesium and so on. Soil acidification affects tree development due to the lack of nutrients and toxic effects, prolonged restriction of root expansion, particularly into the depth of soil (Finlay, 1995; Manrique, 1985), leading to the reduction of fine roots and root distribution, and affecting nutrient and water absorption. Soil acidification can also lead to changes in soil microflora (Bryant et al., 1979). There is a serious problem of orchard soil acidification in China, for instance in the Shandong Province (Wang et al., 2010). Soil acidification of 268 orchards in Jiaodong, Shandong Province, is very serious, accounting for 60.5% of the total production area (Yu et al., 2010). Deterioration in the soil physical and chemical properties reduces root development and changes in soil microflora structure that results in soil acidification affecting tree growth in replant orchards.

2.2 Allelopathic autotoxicity

Earlier studies suggested that the allelochemicals produced by plant root exudation and decomposition of remnants are important factors that may cause replant disease (Börner, 1959; Guenzi and McCalla, 1962; Nicola et al., 2017; Yin et al., 2016c; Zhen et al., 2004). Plants can release some allelochemicals by various means such as aboveground leaching, root exudation and decomposition of plant waste material that may inhibit growth of the same or next crop of plants in the same species or family. This phenomenon is called autotoxicity (Singh et al., 1999; Sun et al., 2015). All plant parts, including leaves, shoots, roots, seeds and fruits, may have autotoxicity (Singh et al., 1999). These autotoxic chemicals are some macromolecular substances that are mainly composed of phenolic compounds, including phenolic acids, organic acids, linear alcohols, tannins, aldehydes, terpenes, amino acids and alkaloids. Phenolic acids include phlorizin, phloretin, benzoic acid, phloroglucinol, ferulic acid, *p*-hydroxybenzoic acid, vanillin, syringic acid and caffeic acid (Yin et al., 2016c; Zhang et al., 2009).

Many studies suggested that phenolic acids are an important cause of ARD (Bai et al., 2009; Wang et al., 2015; Yin et al., 2016c; Zhang et al., 2009). Autotoxicity of phenolic acids vary with the type of phenolic acids and increases with increasing concentration. Root basal respiration rate of *Malus hupehensis* Rhed. seedlings may be induced by a high concentration of cinnamic acid (Gao et al., 2009b). Similarly, a high concentration of benzoic acid inhibits antioxidant enzyme activities and mitochondrial function of *M. hupehensis* Rhed. roots (Gao et al., 2009a; Zhang et al., 2011). Phthalic acid increases the accumulation of MDA in apple seedlings with the associated increase in the permeability of the cell membrane, reducing height, fresh weight and dry weight of young seedlings (Bai et al., 2009). Phlorizin over a certain threshold can reduce the activity of related enzymes in the TCA cycle, resulting in decreased respiration rate in the roots of apple rootstock *M. hupehensis* Rhed. (Wang et al., 2012). Phenolic acids increase membrane permeability, cause outflow of intracellular solutes, induce lipid peroxidation, impede growth and development of the overground parts and ultimately inhibit plant growth or cause tissue necrosis (Li et al., 2010b). Accumulation of soil phenolic acids above a certain concentration can cause stress on plant roots (Han et al., 2012). Phenolic acids inhibit growth and development of the underground parts of a crop (Asao et al., 2003; Blok and Bollen, 1993). The concentrations of phlorizin, phloroglucinol, phloroglucinol, *p*-hydroxybenzoic acid and cinnamic acid in soils with two continuous cropping apple years were sufficiently high to inhibit growth of *M. hupehensis* Rhed. seedlings, particularly root system (Wang et al., 2015; Yin et al., 2016c). A high concentration of fatty acids (myristic acid, palmitic acid and stearic acid) may cause replant disease (Liu et al., 2013). A low concentration of organic acids in continuous cropping soil is favourable for growth of *M. hupehensis* Rhed. seedlings, but a high concentration inhibits seedling growth (Li et al., 2014).

2.3 Soil microflora

Many recent studies support that biotic factors are the major cause of ARD (Mazzola and Manici, 2012). Candidate causal agents of ARD include nematodes, oomycetes, bacteria and fungi (Utkhede et al., 1992), especially fungal and oomycete species. Common pathogens include *Fusarium*, *Rhizoctonia*, *Cylindrocarpon*, *Phytophthora* and *Pythium* (Manici et al., 2013; Tewoldemedhin et al., 2011a; Van Schoor et al., 2009). ARD occurrence is closely related to the changes in soil microflora composition; long-term continuous cropping of one plant species can lead to considerable changes in the soil microflora compositi.on, such as reduced beneficial microorganisms and increased plant pathogens (Franke-Whittle et al., 2015; Li et al., 2010a; Mazzola and Manici, 2012; Tewoldemedhin et al., 2011b; Yim et al., 2013). However, the exact organisms responsible for causing ARD vary from region to region. In Washington, *Cylindrocarpon destructans*, *Phytophthora cactorum*, *Pythium* spp. and *Rhizoctonia solani* are consistently isolated from trees with ARD symptoms, and these are well-known apple pathogens (Mazzola, 1998); In South Africa, *Pythium*, *Cylindrocarpon* and *Fusarium* spp., and, to a lesser extent, *Rhizoctonia* spp. are consistently associated with replant soils (Van Schoor et al., 2009). In Central Europe (Germany, Austria and Italy), *Cylindrocarpon*-like fungi, *Ilyonectria* spp. and *Thelonectria* sp. are associated with ARD (Manici et al., 2013). In Canada, *Cylindrocarpon lucidum* and *Pythium irregulare* are associated with ARD (Braun, 1991). In China, a number of *Fusarium* species, such as *F. oxysporum*, *F. moniliforme*, *F. solani* and *F. proliferatum*, were isolated in old apple orchards in the Bohai Bay region and shown to be strongly pathogenic against the apple rootstock *M. hupehensis* Rhed. seedlings (Liu, 2013; Xu, 2011).

2.4 Rhizosphere micro-ecology

Rhizosphere, as an important interface between plant, soil and microorganism, is the most active ecological niche in soil where material and energy exchanges take place. On the one hand, ARD can lead to the accumulation of plant root exudates, which may, in turn, cause stress and autotoxicity on the root system (Han et al., 2012; Yin et al., 2016a). On the other hand, root exudates can affect the microflora composition in the rhizosphere soil (Landi et al., 2006; Li et al., 2017; Yin et al., 2017). Root exudates provide a rich carbon source and energy for microorganisms, resulting in a greater abundance of specific microorganisms and a higher diversity (Broeckling et al., 2008; Zhao et al., 2015). An increase in the amount of root exudates after continuous cropping can induce an increase in the number of specific fungal species (Broeckling et al., 2008; Zhang et al., 2000): most of the increased fungi are pathogens associated with ARD (Wang et al., 2007). Studies conducted by Yin et al. (2017) found that phloridzin at 0.5 mmol/L and 1.0 mmol/L significantly promoted the reproductive growth of *F. moniliforme* under the *in vitro* culture conditions, and the promotion effect of phlorizin at 1.0 mmol/L is greater. Phenolic acids such as phlorizin at the measured concentration in soil are also found to promote the growth of *F. moniliforme*, *F. oxysporum*, *F. solani* and *F. proliferatum*. Therefore, research on the mechanism of ARD must be based on multiple factors. If we discuss ARD only from one specific aspect such as apple plant, soil microorganisms or root exudates, it may be difficult to reflect the true causes of ARD. The relationship between various factors and ARD is shown in Fig. 1. ARD results from complex interactions between various factors, including apple roots, soil microorganisms and climatic conditions. Understanding biological and ecological interactions in the apple rhizosphere microecosystem is critical to identifying ARD cause(s) at specific sites.

3 Management of ARD

3.1 Rotation, intercropping, interplanting and mixed cropping

In many cases, apple replanting cannot be avoided due to the limited land resources, and the occurrence of ARD is thus expected (Yin et al., 2016c). However, it is feasible to adopt rotation, intercropping, interplanting or mixed cropping methods to manage ARD (Li et al., 2016; Lv et al., 2014; Mazzola and Gu, 2000; Zhang et al., 2013). Rotating wheat with red clover or field peas can increase diversity of soil bacterial community (Lupwayi et al., 1998). Use of short-term wheat cropping sequences during the old orchard renovation

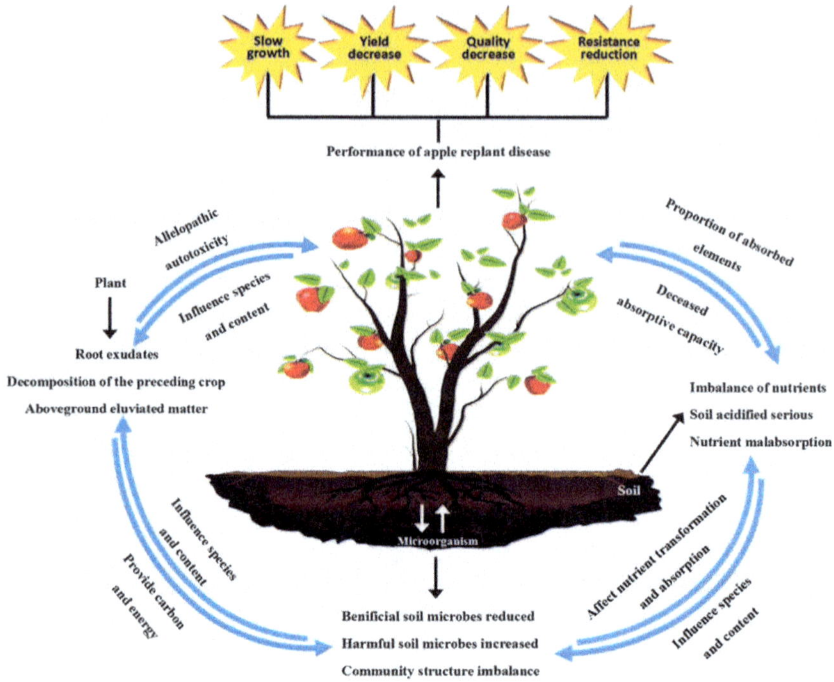

Figure 1 The relationship between factors and replant disease.

period could be useful in the management of ARD (Mazzola and Gu, 2000). Most pathogens are specific to plant species, and thus rotation with non-host plants of the pathogens can effectively reduce the incidence of soil-borne diseases (Li et al., 2016; Mazzola and Gu, 2000).

Results from our laboratory showed that rotation of apple with different crops (spring onion, alfalfa, wheat, potato) has different effects on ARD: apple-spring onion rotation is more effective in reducing ARD development (Lv et al., 2014). Good rotation is one of the best prevention measures to reduce or avoid ARD, but the disadvantage is that it takes a long time, a minimum of 3 years is needed. We have thus adopted a mixed cropping system (Fig. 2): *M. hupehensis* Rhed. seedlings mixed with spring onions leads to reduced fungal populations in the soil, particularly *F. oxysporum,* and increases bacterial populations, ultimately resulting in significant reductions in ARD (Li et al., 2016).

3.2 Deep plough and soil replacement

Deep ploughing leads to soil replacement and is an effective measure to overcome ARD. But this method is both time-consuming and labour-intensive; it is not easily adopted on a large scale. Plant residues may contain pathogens

Figure 2 Mixing the *Malus hupehensis* Rhed. seedlings with spring onions.

and autotoxic substances may be produced after decomposition of plant residues. Therefore we need to eliminate residual roots and leaves as much as possible when using this method (Sun et al., 2015). In the case of continuous apple planting, it is better to conduct deep plough and avoid previous tree stations; thus new apple trees should be replanted in the 'inter-rows' (Kelderer et al., 2012; Yin et al., 2016c), combining the soil treatment method of 'ditching before winter, air-drying freezing-thawing in the winter, backfilling in spring, with soil layer replacement' with other measures is another effective means to manage ARD (Mao and Shen, 2016).

3.3 Application of organic materials

Biochar is produced from thermal degradation of biomass under anaerobic conditions. Biochar can adsorb harmful substances in soil, inhibit bacteria and may reduce replant disease (Atucha and Litus, 2015; Jaiswal et al., 2014; Wang et al., 2014). Under the replanting condition, application of seaweed organic fertilizers to *M. hupehensis* Rhed. can promote the growth of apple seedlings, increase soil enzyme activities and alleviate ARD (Wang et al., 2016). Carbendazim combined with microbial organic fertilizers can reduce ARD (Fu et al., 2016). Applying fluid fermentation of organic material to replanted *M. hupehensis* Rhed. seedlings led to significant reductions in ARD severity (Zhang et al., 2012, 2015). However, the amount of organic material applied must be sufficient to achieve good results, usually 150–450 m^3/ha of organic materials are required (Yin et al., 2013).

3.4 Chemical and physical controls

Soil chemical fumigation is an effective measure to overcome replant diseases (Le et al., 2007; Mazzola and Brown, 2010). Chemical fumigants such as methyl bromide, metalatyl, chloropicrin and 1,3-dichloropropene are often used to sterilize soil to control soil-borne diseases (Jensen and Buszard, 1988; Le et al., 2007; Ross et al., 1983). Although this method is effective, it has many drawbacks, including high cost and pollution to the general environment. As methyl bromide has destructive effects on the atmospheric ozone layer, its large-scale agricultural use is thus limited according to the international conventions, and it has been phased out (Duniway, 2002). Treating replant soil with mylone changes microbial community, increases the ratio of bacteria and fungi and promotes the growth of *M. hupehensis* Rhed. seedlings (Liu et al., 2014). Soil fumigation with dimethyl disulphide (DMDS) can alleviate replant disease of *M. hupehensis* Rhed. seedlings (Wang et al., 2011a). Of carbendazim, thiram, fludioxonil and mancozeb, applied to control *F. oxysporum* in apple replant soil, fludioxonil has the best control effect on *F. moniliforme* in replant soil, and can reduce ARD (Liu et al., 2015). Because of the high toxicity of soil fumigants and the high cost of field application, control by chemical fumigation is gradually replaced by other control measures.

Physical measures such as solar radiation, pasteurization and anaerobic soil disinfestation can also achieve soil disinfection (Li, 1994; Shennan et al., 2014; Zhang et al., 2010). But these physical measures also have limitations. The use of solar disinfection of continuous cropping soil reduced the amount of pathogens in soil (Zhang et al., 2010), but its efficacy is greatly influenced by temperature. Steam treatment of replant apple orchard soil in 50–100°C for a period of time can promote the growth of newly planted apple trees (Yim et al., 2013), and gamma ray-sterilized soil can also reduce ARD, but both methods are too costly to apply (Manici et al., 2013).

Certain plant species can effectively inhibit pathogens associated with replant diseases (Koron et al., 2014; Mazzola and Brown, 2010; Mazzola et al., 2015; Mazzola and Zhao, 2010; Yim et al., 2016). Fumigation of these plants (biofumigation) in soil produces volatile biocidal gases that inhibit or kill pathogens in soil. *Brassica*, *Compositae* and *Allium* plants, and poultry manure are used as biofumigation materials to control soil-borne diseases and plant root-knot nematodes (Koron et al., 2014; Li et al., 2016; Mazzola et al., 2016; Yim et al., 2017). Some studies showed that adding biofumigants such as *Brassica* seed powder and white *Brassica* powder to the continuous cropping soil of apple can significantly reduce ARD (Mazzola and Brown, 2010; Mazzola et al., 2015; Mazzola and Zhao, 2010; Yim et al., 2016).

3.5 Resistant rootstocks

The response of different apple rootstocks to ARD differs significantly among different genotypes (Guo et al., 2015; Rumberger et al., 2004). Rootstock resistance is closely related to the occurrence of replant disease. Breeding rootstocks with strong resistance against the replant disease is an effective way to combat the problem. Rootstocks M7 and M26 were compared to three

Cornell-Geneva series (CG16, CG30 and CG210) in New York. Tree growth on M7, M26 and CG16 was suppressed in the old tree rows, whereas tree growth on CG30 and CG210 rootstocks was not suppressed (Rumberger et al., 2004). When studying the effects of continuous cropping on root vigour and endogenous hormonal levels of apple seedlings on different rootstocks, we found that *M. hupehensis* Rhed. has higher resistance against ARD than *M. Robusta* and *M. sieversii* (Guo et al., 2015). Under continuous cropping condition, *M. hupehensis* Rhed. has the highest tolerance to replant disease among five rootstocks studied (*M. hupehensis* Rehd., *M. prunifolia* (Willd) Borkh, *M. micromalus* Makino, *M. sieversii* (Le deb.) Roem. and *M. baccata* Borkh) (Wang et al., 2011b; Yin et al., 2016b).

3.6 Biological control

When apple orchards are replanted, some antagonistic bacteria may be applied into the soil, which have a certain control effect on soil-borne pathogens. Among the antagonistic bacteria that have been discovered so far, *Bacillus* spp., *Pseudomonas* spp. and *Agrobacterium radiobacter* are frequently reported to have control effects against plant soil-borne pathogenic fungi (Catska and Hudska, 1993; Hyakumachi et al., 2013; Mazzola et al., 2002). A strain of *B. subtilis* named Y-1 can inhibit the *Fusarium* sp. And *R. solani* in apple seedlings and reduce ARD (Ju et al., 2014). *Bacillus subtilis* TS06 and *B. thuringiensis* have some control effects against soil-borne pathogens (Hyakumachi et al., 2013; Zhang, 2012).

Inoculation of arbuscular mycorrhiza fungi (AMF) can greatly reduce ARD development (Čatská, 1994), and may also enhance plant uptake of mineral nutrients (Gastoł and Domagałaświatkiewicz, 2015). *Trichoderma harzianum* bio-organic fertilizer and mycorrhizal fungi *Glomus mosseae* can inhibit *Fusarium* wilt in tobacco (Yuan et al., 2016). An endophytic *Phomopsis liquidambari* strain B3 and a formulated *Atractylodes lancea* powder product significantly reduced peanut replant disease (Su et al., 2016). The addition of *Chaetomium globosum* ND35 fungal fertilizer and *Penicillium oxalicum* A1 strain fertilizer in apple replant soil improves soil microflora in the replant soil, leading to increased biomass of *M. hupehensis* Rhed. seedlings and reduced ARD to a varying degree (Song et al., 2015; Zhang et al., 2016).

4 Problems to be resolved

4.1 Further definition of pathogenesis

Although there are many research studies on replant disease worldwide, they mostly focus on investigating the continuous cropping-driven changes of soil microflora, qualitative and quantitative analysis of the composition of allelochemicals and autotoxins in soil or analysis of changes in the soil physico-chemical properties.

Currently, some progress has been made in understanding soil microflora from replant apple orchards, and several pathogenic microorganisms have been isolated. There are many microorganisms that could cause ARD, but the mechanism of pathogenesis is not clear.

Modern molecular biology techniques such as genomics, metabolomics, transcriptomics and proteomics need to be further applied to study the pathogenic mechanism of replant pathogens, the colonization process of apple rhizosphere and the mechanisms by which beneficial organisms reduce ARD development. Effective measures that alleviate ARD are likely to come from our improved understanding of microbial interactions in apple rhizosphere.

Phenolic acid allelochemicals proved to be important in continuous cropping systems of single crops. Whether phenolic acids are partially responsible for causing ARD is still open to debate. Issues, such as the dynamics of phenolic acids in soil, what phenolic acids may cause pathogenic proliferation and the relationships of various ARD factors with phenolic acids in causing ARD need further research.

4.2 Further integration of prevention and control measures

The occurrence of ARD involves many biological and non-biological factors such as crop, soil and environment, as well as interaction among these factors. Given this complexity, it is unlikely that a single measure can effectively manage ARD in practice. Different management strategies should be combined to manage ARD. At the same time, we also should consider management methods in terms of both control efficacy and potential side effects to the general environment.

ARD management strategies include production of various biological regulators and soil additives based on the principle of allelopathy between organisms, breeding of resistant varieties and stocks, alternative soil disinfection and cultural practices. With national funding, the Apple Replant Disease Prevention and Control Technology Research Group of Shandong Agricultural University has already made significant progress in developing integrated management strategies against ARD in China (Mao and Shen, 2016). However, it should be stressed that the exact implementation of ARD management strategies depends on many factors and is expected to differ among geographic regions.

5 Future trends and conclusion

With many factors that affect ARD, further research is needed to understand ARD complex and develop management strategies. Further research needs to focus on the following areas: (1) breeding of apple rootstocks that possess effective resistance against ARD: this process can be effectively implemented with the aid of molecular tools. (2) Isolation, identification and application of antagonistic bacteria and fungi: in particular, we need to pay attention to consortia of microbial organisms for effective ARD management. And, (3) use of biofumigation: developing environment-friendly soil disinfection technology in combination with other farming management systems and leveraging the diversity of plants.

6 Acknowledgements

The research was supported by the earmarked fund for National Modern Agro-industry Technology Research System of China (CARS-28), the Fruit innovation team in Shandong Province, China (SDAIT-06-07) and the National Natural Science Foundation of China (31672104).

7 Where to look for further information

A good introduction to the subject of replant disease is B. M. Savory, *Specific Replant Diseases*, which can help someone new to the subject to get started (Savory, 1966).

There is a lot of research on the subject, you can keep up with the reports from Mark Mazzola and Zhiquan Mao (most papers were written in Chinese), especially for ARD.

You can also get more information from the journals (e.g. *Phytopathology*, *Soil Biology and Biochemistry*, *Plant and Soil*, *Plant Disease*, *European Journal of Plant Pathology*, *Scientia Horticulturae* etc.).

8 References

Asao, T., Hasegawa, K., Sueda, Y., Tomita, K., Taniguchi, K., Hosoki, T., Pramanik, M. H. R. and Matsui, Y. 2003. Autotoxicity of root exudates from taro. *Scientia Horticulturae* 97 (3–4), 389–96. doi:10.1016/S0304-4238(02)00197-8.

Atucha, A. and Litus, G. 2015. Effect of biochar amendments on peach replant disease. *HortScience* 50 (6), 863–8. doi:10.21273/HORTSCI.50.6.863.

Bai, R., Ma, F., Liang, D. and Zhao, X. 2009. Phthalic acid induces oxidative stress and alters the activity of some antioxidant enzymes in roots of *Malus prunifolia*. *Journal of Chemical Ecology* 35 (4), 488–94. doi:10.1007/s10886-009-9615-7.

Benizri, E., Piutti, S., Verger, S., Pagès, L., Vercambre, G., Poessel, J. L. and Michelot, P. 2005. Replant diseases: bacterial community structure and diversity in peach rhizosphere as determined by metabolic and genetic fingerprinting. *Soil Biology and Biochemistry* 37 (9), 1738–46. doi:10.1016/j.soilbio.2005.02.009.

Blok, W. J. and Bollen, G. J. 1993. The role of autotoxins from root residues of the previous crop in the replant disease of asparagus. *Netherlands Journal of Plant Pathology* 99 (S3), 29–40. doi:10.1007/BF03041393.

Börner, H. 1959. The apple replant problem. I. The excretion of phlorizin from apple root residues. *Contributions of the Boyce Thompson Institute* 20, 39–56.

Braun, P. G. 1991. The combination of *Cylindrocarpon lucidum* and *Pythium irregulare* as a possible cause of apple replant disease in Nova Scotia. *Canadian Journal of Plant Pathology* 13 (4), 291–7. doi:10.1080/07060669109500914.

Broeckling, C. D., Broz, A. K., Bergelson, J., Manter, D. K. and Vivanco, J. M. 2008. Root exudates regulate soil fungal community composition and diversity. *Applied and Environmental Microbiology* 74 (3), 738–44. doi:10.1128/AEM.02188-07.

Brown, G. S. and Koutoulis, L. J. 2008. Overcoming apple replant disease – treatment effects over the first 7 years of orchard life. *Acta Horticulturae* 772, 121–5. doi:10.17660/ActaHortic.2008.772.14.

Bryant, R. D., Gordy, E. A. and Laishley, E. J. 1979. Effect of soil acidification on the soil microflora. *Water, Air and Soil Pollution* 11 (4), 437–45. doi:10.1007/BF00283435.

Čatská, V. 1994. Interrelationships between vesicular-arbuscular mycorrhiza and rhizosphere microflora in apple replant disease. *Biologia Plantarum* 36 (1), 99–104. doi:10.1007/BF02921276.

Catska, V. and Hudska, G. 1993. Use of *Agrobacterium radiobacter* for biological control of apple replant disease. *Acta Horticulturae* 324, 67–72. doi:10.17660/ActaHortic.1993.324.7.

Chen, X., Han, M., Su, G., Liu, F., Guo, G., Jiang, Y., Mao, Z., Peng, F. and Shu, H. 2010. Discussion on today's world apple industry trends and the suggestions on sustainable and efficient development of apple industry in China. *Journal of Fruit Science* 27, 598–604.

Chen, X., Mao, Z., Jiang, Y. and Wang, Z. 2015. Apple industry transformation and upgrading of the proposal of Shandong Province- a policy, two by science and technology, innovation driven, transformation and upgrading, quality and efficiency. *Deciduous Fruits* 47, 2–4.

Duniway, J. M. 2002. Status of chemical alternatives to methyl bromide for pre-plant fumigation of soil. *Phytopathology* 92(12), 1337–43. doi:10.1094/PHYTO.2002.92.12.1337.

Finlay, R. D. 1995. Interactions between soil acidification, plant growth and nutrient uptake in ectomycorrhizal associations of forest trees. *Ecological Bulletins* 44, 197–214.

Franke-Whittle, I. H., Manici, L. M., Insam, H. and Stres, B. 2015. Rhizosphere bacteria and fungi associated with plant growth in soils of three replanted apple orchards. *Plant and Soil* 395 (1–2), 317–33. doi:10.1007/s11104-015-2562-x.

Fu, F., Xiang, L., Xu, S., Liu, X., Shen, X., Chen, X., Yin, C. and Mao, Z. 2016. Effects of carbendazim and bio-organic fertilizer on *Malus hupehensis* seedlings and soil under replant condition. *Acta Horticulturae Sinica* 43, 1452–62.

Gao, X., Hu, Y., Fengxia, Z., Zhiquan, M., Shen, X., Yang, S. and Su, L. 2009a. Effects of benzoic acid application on membrane lipid peroxidation in roots of *Malus hupehensis* seedlings. *Scientia Silvae Sinica* 45, 129–34.

Gao, X., Zhao, F., Shen, X., Hu, Y., Hao, Y., Yang, S., Su, L. and Mao, Z. 2009b. Effects of cinnamon acid on respiratory rate and its related enzymes activity in roots of seedlings of *Malus hupehensis* Rehd. *Scientia Agricultura Sinica* 42, 4308–14.

Gastoł, M. and Domagałaświatkiewicz, I. 2015. Mycorrhizal inoculation of apple in replant soils – enhanced tree growth and mineral nutrient status. *Acta Scientiarum Polonorum. Hortorum Cultus = Ogrodnictwo* 14, 17–37.

Guenzi, W. D. and McCalla, T. M. 1962. Inhibition of germination and seedling development by crop residues. *Soil Science Society of America Journal* 26 (5), 456–8. doi:10.2136/sssaj1962.03615995002600050015x.

Guo, X. J., Wang, R., Zhang, Z., Ge, R., Mao, Z. Q., Guo, L. and Shen, X. 2015. The influence of replant stress on hormone and absorption of root for different apple rootstocks. *Advanced Materials Research* 1120–1121, 926–31. doi:10.4028/www.scientific.net/AMR.1120-1121.926.

Han, C. M., Li, C. L., Ye, S. P., Wang, H., Pan, K. W., Wu, N., Wang, Y. J., Li, W. and Zhang, L. 2012. Autotoxic effects of aqueous extracts of ginger on growth of ginger seedlings and on antioxidant enzymes, membrane permeability and lipid peroxidation in leaves. *Allelopathy Journal* 30, 259–70.

Hyakumachi, M., Nishimura, M., Arakawa, T., Asano, S., Yoshida, S., Tsushima, S. and Takahashi, H. 2013. Bacillus thuringiensis suppresses bacterial wilt disease caused by *Ralstonia solanacearum* with systemic induction of defense-related gene expression in tomato. *Microbes and Environments* 28 (1), 128–34. doi:10.1264/jsme2.ME12162.

Jaiswal, A. K., Elad, Y., Graber, E. R. and Frenkel, O. 2014. *Rhizoctonia solani* suppression and plant growth promotion in cucumber as affected by biochar pyrolysis temperature, feedstock and concentration. *Soil Biology and Biochemistry* 69, 110–8. doi:10.1016/j.soilbio.2013.10.051.

Jensen, P. and Buszard, D. 1988. The effects of chemical fumigants, nitrogen, plastic mulch and metalaxyl on the establishment of young apple trees in apple replant disease soil. *Canadian Journal of Plant Science = Revue Canadienne De Phytotechnie* 68, 255–60. doi:10.4141/cjps88-031.

Ju, R., Zhao, Y., Li, J., Jiang, H., Liu, P., Yang, T., Bao, Z., Zhou, B., Zhou, X. and Liu, X. 2014. Identification and evaluation of a potential biocontrol agent, *Bacillus subtilis*, against *Fusarium* sp. in apple seedlings. *Annals of Microbiology* 64(1), 377–83. doi:10.1007/s13213-013-0672-3.

Kelderer, M., Manici, L. M., Caputo, F. and Thalheimer, M. 2012. Planting in the 'inter-row' to overcome replant disease in apple orchards: a study on the effectiveness of the practice based on microbial indicators. *Plant and Soil* 357 (1–2), 381–93. doi:10.1007/s11104-012-1172-0.

Koron, D., Sonjak, S. and Regvar, M. 2014. Effects of non-chemical soil fumigant treatments on root colonisation with arbuscular mycorrhizal fungi and strawberry fruit production. *Crop Protection* 55, 35–41. doi:10.1016/j.cropro.2013.09.009.

Landi, L., Valori, F., Ascher, J., Renella, G., Falchini, L. and Nannipieri, P. 2006. Root exudate effects on the bacterial communities, CO_2 evolution, nitrogen transformations and ATP content of rhizosphere and bulk soils. *Soil Biology and Biochemistry* 38 (3), 509–16. doi:10.1016/j.soilbio.2005.05.021.

Le, R. H., Ware, A. B. and Pretorius, M. C. 2007. Comparative efficacy of preplant fumigation and postplant chemical treatment of replant citrus trees in an orchard infested with *Tylenchulus semipenetrans*. *Plant Disease* 82, 1323–7.

Li, T. S. C. 1994. Evaluation of chemical and non-chemical treatments for the control of ginseng replant disease. *Acta Horticulturae* 11(363), 141–6. doi:10.17660/ActaHortic.1994.363.19.

Li, C., Li, X., Kong, W., Wu, Y. and Wang, J. 2010a. Effect of monoculture soybean on soil microbial community in the Northeast China. *Plant and Soil* 330 (1–2), 423–33. doi:10.1007/s11104-009-0216-6.

Li, Z. H., Wang, Q., Ruan, X., Pan, C. D. and Jiang, D. A. 2010b. Phenolics and plant allelopathy. *Molecules* 15 (12), 8933–52. doi:10.3390/molecules15128933.

Li, J., Wang, G., Hu, Y., Chen, X. and Mao, Z. 2014. Effect of organic acids on biomass and physiological characteristics of *Malus hupehensis* seedlings under continuous cropping. *Acta Ecologica Sinica* 41, 2489–96.

Li, J., Xiang, L., Pan, F., Chen, X., She, n. X., Yin, C. and Mao, Z. 2016. Effects of *Malus hupehensis* seedlings and *Allium fistulosum* mixed cropping on replanted soil environment. *Acta Ecologica Sinica* 43, 1853–62.

Li, S., Xu, C., Wang, J., Guo, B., Yang, L., Chen, J. and Ding, W. 2017. Cinnamic, myristic and fumaric acids in tobacco root exudates induce the infection of plants by *Ralstonia solanacearum*. *Plant and Soil* 412(1–2), 381–95. doi:10.1007/s11104-016-3060-5.

Liang, L. Z., Zhao, X. Q., Yi, X. Y., Chen, Z. C., Dong, X. Y., Chen, R. F. and Shen, R. F. 2013. Excessive application of nitrogen and phosphorus fertilizers induces soil acidification and phosphorus enrichment during vegetable production in Yangtze River Delta, China. *Soil Use and Management* 29 (2), 161–8. doi:10.1111/sum.12035.

Liu, Z. 2013. The isolation and identification of fungi pathogen from apple replant disease in Bohai Bay region and the screening of the antagonistic *Trichoderma* strains. Master's Thesis. Shandong Agricultural University, China. Available at: http://www.dissertationtopic.net/doc/2147432.

Liu, P., Zhao, H.-J., Zhong, Z., Sun, M., Pang, Y., Ma, Z. and Wan, S. 2013. The effects of three root exudated fatty acids on peanut (*Arachis hypogaea* L.) growth and soil enzymes activities. *Acta Ecologica Sinica* 33 (11), 3332–9. doi:10.5846/stxb201203210382.

Liu, E., Wang, G., Li, Y., Shen, X., Chen, X., Song, F., Wu, S., Chen, Q. and Mao, Z. 2014. Replanting affects the tree growth and fruit quality of gala apple. *Journal of Integrative Agriculture* 13 (8), 1699–706. doi:10.1016/S2095-3119(13)60620-6.

Liu, Y., Bao, J., Jiang, X., Li, P. and Mao, Z. 2015. Effects of four fungicides on soil *Fusarium oxysporum* population and enzymes activity in successive cropping apple orchard. *Journal of Fruit Science* 32, 115–22.

Lupwayi, N. Z., Rice, W. A. and Clayton, G. W. 1998. Soil microbial diversity and community structure under wheat as influenced by tillage and crop rotation. *Soil Biology and Biochemistry* 30 (13), 1733–41. doi:10.1016/S0038-0717(98)00025-X.

Lv, W., Yu, T., Zhu, H., Shen, Q. and Zhang, C. 2006. Effects of cucumber continuous cropping on the soil physico-chemical characters and biological activities. *Chinese Journal of Eco-Agriculture* 14, 119–21.

Lv, Y., Song, F., Li, Y., Shen, X., Chen, X., Wu, S. and Mao, Z. 2014. The influence of different crops rotation on the environment of soil and physiological characteristics of *Malus hupehensis* Rehd. seedlings. *Scientia Agricultura Sinica* 47, 2830–9.

Manici, L. M., Kelderer, M., Franke-Whittle, I. H., Rühmer, T., Baab, G., Nicoletti, F., Caputo, F., Topp, A., Insam, H. and Naef, A. 2013. Relationship between root-endophytic microbial communities and replant disease in specialized apple growing areas in Europe. *Applied Soil Ecology* 72, 207–14. doi:10.1016/j.apsoil.2013.07.011.

Manrique, L. A. 1985. Effect of extreme soil acidity conditions on plant growth and yield of cassava. *Communications in Soil Science and Plant Analysis* 16 (9), 959–70. doi:10.1080/00103628509367658.

Mao, Z.-q. and Shen, X. 2016. Apple replant (cropping) disorder prevention and control technology. *Yantai Fruits* 4, 26–7.

Mazzola, M. 1998. Elucidation of the microbial complex having a causal role in the development of apple replant disease in Washington. *Phytopathology* 88 (9), 930–8. doi:10.1094/PHYTO.1998.88.9.930.

Mazzola, M. and Brown, J. 2010. Efficacy of brassicaceous seed meal formulations for the control of apple replant disease in conventional and organic production systems. *Plant Disease* 94 (7), 835–42. doi:10.1094/PDIS-94-7-0835.

Mazzola, M. and Gu, Y. H. 2000. Impact of wheat cultivation on microbial communities from replant soils and apple growth in greenhouse trials. *Phytopathology* 90 (2), 114–9. doi:10.1094/PHYTO.2000.90.2.114.

Mazzola, M. and Manici, L. M. 2012. Apple replant disease: role of microbial ecology in cause and control. *Annual Review of Phytopathology* 50, 45–65. doi:10.1146/annurev-phyto-081211-173005.

Mazzola, M. and Zhao, X. 2010. *Brassica juncea* seed meal particle size influences chemistry but not soil biology-based suppression of individual agents inciting apple replant disease. *Plant and Soil* 337 (1–2), 313–24. doi:10.1007/s11104-010-0529-5.

Mazzola, M., Granatstein, D. M., Elfving, D. C., Mullinix, K. and Gu, Y. H. 2002. Cultural management of microbial community structure to enhance growth of apple in replant soils. *Phytopathology* 92, 1363–6. doi:10.1094/PHYTO.2002.92.12.1363.

Mazzola, M., Hewavitharana, S. S. and Strauss, S. L. 2015. *Brassica* seed meal soil amendments transform the rhizosphere microbiome and improve apple production through resistance to pathogen re-infestation. *Phytopathology* 105 (4), 460–9. doi:10.1094/PHYTO-09-14-0247-R.

Mazzola, M., Hewavitharana, S. S., Strauss, S. L., Shennan, C. and Muramoto, J. 2016. Anaerobic soil disinfestation and *Brassica* seed meal amendment alter soil microbiology and system resistance. *International Journal of Fruit Science* 16 (suppl. 1), 47–58. doi:10.1080/15538362.2016.1195310.

Neilsen, G. H. and Yorston, J. 1991. Soil disinfection and monoammonium phosphate fertilization increase precocity of apples on replant problem soils. *Journal of the American Society for Horticultural Science* 116 (4), 651–4. doi:10.21273/JASHS.116.4.651.

Nicola, L., Vrhovsek, U., Soini, E., Insam, H. and Pertot, I. 2017. Phlorizin released by apple root debris is related to apple replant disease. *Phytopathologia Mediterranea* 55, 432–42.

Parchomchuk, P., Neilsen, G. H. and Hogue, E. J. 1994. Apple tree growth response to N fertigation of replant soil: comparison of greenhouse pot tests and field results. *Acta Horticulturae* 18 (363), 65–74. doi:10.17660/ActaHortic.1994.363.9.

Ross, R. G., Delbridge, R. W., Kimpinski, J. and Mcrae, K. B. 1983. Control of apple replant disease in Nova Scotia by soil fumigation with vorlex and chloropicrin. *Canadian Journal of Plant Pathology* 5 (3), 177–80. doi:10.1080/07060668309501622.

Rumberger, A., Yao, S., Merwin, I. A., Nelson, E. B. and Thies, J. E. 2004. Rootstock genotype and orchard replant position rather than soil fumigation or compost amendment determine tree growth and rhizosphere bacterial community composition in an apple replant soil. *Plant and Soil* 264 (1/2), 247–60. doi:10.1023/B:PLSO.0000047760.13004.94.

Rumberger, A., Merwin, I. A. and Thies, J. E. 2007. Microbial community development in the rhizosphere of apple trees at a replant disease site. *Soil Biology and Biochemistry* 39 (7), 1645–54. doi:10.1016/j.soilbio.2007.01.023.

Savory, B. M. 1966. *Specific Replant Diseases*. Commonwealth Agricultural Bureaux, England.

Shennan, C., Muramoto, J., Lamers, J., Mazzola, M., Rosskopf, E. N., Kokalisburelle, N., Momma, N., Butler, D. M. and Kobara, Y. 2014. Anaerobic soil disinfestation for soil borne disease control in strawberry and vegetable systems: current knowledge and future directions. *Acta Horticulturae* 1044 (1044), 165–75. doi:10.17660/ActaHortic.2014.1044.20.

Singh, H. P., Batish, D. R. and Kohli, R. K. 1999. Autotoxicity: concept, organisms, and ecological significance. *Critical Reviews in Plant Sciences* 18 (6), 757–72. doi:10.1080/07352689991309478.

Song, F., Wang, S., Zhang, X., Gao, K., Yin, C., Chen, X. and Mao, Z. 2015. Effects of *Chaetomium globosum* ND35 fungal fertilizer on continuous cropping soil microorganism and *Malus hupehensis* seedling biomass. *Acta Horticulturae Sinica* 42, 205–13.

Spath, M., Insam, H., Peintner, U., Kelderer, M., Kuhnert, R. and Franke-Whittle, I. H. 2015. Linking soil biotic and abiotic factors to apple replant disease: a greenhouse approach. *Journal of Phytopathology* 163 (4), 287–99. doi:10.1111/jph.12318.

Su, C., Wang, H., Xie, X., Zhang, W., Li, X., Wang, X. and Dai, C. 2016. Effects of endophytic fungi and *Atractylodes lancea* powder on rhizosphere microflora and trace elements during continuous peanut cropping. *Acta Ecologica Sinica* 36, 2052–65.

Sun, B., Wang, Y., Zhang, X., Shen, X., Chen, X. and Mao, Z. 2015. Effect of residual roots on the biomass of *Malus hupehensis* seedlings, phenolic acids and microbiology in the soil of continuous cropping. *Acta Horticulturae Sinica* 42, 131–9.

Sun, J., Zhang, Q., Li, X., Zhou, B. and Wei, Q. 2017. Apple replant disorder of Pingyitiancha rootstock is closely associated with rhizosphere fungal community development. *Journal of Phytopathology* 165 (3), 162–73. doi:10.1111/jph.12547.

Tewoldemedhin, Y. T., Mazzola, M., Botha, W. J., Spies, C. F. J. and McLeod, A. 2011a. Characterization of fungi (*Fusarium and Rhizoctonia*) and oomycetes (*Phytophthora and Pythium*) associated with apple orchards in South Africa. *European Journal of Plant Pathology* 130 (2), 215–29. doi:10.1007/s10658-011-9747-9.

Tewoldemedhin, Y. T., Mazzola, M., Labuschagne, I. and McLeod, A. 2011b. A multi-phasic approach reveals that apple replant disease is caused by multiple biological agents, with some agents acting synergistically. *Soil Biology and Biochemistry* 43 (9), 1917–27. doi:10.1016/j.soilbio.2011.05.014.

Utkhede, R. S. 2006. Soil sickness, replant problem or replant disease and its integrated control. *Allelopathy Journal* 18, 23–38.

Utkhede, R. S. and Smith, E. M. 2000. Impact of chemical, biological and cultural treatments on the growth and yield of apple in replant-disease soil. *Australasian Plant Pathology* 29 (2), 129–36. doi:10.1071/AP00021.

Utkhede, R. S., Vrain, T. C. and Yorston, J. M. 1992. Effects of nematodes, fungi and bacteria on the growth of young apple trees grown in apple replant disease soil. *Plant and Soil* 139 (1), 1–6. doi:10.1007/BF00012835.

Van Schoor, L., Denman, S. and Cook, N. C. 2009. Characterisation of apple replant disease under South African conditions and potential biological management strategies. *Scientia Horticulturae* 119 (2), 153–62. doi:10.1016/j.scienta.2008.07.032.

Wang, S., Han, X. and Qiao, Y. 2007. Allelopathy of root exudates and their effects on rhizosphere microorganism. *Chinese Journal of Soil Science* 38, 1219–26.

Wang, J., Liu, Q., Li, J., Jin, S. and Yuan, Y. 2010. Analysis on the characteristic and cause of orchard soil acidification in the area of Shandong Peninsula. *Chinese Agricultural Science Bulletin* 26, 164–9.

Wang, F., Wang, Q., Yan, D., Mao, L., Guo, M., Yan, P. and Cao, A. 2011a. Effects of dimethyl disulfide on microbial communities in protectorate soils under continuous cropping. *Chinese Journal of Eco-Agriculture* 19 (4), 890–6. doi:10.3724/SP.J.1011.2011.00890.

Wang, Y., Yin, C., Chen, Q., Shen, X., Jiang, Z. and Mao, Z. 2011b. Study on the difference of adaptability to replant soil in five apple rootstock seedlings. *Acta Horticulturae Sinica* 38, 1955–62.

Wang, Q., Hu, Y., Zhou, H., Zhan, X., Mao, Z. and Zhu, S. 2012. Effects of phloridzin on the tricarboxylic acid cycle enzymes of roots of Malus hupehensis Rehd. *Scientia Agricultura Sinica* 45, 3108–14.

Wang, Y., Pan, F., Wang, G., Zhang, G., Wang, Y., Chen, X. and Mao, Z. 2014. Effects of biochar on photosynthesis and antioxidative system of *Malus hupehensis* Rehd. seedlings under replant conditions. *Scientia Horticulturae* 175, 9–15. doi:10.1016/j.scienta.2014.05.029.

Wang, Y., Pan, F.-b., Xing, Z., Wang, G., Zhang, G., Yanli, H., Chen, X. and Zhiquan, M. 2015. Effects of five kinds of phenolic acid on the function of mitochondria and antioxidant systems in roots of Malus hupehensis Rehd. *Acta Ecologica Sinica* 35, 6566–73.

Wang, Y., Fu, F., Li, J., Wang, G., Wu, M., Zhan, J., Chen, X. and Mao, Z. 2016. Effects of seaweed fertilizer on the growth of *Malus hupehensis* Rehd. seedlings, soil enzyme activities and fungal communities under replant condition. *European Journal of Soil Biology* 75, 1–7. doi:10.1016/j.ejsobi.2016.04.003.

Wilson, S., Andrews, P. and Nair, T. S. 2004. Non-fumigant management of apple replant disease. *Scientia Horticulturae* 102 (2), 221–31. doi:10.1016/j.scienta.2004.01.001.

Wu, F., Zhao, F. and Gu, S. 2002. Effect of the continuous cultivating cucumber on the bio-chemical properties of soil in the plastic greenhouse. *System Sciences and Comprehensive Studies in Agriculture* 18, 20–2.

Xu, W. 2011. Diversity analysis of soil fungi from bohai bay apple replanted orchard and the screening of the antagonistic fungi. Master's Thesis. Shandong Agricultural University, China. Available at: http://www.dissertationtopic.net/doc/410696.

Yim, B., Smalla, K. and Winkelmann, T. 2013. Evaluation of apple replant problems based on different soil disinfection treatments – links to soil microbial community structure? *Plant and Soil* 366 (1–2), 617–31. doi:10.1007/s11104-012-1454-6.

Yim, B., Hanschen, F. S., Wrede, A., Nitt, H., Schreiner, M., Smalla, K. and Winkelmann, T. 2016. Effects of biofumigation using *Brassica juncea* and *Raphanus sativus* in comparison to disinfection using Basamid on apple plant growth and soil microbial communities at three field sites with replant disease. *Plant and Soil* 406 (1-2), 389-408. doi:10.1007/s11104-016-2876-3.

Yim, B., Nitt, H., Wrede, A., Jacquiod, S., Sørensen, S. J., Winkelmann, T. and Smalla, K. 2017. Effects of soil pre-treatment with Basamid® granules, *Brassica juncea*, *Raphanus sativus*, and *Tagetes patula* on bacterial and fungal communities at two apple replant disease sites. *Frontiers in Microbiology* 8, 1604. doi:10.3389/fmicb.2017.01604.

Yin, C., Chen, X., Shen, X., Zhang, Z., Sun, H. and Mao, Z. 2013. Effect of different concentration of organic matter fermentation fluid on the young apple tree biomass and soil properties under replant conditions. *Journal of Plant Nutrition and Fertilizer* 19, 1450–8.

Yin, C., Hu, Y., GongshuaiWand, Zhang, X., Zhou, H., Shen, X., Chen, X. and Mao, Z. 2016a. Effect of main phenolic acids of the apple replanted soil on the roots of *Malus hupehensis* Rehd. seedlings. *Scientia Horticulturae* 49, 961–9.

Yin, C., Xiang, L., Sun, C., Shen, X., Chen, X., Zhou, H. and Mao, Z. 2016b. Effects of different apple rootstocks on the soil microbial quantity and enzyme activity of apple replanted orchard soil. *Acta Horticulturae Sinica* 43, 2423–30.

Yin, C., Xiang, L., Wang, G., Wang, Y., Shen, X., Chen, X. and Mao, Z. 2016c. How to plant apple trees to reduce replant disease in apple orchard: a study on the phenolic acid of the replanted apple orchard. *PLoS One* 11(12), 1–17. doi:10.1371/journal.pone.0167347.

Yin, C., Xiang, L., Wang, G., Wang, Y., Shen, X., Chen, X. and Mao, Z. 2017. Phloridzin promotes the growth of *Fusarium moniliforme* (*Fusarium verticillioides*). *Scientia Horticulturae* 214, 187–94. doi:10.1016/j.scienta.2016.11.035.

Yu, Z., Zhang, Z., Wang, P. and Cong, J. 2010. Analysis on the present situation and cause of orchard soil acidification in the area of Jiaodong. *Yantai Fruits* 2, 31–2.

Yuan, S., Li, M., Fang, Z., Yan, L., Wen, S., Bing, P., Kai, W., Shi, J., Shen, B. and Shen, Q. 2016. Biological control of tobacco bacterial wilt using *Trichoderma harzianum* amended bioorganic fertilizer and the arbuscular mycorrhizal fungi *Glomus mosseae*. *Biological Control* 92, 164–71. doi:10.1016/j.biocontrol.2015.10.013.

Zhang, Y. 2012. Identification and characterization of a *Bacillus subtilis* strain TS06 as bio-control agent of strawberry replant disease (*Fusarium* and *Verticilium* wilts). *African Journal of Biotechnology* 11, 570–80.

Zhang, S., Gao, Z. and Liu, H. 2000. Continuous cropping obstacle and rhizospheric microecology III. Soil phenolic acids and their biological effect. *Chinese Journal of Applied Ecology* 11, 741–4.

Zhang, J., Zhang, D., Mao, Z., Li, Z. and Zhao., M. 2009. Isolation and identification of root exudates in different kinds of apple rootstock seedlings. *Journal of Agricultural University of Hebei* 32, 29–32.

Zhang, L., Li, H., Zhai, S., Zhang, X. and Xie, X. 2010. Effects of solar soil disinfection technology on strawberry protected planting. *Northern Horticulture* 14, 67–8.

Zhang, Z., Mao, Z. and Zhu, S. 2011. Effects of phenolic acids on mitochondria and the activity of antioxidant enzymes in roots of seedlings of *Malus hupehensis* Rehd. *Scientia Agricultura Sinica* 44, 3177–84.

Zhang, Z., Chen, Q., Yin, C., Shen, X., Chen, X., Sun, H., Mao, A. and Mao, Z. 2012. The effects of organic matter on the physiological features of *Malus hupehensis* seedlings and soil properties under replant conditions. *Scientia Horticulturae* 146, 52–8. doi:10.1016/j.scienta.2012.08.011.

Zhang, H., Mallik, A. and Zeng, R. S. 2013. Control of panama disease of banana by rotating and intercropping with Chinese chive (*Allium tuberosum* Rottler): role of plant volatiles. *Journal of Chemical Ecology* 39 (2), 243–52. doi:10.1007/s10886-013-0243-x.

Zhang, G., Zhan, X., Li, Y., Shen, X., Chen, X., Wu, S. and Mao, Z. 2015. Effects of anaerobic and aerobic fermentation of organic materials on continuous cropping soil environment and *Malus hupehensis* seedling biomass. *Acta Horticulturae Sinica* 35, 3663–73.

Zhang, X., Xiang, L., Wang, Y., Wang, G., Liu, H., Sun, G., Shen, X., Chen, X., Zhou, H., Yin, C., et al. 2016. Identification of *Penicillium oxalicum* A1 strain and antagonistic effects on four species of *Fusarium* pathogen of apple. *Acta Horticulturae Sinica* 43, 841–52.

Zhao, X., Zhen, W., Qi, Y., Liu, X. and Yin, B. 2009. Coordinated effects of root autotoxic substances and *Fusarium oxysporum* Schl. f. sp. *fragariae* on the growth and replant disease of strawberry. *Frontiers of Agriculture in China* 3 (1), 34–9. doi:10.1007/s11703-009-0006-1.

Zhao, Y., Wu, L., Chu, L., Yang, Y., Li, Z., Azeem, S., Zhang, Z., Fang, C. and Lin, W. 2015. Interaction of *Pseudostellaria heterophylla* with *Fusarium oxysporum* f.sp. *heterophylla* mediated by its root exudates in a consecutive monoculture system. *Scientific Reports* 5 (1). doi:10.1038/srep08197.

Zhen, W., Cao, K., Dai, L. and Zhang, X. 2004. Simulation of autotoxicity of strawberry root exudates under continuous cropping. *Acta Phytoecol. Sinica* 28, 828–32.

Chapter 4

Fungal diseases of fruit: apple cankers in Europe

Robert Saville and Leone Olivieri, NIAB EMR, UK

1 Introduction

The cultivated apple, *Malus* x *domestica* (syn. *Malus pumila*), is prone to a plethora of fungal diseases. In this chapter, we will explore those that cause canker diseases of apple in Europe with particular focus on European apple canker caused by *Neonectria ditissima*.

Canker is a general term used to describe a disease symptom of the wood defined by necrotic often sunken lesions in the bark, which can eventually lead to the death of the branch. Cankers can be caused by several plant pathogens and pests; the main fungal pathogens causing cankers on apple in Europe are *Neonectria*, *Neofabraea*, *Monilinia* and *Chondrostereum*. The epidemiology of canker pathogens varies. *Neonectria*, *Neofabraea* (*perennans*) and *Chondrostereum* enter through wounds with leaf scar wounds in autumn and pruning cuts being major points of entry; while *Monilinia* cankers result from the infection in the blossom or fruit progressing down through the spur into branches. Table 1 provides a brief summary of each of the cankers caused by the main fungal pathogens of apple in Europe.

Canker diseases cause both direct and indirect economic losses with significance dependant on the disease-causing agent. The cankers themselves lead to direct losses due to the loss of fruiting wood resulting from branch death, the extra cost of pruning out cankered branches, and in severe cases

http://dx.doi.org/10.19103/AS.2019.0046.04

Table 1 The main fungal cankers of apple occurring in Europe. A brief summary of symptoms and epidemiology is presented together with areas of further reading

Disease	Pathogen	Symptoms	Disease cycle and epidemiology	Further reading
Blossom wilt	*Monilinia laxa f.sp. mali*	Brown cracked and distinct cankers forming watermarking (alternating dark and light brown striping), which occur on the fruiting spur, at the base of fruiting spurs and on branches. Forms grey pustules during the spring	This fungus overwinters in infected cankers, dead blossoms on branches or as mummified fruit. Conidia produced on diseased tissues in early spring are disseminated by wind and rain and infect new blossoms. Fungal hyphae then spread from newly infected blossoms to twigs/branches forming new cankers. The disease is favoured by cool moist weather.	Byrde and Willets (1977)
Brown rot	*Monilinia fructigena*	Brown cracked and distinct cankers which occur on fruiting spurs, at the base of fruiting spurs and on branches. Forms buff pustules in the summer	This fungus overwinters as spur cankers or as mummified fruit on the tree or orchard floor and these sporulate in early summer after rain. The spores are spread by wind to infect young fruitlets through wounds. Hyphae can spread to twigs and branches from infected fruit to develop new cankers.	Byrde and Willets (1977)
European apple canker	*Neonectria ditissima*	Cankers appear initially as sunken lesions sometimes forming a papery bark. Cankers become flaky and surrounded by swollen tissue as they mature. Occur on all wood tissue where wounds occur. Fruiting bodies occur in two forms: white, creamy pustules in the summer and red pin-head sized fruiting bodies in the autumn/winter	This fungus overwinters as cankers on the tree. The infection route of this fungus is through wounds. The fungus is capable of producing spores all year round and suitable entry points for infection are available on the host year round meaning the only limiting factor is rain.	Swinburne (1975) Weber (2014)

Perennial canker	*Neofabraea perennans*	Cankers appear as elliptical, sunken and orange, purple or brown lesions in the first year. Cankers continue to expand year on year marked by host produced layers of callus tissue	This fungus infects through wounds associated with pruning, frost damage and wounds resulting from woolly apple aphid (*Eriosoma lanigerum*) infestations.	Webster et al. (2001) Saville (2015)
Silver leaf	*Chondrostereum purpureum*	Blistered and papery barks associated with a pruning wound. Other symptoms include silvered foliage and internal discolouring of affected wood	Fruiting bodies (bracket fungi with smooth purple lower surface and pale brown upper surface) are produced on dead branches in autumn. During wet weather, spores are released from the fruiting bodies from autumn to the following June. Spores infect trees through wounds (generally large pruning wounds).	Webster et al. (2001)

tree death. Indirect losses due to canker pathogens can arise as a result of the pathogens infecting other plant parts during the growing season causing diseases such as blossom blight and fruit rots.

N. ditissima is the most-important canker pathogen of apple in Europe and as such the remainder of this chapter will focus on European apple canker. We will discuss *N. ditissima* in the context of the tree production cycle, introducing the aetiology and epidemiology of the pathogen, reviewing the key control measures, and identifying a number of future research areas. Although largely focused on *N. ditissima*, the most widely spread and devastating of the canker-causing diseases in Europe, certain aspects are applicable across all canker diseases (see Table 1 for further, more specific, reading on each of the canker pathogens). With changes in the plant protection products available and in the methods of tree production and varietal susceptibility, finding solutions for canker management remains an important research target.

2 Apple growing

Infection risk from *N. ditissima* is present through the whole chain of apple production from the production of the tree in the nursery to the storage of the fruit in the store. In order to understand these risk periods, it is important to set the context of apple production.

As a clonally propagated crop, an apple tree is produced by grafting or budding a scion cultivar on to a rootstock. The scion, collected from a mother tree, is a cultivar which has the desired traits for producing fruit. While the rootstock, produced in rootstock beds or layers, provides traits such as control of vigour and may provide pest and disease resistance. Interstocks are sometimes used in which a third genotype, conferring additional desired traits, is grafted between rootstock and scion.

Grafted plants are grown on in a nursery for up to 3 years during which time the plants are irrigated, provided with nutrition and pruned to produce a tree of the desired form as required. During this stage, canker infection risk is high due to: (1) a high number of natural (leaf scars) and artificial (cuts) wounds on the main structural branches of the tree and (2) high inputs of nitrogen as soil-applied fertilizers are used to increase tree vigour but in turn increase susceptibility of the tree to infection by *N. ditissima* (Swinburne et al., 1975a). High inputs of fungicides are thought to delay the expression of canker symptoms but do not eradicate the infection, leading to asymptomatic infections. These asymptomatic infections may develop into visible cankers within the first few years of orchard planting.

Different tree forms lend themselves to different growing systems in orchards. Increasingly, modern orchards are being grown in intensive fruit wall systems (0.5-1 m tree spacing at 3000 to 6000 trees/ha) in contrast to a less intensive standard orchard (> 2 m tree spacing at <1500 trees/ha). The increased tree density of intensive orchards has fuelled the popularity of the spindle or central leader trees, which are generally accepted as less robust than the traditional tree forms, and increases competition for resources, making the intensive orchard system more prone to canker symptom development from asymptomatic (latent) infections.

Once planted and established, ongoing disease management needs to take into account three factors to determine the risk of infection throughout the growing season: wound availability, production of spores and climatic conditions. Each of these factors will be discussed further in the epidemiology section below.

3 Description of symptoms

The canker symptoms of *N. ditissima* (Fig. 1) have been elegantly described by Crowdy (1949). The old/main stem cankers (>2 years) caused by *N. ditissima* are oyster shell like in appearance as lesions expand more rapidly longitudinally than transversely (Swinburne, 1975) with radial layers of host-produced

Figure 1 Symptoms of *N. ditissima* on plant tissues. Cankers on branches (a, b) and trunks (c, d); close up of perithecia on a trunk canker (e) and sporodochia on a branch canker (f); blossom-end rot ('eye rot') on Royal Gala in the orchard (g) and post-harvest (h). Photo credits: L. Olivieri (a-c); M. Papp-Rupar (d); R. J. Saville (e); L. A. Shuttleworth (f, g); A. Berrie (h).

phellogen radiating from the initial point of infection in the centre of the canker. Younger/peripheral cankers (<2 years) develop from infected leaf scars, bud-scale scars or other wounds (Swinburne, 1975). Such cankers, present on 1–2 year growth, tend to girdle the vascular system, particularly on susceptible cultivars, resulting in branch death. It can be difficult to differentiate the causative agents of a canker in the field; however, the sporulating structures, if present, can provide a key diagnostic feature. In the case of *N. ditissima*, the conidia-producing structures, sporodochia, are evident as whitish pustules forming narrow, roughly parallel lines. The ascospore-producing perithecia form in clusters of bright red, pear-shaped bodies. New cankers (<2 years) produce conidia but seldom perithecia. Although not infected directly, leaf symptoms can also be indicative of the presence of *Neonectria* canker. Staining of internal wood tissue and the death of distal leaves (Wormald, 1935) are speculated to result from the production of yet to be identified diffusible toxin(s) produced by the pathogen at the site of infection (Crowdy, 1949; Wiltshire, 1922; Zeller, 1926).

N. ditissima also infects apple fruits resulting in a fruit rot. Infection can be evident in orchards in the form of eye rots although infection can remain latent and develop into a rot during long-term storage (Xu and Robinson, 2010). The fruit rot occurs on the eye, the stalk end or on the cheek dependent on the site of infection which is in turn influenced by the timing of infection. The rots are soft, slightly sunken, with the rotted part easily scooped out from the sound flesh.

Previous observational studies of natural infections at the microscopic level have reported mycelia growing through the phloem and entering the xylem through medullary rays (Goethe, 1877; Zeller, 1926) reaching the lumen of vessels and tracheids via cell pits. Crowdy (1949) described how *N. ditissima* colonizes the cortex and the phloem initially as intracellular hyphae and in latter stages through cell invasion. More recently, Ghasemkhani et al. (2016a) used light microscopy to study infection spread on artificially inoculated young trees and detached shoots of susceptible 'Cox's Orange Pippin' and partially resistant (Santana) cultivars. *N. ditissima* was shown to colonize all plant tissues from the bark to the pith. There is no evidence of systematic deposition of physical barriers in either cultivar, suggesting that the mechanical exclusion of the pathogen does not play a major role in defence responses against canker.

N. ditissima can grow in the xylem beyond the limits of the visible canker on bark. Using histological methods, Crowdy (1949) traced the pathogen in the xylem fibres up to 14 mm beyond the edge of the lesion. Growth of *N. ditissima* in the xylem beyond the leading edge of the canker lesion was later confirmed with microbiological methods (Weber and Hahn, 2013) and immunolocalization (Dewey et al., 1995). Weber and Hahn (2013) showed that *N. ditissima* can spread vertically in the xylem of young apple trees by up to 20 cm beyond the edges of the canker lesion, within 6 months of the first symptoms becoming visible on bark. Dewey et al. (1995) developed an immunofluorescence assay, using a monoclonal antibody, which allows visualization of the fungal hyphae within the xylem vessels in regions remote from the canker lesion. These studies demonstrate that local, non-symptomatic spread can occur from existing symptomatic cankers, which has implications for controlling canker by pruning.

4 Epidemiology

The three factors required for disease initiation for any plant pathogen - inoculum, host and environment - are referred to as the disease triangle (Fig. 2). Through increased understanding of these three factors, infection risk can be determined and disease management plans can be designed and implemented.

4.1 Host factors

Many types of wounds, both natural and artificial, are available on the host throughout the season and their significance with regard to infection by *N. ditissima* varies depending on region. Autumnal leaf scars are generally considered the main entry points for the pathogen in European apple-growing areas (Graf, 1976; Swinburne, 1975); while in New Zealand, wounds produced by fruit thinning and picking are considered the most important entry sites (Amponsah et al., 2015). In Brazil, pruning wounds have been reported to be the most significant entry point during the season (Alves and Nunes, 2017). The regional differences in the relative significance of wound types is a result of numerous factors including differences in agronomy, the availability and regulation of different plant protection products, and the local climatic

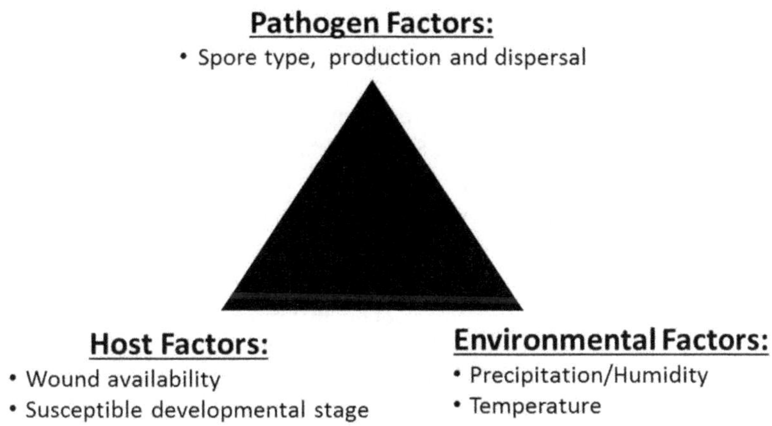

Figure 2 Disease epidemiology, the factors which lead to disease.

conditions and inoculum strength at the times when susceptible wounds are available.

Additional wounds providing entry points are present in early spring as a result of bud break and late spring when rosette leaves detach (Swinburne et al., 1975b), bark cracks caused by frost in winter or by growth cracks and wounds caused by insects or hail in summer (Weber, 2014). Woolly apple aphid (*Eriosoma lanigerum*) colonies have also been implicated in creating entry points for infection (Wiltshire, 1919); however, their significance is not clearly established. During the nursery phase, trees are pruned to shape and promote feathering, both of which create potential entry sites in the trunk and scaffold branches of the tree, which can lead to latent infections.

Host susceptibility varies with tissue types, wound age and developmental stage. The incidence of infection is reduced as pruning wounds age (Xu et al., 1998). There are significant interactions between cultivar and age of pruning wounds, suggesting that this resistance is under genetic control of wound-healing-related processes. The age of the wood also has an effect on susceptibility (Amponsah et al., 2017a) with 3-year-old wood showing increased susceptibility, both in incidence and severity, compared to 1- and 2-year-old wood following artificial inoculation of wounds on the branch. The reason that older wood is more susceptible may result from wound healing taking longer compared to younger wood. In the case of fruit, fruitlets (less than 4 weeks old) and mature fruit (2 weeks prior to harvest) are most susceptible (Xu and Robinson, 2010). Abiotic factors, such as excessive nitrogen input, have been shown to increase the hosts susceptibility to canker development, potentially as a result of trees producing 'softer' wood (Weber, 2014).

4.2 Pathogen factors

Cankers serve as the overwintering form of the disease with the wood providing a perennial substrate on which the pathogen can survive throughout the year (Fig. 3). *N. ditissima* survives primarily as mycelia, which develop sporulating structures from which spores are produced during periods when climatic conditions are favourable. Sporodochia produce conidia, commonly early in the season, but also in the summer and early autumn. Perithecia produce ascospores, which erupt from the canker in late summer and autumn. Ascospores are either forcibly discharged from the perithecium and dispersed by wind or they ooze from the perithecium during periods of high humidity and are washed-off by rain or carried by insects (Swinburne, 1975; Weber, 2014).

Regional differences in spore production and maturation have been reported among the UK (Munson, 1939), NZ (Amponsah et al., 2017b), Germany (Weber, 2014), and the USA (Dubin and English, 1975). The regional and annual variability in spore release highlights the need to understand the dynamics of

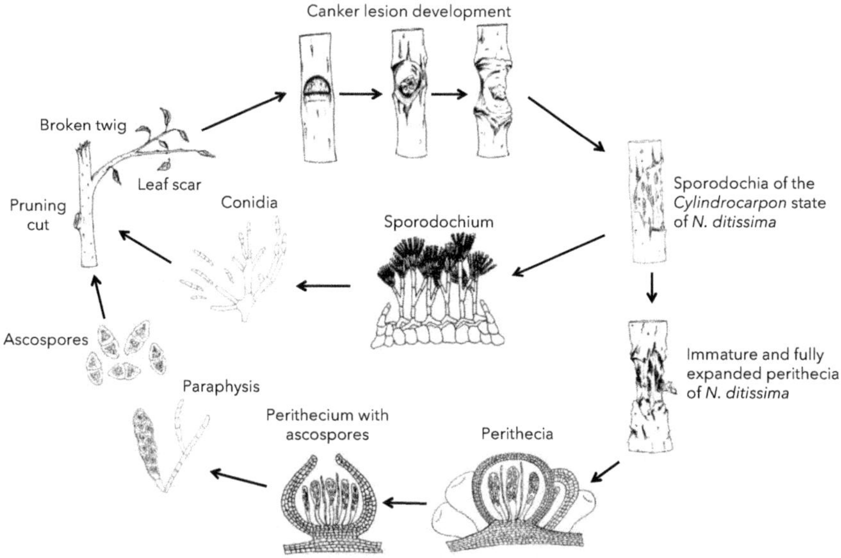

Figure 3 Life cycle of *N. ditissima*. Source: reproduced by kind permission of A. Gomez and co-authors of Gómez-Cortecero et al. (2016).

spore production, and subsequent release and infection to assist with disease forecasting. However, provided there is rainfall, *N. ditissima* inoculum can be considered available all year round.

Airborne ascospores are thought to be responsible for the long-distance dispersal of the disease (Swinburne, 1971) and distances of up to 125 m have been recorded in windy conditions. The dispersal distance of viable spores is likely to be limited by humidity. The localized spread of spores is aided by rain splash. Dispersal distances of spores due to rain splash is generally very short (<15 cm per splash event); however, continual re-splashing of spore-carrying droplets can move spores effectively throughout the canopy (Madden, 1997).

4.3 Environmental factors

European apple canker occurs in temperate apple-growing regions around the world (CABI, 2005) as wet climates and mild temperature favour canker development. Disease risk is considered high in areas which experience >30% rain days per month and an average temperature of 11°C–16°C for >8 h per day (Beresford and Kim, 2011).

A wide range of temperatures for the *in vitro* germination of *N. ditissima* spores has been reported (e.g. 5°C–32°C; Latorre et al., 2002). In the field, the optimum temperature for infection of leaf scars following inoculation is reported as around 15°C (Latorre et al., 2002). Compared to *in vitro* studies, field infection appears to occur over a narrower range (Dubin and English, 1975) suggesting other factors are influencing infection. Leaf scar infection is associated with the number of hours per day with the temperature between 11°C and 16°C and also with the number of ascospores in the spore traps (Dubin and English, 1975). Spore release is also significantly associated with the number of hours per day with the temperature between 5°C and 10°C.

Precipitation and high humidity are essential for infection as water plays an important role in sporulation, spore dispersal, spore germination and infection. Dubin and English (1975) considered frequency and duration, rather than amount, of rainfall more important for all of the above processes. The periods of wetness required for spore release have been quantified by Butt et al. (1994). The number of ascospores discharged from individual cankers within 6 h following 30 min wetting varies greatly, ranging from *c.* 4000 to *c.* 45 000; the number of conidia, released from individual cankers within 6 h after 12 h wetting, ranges from *c.* 17 000 to *c.* 35 000. The importance of the duration of surface wetness on leaf scar (Dubin and English, 1974), pruning wound (Xu et al., 1998) and fruit (Xu and Robinson, 2010) on infection has been demonstrated. A positive correlation of the duration of surface wetness is found with disease incidence when the infection occurred via leaf scar, but not with pruning cut infection. Fruit infection is affected by surface wetness only during the early stages of fruit development.

4.4 Forecasting infection risks

Modelling canker development using relevant experimental data can help growers adopt a rational decision-based strategy for control options. Such epidemiological models have been built for apple scab (Xu et al., 1995), apple powdery mildew (Xu, 1999) and Nectria canker and fruit rot (Xu and Butt, 1994) into a programme called Apple Disease East Malling (ADEM; Xu and Butt, 1996). Patfrut©, developed by Latorre et al. (2002), has been made available commercially through RIMpro®. Field validation of Patfrut© demonstrated that the severity, but not the incidence of cankers, can be reduced when fungicide treatments are applied in accordance with the warnings from the model.

4.5 Nursery infection and detection of latent infection

Understanding the source and timing of infection is important when devising an integrated disease management programme. McCracken et al. (2003) conducted an epidemiological study in the UK with the principal objective of determining the relative significance of nursery versus orchard infection of *N. ditissima*. A total of 27 rootstock and scion combinations from three nurseries were planted out at three locations and canker development was recorded.

The isolates collected from peripheral cankers in the different geographical areas are from different populations suggesting that these infections occurred locally, that is in the orchard. In contrast, the isolates collected from the main stem cankers across the three sites are genetically similar suggesting that these infections originated from a common source, that is the nursery, but remained symptomless until planting in the orchard. Inoculation at key stages (defeathering, budding and heading back) through the propagation phase has demonstrated how latency periods of 3 months to 3 years are possible following inoculation. Although less significant in incidence, main stem cankers resulting from infection occurring during propagation represent a serious loss to growers as these trees act as a source of inoculum to surrounding trees and have to be grubbed out. In contrast, when cankers form on side shoots or minor branches (i.e. orchard-borne infections), the damage can be relatively less severe (dependent on cultivar tolerance), and infections can be pruned out and controlled.

The significance of nursery-borne infection can change from year-to-year and apple growers have reported years when they have experienced devastating losses (>50%) of young trees caused by cankers at or near the base of the main stem (i.e. most likely to originate from nursery infection). The cultural practices of propagation and climatic conditions to which trees are exposed to during the propagation phase may have influenced canker development.

In order to better understand and control nursery-borne infection, there is a need to develop diagnostic tools capable of detecting latent infection. Visual inspection is currently the most widely used strategy for monitoring disease in the orchard or nursery. However, this method only detects symptomatic disease and is labour-intensive. The development and implementation of diagnostic tests for viruses (Maliogka et al., 2018; Torrance and Jones, 1981) of fruit trees at the propagation stage have largely eradicated economically important viruses in the industry, but no such a test currently exists for *N. ditissima*.

Direct isolation on selective artificial media has been used to determine asymptomatic infections (Amponsah et al., 2014; McCracken et al., 2003). However, these methods require specialist knowledge, being better suited for research purposes. In a number of fresh produce crops, in order to predict the risk of rot developing in store or prior to export, a sample of each consignment is exposed to conditions conducive to disease development and assessed for rot development. A similar approach has been adopted to detect asymptomatic infections in nursery trees (Wenneker et al., 2017); this method involves transferring nursery trees into a climate chamber to encourage the development of asymptomatic cankers, requiring long times and weakens the otherwise healthy trees prior to planting.

Molecular and serological techniques are used extensively in diagnostics of plant pathogens (Fang and Ramasamy, 2015), offering the advantage of fast and accurate detection. For *N. ditissima*, the development of such diagnostics has presented some issues, including specificity, tissue selection for sampling and sample preparation. Although so far limited to laboratory use, assays based on polymerase chain reaction (PCR) can be highly specific, quantitative and high throughput. A qPCR assay has been developed enabling the quantification of *N. ditissima* biomass within a plant sample (Garkava-Gustavsson et al., 2013; Ghasemkhani et al., 2016a,b). However, these methods failed to detect the fungus in inoculated but symptomless wood (Ghasemkhani et al., 2016a), indicating that a much higher sensitivity is needed. Monoclonal antibodies have been developed in the past for *N. ditissima* (Dewey et al., 1995) and successfully used in Enzyme-Linked Immunosorbent Assays (ELISA) and immunofluorescence assays to visualize *N. ditissima* in inoculated woody tissue; but this method has not yet been further developed and validated with field samples.

5 Identifying host resistance and pathogen virulence factors

Variations in susceptibility to apple canker have been observed in *Malus* species, apple rootstocks and in different varieties of *M.* x *domestica* (Borecki and Czynczyk, 1985; Gelvonauskienė et al., 2006; Ghasemkhani et al., 2015; Gomez-Cortecero et al., 2016; van de Weg, 1989). Although there is evidence of genetic differences in host responses to *N. ditissima*, little is known of the mechanisms underlying such differences. Canker resistance is quantitative (Gomez-Cortecero et al., 2016; van de Weg and Jansen, 1989); genotypic differences in the susceptibility may partly result from disease escape, for example genotypic differences in the speed of wound healing in relation to *N. ditissima* colonization (Xu et al., 1998). Ghasemkhani (2015) reported differential expression of genes involved in the biosynthesis of lignin, apoptosis and resistance responses between a partially resistant ('Jonathan') and susceptible ('Prima') apple variety during infection by *N. ditissima.* Expression of genes encoding various components of defence signalling pathways increases in response to infection in cv. Jonathan, suggesting that effector triggered immunity (ETI) may be involved in the resistance towards *N. ditissima*. The Rnd1 (Resistance to *N. ditissima* 1, Bus et al., 2019) QTL, identified from seedlings from the cross between 'M9' and Robusta 5, is the first named QTL of major effect for resistance against *N. ditissima*; the underlying genes are expected to be related to establishment of infection rather than disease progression.

While some mechanisms of resistance are constitutively expressed by the host, others are induced as a response to plant recognition of the pathogen. Constitutive traits responsible for varietal differences in resistance to infection (e.g. expedited wound healing, Xu et al., 1998) and tolerance to canker spread (e.g. the production of tyloses and gums in the xylem vessels and phellogen in the peripheral tissue corresponding to bark and vascular cambium) are additional important breeding targets.

In the fruit, the production of the phytoalexin, benzoic acid, is induced after inoculation of immature fruits with *N. ditissima* (Seng et al., 1985); the production of benzoic acid is reduced at maturity and further still in fruits through cold storage, correlating with increases in susceptibility to *N. ditissima*. Noble and Drysdale (1983) reported varietal differences in the production of benzoic acid after *N. ditissima* infection.

Very little is known about the virulence factors of the pathogen. *N. ditissima* isolates are not thought to be specialized to particular host species; for example, common isolates have been isolated from cankers in neighbouring pear and apple orchards (Langrell and Barbara, 2001). Furthermore, Walter et al. (2015) have reported that isolates of *N. ditissima* from apple are able to infect hardwood hosts commonly found in windbreaks. Together this suggests that the pathogen is a general pathogen of hardwood hosts and is unlikely to have a race structure. Gomez-Cortecero et al. (2016) conducted a population analysis of *N. ditissima* using isolates taken from Europe, New Zealand and Brazil and found that there is only slight evidence for geographically structured populations; this finding has major implications for breeding resistance to European apple canker as it means resistance may be deployed worldwide. There is still much to discover about the mechanisms of pathogenicity factors of *N. ditissima* which in turn will inform breeding programmes.

6 Integrated canker control

While breeding resistant varieties remains a long-term goal, in the short-to-medium term, growers need to continue to grow varieties that are highly susceptible to *N. ditissima*. Implementation of an integrated canker management programme is crucial for profitable apple production.

6.1 Cultural control

Cultural control of canker, through pruning out canker lesions and prompt removal of heavily infected trees within the orchard, is currently the most effective and widely adopted practice due to the lack of alternative control measures against *N. ditissima*.

Growth cracks are a site of entry for *N. ditissima* and successful infection leads to cankers localized in the crotches of branches and junctions of different seasons' growth (Swinburne, 1975). Growth cracks are a result of rapid growth that, in turn, is a consequence of high nitrogen inputs. Tree growth management, such as adapting nitrogen and irrigation in young orchards and the use of growth regulators to manipulate tree growth, can reduce the incidence of cankers resulting from infection of growth cracks (Weber, 2014).

The propagation phase is a high risk period for *N. ditissima* infection and can lead to high losses in the orchard following the planting of trees with latent infection of *N. ditissima*. Minimizing wounding during the propagation phase can reduce the risk of *N. ditissima* infection. The current industry standard is the Knipboom method of propagation in which the first year's growth is cut down to 80 cm producing a 1-year-old feathered leader the following year. Some nurseries are reverting to producing maiden trees, the original technique, for the propagation of particularly susceptible varieties (Pers. comm., Henk Nooteboom, Fruitmasters) to minimize wounding during the propagation phase.

Trials evaluating rootstocks for susceptibility to *N. ditissima* have been conducted to inform growers of rootstock choice. There are interactions between rootstock and scion genotypes: the rootstock choice has affected 'Royal Gala' susceptibility but not 'Golden Delicious' (Scheper et al., 2017). Whether the variation in canker susceptibility is inherent or a result of differences in tree vigour conferred by the rootstock is an active area of research.

The apple rot risk assessment concept (Webster et al., 2001) is the culmination of a meta-study identifying the orchard factors that may contribute to infection of fruit by *N. ditissima*. This rot risk assessment forecasts risks of rot development in storage and provides advice on control options to mitigate losses caused by the main apple rots in store. For *Neonectria* fruit rot, the main orchard factor affecting rotting in store is the canker incidence in the orchard. The risk can then be used to determine the requirement of treatment such as the need for pre-harvest fungicide applications and/or post-harvest treatments such as hot water treatment. The outcome of the rot risk assessment also informs the storage potential of the fruit and can be used to determine marketing schedule.

6.2 Chemical control

The primary objective of the chemical control of *N. ditissima* is to reduce infection of new woody tissues and reduce sporulation of existing cankers. In-season control can be largely achieved through programmes targeted primarily for apple scab and powdery mildew control. Fungicides that are mainly active against apple scab such as dodine, dithianon and captan also have some protectant action against *N. ditissima* (Berrie, 2016; Walter et al., 2017a). Carbendazim used to be a key fungicide for managing *N. ditissima*; however, this fungicide is no longer permitted for commercial use in apple production in Europe. Tebuconazole has been identified as a possible alternative product to carbendazim but its use in Europe is restricted to post-harvest, during leaf fall.

SDHI chemicals have been shown to be effective against *N. ditissima* (Berrie, 2016). Copper fungicides give prolonged protection of leaf scars and bud-scale scars against *N. ditissima*, but can be phytotoxic and only used pre-bud burst and post-harvest; the use of copper as a fungicide is becoming increasingly restricted. Low-dose copper products applied throughout the growing season as foliar nutrients have the additional benefit of reducing inoculum. Phosphite, inducing plant natural defences, has been shown to be effective in trials as part of a managed programme (Berrie, 2016). A new product, combining dithianon with potassium phosphonates, provides both curative and protectant properties of dithianon together with induction of the plant's natural defence activity conferred by phosphite.

Timing of treatment application is crucial to make best use of the limited chemical treatments available. In order to protect against wood infection, fungicides are applied at several key growth stages, coinciding with wound availability. Bud-break, petal-fall and leaf-fall (10% and 50%) are the key times for targeting applications (Webster et al., 2001). In orchards with moderate-to-high canker incidence, an extra application of fungicides at 90% leaf fall may be necessary. Infection models have been developed to determine infection risk during these periods (Latorre et al., 2002) and target treatment applications accordingly. For fruit rot, the blossom period and up to 4 weeks following pollination are important risk periods for fruit infection (Xu and Robinson, 2010). In addition, apple fruit becomes more susceptible again as they reach maturity (Xu and Robinson, 2010), as such pre-harvest treatments such as captan, switch (cyprodinil and fludioxonil), geoxe (fludioxonil) and bellis (boscalid and pyraclostrobin) are also recommended if high risk for *Neonectria* rot development in store is predicted.

Spray application technologies have been developed and trialled at Wageningen University in the Netherlands to better target the treatment to infection sites, for example leaf scar wounds, and adaptations have been made to sprayers in the nursery sector to improve coverage in such intensively grown systems (Pers. Comm., M. Wenneker, WUR). Tree injection systems increasingly used to control pests and diseases of forestry and amenity trees (Berger and Laurent, 2019) may provide a useful tool for a more targeted delivery of the treatments to eradicate the disease and/or protect the wounds. Systems, such as the Felco19 (a modified type of secateurs that sprays product on to the exposed pruning wound), may also provide protection from wounds from *N. ditissima*.

Management programmes also need to consider chemical inputs that can be inadvertently beneficial to the pathogen. One such chemical input is urea, used as a foliar feed in season and for reducing overwintering scab inoculum in leaf litter post-harvest (Burchill, 1968). Application of urea may increase leaf scar infection by *N. ditissima* (Dryden et al., 2016), possibly as a result of the nitrogen in the urea acting as a food source for the invading pathogen.

6.3 Biological control

Biological control agents (BCAs) have long been recognized as a method to manage *N. ditissima*. Microbial communities isolated from leaf scars by Swinburne (1973), Swinburne et al. (1975a), and Swinburne and Brown (1976) included a *Bacillus subtilis* strain found to be highly antagonistic to *N. ditissima in vitro*. The use of biocontrol treatments as part of an integrated approach has great potential, for example during picking when harvest intervals of most conventional plant protection products need to be observed. However, so far there have been limited successes in using biocontrol; furthermore, inconsistencies in biocontrol efficacies under field conditions exist (Berrie, 2016; Walter et al., 2017b), especially when targeting leaf scars at leaf fall, when climatic conditions are expected to be less conducive to BCA establishment. Therefore, commercial potential of augmenting the epiphytic communities of fruit/leaf scars with introduced microorganisms has not been widely exploited in the orchard to date. The post-harvest phase may provide better commercial potential for the deployment of BCAs against *N. ditissima*, as the controlled environmental conditions make it easier for BCAs to establish and maintain the antagonist population over a longer time period (Droby et al., 2016; Wilson and Wisniewski, 1989; Wisniewski and Wilson, 1992).

The use of mycoviruses of *N. ditissima* as a BCA has been investigated (Green, 2004). Biocontrol of pathogens using viruses has been demonstrated for chestnut blight (*Cryphonectria parasitica*). Nuss (1992) has shown that the presence of the virus in *C. parasitica* reduces the fungal virulence on chestnut (*Castanea sativa*). A survey failed to find endogenous virus-like dsRNA elements in *N. ditissima* (Green, 2004). However, transfection with the mycovirus of *C. parasitica* is successful and stable in *N. ditissima*. Infected *N. ditissima* isolates have reduced growth *in vitro*. Although no consistent or appreciable differences in fungal virulence were observed on detached fruit infection assays, further work is required in order to determine the potential of this approach.

Biological soil amendments, such as using arbuscular mycorrhizal fungi (AMF), are being increasingly used commercially to aide tree establishment and maximize resource use efficiency. A study on cider apple (Berdeni et al., 2018) has shown that trees inoculated with AMF have significantly reduced amounts of plant material infected by *N. ditissima,* compared to non-inoculated trees.

Microbial endophytes are fungi and bacteria that spend whole or part of their life cycle colonizing internal plant tissues, without producing any sign of infection (Wilson, 1995). Endophytes have been found in every organ of every plant host studied so far (Ryan et al., 2008; Stone et al., 2000). Currently, fungal and bacterial endophytes are receiving increasing attention for their biocontrol potential, which is supported by numerous studies (Bulgarelli et al., 2013; Busby et al., 2016; Compant et al., 2005; Ryan et al., 2008). Unlike the traditional epiphytic BCAs, endophytes are located inside the plant, which protects them against unfavourable environmental conditions.

7 Future trends

7.1 Detection and diagnostics

A key research target for *N. ditissima* control is to develop a method for sensitive, reliable and high throughput detection of the pathogen, particularly prior to planting of new orchards. Monoclonal antibodies have been developed in the past for *N. ditissima* (Dewey et al., 1995) and successfully used in ELISA and immunofluorescence assays to visualize *N. ditissima* in inoculated woody tissues. At NIAB EMR, we have developed a new antibody against *N. ditissima* and currently are evaluating this antibody, together with a qPCR approach, for detecting *N. ditissima* in apple tissues with latent fungal infection.

Developing a sampling strategy represents a major challenge if a diagnostic test is to be implemented commercially. While the systemic nature of most of the major apple viruses makes sampling relatively simple, detecting localized latent infections by a fungal pathogen such as *N. ditissima* needs to consider the locality of the infection in order to sample the tissues to maximize the probability of detection.

Technologies which collect and process image and olfactory data offer further opportunities to develop high throughput disease detection systems for both trees and fruit (Sankaran et al., 2010). Cankers visibly change the surface properties of bark and indirectly have an effect on leaf properties. These changes can be exploited to train imaging software to identify trees with canker within a sample. Use of wavelengths beyond the visible spectrum has been shown to effectively identify latent/early stages of disease development, for example citrus canker (Qin et al., 2009) and wheat scab (Delwiche and Kim, 2000). Electronic noses have enabled in-field applications for the determination of volatile organic compounds (VOCs) which discriminate against healthy and diseased tissues, for example, mango anthracnose (Moalemiyan et al., 2007) and tuber diseases of potato (Lui et al., 2005).

7.2 Endophytes

Endophyte diversity can be studied with culture-dependent methods, which involve the isolation of living endophytes in pure culture and their subsequent identification based either on morphological features or DNA sequences (Cátia Pinto and Gomes, 2016). On the other hand, culture-independent methods, such as meta-barcoding, are based on the isolation of total DNA from plant samples and the analysis of taxon-specific prokaryotic and eukaryotic DNA sequences to dissect microbial community structure (Müller and Ruppel, 2014; Pinto et al., 2014).

Endophytes are receiving increasing attention because of their ability to influence important agronomic traits (Lugtenberg et al., 2016; Ryan et al., 2008). They can promote plant growth (Bulgarelli et al., 2013), improve resistance to abiotic stresses (Bacon, 1993), increase plant biomass and fitness (Clay and Holah, 1999; Marks et al., 1991) and confer disease resistance (Busby et al., 2016; Compant et al., 2005). Endophytes can be manipulated to harness their biocontrol potential. In a plant microbiome manipulation experiment, Zahn and Amend (2017) managed to improve powdery mildew (*Neoerysiphe galeopsidis*) resistance in an endangered wild grass species (*Phyllostegia kaalaensis*). Interest has grown in the commercial exploitation of endophytes to improve plant tolerance to biotic and abiotic stresses (Schneider et al., 2017), which is demonstrated by the increasing number of patents granted on endophytic fungi (Gokhale et al., 2017).

Meta-barcoding has been used to profile microbial community structure associated with internal and external tissues of the apple tree rhizosphere (Mazzola et al., 2015; Zhang et al., 2013), carposphere (Abdelfattah et al., 2016), bark (Arrigoni et al., 2018), flowers (Shade et al., 2013) and xylem tissues (Liu et al., 2018). The composition of fungal and bacterial communities was found to be influenced by a number of factors, including the plant genotype (Arrigoni et al., 2018; Liu et al., 2018). Endophytes associated with specific apple cultivars may be an important component affecting latent canker development, thereby contributing to field resistance. In a pilot study, we have demonstrated that specific endophytes of different apple cultivars are associated with the degree of cultivar susceptibility to *N. ditissima* (Saville and Xu, unpublished data; Fig. 4).

Environmental and orchard-specific factors (such as soil and management) can greatly influence relative abundance of endophytes, which, in turn, may affect canker symptom expression. We are currently testing the hypothesis that endophytes play a role in affecting canker expression of latent infections, and analysing the impact of plant genotypes and environmental conditions on the composition of the endophytic communities. Recently, some positive results have been already obtained with soil treatments with arbuscular mycorrhizal fungi (AMF; *Funneliformis mosseae* and *Rhizophagus irregularis*), which appears to increase apple tree resistance to *N. ditissima* (Berdeni et al., 2018).

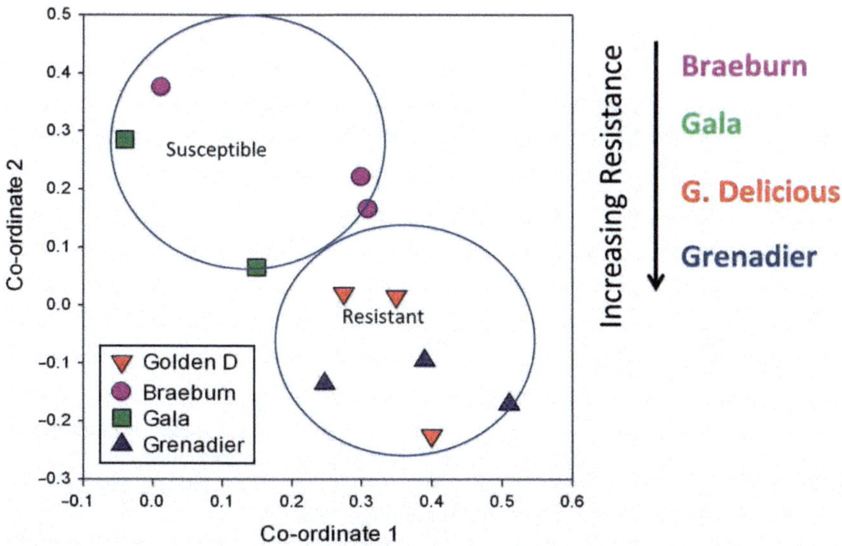

Figure 4 Principle component analysis showing endophyte profiles of different apple cultivars, determined by Illumina Next Generation Sequencing, clustering based on the degree of susceptibility to *N. ditissima*.

8 References

Abdelfattah, A., Wisniewski, M., Droby, S. and Schena, L. 2016. Spatial and compositional variation in the fungal coomunities of organic and conventionally grown apple fruit at the consumer point-of-purchase. *Horticulture Research* 3, 16047. doi:10.1038/hortres.2016.47.

Alves, S. A. M. and Nunes, C. C. 2017. Seasonal susceptibility of apple trees to *Neonectria ditissima* wound infections. *New Zealand Plant Protection* 70, 73–7.

Amponsah, N. T., Walter, M. and Scheper, R. W. A. 2014. Agar media for isolation of *Neonectria ditissima* from symptomatic and asymptomatic apple tissues and production of infective conidia. *New Zealand Plant Protection* 67, 116–22.

Amponsah, N. T., Walter, M., Beresford, R. M. and Scheper, R. W. A. 2015. Seasonal wound presence and susceptibility to *Neonectria ditissima* infection in New Zealand apple trees. *New Zealand Plant Protection* 68, 250–6.

Amponsah, N. T., Scheper, R. W. A., Fisher, B. M., Walter, M., Smits, J. M. and Jesson, L. K. 2017a. The effect of wood age on infection by *Neonectria ditissima* through artificial wounds on different apple cultivars. *New Zealand Plant Protection* 70, 97–105. doi:10.30843/nzpp.2017.70.34.

Amponsah, N. T., Walter, M., Scheper, R. W. A. and Beresford, R. M. 2017b. *Neonectria ditissima* spore release and availability in New Zealand apple orchards. *New Zealand Plant Protection* 70, 78–86. doi:10.30843/nzpp.2017.70.32.

Arrigoni, E., Antonielli, L., Pindo, M., Pertot, I. and Perazzolli, M. 2018. Tissue age and plant genotype affect the microbiota of apple and pear bark. *Microbiological Research* 211(April), 57–68. doi:10.1016/j.micres.2018.04.002.

Bacon, C. W. 1993. Abiotic stress tolerances (moisture, nutrients) and photosynthesis in endophyte-infected tall fescue. *Agriculture, Ecosystems and Environment* 44(1-4), 123-41. doi:10.1016/0167-8809(93)90042-N.

Berdeni, D., Cotton, T. E. A., Daniell, T. J., Bidartondo, M. I., Cameron, D. D. and Evans, K. L. 2018. The effect of arbuscular mycorrhizal colonisation on nutrient status, growth, productivity, and canker resistance of apple (*Malus pumila*). *Frontiers in Microbiology* 9, 1461. doi:10.3389/fmicb.2018.01461.

Beresford, R. M. and Kim, K. S. 2011. Identification of regional climatic conditions favorable for development of European canker of apple. *Phytopathology* 101(1), 135-46. doi:10.1094/PHYTO-05-10-0137.

Berger, C. and Laurent, F. 2019. Trunk injection of plant protection products to protect trees from pests and diseases. *Crop Protection* 124. doi:10.1016/J.CROPRO.2019.05.025.

Berrie, A. 2016. Evaluation of products for control of Nectria galligena on apples. *AHDB Horticulture*. Available at: https://horticulture.ahdb.org.uk/project/evaluation-products-control-nectria-galligena-apples-0 (accessed on 20 September 2005).

Borecki, Z. and Czynczyk, A. 1985. Susceptibility of apple cultivars to bark canker diseases. *Acta Agrobotanica* 38(1), 49-59. doi:10.5586/aa.1985.006.

Bulgarelli, D., Schlaeppi, K., Spaepen, S., van Themaat, E. V. L. and Schulze-Lefert, P. 2013. Structure and functions of the bacterial microbiota of plants. *Annual Review of Plant Biology* 64(1), 807-38. doi:10.1146/annurev-arplant-050312-120106.

Burchill, R. T. 1968. Field and laboratory studies of the effect of urea on ascospore production of *Venturia inaequalis* (Cke.) Wint. *Annals of Applied Biology* 62(2), 297-307. doi:10.1111/j.1744-7348.1968.tb02825.x.

Bus, V. G. M., Scheper, R. W. A., Walter, M., Campbell, R. E., Kitson, B., Turner, L., Fisher, B. M., Johnston, S. L., Wu, C., Deng, C. H., Singla, G., Bowatte, D., Jesson, L. K., Hedderley, D. I., Volz, R. K., Chagné, D. and Gardiner, S. E. 2019. Genetic mapping of the European canker (*Neonectria ditissima*) resistance locus *Rnd1* from *Malus* 'Robusta 5.' *Tree Genetics and Genomes* 15(2). doi:10.1007/s11295-019-1332-y.

Busby, P. E., Ridout, M. and Newcombe, G. 2016. Fungal endophytes: modifiers of plant disease. *Plant Molecular Biology* 90(6), 645-55. doi:10.1007/s11103-015-0412-0.

Butt, D. J., Gunn, N. D. and Xu, X. 1994. Release of *Nectria galligena* spores from apple cankers. *Norwegian Journal of Agricultural Sciences* 17(Suppl.), 309-16.

Byrde, R. J. W. and Willets, H. J. 1977. *The Brown Rot Fungi of Fruit: Their Biology and Control*. Permagon Press, Oxford, UK.

CABI. 2005. CAB International crop protection compendium globule module. CAB International, Wallingford, UK. Available at: https://www.cabi.org/cpc/ (accessed on 20 September 2005).

Clay, K. and Holah, J. 1999. Fungal endophyte symbiosis and plant diversity in successional fields. *Science* 285(5434), 1742-5. doi:10.1126/science.285.5434.1742.

Compant, S., Duffy, B., Nowak, J., Clément, C. and Barka, E. A. 2005. Use of plant growth-promoting bacteria for biocontrol of plant diseases: principles, mechanisms of action, and future prospects. *Applied and Environmental Microbiology* 71(9), 4951-9. doi:10.1128/AEM.71.9.4951-4959.2005.

Crowdy, S. H. 1949. Observations on apple canker. III the anatomy of the stem canker. *The Annals of Applied Biology* 36(4), 483-95. doi:10.1111/j.1744-7348.1949.tb06944.x.

Delwiche, S. R. and Kim, M. S. 2000. Hyperspectral imaging for detection of scab in wheat. *Proceedings of the SPIE* 4203, 13-20.

Dewey, F. M., Li, R. and Swinburne, T. 1995. A monoclonal antibody immunoassay for the detection of *Nectria galligena* in apple fruit and woody tissues. *EPPO Bulletin* 25(1-2), 65–73. doi:10.1111/j.1365-2338.1995.tb01439.x.

Droby, S., Wisniewski, M., Teixidó, N., Spadaro, D. and Jijakli, M. H. 2016. The science, development, and commercialization of postharvest biocontrol products. *Postharvest Biology and Technology* 122, 22-9. doi:10.1016/j.postharvbio.2016.04.006.

Dryden, G. H., Nelson, M. A., Smith, J. T. and Walter, M. 2016. Postharvest foliar nitrogen applications increase *Neonectria ditissima* leaf scar infection in apple trees. *New Zealand Plant Protection* 69, 230–7.

Dubin, H. J. and English, H. 1974. Factors affecting apple leaf scar infection by *Nectria galligena* Conidia. *Phytopathology* 64(9), 1201–3. doi:10.1094/Phyto-64-1201.

Dubin, H. J. and English, H. 1975. Epidemiology of European apple canker in California. *Phytopathology* 65(5), 542–50. doi:10.1094/Phyto-65-542.

Fang, Y. and Ramasamy, R. P. 2015. Current and prospective methods for plant disease detection. *Biosensors* 5(3), 537–61. doi:10.3390/bios5030537.

Garkava-Gustavsson, L., Zborowska, A., Sehic, J., Rur, M., Nybom, H., Lateur, M. and Holefors, A. 2013. Screening of apple cultivars for resistance to European canker, *Nectria galligena*. (Evans, K. M., Lata, B. and Kellerhals, M., Eds) *Acta Horticulturae* 976, 529–36. doi:10.17660/ActaHortic.2013.976.75.

Gelvonauskienė, D., Gelvonauskis, B. and Sasnauskas, A. 2006. Impact of rootstocks on columnar apple tree growth in nursery. *Scientific Works of the Lithuanian Institute of Horticulture and Lithuanian University of Agriculture. Sodininkystė ir Daržininkystė* 25(3), 51–6.

Ghasemkhani, M. 2015. Resistance against fruit tree canker in apple. Swedish University of Agricultural Sciences, Alnarp. Available at: https://pub.epsilon.slu.se/12495/.

Ghasemkhani, M., Sehic, J., Ahmadi-Afzadi, M., Nybom, H. and Garkava-Gustavsson, L. 2015. Screening for partial resistance to fruit tree canker in apple cultivars. *Acta Horticulturae* 1099(1099), 687–90. doi:10.17660/ActaHortic.2015.1099.84.

Ghasemkhani, M., Holefors, A., Marttila, S., Dalman, K., Zborowska, A., Rur, M., Rees-George, J., Nybom, H., Everett, K. R., Scheper, R. W. A. and Garkava-Gustavsson, L. 2016a. Real-time PCR for detection and quantification, and histological characterization of *Neonectria ditissima* in apple trees. *Trees* 30(4), 1111–25. doi:10.1007/s00468-015-1350-9.

Ghasemkhani, M., Zborowska, A., Garkava-Gustavsson, L., Holefors, A., Scheper, R., Everett, K. and Nybom, H. 2016b. Development of a qPCR detection procedure for fruit tree canker caused by *Neonectria ditissima*. *Acta Horticulturae* 1127(1127), 259–64. doi:10.17660/ActaHortic.2016.1127.40.

Goethe, R. 1877. *Mittheilung über den Krebs der Appelbaume*. Leipzig.

Gokhale, M., Gupta, D., Gupta, U., Faraz, R. and Sandhu, S. S. 2017. Patents on endophytic fungi. *Recent Patents on Biotechnology* 11(2), 120–40. doi:10.2174/1872208311666170215151834.

Gomez-Cortecero, A., Saville, R. J., Scheper, R. W. A., Bowen, J. K., de Medeiros, H. A., Kingsnorth, J., Xu, X. and Harrison, R. J. 2016. Variation in host and pathogen in the *Neonectria/Malus* interaction; towards an understanding of the genetic basis of resistance to European canker. *Frontiers in Plant Science* 7, 1365. doi:10.3389/FPLS.2016.01365.

Graf, H. 1976. Die Biologie des Obstbaumkrebses (*Nectria galligena* Bres.) als Grundlage seiner gezielten Bekämpfung. *Mitteilungen des Obstbauversuchsringes des Alten Landes* 31, 68–78.

Green, C. 2004. Hypovirus infection of Neonectria galligena and its potential as a biological control agent for apple canker. Imperial College, University of London.

Langrell, S. R. H. and Barbara, D. J. 2001. Magnetic capture hybridisation for improved PCR detection of *Nectria galligena* from lignified apple extracts. *Plant Molecular Biology Reporter* 19(1), 5–11. doi:10.1007/Bf02824073.

Latorre, B. A., Rioja, M. E., Lillo, C. and Muñoz, M. 2002. The effect of temperature and wetness duration on infection and a warning system for European canker (*Nectria galligena*) of apple in Chile. *Crop Protection* 21(4), 285–91. doi:10.1016/S0261-2194(01)00099-0.

Liu, J., Abdelfattah, A., Norelli, J., Burchard, E., Schena, L., Droby, S. and Wisniewski, M. 2018. Apple endophytic microbiota of different rootstock/scion combinations suggests a genotype-specific influence. *Microbiome* 6(1), 18. doi:10.1186/s40168-018-0403-x.

Lugtenberg, B. J. J., Caradus, J. R. and Johnson, L. J. 2016. Fungal endophytes for sustainable crop production. *FEMS Microbiology Ecology* 92(12), fiw194. doi:10.1093/femsec/fiw194.

Lui, L., Vikram, A., Hamzehzarghani, H. and Kushalappa, A. C. 2005. Discrimination of three fungal diseases of potato tubers based on volatile metabolic profiles developed using GC/MS. *Potato Research* 48(1–2), 85–96. doi:10.1007/BF02733684.

Madden, L. V. 1997. Effects of rain on splash dispersal of fungal pathogens. *Canadian Journal of Plant Pathology* 19(2), 225–30. doi:10.1080/07060669709500557.

Maliogka, V. I., Minafra, A., Saldarelli, P., Ruiz-García, A. B., Glasa, M., Katis, N. and Olmos, A. 2018. Recent advances on detection and characterization of fruit tree viruses using high-throughput sequencing technologies. *Viruses* 10(8), 436. doi:10.3390/v10080436.

Marks, S., Clay, K. and Cheplick, G. P. 1991. Effects of fungal endophytes on interspecific and intraspecific competition in the grasses *Festuca arundinacea* and *Lolium perenne*. *The Journal of Applied Ecology* 28(1), 194–204. doi:10.2307/2404125.

Mazzola, M., Hewavitharana, S. S. and Strauss, S. L. 2015. Brassica seed meal soil amendments transform the rhizosphere microbiome and improve apple production through resistance to pathogen reinfestation. *Phytopathology* 105(4), 460–9. doi:10.1094/PHYTO-09-14-0247-R.

McCracken, A. R., Berrie, A., Barbara, D. J., Locke, T., Cooke, L. R., Phelps, K., Swinburne, T. R., Brown, A. E., Ellerker, B. and Langrell, S. R. H. 2003. Relative significance of nursery infections and orchard inoculum in the development and spread of apple canker (*Nectria galligena*) in young orchards. *Plant Pathology* 52(5), 553–66. doi:10.1046/j.1365-3059.2003.00924.x.

Moalemiyan, M., Vikram, A. and Kushalappa, A. C. 2007. Detection and discrimination of two fungal diseases of mango (cv. Keitt) fruits based on volatile metabolite profiles using GC/MS. *Postharvest Biology and Technology* 45(1), 117–25. doi:10.1016/j.postharvbio.2006.08.020.

Müller, T. and Ruppel, S. 2014. Progress in cultivation-independent phyllosphere microbiology. *FEMS Microbiology Ecology* 87(1), 2–17. doi:10.1111/1574-6941.12198.

Munson, R. G. 1939. Observations on apple canker I. The discharge and germination of spores of *Nectria galligena* Bres. *Annals of Applied Biology* 26(3), 440–57. doi:10.1111/j.1744-7348.1939.tb06982.x.

Noble, J. P. and Drysdale, R. B. 1983. The role of benzoic acid and phenolic compounds in latency in fruits of two apple cultivars infected with *Pezicula malicorticis* or *Nectria galligena*. *Physiological Plant Pathology* 23(2), 207–16. doi:10.1016/0048-4059(83)90003-6.

Nuss, D. L. 1992. Biological control of chestnut blight: an example of virus-mediated attenuation of fungal pathogenesis. *Microbiological Reviews* 56(4), 561–76.

Pinto, C. and Gomes, A. C. 2016. Vitis vinifera microbiome: from basic research to technological development. *BioControl* 61(3), 243–56. doi:10.1007/s10526-016-9725-4.

Pinto, C., Pinho, D., Sousa, S., Pinheiro, M., Egas, C. and Gomes, A. C. 2014. Unravelling the diversity of grapevine microbiome. *PLoS ONE* 9(1), e85622. doi:10.1371/journal.pone.0085622.

Qin, J., Burks, T. F., Ritenour, M. A. and Bonn, W. G. 2009. Detection of citrus canker using hyperspectral reflectance imaging with spectral information divergence. *Journal of Food Engineering* 93(2), 183–91. doi:10.1016/j.jfoodeng.2009.01.014.

Ryan, R. P., Germaine, K., Franks, A., Ryan, D. J. and Dowling, D. N. 2008. Bacterial endophytes: recent developments and applications. *FEMS Microbiology Letters* 278(1), 1–9. doi:10.1111/j.1574-6968.2007.00918.x.

Sankaran, S., Mishra, A., Ehsani, R. and Davis, C. E. 2010. A review of advanced techniques for detecting plant diseases. *Computers and Electronics in Agriculture* 72(1), 1–13. doi:10.1016/j.compag.2010.02.007.

Saville, R. J. 2015. TF-003 A review of the literature of the *Neofabraea* species complex, causative agents of *Gloeosporium* rot in stored apple. Available at: https://horticulture.ahdb.org.uk/publication/tr-tf-003-review-literature-neofabraea-species-complex-causative-agents-gloeosporium-rot.

Scheper, R. W. A., Walter, M., Fisher, B. M., Johnston, S., Curnow, T., Amponsah, N. T., Alspach, P. and Hedderley, D. I. 2017. Resistance of apple and pear rootstocks to *Neonectria ditissima* and their effect on scion susceptibility. *New Zealand Plant Protection* 70, 324. doi:10.30843/nzpp.2017.70.92.

Schneider, C., Hutter, I. and Döring, M. 2017. Commercial use of endophytes in micropropagation. *Acta Horticulturae* 1155(1155), 483–90. doi:10.17660/ActaHortic.2017.1155.71.

Seng, J.-M., Saindrenan, P. and Bompeix, G. 1985. Induction of *Nectria galligena* mutants resistant to benzoic acid and study of their aggressiveness towards immature apples. *Microbiology* 131(8), 1863–6. doi:10.1099/00221287-131-8-1863.

Shade, A., McManus, P. S. and Handelsman, J. 2013. Unexpected diversity during community succession in the apple flower microbiome. *mBio* 4(2), e00602-12. doi:10.1128/mBio.00602-12.

Stone, J., Bacon, C. and White, J. 2000. An overview of endopytic microbes: endophytism defined. In: Bacon, C. and White, J. (Eds), *Microbial Endophytes*. Marcel Dekker, New York, NY, pp. 3–29.

Swinburne, T. R. 1971. The infection of apples, cv. Bramley's Seedling, by *Nectria galligena* Bres. *Annals of Applied Biology* 68(3), 253–62. doi:10.1111/j.1744-7348.1971.tb04644.x.

Swinburne, T. R. 1973. Microflora of apple leaf scars in relation to infection by *Nectria galligena*. *Transactions of the British Mycological Society* 60(3), 389–403. doi:10.1016/S0007-1536(73)80024-5.

Swinburne, T. R. 1975. European canker of apple. *Review of Plant Pathology* 54(10), 787–99.

Swinburne, T. R. and Brown, A. E. 1976. A comparison of the use of *Bacillus subtilis* with conventional fungicides for the control of apple canker (*Nectria galligena*). *Annals of Applied Biology* 82(2), 365–8. doi:10.1111/j.1744-7348.1976.tb00573.x.

Swinburne, T. R., Barr, J. G. and Brown, A. E. 1975a. Production of antibiotics by *Bacillus subtilis* and their effect on fungal colonists of apple leaf scars. *Transactions of the British Mycological Society* 65(2), 211–7. doi:10.1016/S0007-1536(75)80004-0.

Swinburne, T. R., Cartwright, J., Flack, N. J. and Brown, A. E. 1975b. The control of apple canker (*Nectria galligena*) in a young orchard with established infections. *Annals of Applied Biology* 81(1), 61–73. doi:10.1111/j.1744-7348.1975.tb00495.x.

Torrance, L. and Jones, R. A. C. 1981. Recent developments in serological methods suited for use in routine testing for plant viruses. *Plant Pathology* 30(1), 1–24. doi:10.1111/j.1365-3059.1981.tb01218.x.

van de Weg, W. E. 1989. Screening for resistance to *Nectria galligena* Bres. in cut shoots of apple. *Euphytica* 42(3), 233–40. doi:10.1007/BF00034459.

van de Weg, W. E. and Jansen, R. C. 1989. Breeding for resistance to *Nectria galligena*; differences in resistance between seedling populations. Integrated control of pome fruit diseases. *IOBC/WPRS Bulletin* 12(6), 137–45.

Walter, M., Glaister, M. K., Clarke, N. R., Von Lutz, H., Eld, Z., Amponsah, N. T. and Shaw, N. F. 2015. Are shelter belts potential inoculum sources for *Neonectria ditissima* apple tree infections? *New Zealand Plant Protection* 68(1975), 227–40.

Walter, M., Campbell, R. E., Amponsah, N. T., Scheper, R. W. A. and Butler, R. C. 2017a. Evaluation of biological and agrichemical products for control of *Neonectria ditissima* conidia production. *New Zealand Plant Protection* 70, 87–96. doi:10.30843/nzpp.2017.70.33.

Walter, M., Campbell, R. E., Amponsah, N. T., Turner, L., Rainham, D., Kerer, U. and Butler, R. C. 2017b. Can biological products control *Neonectria ditissima* picking wound and leaf scar infections in apples? *New Zealand Plant Protection* 70, 63–72. doi:10.30843/nzpp.2017.70.29.

Weber, R. W. S. 2014. Biology and control of the apple canker fungus *Neonectria ditissima* (syn. *N. galligena*) from a Northwestern European perspective. *Erwerbs-Obstbau* 56(3), 95–107. doi:10.1007/s10341-014-0210-x.

Weber, R. W. S. and Hahn, A. 2013. Obstbaumkrebs (*Neonectria galligena*) und die Apfelsorte "Nicoter" (Kanzi ®) an der Niederelbe. *Mitteilungen des Obstbauversuchsringes des Alten Landes* 68, 247–56.

Webster, A., Cross, J., Berrie, A., Johnston, D., Biddelcombe, C. and Pennell, D. 2001. *The Best Practice Guide for UK Apple Production*. London.

Wenneker, M., de Jong, P. F., Joosten, N. N., Goedhart, P. W. and Thomma, B. P. H. J. 2017. Development of a method for detection of latent European fruit tree canker (*Neonectria ditissima*) infections in apple and pear nurseries. *European Journal of Plant Pathology* 148(3), 631–5. doi:10.1007/s10658-016-1115-3.

Wilson, D. 1995. Endophyte: the evolution of a term and a clarification of its use and definition. *Oikos* 73(2), 274–6. doi:10.2307/3545919.

Wilson, C. L. and Wisniewski, M. E. 1989. Biological control of postharvest diseases of fruits and vegetables: an emerging technology. *Annual Review of Phytopathology* 27(1), 425–41. doi:10.1146/annurev.py.27.090189.002233.

Wiltshire, S. P. 1919. The apple canker fungus. Annual report. Long Ashton Research Station.

Wiltshire, S. P. 1922. Studies on the apple canker fungus. II. Canker infection of apple trees through scab wounds. *Annals of Applied Biology* 9(3–4), 275–81. doi:10.1111/j.1744-7348.1922.tb05959.x.

Wisniewski, M. E. and Wilson, C. L. 1992. Biological control of postharvest diseases of fruits and vegetables: recent advances. *HortScience* 27(2), 94–8. doi:10.21273/HORTSCI.27.2.94.

Wormald, H. 1935. Annual report of East Malling Research Station for 1934.

Xu, X.-M. 1999. Modelling and forecasting epidemics of apple powdery mildew (*Podosphaera leucotricha*). *Plant Pathology* 48(4), 462–71. doi:10.1046/j.1365-3059.1999.00371.x.

Xu, X.-M. and Butt, D. J. 1994. The biology and epidemiology of *Nectria galligena* and an infection warning system. *Norwegian Journal of Agricultural Sciences, Supplement* 17, 317–24.

Xu, X.-M. and Butt, D. J. 1996. ADEM™ a PC-based multiple disease warning system for use in the cultivation of apples . *Acta Horticulturae* 416, 293–6.

Xu, X.-M. and Robinson, J. D. 2010. Effects of fruit maturity and wetness on the infection of apple fruit by *Neonectria galligena*. *Plant Pathology* 59(3), 542–7. doi:10.1111/j.1365-3059.2009.02232.x.

Xu, X.-M., Butt, D. J. and Santen, G. Van. 1995. A dynamic model simulating infection of apple leaves by *Venturia inaequalis*. *Plant Pathology* 44(5), 865–76. doi:10.1111/j.1365-3059.1995.tb02746.x.

Xu, X.-M., Butt, D. J. and Ridout, M. S. 1998. The effects of inoculum dose, duration of wet period, temperature and wound age on infection by *Nectria galligena* of pruning wounds on apple. *European Journal of Plant Pathology* 104(5), 511–9. doi:10.1023/A:1008689406350.

Zahn, G. and Amend, A. S. 2017. Foliar microbiome transplants confer disease resistance in a critically endangered plant. *Peer Journal* 5, e4020. doi:10.7717/peerj.4020.

Zeller, S. M. 1926. European canker of pomaceous fruit trees. Bulletin 222. Bulletin of the Oregon Agricultural Experiment Station, pp. 1–52.

Zhang, Q., Sun, J., Liu, S. and Wei, Q. 2013. Manure refinement affects apple rhizosphere bacterial community structure: a study in sandy soil. *PLoS ONE* 8(10), e76937. doi:10.1371/journal.pone.0076937.

Chapter 5

Fungal diseases of fruit: apple canker in Asia

Baohua Li, Qingdao Agricultural University, China

1 Introduction

Apple cankers in Asia are caused by two main pathogens. In East Asia, the disease is caused by *Valsa* spp., hence the name apple valsa canker; the other is caused by *Botryosphaeria dothidea*. The latter is also present in Europe, North America, South America, and Australia. This chapter focusses on apple canker caused by *Valsa* spp.

Apple valsa canker is mainly distributed in China, Japan, South Korea, and North Korea. It occurs in all apple production areas in China, particularly in the northern cold areas. Apple valsa canker mainly damages the cortex of apple branches, causing the cortex to rot. In the initial stage of disease development, the fungus grows slowly from the entry points under the epidermis, forming reddish brown rotting spots and then becoming dark brown dry tissues after water loss. Canker lesions are confined to the superficial layer of the tissue and can be detected when the epidermis is peeled off in the late autumn and winter. In the early spring, the fungus grows and expands rapidly in the cortex, leading to the rapid decomposition of the cortex and formation of canker

http://dx.doi.org/10.19103/AS.2019.0046.05

Figure 1 Symptoms of apple valsa canker caused by *Valsa mali*: (a) a newly developed canker lesion re-occurring from an old canker lesion; (b) an old canker lesion with orange conidia mass newly released in rainfall. Source: photos from Baohua Li.

lesions. Removing the epidermis of the lesion reveals decayed tissues, which are reddish brown and unorganized, and occasionally microtubule bundles from incompletely decomposed tissues (Fig. 1a). Valsa canker lesions expand quickly in the early spring, up to 10 cm in a week. If it is not controlled in time, it will girdle the branch, causing branch death, and potentially tree death if the canker lesion is on the main trunk. Differential rates of fungal expansion over time result in the light and dark rings. On old canker lesions, there are a large number of fruiting bodies, the orifices of pycnidia (Fig. 1b).

Apple valsa canker has caused devastating damage in Japan and China. For example, about 3.3 million trees were removed because of severe canker development in five Japanese counties, including Hokkaido, Aomori, and Akita. Apple valsa canker was first reported in Xiongyue City, Liaoning Province, China in 1916. Since the late 1940s, there have been four major epidemics of apple valsa canker in China. For example, almost all apple orchards established since 1949 were destroyed in the 1980s because of the canker problem. In the main apple production areas of China in 2008, the overall incidence of trees with valsa canker was 52.7%, with some regions exceeding 85% (Cao et al., 2009).

2 The pathogen

Apple canker pathogen was first named as *Valsa mali* Miyabe et Yamada (Tanaka et al., 1919). In the following two decades, Japanese researchers studied morphological characteristics, culture traits, and pathogenicity of canker pathogens, and confirmed that the pathogen was *V. mali*. In the

1970s, the finding that the morphological characteristics of the apple canker pathogen were the same as those of *Valsa ceratosperma* led to the suggestion that *V. mali* should not be treated as a single species but as a homonym of *V. ceratosperma* (Kobayashi, 1970). Thereafter, these two pathogens have been described as the pathogens causing apple valsa canker, leading to some confusion.

Valsa ceratosperma (Tode) Maire was established in 1818 with its asexual stage *Cytospora sacculus*. Since *V. ceratosperma* was established earlier, many pathogens with the same or similar morphological characteristics as *V. ceratosperma* and *C. sacculus* were named as *V. ceratosperma*. Therefore, *V. ceratosperma* covers many fungal groups with small asci and ascospores in the *Valsa* genus and has a wide range of host species, including more than 20 woody plant species that are distributed around the world (Adams et al., 2005). Strains from *V. ceratosperma* and *C. sacculus* have very different internal transcribed spacer (*ITS*) homology (Adams et al., 2005). ITS sequences showed that these strains belong to three branches: *V. ceratosperma sensu stricto*, *V. ceratosperma sensu Kobayashi*, and *V. ceratosperma sensu KM Old*. The standard strain from the Japanese *V. ceratosperma* (CBS117012, ITS SEQ ID NO: 192192 in GenBank) is in the *V. ceratosperma sensu Kobayashi* branch. This branch differs greatly from *V. ceratosperma sensu stricto* in homology and *V. ceratosperma sensu KM Old* from Australia (Adams et al., 2005).

Taxonomy of the apple valsa canker pathogen was recently studied based on both morphological characteristics and ITS sequences (Wang et al., 2011). Thirteen different rDNA-ITS sequence types were obtained in 159 strains (150 from apple and 9 from pear collected in China); these 13 rDNA-ITS types were divided into three groups. Nine sequence types from 140 apple strains have high homology with the rDNA-ITS sequences from Japanese apples strains (*V. ceratosperma sensu Kobayashi*, AF192326). Two sequence types from 14 strains (5 from apple and 9 from pear) were assigned to the same branch as for the strain *V. ceratosperma* VC38, isolated from pear in Italy. The remaining two sequence types from five apple strains are highly homologous to the ITS sequence of *V. malicola* and assigned to another separate branch (Fig. 2). It was proposed that branches A and B should be treated as *V. mali*. However, these two groups have large differences in colony morphology, ecology, and pathogenicity; thus branch A is named *V. mali* var. *mali* (Vmm) and branch B, *V. mali* var. *piry* (Vmp) (Wang et al., 2011).

Further phylogenetic studies indicated that the difference between Vmm and Vmp is sufficiently large to justify classifying them into two different species (Wang et al., 2014c): Vmm as *V. mali* and Vmp as *V. piry*. Therefore, three pathogens cause apple canker in China, namely *V. mali*, *V. piry*, and *V. malicola,* with *V. mali* accounting for >90% of apple canker in China. *V. malicola* is weakly pathogenic in apple (Wang et al., 2014c).

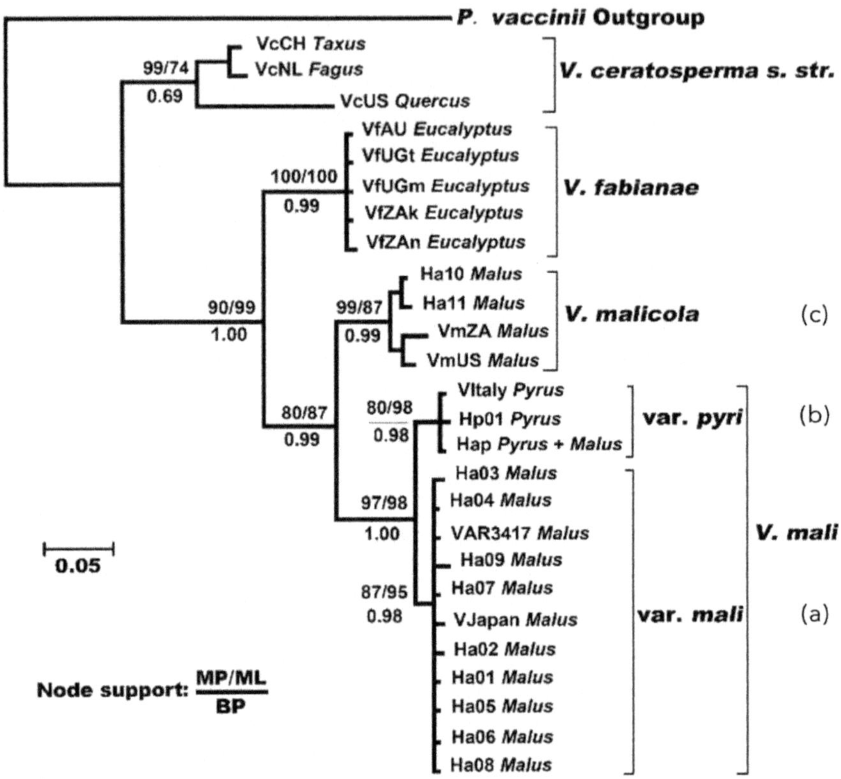

Figure 2 A Bayesian phylogenetic tree constructed from rDNA-ITS sequences showing the interrelationships among *Valsa ceratosperma* sensu stricto strains and samples from China. Host plants (genus) were given after the fungal taxon/haplotype names. Numbers at the nodes refer to the support values (MP/ML/BP): MP and ML representing bootstrapping support, and BP referring to Bayesian posterior probability. Source: adapted from Wang et al. (2011).

There are large genetic variabilities in the canker pathogen (*Valsa spp.*). However, there is no direct correspondence between such variabilities and geographical locations (Zang et al., 2012a), which may be due to the large volume of international trading in apple planting material (Zang et al., 2012a).

3 Biology of *V. mali*

Valsa mali is a typical necrotrophic fungus and grows fast in media rich in nutrients. When *V. mali* grows on potato dextrose agar (PDA) media, the colony is initially white before aerial hyphae have developed; the edge is radial or feathery with daily growth up to 2.5 cm. Gradually, colonies become milky white, yellowish brown, or dark grey, and produce white pycnidia. It can grow on PDA in the range from 5 to 40°C with an optimum at 28–29°C; similarly, pH range is 4–9 with an optimum of 5–6. Colonies on PDA and apple bark decoction media (ABA) can form pycnidia and produce conidia at 5–40°C with an optimum of 25°C. It takes about 20 days for colonies to produce conidia; illumination can shorten this interval and also increase the number of pycnidia. Ascospores can be collated from asci on canker lesions under natural conditions.

Maltose is the most suitable carbon source for *V. mali*; colonies grow best in the maltose (2%) medium. Conidia germinate best in 1% maltose solution, followed by sucrose and glucose. Of vitamins tested, thiamine (1 mg/L) promoted the fungal growth. Among the 16 nitrogen nutrients tested, calcium nitrate, potassium nitrate, and sodium nitrate promoted mycelial growth (Anon., 1971).

V. mali can utilize soluble nutrients in the xylem of apple branches. In a medium prepared with xylem extracts and residues of apple branches, colonies grow at a faster rate than in PDA amended with phloem extracts; the opposite is true for aerial hyphae. Canker pathogens grow faster in the xylem than in the cortex of apple branches (Wang et al., 2018); they do not destroy the xylem.

Conidia usually cannot germinate in pure water but germinate well in apple bark, xylem, leaves, fruit extract, maltose, glucose, and sucrose solution. When conidia germinate, they first expand into a spherical shape and produce germ tubes. Temperature required for conidial germination ranges from 10 to 35°C with an optimum of 25°C. At 25°C, it takes 12 h for conidia to produce germ tubes *in vitro*. Even at 0°C, more than 64% of conidia may germinate after 18 days. Conidial germination requires free water or a relative humidity >97%. Under natural conditions, conidia may survive for a short period of time, usually 2–4 days, but could be as long as 2 weeks (Anon., 1971; Liu et al., 2011a).

Germination of *V. mali* ascospores does not require nutrients and can readily occur in pure water. Nevertheless, nutrients such as apple juice, apple bark juice, and maltose can promote ascospore germination. Temperature and humidity required for ascospore germination are less stringent than for conidia. Ascospore germination can take place at 5 to 40°C with an optimum of 26°C, at which it takes 12 h for *in vitro* germination. Ascospores can germinate with relative humidity >90%. When germinating, ascospores do not swell but immediately produce germ tubes. Under natural conditions, ascospores can survive for 13–20 days (Anon., 1971).

There are 13 chromosomes in *V. mali* with an estimated genome size of 45 Mb. The genome is predicted to encode 11 261 proteins, mainly related to infection and colonization such as cell wall degrading enzymes, secreted proteases, enzymes related to secondary metabolite synthesis, and transport proteins (Yin et al., 2015). Those pathogenicity-related genes predicted for *V. mali* are similar to those involved in the pathogenicity of necrotrophic fungi in general. The *V. mali* genome has a large number of predicted pectinase genes that are needed to decompose pectin, and a limited number of enzyme genes needed for breaking down lignocellulose and keratin (Ke et al., 2013).

4 Epidemiology of apple valsa canker

Apple valsa canker is a typical polycyclic disease with the pathogen overwintering in canker lesions on trees. A single canker lesion can sporulate for 2 years. After sporulation, the cankered tissues are gradually decomposed. However, the pathogen can invade xylem near the lesion without causing apparent visual symptoms and may survive for a long time. For example, canker lesions may re-occur every year or every 2–3 years and last for more than 5 years in the same tissues, indicating that the pathogen can survive for more than 5 years in the xylem. The canker pathogen can penetrate xylem through wounding and may colonize dead wood tissues and wood surface to overwinter.

The exact survival ability of *V. mali* in diseased or colonized tissues in relation to environmental conditions is not yet fully understood. We may speculate that the pathogen can survive as long as it can obtain sufficient nutrients from the host tissues to support essential fungal metabolic functions. The pathogen in xylem can grow and is expected to survive for a long time, potentially throughout the entire tree lifetime. Trees form cortex every year that provides a new vegetative tissue for the fungus to colonize and hence survive in old branches. On the other hand, if the pathogen only colonizes wounds without entering xylem, it will not able to survive for long since dead wood tissues around wounds are gradually decomposed.

Wounds are an important site for colonization, saprophytic survival, and infection of *V. mali*. Subject to conducive temperature and humidity conditions, spores can germinate on and colonize wound surfaces. Pruning wounds are an important entry site for *V. mali*. In the Yantai apple production region, nearly 80% of cankers originate from pruning wounds (Wang et al., 2012). Fresh pruning wounds are particularly susceptible to the pathogen. Pruning wounds made in March remain susceptible to the pathogens until May. Any damage to the cork layer can aggravate canker development, including freezing, sunburn, insect damage, cracks, mechanical injuries and so on. A sudden drop in temperature in early winter can have a significant impact on canker development because of freezing damages to trees. For instance, a period of severe cold weather in the winter of 1976 led not only to a high incidence of shoot and tree injuries, but also to increased canker incidence in the following 2 years. Early defoliation, excessive loading of fruits, drought, and waterlogging can lead to poor development of the cork layer, which can then be easily penetrated by the pathogen.

Conidia are released throughout the year in Xianyang, Shaanxi Province, China, with the maximum and minimum releases in the February to May and November to December periods, respectively (Du et al., 2013). The discharge of

V. mali spores requires high relative humidity (>95%). Under natural conditions, rainfall of 2–5mm can wet cankers sufficiently to release spores. Canker lesions with mature conidia can extrude a large number of conidia within 1–2 minutes after being wetted by rain (Fig. 1b). Temperature is not important for the release of *V. mali* spores. Mature perithecia and pycnidia may also discharge spores under snowing conditions.

The canker pathogen can spread with rain, pruning tools, insect vectors, planting material and so on. Both conidia and ascospores are primarily water-borne, that is dispersed with water (including rain splash and water droplets), but the spore dispersal potential (distance) is not clear. Whether ascospores can be dispersed with air (i.e. air-borne) is yet unclear. Pruning tools can be contaminated with the canker pathogen, and hence may spread the pathogen among trees during pruning, particularly during the winter period (Wang et al., 2016). Insects can be important for spreading the canker pathogens. In Xinjiang, China, the apple branch borer (*Agrilus mali* Matsumura) has apparently caused a large number of apple trees death in wild apple forests. Further investigation, however, showed that the direct cause of tree death is apple valsa cankers, not the apple branch borer. The apple branch borer not only spreads pathogen spores, but also causes many wounds for the canker pathogen to infect. As for many other pathogens, such as *Phytophthora* spp., *V. mali* can be easily spread from one region to another via the trade in planting material, since can survive long-distance travel.

Germination of conidia and ascospores requires free water or high humidity. In free water and relative humidity of 100%, 97%, and 95%, conidial germination can reach 48%, 35%, 18%, and 0%, respectively, when incubated at 0–35°C for 48 h. The corresponding values for ascospores are 69%, 62%, 57%, and 48%. Conidia can germinate on fresh pruning wounds and grow into xylem without the need for additional free water; after 1 week, the pathogen can be isolated in the xylem at a distance of 3 cm from the pruning wound (Chen et al., 2016). Additional free water may lead to more extensive internal colonization. Dead tissues (including xylem) may absorb water and retain water; thus, once spores have landed on such tissues, no additional water is needed for the spores to germinate and colonize at these sites. In the winter and early spring, conidia can still germinate but take longer. For instance, in Huludao, Liaoning Province, China, conidia need 18 days to germinate in late February when the temperature is around –5°C but only 2–3 days in early April. The pathogen can form stroma at 10–35°C with fewer stroma when temperature is below 20°C (Anon., 1971).

Apple valsa canker may remain latent inside host tissues for a long period. The length of the incubation period depends mainly on whether the pathogen can break through host defense and successfully invade living tissues. Wounds on apple shoots when inoculated with fungal mycelium plugs produce typical canker symptoms within 2–3 days, compared with 7–12 days when inoculated with conidia. After lesions become visible, lesions expand rapidly within the first 60 days. Canker lesions resume fast growth in the autumn and spring (Wang et al., 2016).

Under natural conditions, it takes as little as 34 days from initial infection to production of pycnidia and release of conidia (Anon., 1971). However, under artificial inoculations, pycnidia can be formed with 11 days of inoculation and release of conidia within 30 days. Canker lesions formed before October can produce and release conidia in the same calendar year; whereas lesions formed from October onwards usually can only produce conidia from April onwards the following year (Du et al., 2013). Ascospores are produced from July and perithecia are visible on large canker lesions in the autumn and winter.

Given the right conditions, canker lesions can release conidia throughout the growing season (from bud-break to leaf-fall) and colonize or infect branches. New canker lesions may produce spores in the same year, leading to new infections.

Canker development is greatly affected by temperature. Once infected, canker lesions can expand in the range 5–35°C with an optimum around 27°C. In the early spring, late autumn, and winter, although air temperature is low, apple shoots exposed to sunlight can have a surface temperature >5°C, allowing lesion expansion. During the growing season, the rate of canker expansion is usually far less affected by air temperature than by host resistance.

Orchard management practices can greatly affect apple canker development, primarily because these management practices affect tree growth potential (TGP).

There have been four major epidemics of apple canker in China since the late 1940s, all resulting from orchard mismanagement: (1) In the late 1940s, orchard management was neglected because of the civil war. Within a few years, about 1.4 million trees died in Liaoning Province, accounting for about 30% of total production at the time; (2) During the 'Great Leap Forward' movement around 1960, the blind pursuit of high yield led to a severe decline in TGP accompanied with another major canker outbreak; (3) In addition to freezing injury, orchard management was again neglected under the policy of 'Grain as the Guide,' resulting in the third major canker epidemic; (4) In the mid-1980s, most orchards were contracted out to growers/companies with insufficient knowledge of orchard management, leading to the fourth major canker epidemic.

Currently, there is no effective method for accurate and efficient canker monitoring in both nurseries and commercial orchards. A nested PCR method based on eight pairs of specific primers was developed to detect the latent canker pathogen in shoots (Zang et al., 2012b). However, this PCR-based detection method requires high technical skills and stringent testing conditions, and hence is difficult to apply in commercial practice. Due to the complexity of apple canker and the lack of in-depth research, accurate prediction of canker development is not yet possible.

5 Infection process of *V. mali*

Valsa mali is usually considered as a weak pathogen; its infection process can be divided into three stages: (1) spore germination, (2) colonization of dead tissues of wounds and branch surface or xylem, and (3) invasion into the living tissues of cortex (resulting in visible symptoms). After spore germination, the pathogen can also directly invade host living tissues although the probability of such an event occurring is low.

If wounds fail to heal in time, the pathogen can quickly penetrate the living tissues and cause canker lesions. When wound cortex of apple shoots was inoculated *in vitro*, conidia began to swell, then produce germ tubes, and finally penetrate host tissues 6, 16, and 20 h after inoculation (Ke et al., 2013). After 4 days, mycelia invade into the cortex and phloem and canker symptoms become visible on the bark surface. Two weeks after inoculation, mycelia penetrate into xylem vessels, producing pycnidia. *V. mali* mycelia mainly spread between host cell walls, leading to tissue infiltration and cell necrosis.

V. mali can also colonize and survive as a saprophyte in dead tissues on the surface of apple shoots; spores can germinate and grow utilizing nutrients in dead tissues. Because of the large number of spores produced and the ease with which spores can germinate and grow on the shoot surface, the incidence of shoots carrying the pathogen asymptomatically in diseased orchards can be very high. For instance, the incidence of one-year-old shoots with asymptomatic pathogen colonies may reach 100%. The higher the canker incidence, the higher the incidence of shoots with asymptomatic pathogen colonies on the shoot surface (Liu et al., 1979). At sites where apple trees are prone to water logging, dead wood tissues are more likely to harbour asymptomatic pathogen colonies. Since apple shoots form a cork layer annually from June to July, several layers of living cortical cells could provide nutrients for the pathogen, enabling pathogen survival for a long time in the dead layer of shoots. As a saprophyte, whether *V. mali* can invade the living cortical tissues to cause cankers depends on whether it could penetrate the cork layer. Artificial inoculation revealed that mycelia can grow and expand in the dead tissues of the wound but are not able to penetrate the cork layer formed on the wound periphery.

V. mali mainly penetrates shoots from the xylem of pruning wounds and uses the soluble nutrients of the xylem to grow along the gaps in the vessel, the medullary ray, the micropores and so on. However, the pathogen does not destroy the xylem (Fig. 3). Usually, infection via the xylem of pruning wounds does not lead to visible canker lesions until the spring the following year when

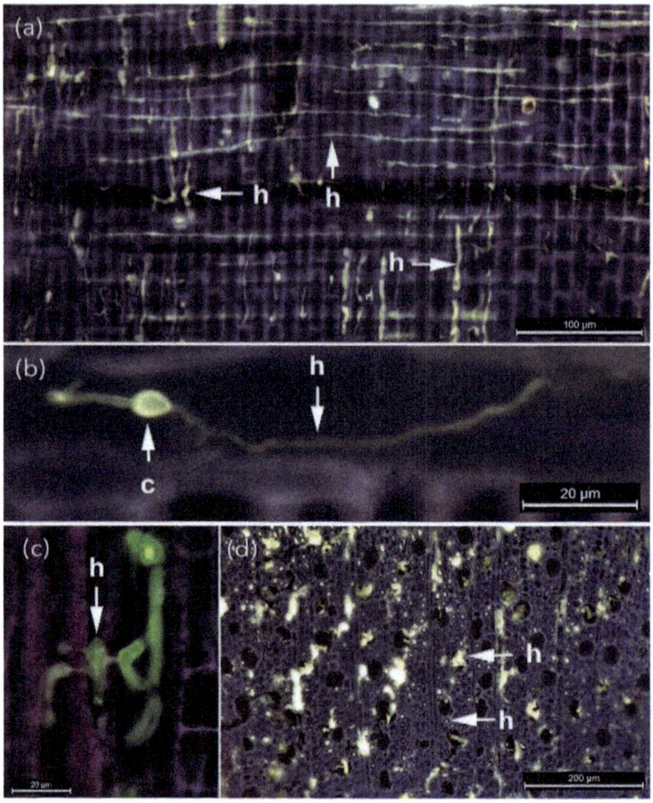

Figure 3 Hyphae (h) and conidia (c) of *Valsa mali* in the xylem tissue of detached 1-year-old 'Fuji' apple shoots inoculated with *V. mali* conidia suspension at pruning wounds. (a) A longitudinal section of xylem tissues showing *V. mali* hyphae colonizing the xylem vessels and tracheids; (b) A longitudinal section of xylem tissues showing germinated conidia within the xylem vessels; (c) A longitudinal section of xylem tissues showing *V. mali* hyphae passing through micropores on the sidewall of tracheids to enter into the adjacent tracheid; (d) A cross-section of the xylem tissues showing *V. mali* hyphal growth and expansion in the xylem vessels, tracheids, and rays. Source: adapted from Chen et al. (2016).

the pathogen reaches the cortex. Inoculation of pruning wounds of one-year-old branches led to 53%, 18%, and 16% of cankers in the spring and autumn the following year, and the spring in the further year after, respectively (Chen et al., 2016).

Pathogen reactivation in old lesions is an important mechanism for valsa canker development, particularly those lesions on pruning wounds. If hyphae in the xylem are not fully removed/killed, the pathogen can grow and expand in the xylem and persist for a long time. When conditions are appropriate, old canker lesions can then resume active development and produce spores.

6 Pathogenesis of *V. mali*

Valsa mali mainly causes the shoot cortex to rot. Mycelia predominantly spread in the intercellular space, first invading the parenchyma of the cortex and then the thick-walled tissue. Plasmolysis and protoplast coagulation occur in the peripheral host cells and the electron density in the cell wall decreases, indicating that peripheral cells have been affected by the exocrine of the pathogen. Hyphae can then invade and expand within the host cell wall, degrading and digesting host cell walls. The cortical tissues are largely digested whereas those cells that have not been digested are loosely arranged. Mycelia can then further invade into the xylem, leading to digestion of the cambia and cortical tissues around the xylem (Ke et al., 2013).

V. mali produces various enzymes, toxins, and pathogenicity factors that interfere with normal tree physiological activities. A total of 585 pathogenicity-related proteins and 215 candidate effector proteins have been identified in pathogen transcriptome during infection (Li et al., 2015). However, no key virulence factors have been identified yet, suggesting that *V. mali* pathogenesis may result from multiple factors.

V. mali can produce five types of cell wall degrading enzymes, namely xylanase, carboxymethyl-cellulase, β-glucosidase, pectin-methyl-polygalacturonase and polygalacturonase. Among them, xylanase and cellulase have the highest and lowest activity, respectively. In an *in vitro* inoculation study, canker symptoms were observed 3–5 days after inoculation, but the enzyme activity was significantly increased 5 days after inoculation and reached the peak 9–13 days after inoculation (Chen et al., 2012a), indicating that these enzymes may play different roles in the infection and pathogenesis of the pathogen.

Pectinase is an important cell wall degrading enzyme produced during infection of *V. mali*. Transcriptomic analysis revealed that the pectinase gene is significantly upregulated during the infection, which is supported by research with electron microscopic and immuno-colloidal gold cytochemical labelling techniques showing pectin decomposition during infection. The pectinase activity is positively correlated with the pathogenicity during the infection (Yin et al., 2015). This is further supported by gene knock-out studies (Wu et al., 2017). In the inoculated apple shoots, pectin-methyl-galacturonase activity increased significantly within 1–2 days after the onset of visible lesions, whereas polygalacturonase (PG) activity increased significantly after 4–6 days (Chen et al., 2012b).

The PG genes *Vmpg2*, *Vmpg3*, *Vmpg7*, and *Vmpg8* are upregulated during infection and colonization. Their role in pathogenicity is also confirmed in gene knock-out studies, which also indicated that these genes are not essential for pathogenicity (Xu et al., 2016). Knocking out other PG genes (*Vmpgx-1, Vmpgx-2, Vmpgx-3, and Vmpgx-4*) had similar effects. The pectate lyase (PL) gene *Vmpl4* is upregulated by 10-fold during infection. Knocking out this gene leads to reduced pathogenicity and upregulation of the other four genes in the same family (Xu et al., 2017). Other cell wall degrading enzymes also play an important role in the pathogenesis of *V. mali*, including xylem (Yu et al., 2018), and β-glucosidase gene (Li et al., 2017b).

A number of genes have been studied for their pathogenicity effects in *V. mali*, including: (1) fungi-specific regulatory factors, velvet protein family genes (Wu et al., 2017); (2) gtp-cyclophosphamide II gene, *Vmgtp1* (Song et al., 2014); (3) high mobility protein, HMG (Jia et al., 2018); (4) non-ribosomal polypeptide synthetase gene, *VmNRPS12* (Ma et al., 2016); (5) ferulic acid esterase B precursor gene; (6) transcription factor; (7) hexokinase genes, *Vmgxk1* and *Vmhxk1*; and (8) zinc-related genes, *Vmzfp3* and *Vntme42* (Li et al., 2017c). These studies showed that all the genes are involved in the metabolism or metabolic regulation of pathogens and affect pathogen growth, reproduction, and pathogenicity.

V. mali produces a variety of candidate effector proteins, which play important roles in pathogenesis, either by inhibiting host resistance responses or by inducing host cell death. There are 193 unidentified proteins in the exocrine proteins of *V. mali*, 101 of which are unique to *V. mali* (Li et al., 2015). These proteins or polypeptides have short amino acid sequences and contain abundant cysteine residues, which may interact with host cells and could be considered as candidate effectors. Among those 70 effectors tested, 7 of them inhibit programmed cell death in *N. benthamiana* induced by the mouse apoptotic protein BAX (Li et al., 2015). Cytochrome P450 (CYP) plays an important role in the toxin anabolism of dead saprotrophic fungi. Toxin synthesis gene (*Vmcyp5*) is upregulated during *V. mali* infection, and is related to melanin synthesis, fruit body production, and pathogenicity (Gao et al., 2018).

V. mali can degrade phlorizin in apple stem cortex during infection and produces five intermediates: protocatechuic acid, *p*-hydroxybenzoic acid, *p*-hydroxyacetophenone, *p*-hydroxyphenylpropionic acid, and phloroglucinol. These intermediates are toxic to apple cortical tissues, particularly protocatechuic and followed by *p*-hydroxyacetophenone and *p*-hydroxyphenylpropionic acid. *V. mali* pathogenicity is positively correlated with the concentration of these toxins (Wang et al., 2014b). Phlorizin is considered as one characteristic substance in apple; apple leaves, shoots, and bark tissues all contain phlorizin, particularly in bark tissues.

7 Resistance to *V. mali*

There are no apple genotypes in the Chinese breeding germplasm that are completely resistant to *V. mali*. Nevertheless, both rootstock and scion genotypes differ greatly in the size of canker lesions, indicating presence of some resistance against pathogen colonization in the cortex. Western apple germplasm is more resistant than wild *Malus* species but the level of resistance is not high (Abe et al., 2010; Liu et al., 2011b). There are four types of resistance identified in 36 apple cultivars commonly used in China (Yin et al., 2017). Resistance of 1150 hybrid seedlings from the 'Purple Pearl' × 'Fuji' cross was evaluated through conidial inoculation of detached shoots *in vitro* (Gao et al., 2012); results suggest that presence or absence of canker lesions is controlled by 4–5 major genes with the susceptible alleles as dominant. In contrast, lesion length (fungal colonization) is polygenically controlled and several QTLs have been identified (Tan et al., 2017).

The occurrence of valsa canker is closely associated with tree growth: new lesions become visible mostly in the early spring. This growth-related resistance may be considered as one specific type of ontogenetic resistance. When water and nutrients required for tree growth are met, TGP is considered to be high and hence increased resistance to the canker pathogen, resulting in fewer and smaller lesions. Ontogenetic resistance against valsa canker could be due to the antitoxin reactions, which is characterized by the formation of cell wall lignification callus pericarp. The ability of forming callus is closely related to TGP, temperature, shoot moisture content, and tree fruit loading and nutrient status (Chen et al., 1982).

Host resistance is related to the activity of host defense enzymes and malondialdehyde (MDA) content. Infection by *V. mali* can induce activity of defense-related enzymes, including peroxidase (POD), polyphenol oxidase (PPO), superoxide dismutase (SOD), and phenylalanine ammonia lyase (PAL). The activities of these enzymes in the callus of resistant rootstock genotypes are maintained at a high level throughout the infection process, whereas in the callus of the susceptible variety 'Fuji,' the enzymatic activities quickly decrease to the original level (Wang et al., 2014a).

There is limited understanding of molecular mechanisms unpinning apple resistance to *V. mali*. Upregulated genes following inoculation are mainly those involved in metabolic activities such as chitin, hormones (e.g. jasmonic and salicylic acid), and cell death. Downregulated genes are mainly involved in the light response and chlorophyll biosynthesis. After infection of susceptible apple cultivars, auxin and abscisic acid (ABA) signal-related genes are downregulated (Yin et al., 2016). A large number of microRNAs (miRNAs) are differentially expressed in infection of 'Fuji'; their target genes are mainly those involved in metabolic activities, such as apoptosis, transcriptional regulation, auxin signalling pathway, ATP binding, DNA binding, and protein binding activity (Feng et al., 2017).

8 Control strategies against apple valsa canker

In the early 1900s, one key control measure was to remove diseased woods or scrap off canker lesions in the early stage of canker development. In Japan, in the 1920s as well as in China in the 1950s, extensive research studies were conducted on how to remove canker lesions and to protect subsequent wounds from canker infection. Even now, scraping off lesions is still an important canker management measure in commercial apple production. Another important control method is to apply protectant agents to protect wounds from infection by the pathogen. In recent times, protectant agents have been changed from inorganic fungicides, such as Bordeaux pulp, to modern fungicides and biopesticides. The matrix of wound protectants is also changed from water and oil to polymer compounds with film-forming capability. Some protectant products also include plant hormones to promote wound healing.

Eliminating latent infections on shoots is becoming the main measure in managing apple canker. Prior to the 1970s, canker management had primarily relied on applying protectant fungicides although this still failed to achieve satisfactory canker control. Since the finding of latent infections on shoots in the 1960s (Liu et al., 1979), the focus in canker management has been gradually moving to eliminating these latent infections. Various measures have been proposed to eradicate latent infections, including (1) scraping off dead bark tissues during the tree dormant period, (2) scraping off shoot surfaces during the period of vigorous tree growth, and (3) spraying or coating shoots with 40% Asomate WP solution just before bud-break and also in early July. As Asomate was banned in 2002 in China, humic acid copper, hydrogen peroxide, thiophanate-methyl, and flusilazole are now used instead to eradicate latent infections nut are not as effective as Asomate.

A set of prevention and control strategies have been proposed and implemented in China since the 1980s. This includes: (1) highlighting crop husbandry management as the key measure to minimize canker development through increased TGP; (2) eliminating/reducing pathogen inoculum in orchards; (3) treating canker lesions in a timely manner and minimizing shoot and tree death; and (4) protecting wounds from canker infection and preventing trees from freezing damage.

A foundation for minimizing canker development is by increasing TGP through crop management. This requires diverse crop husbandry practices. Current crop management usually focusses on increasing fruit yield and quality, and rarely considers disease management. Increasing potassium content in leaves to 1.3% and ensuring nitrogen: potassium within 2:1 can improve tree resistance to apple canker (Peng et al., 2016). However, in commercial apple production, application of nitrogen and potassium fertilizers is usually to improve tree growth and increase fruit yield and quality without consideration of disease development.

Eliminating latent infection in shoots is a key measure to eliminating pathogen inoculum. However, currently there are no reliable and easy methods to detect latent infections in shoots, hence making elimination of latent infection virtually impossible. Furthermore, there are also an insufficient number of effective eradication agents to eliminate latent infections. Scraping off canker lesions in a timely manner is important for minimizing canker development in orchards. Indeed, this may be the only measure to manage canker disease in some regions. Reducing canker entry sites is primarily achieved through protection of wounds via wound coating with macromolecular film-forming agents containing fungicides, phytohormones, or beneficial microbes.

Recently, Li et al. (2017a) proposed that canker disease management should focus on eliminating/minimizing primary inoculum, reducing infectious lesions, and minimizing the risk of accumulation of canker lesions in orchards. This means disease management in nurseries as well as in commercial orchards. Disease management measures in the orchard phase are similar to those previously proposed with the focus on inoculum reduction (Li et al., 2017a; Zhang et al., 2017). The new strategy of managing canker disease has been implemented and is under evaluation in Yantai, Shandong Province, China.

9 Conclusion

Apple valsa canker is the most important disease of apple in East Asia, and the main cause agent is now confirmed as *Valsa mali*. *V. mali* is a typical necrotrophic fungus, which can colonizes and survives on dead tissues or xylem of apple branches. Once colonized, the pathogen invades living tissue and causes canker where it can find an entry. No germplasm has been found to be completely resistant to *V. mali*. The occurrence of apple valsa canker is closely associated with tree growth. The genome of *V. mali* is predicted to encode 11261 proteins, mainly related to infection, colonization, secreted proteases, enzymes related to secondary metabolite synthesis, and transport proteins. No key virulence factors have been identified so far, suggesting that *V. mali* pathogenesis may result from multiple factors. There are no effective measures to control apple valsa canker, and the best strategy is to try to prevent accumulation of the inoculum in the orchards.

10 Where to look for further information

Most research work on apple valsa canker has been carried out in China and Japan. Chinese publications on apple valsa canker can be found on the China Knowledge Resource Integrated Database (CNKI), which is available at: http://www.cnki.net/, Wanfang Data (http://www.wanfangdata.com.cn/), or Baidu Xueshu (http://xueshu.baidu.com/). Japanese publications, mostly before the 1980s, can be found on Japanese Journal of Phytopathology (https://www.ppsj.org/).

The book edited by Ce Cheng, 'Study on occurrence and control of apple valsa canker' (in Chinese), compiled reports, papers, and reviews on apple valsa canker from the 1950s to 2010s, and is available at: http://www.dangdang.com/.

The China Agriculture Research System on Apple conducted most research works on apple valsa canker in China.

11 References

Abe, K., Kotoda, N., Kato, H. and Soejima, J. 2010. Resistance sources to Valsa canker (*Valsa ceratosperma*) in a germplasm collection of diverse *Malus* species. *Plant Breeding* 126, 449–53.

Adams, G. C., Wingfield, M. J., Common, R. and Roux, J. 2005. Phylogenetic relationships and morphology of *Cytospora* species and related teleomorphs (Ascomycota, Diaporthales, Yalsaceae) from Eucalyptus – Preface. *Studies in Mycology* 52, 1–144.

Anon. 1971. A report on the occurrence, infection and pathogen biological characteristics of valsa apple cankers. *Liaoning Agricultural Sciences* 1971, 2–20.

Cao, K., Guo, L., Li, B., Sun, G. and Chen, H. 2009. Investigations on the occurrence and control of apple canker in China. *Plant Protection* 35, 114–6.

Chen, C., Liu, F., Xing, Z., Zhang, X., Shi, X., Guo, J. and Chen, Y. 1982. On the resistance of apple trees to the invasion by *Valsa mali* Miyabe et Yamada: observation on the relationship between wound healing capaty of apple bark and the level of resistance. *Acta Phytopathologyca Sinica* 12, 10–20.

Chen, C., Li, B. H., Dong, X. L., Wang, C. X., Lian, S. and Liang, W. X. 2016. Effects of temperature, humidity, and wound age on *Valsa mali* infection of apple shoot pruning wounds. *Plant Disease* 100(12), 2394–401. doi:10.1094/PDIS-05-16-0625-RE.

Chen, X., Niu, C., Li, B., Li, G. and Wang, C. 2012a. The kinds and activities of cell wall-degrading enzymes produced by *Valsa ceratosperma*. *Acta Agriculturae Boreali-Sinica* 27, 207–12.

Chen, X. L., Niu, C. W., Bao-Hua, L. I., Gui-Fang, L. I. and Wang, C. X. 2012b. The kinds and activities of cell wall-degrading enzymes produced by *Valsa ceratosperma*. *Acta Agriculturae Boreali-Sinica* 27, 207–12.

Du, Z., Li, Z., Gao, X., Huang, L. and Han, Q. 2013. Study on the conidia dispersal and the disease dynamics of apple tree canker caused by *Valsa mali* var. *mali* in Shaanxi. *Journal of Fruit Science* 30, 819–22.

Feng, H., Xu, M., Zheng, X., Zhu, T., Gao, X. and Huang, L. 2017. microRNAs and their targets in apple (*Malus domestica* cv. "Fuji") involved in response to infection of pathogen *Valsa mali*. *Frontiers in Plant Science* 8, 2081. doi:10.3389/fpls.2017.02081.

Gao, T., Liu, X., Zhang, X., Wang, Y., Li, T. and Han, Z. 2012. Inheritance of susceptibility to Valsa canker in *Malus* interspecific hybrids. *Journal of Fruit Science* 29, 717–21.

Gao, M., Wu, Y., Zhu, B., Gao, X., Hao, F. and Huang, L. 2018. Characterization of cytochrome P450 gene *Vmcyp5* in *Valsa mali*. *Acta Microbiologica Sinica* 58, 274–83.

Jia, X., Wang, Y., Sun, G., Sun, G., He, B., Ge, S., Wang, Y. and Liu, H. 2018. Production and phenotypic analysis of mutants with knockout of *VmHMG* gene in *Valsa mali* var. *mali* causing apple canker. *Journal of Nuclear Agricultural Sciences* 32, 448–54.

Ke, X., Huang, L., Han, Q., Gao, X. and Kang, Z. 2013. Histological and cytological investigations of the infection and colonization of apple bark by *Valsa mali* var. *mali*. *Australasian Plant Pathology* 42(1), 85–93. doi:10.1007/s13313-012-0158-y.

Kobayashi, T. 1970. Taxonomic studies of Japanese Diaportliaccac with special reference to their life-histories. *Bulletin of the Government Forest Experiment Station = Ringyo Shikenjo kenkyu hokoku* 226, 242.

Li, Z., Yin, Z., Fan, Y., Xu, M., Kang, Z. and Huang, L. 2015. Candidate effector proteins of the necrotrophic apple canker pathogen *Valsa mali* can suppress BAX-induced PCD. *Frontiers in Plant Science* 6, 579. doi:10.3389/fpls.2015.00579.

Li, B., Lian, S., Wang, C., Dong, X. and Zhang, Z. 2017a. To dicusse again the prevention and treatment of apple valsa canker from the infection by the pathogen from the xylem. *China Fruits* 2017, 88–90.

Li, T., Lian, S., Li, B., Liang, W. and Wang, C. 2017b. Molecular cloning and expression analysis of β-glucosidase gene from *Valsa mali* var. *mali*. *Acta Agriculturae Boreali-Sinica* 32, 78–84.

Li, T., Wu, Y., Cao, K., Wang, S., Huang, L. and Feng, H. 2017c. Screening of *Valsa mali* T-DNA inserted mutants and gene function analysis of *Vmzfp3*. *Journal of Agricultural University of Hebei* 40, 73–8.

Liu, F., Chen, C., Shi, X., Guo, J. and Xing, Z. 1979. Studies on the latent infection of the causal organism of valsa canker of apple. *Journal of Plant Protection* 6, 1–8.

Liu, R., Wang, S., Si, L., Cheng, C., Hu, T. and Cao, K. 2011a. Effect of environmental factor on the conidiospore germination and survival of apple tree canker pathogen. *Journal of Anhui Agricultural Sciences* 2011, 16009–12.

Liu, X., Lv, S., Wang, Y., Wang, K., Li, T., Han, Z. and Zhang, X. 2011b. Evaluation of resistance of malus germplasms to apple canker (*Valsa ceratosperma*). *Journal of Fruit Science* 28, 843–8.

Ma, C., Li, Z., Dai, Q., Han, Q. and Huang, L. 2016. Function of nonribosomal peptide synthetase gene *VmNRPS12* of *Valsa mali*. *Wei Sheng Wu Xue Bao = Acta Microbiologica Sinica* 56(8), 1273–81.

Peng, H. X., Wei, X. Y., Xiao, Y. X., Sun, Y., Biggs, A. R., Gleason, M. L., Shang, S. P., Zhu, M. Q., Guo, Y. Z. and Sun, G. Y. 2016. Management of Valsa canker on apple with adjustments to potassium nutrition. *Plant Disease* 100(5), 884–9. doi:10.1094/PDIS-09-15-0970-RE.

Song, N., Dai, Q., Song, N., Huang, L. and Han, Q. 2014. Construction of knockout vector of GTP cyclohydrolase II gene and mutant's biological characteristics of *Valsa mali*. *Scientia Agricultura Sinica* 47, 2980–9.

Tan, Y., Shen, F., Chen, R., Wang, Y., Wu, T., Li, T., Xu, X., Han, Z. and Zhang, X. 2017. Candidate genes associated with resistance to Valsa canker identified via quantitative trait loci in apple. *Journal of Phytopathology* 165(11–12), 848–57. doi:10.1111/jph.12658.

Tanaka, T. 1919. New Japanese fungi notes and translations: VII. *Mycologia* 11, 148–54.

Wang, X., Wei, J., Huang, L. and Kang, Z. 2011. Re-evaluation of pathogens causing Valsa canker on apple in China. *Mycologia* 103(2), 317–24. doi:10.3852/09-165.

Wang, C., Dong, X., Zhang, Z., Li, G. and Li, B. 2012. Outbreak and the reasons of apple valsa canker in Yantai apple production area in 2011. *Plant Protection* 3, 035.

Wang, C., Chen, X. and Li, B. 2014a. Effects of *Valsa mali* var. *mali* infection on defense enzymes activity and MDA content in apple callus. *Plant Physiology Journal* 50, 909–16.

Wang, C., Li, C., Li, B., Li, G., Dong, X., Wang, G. and Zhang, Q. 2014b. Toxins produced by *Valsa mali* var. *mali* and their relationship with pathogenicity. *Toxins* 6(3), 1139–54. doi:10.3390/toxins6031139.

Wang, X., Zang, R., Yin, Z., Kang, Z. and Huang, L. 2014c. Delimiting cryptic pathogen species causing apple Valsa canker with multilocus data. *Ecology and Evolution* 4(8), 1369–80. doi:10.1002/ece3.1030.

Wang, S., Hu, T., Wang, Y., Luo, Y., Michailides, T. J. and Cao, K. 2016. New understanding on infection processes of Valsa canker of apple in China. *European Journal of Plant Pathology* 146(3), 531–40. doi:10.1007/s10658-016-0937-3.

Wang, X., Pan, T., Lian, S., Wang, C. and Li, B. 2018. Effects of environmental factors on the growth and extension of *Valsa mali* in the xylem of apple branches. *Scientia Agricultura Sinica* 52, 3291–301.

Wu, Y., Xu, L., Yin, Z., Dai, Q., Gao, X., Feng, H., Voegele, R. and Huang, L. 2017. Two members of the velvet family, Vm VeA and Vm VelB, affect conidiation, virulence and pectinase expression in *Valsa mali*: Vm VeA and Vm VelB in *Valsa mali*. *Molecular Plant Pathology* 19(7).

Xu, C., Wu, Y., Dai, Q., Li, Z., Xiaoning, G. and Huang, L. 2016. Function of polygalacturonase genes *Vmpg7* and *Vmpg8* of *Valsa mali*. *Scientia Agricultura Sinica* 49, 1489–98.

Xu, C., Sun, Y., Yuxing, W., Feng, H., Gao, X. and Huang, L. 2017. Pathogenic function of pectate lyase gene *Vmpl4* of *Valsa mali* in apple. *Journal of Fruit Science* 34, 19–25.

Yin, Z., Liu, H., Li, Z., Ke, X., Dou, D., Gao, X., Song, N., Dai, Q., Wu, Y., Xu, J. R., Kang, Z. and Huang, L. 2015. Genome sequence of Valsa canker pathogens uncovers a potential adaptation of colonization of woody bark. *The New Phytologist* 208(4), 1202–16. doi:10.1111/nph.13544.

Yin, Z., Ke, X., Kang, Z. and Huang, L. 2016. Apple resistance responses against *Valsa mali* revealed by transcriptomics analyses. *Physiological and Molecular Plant Pathology* 93, 85–92. doi:10.1016/j.pmpp.2016.01.004.

Yin, H., Zhou, J., Lv, H., Zhang, Z., Qin, N., Niu, G. and Zhao, X. 2017. Evaluation of resistance of 36 apple tree varieties to *Valsa mali*. *Journal of Shanxi Agricultural Sciences* 45, 998–1001.

Yu, C., Li, T., Shi, X., Saleem, M., Li, B., Liang, W. and Wang, C. 2018. Deletion of endo-β-1,4-xylanase *VmXyl1* impacts the virulence of *Valsa mali* in apple tree. *Frontiers in Plant Science* 9, 663. doi:10.3389/fpls.2018.00663.

Zang, R., Huang, L., Li, Z., Gao, X. and Kang, Z. 2012a. Polymorphism analysis of *Valsa mali* isolated from Shaanxi Province using ISSR markers. *Acta Phytophylacica Sinica* 39, 51–7.

Zang, R., Yin, Z., Ke, X., Wang, X., Li, Z., Kang, Z. and Huang, L. 2012b. A nested PCR assay for detecting *Valsa mali* var. *mali* in different tissues of apple trees. *Plant Disease* 96(11), 1645–52. doi:10.1094/PDIS-05-11-0387-RE.

Zhang, Z., Li, B., Lian, S., Wang, C., Dong, X., Li, P., Zhang, M., Xu, Y., Han, W. and Li, S. 2017. Principles and techniques of apple pests and diseases management for saving cost and improving effectiveness. *China Fruits* 2017, 1–7.

Chapter 6

Brown rot: causes, detection and control of *Monilinia* spp. affecting tree fruit

Imre Holb, University of Debrecen and Hungarian Academy of Sciences, Hungary

1 Introduction

Monilinia spp. cause economically important fungal diseases (brown rot) of pome and stone fruits. Here we focus on four *Monilinia/Monilia* species which are *Monilinia fructicola* (Winter) Honey, *Monilinia laxa* (Aderhold & Ruhland) Honey, *Monilinia fructigena* Honey ex Whetzel, and *Monilia polystroma* van Leeuwen, all of which cause significant yield losses around the world.

2 Yield loss and impact on crops

Yield losses caused by brown rot fungi can take place at three stages: (1) pre-harvest in the orchard during blossom, (2) pre-harvest in the orchard during fruit swelling and ripening, and (3) post-harvest during storage and shelf life (Byrde and Willetts, 1977; Holb, 2004a). The level of yield losses can vary considerably depending on climatic conditions, fruit species, cultivar, production system and storage conditions. Fruit crops require regular pre- and post-harvest disease management to control brown rot effectively.

http://dx.doi.org/10.19103/AS.2019.0046.06

In the case of pome fruits, incidences of trees with brown rot, and yield losses caused by *Monilinia* spp. (mainly *M. fructigena*) are usually below 10% in well-managed orchards (e.g. Moore, 1950; van Leeuwen et al., 2000; Holb and Scherm, 2007; Holb, 2008; Xu et al., 2001b; Berrie and Holb, 2014). However, considerable pre-harvest yield losses of up to 45% have been reported from organic pome fruit orchards (Holb, 2004a; Holb and Scherm, 2007; Holb et al., 2011; Berrie and Holb, 2014). Yield loss in unsprayed orchards can range between 50% and 90% (Burchill and Edney, 1972; Garic et al., 1990; Holb and Scherm, 2008). Post-harvest fruit rot of pome fruit, particularly where losses are caused by *M. fructigena*, can be up to 22%, but the loss is usually below 1% in cold storage under CA and ULO storage systems (Falconi and Mendgen, 1994; Holb et al., 2012; Berrie and Holb, 2014).

Monilinia fructicola is considered the most damaging pathogen of the brown rot species, due to the severe losses (up to 90%) in stone fruits, especially in peaches, nectarines and plums in the United States, Canada and Australia (e.g. Byrde and Willetts, 1977; Ogawa and English, 1991; Emery et al., 2000; Luo and Michailides, 2001, 2003). Post-harvest losses of peaches and nectarines can exceed 50% (Janisiewicz and Korsten, 2002), and approach 100% under conducive environmental conditions and high disease pressure (Martínez-García et al., 2013).

Monilinia laxa causes pre-harvest yield loss during the blossom stage due to blossom and shoot blight. In well-managed orchards, blossom and shoot blight incidences usually do not exceed 5%; while blossom damage can range between 10% and 35% in organic stone fruit orchards (e.g. Holb and Schnabel, 2005; Everhart et al., 2011; Holb et al., 2013). Pre-harvest fruit rot is below 10% in well-managed stone fruit orchards (Holb et al., 2013); however, it can be up to 25% in organic production systems (Holb et al., 2013).

3 Causal organisms of brown rot

The genus *Monilinia* belongs to the family Sclerotiniaceae of the Discomycetes, a subdivision of the Ascomycotina. The genus is characterized by the production of conidial and stromatal anamorphs, apothecial ascomata and ascospores (Byrde and Willets, 1977). Honey (1936) subdivided the genus of *Monilinia* into two sections: Junctoriae and Disjunctoriae. In the Disjunctoriae, macroconidia in the conidial chain are separated by a small structure (disjunctor), while in the Junctoriae these structures are absent and are known as the non-disjunctor *Monilinia* species. The four main causal agents of brown rot diseases discussed in this chapter belong to the section Junctoriae, which are polytrophs, not well-specialized pathogens, and have relatively wide geographical distributions (Byrde and Willets, 1977; Holb 2003). *Monilinia fructicola* (G. Winter) Honey is referred to as American brown rot. *Monilinia laxa* (Aderhold & Ruhland) Honey and *M. fructigena* Honey ex Whetzel are referred to as European brown rot. Recently, a new brown rot species was found. *Monilia polystroma* van Leeuwen, referred to as Asiatic brown rot, is closely related to *M. fructigena* (van Leeuwen et al., 2002a). The term 'brown rot fungi' is a common name referring to these four non-disjunctor species.

4 Major fruit hosts and host resistance to brown rot

Brown rot fungi attack fruit trees of the plant family *Rosaceae* in particular the *Prunoideae* and *Pomoideae* sub-family. These groups contain several important orchard crops, which gives worldwide importance to these pathogens. Here, only the economically relevant fruit hosts are given, other host plants for the brown rot species are given by Batra (1991) and Holb (2003). *Monilinia fructicola* has major hosts in the *Prunoideae* including peach (*Prunus persica* (L.) Batsch), apricot (*P. armeniaca* L.), plum (*P. domestica* L.), sweet cherry (*P. avium* L.), sour cherry (*P. cerasus* L.) and almond (*P. dulcis* L.). However, this fungus also attacks quince (*Cydonia oblonga* Mill.), apple (*Malus domestica* Borkh.) and pear (*Pyrus communis* L.) from the *Pomoideae* sub-family. *Monilinia laxa* also has the same major hosts of stone fruit species as *M. fructicola*. This fungus can also attack species of pome fruits such as quince, apple, pear and medlar (*Eriobotrya japonica* (Thunb.)). Apple, pear and quince are the major host for *M. fructigena*. However, this fungus may also attack medlar, peach, apricot, plum, sweet cherry, sour cherry, almond, and hazelnut (*Corylus avellana* L.). *Monilia polystroma* is assumed to have a similar range of hosts to *M. fructigena* as they are closely related to *M. fructigena*.

Host resistance to brown rot fungi greatly differs among and within fruit species (Byrde and Willetts, 1977; Holb, 2004a). Resistance features are associated with some physiological characteristics of the fruit such as thick skin, high acidity and high phenolic content (Byrde and Willetts, 1977), but these have been negatively correlated the marketing quality of such cultivars (Holb, 2004a). Brown rot resistance is also dependent upon fruit maturity stages: mature fruit is more susceptible to brown rot fungi compared to immature fruit (van Leeuwen et al., 2000; Xu et al., 2000).

Several cultivars were evaluated and bred for brown rot resistance (Walter et al., 2004) and host-cultivar examples are given for both low and high levels of resistance to brown rot (Table 1).

In the case of sweet cherry, fruit rot and blossom blight are the major symptoms caused by brown rot fungi. Fruit cracking or other types of wounding are needed for the infection caused by *M. fructigena*. Bargioni (1982) demonstrated that sweet cherry cultivars with thin fruit skins are more susceptible to brown rot than those with thicker skins. In a Canadian breeding program for resistance to brown rot caused by *M. fructicola*, partial resistance

Table 1 Examples of resistant and susceptible fruit cultivars to brown rot caused by *Monilinia* spp.

Fruit	Host resistance	Plant organ	Cultivar, genotype	References
Sweet cherry	Low susceptibility	Fruit	Cultivars with thick fruit skins	Bargioni (1982)
Sweet cherry	Low susceptibility	Fruit	Staccato TM, Stardust, Sweetheart, Santina, Skeena and 13N-07-32	Kappeland and Sholberg (2008)
Sour cherry	Low susceptibility	Blossom, twig	Csengődi, Akasztói, Cigánymeggy 59	Apostol (2005), Rozsnyay and Apostol (2001)
Sour cherry	Low susceptibility	Blossom, twig	Lativiszkaja Nizkaja, Nagy Angol, Mocanesti, Ljubszkaja, Sirpotreb, Oblacsinszkaja, Cigánymeggy 3, Maraska Savena, Mettar, Elegija	Soltész (1997)
Apricot	High susceptibility	Fruit	Royal, Blenheim, Perfection, Derby Royal	Hesse (1938)
Apricot	Tolerant	Blossom, twig	Neptun, Mamaia, Silvana, Sulina, Sirena	Cociu cit. Soltész (1997)
Apricot	High susceptibility	Blossom, twig	Budapest, Mandulakajszi	Soltész (1997)
Apricot	Moderate susceptibility	Blossom, twig	Ceglédi óriás, Liget óriás, Polonais	Soltész (1997)
Apricot	Low susceptibility	Blossom, twig	Borsi-féle kései rózsa, Piroska, Pannónia, Ceglédi bíborkajszi, Magyar kajszi, Rakovszky	Soltész (1997)
Peach	High susceptibility	Fruit	Shipley	Koroknai (1971)
Peach	Moderate susceptibility	Fruit	Alexander, Amsden, Champion, Ford, Győztes, Mayflower	Koroknai (1971)
Peach	Low susceptibility	Fruit	Canada, Carman, Elberta, J.H. Hale, Incrotio Pieri, Magyar arany duránci	Koroknai (1971)
Peach	Tolerant	Blossom	Bolinha	Feliciano et al. (1987), Ogawa and English (1991), Gradziel and Wang (1993), Gradziel et al. (1998)
Plum	High susceptibility	Blossom	Santa Rosa, Wickson, Imperial, French	Ogawa and English (1991)

Fruit	Host resistance	Plant organ	Cultivar, genotype	References
Peach	High susceptibility	Fruit	J.H. Hale, Champion	Soltész (1997)
Peach	Resistant	Fruit	Resistant introgression line F8,1-42	Martínez-García et al. (2013)
Plum	High susceptibility	Fruit	Bluefre, President, Stanley	Soltész (1997)
Plum	Moderate susceptibility	Fruit	Cacanska najbolja, President	Soltész (1997)
Plum	Low susceptibility	Fruit	Besztercei, Silvia, Tuleu gras	Soltész (1997)
Almond	High susceptibility	Blossom, twig	Drake, Jordanolo	Ogawa et al. (1986)
Almond	Moderate susceptibility	Blossom, twig	Ne Plus Ultra, Texas	Ogawa et al. (1986)
Apple	Partial susceptibility	Fruit rot	Early ripening cultivars with low cracking susceptibility	Holb and Scherm (2007, 2008)

was observed with the sweet cherry cultivars 'Staccato™', 'Stardust', 'Sweetheart', 'Santina', 'Skeena' and '13N-07-32' (Kappeland and Scholberg, 2008).

In the case of sour cherry, blossom and twig blight are the most important symptoms caused by brown rot fungi. There are some sour cherry cultivars with low susceptibility or with disease tolerance, such as 'Lativiszkaja Nizkaja', 'Nagy Angol', 'Mocanesti', 'Ljubszkaja', 'Sirpotreb', 'Oblacsinszkaja', 'Cigánymeggy 3', 'Maraska Savena', 'Mettar' and 'Elegija' (Cociu cit Soltész, 1997). Moreover, Rozsnyay and Apostol (2001) and Apostol (2005) indicated that cvs. 'Csengődi', 'Akasztói' and 'Cigánymeggy 59' were partially resistant to *M. laxa*. Cultivars 'Újfehértói fürtös' ('Balaton') and 'Édri bőtermő' ('Danube') are considered to be highly susceptible to *M. laxa* during the blossom stage (Holb and Schnabel, 2005; Holb et al., 2013).

In the case of apricot, both fruit rot and blossom blight caused by *M. fructigena* or *M. laxa,* respectively, can be severe. High susceptibility of cultivars is associated with late blooming and with fruit vulnerability at the fruit ripening stage. Hesse (1938) demonstrated that apricot cvs. 'Royal', 'Blenheim', 'Perfection' and 'Derby Royal' were highly susceptible to blossom infection, whereas 'Tilton' was noticeably less susceptible. In the 1980s, the level of cultivar susceptibility was characterized according to Guerriero and Watkins (1984). In Romania, several apricot cultivars tolerant to brown rot were bred, such as 'Neptun', 'Mamaia', 'Silvana', 'Sulina' and 'Sirena' (Cociu cit Soltész, 1997). Soltész (1997) classified several apricot cultivars into groups based on their susceptibility to brown rot fungi. He suggested that cvs. 'Budapest' and 'Mandulakajszi' are highly susceptible, 'Ceglédi óriás', 'Liget óriás' and 'Polonais' moderately susceptible and 'Borsi-féle kései rózsa', 'Piroska', 'Pannónia', 'Ceglédi bíborkajszi', 'Magyar kajszi' and 'Rakovszky' less susceptible to blossom and twig blight caused by *M. laxa*.

In the case of peach, Koroknai (1971) demonstrated that peach cv. 'Shipley' was highly susceptible, cvs. 'Alexander', 'Amsden', 'Champion', 'Ford', 'Győztes' and 'Mayflower' were moderately susceptible, while 'Canada', 'Carman', 'Elberta', 'J.H. Hale', 'Incrotio Pieri' and 'Magyar arany duránci' had reduced susceptibility to brown rot. In contrast, Soltész (1997) noted that cvs. 'J.H. Hale' and 'Champion' were highly susceptible to *Monilinia* spp. Both authors mentioned that most nectarine are highly susceptible to brown rot because the peels of nectarine species can be damaged easily, leading to multiple wounds on the fruit surface (Soltész, 1997). Adaskaveg et al. (1992) noted that thicker fruit skin and higher phenolic content were found in resistant peach genotypes. Feliciano et al. (1987) and Ogawa and English (1991) noted that peach cv. 'Bolinha' showed moderate resistance to blossom brown rot caused by *M. fructicola*, and a breeding program was started based on the features of this cultivar (Gradziel and Wang, 1993; Gradziel et al., 1998). Other resistant introgression lines were also identified such as 'F8,1-42' (Martínez-García et al., 2013), and genomic and quantitative genetic tools were used to identify genes for brown rot resistance in peach (Martínez-García et al., 2013).

In the case of plum, both fruit rot and blossom infection can occur. Features such as vulnerable fruit peeling, long, wet weather periods during fruit ripening, and clustering of fruits are the main factors responsible for high susceptibility to brown rot infection (Soltész, 1997). Soltész (1997) classified several European plum cultivars into a number of groups based on their susceptibility to brown rot fungi. Cultivars 'Bluefre', 'President' and 'Stanley' are highly susceptible, 'Cacanska najbolja' and 'President' moderately susceptible, and 'Besztercei', 'Silvia' and 'Tuleu gras' less susceptible to fruit rot caused by *M. laxa*. In California, plum cvs. 'Santa Rosa', 'Wickson', 'Imperial' and 'French' suffered from severe blossom infection by *M. laxa,* and in Oregon, Italian prune was found to be susceptible to sporadic blossom infection by both *M. laxa* and *M. fructicola* (Ogawa and English, 1991).

In the case of almond, Ogawa et al. (1986) noted that among almond cultivars, 'Drake' and 'Jordanolo' were highly susceptible to blossom infection by *M. laxa*, and 'Ne Plus Ultra' and 'Texas' were moderately susceptible. Severe blossom infection was rarely observed in cultivars 'Nonpareil', 'Peerless' and 'Davey'.

In the case of apple and pear, fruit rot is mainly dependent upon the presence of wounds caused by biotic and abiotic wounding agents. Therefore, incidence of the disease was significantly higher in organic orchards due to severe injuries by pests (Holb and Scherm, 2007, 2008). Also, late ripening cultivars showed more severe brown rot than early ripening cultivars (Holb and Scherm, 2007).

5 Symptoms

The brown rot fungi infect different aerial parts of the host plant and cause symptoms such as blight affecting blossom, shoot, leaf, twig and fruit, as well as stem canker and fruit rot.

The first symptom caused by brown rot fungi in spring is blossom blight. First, blossom parts (petals, stamens and pistils) show brown lesions, then the whole blossom becomes brown and blighted (Fig. 1). Under humid weather conditions, the infected parts become softened and rotten, while in dry conditions, they are discoloured and brittle (Byrde and Willetts, 1977; Weaver, 1950). Some infected blossoms fall onto the ground, but others can remain attached to the tree showing the typical blossom blight appearance (Batra, 1991; Fig. 1). Blossom blight is caused by *M. fructicola* and *M. laxa*, and occurs mainly on stone fruits (Weaver, 1950; Holb et al., 2013), such as sweet and sour cherries (Wilcox, 1989; Tamm et al., 1995; Holb and Schnabel, 2005), plum (Schlagbauer and Holz, 1990), peach (Sutton and Clayton, 1972), nectarine (Ogawa et al., 1980), apricot and almond (Rudolph, 1925; Osirio et al., 1994). However, it can also occur on apple (Sharma and Kaul, 1987), pear and quince (Byrde and Willetts, 1977).

A secondary effect of blossom blight is shoot and twig infections, although twig infections may also result from infected cankers and blighted leaf tissue (Ogawa et al., 1980). When the blossom is infected, the fungus grows from the blossom parts through the peduncle into the shoots, leaves and twigs. Brown discolouration spreads through the infected tissues, and then the whole twig becomes blighted (Fig. 1). Infected twigs are usually killed by the fungus. With

Figure 1 A dead twig with blighted leaves, blossoms and young fruits caused by *Monilinia laxa* (Aderhold & Ruhland) Honey on sour cherry cv. Cigánymeggy. Photo by I. J. Holb.

severe infections, gummy exudate may also appear on the twig at the base of the diseased blossoms, especially on cherry, apricot and plum (Weaver, 1950; Ogawa et al., 1980). The fungi can sporulate on the basal section of the blighted twig in the following spring (Kable, 1965a).

Fruit can also produce blight symptoms similar to the blossom and twig blight on sour cherry (Holb et al., 2013). Fruit blight occurs on green fruits with sizes <10 mm. Blighted fruits do not rot, but die suddenly with their peduncle attached to the shoot and become firm, like blighted flowers or leaves (Fig. 1). Fruit blight occurs mainly as a result of blossom and/or twig blight. Fruit blight often appears on the same shoots that have already shown blossom and/or shoot blight previously, but it also occurs separately from these symptoms. In rainy periods, blighted fruits may also produce sporodochia and conidia on the dead tissues during the summer, and therefore, may serve as a potential inoculum source for fruit rot (Holb et al., 2013).

Stem cankers are necroses of cortical tissues invaded by mycelia. Cankers seldom develop from direct penetration through the external surfaces of the plants. The infection starts from twig and blossom blight or fruit rot. In the beginning of canker formation, the infected bark dies, then tissues beneath the bark become sunken and discoloured and an open wound develops. Often gum is secreted in the diseased area (Ogawa and English, 1960). If environmental conditions are favourable, the fungus can produce conidia on the active canker and, therefore, stem canker can serve as a source of inoculum. Stem cankers observed on stone fruit species are frequently caused by *M. laxa* and *M. fructicola*, while *M. fructigena* has a mainly affects apple.

Fruit rot is one of the major symptoms of brown rot fungi. Fruit infection starts after fruit set, and it continues during fruit swelling and ripening. Post-harvest damage is also important. Green fruits are more resistant to infection and, as the fruits are ripening, they become increasingly susceptible (Moore, 1950; Falconi and Mendgen, 1994; van Leeuwen et al., 2000; Xu and Robinson, 2000). Young green fruits may be infected, but symptoms may become visible only when the fruits are mature. Batra (1991) and Willetts and Bullock (1993) demonstrated that infection by *M. fructicola* could occur in young fruit, and remain dormant or quiescent until the beginning of fruit ripening, at which stage visual symptoms would begin to appear.

Fruit infection starts with a small, circular and superficial brown spot which progressively increases (Fig. 2). At the same time, conidiophores rupture the epidermis of the fruit and form small tufts over the surface (Fig. 2). Conidia are formed on sporogenous hyphae. When the conditions are very humid, almost the whole fruit surface is covered with conidial tufts or vegetative mycelium. The point of infection is usually a mechanical or physiological injury, particularly wounds caused by insects, birds, growth cracks, mechanical injury

Figure 2 Fruit rot caused by *Monilinia fructigena* (Aderhold & Ruhland) Honey on apple. Conidiophores of the fungus rupture the epidermis of the fruit and form concentric rings. Photo by I. J. Holb.

and other injuries (e.g. Croxall et al., 1951; Tate and Ogawa, 1975; Lack, 1989; van Leeuwen et al., 2000; Xu and Robinson, 2000; Holb and Scherm, 2008). However, both *M. fructicola* and *M. laxa* can infect healthy fruit which are not wounded. Within a few days of infection, if the weather is humid fruit become rotten and usually fall to the ground. However, some fruits lose their water content and this results in the formation of fruit mummies. A specific type of fruit rot is also known as black apple (Fig. 3), which commonly occurs on apples during post-harvest. In the beginning, the colour of infected fruits is brown, but it becomes black during the rotting process. The surface of the black apple is smooth and unruptured by the conidiophores and, in fact, the whole fruit becomes a stromatic mass, and its tissues are completely impregnated by mycelium (Byrde and Willetts, 1977).

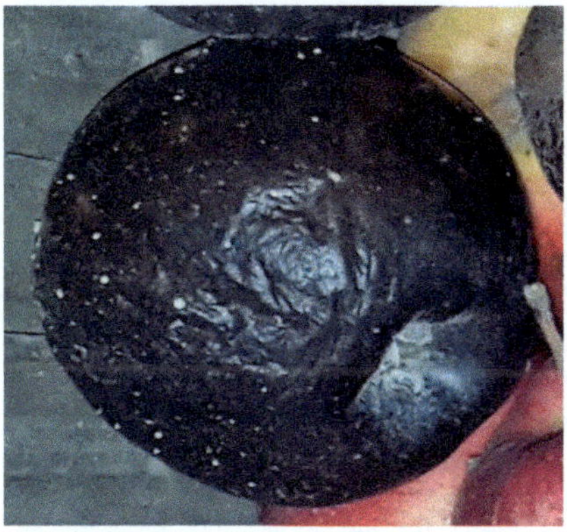

Figure 3 'Black apple': an apple fruit of cv. Idared after the mummification process during storage. Photo by I. J. Holb.

The occurrence and importance of different types of brown rot symptoms vary among fungal species and hosts (Table 2). *Monilinia fructicola* can cause severe fruit rot, blossom and twig blight, and cankers, primarily in stone fruit species and less frequently in pome fruits (Batra, 1991). *Monilinia laxa* attacks mainly stone fruits and causes primarily blossom, shoot and twig blight, and twig cankers. Fruit rot damage is less frequently due to *M. laxa*. *Monilinia fructigena* cause fruit rot primarily in apple, pear and quince, and less frequently in stone fruits. *M. fructigena* rarely causes other symptoms such as twig blight or cankers (Batra, 1991; Willetts and Bullock, 1993). Symptoms caused by *M. polystroma* are very similar to those of *M. fructigena* but lesion expansion rates on fruit are greater, and stroma formation is more abundant in *M. polystroma* than *M. fructigena* (van Leeuwen et al., 2002a; Vasic et al., 2016). Therefore, fruits infected in the field might compose more abundant stroma by *M. polystroma* than *M. fructigena*.

Table 2 Occurrence of major symptom types caused by *Monilinia fructicola, M. laxa* and *M. fructigena* in major host fruit crops

Host	*M. fructicola*	*M. laxa*	*M. fructigena*
Apple	Fruit rot (not serious); spur blight (sometimes); black apple (sometimes); canker	Fruit rot (rare); black apple (rare); canker; *M. laxa* f.sp. *mali* causes blossom wilt	Fruit rot (very common and destructive); black apple; cankers
Pear	Fruit rot	Fruit rot (rare); *M. laxa* f.sp. *mali* causes blossom wilt	Fruit rot
Quince	Fruit rot (sometimes)	Rare	Fruit rot
Peach and nectarine	Fruit rot; blossom and twig blight; canker	Fruit rot; blossom and twig blight; canker	Fruit rot
Apricot	Fruit rot; canker	Fruit rot; blossom and twig blight; canker	Fruit rot (not too significant)
Plum and prune	Fruit rot; blossom blight; canker	Fruit rot; blossom and twig blight; shoot blight; canker	Fruit rot
Sour and sweet cherry	Blossom blight	Fruit rot; blossom and twig blight, shoot blight; fruit blight	Fruit rot
Almond	Fruit rot; blossom and twig blight; canker	Fruit rot; blossom and twig blight; canker	-
Hazelnut	-	-	Nut drop

Monilia polystroma is not included in the table as its symptoms are similar to those of *M. fructigena*. Prepared by combining the study of Byrde and Willetts (1977), Willetts and Bullock (1993) and Holb (2003).

6 Identification and detection

Identification of brown rot fungi can be based on culture (*in vitro*) characteristics and/or on molecular markers for specific genes (Byrde and Willetts, 1977; van Leeuwen and van Kesteren, 1998; van Leeuwen et al., 2002a).

Cultural (*in vitro*) characteristics have been used for a long time to distinguish between the species of brown rot fungi. These culture characteristics are (1) colony growth rate, (2) colony form, (3) colony margins, (4) colony texture and intensity of pigmentation, (5) the production mode of stromata and (6) the zonation around the inoculum plug, that is alternating bands of sporogenous and non-fertile tissue (Byrde and Willetts, 1977; Batra, 1991). Cultural characteristics of four brown rot fungi were summarized by van Leeuwen and van Kesteren (1998) and van Leeuwen et al. (2002a). Some of these characteristics are summarized in Table 3.

On potato dextrose agar (PDA), colonies of the four species are filamentous and the aerial growth is cottony (Batra, 1991). Growth rate is low, low-moderate,

Table 3 Cultural characteristics of brown rot fungi of fruit crops

Monilinia spp.		
Colonies on PDA (22°C, 12 h light/12 h dark regime)		
M. fructicola	High growth rate; intensive sporulation (also continuous in darkness); brownish to ash-grey colour of sporogenous tissue	
M. laxa	Low growth rate, renewed outgrowth of mycelium at the edge of the colony; sparse sporulation, hazel/isabelline colour of sporogenous tissue	
M. fructigena	Low-moderate growth rate; sparse sporulation, distinct tufts/rings of aerial mycelium, buff/pale luteous colour of sporogenous tissue	
M. polystroma	Moderate growth rate; sparse sporulation; intense formation of black stromatal plates initiated after 10-12 days incubation, buff/pale luteous colour of sporogenous tissue	
Conidial dimensions (mean and range)[a]		
M. fructicola	13.1 × 9.1 μm (8)[b]	12.4-14.4 × 8.1-9.9 μm
M. laxa	12.0 × 8.7 μm (4)	11.2-12.8 × 8.2-9.3 μm
M. fructigena	19.2 × 11.3 μm (6)	17.7-20.6 × 10.3-12.3 μm
M. polystroma	15.9 × 9.8 μm (6)	12.8-17.1 × 8.8-10.7 μm
Germ tube morphology[c]		
M. fructicola	Long, straight, 750-900 μm, one germ tube per conidium	
M. laxa	Short, twisted, 150-350 μm, one germ tube per conidium	
M. fructigena	Long, straight, 600-900 μm, often two germ tubes per conidium	
M. polystroma	Long, straight, 700-1000 μm, often two germ tubes per conidium	

[a] Conidia grown on cherry decoction agar at 22°C under NUV light.
[b] 35 conidia measured per isolate in distilled water, number of isolates in parentheses.
[c] After 18 h at 22°C on water agar plates.
Source: Adapted from van Leeuwen et al. (2002).

moderate, and high for *M. laxa*, *M. fructigena*, *M. polystroma* and *M. fructicola*, respectively. On PDA, *M. fructigena* is distinctly buff-coloured, *M. laxa* is hazel or isabelline, while *M. fructicola* has a brownish to ash-grey colour. The margins of the colonies of *M. fructigena* and *M. fructicola* are entire, while in *M. laxa* they are lobed. Via this difference, *M. laxa* can be distinguished from the other two species. *Monilinia fructicola* often forms concentric rings of conidiophores where the conidia are abundant in dark or light/dark cycles. These rings are not so frequently observed in *M. fructigena*. In *M. laxa*, there is a sparse development of conidia over the surface of the colony, and the mycelium sometimes becomes pigmented in the centre of the colonies (Byrde and Willetts, 1977). Mean conidium size of *M. fructigena* and *M. polystroma* was greater than that of *M. fructicola* and *M. laxa* (van Leeuwen et al., 2002a). Features of the germ tubes sprouting from conidia are also different. *M. fructicola*, *M. fructigena* and *M. polystroma* produce long, straight germ tubes before branching, while in *M. laxa* branching starts close to the conidium (Schlagbauer and Holz, 1987). The organized stromata are often not produced in *M. laxa*. In *M. fructigena*, the production of these structures is also irregular, but these structures are more likely to form in the dark. In the case of *M. fructicola*, the stroma formation may occur on the whole surface, or more often beneath the sporulating areas of the colony (Willetts, 1968). *Monilia polystroma* is quite similar to *M. fructigena* in its features, but it produces more stromata and has a colony growth rate greater than *M. fructigena*, while the length and width of conidia are greater for *M. fructigena* compared to *M. polystroma* (van Leeuwen et al., 2002a).

Identification based on culture characteristics can be difficult as there is a lot of variation between isolates of the same species, which are not suitable for a rapid diagnosis when the pathogens occur simultaneously in the same region (Byrde and Willetts, 1977; van Leeuwen and van Kesteren, 1998). Therefore, recent identification and detection research has been focused on molecular methods. Early studies showed that some proteins (enzymes) can be used to distinguish between brown rot species. For instance, arylesterase and acid phosphatase (Penrose et al., 1976), as well as extracellular cell wall degrading enzymes such as pectin esterase and polygalacturonase (Willetts et al., 1977), proved useful in separating brown rot species, but these methods were still laborious. Nowadays, molecular markers are used to distinguish between four brown rot species. Molecular methods allow fruit batches to be tested for pathogen status before long-distance transport. This enables the detection of pathogens requiring quarantine measures, such as *M. fructicola* in Europe (van Leeuwen et al., 2002a). The first molecular diagnosis was based on the presence of a 418-bp group-I intron located in the SSR DNA gene of *M. fructicola*. This intron was absent in *M. laxa* and *M. fructicola* (Fulton and Brown, 1997). As the intron was absent in some *M. fructicola* isolates (Förster and Adaskaveg, 2000; Hughes et al., 2000) the method was not reliable to distinguish species of brown rot fungi. Ioos and Frey (2000) developed a new method based on the ITS between 18S and 28S ribosomal RNA genes, but the differences in the ITS region were not enough to separate *M. fructigena* and *M. polystroma* (van Leeuwen et al., 2002a). Multiplex PCR was developed in order to differentiate *M. fructigena, M. fructicola, M. laxa* and *M. polystroma* (Cote et al., 2004). Another diagnostic assay was based on a conserved region within the *cytb* gene intron located at codon 143 of the gene (Miessner and Stammler, 2010; Hily et al., 2011). A multiplex real-time PCR was also developed to distinguish between four brown rot fungi (van Brouwershaven et al., 2010). This method was further improved and a one-step detection of brown rot fungi was developed by the multiplex real-time PCR assay (Guinet et al., 2016). A real-time PCR method was also developed for detecting latent *Monilinia* infections in nectarine flowers and fruit (Garcia-Benitez et al., 2017), and for quantifying airborne spores of *M. fructicola* from Burkard spore trap samples (Luo et al., 2007). Other methods such as the high-resolution melting (HRM) analysis (Papavasileiou et al., 2016) and volatile organic compounds (VOCs), were also used to differentiate *M. laxa, M. fructigena* and *M. fructicola* (Mang et al., 2015).

7 Disease cycle and epidemiology

Disease cycles of all brown rot fungi rely only on the asexual life stage, except for *M. fructicola* where the sexual stage plays a significant role (Fig. 4). The perfect stage, that is apothecia, rarely form in *M. fructigena* and *M. laxa*. Apothecia of *M. fructigena* were successfully produced in the laboratory (Harada, 1977; Batra and Harada, 1986) and also in the field, but they only form during the second spring after infection (Harada, 1977; Willetts and Harada, 1984). The perfect stage of *M. laxa* was only detected in a few cases in the field (Willetts and Harada, 1984). Due to the rare occurrence of apothecia of *M. fructigena* and *M. laxa*, the perfect stage appears to play an insignificant role in the disease cycle of these two species (Byrde and Willetts, 1977; Willetts and Harada, 1984). The perfect stage of *M. polystroma* is not yet known, but its role in the disease cycle is believed to be very similar to *M. fructigena* (Holb, 2004b). The only brown rot species in which the perfect stage plays a role in the natural disease cycle is *M. fructicola*, although the occurrence of apothecia varies greatly with region (Byrde and Willetts, 1977; Landgraf and Zehr, 1982).

The disease cycle of brown rot fungi in the field, starts from the overwintering forms that are from (1) mummified fruits, (2) cankers on branches, and (3) blighted blossoms, fruits and twigs from previous years (Fig. 4). Spores that emerge from these overwintering forms causing primary infections in the spring are (1) conidia, produced on fruit mummies, blighted blossoms, fruits, twigs, and cankers on branches in the spring and (2) ascospores, produced

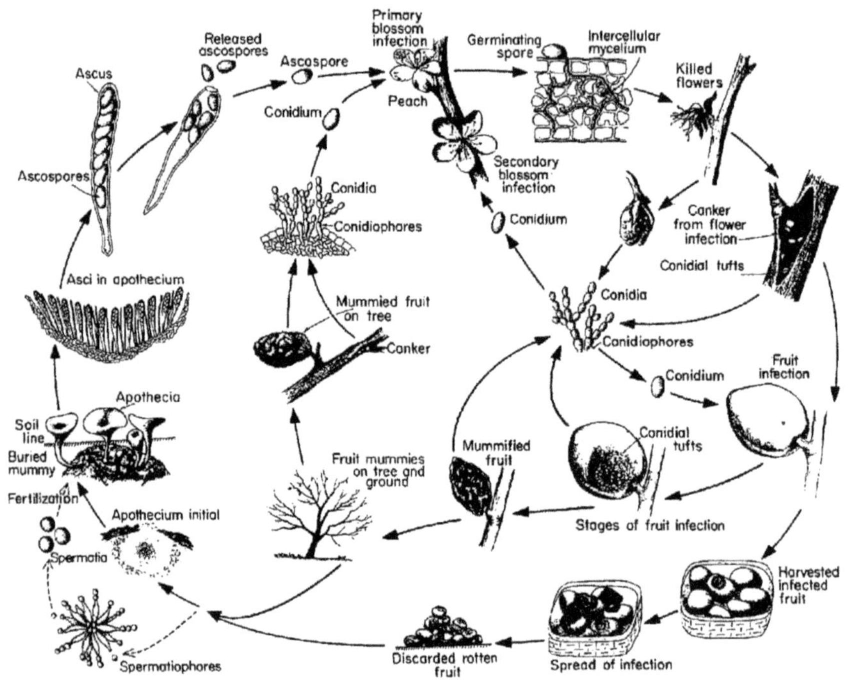

Figure 4 General life cycle of *Monilinia* spp. Source: Adapted from Agrios (1997).

in apothecia on mummified fruits on the ground (Byrde and Willetts, 1977; Holb et al., 2013; Fig. 4). After successful primary infection on blossom, shoots and young fruits, cycles of secondary infection can occur on shoots and twigs. Wounding has a major importance in brown rot epidemiology, as secondary infections and contact spreads in orchards usually occur when the infection enters via wounds (Xu et al., 2007). Mycelium of the fungus spreads via blossoms to shoots and woody tissues as well as via fruit-to-fruit contact. Any infected hosts including ornamental rosaceous trees near the cultivated orchards may serve as a source of inoculum (Byrde and Willetts, 1977; Batra, 1991). Conidia are dispersed by wind, water splash or vectors such as insects and birds (e.g. Lack, 1989). Latent infections of immature fruit may also occur for some brown rot fungi (Emery et al., 2000; Luo and Michailides, 2001; Gell et al., 2008). During, or at the end of the season, some infected fruits form a stroma and mummify (Byrde and Willetts, 1977). These fruit mummies, blighted plant parts and cankers on the branches can overwinter on the tree (Willetts, 1968). Fruits may become infected at harvest time, but fruit rot develops only during storage (Byrde and Willetts, 1977) and spreads quickly by mycelial contact between infected and healthy fruits (Willetts and Bullock, 1993; Holb et al., 2012).

7.1 Specific features of Monilinia fructigena *and* Monilia polystroma

The main overwintering form is the mummified fruits which mainly overwinter on the tree. In springtime, rainfall stimulates the uptake of moisture by the dried-out, mummified fruits. Subsequently, these mummies produce a new crop of conidia (Byrde and Willetts, 1977) as the primary inoculum in order to start an epidemic the following year (Knoche et al., 2000; Leeuwen et al., 2002b). Humid weather conditions and temperatures between 15°C and 20°C in the spring, induce asexual sporulation of *M. fructigena* on the surface of overwintered mummified fruit (Byrde and Willetts, 1977). van Leeuwen et al. (2002b) demonstrated that all infected fruits (mummified and not mummified) can be a source of inoculum next spring if they have overwintered. In most European countries, conidia from fruit mummies are present by late May (Xu et al., 2001b). However, the first infected fruit in the apple or pear tree do not occur before the end of July or the beginning of August (van Leeuwen et al., 2000; Xu et al., 2001b; Holb, 2003; Holb and Scherm, 2007). The 2-month period between the end of May and the end of July is too long for survival of conidia produced on overwintered mummies (Xu et al., 2001a); hence, other sources of inoculum are needed to provide epidemic continuity in early summer (Xu et al., 2001b). In apple, fruit drop occurs as a result of natural and mechanical thinning in June or early July. Conidia from overwintered mummies could infect these fruit on the ground, which then may serve as a bridge, providing inoculum for subsequent infection of fruit on the tree (Holb and Scherm, 2007).

Conidia can germinate over a wide range of temperature when humidity is high (Xu et al., 2001a). However, *M. fructigena*, in contrast to *M. laxa* or *M. fructicola*, can infect fruit only via fresh or partially healed wounds (Byrde and Willetts, 1977; Xu and Robinson, 2000). In general, infection of ripe apples via lenticels (e.g. Sharma and Kaul, 1987) or fruit-to-fruit contact (van Leeuwen et al., 2000) occurs only at low frequencies, although a study by Xu et al. (2001b) attributed up to 44.4% of all brown rot infections on Cox apples to contact spread. When a mature fruit is wounded, conidia can germinate and infect the fruit within as little as 2 h, and the pathogen may be able to infect without additional fruit wetness, because of the presence of free moisture on the fresh wound surface (Xu et al., 2001a). Latent infections of immature fruit as described for *M. fructicola* were not widely investigated for *M. fructigena*. Successful fruit infection caused by *M. fructigena* is dependent on fruit skin injuries. Xu and Robinson (2000) showed that no brown rot developed after inoculation of non-wounded apple fruits with a concentration of 7.5×10^5 conidia mL^{-1}. In field conditions, wounds penetrating the fruit skin can be caused by various abiotic or biotic factors, including insects such as the codling moth (*Cydia pomonella*) (Holb, 2004a; Moore, 1950; Xu et al., 2001b; Holb and Scherm,

2008), the Oriental fruit moth (*Cydia molesta*) (Garic et al., 1990), tortrix moths (*Pandemis* and *Adoxophyes* spp.) (Xu et al., 2001b), the European earwig (*Forficula auricularia*) (Croxall et al., 1951) and various species of wasps (Moore, 1950; Xu et al., 2001b). Other causes of fruit skin wounds include birds such as starlings (*Sturnus vulgaris*), blackbirds (*Turdus merula*) and crows (*Corvus* spp.) (van Leeuwen et al., 2000; Xu et al., 2001b); growth cracks (Byrde and Willetts, 1977; Xu et al., 2001b; Holb and Scherm, 2008), especially in years with late frosts or excessive rainfall, and mechanical injury due to wind storms or hail (Xu et al., 2001b). A few studies also showed that infection can take place via fruit-to-fruit contact (van Leeuwen et al., 2000; Holb and Scherm, 2008) or via lenticels. However, the roles of injury factors vary with seasons and among management strategies. In Britain, the relative importance of different injury types (growth cracks and wounds caused by insects and birds) were established for *M. fructigena* in integrated and conventionally managed apple orchards (Moore, 1950; Xu et al., 2001b). However, fruit rot incidence was low in these well-managed orchards, which limited the ability to quantify the relative importance of wounding agents for brown rot development. In Hungary, insect and bird injuries were the main injury factors by harvest, accounting for 36.6% and 32.2% respectively, in the well-managed integrated blocks. While insect injuries were the leading wound type accounting for 79.4% of brown rot infection in the organic orchards, the major wounding agent was *Cydia pomonella* for both management systems (Holb and Scherm, 2008).

After primary infection on dropped fruits, cycles of secondary infection can occur if there are wounds available on fruit over a long period of time. After infection, a brown lesion develops on the fruit, which is later ruptured by numerous sporodichia (conidial pustules) bursting through the cuticle of the fruit. In the first crop of conidia, millions of spores are produced per fruit, which are subsequently dispersed. The first infected fruitlets in pome fruit orchards usually appear 5–7 weeks after the bloom on the ground, and subsequently infection of healthy fruit occurs continuously up to harvest time (van Leeuwen et al., 2000; Holb, 2008; Holb and Scherm, 2008; Holb et al., 2011). Diseased fruits either fall to the ground or stay on the tree. Most fruits on the ground are degraded by microbiological processes (van Leeuwen et al., 2002a) but in the tree canopy, diseased fruits gradually dry out, shrivel, and turn into firm structures (mummified fruits). These mummified fruits are the main survival structures of the pathogen during the winter, although it has also been observed that the pathogen can survive as mycelium in infected fruits, twigs and branches (Byrde and Willetts, 1977; van Leeuwen et al., 2002a). For example, van Leeuwen et al. (2002b) demonstrated that fruits infected late in the season (e.g. at the beginning of October) had no time to mummify, but if they had managed to overwinter, they then could sporulate profusely under optimal conditions in the following spring.

The life cycle of *M. polystroma* is not well understood yet; however, it is believed that this fungus has a very similar life cycle to *M. fructigena* (van Leeuwen et al., 2002a).

7.2 Specific features of Monilinia laxa

The main feature of the *M. laxa* life cycle is that *M. laxa* is more virulent toward flowers, and therefore, causes more blossom infections compared to either *M. fructigena* or *M. fructicola*. In addition *M. laxa* causes less damage to fruit than *M. fructigena* or *M. fructicola*. *Monilinia laxa* overwinters in twig cankers, as well as in blossom, shoots or fruit affected by blight, and to a lesser extent mummified fruits on the tree (Villarino et al., 2010). Conidia begin to develop on these parts in spring. The peak of conidial production on the overwintered plant parts is between the beginning of flowering and petal fall (Stensvand et al., 2001). These conidia are dispersed by wind (Corbin and Ogawa, 1974), and by rain splash (Corbin, 1963). When they land on susceptible tissues, they germinate and infect within 2–4 h, if free moisture is present and the temperature is favourable (Corbin and Ogawa, 1974). The most critical period for blossom infection is from white flower bud until petal fall. The most frequent infection sites differ on fruit species. Anthers, stigmas and petals are important for cherry. In the case of apricot and prune, the sepals are important; while in almond the stigma is considered as the most susceptible blossom part (Ogawa and English, 1991). Generally, first symptoms appear 3–6 days after infection. Then a rapid necrosis can be seen on the entire blossom. Infections and subsequent symptoms can occur at a wide range of temperatures between 4°C and 30°C, but the fungus is very sensitive to low-moisture conditions, which can strongly limit infections (Weaver, 1950). Between petal fall and pit hardening stage, most stone fruits are resistant to *M. laxa* infection (Fourie and Holz, 2003) but this resistance rapidly decreases with increasing fruit maturity (Fourie and Holz, 2003). Multiple cycles of infections can occur if infection is via a wound, allowing spore production for subsequent infections. Latent infection by *M. laxa* also occurs during the season when healthy intact fruit have been infected by conidia (Schlagbauer, 1989; Xu et al., 2007) and this has a strong correlation with disease incidence at harvest and with brown rot decay during the post-harvest (Gell et al., 2008, 2009). During the season, mummified fruits are formed from the infected fruits for winter survival.

7.3 Specific features of Monilinia fructicola

Monilinia fructicola has the sexual stage which does not occur in either *M. fructigena* or *M. laxa*. Similarly to other brown rot fungi, *M. fructicola* can survive the winter in several ways such as (1) mummified fruits, (2) blighted blossom, and (3) twig canker. In springtime, primary inoculum sources are (1) ascospores from apothecia in the overwintered pseudosclerotial mat (stroma) of mummified fruits on the ground (Holtz et al., 1998), (2) conidia produced on the surface of mummified fruits, (3) conidia produced on blighted blossoms and twig cankers (Sutton and Clayton, 1972).

Apothecia develop in areas such as California where the soil is moist in the spring, but not in areas where springtime soil is dry. Ascospores are forcibly ejected into the air and are carried by air currents. Slight disturbances in the air (often associated with changes in the humidity) initiate ascospore discharge. The liberation of ascospores normally coincides with the emergence of young shoots and blossoms. The primary cycle can also begin with the conidia formed on mummified fruits, and on the above mentioned infected host tissues.

Ascospores or conidia produced from mummified fruit cause blossom blight in the spring under favourable conditions (Byrde and Willetts, 1977; Landgraf and Zehr, 1982). Primary infections of blossoms often provide inoculum for infection of fruit (Jenkins and Reinganum, 1965; Cruickshank and Wade, 1992). When conditions are unfavourable, these fruit infections can remain latent until conditions are unsuitable for fruit rot development. The level of latent infection in fruit is influenced by both primary and secondary infections (Cruickshank and Wade, 1992; Luo et al., 2001). The latent infection can occur over the whole season under favourable conditions.

Blossom blight reduces the number of flowers and twigs, and these infected blossoms and twigs can sporulate and serve as secondary inoculum sources for infection of fruit (Landgraf and Zehr, 1982). Thinned infected fruits on the orchard floor can also be a secondary source of inoculum (Hong et al., 1997). Biggs and Northover (1988) demonstrated that young peach fruits are highly susceptible to *M. fructicola* infection, but the fruits become resistant at pit hardening. The fruits become increasingly susceptible at 2–3 weeks before harvest, therefore, most fruit infections occur during the last few weeks before harvest. These fruit profusely sprorulate and can provide a mass secondary inoculum depending on subsequent infections via wounds. The fungus then survives on twig cankers, blighted blossoms, peduncles and mummified fruits.

8 Disease warning for brown rot

A disease warning system/model defines the conditions under which the pathogen can infect the host (Zadoks and Schein, 1979). In the case of brown rot fungi, such disease warning systems have been developed for predicting infections caused by *M. fructicola, M. laxa* and *M. fructigena*, in order to time fungicide applications during the growing season. However, no disease warning model has been constructed for *M. polystroma*. A preliminary disease warning was established for *M. laxa* on sweet cherry by Tamm et al. (1995). They developed a nonlinear model to describe the incidence of blossom infection in relation to temperature and wetness duration. Luo et al. (2001b) developed a risk analysis system for blossom blight of prune caused by *M. fructicola*. This risk system predicts the extent of blossom blight using infection at bloom stage, inoculum potential, temperature and wetness duration. Four risk levels are identified: no risk, low, moderate and high risk. There is no risk of blossom blight at early bloom stage if the temperature is below 10°C and wetness duration is shorter than 4 h. If wetness duration reaches 24 h, a moderate to high relative risk of blossom blight occurs between 10°C and 25°C. Most conditions result in low to moderate risk of blossom blight. Luo and Michailides (2001) developed a similar risk analysis system for latent infection of prune caused by *M. fructicola*. Based on the above two studies, Luo and Michailides (2003) developed a decision support model to guide fungicide application against prune fruit rot caused by *M. fructicola*. The estimated percentage of branches with fruit rot (<1%, 1–5%, 5–10%, >10%) and relative probability of latent infection leading to fruit rot (<50%, 50–80%, 80–90%, >90%) were used to develop management decisions: (1) safe, no need of fungicide application in the season, (2) wait, continue to investigate latent infection, (3) check historical weather to decide if fungicide application is needed; and (4) spray fungicide immediately.

A disease warning system for *M. fructigena* in organic apple production has been developed, as brown rot can severely affect this production regime. The disease warning model was coupled with a disease management strategy together with insect control (Holb, 2009; Holb et al., 2011).

9 Implications for prevention and brown rot management

According to the life cycle of brown rot fungi, there are at least three essential phases where the disease control measures can be applied: (1) during blossom to reduce blossom blight and twig infection, (2) during fruit swelling and ripening to reduce pre-harvest fruit infection (covering both latent and symptomatic infection), and (3) during storage to reduce post-harvest fruit rot.

In conventional and integrated management, disease control is not usually implemented specifically against *M. fructigena*. This fungus is mainly a wound-pathogen; the most important wounding agents are insects, therefore, brown rot management is based on insect control (Holb, 2004a; Holb and Scherm, 2008). The successive control of these pests significantly influences the incidence of fruit rot caused by *M. fructigena*. Most control measures are performed during blossom development against *M. laxa* in stone fruits (mainly cherries, plums and apricots). *Monilinia fructicola* is controlled mainly during fruit development and ripening, but control is also needed during blossom and storage in most stone fruits (Ogawa and English, 1991). There are only legislative control measures suggested against *M. polystroma* (van Leeuwen et al., 2002a).

Five disease control management practices (legislative control measures, cultural control, physical control, biological control and chemical control methods) are discussed below.

9.1 Legislative measures

Monilinia fructicola is a quarantine organism for Europe (EPPO, 2015) but it has been detected in several European countries (e.g. Bosshard et al., 2006; De Cal et al., 2009; Pellegrino et al., 2009; Duchoslavová et al., 2007; Poniatowska et al., 2013; Papavasileiou et al., 2015). The probability of this pathogen spreading to other EU countries is high. In Spain and Serbia, establishment of this pathogen has changed the population structure of the brown rot fungi in orchards. *Monilinia fructicola* has become more dominant over *M. laxa* and *M. fructigena*, and this change is explained by the multiple fungicide resistance of *M. fructicola*. Import of stone fruits and nursery trees is a common route for spread of the pathogen.

Monilinia fructigena is a quarantine organism in the United States, although it has already been found in the field (Batra, 1979). Authorities in the United States claim that the pathogen has been eradicated from the country.

Monilia polystroma is known to exist in Japan (van Leeuwen et al., 2002a) and its occurrence has been reported in other countries (e.g. Holb, 2004b; Zhu and Guo, 2010; Hilber-Bodmer et al., 2012; Vasić et al., 2013; Di Francesco et al., 2015; Munda, 2016; Poniatowska et al., 2016). The same legislative control measures against *M. polystroma* are proposed as against *M. fructicola* (van Leeuwen et al., 2002a).

9.2 Cultural control

Sanitary measures are effective in reducing or eliminating inoculum sources of brown rot fungi, as primary infections result from overwintered blossoms, twigs, fruits, spurs and branches containing the brown rot fungi (Byrde and Willetts, 1977; Batra, 1991). Removal of infected twigs and branches should be performed immediately when the disease is recognized (Table 4). During winter, when general pruning is performed, some infected twigs and spurs may not be easily recognized and, consequently, are left behind. Mummified fruits are usually visible and therefore can be removed and destroyed easily. This action is especially important in those areas where *M. fructicola* is present, to avoid the formation of the sexual stage. In addition, wild hosts near the orchard should be removed to prevent introduction of inoculum from outside sources (Table 4). Insect control is another prevention step that should be performed due to the insects' importance

Table 4 Cultural and physical control methods against brown rot disease of fruits caused by *Monilinia* spp.

Target organism	Target site	Control method	References
Cultural control			
Monilinia spp.	Infected blossom, twig, fruit	Pruning	Byrde and Willetts (1977), Batra (1991)
Monilinia spp.	Wild host	Removal and eradication	Byrde and Willetts (1977)
Monilinia spp.	Fruit	Preventing insect damage	Byrde and Willetts (1977)
M. laxa	Cherry fruit	Rain shield	Borve and Stensvand (2003), Thomidis and Exadaktylou (2013)
Monilinia spp.	Peach fruit	Irrigation and soil management with clover	Bussi et al. (2016)
Physical control			
M. fructicola	Nectarine, peach, cherry	Elevated CO_2	Ahmadi et al. (1999), Tian et al. (2001), Akbudak et al. (2008)
M. fructigena	Apple	Ultra-low O_2	Balla and Holb (2007), Holb et al. (2012)
M. fructicola	Peach fruit	Cooling of heated and non-heated fruits	Smith and Redit (1966)
M. fructicola, *M. laxa*	Peach fruit	Hydro-cooling, hydrair-cooling	Wells and Bennett (1976)
M. fructicola	Peach, nectarine, plum	Hot water dipping	Margosan et al. (1997), Karabulut et al. 2002, Jemric et al. (2010), Spadoni et al. (2013), Martini and Mari (2014)
Monilinia sp.	Peach, nectarine	Curing treatment	Casals et al. (2010a,b,d)
M. fructicola	Peach and nectarine fruits	Hot water brushing	Karabulut et al. (2002)
M. fructicola	Peach fruit	Ultraviolet light	Stevens et al. (1996)
M. fructicola	Peach fruit	Ultraviolet light with yeast antagonist	Stevens et al. (1998)
M. fructigena	Cherry, peach	Ultraviolet light, heat treatment	Marquenie et al. (2002), Sautter et al. (2011)
M. laxa	Sweet cherry fruit	Sub-atmospheric pressure	Romanazzi et al. (2001)
M. fructicola	Peach fruit	Electrolysed oxidizing water	Al-Haq et al. (2001), Fallanaj et al. (2013)
M. fructicola	Peach	Radiation	Casals et al. (2010e)
M. fructicola	Peach, nectarine	Radiation combined with hot water immerse	Sisquella et al. (2013b)

as wounding agents (Byrde and Willetts, 1977). Covering sweet cherry trees with rain shields made of polyethylene or other waterproof material prior to harvest, can prevent fruit cracking and, therefore, reduce brown rot on fruit (Borve and Stensvand, 2003; Thomidis and Exadaktylou, 2013). The use of plastic shields during rainy periods can therefore reduce the number of fungicide applications from bloom until the end of harvest (Borve and Stensvand, 2003). Combinations of irrigation and soil management treatments with ground cover of *Trifolium repens* in the tree row, reduced brown rot incidence in peach (Bussi et al., 2016).

9.3 Physical control

In recent years, there has been an increased interest in physical methods to control many post-harvest diseases, because no chemical fungicides are allowed in the European Union on stone fruit post-harvest (Sisquella et al., 2013a). Several physical control methods are used to reduce the decay of fruit caused by brown rot fungi, mainly during post-harvest storage conditions, such as controlled atmosphere storage, hydro-cooling, hydrair-cooling, heat, hot water brushing, UV-light, electrolysed oxidizing water and radiation (Table 4).

Controlled atmosphere storage with low or no O_2 and/or high CO_2 concentration has been investigated for controlling brown rot fungi. Brown rot incidence of the nectarine cv. 'Summer Red' significantly decreased when the air in the storage was enriched with 15% CO_2 at 5°C (Ahmadi et al., 1999). Increasing CO_2 concentration further decreased growth of *M. fructicola* and CO_2 concentrations of 15–25% resulted in a significant reduction in lesion size, while a 30% CO_2 concentration completely prevented lesion formation of the fungus (Tian et al., 2001). Storage conditions of 20% CO_2 and 5% O_2 significantly reduced the incidence of *M. fructicola* from 24% to 4% on cherry fruit (Akbudak et al., 2008). Ultra-low oxygen conditions, below 1% O_2 levels, nearly eliminated brown rot on apples caused by *M. fructigena*, compared to the incidence of 18–35% under conventional storage conditions (Holb et al., 2012).

After harvest, fruits are usually hydro-cooled at 0–3°C before shipment. In addition, hot water treatments can precede hydro-cooling, which could reduce post-harvest fruit decay. Cooling fruits before shipment can also be performed by hydrair-cooling, where fruits are cooled with cold air and not with cold water. Hydrair-cooling treatments reduced lesion development as effectively as hydro-cooling treatments (Smith and Redit, 1966; Wells and Bennett, 1976). However, water used for hydro-cooling can be contaminated with pathogen's spores, so it may facilitate spread of post-harvest decay during storage. Therefore, hydrair-cooling is proposed to replace hydro-cooling as this might prevent post-harvest contamination by pathogen inoculum.

Heat treatment is one of the most effective and promising physical methods against brown rot. It can be applied to fruit in the form of hot water, vapour heat, hot air and hot water rinse-brushing. Hot water dipping at 48°C for 12 min, at

55°C for 60 s and at 60°C for 30 or 60 s, significantly reduced disease incidence and lesion size of *M. fructicola* on peaches, nectarines and plums (Karabulut et al., 2002; Jemric et al., 2011; Spadoni et al., 2013). The heat can also stimulate certain host defence mechanisms which can be activated against latent infections of brown rot fungi (Martini and Mari, 2014). However, conventional heat treatments are limited by the low thermal conductivity of fruit. Casals et al. (2010a,b,d) reported a curing treatment at 50°C for 2 h for controlling brown rot in peaches and nectarines, but a shorter period of treatment is needed for commercial applications (Ikediala et al., 2002). Therefore, hot water treatments are suggested using exposure times between 20 s and 2.5 min at temperatures between 45°C and 60°C (Margosan et al., 1997). However, these treatments may need to be combined with other treatments in order to enhance their effectiveness (Casals et al., 2010c; Sisquella et al., 2013b).

Ultraviolet light-C (254 nm - UV-C) was able to reduce post-harvest rot of peaches caused by *M. fructicola* (Stevens et al., 1996). Low doses of UV reduced post-harvest disease incidence by 50–90% but the effectiveness was still reduced when compared to pesticides. UV alone failed to control brown rot when applied to cherries infected with *M. fructigena* (Marquenie et al., 2002) and peaches infected with *M. fructicola* (Sautter et al., 2011). A better reduction in fruit rotting was achieved when UV irradiation was combined with the application of a yeast antagonist (*Debaryomyces hansenii*) (Stevens et al., 1998). UV light is effective for controlling latent infections while the yeast only controlled superficial infections. UV-C light was also successfully combined with heat treatments to kill conidia of *M. fructigena* (Marquenie et al., 2002).

Electrolysed oxidizing water was an effective surface disinfectant on peach fruit, but was not able to reduce disease development of *M. fructicola* (Al-Haq et al., 2001). Electrolysed water can only be a useful physical method if it is free of chlorine and harmful salts (Fallanaj et al., 2013).

A recently applied physical control method against brown rot fungi is radiation. This method does not only heat fruit, but also disinfects and controls post-harvest diseases. A radiation treatment for 18 min reduced brown rot incidence on peaches inoculated with *M. fructicola*, but not on nectarines (Casals et al., 2010e). However, effective brown rot control was achieved on both peach and nectarine fruits when fruits were first immersed in water at 40°C and then treated with radiation treatment for 4.5 min (Sisquella et al., 2013b). Microwave treatments proved to be an unsuitable physical control method as they can cause internal fruit damage (Sisquella et al., 2013b).

9.4 Biological control

The control of brown rot fungi with biological control agents (BCAs) is an environmentally friendly method to replace pesticides, especially during post-harvest storage. Nowadays, a number of biological fungicides targeting brown rot fungi are available commercially, but they are still less effective and more expensive than conventional fungicides.

BCAs against brown rot fungi include filamentous fungi, bacteria, yeast and yeast-like microorganisms. The first studies on the antagonistic effect of microorganisms on brown rot fungi were performed in the 1970s. Heat-killed spores of *Trichoderma viridae* inhibited the mycelial growth of *M. laxa* and *M. fructigena* (Ale-Agha et al., 1974). In 1980 and 1981, microflora of peach twigs and flowers was monitored and the most frequent genera were *Penicillium, Alternaria, Aspergillus* and *Cladosporium* spp. (Melgarejo et al., 1985). Five species (*Aspergillus flavus, Epicoccum nigrum, Penicillium chrysogenum, P. frequentans* and *P. purpurogenum*) inhibited the spore germination and hyphal growth of *M. laxa*. Among these potential microorganisms, *E. nigrum, P. frequentans* and *P. purpurogenum* significantly reduced *M. laxa* infection in springtime (Melgarejo et al., 1986). Further studies revealed that *P. frequentans* produces antifungal compounds that are active against *M. laxa*, inhibiting the germination of *M. laxa* conidia on peach twigs (De Cal et al., 1988), reducing production of stromata (Melgarejo et al., 1989), and leading to unmelanized hyphae of *M. laxa* (De Cal and Melgarejo, 1994). In a field study, *P. frequentans* was applied alone or in alternation with captan to successfully control peach twig blight (De Cal et al., 1990). A mass conidial production of *P. frequentans* was achieved (De Cal et al., 2002) for obtaining self-stable material against *M. laxa*. Post-harvest treatment with conidia of *P. frequentans* reduced the incidence and lesion diameter of brown rot caused by *M. laxa* (De Cal et al., 2002). However, better control was achieved when the *P. frequentans* was applied with additives such as sodium alginate or carboxymethyl cellulose in order to increase adhesion of BCA conidia to fruit surfaces (Guijarro et al., 2008a). Further research revealed that *P. frequentans* should be applied at bloom to enhance its populations on fruit (Guijarro et al., 2008b).

A similar study with *E. nigrum* on peach trees, showed that the most successful treatment was the combined use of *E. nigrum* and captan (Madrigal et al., 1991). Further examination of *E. nigrum* showed that the fungus produces an antifungal compound, flavipin, which is toxic to *M. laxa* and inhibits the respiration process of the fungus (Madrigal and Melgarejo, 1994). In another study, the lytic enzyme preparations of *P. purpurogenum* cultures produced lysis of the hyphae and spores of *M. laxa* (Larena and Melgarejo, 1996). Further studies investigated the production of *E. nigrum* by substrate fermentation in order to test the effect of water activity on accumulation of compatible solutes (Pascual et al., 1999). In addition, accumulation of compatible solutes was also studied in *P. frequentans* grown at reduced water activity (Pascual et al., 2000). Dried *E. nigrum* conidia were produced for obtaining self-stable material against *M. laxa* (Larena et al., 2003). Twig canker and fruit rot of peaches caused by *M. laxa* were successfully controlled by conidial suspensions of *E. nigrum* (De

Cal et al., 2009). When additives of methylcellulose were incorporated into the conidial formulation (Larena et al., 2010), the biocontrol efficiency against *M. laxa* increased. Pre-harvest applications of *E. nigrum* alone, or in combination with fungicides on peaches (Larena et al., 2005) and post-harvest treatments of nectarines with *E. nigrum*, reduced the incidence of brown rot (Mari et al., 2007).

In the case of bacterial antagonists, the *Bacillus subtilis* B-3 strain successfully controlled post-harvest brown rot caused by *M. fructicola* (Pusey and Wilson, 1984). The mechanism of the bacterium appeared to involve production of antifungal substances (iturins and fengycin, antifungal lipopeptides) and showed almost complete suppression of the brown rot at 1 mg mL^{-1} concentration (McKeen et al., 1986; Gueldner et al., 1988). The *B. subtilis* B-3 strain was commercialized for post-harvest control of peach brown rot (Pusey et al., 1988). Since this early work, other BCAs have been developed into commercial products, such as Serenade, the trade name of formulated *B. subtilis* QST713, recommended for use against many plant diseases, including brown rot on stone fruit, and marketed worldwide (AgraQuest, 2009). Another strain of *B. subtilis* (B-192) was also found to be effective against post-harvest brown rot in peach and nectarine. This bacterium strain reduced brown rot infection by 20% on peach, and 40% on nectarine when fruits were inoculated with *M. fructicola* (Fan et al., 2000). *Bacillus* sp. OSU142 reduced infection of apricot blossoms caused by *M. laxa* in an artificial inoculation study (Altindag et al., 2006). For the *B. subtilis* strain CPA-8, fengycin was shown to be a major contributor of antifungal activity against *M. fructicola* and *M. laxa* (Casals et al., 2010c; Yanez-Mendizabal et al., 2012). The *B. amyloliquefaciens* C06 strain also inhibited *M. fructicola* on peaches (Zhou et al., 2008). Cell-free culture filtrates of *B. amyloliquefaciens* C06 inhibited *in vitro* conidial germination and mycelium growth of *M. fructicola* (Liu et al., 2011), and inhibition was due to bacillomycin D and fengycin produced by the bacterium (Liu et al., 2010, 2011). Two other antibiotic-producing bacteria (*Pseudomonas corrugate* and *P. capacia*) were reported as BCAs against post-harvest brown rot caused by *M. fructicola* (Smilanick et al., 1993). Both bacteria significantly reduced the incidence of post-harvest brown rot decay when applied up to 12 h after inoculation with *M. fructicola*.

Epiphytic yeasts were also reported as antagonists of brown rot fungi. The epiphytic yeast, *Metschnikowia pulcherrima* reduced the incidence of *M. laxa* in peaches by 49–89% depending on the strain and fruit species (DeCurtis et al., 1996). Other epiphytic yeasts (*Aureobasidium pullulans* and *Gliocladium roseum*) showed successful control against *M. fructicola* on blossom of cherry cv. 'Royal Anne' under field conditions (Wittig et al., 1997). Several isolates of *A. pullulans* from sweet cherry showed high efficacy against *M. laxa* on artificially wounded sweet cherries (Schena et al., 2003). *Aureobasidium pullulans* strain PL5 reduced *M. laxa* rot development on plums when stored at 1.2°C for 28 days, and on peaches when stored at 1.0°C for 21 days (Zhang et al., 2010b). This isolate also reduced the incidence of brown rot caused by *M. laxa* on peaches when it was combined with HWD (Zhang et al., 2010a). Two *A. pullulans* strains completely inhibited disease development caused by *M. laxa* and *M. fructicola* on peaches and nectarines (Mari et al., 2012). There are two formulated products of an *A. pullulans* strain on the market, 'BoniProtect' and 'BoniProtect forte' (Bio-ferm, 2011), specifically against *Monilinia* species. *A. pullulans* strain in combination with copper, lime sulphur and potassium carbonate successfully controlled blossom blight caused by *M. laxa* in an organic cherry orchard (Holb and Kunz, 2013). *Aureobasidium pullulans* strain (Y126) was able to reduce sporulation on mummified plum fruit, and reduce brown rot development on cherry and plum via nutrient competition (Rungjindamai et al., 2014). Recently, Grzegorczyk et al. (2017) reported post-harvest biocontrol ability of killer yeasts (*Debaryomyces hansenii* KI2a, *D. hansenii* MI1a and *Wickerhamomyces anomalus* BS91) against *M. fructigena* and *M. fructicola* on peach and plum fruits. The production of hydrolytic enzymes, killer toxins and VOCs were hypothesized as the main mechanisms of action against pathogens.

9.5 Chemical control in the field

Sulphur was the first pesticide used against brown rot. Sulphur was applied in some regions every 7 or 14 days from blossom until fruit maturity. This control measure was responsible for a substantial reduction in fruit losses, although the results were economically non-viable (Byrde and Willetts, 1977). This compound still remains the major fungicide against blossom blight and fruit rot in organic production (Holb and Schnabel, 2005; Holb et al., 2011, 2013).

During the 1950s, protective fungicides were introduced, such as captan and dichloran, with better efficacy than sulphur. Captan was superior to dichloran for brown rot control. Captan, for control of blossom blight of stone fruits, requires a minimum of two applications (one at pink bud and the second at full bloom) to provide protection for blossom tissues. These treatments did not control blossom blight completely, but even so, farmers used them often during the blossom period. A mixture of captan and dichloran was also used to control brown rot during post-harvest (Byrde and Willetts, 1977).

In the 1970s, the benzomidazole fungicide group was introduced with a very good efficacy against brown rot fungi. Two applications of benzimidazole at pink bud stage and at full bloom, provided good control due to its curative and protective mode of action. Moreover, benomyl translocated systemically into the non-exposed internal blossom tissues (pistils and stamens), and therefore, only one application of this fungicide at pink bud stage could reduce blossom blight by 92% (Osirio et al., 1994). However, extensive applications of benzimidazole resulted in fungicide resistance by the early 1980s (Zehr, 1982).

In the 1980s, dicarboximides were registered to control brown rot with the active ingredients of vinclozolin, iprodione and prochloraz. However, after a few years of successful commercial use, resistance against this fungicide group was also reported (Zehr, 1982).

In the 1990s, the use of sterol biosynthesis inhibitor (SBI) fungicides was common in fruit orchards against several fungal diseases including brown rot. Wilcox (1990) studied the post-infection and anti-sporulant activities of several SBI fungicides (tebuconazole, propiconazole, myclobutanil, fluzilazole, triforine and fenarimol) against *M. fructicola* on sour cherry. The control efficacy ranged from 97% to 100% when SBI fungicides were applied 24 h after inoculation. Tebuconazole and propiconazole had the best post-infection and anti-sporulant activities; however, when applications were made 72 h after inoculation, almost no control was achieved.

During the growing season, two to four spray applications are performed against brown rot during the blossom period. Later during fruit ripening, two or three other sprays may be applied depending on weather conditions and inoculum pressure (Zehr, 1982).

9.5.1 Pre-harvest chemical control of Monilinia laxa

The first protective spray schedule against *M. laxa* was developed by Rudolph (1925) in California. Under low-to-moderate disease pressure, apricot trees were sprayed once with Bordeaux mixtures at the pink bud stage. However, if the disease was severe, two sprays were advised, one at the tight cluster stage and the other at full bloom. These sprays were phytotoxic to the blossoms.

Early dormant benomyl sprays markedly reduced the number and size of sporodochia of *M. laxa* on blighted almond twigs (e.g. Ramsdell and Ogawa, 1973). The addition of oil increased the penetration of benomyl into the bark and provided additional activity against the fungus (Ramsdell and Ogawa, 1973). By the early 1980s, benomyl-resistant *M. laxa* was detected in apricot orchards (Ogawa et al., 1984). This led to the withdrawal of benomyl a few years later, and other compounds, such as the SBI and the dicarboximide fungicides, were evaluated against *M. laxa*. Most SBI and dicarboximide fungicides (e.g. triazole, piperazine, pyrimidine and imidazole) showed good efficacy against *M. laxa* on several stone fruit crops (Latorre and Lolas, 1986; Ogawa et al., 1988). Cyproconazole and difenoconazole strongly inhibited the mycelial growth of isolates of *M. laxa* and *M. fructigena*, and the minimum inhibitory concentration was more than 30 times lower for these two SBI fungicides than for myclobutanil and triadimenol (Zhang et al., 1991). Iprodione, the dicarboximide fungicide, was effective against both the benomyl-sensitive and benomyl-resistant isolates of *M. laxa* and *M. fructicola*. In the field, applications of iprodione at pink bud (closed blossom) and full bloom (opened blossoms) effectively reduced brown rot twig blight of almond (Osirio et al., 1994). Benzimidazoles, dicarboximides, demethylation inhibitors and strobilurin-type fungicides are the main group of fungicides used to control brown rot in most European countries (Fazekas et al., 2014; Egüen et al., 2016). Fenhexamid, a hydroxyanilide fungicide, was shown to be highly effective against *M. laxa* (Malandrakis et al., 2013). Recently, several fungicides including sulphur, captan, mancozeb, thiophanate-methyl, cyprodinil, myclobutanil, prochloraz, penconazole, fenhexamid and boscalid + piraclostrobin were tested *in vitro* against *M. laxa* isolates. Results showed that 5 out of 12 isolates tested were partially insensitive to boscalid + piraclostrobin, cyprodinil, fenhexamid or prochloraz (Fazekas et al., 2014). Fungicide evaluation studies emphasize that it is important to apply resistance management strategies for all effective fungicide compounds in order to delay the onset and subsequent spread of fungal strains resistant to these products (Rungjindamai et al., 2014).

9.5.2 Pre-harvest chemical control of Monilinia fructicola

Controls against *M. fructicola* focuses on both blossom blight and fruit rot. Applications of benomyl and thiophanate-methyl more effectively controlled the fungus than earlier fungicides (Gilpatrick, 1973; Tate et al., 1974). Excellent protective and after-infection activity of sterol inhibitors and dicarboximides were shown against *M. fructicola* (Szkolnik, 1981). Post-infection applications (18 or 24 h after inoculation) of sterol inhibitors, prochloraz, triadimefon, triforine and iprodione gave excellent blossom blight control. Sterol inhibitors provided effective disease control under heavy disease pressure when the fungicide was applied up to 5 days prior to bloom (van Geluwe et al., 1981). The germ tube tips of *M. fructicola* rupture 2–4 h after the application of SBI fungicides. Propiconazole and penconazole caused the greatest damage in the growth of *M. fructicola* (Dahmen et al., 1988). Manji and Ogawa (1987) demonstrated that if green cherry fruits had quiescent infection of *M. fructicola*, applications of iprodione, triforine and benomyl reduced the incidence of fruit rot at harvest. Among SBI fungicides, tebuconazole gave the best control on immature fruits that have a high incidence of latent infection of *M. fructicola* (Northover and Cerkauskas, 1998). Interactions between components of fungicide mixtures (propiconazole with either benomyl, captan, chlorothalonil, cyprodinil or vinclozolin) were evaluated against *M. fructicola* (Emery et al., 2002) suggesting the absence of synergy between active ingredients. Furthermore, propiconazole-based fungicide mixtures had no advantage over a rotating schedule of fungicides for control of and resistance management in *M. fructicola*. More recently, quinone outside inhibitor (QoI) fungicides have been used against *M. fructicola*. These fungicides have protectant activity and varying levels of curative activity against *M. fructicola* (Burnett et al., 2010). Azoxystrobin, trifloxystrobin and pyraclostrobin + boscalid reduced sporulating area on peach by 15.9%, 42.4% and 0.4%, respectively, while azoxystrobin, trifloxystrobin and pyraclostrobin + boscalid reduced colony growth by 12.3%, 7.5% and 7.4%, respectively (Burnett et al., 2010). Trifloxystrobin had good anti-sporulant activity on both fruit infections and cankers.

9.6 Fungicide resistance

Application of fungicides to blossom and fruit is still the main method for managing brown rot during the growing season. However fungicide-based methods are not sustainable in the longer term, because of the emergence and subsequent spread of *Monilinia* sp. strains with reduced sensitivity to fungicides (e.g. Ma et al., 2005; Thomidis et al., 2009; Malandrakis et al., 2013; Chen et al., 2014). Brown rot resistance to benzimidazole, dicarboximide, SBI, DMI, strobilurin and SDHI fungicides has been reported, and strategies against resistance developed. We only use a few examples to illustrate the problem and possible management strategies.

A few years after the introduction of benzimidazole, benomyl-tolerant isolates of *M. fructicola* and *M. laxa* were found in 1974 in stone fruits orchards in California (Tate et al., 1974). Repeated applications of benomyl during bloom and pre-harvest resulted in the selection of benomyl-resistant *M. fructicola* in Australia (Whan, 1976), Michigan (Jones and Ehret, 1976) and California (Ogawa et al., 1988), and benomyl-resistant *M. laxa* in California (Ogawa et al., 1984). Jones and Ehret (1976) demonstrated that the tolerant isolates grew more slowly, produced fewer conidia and were less virulent than the sensitive ones. However, benomyl at concentrations of 150–300 μg mL^{-1} were ineffective against the benomyl-tolerant isolates, and these isolates were also tolerant to other benzimidazole fungicides such as thiophanate-methyl. As a fungicide resistance management strategy, monitoring benomyl-resistant *M. fructicola* strains and applications of benomyl-captan combination applications were suggested in orchards with low populations of resistant *M. fructicola* (Ogawa et al., 1981). In addition, non-benzimidazole fungicides should be used in orchards with high populations of resistant *M. fructicola*. Benomyl-resistant isolates were also found in *M. laxa* populations in California (Cañez and Ogawa, 1985). Fitness of resistant *M. laxa* isolates was reduced, causing less blossom blight, and resistant isolates had a longer latent period and lower sporulation compared to sensitive ones (Cañez and Ogawa, 1985). Michailides et al. (1986) detected benomyl-resistant isolates of *M. laxa* and *M. fructicola* in several prune and apricot orchards in California. Ogawa et al. (1988) reported that the frequency of benomyl-resistant isolates increased from 20% to nearly

90% after a single benomyl application. By the early 1990s, benzimidazol-resistant isolates of *M. fructicola* persisted in field populations in Australia (Penrose, 1990), New Zealand (Elmer and Gaunt, 1994) and South Carolina (Zehr et al., 1991). Resistance to the benzimidazole fungicides, benomyl and thiophanate-methyl is shown to be due to a point mutation in the beta-tubulin gene at codon 6, causing a replacement of the amino acid histidine by tyrosine, and at codon 198, causing a replacement of the amino acid glutamic acid by alanine in high-resistant isolates. Based on these point mutations allele-specific PCR assays were developed for detection of benzimidazole-resistant isolates of *M. fructicola* (Ma et al., 2003). Yoshimura et al. (2004) found that the highly resistant isolates to benzimidazole thiophanate-methyl caused significantly greater blossom blight than less resistant isolates. Growers then switched to thiophanate-methyl and *M. laxa* strains resistant to this fungicide have been reported on nectarines, peaches, prunes and almonds in California (Ma et al., 2005) and Greece (Thomidis et al., 2009). Recently, Egüen et al. (2015) reported that more than 95% of *M. fructicola* isolates in Spain were resistant to thiophanate-methyl. In addition, Martini et al. (2016) showed that 46 out of 63 *M. fructicola* isolates were resistant to thiophanate-methyl in Italy. All resistant isolates tested showed a point mutation at codons 198 and 83 in the β-tubulin gene, whereas in the latter one the arginine was converted to glutamine.

Resistance in *M. fructicola* to dicarboximide fungicides, iprodione, procymidone and vinclozolin, has been reported in the 1980s (Sztejnberg and Jones, 1978; Ritchie, 1983; Penrose et al., 1985). Strains of *M. fructicola* resistant to iprodione, procymidone and vinclozolin produced smaller lesions or sporulated less than sensitive strains, and were often associated with reduced parasitic fitness (Ritchie, 1983; Elmer and Gaunt, 1994; Sanoamuang and Gaunt, 1995). Sanoamuang and Gaunt (1995) indicated that dicarboximide-resistant *M. fructicola* strains may not overwinter as effectively on mummified fruits and twig cankers as the wild-type strains. These strains, in contrast with benzimidazole-resistant strains, did not persist in field populations unless dicarboximide fungicides were used regularly during the season (Elmer and Gaunt, 1994). *Monilinia fructicola* strains resistant to a dicarboximide fungicide were also cross-resistant to other members of this group (Sztejnberg and Jones, 1978; Ritchie, 1983; Rosenberger and Meyer, 1981). Penrose et al. (1985) demonstrated that vinclozolin-resistant isolates were resistant *in vitro* to two other dicarboximide fungicides, iprodione and procymidone. Egüen et al. (2015) reported that more than 50% of *M. fructicola* isolates in the Ebro Valley in Spain are iprodione-resistant. More recently, Hrustic et al. (2018) showed that isolates of *M. fructicola*, when compared to *M. laxa*, were significantly less sensitive to iprodione in Serbia.

SBI resistance in *M. fructicola* can be induced through ultraviolet radiation of spores, but these isolates lost their resistance after three to nine subculturings on a fungicide-free medium (Reese and Moore, 1982). The potential of *M. fructicola* to build up resistance against SBI fungicides *in vitro* was shown by Nunninger-Ney (1988). To minimize the risk of such a resistance build-up Zhang et al. (1991) recommended using SBI fungicides only in combination with a protective fungicide such as captan or dithianon. Several studies reported DMI resistance in *M. fructicola* in the United States (Zehr et al., 1999; Schnabel et al., 2004; Holb and Schnabel, 2007; Luo and Schnabel, 2008; Villani and Cox, 2011; Chen et al., 2013) and Brazil (May-De Mio et al., 2011). The development of DMI resistance on peach can be delayed when DMI fungicides are used in combination with sulphur fungicide (Holb and Schnabel, 2008). Villani and Cox (2011) demonstrated that *M. fructicola* isolates from 9 out of 18 orchards showed either resistance or reduced sensitivity to fenbuconazole and propiconazole; however, the DMI resistance-associated genetic element 'Mona' was only found in five orchards and not always present in resistant isolates. Resistance to tebuconazole was found in 15.8% of *M. fructicola* isolates collected from São Paulo State in Brazil (May-De Mio et al., 2011) and resistance corresponded to the historic applications of DMI fungicides in the orchard. Chen et al. (2013) reported that 5 out of 22 isolates were resistant to DMI fungicide propiconazole in South Carolina and propiconazole resistance was always associated with the presence of the 'Mona' element located on the *MfCYP51* gene. More recently, Hrustic et al. (2018) showed that isolates of *M. fructicola*, compared to *M. laxa*, are significantly less sensitive to tebucanozole in Serbia.

Partial resistance to strobilurin and SDHI fungicides has been recently reported in *Monilinia* spp. With respect to QoI fungicides, Amiri et al. (2010) observed a reduced sensitivity to azoxystrobin in *M. fructicola*. May-De Mio et al. (2011) reported that all *M. fructicola* isolates were sensitive to azoxystrobin but their EC_{50} increased between 2002 and 2008. Chen et al. (2014) reported azoxystrobin resistance in *M. fructicola* in China, but the resistance was unstable and associated with a fitness penalty, but was not associated with gene mutations. Some *M. fructicola* strains were resistant to both propiconazole and boscalid (Chen et al., 2013). When fruits were treated with propiconazole or boscalid, brown rot incidences were significantly higher on the dual-resistant isolates compared to the sensitive control ones. Isolates of *M. fructicola*, compared to *M. laxa*, were significantly less sensitive to azoxystrobin and boscalid in Serbia (Hrustic et al., 2018).

Several anti-resistance management strategies have been developed by the Fungicide Resistance Action Committee (FRAC) (Staub, 1991). In the case of benzimidazoles, dicarboximides, phenylamides, SBI fungicides and QoI (strobilurine), these fungicides have to be used according to specific anti-resistance strategies against *Monilinia* species. Essential strategies include the following aspects:

1 Apply fungicides according to the manufacturer's recommendations.
2 The number of applications of fungicides must be limited, whether applied alone or in mixtures with other fungicides. Repeated application of fungicides alone should not be practised on the same crop in one season against brown rot.
3 Alternation (block sprays or in sequence) or mixtures with an effective non cross-resistant fungicide are recommended.
4 The mixture partner must have a different mode of action; in addition, it may increase the spectrum of activity or provide needed curative activity.
5 Fungicide input is only one aspect of disease management. Fungicide use does not replace the need for non-chemical control options, including use of resistant cultivars, good agronomic practice, sanitation and all other elements of integrated disease management.

10 Conclusion and future trends

Further research and innovations are needed to implement an IPM strategy against brown rot fungi.

The first basic step is to establish a control threshold in disease forecasting. Growers need to know the action threshold of the brown rot infection, coupled with insect population dynamics and accurate fungicide-resistance status of the orchards. Growers need disease warnings and expert systems, such as those developed by Luo et al. (2001) against *M. fructicola*.

The second step is that more attention needs to be focused on integrated control measures against brown rot fungi. Due to the increasing importance of fungicide resistance, growers have to combine different control methods in order to reduce or replace chemical methods both in pre-harvest and post-harvest brown rot management. This is essential in organic fruit production, where only copper, sulphur, powder rock, botanical fungicides and some microbial products are approved for disease control (Holb, 2009). Also in storage, where the use of non-chemical control methods has high priority and the post-harvest use of fungicides is not permitted. However, there is a need for increased understanding regarding possible synergy of control methods against brown rot fungi.

The third step is that the role of host resistance and the use of more environmentally safe, biological fungicides has to increase. The wide use of highly resistant cultivars and effective non-chemical methods can be effective against blossom, fruit and twig blight. More resistant or tolerant cultivars are needed against brown rot fungi, mainly against *M. fructicola* or *M. laxa*. An increase in commercially available biological products is needed in the future for both pre-harvest and post-harvest brown rot control. These products have to be suitable for combining with other non-chemical control options, and there is a need for sufficient ecological knowledge for their effective use in commercial fruit production (Xu et al., 2011; Xu and Jeger, 2013a,b). In addition, there is an urgent requirement for effective collaboration among commercial enterprises and with research scientists to develop effective BCA products (Rungjindamai et al., 2014).

11 Acknowledgements

This work was partly supported by the Hungarian Scientific Research Fund (K108333) and the State of Hungary, co-financed by the European Social Fund in the framework of TÁMOP-4.2.4.A/2-11/1-2012-0001 'National Excellence Program' under project number A2-SZJ-TOK-13-0061. The work was also supported by the EFOP 3.6.1-16-2016-00022 project, co-financed by the European Union and the European Social Fund.

12 Where to look for further information

Good basic introductions to the subject are Byrde and Willetts, *The Brown Rot Fungi of Fruit*, and Batra, *World Species of Monilinia (Fungi): Their Ecology, Biosystematics and Control* (see Byrde and Willetts (1977) and Batra (1991)). Updated web sources for brown rot fungi are:

- *Monilinia fructigena*: https://www.cabi.org/isc/datasheet/34747
- *Monilinia laxa*: https://www.cabi.org/isc/datasheet/34749
- *Monilinia fructicola*: https://www.cabi.org/isc/datasheet/34746
- *Monilia polystroma*: https://www.cabi.org/isc/datasheet/34755

13 References

Adaskaveg, J. E., Ogawa, J. M., Feliciano, A. J. and Dunlap, M. (1992). Role of the cuticle and phenolic content in resistant and susceptible peach genotypes to brown rot caused by *Monilinia fructicola*. *Phytopathology*, 82, 1093. (Abstract).

AgraQuest. (2009). SERENADE Biofungicide. http://www.marketwired.com/press-release/serenade-biofungicide-included-on-fungicide-resistance-action-committees-2009-code-list-1223821.htm.

Agrios, G. N. (1997). *Plant Pathology*. Academic Press, New York, NY, 635pp.

Ahmadi, H., Biasi, W. V. and Mitcham, E. J. (1999). Control of brown rot decay of nectarines with 15% carbon dioxide atmospheres. *Journal of the American Society for Horticultural Science*, 124, 708–12.

Akbudak, B., Tezcan, H. and Eris, A. (2008). Effect of low-dose gamma irradiation on the quality of sweet cherry during storage. *Italian Journal of Food Science*, 3(20), 381–92.

Al-Haq, M. I., Seo, Y., Oshita, S. and Kawagoe, Y. (2001). Fungicidal effectiveness of electrolyzed oxidizing water on post-harvest brown rot of peach. *HortScience*, 36, 1310–14.

Ale-Agha, N., Dubos, B., Grosclaude, C. and Richard, J. L. (1974). Antagonism between nongerminated spores of *Trichoderma viride* and *Botrytis cinerea*, *Monilia laxa*, *M. fructigena* and *Phomopsis viticola*. *Plant Disease Reporter*, 58, 915–17.

Altindag, M., Sahin, M., Esitken, A., Ercisli, S., Guleryuz, M., Donmez, M. F. and Sahin, F. (2006). Biological control of brown rot (*Moniliana laxa* Ehr.) on apricot (*Prunus armeniaca* L. cv. Hacıhaliloğlu) by *Bacillus*, *Burkholdria*, and *Pseudomonas* application under *in vitro* and *in vivo* conditions. *Biological Control*, 38(3), 369–72.

Amiri, A., Brannen, P. M. and Schnabel, G. (2010). Reduced sensitivity in *Monilinia fructicola* field isolates from South Carolina and Georgia to respiration inhibitor fungicides. *Plant Disease*, 94(6), 737–43.

Apostol, J. (2005). New sweet and sour cherry selections in Hungary. In: *V International Cherry Symposium 795* (pp. 75–8).

Bargioni, G. (1982). *Il ciliegio dolce*. Edagricole, Bologna.

Batra, L. R. (1979). First authenticated North American record of *Monilinia fructigena*, with notes on related species. *Mycotaxon*, 8, 476–84.

Batra, L. R. (1991). World species of *Monilinia* (Fungi): Their ecology, biosystematics and control. Mycologia Memoir No. 16, J. Cramer, Berlin, 246pp.

Batra, L. R. and Harada, Y. (1986). A field record of apothecia of *Monilinia fructigena* in Japan and its significance. *Mycologia*, 78, 913–17.

Berrie, A. M. and Holb, I. (2014). Brown rot diseases. In: Sutton, T. B., Aldwinckle, H. S., Agnello, A. M. and Walgenbach, J. F. (Eds), *Compendium of Apple and Pear Diseases and Pests* (224pp). American Phytopathological Society, St. Paul, MN, pp. 43–5.

Biggs, A. R. and Northover, J. (1988). Early and late-season susceptibility of peach fruits to *Monilinia fructicola*. *Plant Disease*, 72, 1070–4.

Bio-ferm (2011). Safety data sheet Boni Protect – Bio-ferm, pp. 1–10. https://www.yumpu.com/en/document/read/2239488/safety-data-sheet-boni-protect-bio-ferm.

Borve, J. and Stensvand, A. (2003). Use of a plastic rain shield reduces fruit decay and need for fungicides in sweet cherry. *Plant Disease*, 87, 523–8.

Bosshard, E. H. M. S., Hilber-Bodmer, M., Schärer, H. J., Bünter, M. and Duffy, B. (2006). First report of the quarantine brown rot pathogen *Monilinia fructicola* on imported stone fruits in Switzerland. *Plant Disease*, 90, 1554.

Burchill, R. T. and Edney, K. L. (1972). An assessment of some new treatments for control of rotting of stored apples. *Annals of Applied Biology*, 72, 249–55.

Burnett, A. L., Lalancette, N. and McFarland, K. A. (2010). Effect of QoI fungicides on colonization and sporulation of *Monilinia fructicola* on peach fruit and blossom blight cankers. *Plant Disease*, 94, 1000–8.

Bussi, C., Parveaud, C. E., Mercier, V. and Lescourret, F. (2016). Effects of irrigation deprivation and ground cover (*Trifolium repens*) in the tree row on brown rot incidence in peach. *Crop Protection*, 88, 37–44.

Byrde, R. J. W. and Willetts, H. J. (1977). *The Brown Rot Fungi of Fruit: Their Biology and Control*. Pergamon Press, Oxford, 171pp.

Cañez, V. M. and Ogawa, J. M. (1985). Parasitic fitness of benomyl-resistant and benomyl-sensitive *Monilinia laxa*. *Phytopathology*, 75, 1329. (Abstract).

Casals, C., Teixidó, N., Viñas, I., Llauradó, S. and Usall, J. (2010a). Control of *Monilinia* spp. on stone fruit by curing treatments: I. The effect of temperature, exposure time and relative humidity on curing efficacy. *Postharvest Biology and Technology*, 56, 19–25.

Casals, C., Teixidó, N., Viñas, I., Cambray, J. and Usall, J. (2010b). Control of *Monilinia* spp. on stone fruit by curing treatments. II: The effect of host and *Monilinia* spp. variables on curing efficacy. *Postharvest Biology and Technology*, 56, 26–30.

Casals, C., Teixidó, N., Viñas, I., Silvera, E., Lamarca, N. and Usall, J. (2010c). Combination of hot water, *Bacillus subtilis* CPA-8 and sodium bicarbonate treatments to control postharvest brown rot on peaches and nectarines. *European Journal of Plant Pathology*, 128, 51–63.

Casals, C., Viñas, I., Torres, R., Griera, C. and Usall, J. (2010d). Effect of temperature and water activity on in vitro germination of *Monilinia* spp. *Journal of Applied Microbiology*, 108(1), 47–54.

Casals, C., Viñas, I., Landl, A., Picouet, P., Torres, R. and Usall, J. (2010e). Application of radio frequency heating to control brown rot on peaches and nectarines. *Postharvest Biology and Technology*, 58, 218–24.

Chen, F., Liu, X., Chen, S., Schnabel, E. and Schnabel, G. (2013). Characterization of *Monilinia fructicola* strains resistant to both propiconazole and boscalid. *Plant Disease*, 97, 645–51.

Chen, S. N., Shang, Y., Wang, Y., Schnabel, G., Lin, Y., Yin, L. F. and Luo, C. X. (2014). Sensitivity of *Monilinia fructicola* from peach farms in China to four fungicides and characterization of isolates resistant to carbendazim and azoxystrobin. *Plant Disease*, 98, 1555–60.

Corbin, J. B. (1963). Factors determining the length of the incubation period of *Monilinia fructicola* (Wint.) Honey in fruits of *Prunus* spp. *Australian Journal of Agricultural Research*, 14, 51–60.

Corbin, J. B. and Ogawa, J. M. (1974). Springtime dispersal patterns of *Monilinia laxa* conidia in apricot, peach, prune, and almond trees. *Canadian Journal of Botany*, 52, 167–76.

Côté, M. J., Tardif, M. C. and Meldrum, A. J. (2004). Identification of *Monilinia fructigena*, *M. fructicola*, *M. laxa*, and *Monilia polystroma* on inoculated and naturally infected fruit using multiplex PCR. *Plant Disease*, 88, 1219–25.

Croxall, H. E., Collingwood, C. A. and Jenkins, J. E. E. (1951). Observation on brown rot (*Sclerotinia fructigena*) of apples in relation to injury caused by earwigs (*Forficula auricularia*). *Annals of Applied Biology*, 38, 833–43.

Cruickshank, R. H. and Wade, G. C. (1992). The activation of latent infections of *Monilinia fructicola* on apricots by volatiles from the ripening fruit. *Journal of Phytopathology*, 136, 107–12.

Dahmen, H., Hoch, H. C. and Staub, T. (1988). Differential effects of sterol inhibitors on growth, cell membrane permeability, and ultrastructure of two target fungi. *Phytopathology*, 78, 1033–42.

De Cal, A. and Melgarejo, P. (1994). Effects of *Penicillium frequentans* and its antibiotics on unmelanized hyphae of *Monilinia laxa*. *Phytopathology*, 84, 1010–14.

De Cal, A., M.-Sagasta, E. and Melgarejo, P. (1988). Antifungal substances produced by *Penicillium frequentans* and their relationship to the biocontrol of *Monilinia laxa*. *Phytopathology*, 78, 888–93.

De Cal, A., M.-Sagasta, E. and Melgarejo, P. (1990). Biological control of peach twig blight (*Monilinia laxa*) with *Penicillium frequentans*. *Plant Pathology*, 39, 612–18.

De Cal, A., Larena, I., Guijarro, B. and Melgarejo, P. (2002). Mass production of conidia of *Penicillium frequentans*, a biocontrol agent against brown rot of stone fruits. *Biocontrol Science and Technology*, 12, 715–25.

De Cal, A., Gell, I., Usall, J., Viñas, I. and Melgarejo, P. (2009). First report of brown rot caused by *Monilinia fructicola* in peach orchards in Ebro Valley, Spain. *Plant Disease*, 93, 763.

DeCurtis, F., Torriani, S., Rossi, F. and DeCicco, V. (1996). Selection and use of *Metschnikowia pulcherrima* as a biological control agent for postharvest rots of peaches and table grapes. *Annali di Microbiologia ed Enzimologia*, 46, 45–55.

Di Francesco, A., Fruk, M., Martini, C., Jemric, T. and Mari, M. (2015). First report of Asiatic brown rot (*Monilinia polystroma*) on apple in Croatia. *Plant Disease*, 99, 1181.

Duchoslavová, J., Širučková, I., Zapletalová, E., Navratil, M. and Šafářová, D. (2007). First report of brown rot caused by *Monilinia fructicola* on various stone and pome fruits in the Czech Republic. *Plant Disease*, 91, 907.

Egüen, B., Melgarejo, P. and De Cal, A. (2015). Sensitivity of *Monilinia fructicola* from Spanish peach orchards to thiophanate-methyl, iprodione, and cyproconazole: Fitness analysis and competitiveness. *European Journal of Plant Pathology*, 141, 789–801.

Egüen, B., Melgarejo, P. and De Cal, A. (2016). The effect of fungicide resistance on the structure of *Monilinia laxa* populations in Spanish peach and nectarine orchards. *European Journal of Plant Pathology*, 145, 815–27.

Elmer, P. A. G. and Gaunt, R. E. (1994). The biological characteristics of dicarboximide-resistant isolates of *Monilinia fructicola* from New Zealand stone-fruit orchards. *Plant Pathology*, 43, 130–7.

Emery, K. M., Michailides, T. J. and Scherm, H. (2000). Incidence of latent infection of immature peach fruit by *Monilinia fructicola* and relationship to brown rot in Georgia. *Plant Disease*, 84, 853–7.

Emery, K. M., Scherm, H. and Savelle, A. T. (2002). Assessment of interactions between components of fungicide mixtures against *Monilinia fructicola*. *Crop Protection*, 21, 41–7.

EPPO. (2015). List of pests recommended for regulation as quarantine pests. Online publication. European and Mediterranean Plant Protection Organization. http://www.eppo.int/QUARANTINE/listA2.htm

Everhart, S. E., Askew, A., Seymour, L., Holb, I. J. and Scherm, H. (2011). Characterization of three-dimensional spatial aggregation and association patterns of brown rot symptoms within intensively mapped sour cherry trees. *Annals of Botany*, 108, 1195–202.

Falconi, G. J. and Mendegen, D. H. (1994). Epiphytic fungi on apple leaves and their value for control of the post-harvest pathogens *Botrytis cinerea*, *Monilinia fructigena* and *Penicillium expansum*. *Journal of Plant Disease and Protection*, 101(1), 38–47.

Fallanaj, F., Sanzani, S. M., Zavanella, C. and Ippolito, A. (2013). Salt addition improves the control of citrus postharvest diseases using electrolysis with conductive diamond electrodes. *Journal of Plant Pathology*, 95(2), 373–83.

Fan, Q., Tian, S. P., Li, Y. X., Xu, Y. and Wang, Y. (2000). Biological control of postharvest brown rot in peach and nectarine fruits by *Bacillus subtilis* (B-912). *Acta Botanica Sinica*, 42, 1137–43.

Fazekas, M., Madar, A., Sipiczki, M., Miklós, I. and Holb, I. J. (2014). Genetic diversity in *Monilinia laxa* populations in stone fruit species in Hungary. *World Journal of Microbiology and Biotechnology*, 30(6), 1879–92.

Feliciano, A., Feliciano, A. J. and Ogawa, J. M. (1987). *Monilinia fructicola* resistance in peach cultivar Bolinha. *Phytopathology*, 77, 776–80.

Förster, H. and Adaskaveg, J. E. (2000). Early brown rot infections in sweet cherry fruit are detected by Monilinia-specific DNA primers. *Phytopathology*, 90, 171–8.

Fourie, P. H. and Holz, G. (2003). Germination of dry, airborne conidia of *Monilinia laxa* and disease expression on nectarine fruit. *Australasian Plant Pathology*, 32(1), 9–18.

Fulton, C. E. and Brown, A. E. (1997). Use of SSU rDNA group-I intron to distinguish *Monilinia fructicola* from *M. laxa* and *M. fructigena*. *FEMS Microbiology Letters*, 157, 307–12.

Garcia-Benitez, C., Melgarejo, P. and De Cal, A. (2017). Detection of latent *Monilinia* infections in nectarine flowers and fruit by qPCR. *Plant Disease*, 101, 1002–8.

Garic, R., Stamenkovic, S. and Canak, M. (1990). The effect of *Cydia molesta* on the incidence of *Monilinia* spp. on quince fruits. *EPPO Bulletin*, 20, 603–6.

Gell, I., De Cal, A., Torres, R., Usall, J. and Melgarejo, P. (2008). Relationship between the incidence of latent infections caused by *Monilinia* spp. and the incidence of brown rot of peach fruit: Factors affecting latent infection. *European Journal of Plant Pathology*, 121, 487–98.

Gell, I., De Cal, A., Torres, R., Usall, J. and Melgarejo, P. (2009). Conidial density of *Monilinia* spp. on peach fruit surfaces in relation to the incidences of latent infections and brown rot. *European Journal of Plant Pathology*, 123, 415–24.

Gilpatrick, J. D. (1973). Control of brown rot of stone fruits with thiophanate-methyl and a piperazine derivative fungicide. *Plant Disease Reporter*, 57, 457–9.

Gradziel, T. M. and Wang, D. (1993). Evaluation of brown rot resistance and its relation to enzymatic browning in clingstone peach germplasm. *Journal of the American Society for Horticultural Science*, 118(5), 675–9.

Gradziel, T. M., Thorpe, M. A., Bostock, R. M. and Wilcox, S. (1998). Breeding for brown rot (*Monilinia fructicola*) resistance in clingstone peach with emphasis on the role of fruit phenolics. *Acta Horticulturae*, 465, 161–70.

Grzegorczyk, M., Żarowska, B., Restuccia, C. and Cirvilleri, G. (2017). Postharvest biocontrol ability of killer yeasts against *Monilinia fructigena* and *Monilinia fructicola* on stone fruit. *Food Microbiology*, 61, 93–101.

Gueldner, R. C., Reilly, C. C., Pusey, L. P., Costello, C. E., Arrendale, R. F., Cox, R. H., Himmelsbach, D. S., Crumley, F. G. and Cutler, H. G. (1988). Isolation and identification of iturins as antifungal peptides in biological control of peach brown rot with *Bacillus subtilis*. *Journal of Agricultural Food Chemistry*, 36, 366–70.

Guerriero, R. and Watkins, R. (1984). Revised descriptor list for apricot (*Prunus armeniaca*). IBPGR Secretariat, Rome and CEC Secretariat, Brussels, pp. 1–33.

Guijarro, B., Melgarejo, P. and De Cal, A. (2008a). Influence of additives on adhesion of *Penicillium frequentans* conidia to peach fruit surfaces and relationship to the biocontrol of brown rot caused by *Monilinia laxa*. *International Journal of Food Microbiology*, 126(1–2), 24–9.

Guijarro, B., Melgarejo, P., Torres, R., Lamarca, N., Usall, J. and De Cal, A. (2008b). *Penicillium frequentans* population dynamics on peach fruits after its applications against brown rot in orchards. *Journal of Applied Microbiology*, 104(3), 659–71.

Guinet, C., Fourrier-Jeandel, C., Cerf-Wendling, I. and Ioos, R. (2016). One-step detection of *Monilinia fructicola, M. fructigena*, and *M. laxa* on *Prunus* and *Malus* by a multiplex real-time PCR assay. *Plant Disease*, 100, 2465–74.

Harada, Y. (1977). Studies on the Japanese species of *Monilinia* (*Sclerotiniaceae*). *Bulletin of Faculty of Agriculture Hirosaki University*, 27, 30–109.

Hesse, C. O. (1938). Variation in resistance to brown-rot in apricot varieties and seedling progenies. *Proceedings of the American Society for Horticultural Science*, 36, 266–8.

Hilber-Bodmer, M., Knorst, V., Smits, T. H. M. and Patocchi, A. (2012). First report of Asian brown rot caused by *Monilia polystroma* on apricot in Switzerland. *Plant Disease*, 96, 146.

Hily, J. M., Singer, S. D., Villani, S. M. and Cox, K. D. (2011). Characterization of the cytochrome b (cyt b) gene from *Monilinia* species causing brown rot of stone and pome fruit and its significance in the development of QoI resistance. *Pest Management Science*, 67(4), 385–96.

Holb, I. J. (2003). The brown rot fungi of fruit crops (*Monilinia* spp.): I. Important features of their biology. *International Journal of Horticultural Science*, 9(3–4), 23–36.

Holb, I. J. (2004a). The brown rot fungi of fruit crops (*Monilinia* spp.): III. Important features of their disease control. *International Journal of Horticultural Science*, 10(4), 31–48.

Holb, I. J. (2004b). Effect of acidity on growth rate and stroma formation of *Monilia fructigena* and *M. polystroma* isolates. *International Journal of Horticultural Science*, 10(1), 63–7.

Holb, I. J. (2008). Monitoring conidial density of *Monilinia fructigena* in the air in relation to brown rot development in integrated and organic apple orchards. *European Journal of Plant Pathology*, 120, 397–408.

Holb, I. J. (2009). Fungal disease management in environmentally friendly apple production. In: Lichtfouse, E. (Ed.), *Climate Change, Intercropping, Pest Control and Beneficial Microorganisms: Sustainable Agriculture Reviews 2*. Springer Science+Business Media B.V., Dordrecht, pp. 219–93.

Holb, I. J. and Kunz, S. (2013). Integrated control of brown rot blossom blight by combining approved chemical control options with *Aureobasidium pullulans* in organic cherry production. *Crop Protection*, 54, 114–20.

Holb, I. J. and Scherm, H. (2007). Temporal dynamics of brown rot in different apple management systems and importance of dropped fruit for disease development. *Phytopathology*, 97, 1104–11.

Holb, I. J. and Scherm, H. (2008). Quantitative relationships between different injury factors and development of brown rot caused by *Monilinia fructigena* in integrated and organic apple orchards. *Phytopathology*, 98, 79–86.

Holb, I. J. and Schnabel, G. (2005). Effect of fungicide treatments and sanitation practices on brown rot blossom blight incidence, phytotoxicity, and yield for organic sour cherry production. *Plant Disease*, 89, 1164–70.

Holb, I. J. and Schnabel, G. (2007). Differential effect of triazoles on mycelial growth and disease measurements of *Monilinia fructicola* isolates with reduced sensitivity to DMI fungicides. *Crop Protection*, 26, 753–9.

Holb, I. J. and Schnabel, G. (2008). The benefits of combining elemental sulfur with a DMI fungicide to control *Monilinia fructicola* isolates resistant to propiconazole. *Pest Management Science*, 64, 156–64.

Holb, I. J., Balla, B., Abonyi, F., Fazekas, M., Lakatos, P. and Gáll, J. M. (2011). Development and evaluation of a model for management of brown rot in organic apple orchards. *European Journal of Plant Pathology*, 129(3), 469–83.

Holb, I. J., Balla, B., Vámos, A. and Gáll, J. M. (2012). Influence of preharvest calcium applications, fruit injury, and storage atmospheres on postharvest brown rot of apple. *Postharvest Biology and Technology*, 67, 29–36.

Holb, I. J., Szőke, S. and Abonyi, F. (2013). Temporal development and relationship amongst brown rot blossom blight, fruit blight and fruit rot in integrated and organic sour cherry orchards. *Plant Pathology*, 62(4), 799–808.

Holtz, B. A., Michailides, T. J. and Hong, C. X. (1998). Development of apothecia from stone fruit infected and stromatized by *Monilinia fructicola* in California. *Plant Disease*, 82, 1375–80.

Honey, E. E. (1936). North American species of *Monilinia*. I. Occurrence, grouping, and life-histories. *American Journal of Botany*, 23, 100–6.

Hong, C. X., Holtz, B. A., Morgan, D. P. and Michailides, T. J. (1997). Significance of thinned fruit as a source of the secondary inoculum of *Monilinia fructicola* in California nectarine orchards. *Plant Disease*, 81, 519–24.

Hrustić, J., Mihajlović, M., Grahovac, M., Delibašić, G. and Tanović, B. (2018). Fungicide sensitivity, growth rate, aggressiveness and frost hardiness of *Monilinia fructicola* and *Monilinia laxa* isolates. *European Journal of Plant Pathology*, 151, 389–400.

Hughes, K. J. D., Fulton, C. E., McReynolds, D. and Lane, C. R. (2000). Development of new PCR primers for identification of *Monilinia* species. *EPPO Bulletin*, 30(3–4), 507–11.

Ikediala, J. N., Hansen, J. D., Tang, J., Drake, S. R. and Wang, S. (2002). Development of a saline water immersion technique with RF energy as a postharvest treatment against codling moth in cherries. *Postharvest Biology and Technology*, 24, 25–37.

Ioos, R. and Frey, P. (2000). Genomic variation within *Monilinia laxa*, *M. fructigena* and *M. fructicola*, and application to species identification by PCR. *European Journal of Plant Pathology*, 106, 373–8.

Janisiewicz, W. J. and Korsten, L. (2002). Biological control of postharvest diseases of fruits. *Annual Review of Phytopathology*, 40, 411–41.

Jemric, T., Ivic, D., Fruk, G., Matijas, H. S., Cvjetkovic, B., Bupic, M. and Pavkovic, B. (2011). Reduction of postharvest decay of peach and nectarine caused by *Monilinia laxa* using hot water dipping. *Food and Bioprocess Technology*, 4(1), 149–54.

Jenkins, P. T. and Reinganum, C. (1965). The occurrence of quiescent infection of stone fruits caused by *Sclerotinia fructicola* (Wint.) Rehm. *Australian Journal of Agricultural Research*, 16, 131–40.

Jones, A. L. and Ehret, G. R. (1976). Isolation and characterisation of benomyl-tolerant strains of *Monilinia fructicola*. *Plant Disease Reporter*, 60, 765–9.

Kable, P. F. (1965a). Air dispersal of conidia of *Monilinia fructicola* in peach orchards. *Australian Journal of Experimental Agriculture Animal Husbandry*, 5, 166–71.

Kappeland, F. and Sholberg, P. L. (2008). Screening sweet cherry cultivars from the Pacific Agri-Food Research Centre Summerland breeding program for resistance to brown rot (*Monilinia fructicola*). *Canadian Journal of Plant Science*, 88(4), 747–52.

Karabulut, O. A., Cohen, L., Wiess, B., Daus, A., Lurie, S. and Droby, S. (2002). Control of brown rot and blue mold of peach and nectarine by short hot water brushing and yeast antagonists. *Postharvest Biology and Technology*, 24, 103–11.

Knoche, M., Schroder, M. and Hinz, M. (2000). Control of the development of mummified fruit of 'Elstar' apple. *Journal of Horticultural Science & Biotechnology*, 75(3), 328–35.

Koroknai, B. (1971). Őszibarack fajták érzékenysége a moniliás fertőzéssel, levéllikasztó betegséggel és varasodással szemben. In Hungarian (Susceptibility of peach cultivars to brown rot, shot hole, and scab.) *Kertgazdaság*, 3(2), 43–52.

Lack, H. (1989). The spread of apple brown rot (*Monilinia fructigena*) by insects. *Annals of Applied Biology*, 115(2), 221–7.

Landgraf, F. A. and Zehr, E. I. (1982). Inoculum sources for *Monilinia fructicola* in South Carolina peach orchards. *Phytopathology*, 72, 185–90.

Larena, I. and Melgarejo, P. (1996). Biological control of *Monilinia laxa* and *Fusarium oxysporum* f. sp. *lycopersici* by a lytic enzyme-producing *Penicillium purpurogenum*. *Biological Control*, 6, 361–7.

Larena, I., De Cal, A., Linan, M. and Melgarejo, P. (2003). Drying of *Epicoccum nigrum* conidia for obtaining a shelf-stable biological product against brown rot disease. *Journal of Applied Microbiology*, 94(3), 508–14.

Larena, I., Torres, R., De Cal, A., Liñán, M., Melgarejo, P., Domenichini, P., Bellini, A., Mandrin, J. F., Lichou, J., de Eribe, X. O. and Usall, J. (2005). Biological control of postharvest brown rot (*Monilinia* spp.) of peaches by field applications of *Epicoccum nigrum*. *Biological Control*, 32, 305–10.

Larena, I., De Cal, A. and Melgarejo, P. (2010). Enhancing the adhesion of *Epicoccum nigrum* conidia to peach surfaces and its relationship to the biocontrol of brown rot caused by *Monilinia laxa*. *Journal of Applied Microbiology*, 109(2), 583–93.

Latorre, B. A. and Lolas, M. A. (1986). Efficacy of new sterol-inhibiting fungicides for the control of European brown rot. *Phytopathology*, 76, 1107. (Abstract).

Liu, J., He, D., Li, X. Z., Gao, S., Wu, H., Liu, W., Gao, X. and Zhou, T. (2010). γ-Polyglutamic acid (γ-PGA) produced by *Bacillus amyloliquefaciens* C06 promoting its colonization on fruit surface. *International Journal of Food Microbiology*, 142(1–2), 190–7.

Liu, J., Zhou, T., He, D., Li, X. Z., Wu, H., Liu, W. and Gao, X. (2011). Functions of lipopeptides bacillomycin D and fengycin in antagonism of *Bacillus amyloliquefaciens* C06 towards *Monilinia fructicola*. *Journal of Molecular Microbiology and Biotechnology*, 20(1), 43–52.

Luo, Y. and Michailides, T. J. (2001). Factors affecting latent infection of prune fruit by *Monilinia fructicola*. *Phytopathology*, 91, 864–72.

Luo, Y. and Michailides, T. J. (2003). Threshold conditions that lead latent infection to prune fruit rot caused by *Monilinia fructicola*. *Phytopathology*, 93, 102–11.

Luo, C. X. and Schnabel, G. (2008). The cytochrome P450 lanosterol 14α-demethylase gene is a demethylation inhibitor fungicide resistance determinant in *Monilinia fructicola* field isolates from Georgia. *Applied and Environmental Microbiology*, 74(2), 359–66.

Luo, Y., Ma, Z. and Michailides, T. J. (2001a). Analysis of factors affecting latent infection and sporulation of *Monilinia fructicola* on prune fruit. *Plant Disease*, 85, 999–1003.

Luo, Y., Morgan, D. P. and Michailides, T. J. (2001b). Risk analysis of brown rot blossom blight of prune caused by *Monilinia fructicola*. *Phytopathology*, 91, 759–68.

Luo, Y. Z. Ma, Y. H. Reyes, C., Morgan, D. and Michailides, T. J. (2007). Quantification of airborne spores of *Monilinia fructicola* in stone fruit orchards of California using real-time PCR. *European Journal of Plant Pathology*, 118, 145–54.

Ma, Z., Yoshimura, M. A. and Michailides, T. J. (2003). Identification and characterization of benzimidazole resistance in *Monilinia fructicola* from stone fruit orchards in California. *Applied and Environmental Microbiology*, 69(12), 7145–52.

Ma, Z., Yoshimura, M. A., Holtz, B. A. and Michailides, T. J. (2005). Characterization and PCR-based detection of benzimidazole-resistant isolates of *Monilinia laxa* in California. *Pest Management Science*, 61(5), 449–57.

Madrigal, C. and Melgarejo, P. (1994). Mechanisms of action of the antibiotic flavipin on *Monilinia laxa* and *Saccharomyces cerevisiae*. *Mycological Research*, 98, 874–8.

Madrigal, C., Tadeo, J. L. and Melgarejo, P. (1991). Relationship between flavipin production of *Epicoccum nigrum* and antagonism against *Monilinia laxa*. *Mycological Research*, 98, 874–8.

Malandrakis, A., Koukiasas, N., Veloukas, T., Karaoglanidis, G. and Markoglou, A. (2013). Baseline sensitivity of *Monilinia laxa* from Greece to fenhexamid and analysis of fenhexamid-resistant mutants. *Crop Protection*, 46, 13–17.

Mang, S. M., Racioppi, R., Camele, I., Rana G. L. and D'Auria, M. (2015). Use of volatile metabolite profiles to distinguish three *Monilinia* species. *Journal of Plant Pathology*, 97(1), 55–9.

Manji, B. T. and Ogawa, J. M. (1987). Quiescent infections of sweet cherry in California. *Proceedings of 61st Annual Western Orchard Pest Disease Management Conference*, 8. (Abstract).

Margosan, D. A., Smilanick, J. L., Simmons, G. F. and Henson, D. J. (1997). Combination of hot water and ethanol to control postharvest decay of peaches and nectarines. *Plant Disease*, 81, 1405–9.

Mari, M., Torres, R., Casalini, L., Lamarca, N., Mandrin, J. F., Lichou, J., Larena, I., De. Cal, M. A., Melgarejo, P. and Usall, J. (2007). Control of post-harvest brown rot on nectarine by *Epicoccum nigrum* and physico-chemical treatments. *Journal of the Science of Food and Agriculture*, 87(7), 1271–7.

Mari, M., Martini, C., Guidarelli, M. and Neri, F. (2012). Postharvest biocontrol of *Monilinia laxa*, *Monilinia fructicola* and *Monilinia fructigena* on stone fruit by two *Aureobasidium pullulans* strains. *Biological Control*, 60(2), 132–40.

Marquenie, D., Michiels, C. W., Geeraerd, A. H., Schenk, A., Soontjens, C., Van Impe, J. F. and Nicolai, B. M. (2002). Using survival analysis to investigate the effect of UV-C and heat treatment on storage rot of strawberry and sweet cherry. *International Journal of Food Microbiology*, 73(2–3), 187–96.

Martínez-García, P. J., Parfitt, D. E., Bostock, R. M., Fresnedo-Ramírez, J., Vazquez-Lobo, A., Ogundiwin, E. A. and Crisosto, C. H. (2013). Application of genomic and quantitative genetic tools to identify candidate resistance genes for brown rot resistance in peach. *PLoS One*, 8(11), e78634.

Martini, C. and Mari, M. (2014). *Monilinia fructicola*, *Monilinia laxa* (Monilinia rot, brown rot). In: *Postharvest Decay* (pp. 233–65). Academic Press.

Martini, C., Guidarelli, M., Di Francesco, A., Ceredi, G. and Mari, M. (2016). Characterization of thiophanate methyl resistance in Italian *Monilinia fructicola* isolates. *Journal of Plant Pathology*, 98, 453–62.

May-De Mio, L. L., Luo, Y. and Michailides, T. J. (2011). Sensitivity of *Monilinia fructicola* from Brazil to tebuconazole, azoxystrobin, and thiophanate-methyl and implications for disease management. *Plant Disease*, 95, 821–7.

McKeen, C. D., Reilly, C. C. and Pusey, P. L. (1986). Production and partial characterization of antifungal substances antagonistic to *Monilinia fructicola* from *Bacillus subtilis*. *Phytopathology*, 76, 136–9.

Melgarejo, P., Carrillo, R. and M-Sagasta, E. (1985). Mycoflora of peach twigs and flowers and its possible significance in biological control of *Monilinia laxa*. *Transaction for the British Mycological Society*, 85, 313–17.

Melgarejo, P., Carrillo, R. and M.-Sagasta, E. (1986). Potential for biological control of *Monilinia laxa* in peach twigs. *Crop Protection*, 5, 422–6.

Melgarejo, P., De Cal, A. and M.-Sagasta, E. (1989). Influence of *Penicillium frequentans* and two of its antibiotics on production of stromata by *Monilinia laxa* in culture. *Canadian Journal of Botany*, 67, 83–7.

Michailides, T. J. M., Ogawa, J. M. and Opgenorth, D. C. (1986). Distribution of *Monilinia* species and detection of benomyl-resistant isolates in prune and apricot orchards in California. *Phytopathology*, 76, 845. (Abstract).

Miessner, S. and Stammler, G. (2010). *Monilinia laxa, M. fructigena* and M. fructicola: Risk estimation of resistance to QoI fungicides and identification of species with cytochrome *b* gene sequences. *Journal of Plant Diseases and Protection*, 117(4), 162–7.

Moore, M. H. (1950). Brown rot of apples: Fungicide trials and studies of the relative importance of different wound-agents. *Journal of Horticultural Science*, 25, 225–34.

Munda, A. (2015). First report of brown rot on peach caused by *Monilia polystroma* in Slovenia. *Plant Disease*, 99, 1281.

Northover, J. and Cerkauskas, R. F. (1998). Fungicidal suppression of symptomless latent infections of *Monilinia fructicola* in European plums. *Canadian Journal of Plant Pathology*, 20, 234–42.

Nunninger-Ney, C. (1988). Untersuchungen zur Resistenz von *Monilinia fructicola* (Wint.) Honey gegenüber Ergosterol – Biosynthesehemmern. – Inauguraldiss., Universität Basel.

Ogawa, J. M. and English, H. (1960). Relative pathogenicity of two brown rot fungi, *Sclerotinia laxa* and *Sclerotinia fructicola*, on twigs and blossoms. *Phytopathology*, 50, 550–8.

Ogawa J. M. and English, H. (1991). Diseases of temperate zone tree fruit and nut crops. University of California, Division of Agriculture and Natural Resources, Oakland, CA, 461pp.

Ogawa, J. M., English, H., Moller, W. J., Manji, B. T., Rough, D. and Koike, S. T. (1980). Brown rot of stone fruits. Calif. Univ. Div. Agric. Sci. Leaf. 2206, 7pp.

Ogawa, J. M., Manji, B. T., Rough, D. and Sonoda, R. M. (1981). Monitoring and control of benomyl-resistant *Monilinia fructicola. Phytopathology*, 71, 246. (Abstract).

Ogawa, J. M., Manji, B. T., Bostock, R. M., Cañez, V. M. and Bose, E. A. (1984). Detection and characterization of benomyl-resistant *Monilinia laxa* on apricots. *Plant Disease*, 68, 29–31.

Ogawa, J. M., Manji, B. T. and MacSwan, I. C. (1986). Field test procedures for evaluation of fungicides for control of *Monilinia laxa* on stone fruits. In: Hickey, K. D. (Ed.), *Methods for Evaluating Pesticides for Control of Plant Pathogens*. APS Press, St. Paul, MN, pp. 152–4.

Ogawa, J. M., Gubler, W. D. and Manji, B. T. (1988). Effect of sterol biosynthesis inhibitors on diseases of stone fruits and grapes in California. In: Berg, D. and Plempel, M. (Eds), *Sterol Biosynthesis Inhibitors* (583pp). Ellis Horwood Ltd., Chichester, England, pp. 262–87.

Osirio, J. M., Adaskaveg, J. E. and Ogawa, J. M. (1994). Inhibition of mycelial growth of *Monilinia* species and suppression and control of brown rot blossom blight of almond with iprodione and E-0858. *Plant Disease*, 78, 712–16.

Papavasileiou, A., Karaoglanidis, G. S. and Michailides, T. J. (2015). Intraspecific diversity of *Monilinia fructicola* and *M. laxa* populations from blossoms and fruit of different hosts in Greece. *Plant Disease*, 99, 1353–9.

Papavasileiou, A., Madesis, P. B. and Karaoglanidis, G. S. (2016). Identification and differentiation of *Monilinia* species causing brown rot of pome and stone fruit using High-Resolution Melting (HRM) analysis. *Phytopathology*, 106, 1055–64.
Pascual, S., Melgarejo, P. and Magan, N. (1999). Production of the fungal biocontrol agent *Epicoccum nigrum* by solid substrate fermentation: Effect of water activity on accumulation of compatible solutes. *Mycopathologia*, 146(2), 83–9.
Pascual, S., Melgarejo, P. and Magan, N. (2000). Accumulation of compatible solutes in *Penicillium frequentans* grown at reduced water activity and biocontrol of *Monilinia laxa*. *Biocontrol Science and Technology*, 10, 71–80.
Pellegrino, C., Gullino, M. L., Garibaldi, A. and Spadaro, D. (2009). First report of brown rot of stone fruit caused by *Monilinia fructicola* in Italy. *Plant Disease*, 93, 668.
Penrose, L. J. (1990). Prolonged field persistence of resistance to benomyl in *Monilinia fructicola*. *Crop Protection*, 9, 190–2.
Penrose, L. J., Tarran, J. and Wong, A. L. (1976). First record of *Sclerotinia laxa* Aderh. & Ruhl. in New South Wales: Differentiation from *S. fructicola* (Wint.) Rehm. by cultural characteristics and electrophoresis. *Australian Journal of Agricultural Research*, 27, 547–56.
Penrose, L., Koffmann, J. W. and Nicholls, M. R. (1985). Field occurrence of vinclozolin resistance in *Monilinia fructicola*. *Plant Pathology*, 34, 228–34.
Poniatowska, A., Michalecka, M. and Bielenin, A. (2013). Characteristic of *Monilinia* spp. fungi causing brown rot of pome and stone fruits in Poland. *European Journal of Plant Pathology*, 135, 855–65.
Poniatowska, A., Michalecka, M. and Puławska, J. (2016). Genetic diversity and pathogenicity of *Monilinia polystroma* – the new pathogen of cherries. *Plant Pathology*, 65, 723–33.
Pusey, P. L. and Wilson, C. L. (1984). Postharvest biological control of stone fruit brown rot by *Bacillus subtilis*. *Plant Disease*, 68, 753–6.
Pusey, P. L., Hotchkiss, M. W., Dulmage, H. T., Baumgardner R. A., Zehr, E. I., Reilly, C. C. and Wilson, C. L. (1988). Pilot tests for commercial production and application of *Bacillus subtilis* (B-3) for postharvest control of peach brown rot. *Plant Disease*, 72, 622–6.
Ramsdell, D. C. and Ogawa, J. M. (1973). Reduction of *Monilinia laxa* inoculum potential in almond orchards resulting from dormant benomyl sprays. *Phytopathology*, 68, 830–6.
Reese, R. L. and Moore, J. E., Jr. (1982). UV irradiation induced resistance to CGA 64251 in *Monilinia fructicola*. *Phytopathology*, 72, 980. (Abstract).
Ritchie, D. F. (1983). Mycelial growth, peach fruit-rotting capability, and sporulation of strains of *Monilinia fructicola* resistant to dichloran, iprodione, procymidone, and vinclozolin. *Phytopathology*, 73, 44–7.
Romanazzi, G., Nigro, F., Ippolito, A. and Salerno, M. (2001). Effect of short hypobaric treatments on postharvest rots of sweet cherries, strawberries and table grapes. *Postharvest Biology and Technology*, 22, 1–6.
Rosenberger, D. A. and Meyer, F. W. (1981). Postharvest fungicides for apples: Development of resistance to benomyl, vinclozolin and iprodione. *Plant Disease*, 65, 1010–13.
Rozsnyay, Zs. and Apostol, J. (2001). Breeding for sweet and sour cherry disease resistance in Hungary. In *IV International Cherry Symposium 667* (pp. 117–22).
Rudolph, B. A. (1925). Monilinia blossom blight (brown-rot) of apricots. *California University Agricultural Experimental Station Bulletin*, 383.

Rungjindamai, N., Jeffries, P. and Xu, X. M. (2014). Epidemiology and management of brown rot on stone fruit caused by *Monilinia laxa*. *European Journal of Plant Pathology*, 140, 1–17.

Sanoamuang, N. and Gaunt, R. E. (1995). Persistence and fitness of carbendazim- and dicarboximide-resistant isolates of *Monilinia fructicola* (Wint.) Honey in flowers, shoots and fruit of stone fruit. *Plant Pathology*, 44, 448–57.

Sautter, C. K., Brackmann, A., Anese, R. D. O., Weber, A., Rizzatti, M. R. and Pavanello, E. P. (2011). Control of brown rot and physical-chemical characteristics in 'Magnum'peaches postharvest treated with abiotic elicitors. *Revista Ceres*, 58(2), 172-7.

Schena, L., Nigro, F., Pentimone, I., Ligorio, A. and Ippolito, A. (2003). Control of postharvest rots of sweet cherries and table grapes with endophytic isolates of *Aureobasidium pullulans*. *Postharvest Biology and Technology*, 30(3), 209–20.

Schlagbauer, G. (1989). Occurrence of latent *Monilinia laxa* infections on plums, peaches and apricots. *Phytophylactica*, 21(1), 35–8.

Schlagbauer, H. E. and Holz, G. (1987). Blossom blight and brown rot of stone fruit caused by *Monilinia laxa* in the Cape province of South Africa. *Phytophylactica*, 19, 513–14.

Schlagbauer, H. E. and Holz, G. (1990). Infection and colonization of plum blossoms by *Monilinia laxa*. *Phytophylactica*, 22, 419–22.

Schnabel, G., Bryson, P. K., Bridges, W. C. and Brannen, P. M. (2004). Reduced sensitivity in *Monilinia fructicola* to propiconazole in Georgia and implications for disease management. *Plant Disease*, 88, 1000–4.

Sharma, R. L. and Kaul, J. L. (1987). Blossom blight and fruit rot of apple. *Indian Journal of Plant Pathology*, 5, 205–6.

Sisquella, M., Vinas, I., Teixidó, N., Picouet, P. and Usall, J. (2013a). Continuous microwave treatment to control postharvest brown rot in stone fruit. *Postharvest Biology and Technology*, 86, 1–7.

Sisquella, M., Casals, C., Picouet, P., Vinas, I., Torres, R. and Usall, J. (2013b). Immersion of fruit in water to improve radio frequency treatment to control brown rot in stone fruit. *Postharvest Biology and Technology*, 80, 31–6.

Smilanick, J. L., Denis-Arrue, R., Bosch, J. R., Gonzalez, A. R., Henson, D. and Janisiewicz, W. J. (1993). Control of postharvest brown rot of nectarines and peaches by *Pseudomonas* species. *Crop Protection*, 12, 513–20.

Smith, W. L. and Redit, W. H. (1966). Relation of cooling methods to postharvest brown rot of heated and nonheated peaches. *Phytopathology*, 56, 586. (Abstract).

Soltész, M. (1997). Kórokozókkal és kártevőkkel szembeni ellenállóság. (Resistance to diseases and pests). In: Soltész, M. (Ed.), *Integrált gyümölcstermesztés*. (*Integrated Fruit Production*). Mezőgazda Kiadó, Budapest, pp. 71–84.

Spadoni, A., Neri, F., Bertolini, P. and Mari, M. (2013). Control of Monilinia rots on fruit naturally infected by hot water treatment in commercial trials. *Postharvest Biology and Technology*, 86, 280–4.

Staub, T. (1991). Fungicide resistance: Practical experiences with antiresistance strategies and the role of integrated use. *Annual Review of Phytopathology*, 29, 421–42.

Stensvand, A., Talgo, V. and Borve, J. (2001). Seasonal production of conidia of *Monilinia laxa* from mummified fruits, blighted spurs and flowers of sweet cherry. *Gartenbauwissenschaft*, 66(6), 273-81.

Stevens, C., Wilson, C. L., Lu, J. Y., Khan, V. A., Chalutz, E., Droby, S., Kabwe, M. K., Haung, Z., Adeyeye, O., Pusey, L. P., Wisniewski M. E. and West, M. (1996). Plant hormesis induced by ultraviolet light-C for controlling postharvest diseases of tree fruits. *Crop Protection*, 15, 129–34.

Stevens, C., Khan, V. A., Lu, J. Y., Wilson, C. L., Pusey, P. L., Kabwe, M. K., Igwegbe, E. C. K., Chalutz, E. and Droby, S. (1998). The germicidal and hermetic effects of UV-C light on reducing brown rot disease and yeast microflora of peaches. *Crop Protection*, 17, 75–84.

Sutton, T. B. and Clayton, C. N. (1972). Role and survival of *Monilinia fructicola* in blighted peach branches. *Phytopathology*, 62, 1369–73.

Szkolnik, K M. (1981). Protective and after-infection activity of sterol inhibitors against *Monilinia fructicola*. *Phytopathology*, 71, 565. (Abstract).

Sztejnberg, A. and Jones, A. L. (1978): Tolerance of the brown rot fungus *Monilinia fructicola* to iprodione, vinclozolin and procymidone. *Phytopathological News*, 12, 187–8.

Tamm, L., Minder, C. E. and Flückiger, W. (1995). Phenological analysis of brown rot blossom blight of sweet cherry caused by *Monilinia laxa*. *Phytopathology*, 85, 401–8.

Tate, K. G. and Ogawa, J. M. (1975). Nitidulid beetles as vectors of *Monilinia fructicola* in California stone fruits. *Phytopathology*, 65, 977–83.

Tate, K. G., Ogawa, J. M., Manji, B. T. and Bose, E. (1974). Survey of benomyl-tolerant isolates of *Monilinia fructicola* and *Monilinia laxa* in stone fruit orchards of California. *Plant Disease Reporter*, 58, 663–5.

Thomidis, T. and Exadaktylou, E. (2013). Effect of a plastic rain shield on fruit cracking and cherry diseases in Greek orchards. *Crop Protection*, 52, 125–9.

Thomidis, T., Michailides, T. and Exadaktylou, E. (2009). Contribution of pathogens to peach fruit rot in northern Greece and their sensitivity to iprodione, carbendazim, thiophanate-methyl and tebuconazole fungicides. *Journal of Phytopathology*, 157(3), 194–200.

Tian, S. P., Fan, Q., Xu, Y. and Jiang, A. L. (2001). Evaluation of the use of high CO_2 concentrations and cold storage to control *Monilinia fructicola* on sweet cherries. *Postharvest Biology and Technology*, 22, 53–60.

van Brouwershaven, I. R., Bruil, M. L., van Leeuwen, G. C. M. and Kox, L. F. F. (2010). A real-time (TaqMan) PCR assay to differentiate *Monilinia fructicola* from other brown rot fungi of fruit crops. *Plant Pathology*, 59(3), 548–55.

van Geluwe, J., James, J. R. and Davidson, R. (1981). Length of protective of CGA-64251 for control of brown rot blossom blight in stone fruit. *Phytopathology*, 71, 883. (Abstract)

van Leeuwen G. C. M. and van Kesteren H. A. (1998). Delineation of the three brown rot fungi of fruit crops (*Monilinia* spp.) on the basis of quantitative characteristics. *Canadian Journal of Botany*, 76, 2042–50.

van Leeuwen, G. C. M., Stein, A., Holb, I. and Jeger, M. J. (2000). Yield loss caused by *Monilia fructigena* (Aderh. & Ruhl.) Honey, and spatio-temporal dynamics of disease development. *European Journal of Plant Pathology*, 106, 519–28.

van Leeuwen, G. C. M., Baa Yen, R. P., Holb I. J. and Jeger, M. J. (2002a). Distinction of the Asiatic brown rot fungus *Monilia polystroma* sp. nov. from *Monilia fructigena*. *Mycological Research*, 106, 444–51.

van Leeuwen, G. C. M., Holb, I. J. and Jeger, M. J. (2002b). Factors affecting mummification and sporulation of pome fruit infected by *Monilinia fructigena* in Dutch orchards. *Plant Pathology*, 51, 787–93.

Vasić, M., Duduk, N. and Ivanović M. S. (2013). First report of brown rot caused by *Monilia polystroma* on apple in Serbia. *Plant Disease*, 97: 145.

Vasić, M., Duduk, N., Vico, I., Rančić, D., Pajić, V. and Backhouse, D. (2016). Comparative study of *Monilinia fructigena* and *Monilia polystroma* on morphological features, RFLP analysis, pathogenicity and histopathology. *European Journal of Plant Pathology*, 144, 15–30.

Villani, S. M. and Cox, K. D. (2011). Characterizing fenbuconazole and propiconazole sensitivity and prevalence of 'Mona' in isolates of *Monilinia fructicola* from New York. *Plant Disease*, 95, 828–34.

Villarino, M., Melgarejo, P., Usall, J., Segarra, J. and De Cal, A. (2010). Primary inoculum sources of *Monilinia* spp. in Spanish peach orchards and their relative importance in brown rot. *Plant Disease*, 94, 1048–54.

Walter, M. A. D., Gillian, F., McLaren, B., Jenny, A., Fraser, B., Chris, M., Frampton, C., Kirsty, S. H., Boyd-Wilson, A. and Perry J. H. (2004). Methods of screening apricot fruit for resistance to brown rot caused by *Monilinia* spp. *Australasian Plant Pathology*, 33, 541–7.

Weaver, L. O. (1950). Effect of temperature and relative humidity on occurrence of blossom blight of stone fruits. *Phytopathology*, 40, 1136–53.

Wells, J. M. and Bennett, A. H. (1976). Hydro-cooling and hydrair-cooling with fungicides for reduction of postharvest decay of peaches. *Phytopathology*, 66, 801–5.

Whan, J. H. (1976). Tolerance of Sclerotinia fructicola to benomyl. *Plant Disease Reporter*, 60, 200–1.

Wilcox, W. F. (1989). Influence of environment and inoculum density on the incidence of brown rot blossom blight of sour cherry. *Phytopathology*, 79, 530–4.

Wilcox, W. F. (1990). Postinfection and antisporulant activities of selected fungicides in control of blossom blight of sour cherry caused by *Monilinia fructicola*. *Plant Disease*, 74, 808–11.

Willetts, H. J. (1968). Factors influencing the production of stromata and microconidia by *Sclerotinia fructicola* (Wint.) Rehm. *Annals of Botany*, 32, 219–32.

Willetts, H. J. and Burllock, S. (1993). Cytology, histology and histochemistry of fruit infection by *Monilinia* species. In: Biggs, A. R. (Eds), *Handbook of Cytology, Histology and Histochemistry of Fruit Tree Diseases*. CRC Press, Boca Raton, FL, pp. 113–36.

Willetts, H. J. and Harada, Y. (1984). A review of apothecial production by *Monilinia* fungi in Japan. *Mycologia*, 76, 314–25.

Willetts, H. J., Byrde, R. J. W., Fielding, A. H. and Wong, A. L. (1977). The taxonomy of the brown rot fungi (*Monilinia* spp.) related to their extracellular cell wall-degrading enzymes. *Journal of General Microbiology*, 103, 77–83.

Wittig, H. P. P., Johnson, K. B. and Pscheidt, J. W. (1997). Effect of epiphytic fungi on brown rot blossom blight and latent infections in sweet cherry. *Plant Disease*, 81, 383–7.

Xu, X. M. and Jeger, M. J. (2013a). Combined use of two biocontrol agents with different biocontrol mechanisms most likely results in less than expected efficacy in controlling foliar pathogens under fluctuating conditions: A modeling study. *Phytopathology*, 103, 108–16.

Xu, X. M. and Jeger, M. J. (2013b). Theoretical modeling suggests that synergy may result from combined use of two biocontrol agents for controlling foliar pathogens under spatial heterogeneous conditions. *Phytopathology*, 103, 768–75.

Xu, X. M. and Robinson, J. D. (2000). Epidemiology of brown rot (*Monilinia fructigena*) on apple: Infection of fruits by conidia. *Plant Pathology*, 49, 201–6.

Xu, X. M., Guerin, L. and Robinson, J. D. (2001a). Effects of temperature and relative humidity on conidial germination and viability, colonization and sporulation of *Monilinia fructigena*. *Plant Pathology*, 50, 561–8.
Xu, X. M., Robinson, J. D., Berrie, A. M. and Harris, D. C. (2001b). Spatio-temporal dynamics of brown rot (*Monilinia fructigena*) on apple and pear. *Plant Pathology*, 50, 569–78.
Xu, X. M., Bertone, C. and Berrie, A. (2007). Effects of wounding, fruit age and wetness duration on the development of cherry brown rot in the UK. *Plant Pathology*, 56, 114–19.
Xu, X. M., Jeffries, P., Pautasso, M. and Jeger, M. J. (2011). Combined use of biocontrol agents to manage plant diseases in theory and practice. *Phytopathology*, 101, 1024–31.
Yánez-Mendizábal, V., Zeriouh, H., Viñas, I., Torres, R., Usall, J., de Vicente, A., Perez-Garcia, A. and Teixidó, N. (2012). Biological control of peach brown rot (*Monilinia* spp.) by *Bacillus subtilis* CPA-8 is based on production of fengycin-like lipopeptides. *European Journal of Plant Pathology*, 132, 609–19.
Yoshimura, M. A., Luo, Y., Ma, Z. and Michailides, T. J. (2004). Sensitivity of *Monilinia fructicola* from stone fruit to thiophanate-methyl, iprodione, and tebuconazole. *Plant Disease*, 88, 373–8.
Zadoks, J. C. and Schein, R. D. (1979). *Epidemiology and Plant Disease Management*. Oxford University Press, New York, NY, 428pp.
Zehr, E. I. (1982). Control of brown rot in peach orchards. *Plant Disease*, 66, 1101–5.
Zehr, E. I., Toler, J. E. and Luszcz, L. A. (1991). Spread and persistence of benomyl-resistant *Monilinia fructicola* in South Carolina peach orchards. *Plant Disease*, 75, 590–3.
Zehr, E. I., Luszcz, L. A., Olien, W. C., Newall. W. C. and Toler, J. E. (1999). Reduced sensitivity in *Monilinia fructicola* to propiconazole following prolonged exposure in peach orchards. *Plant Disease*, 83, 913–16.
Zhang, Y., Boomer, M. and Rüegg, J. (1991). Variation in sensitivity to sterol biosynthesis inhibitors (SBI) of different isolates of *Monilinia* species. *Journal of Plant Disease and Plant Protection*, 98, 317–22.
Zhang, D., Spadaro, D., Garibaldi, A. and Gullino, M. L. (2010a). Selection and evaluation of new antagonists for their efficacy against postharvest brown rot of peaches. *Postharvest Biology and Technology*, 55, 174–81.
Zhang, D., Lopez-Reyes, J. G., Spadaro, D., Garibaldi, A. and Gullino, M. L. (2010b). Efficacy of yeast antagonists used individually or in combination with hot water dipping for control of postharvest brown rot of peaches. *Journal of Plant Diseases and Protection*, 117, 226–32.
Zhou, T., Schneider, K. E. and Li, X. Z. (2008). Development of biocontrol agents from food microbial isolates for controlling post-harvest peach brown rot caused by *Monilinia fructicola*. *International Journal of Food Microbiology*, 126, 180–5.
Zhu, X. Q. and Guo, L. Y. (2010). First report of brown rot on plum caused by *Monilia polystroma* in China. *Plant Disease*, 94, 478.

Chapter 7

Apple mosaic virus: biology, epidemiology and detection

Karel Petrzik, Biology Centre CAS, Czech Republic

1 Introduction

Apple mosaic virus (ApMV) is a distinct species within the relatively large genus *Ilarvirus*, which comprises 22 different virus species and seven putative species or synonyms (ICTV release 2017, email ratification 2018, MSL32 and GenBank release April 2018) (Fig. 1). Ilarviruses, whose name is an acronym derived from *i*sometric *la*bile *r*ingspot, have been recognized exclusively as plant viruses occurring in various mono- and dicotyledonous plant species, including woody plants as well as herbs. Recently, however, one exception has been identified, as ilarvirus-like sequences were obtained by Illumina sequencing of *Ixodes scapularis* tick. The sequences are part of the genome of the novel virus named blacklegged tick-associated virus, which in this case would be the first identified invertebrate ilarvirus (Tokarz et al. 2018). Related to ilarviruses in taxonomy are plant viruses of the genera *Alfamovirus*, *Anulavirus*, *Bromovirus*, *Cucumovirus* and *Oleavirus* which share similar genome organization but differ in host preference and some also in particle morphology.

http://dx.doi.org/10.19103/AS.2019.0046.09

ApMV has been known since 1928, when it was described in *Rosa* spp. and apple (White 1928). This virus has been also found in birch, hop, plum and hazelnut and a number of synonyms of the virus name have been assigned with time (e.g. Rose mosaic virus, Mild ApMV, Severe ApMV, Birch line pattern virus, Birch ringspot virus, Hop virus A, Hop virus C, European plum line pattern virus, Dutch plum line pattern virus, Hazel mosaic virus and Horse chestnut yellow mosaic virus) (Petrzik and Lenz 2011). ApMV is present often in mixed infections

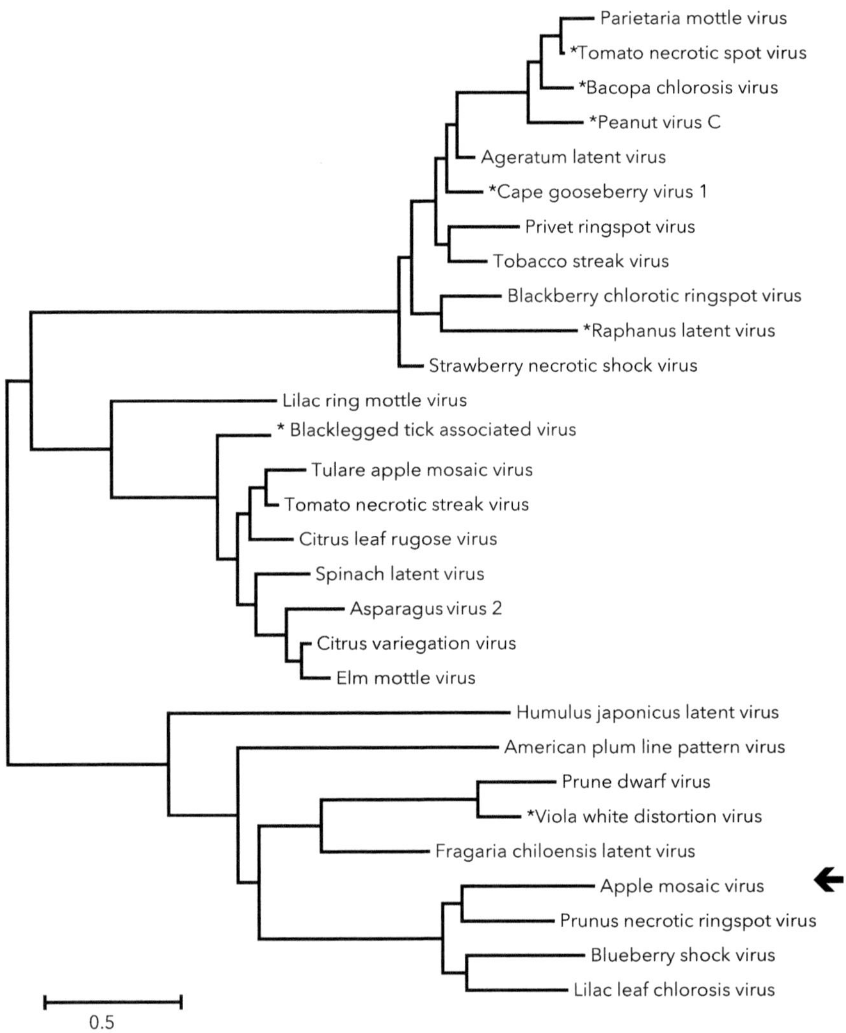

Figure 1 Phylogenic neighbour-joining tree of recently known ilarviruses based on capsid protein amino acid sequences. Viruses not approved by the ICTV authority are marked by an asterisk (*). Apple mosaic virus is marked with an arrow.

with other ilarviruses (*Prunus necrotic ringspot virus* [PNRSV] or *Prune dwarf virus* [PDV]) on their common stone fruit hosts, or with Apple chlorotic leaf spot virus (ACLSV), Apple stem pitting virus (ASPV), Apple stem grooving virus (ASGV) and other apple-infecting viruses on apple host. No recent studies have been made on the effect of ApMV on growth and yield. The observed impact of ApMV in gardens is often influenced by the presence of co-hosting viruses, and there is no reason to suppose that the virus may cause any significant economic impact. Even though this virus occurs worldwide, that occurrence is in fact not common and in apple orchards ApMV is present in the range of 0.2–4%. Mechanical transmission of ApMV isolates from woody hosts to herbaceous hosts is very difficult, but they are readily transmissible from herbaceous hosts. To date, no pathotypes of ApMV have been described despite the fact that length variants in the *capsid protein* gene are known and differences exist among isolates in their ability to infect particular herbaceous species. ApMV is serologically distantly related to PNRSV, and it could have been erroneously identified as PNRSV in serological tests in the past (Fulton 1972).

2 Feature of the genome

The genome of ApMV is composed of single-stranded RNA of positive polarity. About 8500 nt long, it is separated into three segments (Fig. 2) that are encapsidated separately in quasi-spherical particles with a diameter of 25–29 nm. The virus consists of 180 CP subunits. The genomic segments have 5′ terminal cap structures and 3′ termini without the poly(A) tail but forming strong complex secondary structures at the 3′ end. All three segments must be present in the host plant to initiate infection. The RNA1 encodes a polyprotein with the methyltransferase and helicase domains. The RNA2 encodes a polyprotein with RNA-dependent RNA polymerase domains. Both are involved in genome replication and transcription of subgenomic sgRNA4 from RNA3. The RNA3 contains two ORFs, the first with cell-to-cell movement activity and the other CP. The *CP* is translated from subgenomic RNA4 that is derived from RNA3 (Sánchez-Navarro and Pallás 1994) (Fig. 2).

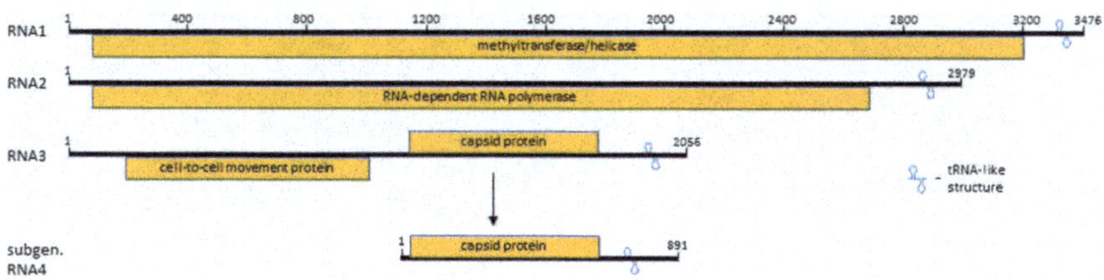

Figure 2 Arrangement of the ApMV genome. Open reading frames are marked as yellow rectangles. All segments are drawn in scale.

Of all the viral genes, the *capsid protein* gene is the most known and there are more than 100 CP sequences from different host plants stored in GenBank. Sequence comparison has identified a mutation hot spot around nt position 141, where the 15 nt deletion was present in all hop-infecting isolates, 9 or 6 nt deletion in most hazelnut isolates and 6 nt deletion in the almond isolate (Petrzik and Lenz 2002; Cieślińska and Valasevich 2016). Close to this site, a frameshift mutation begins in five apple-infecting isolates. Thus, the capsid protein of recently known isolates has a size of 220–223 amino acids (Fig. 3; see also Table 1).

In contrast to point mutations, which slowly change a genome, recombination is a very powerful natural process that can lead to dramatic changes in the biological properties of the virus. These include the acquisition of a different host, as well as serotype and pathotype differentiation. In ApMV, two recombination events have been identified in the *CP* gene (Boulila 2010; Petrzik 2018, unpublished results). These indicate that infection with several different strains has occurred in the past within one host or that recombination has occurred between different quasi-species present in a single host. It has been predicted that two regions were recombinant: the area between nt 182 and 457 in one isolate from pear and the region between nt 318 and 654 in three isolates from pear and two from apple (Fig. 4). Most probably, these changes have no effect on the biological properties of the recombinants.

The *CP* gene is a multifunctional gene that works in replication, in propagating infection and is required for systemic movement and for cell-to-cell spread in plants (van der Vossen et al. 1994). It has been speculated that different sequence variants exist in the most frequent hosts (Grimová et al. 2013). Indeed, *CP* sequence analysis and phylogenetic tree calculations have identified three groups of isolates that strongly correlated with apple, hop and hazelnut/rose hosts (Fig. 5). These seem to be the primary hosts of the virus.

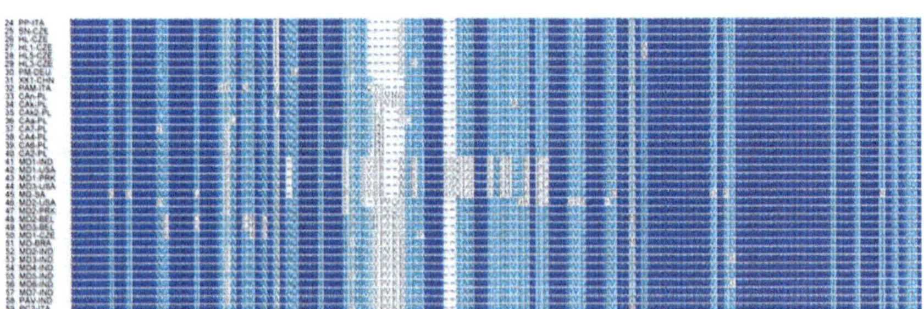

Figure 3 Alignment of CP protein. Alignment of the N-terminal part of the ApMV CP amino acid sequence around the mutation hot spot with deletions and frameshift area (marked red). For isolate acronyms see Table 1.

Table 1 ApMV CP sequences

Isolate	Access. no.	Original host	Geographic origin
HL 05_MB*	AF473580	*H. lupulus*	Australia
HL 42_2001*	AF473581	*H. lupulus*	Australia
HL 16_2001*	AF473582	*H. lupulus*	Australia
HL 20_2001*	AF473583	*H. lupulus*	Australia
HL 28_2001*	AF473584	*H. lupulus*	Australia
HL 26_MB*	AF473585	*H. lupulus*	Australia
HL 22_MB*	AF473586	*H. lupulus*	Australia
HL 34_MB*	AF473587	*H. lupulus*	Australia
HL 27_MB*	AF473588	*H. lupulus*	Australia
HL 17_2001*	AF473589	*H. lupulus*	Australia
HL 35_2001*	AF473590	*H. lupulus*	Australia
HL 37_2001*	AF473591	*H. lupulus*	Australia
HL 10_2001*	AF473592	*H. lupulus*	Australia
HL 11_2001*	AF473593	*H. lupulus*	Australia
HL 24_2001*	AF473594	*H. lupulus*	Australia
HL 25_2001*	AF473595	*H. lupulus*	Australia
HL 17_MB*	AF473596	*H. lupulus*	Australia
MD1-PRK	AF548367	*Malus* ssp.	South Korea
*	AM403478	Unknown	China
	AM490197	Unknown	China
MD2-USA	AMQCOATPA	*Malus* ssp.	USA
MD1-USA	AMU15608	*Malus x domestica*	USA
MD1-CZE	AY054385	*Malus x domestica*	Czech Republic
PD-CZE	AY054386	*Prunus domestica*	Czech Republic
HL-CZE	AY054387	*H. lupulus*	Czech Republic
PAM-ITA	AY054388	*Prunus amygdalus*	Italy
PC1-CZE	AY054389	*Pyrus* ssp.	Czech Republic
MD2-PRK	AY125977	*Malus x domestica* cv. Fuji	South Korea
MD3-BEL	AY542540	*Malus x domestica*	Belgium
MD2-BEL	AY542541	*Malus x domestica*	Belgium
PC6-CZE	AY542542	*Pyrus* ssp.	Czech Republic
PC5-CZE	AY542543	*Pyrus* ssp.	Czech Republic
PC4-CZE	AY542544	*Pyrus* ssp.	Czech Republic
PC3-CZE	AY542545	*Pyrus* ssp.	Czech Republic
PC2-ITA	AY542546	*Pyrus* ssp.	Italy
FRA-USA	AY854050	*Fragaria vesca*	USA
	FJ429309	*Malus x domestica* cv. Golden Delicious	India

(*Continued*)

Table 1 (*Continued*)

Isolate	Access. no.	Original host	Geographic origin
MD2-IND	FJ429311	*Malus x domestica* cv. Golden Delicious	India
	FM178274	*Malus x domestica*	India
MD7-IND	FN435314	*Malus x domestica*	India
MD6-IND	FN435315	*Malus x domestica*	India
MD5-IND	FN435316	*Malus x domestica*	India
MD4-IND	FN435317	*Malus x domestica* cv. Golden Delicious	India
PAV-IND	FN546183	*Prunus avium*	India
MD3-IND	FN547927	*Malus x domestica*	India
	FN564150	*Malus x domestica*	India
*	FR750246	*Malus x domestica*	India
MD-BRA	GQ131805	*Malus x domestica* cv. Fuji	Brazil
MD1-IND	HE574164	*Malus x domestica* cv. Golden Delicious	India
HL-BGR	HE866936	*H. lupulus* Cocnean	Bulgaria
HL-SVN	HE866937	*H. lupulus* Styrie	Slovenia
HL-FRA	HE866938	*H. lupulus* Strieselspalt	France
HL-GBR	HE866939	*H. lupulus* Northern Brewer	Great Britain
HL-DNK	HE866940	*H. lupulus* Nordgaard 978	Denmark
HL-USA	HE866941	*H. lupulus* Sacramento English Cluster	USA
HL-NZL	HE866942	*H. lupulus* Calicross	New Zealand
HL-AUT	HE866943	*H. lupulus* Santaler	Austria
HL-AUS	HE866944	*Humulus lupulus* Ringwood Special	Australia
HL-RUS	HE866945	*H. lupulus* Serebrjanka	Russia
HL-SWE	HE866946	*H. lupulus* Svalof 825-17	Sweden
HL-ZAF	HE866947	*H. lupulus* Quteniqua	South Africa
HL1-CZE	HE866948	*H. lupulus* Kazbek	Czech Republic
HL2-CZE	HE866949	*H. lupulus* Osvaldův klon	Czech Republic
HL3-CZE	HE866950	*H. lupulus* Premiant	Czech Republic
HL4-CZE	HE866951	*H. lupulus* Lučan	Czech Republic
HL5-CZE	HE866952	*H. lupulus* Bor	Czech Republic
HL6-CZE	HE866953	*H. lupulus* Russian hybrid	Czech Republic
HL7-CZE	HE866954	*H. lupulus* Blato	Czech Republic
HL8-CZE	HE866955	*H. lupulus* Blšanka	Czech Republic

Isolate	Access. no.	Original host	Geographic origin
HL9-CZE	HE866956	*H. lupulus* Zlatan	Czech Republic
Hpl1-CZE	HE866957	*H. lupulus*	Czech Republic
Hpl2-CZE	HE866958	*H. lupulus*	Czech Republic
CA-CZE	HE866959	*Corylus avellana*	Czech Republic
SN-CZE	HE866960	*Sambucus nigra*	Czech Republic
PP-ITA	HE866961	*Prunus persica*	Italy
PAR-ITA	HE866962	*Prunus armeniaca*	Italy
MDI2-PL	HG328263	*Malus domestica*	Poland
MD12-PL	HG328264	*Malus domestica*	Poland
MD17-PL	HG328265	*Malus domestica*	Poland
MD3-PL	HG328266	*Malus domestica*	Poland
MD1-PL	HG328267	*Malus domestica*	Poland
MD-SK	HG328268	*Malus domestica*	Slovakia
MDM-BL	HG328269	*Malus domestica*	Belarus
MDP-BL	HG328270	*Malus domestica*	Belarus
RO-PL	HG328271	*Rosa* spp.	Poland
CAn-PL	HG328276	*Corylus avellana*	Poland
CAk2-PL	HG328277	*Corylus avellana*	Poland
CAa-PL	HG328278	*Corylus avellana*	Poland
CAk-PL	HG328279	*Corylus avellana*	Poland
MDz-BL	HG328280	*Malus domestica*	Belarus
MD7-PL	HG328281	*Malus domestica*	Poland
CA7-PL	HG328282	*Corylus avellana*	Poland
CA4-PL	HG328283	*Corylus avellana*	Poland
CA6-PL	HG328284	*Corylus avellana*	Poland
CA2-PL	HG328285	*Corylus avellana*	Poland
MD1-LAT	HQ609545	*Malus* spp.	Latvia
MD2-LAT	HQ609546	*Malus* spp.	Latvia
PC1-LAT	HQ609547	*Pyrus communis*	Latvia
PC2-LAT	HQ609548	*Pyrus communis*	Latvia
MD3-LAT	HQ609549	*Malus* spp.	Latvia
MD4-LAT	HQ609550	*Malus* spp.	Latvia
MD5-LAT	HQ609551	*Malus* spp.	Latvia
LL1-NOR	KC469067	*Lasallia pustulata*	Norway
LL6-AN	KC469068	*Usnea antarctica*	Antarctica
LL2-NOR	KC469069	*Usnea chaetophora*	Norway
LL1-CZ	KC469070	*Xanthoria parietina*	Czech Republic
LL4-CZ	KC469071	*Usnea hirta*	Czech Republic

(*Continued*)

Table 1 (*Continued*)

Isolate	Access. no.	Original host	Geographic origin
LL5-CZ	KC469072	*Pseudevernia furfuracea*	Czech Republic
MDS-BL	KR779895	*Malus domestica*	Belarus
MDK2-BL	KR779896	*Malus domestica*	Belarus
MDK1-BL	KR779897	*Malus domestica*	Belarus
MD2-BL	KR779898	*Malus domestica*	Belarus
MD1-BL	KR779899	*Malus domestica*	Belarus
RO-EC	KT590399	*Rosa* spp.	Ecuador
MD13a-TUR	KU645884	*Malus domestica*	Turkey
MD38a-TUR	KU645885	*Malus domestica*	Turkey
MD44a-TUR	KU645886	*Malus domestica*	Turkey
MD49-TUR	KX034205	*Malus domestica*	Turkey
MD13-TUR	KX034206	*Malus domestica*	Turkey
MD38-TUR	KX034207	*Malus domestica*	Turkey
MD44-TUR	KX034208	*Malus domestica*	Turkey
MD45-TUR	KX034209	*Malus domestica*	Turkey
MD47-TUR	KX034210	*Malus domestica*	Turkey
MD48-XX	KX034211	*Malus domestica*	Turkey
RO5-EQ	KX646545	*Rosa* spp.	Ecuador
RO13-EQ	KX646546	*Rosa* spp.	Ecuador
RO22-EQ	KX646547	*Rosa* spp.	Ecuador
RO23-EQ	KX646548	*Rosa* spp.	Ecuador
MD-SA	KY488189	*Malus domestica*	Saudi Arabia
PR-AUS	KY883320	Prunus	Australia
MD151-CAN	KY971019	*Malus domestica*	Canada
MD157-CAN	KY971020	*Malus domestica*	Canada
MD160-CAN	KY971021	*Malus domestica*	Canada
PE-KO	LC006865	Peach	South Korea
HL3-BL	LT546544	*H. lupulus*	Belarus
HL2-BL	LT546545	*H. lupulus*	Belarus
HL1-BL	LT546546	*H. lupulus*	Belarus
MD1a-CZE	LT574866	*Malus domestica*	Czech Republic
MD2a-CZE	LT574867	*Malus domestica*	Czech Republic
MD3a-CZE	LT574868	*Malus domestica*	Czech Republic
MD4a-CZE	LT574869	*Malus domestica*	Czech Republic
MD5a-CZE	LT574870	*Malus domestica*	Czech Republic
MS6-CZE	LT574871	*Malus silvestris*	Czech Republic
MS7-CZE	LT574872	*Malus silvestris*	Czech Republic
MD3-USA	NC_003480	*Malus x domestica*	USA

Isolate	Access. no.	Original host	Geographic origin
PNRSV	NC_004364	*Prunus persica*	USA
PM-DEU	S78319	*Prunus mahaleb*	Germany

*Partial CP sequence.

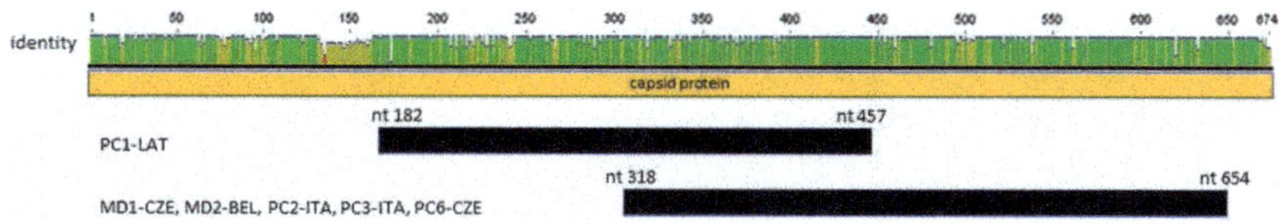

Figure 4 Position of putative recombination in the *capsid protein* gene of ApMV. Nucleotide sequence alignment of the *CP* gene of 96 ApMV isolates revealed mutational hot spot around nt positions 140 and 500. Putative recombinant regions are marked as black rectangles.

Stone fruit seems to be an accidental host of ApMV, as plum, peach, apricot and almond sequences are distributed in all three groups of isolates. Recently, no sequences are known for ApMV from weeds, birches, horse chestnuts and other rare hosts that would enable incorporating them into phylogenetic analysis and to reveal their relationships.

The most significant difference between the apple and hop groups of isolates is the presence of an amino acid change from Gly to Asn in the vicinity of the zinc-finger motif at amino acid position 7 of the *CP* gene. This could affect its nucleic acid- and/or protein-binding activity and, indirectly, its host preference (Grimová et al. 2013). Gly_7 is specific for all isolates from hop and hazelnut hosts, while Asn_7 is specific for apple sequences. No specific amino acid position has been detected that can discriminate the hazelnut/rose group. These data indicate that ApMV has long infected apple trees as well as hop plants and that it has adapted its *CP* accordingly. In all the virus populations analysed, a purifying selection has occurred due to natural selection that conserved protein sequences despite the occurrence of nucleotide polymorphism (Grimová et al. 2013).

3 Host range of ApMV

Because hosts belong to different taxonomic families, there is no clear host preference for ApMV. Most *Rosaceae* plants do, however, seem to be sensitive to ApMV (e.g. apples, pears, quinces, sweet and sour cherries, plum, roses, strawberries, raspberries, hawthorn). Among the *Corylaceae*, hazelnut is often infected in some countries and birches are known hosts. In the *Cannabaceae* family, hop is the only known host. The presence of ApMV in weeds is much

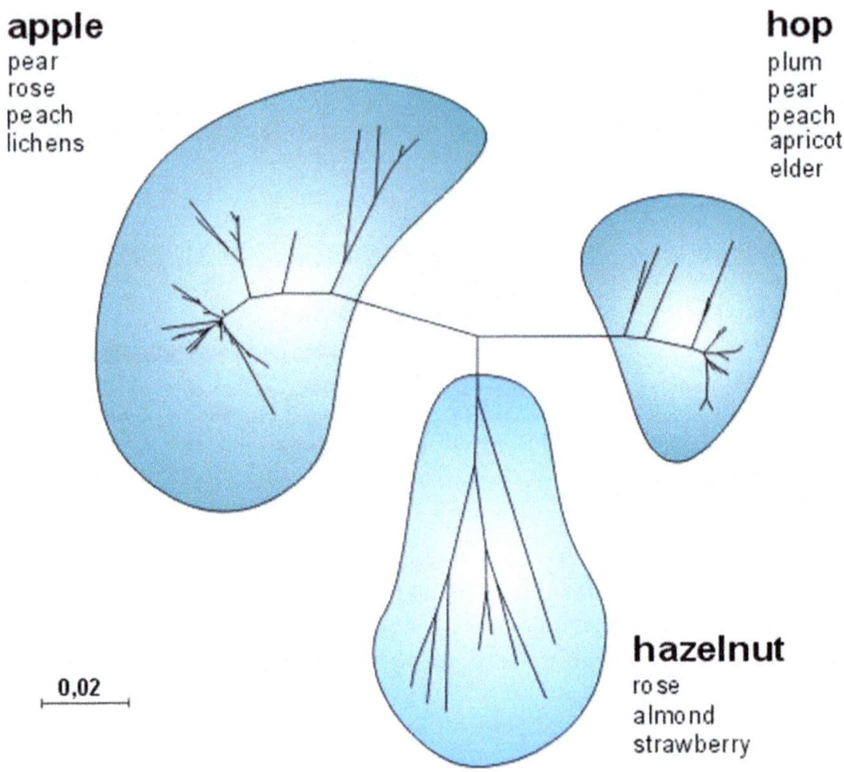

Figure 5 Phylogenic maximum-likelihood tree computed from the *CP* protein sequences of 139 isolates of ApMV. Grouping of ApMV isolates based on the ML phylogenetic tree. The main host in each group is marked with bold font; minority hosts in each group are marked with smaller fonts. The tree was computed from the best alignment of 329 nt of 139 CP sequences without detected recombinants. Bootstrap values were applied using 1000 replicates. Bar = number of substitutions per site. List of sequences with AC numbers is in Table 1.

less known. In recent research within a hazelnut orchard, ApMV was detected in about 18% of tested weed samples, including in the perennials *Artemisia vulgaris*, *Salvia verbenaca*, *Clematis vitalba* and *Rubus canescens*, and the annuals *Campanula* sp. and *Scandix* sp. (Sokmen et al. 2005). No resistance is known for ApMV in hop and other hosts. In the Czech Republic, all 13 of the most commonly grown hop cultivars have tested positive for ApMV (Petrzik and Svoboda 1997).

It should be noted that host-range tests conducted in the 1950s were solely biological tests and that they were carried out with the aim of finding the best host for stepwise characterization of the virus. Successful transmission of the virus was documented by the development of symptoms in these cases and then confirmed by back inoculation to cowpea (*Vigna sinensis*). In the pioneering work of Fulton (1952), the inoculum source was sap from cowpea leaves that had been manually infected by material from a *Rosa setigera* plant. In such trials, *Vinca rosea*, *Convolvulus tricolor*, *Cynara cardunculus*, *Citrullus vulgaris*, *Cucumis sativus*, *Cucurbita maxima*, *Cucurbita pepo*, *Crotalaria intermedia*, *Cyamopsis tetragonolobus*, *Dolichos biflorus*, *Phaseolus vulgaris*, *Phaseolus aureus*, *Vigna sinensis*, *Fragaria vesca*, *Potentilla monspeliensis*, *Potentilla recta*, *Rosa setigera*, *Nicotiana benthamiana*, *Nicotiana forgetiana*, *Nicotiana fragrans*, *Nicotiana glutinosa*, *Nicotiana occidentalis*, *Nicotiana rustica* and *Petunia hybrida* were infected by mechanical inoculation, while *Rosa multiflora*, *Nerium oleander* and *Fragaria* sp. (strawberry) were infected by grafting to infected *Rosa setigera*, *Vinca rosea* and *F. vesca*, respectively (Fulton 1952). The transmission and propagation of ApMV in most of the aforementioned hosts has recently been confirmed by serological or molecular methods. Additional natural hosts as well as laboratory hosts of ApMV were diagnosed and verified serologically by enzyme-linked immunosorbent assay (ELISA) or by polymerase chain reaction (PCR)-specific amplification (Ertunç and Gospodarynk 2016) (Table 2). Through the assays it is concluded that the virus is mechanically transmissible with difficulty from apple tree and rose to herbaceous hosts, but it is readily transmissible from the herbaceous host. Although the virus has been detected on lichens, that situation remains ambiguous: ApMV sequences were amplified from lichen thalli as well as from axenic cultures of lichen algal symbionts and the sequences were closely related to those of ApMV from apple trees and from hop. It has been suggested that lichens could serve as a reservoir for this virus (Petrzik et al. 2013).

4 Symptoms

It seems that ApMV is not tissue-specific. The virus is present in all tissues and has been detected in leaves, floral parts, shoots and roots of infected plants. Although some bright yellow spots and lines are typical in some hosts, colour changes should not be considered as a diagnostic marker (Fig. 6 and 7). Often, infected symptomatic leaves can occur on a single branch of the tree and fall off prematurely. Some apple and pear cultivars can be infected without showing visible symptoms on the leaves (Petrzik 2005). Fruits do not develop diagnostic symptoms. In hops, leaves of sensitive cultivars show a light green pattern, but insensitive cultivars remain without colour changes despite being infected (Fig. 8). In strawberry, ApMV has been associated not with yellow symptoms but with strawberry leafroll disease (Tzanetakis and Martin 2005). *Cyamopsis tetragonoloba* (guar bean) is the recommended assay host, wherein most ApMV isolates induce distinct necrotic lesions after mechanical inoculation (Fulton 1972) (Fig. 9). Cucumber (*Cucumis sativus*), wherein chlorotic lesions

Table 2 Natural and laboratory hosts of ApMV

	Detection method	Reference
Natural host		
Aesculus × carnea	ELISA	Sweet and Barbara (1979)
Aesculus hippocastanum	ELISA	Sweet and Barbara (1979)
Artemisia vulgaris	ELISA	Sokmen et al. (2005)
Berberis thunbergii atropurpurea	RT-PCR	Ertunç and Gospodarynk (2016)
Betula pendula	ELISA	Polák and Zieglerová (1997)
Campanula sp.	ELISA	Sokmen et al. (2005)
Clematis vitalba	ELISA	Sokmen et al. (2005)
Fragaria × ananassa	ELISA, RT-PCR	El-Razik et al. (2017)
Prunella sp.	ELISA	Sokmen et al. (2005)
Prunus spinosa	ELISA	Ilbagi et al. (2008)
Rosa × damascena	DAS-ELISA	Yardimci and Çulal (2009)
Rosa gallica var. *aegyptiaca*	DAS-ELISA	El-Den et al. (2013)
Rubus canescens	ELISA	Sokmen et al. (2005)
Salvia verbenaca	ELISA	Sokmen et al. (2005)
Sambucus nigra	RT-PCR	Grimová et al. (2013)
Scandix sp.	ELISA	Sokmen et al. (2005)
Sorbus aucuparia	ELISA	Polák and Zieglerová (1997)
Laboratory host		
Amaranthus retroflexus	ELISA	Ghanem and Ashour (2002)
Capsicum annuum	ELISA	El-Razik et al. (2017)
Chenopodium amaranticolor	ELISA	El-Razik et al. (2017)
Chenopodium murale	ELISA	Ghanem and Ashour (2002)
Chenopodium rubrum	ELISA	Sweet and Barbara (1979)
Chenopodium quinoa	ELISA	El-Razik et al. (2017)
Cucumis sativus	ELISA	El-Razik et al. (2017)
Cyamopsis tetragonoloba	ELISA	Sweet and Barbara (1979)
Datura metel	ELISA	El-Razik et al. (2017)
Datura stramonium	ELISA	Ghanem and Ashour (2002)
Gomphrena globosa	ELISA	Ghanem and Ashour (2002)
Helianthus annuus	ELISA	Ghanem and Ashour (2002)
Lycopersicon esculentum	ELISA	El-Razik et al. (2017)
Nicotiana benthamiana	ELISA	El-Razik et al. (2017)
Nicotiana glutinosa	ELISA	El-Razik et al. (2017)
Petunia hybrida	ELISA	El-Razik et al. (2017)
Phaseolus vulgaris	ELISA	El-Razik et al. (2017)
Ricinus communis	ELISA	Ghanem and Ashour (2002)
Solanum melongena	ELISA	El-Razik et al. (2017)

	Detection method	Reference
Vicia faba	ELISA	El-Razik et al. (2017)
Vigna unguiculata	ELISA	Ghanem and Ashour (2002)
Vinca minor	ELISA	Sweet and Barbara (1979)

Figure 6 Symptoms associated with ApMV infection visible on the leaves of Granny Smith apple trees. Photo provided by Professor Filiz Ertunç, Ankara University, Turkey.

Figure 7 Symptoms associated with ApMV infection visible on the leaves of hazelnut trees. Photo provided by Professor Filiz Ertunç, Ankara University, Turkey.

Figure 8 Symptoms associated with ApMV on hop. Photo Karel Petrzik.

Figure 9 Necrotic lesions in *Cyamopsis tetragonoloba* induced by ApMV infection. Reproduced from: *AAB Descriptions of Plant Viruses* No. 83 (Fulton 1972).

occur on infected cotyledons along with the plant's extreme stunting, seems to be suitable for ApMV diagnosis also (Fulton 1972).

5 Transmission and vectors

The virus is graft-transmissible and persists in vegetative propagation material, which probably is the main source of virus infection. ApMV's introduction to a garden is linked strongly to infected (not certified as virus-free) planting material, whereas subsequent spreading is attributed to physical contact between plants and/or to mechanical operations of the gardener (Dhingra 1972). Mechanical transmission by such agronomic practices as irrigation, nutrition and pest control are considered the most likely means of spreading ilarviruses (PNRSV and ApMV) between hop plants and within hop gardens.

The most common means by which ApMV spreads between gardens is not yet identified. Although there is suspicion that the virus is transmissible by pollen, experiments in *Prunus* trees and on hops did not support this hypothesis (Fulton 1972; Digiaro et al. 1992). ELISA tests have shown the presence of the virus in anthers from the horse chestnut (*Aesculus carnea*) tree but not in those of *A. hippocastanum* tree (Sweet and Barbara 1979). Transmission by pollen seems unimportant and improbable in hop, because virus spread occurs in the absence of male plants within hop gardens (Pethybridge et al. 2002a). In roses, 651 seedlings produced from seed collected from virus-infected plants were tested for ApMV but no virus was detected. Furthermore, pollen from infected plants was applied for 3 years to uninfected rose plants but for the subsequent 2 years all ELISA tests were negative for ApMV (Golino et al. 2007). While in hazelnut trees generally 8% of seeds from healthy trees tested positively for ApMV, the percentage of infected seeds increased to 54% in ApMV-infected hazelnut trees, and, in a few cases, the virus was also detected in the corresponding involucre (Rovira and Aramburu 2001).

ApMV's slow and random occurrence does not suggest that any insect plays an important role in its transmission. Repeated trials using *Myzus persicae* aphid failed to transmit the virus after the aphids fed on infected *N. rustica* (Fulton 1952). On the other hand, Tobacco streak virus (*Ilarvirus*) is known to be vectored by *Frankliniella schultzei*, *Thrips tabaci* and *Microcephalothrips abdominalis* thrips (Sharman et al. 2015). Mechanical transmission of ApMV by some less common species of thrips, by oribatid mites (Acari) or by other animals feeding on pollen grains containing the virus could also be considered for virus control.

6 Epidemiology and geographical distribution

ApMV is distributed worldwide wherever its hosts are present (Fulton 1972) (Fig. 10). However, in many countries, only the most important ApMV hosts are tested, and especially in apples. ApMV is often included into a battery of tests for the presence of apple-infecting viruses. ApMV generally is present in apples at much lower frequencies than are other apple-infecting viruses, such as ACLSV, ASPV and Apple stem grooving virus (ASGV): From 68 pome fruit trees in a germplasm collection in the Czech Republic, ApMV was present in one apple tree (1.5%), while ACLSV, ASPV and ASGV were present in 96.5%, 89.7% and 34.5% of trees, respectively (Hassan et al. 2006). In a screening of more than 800 apple trees in Latvia, ApMV was present in 2%, ACLSV in 11%, ASGV in 1.6% and ASPV in 3.9% of trees (Pūpola et al. 2011).

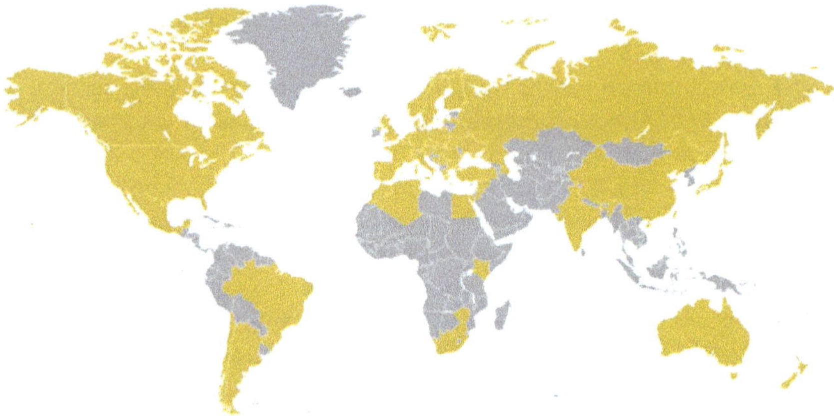

Figure 10 Geographical distribution of ApMV. Countries with ApMV presence are shaded yellow. Data were obtained from EPPO PQR database (Version 09 June 2017) and reports in the literature.

In Tunisia, testing of apple, pear and quince showed that 4% of samples were infected with ApMV while ACLSV occurred in 34% (Mahfoudhi et al. 2013). In many other countries, ApMV is present only sporadically in a range of up to 2-4% on apple trees (see Table 3). Exceptions have been found within apple orchards in India's Kashmir Valley, which showed infection of about 69% as detected by double-antibody sandwich (DAS) ELISA (Verma et al. 2014), and in another region within India where the occurrence was about 15% (Kumar et al. 2012).

The incidence of ApMV in stone fruit trees seems to be similar to that in apple trees in many countries (within the range of 1-4%) (Fiore et al. 2016; Kinoti et al. 2018; Gospodaryk et al. 2013). An exception is that a high (22.2%) incidence of ApMV was observed on *Prunus spinosa* in Turkey (Ilbagi et al. 2008).

ApMV seems to occur very often in hazelnut, where seed transmission has been proven. It was found in 13.6% of samples in Turkey and 15% in Spain. Crop losses ranged from 19.7% to 35.6% (Akbaş and Değirmenci 2009; Rovira and Aramburu 2001). Testing in Poland detected no ApMV in local hazelnut varieties but did find the virus in imported varieties (Kobyłko et al. 2005).

The virus generally is present in hop gardens at low incidence, but those rates could increase with time. One set of observations described an increasing ApMV presence in small plots cultivated from virus-free clones to 6%, 13%, 18% and 23% after 3, 4, 5 and 6 years, respectively (Thresh and Ormerod 1976).

Table 3 ApMV incidence in distinct crops and countries

Country	Incidence (%)	Method	No. of samples positive/total	Reference
Apple, pear, quince				
Syria	0.2	ELISA	2/1145	Ismaeil et al. (2006)
Latvia	2	ELISA	18/870	Pūpola et al. (2011)
Australia	3.5	RT-PCR	6/173	Constable et al. (2007)
Belarus/Poland	1.1-3.7	ELISA	2/173, 5/133	Valasevich et al. (2015)
Tunisia	4	ELISA, RT-PCR	248	Mahfoudhi et al. (2013)
Stone fruit trees				
Chile	1.1	RT-PCR	12/1070	Fiore et al. (2016)
Latvia	2	ELISA, RT-PCR	13/641	Gospodaryk et al. (2013)
Algeria	2.2	ELISA	10/454	Rouag et al. (2008)
Australia	3.9	RT-PCR	4/127	Kinoti et al. (2018)
Turkey	22.2	ELISA	12/54	Ilbagi et al. (2008)
Hazelnut				
Poland	0	ELISA	0/532	Kobyłko et al. (2005)
Turkey	7.9	ELISA	18/229	Sevik (2015)
Poland	10	ELISA, RT-PCR	13/125	Cieślińska and Valasevich (2016)
Turkey	13.6	ELISA	197/1465	Akbaş and Değirmenci (2009)
Spain	15	ELISA	not specified	Rovira and Aramburu (2001)
Hop				
USA	1-6	ELISA	ApMV + PNRSV	Pethybridge et al. (2002a)
USA	38	ELISA	88/234	Hampton (1988)
New Zealand	68	ELISA	41/60 gardens	Pethybridge et al. (2009)
Strawberry				
Egypt	2.8	ELISA	85/3009	El-Razik et al. (2017)
Rose				
Turkey	17.8	ELISA	39/218	Yardimci and Çulal (2009)
Poland	70.8	RT-PCR	17/24	Valasevich et al. (2015)

7 Economic impact

It is not easy to determine the economic impact of ApMV independently, because the virus is often found on apple trees in mixed infections with other apple-infecting viruses or with PDV and PNRSV on stone fruit trees. In general, the effect could be synergistic or antagonistic vis-à-vis other viruses, but there is no precise data on this subject. We can expect that the effect of infection on fruit yields should be in a range similar to or lower than the incidence of the virus in that host.

In apple orchards it seems that ApMV alone has no visible detrimental effect on apple production. Older works from before 1970s had reported that apple bud set is affected, but this effect is not mentioned in more recent works.

In hazelnut, however, the yield from healthy trees has been observed to be 77% greater than that from ApMV-infected trees, due not to differences in nut weight but to the production of more nuts by healthy trees (Rovira and Aramburu 2001). Crop losses ranging from 19.7% to 35.6% have been reported (Akbaş and Değirmenci 2009; Rovira and Aramburu 2001).

ApMV infection causes significant crop yield reduction of as much as 25% in commercial almond trees affected by almond mosaic, a complex disease in which ApMV plays a major aetiological role (Martelli and Savino 1997).

Some ApMV-infected hop varieties have been reported to possess lower alpha and beta bitter acids, but the contents vary by hop cultivar. The reduction could be by as much as 22% in alpha acids and as much as 7% in beta acids (Kremheller et al. 1989; Pethybridge et al. 2002b).

8 Curation

Most probably, viruses have been a part of life on Earth since its inception and could be present in all cellular organisms. Some viruses are constituent parts of their hosts' genomes and are present persistently and at low concentration in all cells without causing visible damage to the host. Harmful viruses, however, are present in much higher concentrations and are unevenly distributed in plants. It is anticipated that at least the apical meristem of a plant should be virus free and that viruses could therefore be eradicated and virus-free plants could be produced using meristem tip culture techniques (Pennazio and Redolfi 1974).

In the case of ApMV, the absolute concentration of virus in the plant is not known. In comparison with ASPV and ASGV, the concentration of ApMV estimated by real-time PCR (RT-PCR) is either similar to or much lower (0.05–0.0001%) than those seen in ASPV and AGSV (Malandraki et al. 2017).

Meristem culture alone or in combination with thermotherapy, cryotherapy or antivirotics is generally used for virus eradication. Thermotherapy (cultivation at elevated temperature for some time) and use of antivirotics (ribavirin, acyclic nucleoside phosphonates) could help to eliminate DNA and ssRNA viruses (Špak et al. 2010; Hu et al. 2012). Cycloheximide is used with varying effectivity in eliminating dsRNA viruses from fungal tissue (Aminian et al. 2011).

Thermotherapy of apple terminal and apical buds carried out at 37–40°C for 4 weeks and followed by culturing of meristems of different sizes has been shown to eliminate ApMV as well as ACLSV, ASGV and ASPV (Vivek and Modgil 2018). Providing about 88% efficiency, meristem 0.5–1 mm high from axillary and apical buds was found to be the best material for meristem culture to eliminate ApMV from red raspberry (*Rubus idaeus*) (Theiler-Hedtrich and Baumann 1989). Starting with a meristem cutting as small as possible is better also for elimination of the virus in apples. Working with three apple varieties, Plopa and Preda found that 0.3 mm explants produced 42–50% virus-free plants while explants 1 mm in size yielded just 17–23% virus-free plants (Plopa and Preda 2012). Hu et al. (2015) found that the efficiency of virus elimination after thermotherapy for 30 days at 37°C and shoot tip grafting from apple was 67%.

Virus-free potato, raspberry, banana and other hosts were obtained by cryotherapy of shoot tips. In this method, plant shoot tips were grown with cold acclimation and pre-cultivated with a mixture of glycerol, ethylene glycol, DMSO and sucrose. The dissected shoot tips are then immersed for some time in liquid nitrogen. After regeneration, the shoots were grown in standard medium. In the case of apple shoot material, ApMV infection was eliminated from 88.9% of shoots (Romadanova et al. 2016). When hop terminal meristems were the starting material, however, only 2% and 4% of samples regrew after cryotherapy and only 6 of 62 samples were found by RT-PCR assay to be free of ApMV (Nukari et al. 2014).

9 Detection techniques

9.1 ELISA

Although ApMV particles are very unstable and lose most of their infectivity in a few minutes, polyclonal antisera are produced by and available from many providers. Double-antibody sandwich (DAS) ELISA seems to be the gold standard as a reliable qualitative assay for detecting ApMV in plant tissue sap from various hosts. Test systems are available from Agdia (US), Loewe (Switzerland), Bioreba (Switzerland), Agritest (Italy), Creative Diagnostics (US), Krishgen BioSystems (India), NeoBiotech (France) and many others. Two serotypes – the ApMV-hop serotype and the ApMV-intermediate serotype have been found infecting commercial Australian hops (Crowle et al. 2003). However, the serotypes are not recognized in current detection systems. If specified, then the antisera are prepared against the ApMV-hop isolate. Most often leaves are used as the testing material, but hop shoots and hazelnut flower and bark tissues also have been used for that purpose (Akbaş and Değirmenci 2010). ELISA is limited to the early spring season for most plants and appears to be inappropriate for use with dormant woody tissue (Schröder 2018; Akbaş et al. 2010). The advantage of the ELISA commonly used for large-scale screening is its lower sensitivity to sequence variability, and its disadvantage is a higher detection limit compared to RT-PCR.

9.2 RT-PCR

Faster and more sensitive than ELISA, but also more laborious is detection by reverse transcription (RT) PCR. This involves amplifying part of the viral genome. Total RNA should be isolated from the tested sample, and a set of primers and enzymes is necessary for this assay. The template RNA isolation is a critical step for routine use of PCR technology. RNA isolation protocol should be optimized for distinct plant material, because not all methods are equally well suited for a given sample type (Akbaş and Değirmenci 2010). The most popular phenol/chloroform, lithium chloride and guanidine thiocyanate extractions are not effective with many materials (Akbaş and Değirmenci 2010). Silica capture-based systems (Rott and Jelkmann 2001) with some improvements (addition of active charcoal) for the detection of plant viruses from woody plants give better quality and increase the integrity of isolated RNA. Alternatives include purifying virus particles by immunocapture utilizing specific antibodies for subsequent use in RT-PCR (Jelkmann and Keim-Konrad 1997; Salem et al. 2004; Moury et al. 2000) or by their non-specific attachment to the walls of the PCR tube (Rowhani et al. 1995; Petrzik and Svoboda 1997). Comparison of the sensitivity showed that non-specific attachment-PCR, immunocapture-PCR and RT-PCR are 100, 10 000 and 10 000 times more sensitive, respectively, than ELISA (Rowhani et al. 1995).

The PCR test should be prepared as a one-step process in a single tube containing also the reverse transcriptase and DNA polymerase enzymes or as a two-step process with separate reverse transcription and amplification reactions. Specificity of the test is influenced by the primers, which are designed from the most conserved parts of the ApMV genome, mostly on the *CP* gene. It is not recommended to direct primers to nucleotide variability hot spots. In the *CP* gene, more than one-third of positions are variable (280/675) and it is difficult to find primers hybridizing with 100% effectivity to all *CP* variants. Table 4 lists some specific detection primers previously used successfully in distinct hosts (Fig. 11).

A cost-effective way for sensitively detecting several related viruses is to use degenerate primers hybridizing the highly conserved motives. A nested PCR system specific for several ilarviruses has been developed by Maliogka et al. (2007) and Untiveros et al. (2010) (Table 5). If necessary, individual viruses could be identified by sequencing or by restriction enzyme digestion of the obtained amplification products.

Multiplex detection of several viruses in a mixture of virus-specific primers has been tested as a means of reducing the cost of virus screening (Hassan et al. 2006; Menzel et al. 2003, 2002; Kumar et al. 2014). In these systems, it was possible to detect 3–8 viruses simultaneously (Saade et al. 2000; Sánchez-Navarro et al. 2005). With an increasing number of primers in multiplex PCR,

Table 4 Specific PCR detection primers

Primers	Product length	Target gene	Used in host	Reference
5′-TTGCGTTAATTGCAAGTGG 5′-TCAAAAGTTGTGTTTGGAG	471	*CP*	Prunus	Kinoti et al. (2018)
5′-CGTAGAGGAGGACAGCTTGG 5′-CCGGTGGTAACTCACTCGTT	431	*CP*	Apple, pear	Hassan et al. (2006)
5′-ATCCGAGTGAACAGTCTATCCTCTAA 5′-GTAACTCACTCGTTATCACGTACAA	262	*CP*	Hazelnut, strawberry	Menzel et al. (2002)
5′-TGGATTGGGTTGGTGGAGGAT 5′-TAGAACATTCGTCGGTATTTG	261	*CP*	Hop	Petrzik and Svoboda (1997)
5′-CAACCGAGAGGTTGGCA 5′-TTCTAGCAGGTCTTCATCGA	161	*CP*	Apple	Candresse et al. (1998)
5′-CAAGCGAACCCGAATAAGG 5′-CTAACCTCCCAAGCTGTCC	87	*CP*	Apple	Winkowska et al. (2016)
5′-AGTTTGTGTGATGTGAGAT 5′-ATTTCTAAGGCGTAACTTC	222	*Met*	Prunus	Kinoti et al. (2018)
5′-TCATTGGATCCCTTTGCTTC 5′-AAACTCGTCGTCCCTATCC	383	*RdRp*	Prunus	Kinoti et al. (2018)
5′-GAGGAGGACAGCTTGGGAGGTTAG 5′-GCAGTCACTTCTCGACTGCTCAGG 5′-CCCGAATGTTGAGCCGAAAATTC-probe	89	*CP*	Apple, pear	Nickel and Fajardo (2014)

however, the sensitivity of the assay eventually drops to a level comparable to that of ELISA.

Other multiplex techniques are based on nucleic acid hybridization of the tested sample with a range of specific probes attached to oriented matrices (arrays). The earliest microarray attempts had the ability to detect a limited number of viruses (Bystricka et al. 2005; Lenz et al. 2008). Recently, a chip containing about 4000 probes for 538 species was prepared (Nam et al. 2014). In this method, the sample nucleic acid labelled with one fluorophore (cy3) and also a healthy control sample labelled with another fluorophore (cy5) are hybridized simultaneously with the probes on the chip. The hybridized area is then scanned by laser and the image is analysed. It should be noted that the advantage of multiplex detection of many viruses

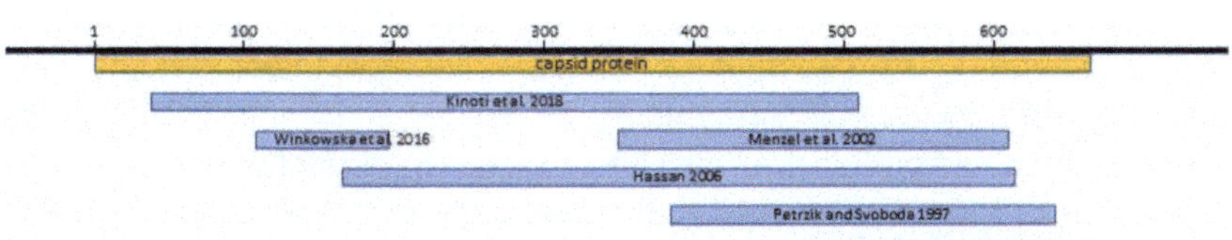

Figure 11 Position of amplified regions on ApMV *CP* gene in specific PCR detection of the virus. For primer sequences and references see Table 4.

Table 5 Ilarvirus universal PCR primers

Primers	Product length	Target gene	Target viruses	Reference
5′-TCRAYRTTYGAYAARTCNCA 5′-GGTTGRTTRTGHGGRAAYTT	~380	*RdRp*	Bromoviruses	Untiveros et al. (2010)
5′-GCNGGWTGYGGDAARWCNAC 5′-AMDGGWAYYTGYTYNGTRTCACC	~300	*Met*	Bromoviruses	Untiveros et al. (2010)
5′-CBATHACHTATCAYRADAARGG 5′-GGTTGRTTRTGHGGRAAYTT	~560	*RdRp*	Bromoviruses	Untiveros et al. (2010)
5′-YTCIAMRTTYGAYAARTC 5′-GGYTGRTTRTGIGGRAA	381	*RdRp*	Ilarviruses	Maliogka et al. (2007)
nested with				
5′-TCGAMRTTYGAYAARTCICA 5′-TGGGGRAAYTTIGYYTCRA	371			

N = any nucleotide, I = Inosine, Y = C or T, R = A or G, W = A or T, M = A or C, H = A or T or C, D = A or G or T, B = C or G or T, H = A or C or T.

is offset by weaker (up to ten times) sensitivity in comparison with specific RT-PCR (Nam et al. 2014).

Although not primarily a detection method, next-generation sequencing (NGS) – also termed high-throughput sequencing (HTS) – has the ability to ascertain minute amounts of any virus without prior limitation (Cho et al. 2016). These methods are laborious and expensive, and thus are not yet expedient for routine use in field detection.

10 Summary and future trends

Despite that this virus's name refers to apple, ApMV appears to be more important in relation to hazelnut production, wherein it spreads through infected seeds and infects the host in higher proportion than its infection levels in apple trees. Deleterious effect of ApMV is also significant in hop production, where the content of bitter acids is reduced in infected plants. Apple, plum, apricot and almond trees seem to be infected less frequently and the virus infection has been observed to have less pronounced effect on fruit yield. The original source of the virus remains in doubt, but the list of natural weed hosts is continuously growing. The virus is vectored by vegetative propagation of infected material and by agricultural practices. No insect vector is known to date. At least in the case of hazelnut, ApMV is seed transmitted. Because no gene resistance to ApMV is presently known, propagation of virus-free seed is a key prerequisite for limiting the further spread of ApMV, and it is a quarantine pest in the United States, Israel and Norway (EPPO 2017).

As is true of other plant viruses, efforts should be directed to understanding ApMV's pathogenicity and its interactions with the host(s) and with co-inhabiting viruses. With the development of next-generation (high-throughput) sequencing (NGS or HTS), new ApMV hosts will be found, especially among weeds and non-cultivated trees, and their roles should be clarified. Routes of ApMV transport to the embryo in hazelnuts and details of vertical transmission there should be clarified.

11 Where to look for further information

Crop Protection Compendium (available online at https://www.cabi.org/cpc/) is a comprehensive site for information on crop pests. It provides access to primary scientific information and contains almost 400 000 abstracts.

Plant Quarantine Data Retrieval system (PQR) is the European and Mediterranean Plant Protection Organization (EPPO) database of quarantine pests available online at http://www.eppo.int. At present, PQR contains information about more than 1600 pests, preferentially that of EPPO/EU region. It gives access for each pest to obtain lists of host plants and geographical distribution. PQR contains nomenclatural and taxonomic details on pests and hosts.

Descriptions of Plant Viruses published by the CMI/AAB (available at http://dpvweb.net/dpv/index.php) provides a source of information about plant viruses, viroids and satellites. This database contains 422 descriptions of plant viruses or virus groups and contains data up to year 2011.

12 References

Akbaş, B. and Değirmenci, K. (2009), Incidence and natural spread of *Apple mosaic virus* on hazelnut in the west Black Sea coast of Turkey and its effect on yield, *J. Plant Pathol.*, 91, 767–71.

Akbaş, B. and Değirmenci, K. (2010), Simultaneous detection of *Apple mosaic virus* in cultivated hazelnuts by one-tube RT-PCR, *Afr. J. Biotechnol.*, 9, 1753–7.

Aminian, P., Azizollah, A., Abbas, S. and Naser, S. (2011), Effect of double-stranded RNAs on virulence and deoxyvalenol production of *Fusarium graminearum* isolates, *J. Plant Prot. Res.*, 51. doi:10.2478/v10045-011-0006-z.

Boulila, M. (2010), Putative recombination events and evolutionary history of five economically important viruses of fruit trees based on coat protein-encoding gene sequence analysis, *Biochem. Genet.*, 48, 357–75.

Bystricka, D., Lenz, O., Mraz, I., Piherova, L., Kmoch, S. and Sip, M. (2005), Oligonucleotide-based microarray: A new improvement in microarray detection of plant viruses, *J. Virol. Methods*, 128, 176–82.

Candresse, T., Kofalvi, S. A., Lanneau, M. and Dunez, J. (1998), A PCR-ELISA procedure for the simultaneous detection and identification of prunus necrotic ringspot (PNRSV) and apple mosaic (ApMV) ilarviruses, *Acta Hortic.*, 472, 219–26.

Cho, I.-S., Igori, D., Lim, S., Choi, G-S., Hammond, J., Lim, H.-S. and Moon, J. S. (2016), Deep sequencing analysis of apple infecting viruses in Korea, *Plant Pathol. J.*, 32, 441–51.

Cieślińska, M. and Valasevich, N. (2016), Characterization of *Apple mosaic virus* isolates detected in hazelnut in Poland, *J. Plant Dis. Prot.*, 123, 187–92.

Constable, F. E., Joyce, P. A. and Rodoni, B. C. (2007), A survey of key Australian pome fruit growing districts for exotic and endemic pathogens, *Australas. Plant Pathol.*, 36, 165–72. doi:10.1071/AP07003.

Crowle, D. R., Pethybridge, S. J., Leggett, G. W., Sherriff, L. J. and Wilson, C. R. (2003), Diversity of the coat protein-coding region among Ilarvirus isolates infecting hop in Australia, *Plant Pathol.*, 52, 655–62.

Dhingra, K. L. (1972), Transmission of apple mosaic by natural root grafting, *Indian J. Hort.*, 29, 348–50.

Digiaro, M., Savino, V. and Terlizzi, B. D. (1992), Ilarviruses in apricot and plum pollen, *Acta Hortic.*, 309, 93–8.

El-Den, K. M. E., El Awady, M. A. M., Said, M. M., El Hallous E. I. and El-Tarras, A. A. (2013), Improvement of serological diagnosis of viruses associated with rose mosaic disease affecting Egyptian- rose growth in Taif, Kingdom of Saudi Arabia, *Life Sci. J.*, 10, 1673–9.

El-Razik, R. A. A., Youssef, S. A., El-Morsy, S., El-Sheikh, M. A. and Shalaby, A. A. (2017), Molecular characterization of apple mosaic virus (ApMV) isolated from strawberry plants in Egypt, *Int. J. Sci. Eng. Res.*, 8, 709–14.

EPPO (2017), PQR- EPPO database on quarantine pests (available online), http://www.eppo.int.

Ertunç, F. and Gospodarynk, A. (2016), A new natural host of apple mosaic virus: Berberis thunbergii atropurpureum, *J. Appl. Biol. Sci.*, 10(2), 26–8.

Fiore, N., Zamorano, A., Pino, A. M., Gonzáles, F., Rosales, I. M., Sánchez-Navarro, J. A. and Pallás, V. (2016), Survey of stone fruit viruses and viroids in Chile, *J. Plant Pathol.*, 98, 631–5. doi:10.4454/JPP.V98I3.032.

Fulton, R. W. (1952), Mechanical transmission and properties of rose mosaic virus, *Phytopathology*, 42, 413–16.

Fulton, R. W. (1972), *Apple Mosaic Virus. CMI/AAB Descriptions of Plant Viruses*, Applied Biologists, Wellesbourne, UK, p. 83.

Ghanem, G. A. M. and Ashour, A. M. A. (2002), Isolation and partial characterization of Apple mosaic ilarvirus (ApMV) in Egypt, *Egypt. J. Phytopathol.*, 30, 14.

Golino, D. A., Sim, S. T., Cunningham, M. and Rowhani, A. (2007), Transmission of rose mosaic viruses, *Acta Hortic.*, 751, 217–24.

Gospodaryk, A., Moročko-Bičevska, I., Pupola, N. and Kale, A. (2013), Occurrence of stone fruit viruses in plum orchards in Latvia, *Proc. Latvian Acad. Sci.*, 67, 116–23. doi:10.2478/prolas-2013-0018.

Grimová, L., Winkowska, L., Ryšánek, P., Svoboda, P. and Petrzik, K. (2013), Reflects the coat protein variability of apple mosaic virus host preference? *Virus Genes*, 47, 119–25. doi:10.1007/s11262-013-0925-z.

Hampton, R. O. (1988), Health status (virus) of native North American *Humulus lupulus* in the natural habitat, *J. Phytopathol.*, 123, 353–62.

Hassan, M., Myrta, A. and Polak, J. (2006), Simultaneous detection and identification of four pome fruit viruses by one-tube pentaplex RT-PCR, *J. Virol. Methods*, 133, 124–9.

Hu, G. J., Hong, N., Wang, L. P., Hu, H. J. and Wang, G. P. (2012), Efficacy of virus elimination from in vitro-cultured sand pear (*Pyrus pyrifolia*) by chemotherapy combined with thermotherapy, *Crop Prot.*, 37, 20–5.

Hu, G. J., Zhang, Z. P., Dong, Y. F., Fan, X. D., Ren, F. and Zhu, H. J. (2015), Efficiency of virus elimination from potted apple plants by thermotherapy coupled with shoot-tip grafting, *Australas. Plant Pathol.*, 44, 167–73.

Ilbagi, H., Citir, A. and Bostan, H. (2008), *Prunus spinosa* L. – A natural wild host of some important fruit viruses in Tekirdag, Turkey, *Acta Hortic.*, 781, 33–6.

Ismaeil, F., Al-Jabor, K., Myrta, A., Mando, M. J., Al-Saadoun, E., Hassan, M. and Al-Chaabi, S. (2006), Viruses of pome fruit trees in Syria, *Bull. OEPP/EPPO Bull.*, 36, 65–8.

Jelkmann, W. and R. Keim-Konrad. (1997), An immunocapture polymerase chain reaction and plate trapped ELISA for the detection of apple stem pitting virus, *J. Phytopath.*, 145, 499–504.

Kinoti, W. M., Constable, F. E., Nancarrow, N., Plummer, K. M. and Rodoni, B. (2018), The incidence and genetic diversity of apple mosaic virus (ApMV) and prune dwarf virus (PDV) in Prunus species in Australia, *Viruses*, 10, 136.

Kobyłko, T., Nowak, B. and Urban, A. (2005), Incidence of Apple mosaic virus (ApMV) on hazelnut in south-east Poland, *Folia Hortic.*, 17, 153–61.

Kremheller, H. T., Rossbauer, G. and Ehrmaier, H. (1989), Reinfection of virus-free planted hop gardens with *Prunus necrotic ringspot* and *hop mosaic virus*. Effects of the virus infection upon the yield, alpha acids, and the disease symptoms of the various hop varieties, *in Proceedings, International Workshop Hop Virus Disease*, Giessen, Germany, pp. 133–6.

Kumar, S., Singh, R. M., Ram, R., Badyal, J., Hallan, V., Zaidi, A. A. and Varma, A. (2012), Determination of major viral and sub viral pathogens incidence in apple orchards in Himachal Pradesh, *Indian J. Virol.*, 23, 75–9.

Kumar, S., Singh, L., Ram, R., Zaidi, A. A. and Hallan, V. (2014), Simultaneous detection of major pome fruit viruses and a viroid, *Indian J. Microbiol.*, 54, 203–10.

Lenz, O., Petrzik, K. and Špak, J. (2008), Investigating the sensitivity of a fluorescence-based microarray for the detection of fruit-tree viruses, *J. Virol. Methods*, 148, 96–105.

Mahfoudhi, N., El Air, M., Moujahed, R., Salleh, W. and Djelouah, K. (2013), Occurrence and distribution of pome fruit viruses in Tunisia, *Phytopathol. Mediterr.*, 52, 136–40. doi:10.14601/Phytopathol_Mediterr-11084.

Malandraki, I., Beris, D., Isaioglu, I., Olmos, A., Varveri, C. and Vassilakos, N. (2017), Simultaneous detection of three pome fruit tree viruses by one-step multiplex quantitative RT-PCR, *PLoS ONE*, 12(7), e0180877.

Maliogka, V. I., Dovas, C. I. and Katis, N. I. (2007), Demarcation of ilarviruses based on the phylogeny of RNA2-encoded RdRp and a generic ramped annealing RT-PCR, *Arch. Virol.*, 152, 1687–98.

Martelli, G. and Savino, V. (1997), Infectious diseases of almond with special reference to the mediterranean area 1, *Bull. OEPP/EPPO Bull., 27*, 525–34.

Menzel, W., Jelkmann, W. and Maiss, E. (2002), Detection of four apple viruses by multiplex RT-PCR assays with coamplification of plant mRNA as internal control, *J. Virol. Methods*, 99, 81–92.

Menzel, W., Zahn, V. and Maiss, E. (2003), Multiplex RT-PCR-ELISA compared with bioassay for the detection of four apple viruses, *J. Virol. Methods*, 110, 153–7.

Moury, B., Cardin, L., Onesto, J.-P., Candresse, T., and Poupet, A. (2000), Enzyme-linked immunosorbent assay testing of shoots grown in vitro and the use of immunocapture-reverse transcription-polymerase chain reaction improve the detection of *Prunus necrotic ringspot virus* in rose, *Phytopathol.*, 90(5), 522–8. doi:10.1094/PHYTO.2000.90.5.522.

Nam, M., Kim, J-S., Lim, S., Park, C. Y., Kim, J-G., Choi, H.-S., Lim, H.-S., Moon, J. S. and Lee, S.-H. (2014), Development of the large-scale oligonucleotide chip for the diagnosis of plant viruses and its practical use, *Plant Pathol. J.*, 30, 51–7.

Nickel, O. and Fajardo, T. V. M. (2014), Detection of viruses in apples and pears by real time RT-PCR using 5′-hydrolysis probes, *J. Plant Pathol.*, 96, 207–13.

Nukari, A., Laamanen, J., Uosukainen, M. and Lemmetty, A. (2014), Comparison of virus eradication of apple mosaic virus from hop by encapsulation-dehydration cryotherapy and meristem culture methods, *Acta Hortic.*, 1039, 113-20.

Pennazio, S. and Redolfi, P. (1974), Potato virus X eradication in cultured potato meristem tip, *Potato Res.*, 17, 333–5.

Pethybridge, S. J., Nelson, M. E., Eastwell, K. C., Klein, R. E., Kenny, S. T. and Wilson, C. R. (2002a), Incidence and spatial distribution of viruses in hop gardens of Washington State, *Plant Dis.*, 86, 661–5.

Pethybridge, S. J., Wilson, C. R., Hay, F. S., Leggett, G. W. and Sherriff, L. J. (2002b), Effect of viruses on agronomic and brewing characteristics of four hop (*Humulus lupulus* L.) cultivars in Australia, *Ann. App. Biol.*, 140, 97–105.

Pethybridge, S. J., Fletcher, J. D., Hay, F. S. and Beatson, R. A. (2009), Prevalence and incidence of viruses in New Zealand hop (*Humulus lupulus*) gardens, *New Zeal. J. Crop. Hort. Sci.*, 37, 235–41.

Petrzik, K. (2005), Capsid protein sequence gene analysis of Apple mosaic virus infecting pears, *Eur. J. Plant Pathol.*, 111, 355–60.

Petrzik, K. and Lenz, O. (2002), Remarkable variability of Apple mosaic virus capsid protein gene after nucleotide position 141, *Arch. Virol.*, 147, 1275–85.

Petrzik, K. and Lenz, O. (2011), Apple mosaic virus in pome fruits, *in Virus and Virus-like Diseases of Pome and Stone Fruits* (A. Hadidi, M. Barba, T. Candresse and W. Jelkmann, eds). American Phytopathological Society, St Paul, MN, pp. 25-9.

Petrzik, K. and Svoboda, P. (1997), Screening of apple mosaic virus in hop cultivars in the Czech Republic by reverse transcription-polymerase chain reaction, *Acta Virol.*, 41, 101-3.

Petrzik, K. Vondrák, J., Barták, M., Peksa, O. and Kubešová, O. (2013), Lichens – a new source or yet unknown host of herbaceous plant viruses? *Eur. J. Plant Pathol.*, 138, 549-59.

Plopa, C. and Preda, S. (2012), Elimination of apple mosaic virus by tissue culture of some infected apple cultivars, *Acta Hortic.*, 981, 517-22. doi:10.17660/ActaHortic.2013.981.83.

Polák, Z. and Zieglerová, J. (1997), Spontaneous occurrence of apple mosaic virus in some forest and ornamental woody species, *XIVth Slovak and Czech Plant Protection Conference Proceedings*, Nitra 1997, pp. 87–8.

Pūpola, N., Moročko-Bičevska, I., Kāle, A. and Zeltiņš, A. (2011), Occurrence and diversity of pome fruit viruses in apple and pear orchards in Latvia, *J. Phytopathol.*, 159, 597–605. doi:10.1111/j.1439-0434.2011.01812.x.

Romadanova, N. V., Mishustina, S. A., Gritsenko, D. A., Omasheva M. Y., Galiakparov, N. N., Reed, B. M. and Kushnarenko, S. V. (2016), Cryotherapy as a method for reducing the virus infection of apples (*Malus* sp.), *Cryo Lett.*, 37, 1–9. doi:10.1016/j.cryobiol.2015.10.093.

Rott, M. E. and Jelkmann, W. (2001), Characterization and detection of several filamentous viruses of cherry: Adaptation of an alternative cloning method (DOP-PCR), and modification of an RNA extraction protocol, *Eur. J. Plant Pathol.*, 107, 411-20.

Rouag, N., Guechi, A., Matic, S. and Myrta, A. (2008), Viruses and viroids of stone fruits in Algeria, *J. Plant Pathol.*, 90, 393–5.

Rovira, M. and Aramburu, J. (2001), Incidence of apple mosaic ilarvirus (ApMV) in Catalonia (Spain) and its effects on 'Negret' hazelnut, *Acta Hortic.*, 556, 509–12.

Rowhani, A., Maningas, M. A., Lile, L. S., Daubert, S. D. and Golino, D. A. (1995), Development of a detection system for viruses of woody plants based on PCR analysis of immobilized virions, *Phytopathol.*, 85, 347–52.

Saade, M., Aparicio, F., Sánchez-Navarro, J. A., Herranz, M. C., Myrta, A., Di Terlizzi, B. and Pallás, V. (2000), Simultaneous detection of three Ilarviruses affecting stone fruit trees by nonisotopic molecular hybridization and multiplex reverse-transcription polymerase chain reaction, *Phytopathology*, 90, 1330–6.

Salem, N., Mansour, A., Al-Musa, A., Al-Nsour, A. and Hammond, R. (2004), Identification and partial characterization of Prunus necrotic ringspot virus on stone fruits in Jordan, *J. Plant Pathol.*, 86, 85–90.

Sánchez-Navarro, J. A. and Pallás, V. (1994), Nucleotide sequence of apple mosaic ilarvirus RNA4, *J. Gen. Virol.*, 75, 1441–5.

Sánchez-Navarro, J. A., Aparicio, F., Herranz, M. C., Minafra, A., Myrta, A. and Pallás, V. (2005), Simultaneous detection and identification of eight stone fruit viruses by one-step RT-PCR, *Eur. J. Plant Pathol.*, 111, 77–84.

Schröder, M. (2018), Seasonal detection of pome fruit viruses by RT-PCR in apple mother trees used for scion production, *J. Kulturpflanzen*, 70, 124–9. doi:10.1399/JKI.2018.04.02.

Sevik, M. A. (2015), Occurrence and distribution of apple mosaic virus on hazelnut cultivars in Giresun, Turkey, *Biosci. J.*, 31, 1307–11.

Sharman, M., Thomas, J. E. and Persley, D. M. (2015), Natural host range, thrips and seed transmission of distinct Tobacco streak virus strains in Queensland, Australia, *Ann. Appl. Biol.*, 167, 197–207.

Sokmen, M. A., Yilmaz, N. D. K., Mennan, H. and Sevik, M. A. (2005), Natural weed hosts of apple mosaic virus in hazelnut orchards in Turkey, *J. Plant Pathol.*, 87, 239–42.

Špak, J., Holý, A., Pavingerová, D., Votruba, I., Špaková, V. and Petrzik, K. (2010), New in vitro method for evaluating antiviral activity of acyclic nucleoside phosphonates against plant viruses, *Antiviral Res.*, 88, 296–303. doi:10.1016/j.antiviral.2010.09.019.

Sweet, J. B. and Barbara, D. J. (1979), A yellow mosaic disease of horse chestnut (*Aesculus* spp.) caused by apple mosaic virus, *Ann. Appl. Biol.*, 92, 335–41.

Theiler-Hedtrich, R. and Baumann, G. (1989), Elimination of Apple mosaic virus and Raspberry bushy dwarf virus from infected red raspberry (*Rubus idaeus* L.) by tissue culture, *J. Phytopathol.*, 127, 193–9.

Thresh, J. M. and Ormerod, P. J. (1976), Virus diseases of hop. Report of the East Malling Research Station 1975, pp. 122–3.

Tokarz, R., Sameroff, S., Tagliafierro, T., Jain, K., Williams, S. H., Cucura, D. M., Rochlin, I., Monzon, J., Carpi, G., Tufts, D., Diuk-Wasser, M., Brinkerhoff, J. and Lipkin, W. I. (2018), Identification of novel viruses in *Amblyomma americanum*, *Dermacentor variabilis*, and *Ixodes scapularis* ticks, *mSphere*, 3, e00614-17. doi:10.1128/mSphere.00614-17.

Tzanetakis, I. E. and Martin, R. (2005), First report of strawberry as a natural host of Apple mosaic virus, *Dis. Note*, 89, 431.

Untiveros, M., Perez-Egusquiza, Z. and Clover, G. (2010), PCR assays for the detection of the genus *Ilarvirus* and family *Bromoviridae*, *J. Virol. Methods*, 165, 97–104.

Valasevich, N., Cieślińska, M. and Kolbanova, E. (2015), Molecular characterization of Apple mosaic virus isolates from apple and rose, *Eur. J. Plant Pathol.*, 141, 839–45.

van der Vossen, E. A. G., Neeleman, L. and Bol, J. F. (1994), Early and late functions of Alfalfa mosaic virus coat protein can be mutated separately, *Virology*, 202, 891–903.

Verma, R. K., Ahmed, N., Mir, J. I., Verma, M. K., Srivastava, K. K. and Focktoo, S. Z. (2014), Detection of apple mosaic and chlorotic leaf spot viruses by DAS-ELISA from farmers orchards of Kashmir valley, *Indian J. Hort.*, 71, 567–70.

Vivek, M. and Modgil, M. (2018), Elimination of viruses through thermotherapy and meristem culture in apple cultivar Oregon Spur-II, *Virus Dis.*, 29, 75–82.

White, R. P. (1928), An infectious chlorosis of roses, *Plant Disease Report*, 12, 33–4. In: Brunt, A. A., Crabtree, K., Dallwitz, M. J., Gibbs, A. J., Watson, L. and Zurcher, E. J. (Eds) (1996 onwards), Plant Viruses Online: Descriptions and Lists from the VIDE Database. Version: 20th August 1996, URL: http://biology.anu.edu.au/Groups/MES/vide/

Winkowska, L., Grimova, L. and Rysanek, P. (2016), Quantitative detection of four pome fruit viruses in apple trees throughout the year, *Phytopathol. Mediterr.*, 55, 207–24. doi:10.14601/Phytopathol_Mediterr-16825.

Yardimci, N. and Çulal, H. (2009), Occurrence and incidence of Prunus necrotic ringspot virus, Arabis mosaic virus, and Apple mosaic virus on oil rose (Rosa damascena) in the Lakes Region of Turkey, *New Zeal. J. Crop Hort. Sci.*, 37, 95–8.

Chapter 8

Plum pox virus: detection and management

Manuel Rubio, Federico Dicenta and Pedro Martínez-Gómez, CEBAS-CSIC, Spain

1 Introduction

Stone fruit trees are affected by a large number of viruses that can cause economic losses to farmers (Rubio et al., 2017). These viruses, alone or in combination, can affect fruit yield, fruit maturity and tree growth. However, Sharka disease, caused by *Plum pox virus* (PPV), is the most important virus inducing extended yield losses in plums (*Prunus domestica* L.), apricots (*P. armeniaca* L.) and peaches [*P. persica* (L.) Batsch]. It results in reduced fruit quality, premature fruit drop and rapid natural virus spread by aphid vectors (Scholthof et al., 2011).

Although the impact of this virus on stone fruit production is huge, it is difficult to accurately evaluate the economic impact it causes. In the best-documented case, the financial loss caused by Sharka disease over 30 years was estimated to be over €10 000 million (Cambra et al., 2006). Furthermore, Sharka was included among the top ten plant viruses in terms of its scientific and economic importance worldwide (Scholthof et al., 2011).

This chapter provides general information on the status of this disease around the world, with emphasis on strains, detection methods, orchard management and breeding for resistance.

http://dx.doi.org/10.19103/AS.2019.0046.10

2 Case study: the *Plum pox virus*

PPV belongs to the genus *Potyvirus* in the family *Potyviridae*. PPV is a RNA virus with flexuous filamentous particles approximately 750 nm in length × 15 nm in diameter. PPV particles are composed of one molecule of RNA (positive-sense, single-stranded RNA) and a protein envelope. The genome is expressed as a 350-kDa polyprotein precursor that is proteolytically processed by viral and host proteases into ten smaller functional proteins including P1, HCPro, P3, 6K1, CI, 6K2, VPg, NIapro, NIb and CP. P1 and HCPro are responsible for processing themselves at their respective C-termini, and NIa performs the rest of the cleavages within the polyprotein. The cylindrical inclusion (CI) protein in potyviruses is involved in virus replication and cell-to-cell movement. Studies of the interaction of the chloroplast PSI-K protein with the potyviral CI protein have found that the PSI-K protein has a negative effect on viral infection (García et al., 2014) (Fig. 1).

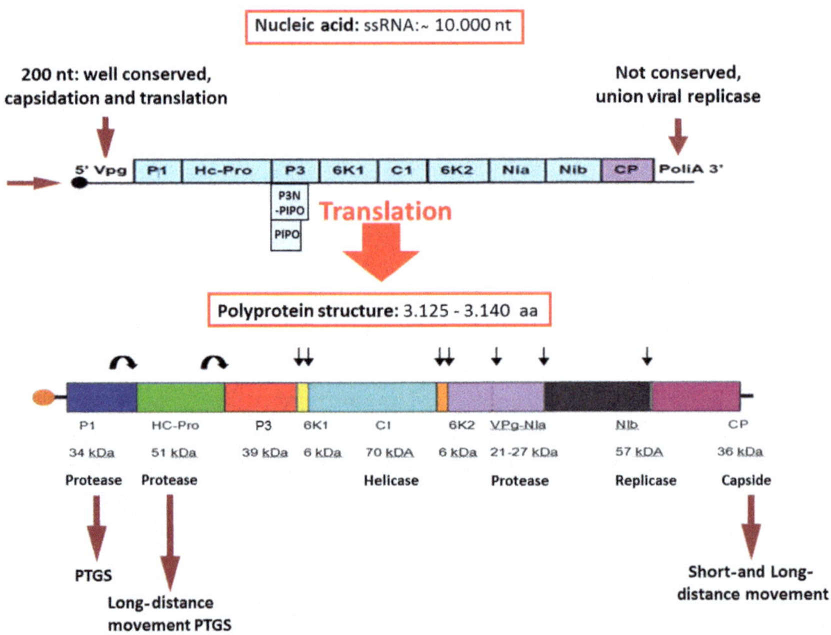

Figure 1 Scheme of the nucleic acid sequence of *Plum pox virus* and structure of the polyprotein generated by the direct translation of this sequence. The different protein regions are represented with the cut-off points that are as follows: P1 first protein, Hc-Pro helper factor protein, P3 third protein including cell to cell movement P3N-PIPO and PIPO, 6K1 first peptide 6K, C1 cell inclusion protein, 6K2 second 6k peptide, NI nuclear inclusion protein 'a', NIb nuclear inclusion protein 'b', CP capsid protein. Source: adapted from Ilardi and Di Nicola-Negri (2011) and García et al. (2014).

PPV symptoms were first observed in Bulgarian plums between 1915 and 1918, although some reports indicate that symptoms were observed in Macedonian plums as early as 1910. However, the first paper describing the viral nature of the disease did not appear until 1932, when Atanosoff (1932) named it 'Sarka po slivite' meaning 'Pox of Plum' (hence Sharka). Since then, PPV has spread throughout Europe, the Mediterranean, the Middle East (Egypt and Syria), India and South and North America (Table 1; Fig. 2). In Europe alone, the European and Mediterranean Plant Protection Organization (EPPO) estimates that more than 100 million stone fruit trees have been infected with PPV (https://www.eppo.int/). The steady progress of Sharka in Europe, and the severity of the disease, led to the development of the Sharka International Working Group within the framework of the EPPO in the 1970s, which allowed for the coordination of research and free flow of information between countries (https://www.eppo.int/).

PPV is similar to other plant viruses, in that it is comprised of several strains based on biology, serological reactions and molecular and biological data. To

Table 1 Distribution, isolate and year of detection of Sharka by countries and different strains declared

Year	Country	Year	Country
1917	Bulgaria (D, M, Rec)	1987	Egypt (EA); Slovenia (D, M)
1935	Serbia (D, M, Rec)	1988	Croatia (D, M)
1938	Hungary (D, M, C)	1992	Chile (D)
1941	Romania (D, M, C, Rec)	1994	Estonia (D); Georgia (D, M)
1947	Albania (D, M, Rec, T)	1995	Lithuania (D); India (D)
1950	Slovakia (D, M, Rec)	1998	Norway (D)
1952	Czech Republic (D, M, Rec)	1999	USA (D)
1956	Germany (D, M, C)	2000	Canada (D, W); China (D), Iran (D, M)
1961	Austria (D); Poland (D, M)	2001	Jordan (M)
1962	Moldavia (C); Russia (D, C, CR, Rec)	2002	Tunisia (D)
1965	Holland (D); United Kingdom (D)	2004	Bosnia-Herzegovina (D, M, Rec); Kazakhstan (D); Argentina (D)
1967	Greece (D, M); Switzerland (D); Ukraine (D, M)	2006	Pakistan (D, Rec)
1968	Turkey (D, M, Rec, T, An)	2007	Montenegro (D, M)
1970	France (D, M, Rec); Sweden (D)	2010	Japan (D)
1973	Italia (D, M, C, Rec)	2011	Byelorussia (D, M, Rec)
1974	Belgium (D)	2015	Finland (D)
1982	Cyprus (M)	2016	South Korea (D)
1984	Portugal (D); Spain (D)	2017	Mexico (D)
1986	Syria (D, M)		

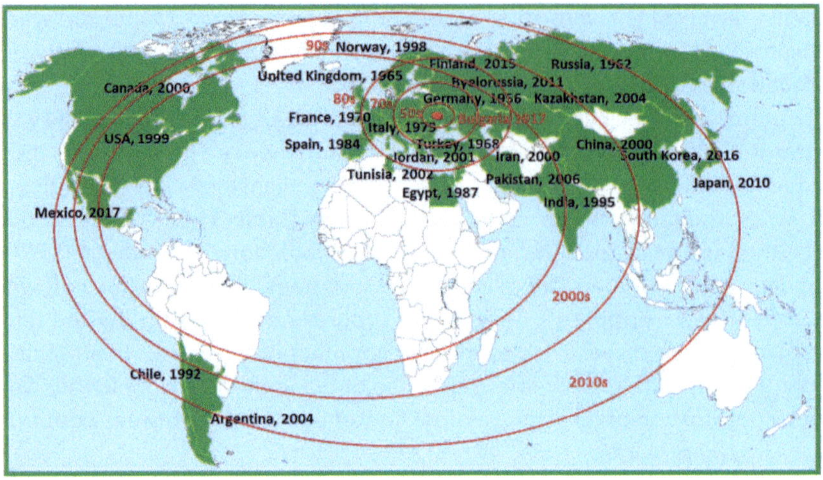

Figure 2 *Plum pox virus* distribution around the world including the year of the first detection in the most important countries and the expansion circles during the century of the presence in the agriculture.

date, nine PPV strains have been recognized that include PPV-D, PPV-M, PPV-Rec, PPV-EA, PPV-C, PPV-T, PPV-W, PPV-CR and PPV-An.

Among these nine strains, M (Marcus) and D (Dideron) are the most extensive with a recombination of both called Rec. These three strains are widely distributed in Europe, with PPV-D being the most common globally. The other groups are less extensive, and mainly classified by their molecular characterization by hosts and geographical distribution. For example, El Amar (PPV-EA) has been reported in Egypt, PPV-C has only been detected on cherry trees in Moldavia and Central Europe, PPV-T was characterized in Turkey and PPV-W (Winona) in Ontario (Canada). PPV-CR is another strain reported in sour cherry trees in Russia, and PPV-An has been detected in Albania, possibly as an ancestor of the Marcus strain. Recently, a new group has been proposed as a new variant of the cherry group called PPV-CV (Cherry Volga) (Chirkov et al., 2018). These new strains demonstrate the dangerous capacity of the virus to mutate. However, it is remarkable that the genomes of the three major strains show a 95% similarity, being only partially differentiable by their biological or epidemiological properties. It is not clear whether different strains prefer different hosts, as it is widely accepted that the three main strains or groups (D, M and Rec) are able to infect peach, plum and apricot. However, in natural conditions some preferences have been observed that include PPV-M/peach, PPV-D/Plum/apricot and PPV-Rec/plum (Sihelska et al., 2017). In addition, PPV-M is much more aggressive in peach and spreads rapidly in orchards via aphids, as well as being the only strain reported to transmit the virus via seeds (García et al., 2014) (Table 2).

Table 2 Main PPV isolates described at this moment

Isolate type	Description
PPV-D (Dideron)	Originally described in apricot in France in the 1970s, it is the most abundant isolate, characterized by a lower aggressiveness and a low speed of diffusion. It affects the apricot tree, plum tree and, to a lesser extent, the peach tree. It is believed to be the original isolate detected in Bulgaria in 1917.
PPV-M (Marcus)	Described in 1979 in a peach orchard in Greece, it presents high aggressiveness and speed of diffusion. Affects peach, apricot and plum.
PPV-EA (El Amar)	Originally described in apricot tree in Egypt in 1987, it has high aggressiveness. Affects peach, apricot and plum.
PPV-C (Cherry)	Originally described in cherry tree in Moldavia in 1994. It is the unique PPV that affects the cherry trees in the nature, although it affects peach, apricot and plum trees. It is located in Central Europe.
PPV-Rec (Recombinant)	Recombinant isolate between PPV-M and PPV-D with a breakpoint at the carboxyl-terminal end of the NIb gene described in 2004.
PPV-T (Turkey)	A strain originated in Turkey in 2009.
PPV-W (Winona)	A strain, apparently originated in Eastern Europe and Russia but previously described in 2005 in Canada.
PPV-An (Ancestor)	A new strain described as a possible ancestor of Marcus strains, being detected in Albania in 2012.
PPV-CR (Cherry Russian)	A new strain, apparently originating in Eastern Europe and Russia in cherry in 2013.
PPV-CV (Cherry Volga)	Isolate propose as a new strain group, as a variable of the cherry group. It is a new strain, apparently originating in Eastern Europe and Russia in sour cherry in 2018.

The efficiency of PPV transmission is dependent on the virus strain, host cultivar, age of the host cultivar, aphid species and time of year. Viral infection can cause considerable losses with characteristic symptoms in the leaves and fruits, and, in the case of apricots, in the stone (Fig. 3). PPV has been transmitted by at least 20 different aphid species, although only 4–6 are considered important vectors. However, its long-distance distribution via infectious biological material transported for horticultural purpose is more significant (García et al., 2014).

3 PPV detection

PPV can be detected using a biological, serological or molecular test; identification requires either a serological or a molecular test. Further tests may also be carried out to identify the strain of PPV present. The simplest method for detection of plum pox is the use of biological index hosts. PPV can be detected in herbaceous indicator hosts by mechanical inoculation to such diagnostic hosts as *Chenopodium foetidum* and several *Nicotiana* species. The virus can

Figure 3 *Plum pox virus* (Sharka) symptoms in fruits, endocarps and leaves of different *Prunus* species including plum, apricot, peach, prune, almond and Pollizo plum.

also be reliably detected in woody indicator plants by chip budding to hosts such as GF305 peach (Sochor et al., 2012).

The first, highly important, step in the process of PPV diagnostics is based on the serological assay, specifically on the enzyme-linked immunosorbent assay (ELISA). PPV detection by ELISA was included in some of the first work done by Clark and Adams (1977) on the application of this technology for the detection of plant viruses. As with other tree fruit viruses, PPV concentration can be low at certain times of the year and in some *Prunus* cultivars. In a 2010 study, plum, peach and nectarine samples showing mosaic, chlorosis, necrosis and ring pattern were collected and evaluated using a DAS-ELISA (double antibody sandwich-ELISA) method and a polyclonal antibody (Martínez-Gómez et al., 2003a). None of the samples showed a positive reaction in the DAS-ELISA test. Since most of the samples were collected from an infected province (Golestan), further studies on a wider range of samples and using more sensitive methods are required (Hosseinzadeh et al., 2012).

Increasingly sensitive and accurate detection of plum pox became possible in the 1980s through the application of cDNA and cRNA probes, which helped to overcome the problem posed by low concentrations of the virus. As plant pathologists applied polymerase chain reaction (PCR) technology to plant virus detection in the early 1990s, PPV was among the first viral targets amplified. Molecular methods using reverse transcription-PCR (RT-PCR) may be more expensive and/or time consuming than serological techniques, especially for large-scale testing. However, molecular methods, especially real-time PCR or quantitative PCR (qPCR), are generally more sensitive than serological techniques. The use of qPCR also avoids the need for any post-amplification processing (e.g. gel electrophoresis) and is therefore quicker with less opportunity for contamination than conventional PCR. With the exception of immunocapture (IC)-RT-PCR (for which RNA isolation is not required), RNA extraction should be done using appropriately validated protocols. To avoid the spread of PPV over long distances by the transportation of plant material, reliable detection methods are needed for accurate detection of the virus in symptomless nursery plants and propagative material (Scholthof et al., 2011; Sochor et al., 2012).

Recently, the wide application of high-throughput sequencing technologies has significantly facilitated virus detection in fruit trees, including PPV. These new technologies allow for detection of multiple viruses and identification of new virus or isolates (Maliogka et al., 2018).

4 Sharka symptoms and transmission

PPV produces significant symptoms in *Prunus* species (Fig. 3). It is particularly detrimental to apricots, European plums, peaches and Japanese plums, because it can seriously reduce yield and fruit quality. Losses in susceptible cultivars may reach 100% in some cases, and PPV symptoms may appear in leaves, shoots, bark, petals, fruits and even stones. The symptoms change according to the *Prunus* species and cultivar, PPV strain, season and location. Infected leaves may show chlorotic spots or lightly pigmented yellow rings or line patterns. Fruits may become deformed or irregular in shape, and may develop brown or necrotic areas under the discoloured rings, patterns and chlorotic bands or blotches (Fig. 3). Some peach cultivars may show colour-breaking symptoms on the flower petals. In addition, some plum trees suffer from premature fruit drop (Mohammadi et al., 2012). Infected almond trees generally show slight symptoms of the virus in the leaves. Generally, the fruits of early maturing cultivars of all susceptible species show more marked symptoms than those of late maturing cultivars (Myrta et al., 2003).

As indicated above, PPV affects most species within the *Prunus* genus. Through germplasm evaluation studies carried out for more than 20 years at CEBAS, we have evaluated about 5000 different genotypes, mainly of apricot but also of 18 different species of *Prunus*. Genomic analysis indicates the MATH gene(s) as a candidate(s) for PPV resistance in apricot (*Prunus armeniaca* L.). Together with results published by different groups, mainly from France and Central Europe, this reveals a great heterogeneity in the intensity of the symptoms of the disease and the specificity of the isolates to infect different species according to their taxonomic classification.

Within the genus *Prunus*, the subgenus *Prunus* (including the apricot and plum) shows the least specificity for PPV isolates that affect these species and greater aggressiveness with greater symptom severity. At the opposite pole is the subgenus *Amygdalus*, including species such as peach and almond (*P. dulcis* Mill., DA Webb), where the intensity of symptoms is lower (especially in almond) but specificity of isolates is greater. In an intermediate situation is the group including cherry trees (*Cerasus* subgenera) (Martínez-Gómez et al., 2000, 2004; Rubio et al., 2003, 2012) (Fig. 4).

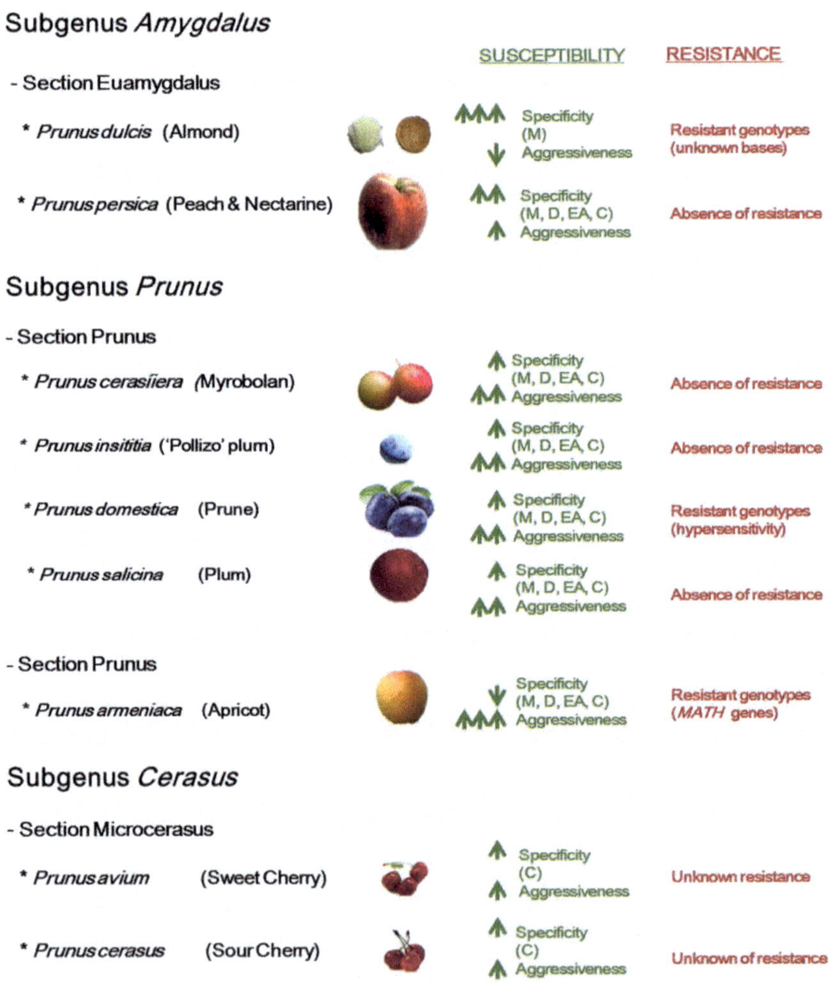

Figure 4 *Plum pox virus* infection in the different species of the *Prunus* genus.

5 Sharka orchard management

When considering orchard management, it is very important to pay attention to the manifestation of possible PPV symptoms in the orchard. Visual observation is crucial as the first method of detecting Sharka disease. Symptoms can be detected in early spring on the leaves of affected plants, and later on the fruits. This method has the advantage of being easy and rapid; however, it also presents significant inconveniences such as possible confusion with symptoms of other diseases and the difficulty of localizing symptoms on adult trees due to the random distribution of PPV. Another problem often faced is the time that Sharka symptoms take to develop: a plant or a tree can be infected but remain symptom-free for a long time, and thus could be erroneously considered a healthy tree. All these issues mean that the use of a reliable method of detecting PPV is necessary. Fortunately, nowadays fast and easy reactive strip tests are available for PPV detection. These methods are based on lateral flow immunochromatography, using specific antibodies against PPV. In any case, if a farmer or producer suspects of the presence of Sharka, we recommend contacting local plant protection services.

If the disease is detected it is important to remove and destroy the infected trees as soon as possible (Fig. 5). A single tree could serve as a focus of dispersion to nearby trees as well as to adjoining orchards. Another important measure is to control sucker shoots, since these could be a good reservoir of PPV and are frequently a refuge for aphids. Controlling aphids with insecticides could help to partially protect the orchards; however, the non-persistent

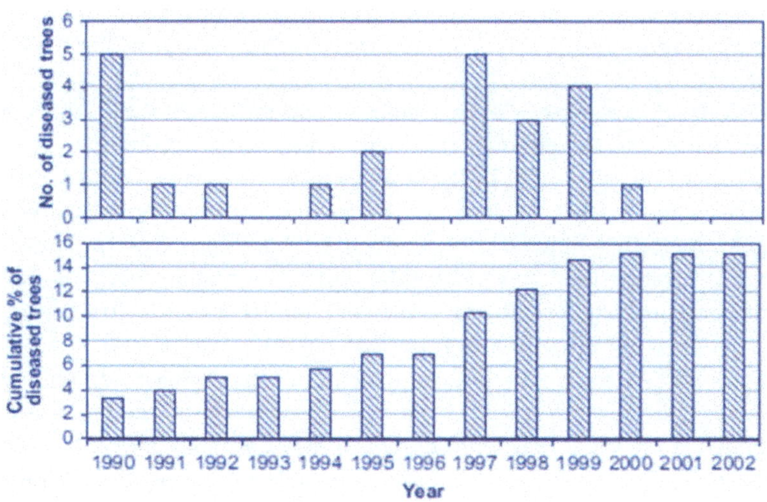

Figure 5 Monitoring *Plum pox virus* infection in an orchard from the Region of Murcia (Southeast of Spain) during the period 1990–2002.

manner in which aphids transmit PPV makes the effective control of the disease very difficult (Martínez-Gómez et al., 2003b). Another measure that can help to prevent Sharka dispersion is weed control, since several weed species are effective hosts of PPV and during the winter can act as reservoirs of the virus as well as hosting numerous aphid species including *Aphis spiraecola*, *Aphis gossypii*, *Aphis fabae*, *Aphis citricola*, *Hyalopterus pruni*, *Aphis craccivora*, *Brachycaudus cardui*, *Brachycaudus prunicola*, *Myzus varians* or *Myzus persicae* (Avinent et al., 1994).

6 PPV control for sustainable cultivation

PPV can be managed by multiple approaches such as certification programmes, vector control and the use of resistant varieties. PPV has proven to be a very difficult disease to control, and there is no antiviral treatment that can be applied to infected trees or orchards. The best method for controlling PPV is to prevent the spread of the virus to new fruit growing areas. Strategies for managing PPV are aimed primarily at preventing its introduction, by the use of virus-tested clean nursery stock. Once PPV is detected, strict quarantine, eradication and ongoing surveys are the only useful management strategies, since a tree, once infected, will never be free of the disease.

Insecticide management strategies that keep aphid populations low may help to slow the movement of the virus in areas where PPV is rare, but may not be a good idea in some situations. Chemical treatments can sometimes cause winged aphids to leave treated areas, taking the virus with them to infect new hosts. An alternative control measure is the removal and destruction of infected trees as quick as possible. As stated above, a single infected tree in an orchard can serve as a source of infection for all surrounding trees and adjacent orchards (Sochor et al., 2012).

Encouraging plant resistance in cultivars and rootstocks is the ideal control strategy for PPV in the different *Prunus* species (Martínez-Gómez et al., 2000, 2004; Rubio et al., 2003, 2012). There are several PPV-resistant apricot cultivars available in the market, all obtained by traditional breeding using resistant genes from North American cultivars such as 'Stark Early Orange', 'Harlayne', 'Goldrich', 'Orange Red' and so on (Martínez-Gómez et al., 2000). So far, the scientific community has made significant efforts towards identifying the genetic control of PPV resistance in apricot (Fig. 4).

The case of peach, where no sources of resistance have been identified so far, requires the adoption of alternative methods to obtain resistance. The peach is the most important *Prunus* species, with an annual production of 21.63 million tons in 2015 (http://faostat.fao.org). As this production is seriously threatened by different strains of PPV, the lack of resistance in this species could potentially be solved by incorporating resistant genes from other related species. To this end, *Prunus ferganensis* (Kostov and Rjabov) and *Prunus davidiana* (Carrière) Franch were the first species to be described as potential sources of resistance to PPV in peach. However, the possible transmission of such resistance has been questioned, as the genetic background is important in the effective transmission of resistance (Fig. 4; Rubio et al., 2010).

Another alternative is the use of the closely related almond species, as almond is a good candidate for a resistance donor by interspecific crosses (Martínez-Gómez et al., 2004). One study demonstrated that grafting the almond cultivar 'Garrigues' onto 'GF305' peach seedlings seriously affected by Sharka can progressively reduce symptoms of the disease as well as virus accumulation (Rubio et al., 2013a). Furthermore, grafting 'Garrigues' onto 'GF305' before PPV inoculation completely prevented viral infection, showing that resistance is constitutive and not induced by the virus (Rubio et al., 2013a).

The case of Japanese plum (*Prunus salicina* L.) is similar to peach, in that there is no PPV resistance within the commercial plums that have been studied (Rubio et al., 2011) (Fig. 4). Again, resistance could be found in related species and interspecific crosses with resistant apricots could produce promising interspecific resistant 'plumcot' (*Prunus salicina* × *Prunus armeniaca*) cultivars (Nicolás-Almansa et al., 2018). In the prune or European plum (*Prunus domestica* L.), resistance is also scarce. Breeding for resistance in European plums is based on their hypersensitive response to PPV consisting of the death of the affected area while the rest of the tree remaining free of the virus (Hartmann and Petruschke, 2000; Neumüller et al., 2009).

Rootstocks can be considered as a special group, as it includes many different species and hybrids. Rootstocks can play an important role in the spread of Sharka, mainly at nursery level, due to their clonal propagation, high planting density and juvenility. At this stage, plants are very susceptible to aphids, the main vector of PPV. Direct infection of rootstocks in the field is unlikely, except for those with a natural tendency to produce suckers (e.g. *Prunus insititia* (L.) C.K. Schneid), which are frequently visited and colonized by aphids and, therefore, likely to be infected with PPV (Rubio et al., 2013b). Data about rootstock susceptibility is scarce, and the studies carried out show a very wide range of behaviours depending on the species, strains and isolates being tested. Both the generalized susceptibility of *P. insititia* and the high level of resistance showed by the interspecific hybrid (almond × peach) 'GF677' are worth noting (Rubio et al., 2005, 2013b). The lack of resistance in most commercial rootstocks used for stone fruit production emphasizes the need for breeding PPV-resistant rootstocks for these species.

The transfer of resistant genes to stone fruit crops by conventional breeding techniques is challenging and time-consuming. Genetic engineering and the use of biotechnology can help to develop Sharka-resistant cultivars (Ilardi and Di Nicola-Negri, 2011). A transgenic European plum called 'HoneySweet' resistant to PPV has been released (Scorza et al., 2013). This was the first genetically engineered PPV-resistant plum to be commercialized, and so far it is only available in the United States. However, there was a delay of 25 years between the engineering of the plum and its commercial release, suggesting that even when an affordable transgenic PPV-resistant *Prunus* is available, there are still many impediments to growing and commercializing transgenic fruit, including consumer acceptance and the legal issues surrounding production in the European Union (Rubio et al., 2017).

7 Genetic and molecular basis of PPV resistance in *Prunus*

The study of the genetics and molecular basis of PPV resistance in *Prunus* has been focussed mainly in the apricot species, where clearly resistant genotypes have been identified (Fig. 4). In this species, different hypotheses have been established on the genetic control of resistance to PPV, including that this oligogenic control is due to a single gene controlled by other two or three (Rubio et al., 2007; Llácer et al., 2008). In the case of the prune or European plum (*Prunus domestica* L.) hypersensitive resistance has been described as monogenic (Hartmann and Neumüller, 2009).

Regarding the genomic bases of this resistance in apricot, a main QTL (quantitative trait loci) has been located in the linkage group (LG) 1 of the apricot tree as responsible for PPV resistance. These results have been observed in the species *Prunus davidiana* (along with other QTLs in other groups) and used as a source of resistance in peach trees, although recent studies suggest that the genetic background of the susceptible parent also strongly influences resistance (Rubio et al., 2010). To date, more than 60 QTLs linked to resistance to PPV have been identified in apricot and *P. davidiana* (Salazar et al., 2014). In addition, a genomic region of 194 kb in LG1 (scaffold 1 of the peach reference genome, positions 8.050.805-8.244.925), where the *PPVres* resistance gene is located, has been described as responsible for PPV resistance in apricot (Zuriaga et al., 2013). However, recent studies indicate the possibility of a second locus being involved (Rubio et al., 2014). Zuriaga et al. (2013) also describe a family of MATH domain homology genes involved in resistance control. On the other hand, other hypotheses have been established about genes that may be related to resistance as the translation initiation factor in eukaryotes eIF4E (Marandel et al., 2009), although at present they seem to be ruled out as a mechanism in *Prunus*, or RTM-type genes [*Restricted Tobacco etch virus* (TEV) movement] (Pagny et al., 2012) that belong to the same MATH gene family described by Zuriaga et al. (2013).

From a transcriptomic point of view, transcriptome sequencing (RNA-Seq) approaches have been assayed for optimal understanding of the molecular processes associated with the expression of the character of susceptibility/ resistance to PPV in fruit trees of the *Prunus* genus (Rubio et al., 2015a,b). These studies showed that the early response to PPV infection in peach and apricot trees is associated with the induction of genes related to resistance to pathogens such as jasmonic acid, resistance proteins, chitinases, cytokinins or Lys-M proteins. On the other hand, when the virus was established, an overexpression of *Dicer protein 2a* genes was observed that could also show a gene-silencing response of the suppressed plant by the PPV proteins HCPro and P1. This phenomenon of gene silencing can be the cause of the elimination of virus symptoms in infected susceptible plants. In studies of resistance to PPV in apricot trees, the results obtained indicate that RTM-type genes [*Restricted Tobacco etch virus (TEV) movement*] within the gene family of MATH domain homology seem to be the main genes involved in the control of this resistance. In addition to these genes, other candidate genes involved in resistance are *Pleiotropic drug resistance 9* gene, *Cysteine-rich secretory* proteins, *Antigen 5* and *Pathogenesis-related 1 protein* (CAP) and late embryogenesis abundant protein (LAE) (Fig. 6). These results have been validated through genomic assays (Zuriaga et al., 2013, 2018).

Regarding the biochemical bases of the PPV/*Prunus* interaction, different studies have shown that in susceptible apricot and peach genotypes the infection produced an oxidative stress with an increase in peroxidation of leaf lipids, oxidation of proteins and accumulation of H_2O_2 together with a decrease in some antioxidant enzymes. These effects were also observed in the chloroplasts (which are the most affected cellular compartment) where the alteration of their ultrastructure and a decrease in the levels of some proteins involved in photosynthesis and carbohydrate metabolism were observed. By contrast, resistant apricot genotypes showed an increase in antioxidant enzymes in the symplast, chloroplast and apoplast (Hernández et al., 2004, 2006). Different results also suggest that the greater antioxidant capacity of resistant varieties could be associated, among other factors, with their resistance. In protein studies by means of 2D electrophoresis, we have found that PPV infection causes the reduction of mandelonitrile lyase (MDL) in peach trees (Díaz-Vivancos et al., 2006). This enzyme catalyses the decomposition of mandelonitrile in benzaldehyde, which can be oxidized to produce the benzoic acid that is precursor of salicylic acid (SA). These data suggest that the decrease in CDM activity may result in a decrease in SA biosynthesis in susceptible plants (Clemente-Moreno et al., 2015).

Finally, in a recent study of the hormonal balance in the control 'GF305' peach and the 'GF305' peach inoculated with a PPV-D isolate, both grafted and not grafted with the almond cultivar 'Garrigues' (Nikbakht-Dehkordi et al., 2018), showed the important differences between treatments. PPV inoculation produced a significant increase in GA3 and ABA and a decrease in

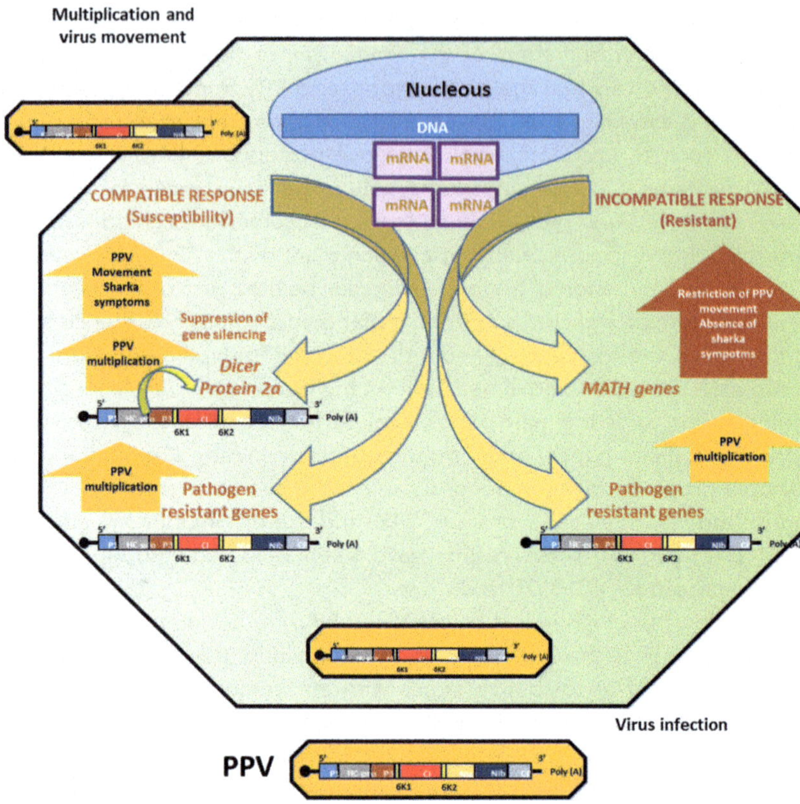

Figure 6 Schematic representation of the PPV/*Prunus* interaction analysed by high-throughput sequencing of different transcriptomes (RNA-Seq) from different apricot and peach genotypes susceptible and resistant to the virus. Source: adaptation from the works of Rubio et al. (2015a,b).

the other phytohormones analysed, including tZ, GA4, ACC, SA and JA. These imbalances were related to the virus infection and the presence of chlorosis symptoms, particularly in the case of the ABA concentration. Additionally, grafting the 'Garrigues' onto 'GF305' produced an increase in GA_3, GA_4, SA and ABA and a decrease in the rest of the phytohormones analysed, tZ, ACC and JA. However, grafting 'Garrigues' almond onto the PPV-inoculated peach 'GF305' produced the opposite effect in some phytohormones, resulting in an increase in tZ, SA and JA. These results showed the significant involvement of SA in the induced resistance response in peach after 'Garrigues' grafting. A significant decrease in ABA and increase in ACC was observed in relationship to the elimination of symptoms. This trend was also detected in the case of the inoculation of 'Garrigues'. These results also show the implication of tZ and JA in virus infection and the induced resistance in peach from almond through a

SA-induced resistance. This SA-induced resistance, consisting of a reduction in symptoms and an absence of other reactions producing necrosis, seems to be different to the SAR and ISR described. SA signalling should activate a cytokinin response to induce PPV resistance linking plant growth and defence. JA also participates in the response against PPV.

8 Future trends and conclusion

Plum pox or Sharka, a viral disease caused by PPV, severely affects the production of *Prunus* species in Europe. This disease is currently globally distributed. Due to the economic impact of this disease in countries where it is prevalent, the latest developments concerning the biology, epidemiology and disease management are provided in this chapter to improve awareness among growers and professionals involved with *Prunus* production. Knowledge and understanding of this virus will be fundamental to success in any attempt to detect and control this disease. After 70 years of research in plum pox, there still exist a lot of unknown facts about the virus.

There are a number of potential future scenarios as regards Sharka disease, depending on the current situation in each producing area of the world. One possible scenario is that those areas not affected by PPV to date will continue free of the virus. Although it is unlikely, it is possible that the disease could be contained by applying the appropriate measures of surveillance and control, especially by strict supervision of plant material imports. The second scenario is one in which PPV is present, and continues its dispersion more or less quickly, but with the risk of new strains emerging. This is the case in countries where PPV-D is present and there is a high risk of the appearance of more aggressive strains in peach trees, such as M or Rec. In those countries where type D, M and Rec strains (France, Italy, Greece etc.) have already been detected, control measures should continue to be applied, since in recent years significant concerns have arisen about the high number of peach plantations seriously affected, and consequent losses in production. Finally, the third scenario would be one in which the affected areas ceased the application of control measures for Sharka containment, which would increase the number of plantations affected and would seriously threaten the continuity of certain stone fruit in countries such as Iran (Nikbakht-Dehkordi et al., 2017). This third scenario will produce the elimination of the stone fruit culture if new resistant cultivars and control measurements are incorporated.

In conclusion, Sharka is still a very important disease at the global level, and despite all the efforts made and resources allocated to date, it remains a threat to stone fruit production. However, with the appropriate measures and collaboration between the different stakeholders (authorities, nurserymen, producers and researchers) the disease can be controlled, although it is likely impossible to eradicate it. Proper management, control of nursery plants and obtaining resistant varieties (when possible) will be the key to keeping the disease under control without causing significant damage.

9 Acknowledgements

This study has been partially supported by the following projects: 'Genetics and molecular basis of multiple resistance to *Plum pox virus* (PPV) and Apple chlorotic leaf spot virus (ACLSV) in apricot' (AGL2015-68021-R) of the Spanish Ministry of Economy and Competiveness and 'Breeding stone fruit species assisted by molecular tools' of the Seneca Foundation of the Region of Murcia (19879/GERM/15).

10 Where to look for further information

PPV research is advancing fast in various sectors including detection and characterization, epidemiology and transmission, control of vector and host resistance. Actually PPV is considered by the American Society of Plant Pathology (https://www.apsnet.org) and the European Society for Virology (http://www.eusv.eu/) as a model potyvirus inside the top ten plant viruses together with a major disease. The current state of the art in PPV research is highlighting the valuable work carried out in different laboratories throughout the world to combat this severe plant disease under the coordination of the European and Mediterranean Plant Protection Organization (EPPO) (https://www.eppo.int). The increasing knowledge about the complex molecular relationships that PPV have with their hosts is building a complete picture of the viral infection cycle and viral genome expression towards a definitive control of the virus using classical breeding and genetic engineering.

11 References

Atanasoff, D. 1932. Plum pox a new virus disease. *Ann. Univ. Sofia, Fac. Agri. Silviculture* 11, 49–69.

Avinent, L., Hermoso de Mendoza, A. and Llácer, G. 1994. Transmission of Plum pox potyvirus in Spain. *EPPO Bull.* 36, 669–74.

Cambra, M., Capote, N., Myrta, A. and Llácer, G. 2006. Plum disease. *EPPO Bull.* 36(2), 202–4. doi:10.1111/j.1365-2338.2006.01027.x.

Chirkov, S., Sheveleva, A., Ivanov, P. and Zakubanskiy, A. 2018. Analysis of genetic diversity of Russian sour cherry *Plum pox virus* isolates provides evidence of a new strain. *Plant Dis.* 102(3), 569–75. doi:10.1094/PDIS-07-17-1104-RE.

Clark, M. F. and Adams, A. N. 1977. Characteristics of the microplate method of enzyme-linked immunosorbent assay for the detection of plant viruses. *J. Gen. Virol.* 34, 51–7.

Clemente-Moreno, M. J., Hernández, J. A. and Díaz-Vivancos, P. 2015. Sharka: how do plants respond to *Plum pox virus* infection? *J. Exp. Bot.* 66(1), 25–35. doi:10.1093/jxb/eru428.

Díaz-Vivancos, P., Rubio, M., Periago, P. M., Martínez-Gómez, P. and Hernández, J. A. 2006. The apoplastic antioxidant system in *Prunus*: response against long-term PPV infection. *J. Exp. Bot.* 57, 3813–24.

García, J. A., Glasa, M., Cambra, M. and Candresse, T. 2014. *Plum pox virus* and sharka: a model potyvirus and a major disease. *Mol. Plant Pathol.* 15(3), 226–41. doi:10.1111/mpp.12083.

Hartmann, W. and Neumuller, M. 2009. Plum breeding. In: Mohan, J. S. and Priyadarshan, P. M. (Eds), *Breeding Plantation Tree Crops: Temperate Species*. Springer, New York, pp. 161–231. doi:10.1007/978-0-387-71203-1_6.

Hartmann, W. and Petruschke, M. 2000. Sharka resistant plums and prunes by utilization of hypersensitivity. *Acta Hortic.* 538, 391–5.

Hernández, J. A., Rubio, M., Olmos, E., Ros-Barceló, A. and Martínez-Gómez, P. 2004. Oxidative stress induced by long-term *Plum pox virus* infection in peach (*Prunus persica*). *Physiol. Plant.* 122, 486–95.

Hernández, J. A., Díaz-Vivancos, P., Rubio, M., Olmos, E., Ros-Barceló, A. and Martínez-Gómez, P. 2006. Long-term *Plum Pox Virus* infection produces an oxidative stress in a susceptible apricot, *Prunus armeniaca*, cultivar but not in a resistant cultivar. *Physiol. Plant.* 126, 140–52.

Hosseinzadeha, H., Nasrollanejada, S., Dordiania, Z. and Jamshidnezhad, V. 2012. Serological detection of sharka quarantine disease (*Plum pox virus*) on stone fruit trees in Golestan Province. *Arch. Phytopathol. Plant Prot.* 45(16), 1879–83. doi:10.1080/03235408.2012.718178.

Ilardi, V. and Di Nicola-Negri, E. D. 2011. Genetically engineered resistance to *Plum pox virus* infection in herbaceous and stone fruit hosts. *GM Crops* 2(1), 24–33. doi:10.4161/gmcr.2.1.15096.

Llácer, G., Badenes, M. L. and Romero, C. 2008. Problems in the determination of inheritance of *Plum pox virus* resistance in apricot. *Acta Hortic.* 781(781), 263–8. doi:10.17660/ActaHortic.2008.781.39.

Malioogka, V. I., Minafra, A., Saldarelli, P., Ruiz-García, A. B., Glasa, M., Katis, N. and Olmos, A. 2018. Recent advances on detection and characterization of fruit tree viruses using high-throughput sequencing technologies. *Viruses* 10(8), 436. doi:10.3390/v10080436.

Marandel, G., Pascal, T., Candresse, T. and Decroocq, V. 2009. Quantitative resistance to *Plum pox virus* in *Prunus davidiana* linked to components of the eukaryotic translation initiation complex. *Plant Pathol.* 58(3), 425–35. doi:10.1111/j.1365-3059.2008.02012.x.

Martínez-Gómez, P., Dicenta, F. and Audergon, J. M. 2000. Behaviour of apricot (*Prunus armeniaca* L.) cultivars in presence of sharka (*Plum pox Potyvirus*): a review. *Agronomie* 20(4), 407–22. doi:10.1051/agro:2000137.

Martínez-Gómez, P., Rubio, M., Aparicio, F., Pallás, V. and Dicenta, F. 2003a. Comparative analysis of three diagnostic methods for the evaluation of PPV resistance in apricot breeding programs. *Acta Hortic.* 662, 397–401.

Martínez-Gómez, P., Dicenta, F. and Egea, J. 2003b. Effect of a traditional control method (tree removal) on the spread of sharka in an apricot orchard in Southeastern Spain. *Phytopathol. Mediterr.* 42, 161–6.

Martínez-Gómez, P., Rubio, M., Dicenta, F. and Gradziel, T. M. 2004. Resistance to *Plum pox virus* (RB3.30 isolate) in a group of California almonds and transfer of resistance to peach. *J. Am. Soci. Hortic. Sci.* 129, 544–8.

Mohammadi, M., Goharzad, F. and Zaghi, A. 2012. Association of *Plum pox virus* strain m with plum fruit dropping in Mazandaran Province. *Appl. Entomol. Phytopathol.* 80, 97–8. (in Persian with English summary).

Myrta, A., Di Terlizzi, B., Savino, V. and Martelli, G. P. 2003. Virus diseases affecting the Mediterranean stone fruit industry: a decade of surveys. In: Myrta, A., Di Terlizzi, B. and Savino, V. (Eds), *Virus and Virus-Like Diseases of Stone Fruits, with Particular Reference to the Mediterranean Region*. CIHEAM, Bari, pp. 15–23. (Options Méditerranéennes: Série B. Etudes et Recherches; n. 45).

Neumuller, M., Lanzl, S., Hartmann, W., Feucht, W. and Treutter, D. 2009. Towards an understanding of the inheritance of hypersensitivity resistance against the sharka virus in European plum (*Prunus domestica* L.): generation of interspecific hybrids with lower ploidy levels. *Acta Hortic.* 814(814), 721–6. doi:10.17660/ActaHortic.2009.814.122.

Nicolás-Almansa, M., Guevara, A., Salazar, J. A., Rubio, M., Martínez-Gómez, P. and Ruiz, D. 2018. Herencia de caracteres agronómicos en ciruelo japonés (*Prunus salicina* Lindl.) y plumcot (*P. salicina* Lindl. x *P. armeniaca* L.). *Acta Hortic.* 80, 312–6.

Nikbakht-Dehkordi, A., Babeian, N., Karimpour, S., Martínez-Gómez, P., Rubio, M. and Bagheri, N. 2017. Sharka (*Plum pox virus*): a forgotten disease in Iran. *Int. J. Hortic. Sci. Technol.* 4, 183–91.

Nikbakht-Dehkordi, A. N., Rubio, M., Babaeian, N., Albacete, A. and Martínez-Gómez, P. 2018. Phytohormone signaling of the resistance to *Plum pox virus* (PPV, Sharka Disease) induced by almond (*Prunus dulcis* (Miller) Webb) grafting to peach (*P. persica* L. Batsch). *Viruses* 10(5), 238. doi:10.3390/v10050238.

Pagny, G., Paulstephenraj, P. S., Poque, S., Sicard, O., Cosson, P., Eyquard, J. P., Caballero, M., Chague, A., Gourdon, G., Negrel, L., et al. 2012. Family-based linkage and association mapping reveals novel genes affecting *Plum pox virus* infection in *Arabidopsis thaliana*. *New Phytol.* 196(3), 873–86. doi:10.1111/j.1469-8137.2012.04289.x.

Rubio, M., Martínez-Gómez, P. and Dicenta, F. 2003. Resistance of almond cultivars to *Plum pox virus* (sharka). *Plant Breed.* 122(5), 462–4. doi:10.1046/j.1439-0523.2003.00872.x.

Rubio, M., Martínez-Gómez, P., Pinochet, J. and Dicenta, F. 2005. Evaluation of resistance to sharka (*Plum pox virus*) of several *Prunus* rootstocks. *Plant Breed.* 124(1), 67–70. doi:10.1111/j.1439-0523.2004.01068.x.

Rubio, M., Martínez-Gómez, P., Audergon, J. M. and Dicenta, F. 2007. Testing genetic control hypotheses for *Plum pox virus* resistance in apricot. *Sci. Hortic.* 112, 361–5.

Rubio, M., Pascal, T., Bachellez, A. and Lambert, P. 2010. Quantitative trait loci analysis of PPV resistance in Prunus davidiana P1908. *Tree Genet. Genomes* 6, 291–304.

Rubio, M., García-Ibarra, A., Dicenta, F. and Martínez-Gómez, P. 2011. *Plum pox virus* (Sharka) sensitivity in *Prunus salicina* and *Prunus cerasifera* cultivars against a Dideron-type isolate. *Plant Breed.* 130(2), 283–6. doi:10.1111/j.1439-0523.2010.01813.x.

Rubio, M., Martínez-Gómez, P., García-Bruntom, J., Pascal, T., García-Ibarra, A. and Dicenta, F. 2012. Sensitivity of peach cultivars against a Dideron isolate of *Plum pox virus*. *Sci. Hortic.* 144, 81–6. doi:10.1016/j.scienta.2012.06.038.

Rubio, M., Martínez-Gómez, P., García, J. A. and Dicenta, F. 2013a. Interspecific transfer of resistance to *Plum pox virus* by grafting. *Ann. Appl. Biol.* 163, 466–74.

Rubio, M., Dicenta, F., Masse, M. and Duval, H. 2013b. Susceptibility of *Prunus* rootstocks against Marcus and Dideron isolates of *Plum pox virus* by graft-inoculation. *Ann. Appl. Biol.* 162(2), 214–20. doi:10.1111/aab.12015.

Rubio, M., Ruiz, D., Egea, J., Martínez-Gómez, P. and Dicenta, F. 2014. Opportunities of marker assisted selection for *Plum pox virus* resistance in apricot breeding. *Tree Genet. Genomes* 10, 513–25.

Rubio, M., Rodríguez-Moreno, L., Ballester, A. R., Castro de Moura, M. C., Bonghi, C., Candresse, T. and Martínez-Gómez, P. 2015a. Analysis of gene expression changes in peach leaves in response to *Plum pox virus* infection using RNA-Seq. *Mol. Plant Pathol.* 16(2), 164–76. doi:10.1111/mpp.12169.

Rubio, M., Ballester, A. R., Olivares, P. M., Castro de Moura, M., Dicenta, F. and Martínez-Gómez, P. 2015b. Gene expression analysis of *Plum pox virus* (Sharka) susceptibility/resistance in apricot (*Prunus armeniaca* L.). *PLoS ONE* 10(12), e0144670. doi:10.1371/journal.pone.0144670.

Rubio, M., Martínez-Gómez, P., Marais, A., Sánchez-Navarro, J. A., Pallás, V. and Candresse, T. 2017. Recent advances and prospects in *Prunus* virology. *Ann. Appl. Biol.* 171(2), 125–38. doi:10.1111/aab.12371.

Salazar, J. A., Ruiz, D., Campoy, J. A., Sánchez-Pérez, R., Crisosto, C. H., Martínez-García, P. J., Blenda, A., Jung, S., Main, D., Martínez-Gómez, P. et al. 2014. Quantitative trait loci and mendelian trait loci analysis in *Prunus*. *Plant Mol. Biol. Rep.* 32, 1–18.

Scholthof, K. B. G., Adkins, S., Czosnek, H., Palukaitis, P., Jacquot, E., Hohn, T., Hohn, B., Saunders, K., Candresse, T., Ahlquist, P., et al. 2011. Top 10 plant viruses in molecular plant pathology. *Mol. Plant Pathol.* 12(9), 938–54. doi:10.1111/j.1364-3703.2011.00752.x.

Scorza, R., Callahan, A., Dardick, C., Ravelonandro, M., Polak, J., Malinowski, T., Zagari, I., Cambra, M. and Kamenova, I. 2013. Genetic engineering of *Plum pox virus* resistance: 'HoneySweet' plum – from concept to product. *Plant Cell Tiss. Organ Cult.* 115(1), 1–12. doi:10.1007/s11240-013-0339-6.

Sihelská, N., Glasa, M. and Šubr, Z. W. 2017. Host preference of the major strains of *Plum pox virus*–opinions based on regional and world-wide sequence data. *J. Integr. Agric.* 16(3), 510–5. doi:10.1016/S2095-3119(16)61356-4.

Sochor, J., Babula, P., Adam, V., Krska, B. and Kizek, R. 2012. Sharka: the past, the present and the future. *Viruses* 4(11), 2853–901. doi:10.3390/v4112853.

Zuriaga, E., Soriano, J. M., Zhebentyayeva, T., Romero, C., Dardick, C., Cañizares, J. and Badenes, M. L. 2013. Genomic analysis reveals MATH gene(s) as candidate(s) for *Plum pox virus* (PPV) resistance in apricot (*Prunus armeniaca* L.). *Mol. Plant Pathol.* 14(7), 663–77. doi:10.1111/mpp.12037.

Zuriaga, E., Romero, C., Blanca, J. M. and Badenes, M. L. 2018. Resistance to *Plum pox virus* (PPV) in apricot (*Prunus armeniaca* L.) is associated with down-regulation of two MATHd genes. *BMC Plant Biol.* 18(1), 25. doi:10.1186/s12870-018-1237-1.

Part 2

Integrated fruit disease management

Chapter 9

Disease monitoring and decision making in integrated fruit disease management

Angela Berrie, NIAB EMR, UK

1 Introduction

Today's consumers expect perfect fruit which are produced without the use of pesticides or at least are free of pesticide residues. Unfortunately, most of the fruit cultivars desired by the market are susceptible to most diseases and require frequent applications of fungicides to produce a marketable crop. Pest control has the advantage that, as the pest is generally visible, sprays can be targeted to the presence or absence or to a particular number threshold, following regular crop inspections. While scouting for disease symptoms in fruit crops is an important part of decision making in integrated disease management, the approach used for pests cannot be applied to many diseases. Often there is a lag between fungal infection and the appearance of visible disease symptoms, so relying on visual assessments alone can be very misleading. For example, infection of apple rosette leaves with apple scab pre-blossom in cold weather often does not appear as visible scab lesions on the leaf until after blossom, some 3 weeks later (Sutton et al., 2013). Regular inspection of fruit crops for diseases is a vital part of integrated disease management. Combining this information with that generated from disease warning systems can play a significant role in achieving perfect fruit with minimal use of fungicides. This chapter will include general discussions on fruit disease monitoring, disease warning systems and decision making and then use examples of specific diseases in apple and pear to illustrate the role of both in integrated disease management.

http://dx.doi.org/10.19103/AS.2019.0046.11

2 General considerations

2.1 Monitoring fruit crops for diseases

Fruit crops are generally monitored for all pests and diseases and agronomic factors such as weeds at the same time by specialist crop walkers or agronomists on a weekly or fortnightly basis (or timed to coincide with decisions on treatments for the next spray round) from the beginning of growth in the spring to the end of the crop at harvest. More limited monitoring in the dormant period can also take place and be important for the overall disease management of the crop. Key requirements for successful crop monitoring are:

- An understanding of the epidemiology of the disease in relation to the growth of the crop so that the correct part of the fruit crop is targeted during the disease searches (see Table 1).
- Confidence in disease recognition and the ability to spot the early signs of the disease.
- Adequate sampling. Fifty trees is considered the ideal sample size in each orchard but this is probably impractical. In practice, a minimum of 10 trees, preferably 20, is likely to be more manageable but it is important to maintain a general vigilance for signs of diseases as the orchard is walked through to compensate for the reduced sample size.
- Ideally, the 10–20 trees will be at random and spread over the whole block of fruit but in practice the sampling route will usually be a compromise between the ideal and the shortest route for covering the ground. Particular attention should be paid to areas of the orchard with higher disease risk, e.g. for apple scab, this would be dips or sheltered areas in the orchard where the trees are likely to remain wetter for longer.
- Monitoring in most cases will be by eye but a x 10 hand lens is essential to check and confirm the early signs of disease.
- Keeping records of the monitoring results is desirable for future reference and for the end-of-season post mortem on the success of the disease control on the farm.

Inspection methods, timing and target diseases have been formalised for some crops especially apples and pears. These will vary according to the country and the range of disease problems present. Monitoring methods for diseases in apples and pears in the UK have been described by Cross and Berrie (1994, 2001) and an example for apples is given in Table 1. Guidelines for disease monitoring for other fruit crops are less well documented but follow a similar pattern to those described for apple and pear.

Table 1 An example of sampling methods, treatment thresholds and actions and their timings if thresholds are exceeded for diseases in Integrated Disease Management used in UK apple orchards, based on an assessment of 10–20 trees/orchard

Time/growth stage Sampling unit per tree	Disease	Threshold	Action and time for treatment
Dormant period			
Whole tree	Wood scab	Presence	Remove during winter pruning
	Apple mildew (silvered shoots)		
	European apple canker		
	Wood-rotting fungi		
	Mummified brown rot fruit		Remove or push into alley way to be macerated by mower
Whole orchard	Overwintering leaf litter	Easily found	Macerate to encourage breakdown prior to bud-burst
Bud-burst to Pink bud			
Whole tree	Apple scab	% trees infected: ≥5% = low 5–20% = moderate >20% = high	Intensify programme of eradicant and protectant fungicide sprays at full dose. Reduce spray interval according to severity and risk of infection
Four blossom trusses	Primary blossom mildew	2% mildewed blossoms	Intensify programme of eradicant and protectant fungicide sprays at full dose and medium volume (500 L/ha). Reduce spray interval
Early blossom			
Whole orchard	Blossom wilt	Presence previous year	Apply suitable fungicide at early blossom and repeat 10 days later
Late blossom to 80% petal fall			
Whole tree	Apple scab	Presence of visible symptoms	Intensify programme of eradicant and protectant fungicide sprays at full dose. Reduce spray interval

(*Continued*)

Table 1 (*Continued*)

Time/growth stage Sampling unit per tree	Disease	Threshold	Action and time for treatment
	Primary vegetative mildew	2% mildewed shoots	Intensify programme of eradicant and protectant fungicide sprays at full dose and medium volume (500 L/ha). Reduce spray interval
	European apple canker/*Neonectria* fruit rot	>25% trees with canker	Indicates high risk of fruit infection and need to apply sprays at petal fall and 14 days later to protect fruit
Mid-May			
Ten trusses on each of four branches	Scab on leaves, flowers or fruitlets	% trusses infected: ≥2.5% = low 2.5–9.0% = moderate >9% = high	Intensify programme of eradicant and protectant fungicide sprays at full dose. Reduce spray interval according to severity and risk of infection
Every 14 days after petal fall (June, July, August and, where necessary, September)			
Whole tree	Blossom wilt (*Monilinia laxa fsp. mali*)	Presence	Prune out. Treat with fungicide in following year
	Canker in 1–2 year old shoots	Presence	Prune out. Treat with fungicide at leaf fall
	Neonectria eye rot	Presence	A moderate to high incidence indicates conditions were favourable for fruit infection at petal fall and a moderate to high risk of rotting in store
Four shoots per tree	Scab on leaves	% shoots infected: ≥2.5% = low 2.5–9.0% = moderate >9% = high	Intensify programme of eradicant and protectant fungicide sprays at full dose. Reduce spray interval according to severity and risk of infection
Top (apical) five leaves in two growing shoots per tree	Secondary apple powdery mildew	8% of leaves or 30% of shoots mildewed	Intensify programme of fungicide sprays at full dose and medium volume. Reduce spray interval

Fruit whole tree	Sooty blotch and Fly speck	Presence	Earmark for treatment next year. Implement cultural control measures
14–7 days before harvest			
Whole orchard – rot risk	Brown rot (*Monilinia fructigena*)	Assess incidence on fruit >1% indicates high risk of rot in store	Selective picking at harvest
	Phytophthora rot (*Phytophthora syringae*)	Assess incidence of bare ground in tree row. 20% or > bare ground = risk % of crop <0.5 m from ground. 15% or > of fruit = risk	20 mm or > rain in 15 days before harvest in at risk orchards = risk of rot in store. Options are to avoid harvesting low fruit for store or market fruit early to limit losses
Post-harvest in orchard			
Whole orchard	Canker	Damaging levels present (25% or > trees with canker)	Apply approved sprays at leaf fall
10–20 leaves/tree	Late scab infection on leaf undersides of leaves (diffuse scab)	Present ≤3% scabbed = low >3% scabbed = high	Spray urea shortly before leaf fall and macerate fallen leaves in winter

2.2 Disease forecasting or warning systems

Disease warning systems usually require input of temperature, relative humidity, leaf surface wetness and rain fall from a weather station to generate disease risk information. In many countries there are networks of weather stations, either government or privately managed, which can provide the weather data for input into the models. How relevant these warnings are to a particular fruit farm will depend on proximity to the weather station and the local topography. Having dedicated weather stations on site can be more advantageous, but these will need to be regularly maintained, particularly checking that the rain gauge is not blocked.

Disease warning systems can be used in a variety of ways in integrated disease management, depending on the fruit crop and the area to be treated, the disease, fungicide availability and the economics. Above all the system in use has to be easy to understand and implement and be part of an integrated approach to disease control. It is important to emphasise on the need to continue regular crop inspections for diseases. It is very easy to assume that once a disease warning system is installed, all decisions on fungicide use can be decided from the comfort of the office. Such a strategy usually ends in disaster. In most cases, as the systems are weather-based, they give details of disease infection that has occurred and are therefore dependent on the availability of control methods that can eradicate infection. As weather forecasts are becoming more reliable, some warning systems incorporate a weather forecast into the system to assist in decision making. Details of some of the main computer-based disease warning systems available for use in fruit disease management are given in Table 2.

2.3 Making decisions on control

Integrated management of diseases requires a good knowledge of the biology and epidemiology of the pathogen and combines various methods of disease control such as use of resistant cultivars, cultural control, biocontrol and use of fungicides. Generally in fruit production use of resistant cultivars is not an option as cultivar choice is dictated by the market and, as stated earlier, the cultivars preferred are often the most disease susceptible. Cultural practices such as the elimination of inoculum should be carried out as routine practices in an integrated disease control programme. So integrated control of apple scab will include methods such as post-harvest urea sprays or leaf mashing to encourage leaf litter breakdown and hence the reduction in overwintering scab inoculum (MacHardy, 1996) and control of brown rot (*Monilinia fructigena*) in apples and stone fruit will include removal of tree cankers and mummified fruit.

Table 2 Summary of some computer-based disease warning systems used in apple disease management

Model/decision system	Target diseases	Weather criteria	Other factors	Output
RIMpro (www.rimpro.eu)	Apple scab	Temperature, relative humidity (RH) and rain, based on Mills and Laplante (1951). Includes forward weather forecast	Ascospore maturity model	1 Ascospore discharge 2 Scab risk periods 3 Forward forecast scab risk
VENTEM (Xu et al., 1995)	Apple scab	Temperature, RH and rain, based on new East Malling model	Cultivar susceptibility Inoculum level	Scab risk which can be orchard specific
PODEM (Xu, 1999)	Apple powdery mildew	Temperature, rain and RH based on East Malling model	Cultivar susceptibility Inoculum level	Current mildew risk and future predicted risk
NECTEM (Xu, 1993)	*Neonectria ditissima*	Temperature, RH, rain and duration of surface wetness	Input of leaf fall dates	Risk of fruit infection Risk of leaf scar infection at leaf fall
Neonectria (Latorre et al., 2002)	*Neonectria ditissima*	Temperature, RH, rain and duration of surface wetness		Risk of leaf scar infection at leaf fall
RIMpro (Trapman, 2004)	Sooty Blotch and Fly Speck	Temperature, rain, RH and duration of surface wetness		Risk of fruit colonisation
BIS95 (Billing, 1996)	Fire blight	Rain and temperature		Geared to fire blight in UK. Timing of orchard searches to find new fire blight infections
Maryblyte (Lightner and Steiner, 1992)	Fire blight	Rain and temperature		Potential fire blight infection periods

Control methods employed in the growing season are generally based on conventional fungicides or biopesticides where effective products exist. More recently, as fungicides become more limited, interest has been growing in the use of biostimulants or elicitors as alternative approaches to disease control with some success (Berrie et al., 2015). These can be used to supplement control programmes based on fungicides or biopesticides, but tend to be applied as routine treatments. Biopesticides are more effective when applied as protectant treatments (Berrie et al., 2015), but are not usually effective as curatives or eradicants. Decisions on treatment of diseases based on crop monitoring and/or warnings from disease forecasting systems are generally fungicide-based. As most disease warning systems are weather-based, the information on infection is historic and requires a fungicide with curative or 'kick-back' action. There is information available on fungicide action (protectant or curative) for apple scab and powdery mildew (Tables 3 and 4), but similar information for other diseases is lacking. Most approaches to integrated disease control rely on a basic protectant programme and information from disease warning systems is used to adjust the programme in terms of fungicide or biopesticide choice and spray interval. The use of disease monitoring and warning systems in relation to the management of some key diseases of apple and pear crops will now be given as examples.

Table 3 Fungicide groups for the control of apple scab protectant and eradicant (kick-back)

Chemical group	Examples	Kick-back (days)	Protectant action (days)	Resistance risk
SDHI	fluopyram fluxapyroxad penthiopyrad	0	7-10	Medium to high
QoI	pyraclostrobin kresoxim-methyl trifloxystrobin	0	7-10	High
anilino-pyrimidines	cyprodinil pyrimethanil	3	10-14	Medium
DMI	difenoconazole fenbuconazole myclobutanil	4	6-9	Medium
guanidine	dodine	1	7-14	Low-medium
dithiocarbamate	mancozeb	0	7-10	Low
phthalimide	captan folpet	1-2 at low temperature	10-14	Low
quinone	dithianon	2	10-14	Low
inorganic	sulphur	0	10	Low

Table 4 Fungicide groups for the control of powdery mildew curative/anti-sporulant/protectant

Chemical group	Examples	Anti-sporulant (AS) action/ curative	Protectant action (days)	Resistance risk
Hydroxy-pirimidine	Bupirimate	AS	7-10	Medium
SDHI	Fluopyram Fluxapyroxad Isopyrazam Penthiopyrad	AS	7-10	Medium-high
Strobilurin (QoI)	Kresoxim-methyl Pyraclostrobin Trifloxystrobin	No	7	Medium-high
Aza-naphthalenes	Proquinazid	AS	7-10	Medium
Dinitrophenyl-crotonates	Meptyldinocap	No	7	Low
DMI	Myclobutanil Penconazole	AS	7	Medium
Phenyl-acetamide	Cyflufenamid	AS	7-10	Medium
Inorganic	Potassium Bicarbonate	AS	0	Low
	Sulphur	No	10	Low

3 Apple disease examples

*3.1 Apple scab (***Venturia inaequalis***)*

Apple scab is the most economically important disease of apple in the world. All above ground parts of the tree are attacked. Scab infection of fruit is most obvious. Most commercial UK apple varieties are susceptible to scab; Gala is the most susceptible. The scab fungus overwinters mainly as the sexual state (pseudothecia) on leaves on the orchard floor, but it can also survive as mycelium in lesions on the tree as wood scab, bud scale scab or shoot base scab. The importance of this source of inoculum varies with cultivar, region and season, but it is important that it is not ignored as a potential inoculum source with all the focus on ascospores. In spring, spores, either ascospores released in rain from overwintering leaves or conidia from tree sources, infect developing leaves and fruit to initiate the new scab epidemic. Wet leaves or fruit are essential for infection. Once new infections are established, conidia are produced and these spread in wind and rain to infect other new leaves and fruit. Infection of fruit near harvest can lead to the development of storage scab on fruit in store. In late summer/autumn, older leaves which had become resistant to scab with maturity become susceptible to scab again as the cuticle cracks with age. This diffuse late scab developing on leaves contributes most to the overwintering scab on the orchard floor.

Successful control of apple scab combines cultural control, orchard monitoring, and disease warning systems in an integrated approach to minimise fungicide inputs to achieve high-quality scab-free fruit.

3.1.1 Disease monitoring - visual assessment

Details of monitoring methods can be found in Table 1 and elsewhere (Cross and Berrie, 1994; Cross and Berrie, 2001). However, it is not possible in the early part of the season to base decisions on fungicide use on monitoring for visual scab because of the long time (up to 3 weeks at low temperatures) between infection and visual scab (Sutton et al., 2013). Decisions on fungicide use are usually based on a combination of routine applications and weather-related risks. Searches for and assessment of scab are still important, to check on inoculum levels at the start of and during the season and to assess the success of control measures such that any modification to spray decisions can be made.

3.1.2 Disease monitoring - scab inoculum

Pseudothecia, overwintering in leaf litter under trees and in the windbreaks, is one of the most important inoculum sources for scab in spring. Formal assessments of the size of the inoculum (potential ascospore dose or PAD) can be made (McHardy, 1996). Noting the amount of scab in the orchard the previous season and the incidence of leaf litter left in orchards in spring will also give a reasonable assessment of scab inoculum. The amount of leaf litter left is dependent on a number of factors but particularly winter conditions (cold and snow cover will preserve leaves; mild wet winters will encourage leaf rotting) and earthworm activity. Apple variety will also have an effect - Cox leaves tend to fall early and rot rapidly, while Bramley and Jonagold leaves can remain on the trees until December and rot slowly and still be in abundance in orchards in spring. Scab ascospore maturity and discharge can also be monitored using spore traps and this information is provided nationally in some countries. Prediction of ascospore discharge - the proportion of the season's ascospores that are mature on a given date and the date from which ascospore discharge is almost complete are also useful for managing apple scab in an orchard and some scab warning systems such as RIMpro (www.rimpro.eu) incorporate this information. However, it is important to remember that firstly this information is based on models and must be interpreted in relation to local conditions and secondly that ascospores are not the only source of inoculum in the spring. The scab fungus can also overwinter on the tree as mycelium, which can produce conidia to infect leaves. This source can be quite important, depending on the cultivar and the scab incidence the previous season. Once new scab lesions are visible on leaves in spring, then conidial sources of inoculum are equally as important as ascospores.

3.1.3 Apple scab warning systems

In many countries in Europe, a network of weather stations measuring temperature, rainfall, relative humidity and leaf surface wetness and linked to computers have been established to monitor infection periods and advise growers on the need to spray. These are usually all based on the Mills or Mills/MacHardy scab infection models (Mills and LaPlante, 1951). Similarly, software provided by companies producing automatic weather stations for use in apple orchards, e.g. Metos, are also all based on the Mills apple scab infection model. In the UK, scab warnings, mainly based on RIMpro, are provided by private companies at a cost. Warnings given by Mills models are based only on weather and give information about scab infections that have occurred. RIMpro also incorporates the weather forecast to assist in making decisions on treatment.

3.1.4 Making decisions on treatment

Apple scab is a serious disease of apples and failure to achieve control results in economic loss. Growers aim for zero fruit scab at harvest and levels below 1% fruit scab at harvest are acceptable. Generally, the most reliable and easiest approach to manage practically, to achieve 0–1% fruit scab, is a routine fungicide programme. Scab warnings can improve scab control, reduce costs and fungicide inputs, all of which are of increasing importance in modern fruit production. Usually apple scab warning systems are associated with curative spraying where fungicides with curative/eradicant action are applied soon after infection periods are detected (Table 3). Such an approach can achieve substantial reductions in fungicide inputs; however, in the UK, curative spraying is also the most risky strategy to adopt – scab warnings have to be reliable and the grower must have the capacity to rapidly respond to warnings. Possible control strategies to be considered are shown in Table 5.

Which of these strategies is adopted depends on several factors. Farm size is critical. Small farms where the orchard spraying can be completed in around an hour could rely on a curative strategy spraying only in response to scab warnings, whereas this approach would be impossible for large farms where several days are required to complete a spray round. On such large farms, the orchards are often split up into high- and low-risk scab orchards so that control measures can be more targeted.

Availability of fungicides is also critical as a curative spraying approach would rely heavily on fungicides with 'kick-back' action (Table 3), which could lead to overuse of some fungicide groups such as DMIs and the risk of fungicide

Table 5 Strategies for making use of apple scab warnings

Strategy	Fungicide use	Scab warning use	Orchard monitoring input	Comments
Routine	Routine at 7–14 day intervals	No	Yes	Routine spraying from bud burst to end of scab risk, usually mid-summer
Curative	Not routine	Yes	Yes	Fungicides applied only in response to scab warning
Mixed Key stage	Routine fungicide at key stages such as bud burst, petal fall	Yes	Yes	Fungicide applied routinely at key stages of bud burst and petal fall and in response to scab warnings and other orchard activities such as pest sprays at other times
Mixed curative	Routine fungicide at 7–14 day intervals	Yes	Yes	Generally a routine fungicide programme from bud burst to mid-summer but fungicide choice and spray interval adjusted for scab warnings

resistance. In recent years, changes in fungicide legislation in some countries have resulted in restrictions on numbers of applications for some fungicide groups, particularly DMIs, which has meant a curative only strategy is less likely.

In trials over several years in the UK, the most practical approach has been the key stage – curative strategy (Berrie, 1994; Berrie and Xu, 2003). This system achieves similar or better scab control compared to the routine system at lower fungicide inputs. It can also be integrated with pest and nutrient sprays into the orchard protection programme.

*3.2 Apple powdery mildew (*Podosphaera leucotricha*)*

Powdery mildew is one of the most important diseases of apple in the UK, reducing yield and fruit quality on susceptible varieties. All the main varieties grown in the UK are susceptible, especially Cox and Jonagold. The fungus overwinters as mycelium in fruit or vegetative buds, which emerge as mildewed blossoms at pink bud (primary blossom mildew) or mildewed shoot tips at petal fall (primary vegetative mildew). During spring and summer, spores from the primary mildew spread to developing shoots to initiate the secondary mildew epidemic. Mildew colonises fruit buds in June/July and vegetative buds at the end of shoot growth in late summer where it remains quiescent until the following spring. The disease is favoured by warm temperatures above 18°C and humid conditions. Water on the leaf surface is not required for spore germination. So, in contrast to apple scab, provided that there is shoot growth and inoculum, and it is not raining, every day in summer is a potential mildew infection day. Under favourable conditions, the fungus can infect leaves and produce sporing mildew colonies in about 4–5 days. Mildew inoculum level is the key factor in determining the seasonal epidemic. Therefore strategies for control depend on maintaining primary mildew at a low level and hence fungicide protection is essential from pink bud to the end of shoot growth, which can amount to 10–15 sprays per season. Managing the mildew epidemic through careful monitoring of disease incidence is essential to rationalise fungicide use and to check that control measures are effective. Once primary mildew levels are high, effective control becomes difficult.

3.2.1 Mildew monitoring

Since the time between mildew infection and visible mildew can be short (4-5 days), regular monitoring of the disease on apple trees in orchards can be used as a basis for decision making on treatment. As the powdery mildew fungus only infects young leaves, the progress of the mildew epidemic in the orchard can be followed by assessing mildew incidence on the shoot tips. Butt and Barlow (1979) developed a system for mildew management which could be used as a basis for decision making on fungicide use. The objective of mildew management is to adopt a flexible approach in which fungicide dose, spray interval and spray volume are adjusted to match the levels of epidemic activity.

Primary mildew is assessed on blossoms at pink bud and vegetative shoots at petal fall (Table 1). Primary mildew levels of <1% mildewed blossoms or shoots are satisfactory. Levels >2% indicate a problem. Assessment of primary mildew provides information of the success of last season's control programme and indicates the problem for the current season and can be used as a guide for any adjustments needed to the current seasons' planned programme.

Secondary mildew incidence (Table 1) is assessed on the youngest five leaves on two actively growing shoots on each of 10–20 trees per orchard. Growth stage and current and forecast weather are also recorded. Mildew incidence is expressed as per cent mildewed leaves or mildewed shoots. The success of a mildew management system and reducing fungicide inputs without jeopardising disease control are dependent on regular monitoring of secondary mildew, so that changes in epidemic can be responded to and primary mildew build-up avoided.

3.2.2 Mildew warning systems

PODEM (Xu, 1999) is an apple powdery mildew model contained in ADEM, which is a PC-based system, which gives warnings of several apple diseases, including models of apple scab, apple canker and fruit rot and fire blight as well as powdery mildew. The models are driven by one or more of the following weather variables *viz* rainfall, surface wetness duration, ambient temperature and ambient relative humidity obtained from local weather stations. The mildew model forecasts the epidemic of secondary mildew by combining the effects of weather on spore production and dispersal, infection and subsequent colony development on young leaves, with varietal susceptibility and the amount of mildew inoculum recorded in the orchard. The index of mildew risk generated can then be used to assist in decision making on fungicide use. The mildew risk generated is predictive as it forecasts the amount of mildew in the orchard, which is likely to be sporing in 3–4 days-time.

Regular monitoring of the mildew incidence on apple shoots post blossom and the current and forecast weather provide sufficient information for decision making for mildew management without the need to consult disease warning systems. However, by using the mildew risk derived from PODEM in ADEM, which gives information on the future progress of the epidemic, the decisions on fungicide use are more robust.

3.2.3 Making decisions on control

Cultural control is based on removal of primary inoculum by pruning out primary mildew to reduce the inoculum at the beginning of the season as soon as it emerges. This must be done rapidly to limit spread of inoculum. An alternative approach is to eliminate overwintering mildew in buds during the dormant period using surfactant-based treatments. However, currently in the UK, there are no products approved for this approach.

Post-blossom, once shoot growth starts, fungicides for powdery mildew control are generally applied routinely at 7–14 day intervals until shoot growth stops. However, decisions on fungicide choice (protectant or anti-sporulant), dose and frequency of use can be based on regular assessment of the mildew incidence on the youngest five leaves of the shoots (per cent secondary mildew) to give the size of the current mildew epidemic, combined with the amount of shoot growth (rapid or slow) and the current and forecast weather (temperature and humidity). Decisions on control options are summarised in Table 6. The activity of various fungicide groups (protectant or anti-sporulant) are given in Table 4.

*3.3 European apple canker (***Neonectria ditissima***)*

Apple canker is one of the most important diseases of apple in the UK and parts of Europe and on susceptible varieties, it can cause serious losses. The fungus causes cankers on trees and also a fruit rot both in the orchard and in store. The fungus produces spores, either ascospores or conidia on cankers, all year round and there are entry points, either natural such as leaf

Table 6 Guidelines for decisions on fungicide use based on secondary mildew assessments

Disease rating	Mildewed leaves (%)	Mildewed shoots (%)	Action after petal fall
Low	<2-3	<10	In cool weather with rainy spells or when shoot growth is slow, opportunity to reduce fungicide by reducing dose (minimum dose = 25%) or extending spray interval.
Moderate	3-9	10-30	Maintain control. Consider improving programme by reducing spray interval or increasing fungicide dose (not exceeding label maximum) especially if weather is warm and humid and shoot growth rapid.
Potentially high	>9	>30	Improve control immediately especially if shoots are growing, irrespective of weather. Shorten spray interval, increase fungicide dose (not exceeding label maximum), possibly increase spray volume. Consider changing fungicide from protectant to anti-sporulant. Check sprayer is working correctly.

or fruit scars or artificial such as pruning cuts, for infection on the apple tree all year round. The limiting factor is rain and wet seasons, particularly wet autumns, usually result in significant canker incidence in orchards and on fruit. Spores (mainly conidia) spread from cankers in wet weather to infect developing fruit. Most infection of fruit that results in rotting in store occurs in the period from late blossom to two weeks after petal fall (Xu and Robinson, 2010). Some of the fruit infection that occurred at this time may appear in the orchard as visible rots at the calyx end of the fruit. The proportion of the infected fruit that appears as orchard rots varies depending on seasonal conditions. In problem orchards, routine treatments to control canker are required every year.

3.3.1 Disease monitoring

Orchard monitoring for *Neonectria* cankers is included throughout the year (Table 1) to enable decisions to be made on the need to protect fruit from infection in countries where *Neonectria* is a problem in stored fruit and to protect wood from infection at key times such as leaf fall. Inspection of orchards for *Neonectria* cankers during winter pruning and for shoot die back in spring/summer due to canker will give an indication of the problem in orchards. Cankers on trees are more easily seen in the dormant period. Cankers resulting from infection of leaf scars the previous autumn will not become visible until the following spring, usually from blossom time onwards, depending on winter temperatures. Orchards where 25% or more of the trees have cankers are high risk orchards. An assessment of *Neonectria* rot incidence during fruit grading from store will also give an indication of canker incidence in the orchard. A high incidence of fruit with *Neonectria* eye rots in early summer is an indication of likely rot problems in stored fruit. The most important weather factor is rain, particularly at key times for infection such as late blossom/petal fall (fruit infection) or autumn leaf fall (wood cankers).

3.3.2 European canker warning systems

NECTEM is a disease forecasting model for *Neonectria* canker and fruit rot contained in ADEM. The models are driven by the weather variables - rainfall, surface wetness duration, ambient temperature and ambient relative humidity. The models use the weather data to determine the favourability of the weather for *Neonectria* infection of fresh leaf scars and near-mature fruits and indicate the incidence of disease likely to occur at these two infection sites. The leaf fall model assumes that the inoculum level is high and varietal susceptibility to *Neonectria* is high. In addition to weather, the system requires inputs on the age of leaf scars based on the date for 10%, 50% and 90% leaf fall.

A warning system for infection of leaf scars has also been developed in Chile (Latorre et al., 2002). Warnings are generated on leaf scar infection from inputs of temperature, rain, leaf surface wetness and relative humidity.

Currently it is difficult to see how these warning systems can be used in practice as the warnings give information on infection that has occurred and fungicides that are effective against *Neonectria* are limited and none are curative in action.

3.3.3 Making decisions on control

Effective control of canker requires an integrated approach, combining cultural methods and fungicide applications at key times. In problem orchards, routine treatments are required every year. In winter, cankers should be pruned out where possible or cut back to healthy tissue. The wounds should be treated with a recommended canker paint immediately after pruning. It is especially important to ensure wounds on young trees are painted. Infected prunings are a source of inoculum and should be removed from the orchard and burnt or macerated up to enhance decay. Autumn leaf scar cankers that appear in summer as shoot die back should be removed as soon as possible to reduce inoculum.

In the UK, effective fungicides such as carbendazim or copper are either no longer permitted or their use restricted. Alternative fungicides are much less effective. Therefore decisions on fungicide use are best timed to key stages to protect against infection, rather than disease warnings generated by models. Key stages for protection against *Neonectria* infection of wood are bud burst (bud scale scars), fruitlet drop (fruit scars), summer leaf fall (leaf scars), harvest (fruit scars) and autumn leaf fall (timed at 10%, 50%, 90% leaf fall to protect leaf scars). Fungicides applied for control of apple scab such as captan, dithianon or dodine also give some protection against *Neonectria* infection.

Orchards at risk from *Neonectria* rot can be identified in spring based on the incidence of cankered trees in the orchard (<5% = low risk, 5–25% = moderate risk, >25% = high risk) and the rot history from pack house records. Orchards at risk should be treated at the end of blossom and 2 weeks later with a further spray before harvest if the summer has been wet. The risk of *Neonectria* rot in store can then be further assessed based on the rainfall between blossom and harvest (Table 7). As *Neonectria* rot does not usually appear in stored fruit until 2–3 months after harvest (usually December), in orchards where a high canker risk has been identified (>25% trees with canker), the best option may be to avoid chemical treatment and schedule the fruit for early marketing before Christmas to minimise losses. The *Neonectria* risk of fruit from lower risk orchards is based on the amount of rainfall from blossom-harvest. In seasons when rainfall is above average (average rain fall per month in the UK is about 40 mm), this fruit may also need to be scheduled for early marketing if sprays have not been applied in blossom.

3.4 Sooty blotch and fly speck

Sooty Blotch and Fly Speck are two distinct diseases of apple and pear fruit but often occur together. Both diseases are caused by fungal disease complexes. The fungi cause characteristic symptoms of sooty or black shiny

Table 7 Orchard canker risk and action for *Neonectria* fruit rot

Orchard canker risk	Rain blossom-harvest	Action
High (>25% of trees cankered)	<Average	Apply sprays in blossom or market fruit pre-Christmas if no sprays applied
	Average	
	>Average	
Moderate (5–25% trees cankered)	<Average	Low risk No special action needed
	Average	Apply sprays in blossom or market pre-Christmas if no sprays applied
	>Average	
Low (<5% trees cankered)	<Average	Low risk No special action needed
	Average	
	>Average	Apply sprays in blossom or market pre-Christmas if no sprays applied

Average monthly rainfall in UK is around 40 mm.

dots on the surface of near mature fruit. Both pathogens colonise the waxy surface of the fruit but do not penetrate lower layers. In the UK, the diseases occur sporadically in well-managed orchards sprayed conventionally, but can cause serious losses in wet seasons in organically managed orchards or where fungicide use is reduced. In other countries such as USA, the diseases are more prevalent in the humid fruit growing regions and can cause significant economic losses due to the unsightly appearance of the fruit that makes them unmarketable. The fungi overwinter on apple twigs and on many hedgerow and wild tree species. In spring, spores produced from these sources spread in wind and rain from late pink bud onwards to colonise developing fruit in the spring and early summer. Secondary spread from these early colonies occurs throughout the summer. Growth of these fungi is favoured by high humidity and temperatures from 18°C to 27°C. However, visual symptoms on fruit do not appear until 3–6 weeks after initial colonisation (Brown and Sutton, 1995).

3.4.1 Disease monitoring

As disease symptoms do not appear until 3–6 weeks after initial colonisation, orchard monitoring as a basis for decision making is not practical. However, identifying the presence of the problem in one year can be used as an alert for treatment for future years.

3.4.2 Warning systems

There are several disease models, which are variations on the original model of Brown and Sutton (1995) to predict sooty blotch and fly speck infection periods. RIMpro includes a model based on leaf wetness, rain and temperature (Trapman, 2004). It numerically grades risk and degree of infection to serve as a warning system and allows sprays to be more targeted.

3.4.3 Making decisions on control

In orchards where the problem has been identified, pruning trees to allow light and air to enter the tree canopy to speed up drying is an important cultural method to improve control. In the UK, in well-managed conventional orchards, which receive frequent fungicide applications for scab control, specific control measures for sooty blotch and fly speck are not usually required. In organically managed orchards or in countries where the disease complex is more prevalent, control is usually achieved by routine fungicide applications from early summer. Use of a warning system allows the sprays to be more targeted.

3.5 Storage rots

Apples are generally cold stored for up to 6 months or more, depending on cultivar, to regulate the supply of fruit on to the market and ensure apples are available for most of the year. Apples in long-term storage are generally stored in controlled atmosphere (low oxygen and carbon dioxide) and this combined with low temperature reduces rot development as well as maintaining fruit quality. Fungal rots are responsible for significant losses in stored apple in most seasons in the UK (Berrie and Lower, 2012). They can be divided into two broad groups: those causing rots primarily after harvest (store diseases) and those that also cause rots in the orchard (orchard diseases). Rotting in store due to the latter group mainly results from fruit infection that occurred before harvest, but remained symptomless and subsequently developed in store. The store diseases are usually wound rots, which gain entry to damaged fruit at harvest. The fungi, and their main characteristics, are given in Tables 8 and 9.

Apple varieties vary considerably in their susceptibility to rotting. In the UK, Cox is most susceptible to rotting. Losses in other varieties such as Bramley, Braeburn, Gala or Jonagold are usually much lower, but occasionally certain fungi, e.g. *Phytophthora syringae*, can cause significant losses in these varieties. Orchards differ considerably in actual losses due to rots and the main fungi responsible. Information on losses in store and the rots responsible for an orchard is important if losses are to be minimised and the appropriate control measures applied. This information can be obtained from assessment of rots in the pack house during grading.

There are several important factors that affect fruit susceptibility to rotting:

Fruit maturity – correct harvest date is important as over-mature fruit are more prone to damage and rotting.

Handling – poor handling of fruit at harvest increases the risk of wound fungi such as *Botrytis*, *Penicillium* and *Mucor*.

Mineral composition – fruit low in calcium and high in potassium and nitrogen are more prone to rots particularly *Gloeosporium* rot in Cox. Therefore, correct mineral composition is important for varieties such as Cox and Bramley.

Weather – rainfall is the most critical factor in determining infection of apples by fungi. Wet summers and harvests usually result in high rot incidence in store.

Until recently, post-harvest fungicide drenches would have been used almost exclusively for control of rotting in store. This approach, though scientifically sound and with minimal risk to the operator and to the environment compared to pre-harvest orchard sprays, is no longer acceptable to the consumer and hence

Table 8 Storage rots (orchard diseases)

Disease	Varieties	Source	Entry	Weather factors	Cultural control	Fungicides
Brown rot (*Monilinia fructigena*)	All varieties	Cankers, mummified fruit on orchard floor and tree	Initially by wounds. Spread by contact	Warm summer + rain	Prune out cankers Remove/collect mummies Good control of pest and disease Avoid fruit damage Close supervision of pickers to ensure no damaged fruit is stored	Pre-harvest spray depending on risk
Black rot (*Botryosphaeria obtusa*)	Cox Fiesta + others	Cankers, mummified fruit, dead twigs, prunings, weeds	Direct and wounds	Warm (20°C) + rain, humid	Prune out cankers and dead wood Remove/collect mummies Avoid piles of prunings in orchard	At present not important enough in UK to require special measures
Gloeosporium rot (*Neofabraea* spp.)	Cox, Gala Jonagold Egremont Russet	Cankers, pruning stubs, mummified fruits, weeds	Direct, through lenticels, eye, stalk, russet cracks	Warm weather + rain	Prune out cankers, dead stubs and die-backs Remove mummies	Pre-harvest sprays July–September
Phytophthora rot (*Phytophthora syringae*)	All varieties	Soil splash	Direct entry, spread by contact	Wet especially near fruit picking	Maximum width grass strip Mulch to cover soil to reduce soil splash Remove low hanging branches Selective fruit picking Not muddying bulk bins Encourage earthworms to remove leaf litter	Post-harvest urea to trees to encourage leaf rotting Pre-harvest sprays

Neonectria rot (*Neonectria ditissima*)	Cox, Bramley, Gala Egremont Russet	Cankers	Direct through lenticels, stalk end, eye end	Wet autumn at leaf fall. Wet spring, summer	Remove cankers and macerate with prunings	Post-harvest copper sprays to protect leaf scars, sprays at bloom and petal fall Pre-harvest sprays
Diaporthe rot (*Diaporthe perniciosa*)	Cox, Bramley, Jonagold, Gala	Cankers, dead wood, mummies	Direct on mature fruit long-term stored	Warm and wet	Remove cankers, dead and weak wood during pruning	Rot usually not important enough to merit special control measures
Botrytis eye and core rot	Cox	Ubiquitous especially dead plant material, for example flowers	Direct from dead petals to eye or to core. May require mature fruit to rot	Humid or wet	None	Possibly sprays during bloom
Fusarium rot, *Fusarium* core rot (*Fusarium* spp)	Cox, Bramley, Cameo	Orchard soil, plant debris, cankers, dirty bulk bins	Wounds Through open calyx at blossom	Rain, warm weather	Prune out old *Neonectria* cankers and pruning stubs.	Usually not important enough to merit special attention on most varieties. On Bramley and Cameo apply fungicide sprays at blossom and petal fall

Table 9 Storage rots (store diseases)

Disease	Varieties	Source	Entry	Weather factors	Cultural control	Fungicides
Botrytis rot (*Botrytis cinerea*)	All varieties	Plant debris, orchard soil, dirty bulk bins	Wounds. Infected stems	Store disease rapid growth at low temperatures	Careful picking to avoid wounds Good control of pest and disease Avoid muddying bulk bins and introducing debris such as leaves Clean bulk bins	Pre-harvest sprays may give some control
Blue Mould (*Penicillium expansum*)	All varieties especially Bramley	Plant debris, orchard soil, dirty bulk bins	Wounds, direct entry on over-mature fruit	Store disease	Careful picking and handling to avoid damage Good control of pest and disease Avoid muddying bins and introducing debris such as leaves Clean bulk bins	Post-harvest dips/ drenches aggravate the problem
Mucor rot (*Mucor* spp)	All varieties	Orchard soil, dirty bulk bins	Wounds or through open calyx in drench solution	Store disease rapid growth at low temperatures. Wet harvesting	Avoid muddying bins and introducing debris Clean bulk bins Remove fallen fruit from orchard after harvest to reduce fungus population in soil Store dry fruit	No fungicides are effective against *Mucor* Post-harvest dips/ drenches aggravate the problem

markets, because of the high residue remaining on fruit after treatment, which does not decline in store. In most countries, there are no longer any approved products for use on apples as post-harvest fungicide treatments. Therefore, the emphasis must be on integrated control and alternative methods, which are applied in the orchard. Research in the UK has shown that apple orchards vary considerably in the actual losses in store and the main fungi responsible, depending on variety and various orchard factors. Orchard monitoring is an important part of storage rot management so that information on losses due to rots and the fungi responsible can be collected for each orchard. This information can then be used to determine risk and provide a basis for decision making.

For some apple varieties, where physiological disorders are important such as bitter pit and scald, the most important pre-harvest factor to assess is the suitability of the fruit for storage. Physiological disorders can result in the loss of the whole store of fruit in extreme cases, whereas losses due to rots are generally much lower than this.

3.5.1 Disease monitoring and rot risk assessment

Rot risk assessment (Berrie, 2000) was developed at NIAB EMR for the UK, (but the principles are applicable to most countries) to (1) identify those orchards with fruit likely to have significant rot in store so that appropriate measures can be taken to minimise losses and (2) to avoid unnecessary treatment on orchards where a minimal risk of rotting has been identified. Rot risk assessment is a whole season system, requiring assessment of factors from bud burst to harvest and post-store in the pack house. The key factors assessed are given in Table 10. Since some fungicide treatments for rot control may need to be applied early in the season, assessment of orchard factors needs to be done initially pre-bloom and repeated nearer harvest. Near harvest assessments of inoculum are particularly important for rots such as brown rot, the incidence of which can change very rapidly as the fruit matures.

A general walk through the orchard should be done examining trees for canker, low hanging fruit, rot incidence etc. A more formal assessment should be included on at least 20 trees selected at random following a W pattern across the orchard if possible. The rots and relevant criteria are given in Table 11. Keeping records for individual orchards, of losses due to rots, and the fungi responsible, when the fruit is being graded from store can provide valuable information on the current problems in the orchard and the ones likely to occur in future. A sample of at least 100 rotted fruit selected at random from the rot bin when the fruit is being graded should be sufficient to give a reasonable assessment. In this way a database can be built up for each orchard providing information to assist in decisions on treatment and storage potential.

Table 10 Rot risk factors

Factor	How assessed	When assessed	Fungal rot affected
Rainfall	Nearest weather station or on farm weather station	Daily rainfall blossom to harvest	*Neonectria* rot - Rain blossom to harvest *Botrytis* rot - Rain blossom to harvest *Phytophthora* rot - Rain in 15 days pre-harvest *Gloeosporium* rot - Rain in month before harvest
Fungal inoculum	Orchard assessment	Bud burst to harvest	*Neonectria* rot - Canker incidence on trees Brown rot - Incidence of mummies in spring and new rots pre-harvest
Orchard type	Incidence of bare ground in herbicide strip under trees and low hanging fruit	One month before harvest	*Phytophthora*
Orchard rot history	Sampling rot bin in pack house	After grading orchard fruit out of store	All rots
Fruit storage potential	Mineral composition, fruit firmness	Random sample of 20 fruit month before harvest	All rots

The suitability of the fruit from an orchard for storage is the most important aspect of rot risk assessment. If the fruit is not suitable for storage, then it will be scheduled for early marketing and storage rot risk is in most cases irrelevant. The mineral composition of fruits is closely correlated to the level of rots found after medium- or long-term Controlled Atmosphere (CA) storage. Increased levels of nitrogen and potassium are associated with higher levels of rots, while increased calcium levels in the fruit tend to reduce rot incidence. The importance of adequate concentrations of calcium and phosphorous in apple fruit has been well established through research during the past 30 years and fruit mineral thresholds have been defined for ensuring a high probability of freedom from both fungal wastage and physiological disorders. Samples of fruit for mineral analysis should be taken from orchards which may be stored, 2–3 weeks prior to harvest. The suitability of fruit for storage is based on the results of the mineral analysis.

3.5.2 Making decisions on control

Once a rot risk has been identified for fruit from an orchard that is intended for medium- to long-term storage, then decisions on treatment or time of marketing

Table 11 Procedures for assessment of orchard factors affecting rotting in store

Fungal rot	Timing	Factor	Assessment procedure	Criteria for risk
Brown rot	July/August	Incidence of brown rot in orchard	Select 20 trees at random and assess incidence of brown rot on tree and on floor to obtain % fruit with brown rot	Brown rot incidence >1% = high risk
Neonectria rot	April	Incidence of canker	Select 20 trees at random and assess incidence of cankered trees	>25% trees with canker = high risk 5–25% = moderate risk <5% = low risk 0 = no risk
Phytophthora rot	July and August/September	(1) % bare ground	Inspect whole orchard Estimate bare ground under trees taking into account mulch and weed cover	1 100% bare ground (overall herbicide) = high risk 2 Overall grass or mulch or weed cover (0% bare ground) = low risk 3 herbicide strip (20% or > bare ground) = risk
		(2) % crop <0.5 m from ground	Select 20 trees at random and assess % crop <0.5 m from ground	15% or > = risk

can be made. This information for *Neonectria* rot, *Phytophthora* rot and brown rot is summarised in Tables 12 and 13. For Gloeosporium (*Neofabraea* spp.) and *Botrytis* (latent) rot, research in the UK did not identify any clear factors on which risk assessment could be based. Decisions on risk are therefore mainly based on rot history and rainfall. For *Penicillium* rot, *Mucor* rot and *Botrytis* (wound) rot, risk assessment is purely dependent on crop damage. Attention to crop handling and hygiene at harvest should avoid these problems (Table 14).

Where decisions based on rot risk are made at harvest, options for control or minimising losses due to rots are based on reducing the storage time (earlier marketing). Where orchard fungicide sprays are considered, decisions will need to be made much earlier, even at the start of the season. In these cases, decisions can be made based on rot history from pack house records, or from orchard types identified as at risk, e.g. for *Phytophthora* – potential low hanging fruit, a high percentage bare ground or for *Neonectria* rot – a high incidence of cankered trees.

3.6 Fire blight

Fire blight is a serious disease of pome fruits, especially pear, quince and apple and a number of ornamentals of which hawthorn is the most significant because of its widespread occurrence in hedgerows around or near to orchards in the UK. The disease is caused by bacteria – *Erwinia amylovora* – and occurs more frequently in warmer climates, such as the USA where the disease is a problem in most seasons. In the UK, the climate is more marginal for fire blight so the disease occurs sporadically with serious outbreaks occurring only once or twice every 5 years or so. Pears and young apple trees are more susceptible and under favourable warm moist conditions mature pear trees can be killed within 3 days of the initial infection.

First obvious symptoms are dead blossoms or dark brown leaves hanging from a truss or branch, resulting from spring or summer infection of blossoms. The bark of diseased areas or branches (cankers) is dark-green or brown, often water-soaked and with an indistinct margin between healthy and infected tissue. Shoot infection, especially on apple is a characteristic; initially the tips wilt and droop with ooze drops on the stem. Later the leaves turn brown or black on pears.

The bacteria overwinter in orchard trees, hedgerow hawthorns and susceptible ornamentals in bark tissue along the edges of cankers formed from the previous year's infections. In spring these bacteria multiply and produce ooze, which is spread by wind-blown rain and insects to infect flowers in spring or shoot tips. The bacteria can also invade via wounds made by insects or hail or wind and through natural openings such as leaf scars or nectathodes, giving typical symptoms in the tissues. Fire blight risk is greatest when temperatures

Table 12 Apple orchard diseases – Rot management options

Fungal rot	Assessment timing	Orchard Risk	Rot history	Rain	Decision
Brown rot	July/August	Orchard incidence high (>1%)	N/A	N/A	• Selective picking at harvest and/or • Market fruit early
		Orchard incidence low	N/A	N/A	Selective picking at harvest
Botrytis (eye rot)	August	N/A	Moderate-high incidence	Rain June to harvest average or >average	Market fruit by January
				Rain June to harvest <average	No specific action needed
		N/A	Low incidence	Rain June-harvest = average or <average	No specific action needed
Neonectria rot	April	Moderate-high incidence cankered trees	Moderate - high incidence	N/A	Apply fungicide spray in blossom and petal fall
		Low incidence cankered trees	Low incidence	N/A	No fungicide treatment in blossom/petal fall needed
	August/September	Low-high incidence cankered trees	N/A	>Average rain blossom-harvest	
		Moderate-high incidence cankered trees	N/A	Average rain blossom-harvest	Market fruit before January if no sprays applied in blossom/petal fall
		High incidence cankered trees	N/A	<Average rain blossom-harvest	

Table 13 Apple orchard diseases – Rot management options for *Phytophthora* rot

		Orchard risk			
Fungal rot	Assessment timing	% bare ground	% crop <0.5 m above ground	Rain in 15 days pre-harvest	Decision
Phytophthora rot	Mid July	0% = overall grass, weed cover or mulch	N/A	N/A	No sprays needed
		20% or > bare ground	>15%	N/A	Apply fungicide 28 and 14 days pre-harvest or delay action until harvest
	August/September/Harvest	0% = overall grass, weed cover or mulch	N/A	N/A	No specific action as risk very low
				Low or no rain = low risk	No specific action as risk very low
		20% or > bare ground	>15%	20 mm or > = high risk	• Selectively pick at harvest or • Market fruit early to minimise losses if no fungicide treatment applied
			<15%	N/A	No specific action as risk very low

Table 14 Apple storage diseases - wound pathogens - rot management options

Fungal rot	Inoculum sources	Control options
Penicillium rot	Dirty fruit bins Damaged fruit	• Bin hygiene • Selective fruit picking - only sound fruit in the bin • Good harvest supervision
Mucor rot	Dirty fruit bins Damaged fruit Soil contamination	
Botrytis rot	Dirty fruit bins Damaged fruit Plant debris	

exceed 18°C and there is rain. Disease development occurs between 5°C and 30°C with an optimum of 27°C.

3.6.1 Disease monitoring

In the UK there are no recommended chemical control measures for fire blight. Control is dependent on understanding the disease, prevention, vigilance and rapid action when symptoms are seen. In other countries where chemical control is an option, the above are still important in an integrated approach but can be combined with the use of chemical sprays.

Routine inspections for fire blight symptoms should be made at key times - soon after leaf fall, during winter pruning, soon after bud break, mid-June and in late July to early August. Young trees should be checked more frequently. Inspections should also include hawthorn hedgerows and any susceptible ornamentals in the area. Additional inspections should be made if there is frost damage to blossom or following damaging storms, fire blight warnings or if fire blight has been spotted nearby. Once spotted, diseased parts should be removed as soon as possible, ensuring the cut is well below the diseased area, and burnt.

3.6.2 Warning systems and their use in practice

Various fire blight forecasting systems have been devised worldwide, e.g. Maryblyte (in the USA) (Lightner and Steiner, 1992) or the Billings Integrated System 1995 (BIS95) (Billing, 1996; Xu and Butt, 1996) in the UK. Both are weather-based warning systems.

3.6.3 Making decisions on control

How warning systems are used as part of a monitoring system is dependent on the country and the frequency with which fire blight occurs. In other countries, the use of antibiotics is permitted for the control of fire blight. Then warning systems can be used to indicate infection days and the timing of sprays for fire blight control. However, in the UK, the use of antibiotics is not permitted and there are no other recommended chemical control methods. BIS95 was developed for fire blight in the UK and, because of the sporadic nature of the disease, risk days identified are potential infection days, not necessarily actual infection days. For each of these risk days, the model also indicates the earliest date when early fire blight symptoms might be seen, which can be used to time field searches in at-risk orchards to spot infection as soon as possible to limit spread. In the UK, this is the key to managing fire blight as growers can become complacent in checking orchards for fire blight if the disease is not often found. Where UK growers have had problems in pear orchards resulting in the grubbing of many trees, monitoring in subsequent seasons is often done more regularly.

4 Conclusion and future trends

The examples given above have demonstrated how in current 'best practice', orchard monitoring and, where appropriate, disease warning systems can be integrated to make decisions on treatment to control diseases to produce quality fruit with minimum input of fungicides. For apple scab, sooty blotch and fly speck, the computer-based warning systems provide key information to assist in decision making as orchard monitoring alone is insufficient. However, for powdery mildew and storage rots, orchard monitoring is sufficient to provide information on which to make decisions. Disease warning systems are available for European apple canker, but available fungicides are only partially effective and not curative so that decisions are based on key growth stages for protectant sprays and orchard monitoring. Two warning systems are used in fire blight management but how they are used in practice differ. Where chemicals are available for control as in USA, the warning system can be used to time sprays. In the UK, where there are no approved products for control, the warning system is used to time orchard searches for symptoms.

In the future the technology available to assist in decision making will increase. Also, more sophisticated warning systems based on weather and smart spore traps (www.agritech-east.co.uk/fungal-spore-trap) are being developed that indicate whether the pathogen is present and active. In addition, drones will be used in crop monitoring as well as orchard scanning systems. However, it is important to emphasise that while these new technologies will provide additional information to assist in decision making, they will not replace orchard visits to monitor for disease presence.

In the future the availability of fungicides will most likely decline through legislation and public pressure so there will be increased use of biopesticides, biostimulants and elicitors for disease control. The new advances in technology will be important in targeting the few fungicides still available for maximum benefit.

5 Where to look for further information

Rossii et al. (2012) published a review article entitled '*Helping farmers face the increasing complexity of decision-making for crop protection*' which provides some interesting discussions. In addition, up-to-date information and its use in practice is available, usually on the internet, from the Extension departments of Agricultural Universities and Agronomy organisations around the world.

6 References

Berrie, A. M. 1994. Practical experience with the VENTEM™ system for control of apple scab in the United Kingdom. *Norwegian Journal of Agricultural Sciences* 17, 295–301.

Berrie, A. M. 2000. Pre-harvest assessment of the risk of storage rots in Cox apples. *IOBC-WPRS Bulletin, Integrated Control of Pome Fruit Diseases* 23, 159–69.

Berrie, A. M. and Lower, K. 2012. Incidence and losses due to fungal rots in cold-stored apples in the UK. In: Creemers, P. (Ed.), *Proceedings of the 9th International IOBC-WPRS Workshop on Pome Fruit Diseases at Hasselt*, Belgium, 29 August – 2 September 2011. *IOBC-WPRS Bulletin* 84, 265–71.

Berrie, A. M. and XU, X. M. 2003. Managing apple scab (*Venturia inaequalis*) and powdery mildew (*Podosphaera leucotricha*) using Adem™. *International Journal of Pest Management* 49(3), 243–9. doi:10.1080/0967087031000101089.

Berrie, A. M., Saville, R. and Xu, X.-M. 2015. Management and control of apple powdery mildew. *IOBC-WPRS Bulletin* 110, 173–85.

Billing, E. 1996. BIS95, an improved approach to fire blight risk assessment. *Acta Horticulturae* 411(411), 121–6. doi:10.17660/ActaHortic.1996.411.26.

Brown, E. M. and Sutton, T. B. 1995. An empirical model for predicting the first symptoms of sooty blotch and fly speck of apples. *Plant Disease* 79(11), 1165–8. doi:10.1094/PD-79-1165.

Butt, D. J. and Barlow, G. P. 1979. The management of apple powdery mildew: a disease assessment method for growers. In: *British Crop Protection Conference – Pests and Diseases*, Brighton, UK. British Crop Protection Council, pp. 77–86.

Cross, J. V. and Berrie, A. M. 1994. Sampling and assessment of pests and diseases as the basis for decision making in orchards in the UK. *Aspects of Applied Biology* 37, 225–36.

Cross, J. V. and Berrie, A. M. 2001. Integrated pest and disease management in apple production. In: *The Best Practice Guide for UK Apple Production*, pp. 2.1-2.94. Available at: https:ahdb.org.uk/sector/tree-fruit.

Latorre, B. A., Rioja, M. E., Lillo, C. and Muñoz, M. 2002. The effect of temperature and wetness duration on infection and a warning system for European canker (*Nectria galligena*) of apple in Chile. *Crop Protection* 21(4), 285–91.

Lightner, G. W. and Steiner, P. W. 1992. Maryblyt™. A computer model for predicting of fire blight disease in apples and pears. *Computers and Electronics in Agriculture* 7(3), 249–60. doi:10.1016/S0168-1699(05)80023-7.

MacHardy, W. E. 1996. *Apple Scab: Biology, Epidemiology and Management*. APS Publishing, MN.

Mills, W. D. and LaPlante, A. A. 1951. Apple scab. Diseases and insects in the orchard. *Cornell Extension Bulletin* no. 711, 20–100.

Rossi, V., Caffi, T. and Salinari, F. 2012. Helping farmers face the increasing complexity of decision-making for crop protection. *Phytopathologia Mediterranea* 51(3), 457–79.

Sutton, T. B., Aldwinckle, H. S., Agnello, A. M. and Walgenbach, J. F. 2013. *Apple Scab in Compendium of Apple and Pear Diseases and Pests* (2nd edn.). The American Phytopathological Society, p. 11.

Trapman, M. 2004. A simulation program for the timing of fungicides to control Sooty Blotch in organic apple growing. In: *Proceedings of the 11th International Conference on Cultivation Techniques and Phytopathological Problems in Organic Fruit Growing*, pp. 56–66.

Xu, X. M. 1993. PC-based disease warning systems for use by apple growers. In: *EPPO Bulletin for Computerised Advisory Systems for Plant Protection*, Eslöv, 23 November 1992, pp. 595–600.

Xu, X. M. 1999. Modelling and forecasting the epidemics of apple powdery mildew (*Podosphaera leucotricha*). *Plant Pathology* 48(4), 462–71. doi:10.1046/j.1365-3059.1999.00371.x.

Xu, X. M. and Butt, D. J. 1996. Adem™ a PC-based multiple disease warning system for use in the cultivation of apples. *Acta Horticulturae* 416, 293–6.

Xu, X. M. and Robinson, J. D. 2010. Effect of fruit maturity and wetness on the infection of apple fruit by *Neonectria galligena*: infection of apple fruit by *N. galligena*. *Plant Pathology* 59(3), 542–7. doi:10.1111/j.1365-3059.2009.02232.x.

Xu, X. M., Butt, D. J. and Van Santen, G. 1995. A dynamic model simulating infection of apple leaves by *Venturia inaequalis*. *Plant Pathology* 44(5), 865–76. doi:10.1111/j.1365-3059.1995.tb02746.x.

Chapter 10

Breeding fruit cultivars with durable disease resistance

Vincent G. M. Bus and Joanna K. Bowen, The New Zealand Institute for Plant and Food Research Limited, New Zealand; Andrea Patocchi, Agroscope, Switzerland; Giovanni A. L. Broggini, ETH Zürich, Switzerland; Satish Kumar, The New Zealand Institute for Plant and Food Research Limited, New Zealand; and François N. D. Laurens, Institut National de Recherche Agronomique, France

1 Introduction

Pest and disease control is a critical aspect of the successful production of most commercial crops, so resistant cultivars can make a significant contribution to an integrated management strategy. In nature, resistance is the rule – in the form of nonhost resistance where an entire plant species is resistant to all genetic variants of most pathogens (Mysore and Ryu, 2004) – and susceptibility is the exception. These exceptions mostly involve host-specific pathogens and insects with highly adapted infection machinery that make monocultures especially vulnerable to attack, and hence, need some form of treatment to maintain pests and diseases below the threshold levels of economic damage. For example, apple (*Malus* × *domestica*) is a host for at least 70 diseases that cause tree and/or postharvest fruit losses (Sutton et al., 2014). While many of these may be relatively benign and only occasionally require intervention, some of the highly specialised pathogens and pests may devastate crop production if not controlled. In this chapter we discuss the basic requirements – germplasm and breeding tools – for a breeding programme to achieve durable resistances in new cultivars as fast as possible, largely focusing on apple as a case study.

http://dx.doi.org/10.19103/AS.2019.0046.12

As in many crops, wild species provide a rich source of resistance genes for backcrossing into the domesticated apple. Crabapple species, such as *M. × floribunda*, *M. micromalus* and *M. baccata*, were used as sources for mainly scab (*Venturia inaequalis*) resistance when the collective Purdue-Rutgers-Illinois (PRI) programme in the United States started a backcross breeding programme in 1945. This followed the discovery by Dr L.F. Hough that scab resistance was segregating in germplasm about 30 years after Dr C.S. Crandall had raised progenies from a number of crabapples (Crosby et al., 1992). Descendants of particularly *M. ×floribunda* clone 821 have since been distributed worldwide to provide the basis of many regional breeding programmes, which in turn have provided advanced breeding germplasm to other breeding programmes, including in New Zealand (White and Bus, 1999). For much of the last century, pesticides were the primary means of pest and disease control and were applied indiscriminately because they were cheap and easy to use. This started to change when the publication of the book *Silent Spring* (Carson, 1962) exposed the negative environmental effects of the extensive use of pesticides and instigated consumer awareness of the risks of pesticide use in food production. In response, integrated pest management (IPM) strategies were developed to reduce the use of pesticides and to favour biological control. IPM has worked well for managing many insect pests (Walker et al., 2017), but this approach still remains a challenge for disease control. Instead, resistance breeding is seen as the main way forward for sustainable disease management and thus is a high priority target in most crop breeding programmes, both in high-value crops such as apple (Laurens, 1999) and particularly in low-input commodity crops, such as cereals. Pest and disease control using chemical intervention forms a major part of production costs of the commodity crops; therefore, major economic gains can be made by growing resistant cultivars. In apple, those costs are relatively minor (Merwin et al., 1994); hence, an assumed consumer preference for residue-free fruit, enhanced by other socio-economic benefits (Gardiner et al., 2007), has been a major driver for the development of resistant cultivars. Concurrently, stringent regulations around pesticide availability and usage, particularly in Europe, combined with fewer new chemistry introductions are further reducing pest and disease control options for growers.

Reduced availability of new chemistry will lead to heavy use of the few remaining products to control pests and diseases: this almost invariably leads to reduced efficacy, with genotypes often developing multiple resistances when exposed to fungicides with different modes of action. The history of pesticide usage shows many examples of plant pathogen resistance, such as *Penicillium* postharvest moulds on citrus, with genotypes resistant to 2-amino butane, sodium orthophenylphanate and the benzimidazoles (Davé et al., 1980) having been reported nearly 40 years ago. At about the same time, resistance to benzimidazoles was identified in *P. expansum* isolated from pear (Wicks,

1977). The issue is also gradually exhibited in the field as resistance to a range of pesticides has been identified here. This possibly, and controversially, may have been enhanced by IPM practices aimed at reducing chemical inputs by, for example, reducing their application rates or applying them curatively (Beckerman et al., 2015). Other examples of resistance to (multiple) active ingredients in pipfruit are brown spot (*Stemphylium vesicarium*) (Alberoni et al., 2010), apple scab (Chapman et al., 2011; Frederick et al., 2014), fire blight (*Erwinia amylovora*) (Mayerhofer et al., 2009; Tancos et al., 2016) and codling moth (*Cydia pomonella*) (Bosch et al., 2018). This does not necessarily mean that a pesticide has been rendered unusable in the presence of resistant pathogen or pest genotypes as, for example, increased dosage may still be a financially viable option in a fungicide resistance management strategy (van den Bosch et al., 2018). While research is under way to improve the durability of pesticides (Baker et al., 2018), resistance breeding remains a significant option, provided that a basic premise, that is the relevant industry being large enough to commit long-term to the development of new cultivars, is met (Wolfe and Gessler, 1992).

2 Pathways to durable resistance

2.1 The plant defence system

The development of fungicide resistance demonstrates the high evolutionary adaptability of plant pathogens. The boom–bust cycles in the use of rust-resistant wheat cultivars (Hulbert and Pumphrey, 2014) are a well-known phenomenon, where the initially functioning resistance of a novel gene in a new cultivar is rapidly reduced. This is the result of selection pressure, with evolving pathogen races overcoming the resistance within a few years of its introduction and rendering it ineffective. Pests and diseases that are most damaging to a crop are often host specific, that is well adapted. They are able to bypass or suppress the host's pathogen-associated molecular pattern (PAMP)-triggered immunity (PTI) effectuated by external pattern-recognition receptors (PRRs) recognising highly conserved PAMPs and initiating a low-level defence response (Dangl and Jones, 2001). Bacterial pathogens such as *E. amylovora* have developed the type III secretion system (T3SS) to inject effectors into host cells to manipulate the cell's metabolism to their advantage (Vleeshouwers and Oliver, 2014). Fungal and oomycete pathogens can also introduce effectors intracellularly to counteract PTI. However, when the host carries resistance genes, an effector-triggered immunity (ETI) may be activated when the resistance protein recognises, directly or indirectly (Kourelis and van der Hoorn, 2018), a cognate effector (Dangl and Jones, 2001). Many of these plant resistance genes belong to the nucleotide-binding site - leucine-rich repeat (NBS-LRR or NLR) class of genes (Borrelli et al., 2018; Lai and Eulgem, 2018), of which apple carries about 1000 (Velasco et al., 2010). One example is the *Rvi15* scab resistance gene (with a Toll/interleukin-1 receptor (TIR) domain at the amino terminus), which recognises the *AvrRvi15* effector from *V. inaequalis* (Schouten et al., 2014). Another is the *Fb_Mr5* gene from 'Robusta 5' that is effective against the pathogen *E. amylovora* (Fahrentrapp et al., 2013), with a coiled-coil (CC) motif at the N terminus. Races of *E. amylovora* able to overcome the resistance controlled by *Fb_Mr5* have a single base pair mutation in the cognate effector gene, $avrRpt2_{EA}$, which leads to the substitution of a cysteine for a serine enabling virulence on *M. ×robusta* 5 carrying *Fb_Mr5* (Vogt et al., 2013). Like in effector genes, small differences in DNA sequences can show differential functionality in resistance genes, too. As few as six amino acid differences were shown to alter recognition specificities of resistance proteins in the flax–flax rust (*Melampsora lini*) system (Ellis et al., 1999; Dodds et al., 2001). This system is the one for which Flor (1956) demonstrated that gene-for-gene relationships are at the base of these differential interactions with the resistance protein specifically recognising the cognate effector protein. Recently, Flor's theory has been shown to be overly simplistic in some instances. For example, in the wheat–wheat blast (*Pyricularia oryzae*) pathosystem, the effector encoded by the *AvrRmg8* gene is recognised by two resistance proteins, encoded by resistance genes *Rmg7* and *Rmg8* located on homologous wheat chromosomes (Anh et al., 2018). From recent research into the cereal powdery mildew fungus *Blumeria graminis*, it has been hypothesised that resistance conferred by the *Pm3f* gene is effectuated by recognition of an effector in the pathogen, provided that the relevant suppressor is turned off (Bourras et al., 2016; Praz et al., 2018). Other variants of resistance and effector protein interactions are the guard and decoy gene concepts, where resistance gene proteins monitor pathogen effector targets, such as the common bacterial PTI-suppressing protein RIN4 or mimic such a target (van der Hoorn and Kamoun, 2008).

A better understanding of host-pathogen interactions at the molecular level (Stuthman, 2002) will be of great help to the breed for durable resistance, particularly if it aids in predicting the durability of resistance genes. Since both host and pathogen genes contribute to durability, the resistance proteins that recognise those effectors will be the most durable if effectors can be identified that are essential for pathogen virulence, and therefore, cannot be lost or mutated without a crippling effect on pathogen fitness.

2.2 Resistance phenotypes

Resistance reactions are often typified by a hypersensitive response (HR) recognisable as pin-point necrotic reactions as the result of apoptosis (programmed cell death) (Heath, 1998; Lam et al., 2001), that is plant cells self-destructing to stop a biotrophic pathogen in its tracks. This may be complemented by the production of anti-microbial phytoalexins (Anderson, 1991). In apple, several scab resistance genes, such as *Rvi4*, *Rvi5* and *Rvi15*, condition HR, but there are two other predominant resistance reactions seen under controlled environmental conditions that are rarely observed in the field: the stellate necrosis (SN) reaction conditioned by for example *Rvi2* and *Rvi8* and partial resistance involving chlorosis often with limited sporulation, conditioned by for example *Rvi6* and *Rvi12* (Bowen et al., 2011). Resistance initiated by different resistance genes may show phenotypic variation, but they all involve the production of polyphenols in many more cells than those in contact with the pathogen spore and mycelium (Bus et al., 2005a, 2010; Galli et al., 2010; Clark et al., 2014) as visualised with fluorescence microscopy (Fig. 1). *V. inaequalis* sometimes shows fluorescence, too, presumably as an indicator of a stress response (Bus et al., 2010) (Fig. 1b).

Polyphenols have long been implicated in disease resistance to apple and pear scab (Kirkham, 1957), and there has been a steady trickle of reports on their role in resistance to apple against scab (Mayr et al., 1997; Usenik et al., 2004; Slatnar et al., 2012) and fire blight (Pontais et al., 2008; Vrancken et al., 2013; Skłodowska et al., 2018). However, the role of polyphenols is difficult to demonstrate conclusively as they are also synthesised during susceptible interactions (Kalisz et al., 2015; Oliveira Lino et al., 2016), but presumably too slowly and/or at levels insufficient to effectuate resistance. This was, for example, shown in the transcriptional reprogramming in *Arabidopsis thaliana* following *Pseudomonas syringae* infection (Mine et al., 2018). In some cases, resistance may be attributable to specific polyphenols, as suggested for the malusfuran aglycone that functions as a phytoalexin against *V. inaequalis* (Hrazdina et al., 1997). A number of polyphenols have been implicated in fireblight resistance in apple (Skłodowska et al., 2018), and specifically biphenyls and dibenzofurans in European pear (Chizzali and Beerhues, 2012). In stone fruit, higher levels of chlorogenic acid and caffeic acid in peach skin were indicated to contribute to brown rot (*Monilinia fructicola*) resistance, but also caused an increased susceptibility to flesh browning (Gradziel and Wang, 1993). The production of pathogen-inhibiting compounds is part of a complex resistance response that can also involve the generation of reactive oxygen species acting as signalling compounds, hormones and pathogenesis-related and ribosome-inactivating proteins (RIPs) (Zhu et al., 2018), resulting in chromatin modifications (Ramirez-Prado et al., 2018), ion influx and cell wall fortification. These resistance responses involve signal transduction cascades (Oliveira Lino et al., 2016) with components shared between PTI and ETI, in particular mitogen-activated protein kinases and transcription factors (Lee et al., 2017; Peng et al., 2018).

2.3 Differential interactions and environmental conditions

In apple, resistance breeding to date has predominantly been based on so-called major genes, which refers to their strong effect, such as induction of an HR or SN. However, the spectrum of a resistance gene, which is the proportion

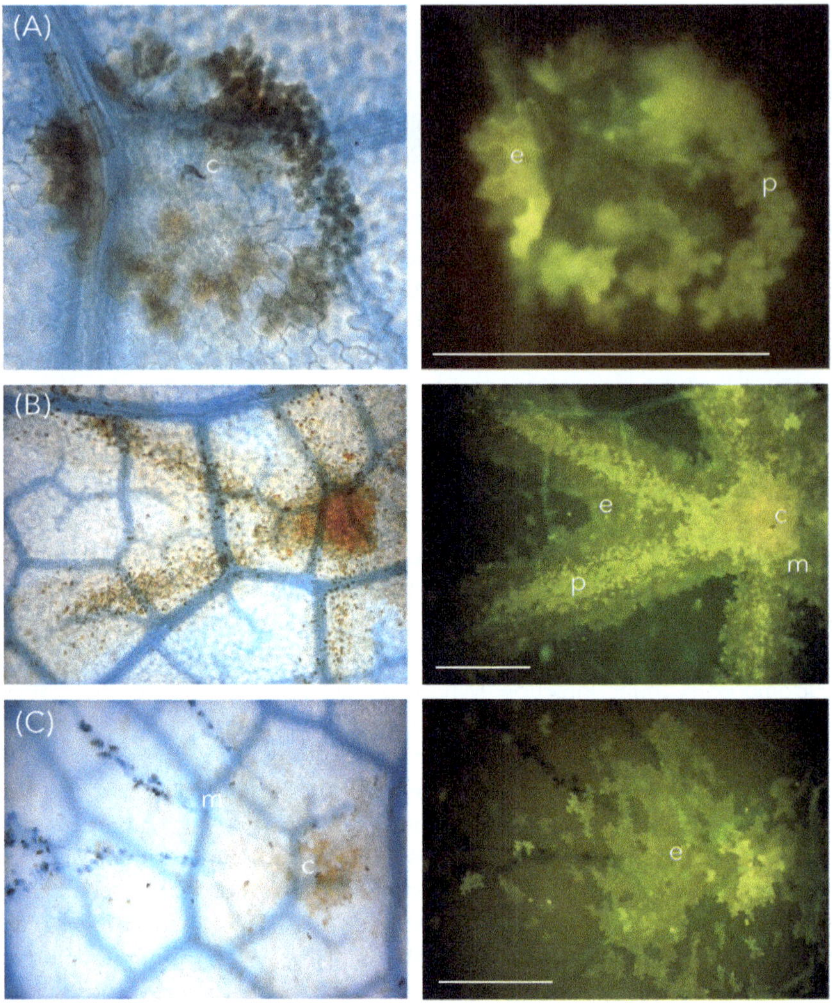

Figure 1 Brightfield (left) and autofluorescence (right) microscopic observations of different types of resistance reactions to the scab pathogen *Venturia inaequalis* on leaves of young apple seedlings: (A) hypersensitive response on progeny BJ55 derived from 'Antonovka' B VIII 33,8; (B) stellate necrosis on progeny BJ927 of *Malus sieversii* GMAL3634; and (C) necrotic/chlorotic reaction on progeny BK758 of *M. baccata jackii*. The resistance reactions resulted from infections by single conidia (c) with the host responding with polyphenol production in the epiderm (e) and palisade parenchym (p). Sometimes the mycelium (m) fluoresces, which indicates of stress in the pathogen, too (Fig. 1b). Scale bars are 1 mm.

of the isolates present in a pathogen population to which the gene is effective, is more relevant for breeding purposes. A broad spectrum, that is effective against most if not all isolates of the pathogen population, is preferred; hence, this was a key criterion in the decision to include resistance genes in the revised scab differential host-pathogen interaction nomenclature (Bus et al., 2011). Nevertheless, the highly susceptible cultivar 'Golden Delicious' was included since the *Rvi1–AvrRvi1* interaction has been a model system for understanding the basis of ephemeral resistance. While it is overcome by the majority of *V. inaequalis* populations (Parisi et al., 2004; Laurens et al., 2004), that is it is a narrow-spectrum gene, it induces strong necrosis in incompatible interactions (Durel et al., 2000), which demonstrates that the effect of a resistance gene is independent of its spectrum. The apple germplasm contains an abundance of narrow-spectrum scab resistance genes in 'susceptible' cultivars, many of which condition strong reactions, including the chlorotic (or 'fleck') type (Boone and Keitt, 1957). The complementary avirulence effector genes for these resistance genes segregate in a Mendelian fashion, showing that they are involved in gene-for-gene relationships. Examples of this are *Vsv* (Bus et al., 2005a) and *Vt57* (Bus et al., 2005b) scab resistance genes in their relationships with *AvrVsv* and *AvrVt57*, respectively (Broggini et al., 2011).

Describing resistance genes in terms of their effect (major or minor) and spectrum (broad or narrow) helps to avoid the confusion that is associated with the terms quantitative or polygenic resistance – typically these are associated with only partial resistance and are assumed to be broad spectrum (Poland et al., 2009). However, many single genes can also confer partial resistance allowing limited pathogen development, while combinations of partial resistance genes can confer complete resistance. The initiation of the resistance response through the recognition of pathogen effectors occurs irrespective of the spectrum of resistance genes. While avirulence to narrow-spectrum resistance genes appears to be maintained in pathogen populations, it is not clear what role they can play in breeding for durable resistances, although there are examples of such genes providing resistance to some multi-virulent races, for example, the *Vd3* gene (Soriano et al., 2009) to *V. inaequalis* genotype EU-NL24 (Caffier et al., 2015).

While some broad-spectrum resistance genes still appear to be completely effective against pathogen or pest populations, differential interactions have been demonstrated for many of those used in apple breeding (Table 1). In other words, races (pathogen) or biotypes (insect) exist that are able to completely overcome a specific resistance gene or combination of genes carried by a host (Fig. 2). However, symptoms are not always clear-cut as resistant trees can show reduced levels of infection by pathogen genotypes that do carry the relevant effector(s) under conditions conducive to disease development, which clouds race determination. Such intermediate responses can be the result of transcriptional dosage variation of both effector genes in the parasite (Tan and Oliver, 2017) and resistance genes in the host (Lai and Eulgem, 2018) under the influence of the environment (i.e. the disease triangle, Agrios, 2005). The environmental effect was nicely demonstrated by the positive correlation

Table 1 Differential interactions reported for major resistance genes/quantitative resistance loci used in apple breeding

Pathogen/pest species	R gene	References
Venturia inaequalis	*Rvi1*	Shay and Hough (1952), Caffier et al. (2015), Masny (2017)
	Rvi2	Shay and Williams (1956), Bus et al. (2005b), Padder et al. (2013), Caffier et al. (2015), Masny (2017)
	Rvi3[a]	Shay and Williams (1956), Bengtsson et al. (2000), Sandskär and Liljeroth (2005), Caffier et al. (2015), Bastiaanse et al. (2015), Masny (2017)
	Rvi4	Shay and Williams (1956), Caffier et al. (2015)
	Rvi5	Williams and Brown (1968), Bengtsson et al. (2000), Caffier et al. (2015)
	Rvi6	Parisi et al. (1993), Roberts and Crute (1994), Lateur et al. (2002), Lefrancq and Lateur (2009), Beckerman et al. (2009), Caffier et al. (2015)
	Rvi7[b]	Roberts and Crute (1994), Parisi et al. (2000), Bengtsson et al. (2000), Lefrancq and Lateur (2009), Beckerman et al. (2009), Caffier et al. (2015)
	Rvi8	Bus et al. (2005a), Caffier et al. (2015), Masny (2017)
	Rvi9	Caffier et al. (2015)
	Rvi10	Caffier et al. (2015), Masny (2017)
	Rvi11	Sandskär and Liljeroth (2005)
	Rvi13	Lefrancq and Lateur (2009), Caffier et al. (2015)
Podosphaera leucotricha	*Pl-m*	Lespinasse (1983)
	Pl1	Urbanietz and Dunemann (2005)
	Pl2	Caffier and Laurens (2005)
Erwinia amylovora	*FBR5*	Norelli et al. (1986)
Eriosoma lanigerum	*Er1*	Giliomee et al. (1968), Rock and Zeiger (1974), Sen Gupta and Miles (1975), Young et al. (1982)
	Er2	Cummins and Aldwinckle (1983)
	Er3	Sandanayaka et al. (2003, 2005)

[a] *Rvi3* is a complex of scab resistance genes, each of them showing differential interactions (Bastiaanse et al., 2015).
[b] In the absence of a specific host (7), host (6,7) *M.* ×*floribunda* 821 was used, therefore the infecting strains are race (6,7) in all studies, but the one by Caffier et al. (2015).

between the incidence of disease symptoms and the total infection pressure over a season in an extended field study on cider apples (Martinez-Bilbao et al., 2012).

Intermediate resistance responses may also be a proxy for the process of race selection, described as an 'adaptive walk' towards the fixing of a mutation into a pathogen population (Frantzeskakis et al., 2018) as race isolates will be preferentially selected and proportionally increase to become the majority in the population. For example, *V. inaequalis* isolate 1638 showed the typical

Figure 2 Differential interactions of four *Venturia inaequalis* genotypes on progeny 2009-077-063 of a 'Honeycrisp' × A248R22T004 progeny carrying one of the 'Honeycrisp' genes (Clark et al., 2014). Genotype EU-NL05 (a) is incompatible, showing a combination of chlorotic and necrotic resistance reactions, and genotypes 1774-1 (b), 1639 (c) and EU-NL24 (d) are compatible, showing extensive sporulation without any macroscopic signs of resistance reactions.

HR conditioned by the *Rvi4* gene in scab differential host (4), but with limited sporulation (Bus et al., 2011), which is very rare as the gene normally stops any infections in its tracks. In orchard studies, *V. inaequalis* adaptation to the host from which they were sampled has been evident in the higher disease incidence in subsequent controlled experiments when compared with other cultivars carrying partial resistances in the same mixed stand orchard (Sierotzki and Gessler, 1998; Barbara et al., 2008). Such resistance erosion can also occur on hosts carrying more than one partial resistance gene (Caffier et al., 2014), including ones for which gene-for-gene relationships have been demonstrated (Caffier et al., 2016). Mutation fixing does not necessarily occur on all resistant hosts (Beckerman et al., 2009) as *V. inaequalis* genotypes do not necessarily accumulate all virulences from the partially resistant cultivars in mixed-cultivar orchards (Passey et al., 2016) into a 'super strain' (Xu, 2012). This may indicate that the accumulation of some virulences are mutually exclusive, for example, by having an effect on pathogen fitness, which at the same time may provide a pathway to durable resistance.

2.4 Durable resistance strategies

Practical definitions for durable resistance are that it 'remains effective during a prolonged and wide spread use in an environment favourable to disease'

(Johnson, 1978) and that it is adequate 'throughout the useful lifetime expected from a variety' (Leach et al., 2001). The perennial habit of most fruit crops therefore provides greater challenges to the breeders than annual crops, since with the latter, responses to disease pressures can be rapidly adjusted through the introduction of cultivars with new resistances.

Some pathogen genotypes can overcome multiple resistance genes (Caffier et al., 2015; Peil et al., 2018) without a significant reduction in pathogenic fitness in spite of these races having to alter or shed effectors to prevent recognition by the cognate resistance genes (Parlevliet, 2002). Nevertheless, some, usually conserved (Zhang and Coaker, 2017) effectors, such as the Pep1 effector protein of the smut pathogen *Ustilago maydis* of maize (Doehlemann et al., 2009), are essential to pathogenicity. Loss of these essential effectors negatively affects pathogen fitness, and hence, identifying them underpins the development of durable resistances in many host-pathogen systems (Leach et al., 2001; Brown, 2015). The so-called 'effectoromics' strategy has become an important tool in resistance gene identification and cultivar breeding (Vleeshouwers and Oliver, 2014). Indeed, this approach is being followed for apple scab resistance. Whilst no effectors from *V. inaequalis* with a role in virulence have yet been molecularly characterised, candidate effectors have been catalogued (Thakur et al., 2013; Deng et al., 2017), including orthologues of known effectors from related pathogens being identified, such as AvrLm6 from *Leptosphaeria maculans* and Ave1 from *Verticillium dahliae* (Shiller et al., 2015; Deng et al., 2017). Numerous small, secreted proteins (SSPs), predicted to be synthesised during infection since their encoding genes have higher expression levels *in planta* than *in vitro*, have been identified, too. A subset of these putative SSPs is predicted to have identical sequence in genotypes capable of causing disease on both apple and loquat. Indeed, several are also predicted to be present in *V. pirina*, the causal agent of European pear scab, albeit with a lower level of similarity. This suggests that these conserved candidate effectors have an essential role in pathogenicity of these related fungi on *Maloideae*, contributing to their ability to colonise the subcuticular space.

A second strategy for durable resistance is the pyramiding of resistance genes involved in ETI through marker-assisted breeding (MAB) (see Section 4.1). As in the major agricultural crops, the strategy in fruit crops typically requires the stacking of genes with both large effect and broad spectrum, but the variant of pyramiding large effect and partial resistance genes is suggested to be at least equally effective (Palloix et al., 2009; Brun et al., 2010; Lasserre-Zuber et al., 2018). There are occasional examples of single-gene resistances having lasted, but the general rule is that they do not. Examples of durable genes are the *Sr2* and *Sr26* black stem rust (*P. graminis*) genes in wheat and the *mlo* powdery mildew (*Blumeria graminis*) gene in barley (Johnson, 2000;

Acevedo-Garcia et al., 2014). A most interesting case is the *Lr34* adult plant resistance gene, proving to be durable against leaf rust (*P. triticina*), yellow/stripe rust (*P. striiformis* f.sp. *tritici*) and powdery mildew simultaneously for over 50 years in wheat (Krattinger et al., 2009). In general terms, at least for yellow/stripe rust, it appears that genes conferring immunity or partial resistance are equally likely to be durable (Johnson, 2000). However, co-evolutionary theory predicts that partial resistances should be more durable (Brown, 2015). It is the spectrum of the resistance gene that is associated with the change in composition of the pathogen population, as was shown for apple scab (Lê Van et al., 2013). The wider the spectrum of a gene, the stronger the reduction in the diversity of the pathogen population will be: if there is no fitness penalty to having acquired the virulence, the virulent genotypes will proliferate and render the resistance ineffective (Brown, 2015).

As shown in Table 1, races/biotypes have been identified for most resistance genes used in apple breeding; hence, gene pyramiding is essential. A guiding principle for maximum durability is that complexity of the resistance mechanisms should match the complexity of specialisation in the pathogen (Stuthman, 2002). As not all genes and quantitative trait loci (QTLs) contribute equally to resistance pyramids (Mundt, 2018), detailed knowledge how they affect different aspects of pathogen development in order to reduce colonisation (e.g. infection rate, latent period) and reproduction rates (e.g. spore production) is useful (Bourget et al., 2015). In apple, for example, the 'Fiesta' and TN10-8 scab QTLs have different effects on pre- and post-penetration development of *V. inaequalis*, with their combined effect virtually inhibiting all development (Laloi et al., 2017). Similarly, different apple accessions were found to affect different biological parameters of woolly apple aphid development, such as settlement, reproduction and survival rates (Sandanayaka et al., 2005).

Increasing the number of genes/QTLs in a pyramid does not necessarily increase the level of resistance (den Boer et al., 2014), accentuating the need to understand which aspect of pathogen development is affected by the resistance gene/QTLs to avoid the hit-and-miss approach in the absence of such knowledge. A successful example is the crown rust (*P. coronata* f.sp. *avenae*)-resistant oat line MN841801, which carries two complementary genes for partial adult plant resistance that has remained effective for over 30 years, although there are indications for other resistance factors being involved (Nazareno et al., 2018). An unsuccessful example is the oat cultivar 'AC Assiniboia' carrying three *Pc* genes to the same disease, where the resistance was overcome 4 years after its introduction (Leonard, 2007). Such findings accentuate the need to understand not only the function of the resistance gene, but also the virulence diversity in the relevant pathogen population (Stam and McDonald, 2018).

Most apple scion cultivar breeding for durable resistance is still at the pre-commercial stage of typically pyramiding only two genes, for example, *Rvi6* with *Rvi2* or *Rvi4* for scab resistance (Kellerhals et al., 2013; Bus et al., 2017). Fruit breeders are looking to the progress made in crops like cereals and potatoes to see how this strategy of gene pyramiding delivers for high-risk pathogens that have large population sizes as well as high reproduction rates and gene flow (McDonald and Linde, 2002). Rice breeders, for example, have been pyramiding at least three genes for bacterial blast (*Xanthomonas oryzae* pv. *oryzae*) resistance for a number of years, including the use of both dominant and recessive genes (Singh et al., 2001). Similarly, an association mapping study of a population of 250 elite wheat breeding lines showed that over half of them comprised at least three black stem rust *Sr* genes, with one line postulated to carry nine genes (Edae et al., 2018).

Virulent genotypes may already exist in pathogen populations following either *de novo* development through mutations or through introduction from non-agricultural disease reservoirs (Lemaire et al., 2016; Sitther et al., 2018), where they will not have been intensively selected for in the absence of temporal and/or spatial exposure to their relevant hosts. Their further dispersal may be accelerated by human-mediated gene flow, which can lead to greater population diversity and therefore increases the pathogen's adaptive capacity (McDonald and Linde, 2002; Meng et al., 2018). Like disease populations, insect populations can be highly dynamic too (Zhang et al., 2018), which shows that breeders need to keep track of changes in the populations in order to be able to respond to them (Nazareno et al., 2018).

The presence of virulent genotypes can be monitored through the use of trap orchards (Lespinasse, 1989) consisting of the differential hosts carrying single genes used in breeding programmes. For apple scab, a differential host set representing mostly major effect resistance genes has been assembled (Bus et al., 2011), with trap orchards having been or being established worldwide (Patocchi et al., 2009) (http://www.vinquest.ch). Even if no virulent genotypes are present, the pyramiding strategy should be applied from the start of a breeding programme. No cultivars with single-gene resistances should be released to prevent the selection of races/biotypes sequentially acquiring additional virulences (Pink, 2002), although commercial demands may overrule that. Naturally, a large supply of resistance genes will provide the breeders with options for pyramiding, which may be facilitated by maintaining natural gene pyramids in germplasm that has emerged as the result of natural selection.

The third approach to durable resistance is the use of nonhost resistance, which has long been recognised as an avenue to achieving durability, although its use in classical breeding is often limited by interspecific sterility (Lee et al., 2017). In its most extreme, and possibly most durable, form, the pathogen does not recognise the host, and hence fails to develop infection structures. At the next level, the pathogen's attempts to infect are prevented by the plant's cytoskeleton providing a physical barrier (Mysore and Ryu, 2004). When the pathogen does establish an interaction with the host, further development may be inhibited by constitutive resistance or through active pathogen recognition of conserved structures by the host (Lee et al., 2017), for example the bacterial flagellum or fungal chitin, resulting in PTI. A very strong resistance response involving callose deposition and tumour-like growths was observed when *E. amylovora* was inoculated on *Arabidopsis thaliana* (Hamdoun et al., 2018). The nonhost resistance response can involve an HR that may involve different resistance cascades than those of ETI (Christopher-Kozjan and Heath, 2003) as well as multiple resistance protein–effector interactions (Lee et al., 2017). An example of the latter is the lettuce species *Lactuca saligna* accession CGN05271, whose pyramid of three moderate level partial resistance genes conferred immunity to downy mildew (*Bremia lactucae*) (Jeuken et al., 2008). Interestingly, when this nonhost resistance was backcrossed to *L. sativa*, recombining the QTLs did not attain immunity, suggesting that the genetics of the nonhost resistance was more complex than first thought (den Boer et al., 2014). Very similar to this lettuce example is the *Venturia* spp. *Pyrus* pathosystem, in which putative nonhost interactions have proven to be very useful for scab resistance breeding in interspecific pear (Brewer et al., 2009). The European pear (*P. communis*) and Asian pear (*P. pyrifolia* and *P. ×bretschneideri*) species commonly used in breeding are mutually exclusive hosts for the Asian (*V. nashicola*) and European (*V. pirina*) scab species, respectively, with the F1 pear generation being virtually immune to both scab species. With seven QTLs identified in a family from a cross between an F2 and an F3 interspecific parent, the resistance involves differential and non-differential genes (Won et al., 2014): further research is required to characterise them as host or nonhost genes. Similarly, both types of genes were also identified in peach with nonhost resistance introduced from almond (Martínez-García et al., 2013). The relative contribution of either PTI or ETI to nonhost resistance is thought to be related to the evolutionary relationship between host and nonhost. If the plant species are phylogenetically divergent, then PTI is thought to predominate, whereas in more closely related species, ETI is predicted to be more important (Schulze-Lefert and Panstruga, 2011). An understanding of the molecular interactions underlying nonhost resistance would facilitate its use in a breeding programme and its subsequent deployment.

3 Resistance breeding targets

Some of the most serious diseases, for example apple scab and powdery mildew (*Podosphaeria leucotricha*), have hitch-hiked with their host to all corners of the world, or were inadvertently introduced at some stage and slowly spread over a country or continent (e.g. fire blight in Europe, Van der Zwet and Keil, 1979) and woolly apple aphid (*Eriosoma lanigerum*) in China

(Zhou et al., 2015). Others have remained relatively localised, for example Valsa canker (*Valsa ceratosperma*, *V. mali*) in China, Japan and Korea (Proffer, 2014) and cedar apple rust (*Gymnosporangium juniper-virginianae*) in eastern North America (Aldwinckle, 2014). However, in the world of free trade, there is the constant risk of pests and diseases spreading, even with stringent border controls in place, as demonstrated by the more recent introductions into Europe and the Americas of the apple leaf drop disease or Marssonina blotch (*Diplocarpon mali*, anamorph *Marssonina coronaria*) from Asia (Takahashi et al., 2014; Hinrichs-Berger and Brüstle, 2015) and *Monilinia fructicola*, which causes brown rot of stone fruit, from North America (Villarino et al., 2012; Jänsch et al., 2012). An example in insects is the highly problematic brown marmorated stink bug (*Halyomorpha halys*), with its indiscriminate polyphagous feeding habits that are damaging to a wide range of crops, from Asia to both the United States and Europe (Kriticos et al., 2017; Valentin et al., 2017). Regional resistance breeding programmes therefore will have their specific as well as internationally shared targets.

The most basic requirement for resistance breeding is the availability of resistant germplasm; hence, its long-term conservation is important to supply breeding programmes with a sustained stream of resistances and other traits (Byrne et al., 2018). The use of in-plant resistance as an alternative to the crude fungicides used in the early days of commercial apple production was considered by Aderhold (1902) who consistently rated 'Antonovka', 'Dean's Codlin', 'Doppelter Holländer', 'Heinemann's Schlotterapfel', 'Parmaine de Pless' and 'Rothgestreifter Sämling' as highly scab resistant in a 5-year study. Ever since, breeders have evaluated apple germplasm apple to identify sources of resistance, sometimes involving hundreds of accessions or seedling populations being assessed for multiple diseases, typically for apple scab, powdery mildew and/or fire blight (Schmidt, 1938; Mihatsch and Mildenberger, 1966; Alston, 1969; Pitera and Bogdanowicz, 1992; Fischer and Dunemann, 2000; Bus et al., 2002; Luby et al., 2002). Examples of extensive evaluations for regional intractable pests and diseases are woolly apple aphid (Cummins et al., 1981), European canker (*Neonectria ditissima*) (Lateur and Populer, 1994) and cedar apple rust and frog eye leaf spot (*Botryosphaeria obtusa*) (den Boer and Green, 1995).

Resistance evaluations will only be meaningful if performed under climatic conditions conducive to the disease on plants that are receptive to pathogen infection. Resistance phenotyping of germplasm for diseases such as scab and mildew in the field is often facilitated by taking advantage of the natural infection period in the disease cycle while withholding the application of fungicides. Being able to replicate those optimal disease conditions in a glasshouse, as is common for *V. inaequalis* and *E. amylovora*, provides more flexibility in timing of resistance evaluation than in traditional germplasm evaluation. For

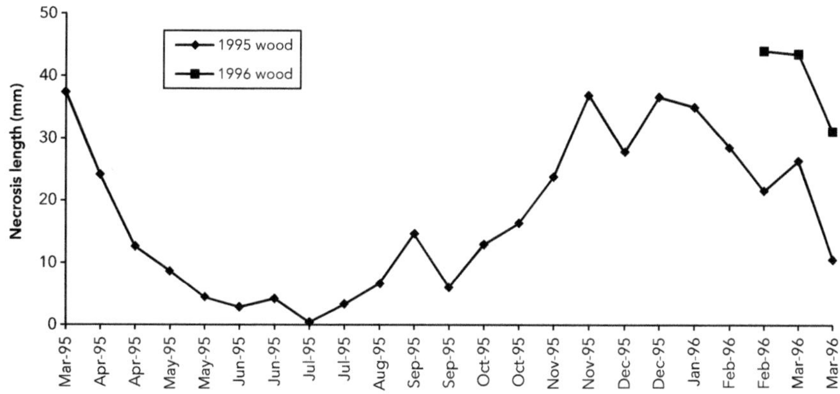

Figure 3 Seasonal variation in shoot susceptibility of 'Cox's Orange Pippin' apple to *Phytophthora cactorum* genotype 133B in an excised shoot test (Jeffers et al., 1981) in Hawke's Bay, New Zealand. The wood is most susceptible in the November–February period, which coincides with shoot extension during spring–summer.

some diseases, laboratory protocols using excised plant material have been developed. For example, a cut-shoot method developed for crown/collar rot (*Phytophthora cactorum*) (Jeffers et al., 1981; Borecki and Millikan, 1969) was used to study host phenology and pathogen activity (Gates and Millikan, 1972; Browne and Mircetich, 1996) (Fig. 3), which helped determine the optimal timing for germplasm screening, as well as germplasm evaluations themselves (Utkhede, 1986; Utkhede and Quamme, 1988). Similarly, an *in vitro* technique was used to study silverleaf (*Chondrostereum purpureum*) disease phenology in a number of host species (Spiers et al., 1998) and resistance in apple germplasm (Bus et al., 1996).

4 Application of breeding tools

4.1 Marker-assisted breeding

One way of achieving durable resistance is to pyramid several major genes or QTLs against the same pathogen in a single genotype. Once the crossing strategy has been implemented, one needs to identify the seedlings carrying these pyramids. Using genotypes of the pathogen with differential interactions for the resistance genes used in the pyramids is laborious and becomes complicated when pyramids carry three or more resistance genes. It is much more efficient to use molecular markers associated with these resistances and perform MAB, which has the added advantage of enabling early selection when seedlings are still very young, hence improving breeding efficiency.

In apple, molecular markers are available for major, both in effect and spectrum, resistance loci against apple scab (e.g. Gygax et al., 2004; Bus et al., 2005b, 2011; Dunemann and Egerer, 2010), powdery mildew (Markussen et al.,

1995; Evans and James, 2003; James et al., 2005; Bus et al., 2010) and fire blight (Peil et al., 2009; Roux et al., 2010; van de Weg et al., 2017) as well as resistance to insects such as rosy apple aphid (*Dysaphis plantaginea*) (Pagliarani et al., 2016), woolly apple aphid (Bus et al., 2008, 2010) and brownheaded leafroller (Bus et al., 2017). In peach, genetic markers have been identified for the intractable brown rot disease (Martínez-García et al., 2013), green peach aphid (*Myzus persicae*) (Lambert et al., 2016) and root-knot nematodes (*Meloidogyne* spp.) (Bus et al., 2009). In pear, the main focus has been on fire blight (Dondini et al., 2006; Montanari, 2015) and the major insect pest pear psylla (*Cacopsylla pyri*) (Montanari et al., 2015; Dondini et al., 2015), while recently resistance QTLs were mapped for two minor pests, sawfly (*Caliroa cerasi*) and blister mite (*Eriophyes pyri*) (Brewer et al., 2018).

For all these genes, the general approach to marker identification has been genetic mapping in single generation crosses with a single resistance source. However, the usefulness of the markers may be low because they are not readily transferable across generations or do not show linkage with the same resistance in other genetic backgrounds. The genome-wide association approach takes the varied genetic background into account as it involves diverse germplasm that represents many generations of hybridisation, resulting in greater linkage disequilibrium and hence stronger marker–trait associations (Oraguzie and Wilcox, 2007). This approach has become much more feasible with the advent of whole genome genotyping, through either sequencing or single nucleotide polymorphism (SNP) arrays as the preferred method of genotyping germplasm since it delivers an abundance of markers (Kumar et al., 2017; Dougherty et al., 2018). Ultimately, the goal is to identify functional markers that are derived from the gene itself, hence having a 100% success rate as demonstrated by the Vfa2 marker (Bus et al., 2009) derived from the *Rvi6* scab resistance gene (Xu and Korban, 2002).

Over time, the type of molecular markers for use in MAB has changed. Initially in the 1990s, SCAR markers were predominantly used, but the large availability of single sequence repeat (SSR) markers mapped over the whole apple genome during the first decade of the new millennium led to a switch to this type of marker. Because of their defined position on the apple genome, the development of an SSR reference map (Liebhard et al., 2002) greatly increased the efficiency of mapping new resistances, and traits in general, by taking a genome scanning approach (GSA) (Patocchi et al., 2005).

More recently, apple genome sequences (Velasco et al., 2010; Daccord et al., 2017) have become available and, combined with the decreasing costs of DNA sequencing, have allowed the development of SNP chips (Khan et al., 2012; Chagné et al., 2012; Bianco et al., 2014, 2016), making genome-wide marker genotyping more affordable and further boosting the mapping of new loci (Guan et al., 2015; van de Weg et al., 2018; Laurens et al., 2018). Compared with SSR and SCAR markers, SNP markers have the advantage of being automatable in modern high-throughput genotyping systems as well as being cheaper. Hence, specific SNP markers have been developed for previously mapped resistance loci (Padmarasu et al., 2014; Cova et al., 2015; Jänsch et al., 2015; Pagliarani et al., 2016). For a number of loci, KASP (competitive allele-specific PCR) genotyping assays were developed, which were validated and finally applied in a pilot study (Baumgartner et al., 2016). Then, in addition to the optimisation of the MAB procedure, an outsourcing genotyping pipeline was developed and tested, hence providing access to MAB for breeding companies without their own DNA analysis capability.

Developing molecular markers is the first step towards increasing breeding efficiency, the next is to determine at which time point(s) to most efficiently and effectively use them in the breeding scheme. The practical implementation of MAB, when stacking multiple disease resistance genes, adds complexity to applied breeding programmes, as the population size needed for maintaining and fixing multiple genes quickly outgrows the reasonable available resources for MAS. Edge-Garza et al. (2015) developed a decision support system 'MASS (Marker-Assisted Seedling Selection) Efficiency calculator V1.0', which can be used to test application scenarios of MASS in order to identify the conditions in which it is expected to increase the cost efficiency.

Traditional MAB remains ineffective when many genes with little effect segregate and reliable markers are missing. In such situations, genomic selection (GS) for complex traits can complement the MAB for 'major' genes.

4.2 Genomic selection

Phenotypic and molecular breeding approaches must be matched to the genetic architecture of disease resistance: resistance to most pests and diseases ranges from single (major effect) gene to multiple genes with small effects. Major effect resistance genes can be mapped and flanking or preferentially functional markers can be developed for breeding purposes. In contrast, resistance conditioned by multiple small effect genes can be approached similarly to other complex agronomic traits, using breeding methodology developed for quantitatively inherited traits.

GS, as proposed by Meuwissen et al. (2001), involves predicting the total breeding values of selection candidates based on genome-wide genetic markers. GS is best suited for quantitative or polygenic traits that are controlled by many genes of small-to-moderate effect. To implement GS, a training population consisting of breeding lines phenotyped for target traits and genotyped with genome-wide markers is used to design a prediction model for a combination of traits. That prediction model is then applied to a new population of selection candidates that have been genotyped with genome-wide markers.

Using only molecular information in GS early in the breeding cycle rather than waiting for trees to mature to be able to phenotype is especially useful for increasing the rate of genetic gain by reducing the breeding cycle time, increasing the selection intensity and possibly increasing selection accuracy. Factors that affect the accuracy of GS include the heritability of the trait, the rate of linkage disequilibrium (LD) decay, the marker density, the number of individuals in the training population and the genetic relationship between training and selection populations (Muranty et al., 2015). It was observed that the cycle time for GS needed to be only about 30% of the phenotypic cycle to achieve superior gains from implementing this approach in apple breeding programmes (Kumar et al., 2012b).

One of the major benefits of applying GS for quantitative disease resistance breeding is that more breeding candidates can be evaluated, leading to higher cost efficiency and selection intensity. A number of studies have demonstrated the effectiveness of the current GS models to capture and predict the genetic variation for disease resistance, particularly quantitative disease resistance in animal (Vallejo et al., 2017; Barría et al., 2018) and plant (Rutkoski et al., 2015; Arruda et al., 2016; Juliana et al., 2017) species. Nevertheless, while the cost/benefit ratio probably stacks up for annual crops, this may not always be the case in fruit crops for breeders to take up the technology.

4.3 Early flowering and fast breeding

Since wild species are often used as donors of resistance genes, the time taken for their deployment in commercial cultivars can be very long, especially in tree fruits. Breeding new tree fruit varieties is time-consuming because of protracted generation cycles. The breeding and selection of a new apple cultivar can be achieved in a single generation of 15–25 years when established cultivars with a good fruit quality are used as parents in crosses, but can take more than 50 years when a trait from a wild apple has to be introduced over multiple generations (Schouten et al., 2006). Generally, at least five generations are required to remove (most of) the linkage drag of deleterious genes affecting horticultural performance that accompany the gene of interest from the wild apple (Le Roux et al., 2012). In some cases, such as the scab resistance gene *Rvi6*, backcrossing has taken up to ten generations without a new cultivar that could compete with the current susceptible commercial cultivars. The long-lasting juvenile period of fruit trees, which can last 4–10 years (Flachowsky et al., 2009), is a major drawback in this process. Fast-tracking breeding by optimising growth conditions (Volz et al., 2009) or using genotypes genetically modified for early flowering (Flachowsky et al., 2011) are two approaches that allow the introgression of new traits from wild germplasm into pre-breeding material (plants of the fifth generation) in a reduced amount of time.

The early flowering approach currently tested in apple is based on the transgenic 'Pinova' line T1190 (Flachowsky et al., 2009), in which the FRUITFULL-homolog *BpMADS4* of silver birch (*Betula pendula*) is overexpressed. Using this approach, one generation per year has been demonstrated to be feasible (Flachowsky et al., 2011; Le Roux et al., 2012; Schlathölter et al., 2018). This is considerably faster than the fast-track approach, but still has a juvenile period of 2–3 years (Baumgartner et al., 2014). In addition to the extreme reduction in generation time, the early flowering approach based on the T1190 line has the added advantage that breeders can select progeny that do not carry the genetic modification in the final generation, provided the genes are not linked. So, in the earlier generations, progeny carrying both the early flowering gene and the gene of interest are selected as parents for the next cross, but in the last generation, only the seedlings carrying the gene of interest without the early flowering cassette (so-called null segregants) are selected (Fig. 4).

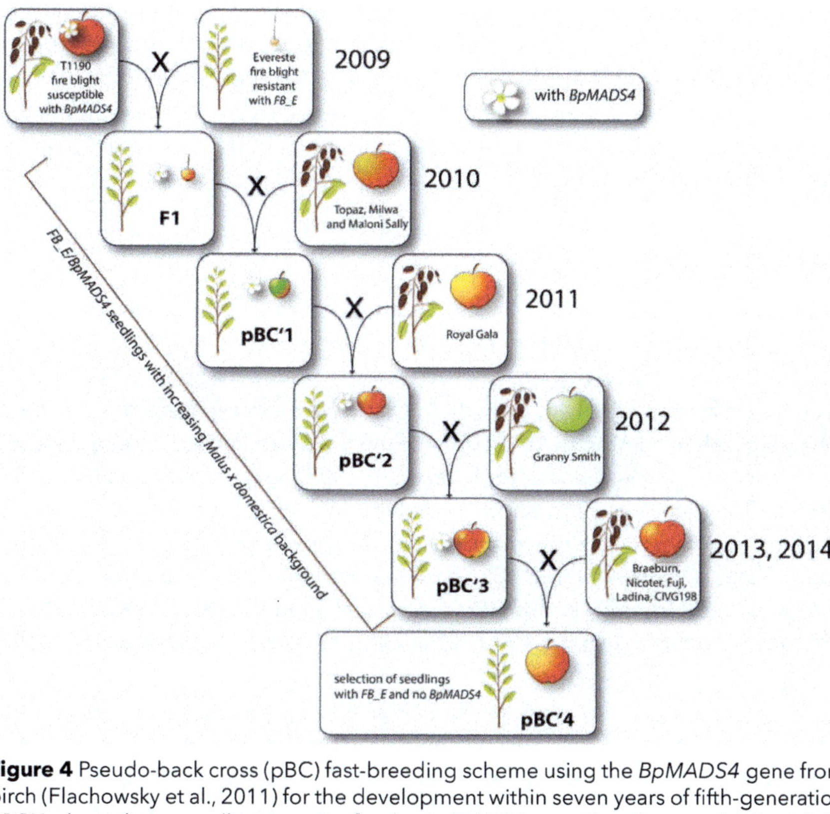

Figure 4 Pseudo-back cross (pBC) fast-breeding scheme using the *BpMADS4* gene from birch (Flachowsky et al., 2011) for the development within seven years of fifth-generation pBC'4 plants that are null segregants for the *BpMADS4* gene, but do carry the fire blight resistance gene *Fb_E* from *Malus* 'Evereste' (Durel et al., 2009). This fast-breeding scheme was described by Schlathölter et al. (2018).

Generally, in both fast-breeding approaches, only foreground selection for the trait of interest is applied at each generation. The undesirable traits from the wild relative introduced into the breeding programme when introgressing the trait of interest are assumed to be progressively eliminated by sexual recombination over the generations by (pseudo-)backcrossing with high-quality apple genotypes as parents. As mentioned above, plants of the fifth generation (i.e. year five from initial cross) are generally considered to have lost most of the unwanted traits; therefore, the best selections can then be used as parents in the final cultivar breeding step. Again, the application of GS can considerably increase the breeding efficiency through the early selection of higher quality parents at each generation (Kumar et al., 2012a,b; Muranty et al., 2015).

The legal status of the final products of the early flowering system (i.e. the null segregants) is not clear everywhere. For instance, in the United States, US Department of Agriculture (USDA) regulators have decided that they will be outside the regulatory authority, as long as these genotypes have been tested for phenotype and molecularly shown to not contain transgenes or pieces of transgenes (USDA, 2011, 2014; McGarry et al., 2017). As a consequence, the method is being used in apple pre-breeding in the United States (Norelli et al., 2015). In contrast, the European Union and other countries have ruled this strategy is not permitted; therefore, this technology remains in the experimental phase in these territories.

4.4 Cisgenesis

Genetic engineering for apple improvement has been investigated extensively in recent years. However, almost all attempts failed to generate products ready to reach the market, with only one exception – Arctic® Apples produced by Okanagan Specialty Fruits in Canada (https://www.arcticapples.com). To date, transgenesis has commonly been used to improve resistance to diseases, such as fire blight in apple (Gessler and Patocchi, 2007), but these plants had to be identified as genetically modified organisms and were often met with caution amongst regulators and the public. Arctic® Apples are also the product of transgenesis, but unlike other programmes, the non-browning apple product was designed to directly benefit the consumer rather than the orchardists. This may be why the product has been accepted in the market place whilst others have failed.

In order to avoid concerns associated with the generation of novel genetic combinations, the exclusive use of genes derived from a sexually compatible donor was proposed (Schouten et al., 2006). This approach is known as 'cisgenesis' and utilises 'cisgenes' that are from the same gene pool as the genes used by breeders in conventional breeding. Cisgenic crops can be considered to be as safe as conventionally bred cultivars, in which the same gene has been introgressed (Andersson et al., 2012). In order to apply cisgenesis for enhancing biotic stress resistance in apple crops, genes with known resistance function are required. Several projects have sought to identify disease resistance genes and the molecular markers linked to favourable traits (as discussed for MAB), determining the position of resistance loci on genetic maps and allowing positional cloning. *Rvi6* was the first apple scab resistance gene cloned (Belfanti et al., 2004), followed by *Rvi15* (Schouten et al., 2014) and *Fb_Mr5* (Broggini et al., 2014). Fine-mapping of other genes, such as *Rvi5* (Cova et al., 2015) and the *Dp-fl* (Pagliarani et al., 2016) as a first step to cloning are in progress in anticipation of cisgenesis becoming a non-regulated technique.

To introgress the cloned gene into a new host, selectable markers aligned with the gene of interest, usually antibiotic resistance genes, are used to screen for genotypes containing the cisgene. However, such markers are unwanted in the final product and therefore different methods to excise them using recombinase activity have been developed, including the pMF1 (Schaart et al., 2004) and p9.FLP vectors. These systems target recombination recognition sites that flank the markers and can be excised (via recombinase activity) once stably transformed plantlets have been identified, leaving the final product containing the desired cisgene but without additional foreign genes. This approach has been used to generate the first *Rvi6*-cisgenic apple lines (Vanblaere et al., 2011; Würdig et al., 2015), which have been assessed in a field trial in The Netherlands (Krens et al., 2015). More recently, cisgenic lines carrying *Rvi15* (Broggini et al., *unpublished data*) or *Fb_Mr5* have been generated (Kost et al., 2015), with the latter being under a second field evaluation in Switzerland (www.protectedsite.ch). Evidence to date reveals that disease resistance can be achieved using cisgenesis, but as *Rvi6*- and *Fb_Mr5*-virulent genotypes have already been identified (Norelli et al., 1986; Parisi et al., 1993), further work is required to develop an enduring disease resistance strategy.

4.5 Genome editing

The genome editing revolution has led researchers to modify genomic sequences using sequence-specific nucleases (SSN), such as zinc-finger nucleases (ZFNs), transcription activator-like effector nucleases (TALENs) and the CRISPR-Cas9 system. ZFNs and TALENs utilise specifically designed modular assemblies of protein elements to target a specified sequence. In contrast, sequence specificity of the Cas9 nuclease is determined by a portion of guide RNA (gRNA) that is taken up by the Cas9 protein to become active. The same Cas9 protein combined with different gRNAs results in different SSNs. When SSNs, such as Cas9, ZFN or TALEN, specifically bind to their target DNA sequences, a double-strand break (DSB) is induced at a specific position and two possible DNA-repairing mechanisms are activated: non-homologous end joining (NHEJ) and homology-directed repair (HDR). NHEJ results in a template-less joining of the DNA ends, which often results in generation of small insertion/deletions (indels) and mutations close to the position of the DSB. In contrast, HDR requires a template-sharing sequence homology to the region flanking the DSB in order to guide the repairing mechanism. HDR occurs at a lower frequency compared with NHEJ (Gao, 2018), and the provided template sequence is copied into the genome replacing the disrupted allele. Amongst pipfruit crops, the CRISPR-Cas9 system is most advanced in apple. For example, Nishitani et al. (2016) stably transformed apple to achieve Cas9-mediated targeted mutagenesis of the phytoene desaturase (*PDS*) gene, which resulted in albino plant lines being produced. A homozygous mutation frequency of 13.6% was achieved for one of the four tested sgRNAs, while two other sgRNA induced no mutations. However, this has only served to demonstrate the possible use of Cas9 nucleases in apple as the resulting product had no improved traits and was still transgenic because of the RNA and Cas9 protein used. There is therefore a drive to utilise Cas9-mediated genome editing in novel systems that do not render the plant transgenic. This is especially important in crops such as apple given the difficulty and time required to breed out foreign DNA without negatively affecting variety traits (Woo et al., 2015). For example, Malnoy et al. (2016) transfected apple and grape protoplasts with pre-assembled ribonucleoprotein complexes (RNPs; *in vitro* assembled sgRNA and Cas9 protein) in order to avoid integration of foreign sequences in the final product. This approach showed that it is possible to obtain nucleotide indels in target genes (e.g. DIPM-1, 2 and 4) without introducing foreign genetic material. However, regeneration of plants from the mutagenised protoplasts was not achieved and highlights a bottleneck in the application of genome editing technology in plant species recalcitrant to regeneration. All experiments showed that it is possible to induce gene disruption via NHEJ, and this mechanism could be used to increase resistance to diseases by disrupting susceptibility genes, such as the apple *MdMLO19* powdery susceptibility gene (Pessina et al., 2016). However, further modification of target sequences utilising HDR and resulting in allele replacement has great potential, for example, to replace non-functional alleles of resistance genes with functional ones. In order to generate superior (and non-transgenic) genome-edited lines, several issues need to be overcome, including the regeneration of whole plants from protoplasts and HDR systems using a suitable (and large) donor template (Gao, 2018). Only then will researchers be able to fully embrace the significant potential of genome editing in apple and other recalcitrant species to achieve durable resistance.

Acceptance of the technology varies between territories. In the United States, where the Arctic apples genetically modified for the non-browning trait (Waltz, 2015) can already be sold, CRISPR/Cas9-edited crops can be cultivated and sold free from regulation as the USDA is giving 'free passes' to gene-edited plants (Waltz, 2018). False flax enhanced for omega-3 oil production and drought- and salt-tolerant soybean are two of the several plant crop examples that are set to reach market without oversight from the USDA. In addition, the USDA has received 57 enquiries relating to editing of crops with regard to requirements for USDA regulation and the majority of these have been granted non-regulated status without the requirement for oversight. In contrast, genome editing continues to be regulated in Europe since it still is regarded as genetic modification (Callaway, 2018).

5 Conclusion and future trends

Durably resistant cultivars are a key element to successful disease management in most crops. Whilst consumers prefer low or even zero pesticide residues on the fruit they eat, they are not necessarily the determining factor in their repeat purchasing decisions, but rather attractiveness and eating quality. The latter two qualities also are the primary breeding objectives for high-value fruit crops in order to meet consumer preferences and expectations, while for growers, high yields and packout rates are important. As genetic gain is still being made in all these traits (R. Volz, *pers. comm.*), introgressing resistances from wild apples has an associated risk that the combined genetic gains in disease resistance and fruit quality may not happen fast enough to catch up and compete directly with high-quality fruit cultivars on the supermarket shelf. This challenge is all the greater when combining multiple resistances – to gain the most benefit – into the one cultivar. The case for using alternative technologies to retrofit resistance gene cassettes into high-quality cultivars rather than through traditional breeding is therefore compelling. Some urgency to improve genetic resistances using new molecular breeding techniques is advocated in order to maintain global food security (Periyannan, 2018), with the threat of climate change and invasive pest species adding to the challenge.

Regardless of the route to durably resistant cultivars, today, and in the longer term, an integrated control strategy will continue to be required, as exemplified by that implemented for high-risk diseases such as late blight in potato (Naerstad et al., 2007) and rusts in the cereals (Burdon et al., 2014), but also for, for example, apple scab (Beckerman, 2015). The application of cultural, biological and chemical methods will continue to play a role in disease control (Mundt et al., 2002; Didelot et al., 2016; Carolan et al., 2017; Xu, 2012) not only to improve efficacy in preventing crop losses, but also to extend the useful lifetime of all pest and disease control strategies. This will include resistance using damage-associated molecular patterns (DAMPs) (Quintana-Rodriguez et al., 2018) and in the case of insect control, herbivore-induced plant volatiles (War et al., 2011).

The breeding for durable resistances based on the active recognition of the pathogen through PTI and ETI inducing a signalling cascade can be complemented by the breeding for additional components of resistance, such as plant architectural and morphological traits that will reduce disease establishment (Niks and Rubiales, 2002). Crop mixture strategies have been shown to have some effect against apple scab (Didelot et al., 2007; Parisi et al., 2013; Xu, 2012). Whilst regarded as less effective when the breeding lines each carry only single-gene resistances (Burdon et al., 2014), they can be made effective if trees of the same cultivar carry different gene pyramids. In outbreeding fruit crops, this is only practical by retro-fitting them with resistance gene cassettes.

An in-depth understanding of the molecular interaction between resistance genes, their cognate effectors and effector targets, underpinned by evolutionary principles applied to host–pathogen interaction research (Upson et al., 2018), opens the intriguing possibility of developing durably resistant cultivars with complementary use of designer resistance genes and traditional breeding strategies. Possible strategies for engineering bespoke resistances involve, for example, the mutation of a resistance gene or effector target, to expand pathogen recognition, or the combination of effector targets, to act as bait, with signalling modules for entirely novel immune receptors (Rodriguez-Moreno et al., 2017; Cesari, 2018).

An integrated approach as outlined above will increase the resilience of fruit production systems and improve environmental, social and economic sustainability.

6 Where to look for further information

Borrelli, G. M., Mazzucotelli, E., Marone, D., Crosatti, C., Michelotti, V., Valè, G. and Mastrangelo, A. M. 2018. Regulation and evolution of NLR genes: A close interconnection for plant immunity. *International Journal of Molecular Sciences,* 19, 1662.

Brown, J. K. M. 2015. Durable resistance of crops to disease: A Darwinian perspective. *Annual Review of Phytopathology,* 53, 513-39.

Lee, H. A., Lee, H. Y., Seo, E., Lee, J., Kim, S. B., Oh, S., Choi, E., Choi, E., Lee, S. E. and Choi, D. 2017. Current understandings of plant nonhost resistance. *Molecular Plant-Microbe Interactions,* 30, 5-15.

Mundt, C. 2018. Pyramiding for resistance durability: Theory and practice. *Phytopathology,* 108, 792-802.

7 References

Acevedo-Garcia, J., Kusch, S. and Panstruga, R. 2014. Magical mystery tour: MLO proteins in plant immunity and beyond. *New Phytologist*, 204, 273–81.

Aderhold, R. 1902. Ein Beitrag zur Frage der Empfänglichkeit der Apfelsorten fur *Fusicladium dendriticum* (Wallr.) Fuck. und deren Beziehungen zum Wetter. *Arbeiten aus der Biologischen Abteilung für Land- und Forstwirtschaft*, 2, 560–6.

Agrios, G. N. 2005. *Plant Pathology*. Burlington, MA: Elsevier Academic Press.

Alberoni, G., Cavallini, D., Collina, M. and Brunelli, A. 2010. Characterisation of the first *Stemphylium vesicarium* isolates resistant to strobilurins in Italian pear orchards. *European Journal of Plant Pathology*, 126, 453–7.

Aldwinckle, H. S. 2014. Cedar apple rust. *In:* Sutton, T. B., Aldwinckle, H. S., Agnello, A. M. and Walgenbach, J. F. (Eds), *Compendium of Apple and Pear Diseases and Pests*. 2nd ed. St. Paul, MN: APS Press, pp. 15–17.

Alston, F. H. 1969. Response of apple cultivars to mildew (*Podosphaera leucotricha*). *Report East Malling Research Station for 1968*, pp. 133–5.

Anderson, A. J. 1991. Phytoalexins and plant resistance. *In:* Sharma, R. P. and Salunkhe, D. K. (Eds), *Mycotoxins and Phytoalexins*. London, UK: CRC Press.

Andersson, H. C., Arpaia, S., Bartsch, D., Casacuberta, J., Davies, H., Du Jardin, P., Flachowsky, G., Herman, L., Jones, H., Kärenlampi, S., Kiss, J., Kleter, G., Kuiper, H., Messean, A., Nielsen, K. M., Perry, J., Pöting, A., Sweet, J., Tebbe, C., Von Wright, A. J., Wal, J. M. and Modif, E. P. G. 2012. Scientific opinion addressing the safety assessment of plants developed through cisgenesis and intragenesis. *EFSA Journal*, 10, 2561.

Anh, V. L., Inoue, Y., Asuke, S., Vy, T. T. P., Anh, N. T., Wang, S. Z., Chuma, I. and Tosa, Y. 2018. *Rmg8* and *Rmg7*, wheat genes for resistance to the wheat blast fungus, recognize the same avirulence gene *AVR-Rmg8*. *Molecular Plant Pathology*, 19, 1252–6.

Arruda, M. P., Lipka, A. E., Brown, P. J., Krill, A. M., Thurber, C., Brown-Guedira, G., Dong, Y., Foresman, B. J. and Kolb, F. L. 2016. Comparing genomic selection and marker-assisted selection for Fusarium head blight resistance in wheat (*Triticum aestivum* L.). *Molecular Breeding*, 36, 84.

Baker, S., Thomson, N., Weill, F.-X. and Holt, K. E. 2018. Genomic insights into the emergence and spread of antimicrobial-resistant bacterial pathogens. *Science*, 360, 733–8.

Barbara, D. J., Roberts, A. L. and Xu, X. M. 2008. Virulence characteristics of apple scab (*Venturia inaequalis*) isolates from monoculture and mixed orchards. *Plant Pathology*, 57, 552–61.

Barría, A., Christensen, K. A., Yoshida, G. M., Correa, K., Jedlicki, A., Lhorente, J. P., Davidson, W. S. and Yáñez, J. M. 2018. Genomic predictions and genome-wide association study of resistance against *Piscirickettsia salmonis* in coho salmon (*Oncorhynchus kisutch*) using ddRAD sequencing. *G3: Genes, Genomes, Genetics*, 8, 1183–94.

Bastiaanse, H., Muhovski, Y., Mingeot, D. and Lateur, M. 2015. Candidate defense genes as predictors of partial resistance in 'President Roulin' against apple scab caused by *Venturia inaequalis*. *Tree Genetics & Genomes*, 11, 125.

Baumgartner, I. O., Patocchi, A., Lussi, L., Kellerhals, M. and Peil, A. 2014. Accelerated introgression of fire blight resistance from *Malus* × *robusta* 5 and other wild germplasm into elite apples. *Acta Horticulturae*, 1056, 281–7.

Baumgartner, I. O., Kellerhals, M., Costa, F., Dondini, L., Pagliarani, G., Gregori, R., Tartarini, S., Leumann, L., Laurens, F. and Patocchi, A. 2016. Development of SNP-based assays for disease resistance and fruit quality traits in apple (*Malus* × *domestica* Borkh.) and validation in breeding pilot studies. *Tree Genetics & Genomes*, 12, 1–21.

Beckerman, J., Chatfield, J. and Draper, E. 2009. A 33-year evaluation of resistance and pathogenicity in the apple scab-crabapples pathosystem. *HortScience*, 44, 599–608.

Beckerman, J. L., Sundin, G. W. and Rosenberger, D. A. 2015. Do some IPM concepts contribute to the development of fungicide resistance? Lessons learned from the apple scab pathosystem in the United States. *Pest Management Science*, 71, 331–42.

Belfanti, E., Silfverberg-Dilworth, E., Tartarini, S., Patocchi, A., Barbieri, M., Zhu, J., Vinatzer, B. A., Gianfranceschi, L., Gessler, C. and Sansavini, S. 2004. The *HcrVf2* gene from a wild apple confers scab resistance to a transgenic cultivated variety. *Proceedings of the National Academy of Sciences*, 101, 886–90.

Bengtsson, M., Lindhard, H. and Grauslund, J. 2000. Occurrence of races of *Venturia inaequalis* in apple scab race screening orchard in Denmark. *IOBC/WPRS Bulletin*, 23(12), 225–9.

Bianco, L., Cestaro, A., Sargent, D. J., Banchi, E., Derdak, S., Di Guardo, M., Salvi, S., Jansen, J., Viola, R., Gut, I., Laurens, F., Chagné, D., Velasco, R., Van De Weg, E. and Troggio, M. 2014. Development and validation of a 20K single nucleotide polymorphism (SNP) whole genome genotyping array for apple (*Malus × domestica* Borkh). *PLoS One*, 9, e110377.

Bianco, L., Cestaro, A., Linsmith, G., Muranty, H., Denancé, C., Theron, A., Poncet, C., Micheletti, D., Kerschbamer, E., Di Pierro, E. A., Larger, S., Pindo, M., Van De Weg, E., Davassi, A., Laurens, F., Velasco, R., Durel, C. E. and Troggio, M. 2016. Development and validation of the Axiom® Apple480K SNP genotyping array. *Plant Journal*, 86, 62–74.

Boone, D. M. and Keitt, G. W. 1957. *Venturia inaequalis* (Cke.) Wint. XII. Genes controlling pathogenicity of wild-type lines. *Phytopathology*, 47, 403–9.

Borecki, Z. and Millikan, D. F. 1969. A rapid method for determining the pathogenicity and factors associated with pathogenicity of *Phytophthora cactorum*. *Phytopathology*, 59, 247–8.

Borrelli, G., Mazzucotelli, E., Marone, D., Crosatti, C., Michelotti, V., Valè, G. and Mastrangelo, A. 2018. Regulation and evolution of NLR genes: A close interconnection for plant immunity. *International Journal of Molecular Sciences*, 19, 1662.

Bosch, D., Rodriguez, M. A. and Avilla, J. 2018. Monitoring resistance of *Cydia pomonella* (L.) Spanish field populations to new chemical insecticides and the mechanisms involved. *Pest Management Science*, 74, 933–43.

Bourget, R., Chaumont, L., Durel, C. E. and Sapoukhina, N. 2015. Sustainable deployment of QTLs conferring quantitative resistance to crops: First lessons from a stochastic model. *New Phytologist*, 206, 1163–71.

Bourras, S., McNally, K. E., Müller, M. C., Wicker, T. and Keller, B. 2016. Avirulence genes in cereal powdery mildews: The gene-for-gene hypothesis 2.0. *Frontiers in Plant Science*, 7, 241.

Bowen, J. K., Mesarich, C. H., Bus, V. G. M., Beresford, R. M., Plummer, K. M. and Templeton, M. D. 2011. *Venturia inaequalis*: The causal agent of apple scab. *Molecular Plant Pathology*, 12, 105–22.

Brewer, L. R., Alspach, P. A., Morgan, C. and Bus, V. G. M. 2009. Resistance to scab caused by *Venturis pirina* in interspecific pear (*Pyrus* spp.) hybrids. *New Zealand Journal of Crop and Horticultural Science*, 37, 211–18.

Brewer, L., Shaw, P., Wallis, R., Alspach, P., Aldworth, M., Orellana-Torrejon, C., Chagné, D. and Bus, V. G. M. 2018. Genetic mapping of pear sawfly (*Caliroa cerasi*) and pear blister mite (*Eriophyes pyri*) resistance in an interspecific pear family. *Tree Genetics & Genomes*, 14, 38.

Broggini, G. A. L., Bus, V. G. M., Parravicini, G., Kumar, S., Groenwold, R. and Gessler, C. 2011. Genetic mapping of 14 avirulence genes in an EU-B04 × 1639 progeny of *Venturia inaequalis*. *Fungal Genetics and Biology*, 48, 166–76.

Broggini, G. A. L., Wöhner, T., Fahrentrapp, J., Kost, T. D., Flachowsky, H., Peil, A., Hanke, M. V., Richter, K., Patocchi, A. and Gessler, C. 2014. Engineering fire blight resistance into the apple cultivar 'Gala' using the *FB_MR5* CC-NBS-LRR resistance gene of *Malus × robusta* 5. *Plant Biotechnology Journal*, 12, 728–33.

Brown, J. K. M. 2015. Durable resistance of crops to disease: A Darwinian perspective. *Annual Review of Phytopathology*, 53, 513–39.

Browne, G. T. and Mircetich, S. M. 1996. Effects of month of inoculation on severity of disease caused by *Phytophthora* spp. in apple root crowns and excised shoots. *Phytopathology*, 86, 290–4.

Brun, H., Chevre, A. M., Fitt, B. D. L., Powers, S., Besnard, A. L., Ermel, M., Huteau, V., Marquer, B., Eber, F., Renard, M. and Andrivon, D. 2010. Quantitative resistance increases the durability of qualitative resistance to *Leptosphaeria maculans* in *Brassica napus*. *New Phytologist*, 185, 285–99.

Burdon, J. J., Barrett, L. G., Rebetzke, G. and Thrall, P. H. 2014. Guiding deployment of resistance in cereals using evolutionary principles. *Evolutionary Applications*, 7, 609–24.

Bus, V. G., Spiers, A. G., Brewster, D. T. and Hofstee, M. E. 1996. Preliminary screening of apple germplasm for resistance to silverleaf infection. *New Zealand Journal of Crop and Horticultural Science*, 24, 1–6.

Bus, V. G. M., Alspach, P. A., Hofstee, M. E. and Brewer, L. R. 2002. Genetic variability and preliminary heritability estimates of resistance to scab (*Venturia inaequalis*) in an apple genetics population. *New Zealand Journal of Crop and Horticultural Science*, 30, 83–92.

Bus, V. G. M., Laurens, F. N. D., Van De Weg, W. E., Rusholme, R. L., Rikkerink, E. H. A., Gardiner, S. E., Bassett, H. C. M., Kodde, L. P. and Plummer, K. M. 2005a. The *Vh8* locus of a new gene-for-gene interaction between *Venturia inaequalis* and the wild apple *Malus sieversii* is closely linked to the *Vh2* locus in *Malus pumila* R12740-7A. *New Phytologist*, 166, 1035–49.

Bus, V. G. M., Rikkerink, E. H. A., Van De Weg, W. E., Rusholme, R. L., Gardiner, S. E., Bassett, H. C. M., Kodde, L. P., Parisi, L., Laurens, F., Meulenbroek, E. J. and Plummer, K. M. 2005b. The *Vh2* and *Vh4* scab resistance genes in two differential hosts derived from Russian apple R12740-7A map to the same linkage group of apple. *Molecular Breeding*, 15, 103–16.

Bus, V. G. M., Chagné, D., Bassett, H. C. M., Bowatte, D., Calenge, F., Celton, J.-M., Durel, C.-E., Malone, M. T., Patocchi, A., Ranatunga, A. C., Rikkerink, E. H. A., Tustin, D. S., Zhou, J. and Gardiner, S. E. 2008. Genome mapping of three major resistance genes to woolly apple aphid (*Eriosoma lanigerum* Hausm.). *Tree Genetics & Genomes*, 4, 223–36.

Bus, V. G. M., Esmenjaud, D., Buck, E. and Laurens, F. 2009. Application of genetic markers in rosaceous crops. *In:* Folta, K. M. and Gardiner, S. E. (Eds), *Genetics and Genomics of Rosaceae*. New York, NY: Springer, pp. 563–99.

Bus, V. G. M., Bassett, H. C. M., Bowatte, D., Chagné, D., Ranatunga, C. A., Ulluwishewa, D., Wiedow, C. and Gardiner, S. E. 2010. Genome mapping of an apple scab, a powdery mildew and a woolly apple aphid resistance gene from open-pollinated Mildew Immune Selection. *Tree Genetics & Genomes*, 6, 477–87.

Bus, V. G. M., Rikkerink, E. H. A., Caffier, V., Durel, C.-E. and Plummer, K. M. 2011. Revision of the nomenclature of the differential host-pathogen interactions of *Venturia inaequalis* and *Malus*. *Annual Review of Phytopathology*, 49, 391–413.

Bus, V., Singla, G., Ward, S., Brewer, L., Morgan, C., Bowatte, D., Bassett, H., Attfield, B., Colhoun, C., Bastiaanse, H., Walter, M., Scheper, R., Fisher, B., Won, K., Montanari, S., Volz, R., Chagné, D. and Gardiner, S. E. 2017. Progress in pipfruit resistance breeding and research at Plant & Food Research. *Acta Horticulturae*, 1172, 7–14.

Byrne, P. F., Volk, G. M., Gardner, C., Gore, M. A., Simon, P. W. and Smith, S. 2018. Sustaining the future of plant breeding: The critical role of the USDA-ARS National Plant Germplasm System. *Crop Science*, 58, 451–68.

Caffier, V. and Laurens, F. 2005. Breakdown of *Pl2*, a major gene of resistance to apple powdery mildew, in a French experimental orchard. *Plant Pathology*, 54, 116–24.

Caffier, V., Lasserre-Zuber, P., Giraud, M., Lascostes, M., Stievenard, R., Lemarquand, A., van de Weg, E., Expert, P., Denancé, C., Didelot, F., Le Cam, B. and Durel, C.-E. 2014. Erosion of quantitative host resistance in the apple-*Venturia inaequalis* pathosystem. *Infection, Genetics and Evolution*, 27, 481–9.

Caffier, V., Patocchi, A., Expert, P., Bellanger, M. N., Durel, C. E., Hilber-Bodmer, M., Broggini, G. A. L., Groenwold, R. and Bus, V. G. M. 2015. Virulence characterization of *Venturia inaequalis* reference isolates on the differential set of *Malus* hosts. *Plant Disease*, 99, 370–5.

Caffier, V., Le Cam, B., Al Rifai, M., Bellanger, M. N., Comby, M., Denancé, C., Didelot, F., Expert, P., Kerdraon, T., Lemarquand, A., Ravon, E. and Durel, C. E. 2016. Slow erosion of a quantitative apple resistance to *Venturia inaequalis* based on an isolate-specific quantitative trait locus. *Infection Genetics and Evolution*, 44, 541–8.

Callaway, E. 2018. CRISPR plants now subject to tough GM laws in European Union. *Nature*. https://www.nature.com/articles/d41586-018-05814-6#correction-0.

Carolan, K., Helps, J., van den Berg, F., Bain, R., Paveley, N. and van den Bosch, F. 2017. Extending the durability of cultivar resistance by limiting epidemic growth rates. *Proceedings of the Royal Society B: Biological Sciences*, 284, 20170828.

Carson, R. 1962. *Silent Spring*. Boston, MA: Houghton Mifflin Co. Inc.

Cesari, S. 2018. Multiple strategies for pathogen perception by plant immune receptors. *New Phytologist*, 219, 17–24.

Chagné, D., Crowhurst, R. N., Troggio, M., Davey, M. W., Gilmore, B., Lawley, C., Vanderzande, S., Hellens, R. P., Kumar, S., Cestaro, A., Velasco, R., Main, D., Rees, J. D., Iezzoni, A., Mockler, T., Wilhelm, L., van de Weg, E., Gardiner, S. E., Bassil, N. and Peace, C. 2012. Genome-wide SNP detection, validation, and development of an 8K SNP array for apple. *PLoS One*, 7, e31745.

Chapman, K. S., Sundin, G. W. and Beckerman, J. L. 2011. Identification of resistance to multiple fungicides in field populations of *Venturia inaequalis*. *Plant Disease*, 95, 921-6.

Chizzali, C. and Beerhues, L. 2012. Phytoalexins of the Pyrinae: Biphenyls and dibenzofurans. *Beilstein Journal of Organic Chemistry*, 8, 613–20.

Christopher-Kozjan, R. and Heath, M. C. 2003. Cytological and pharmacological evidence that biotrophic fungi trigger different cell death execution processes in host and nonhost cells during the hypersensitive response. *Physiological and Molecular Plant Pathology*, 62, 265–75.

Clark, M. D., Bus, V. G. M., Luby, J. J. and Bradeen, J. M. 2014. Characterization of the defence response to *Venturia inaequalis* in 'Honeycrisp' apple, its ancestors, and progeny. *European Journal of Plant Pathology*, 140, 69–81.

Cova, V., Bandara, N., Liang, W., Tartarini, S., Patocchi, A., Troggio, M., Velasco, R. and Komjanc, M. 2015. Fine mapping of the *Rvi5* (*Vm*) apple scab resistance locus in the 'Murray' apple genotype. *Molecular Breeding*, 35, 1–12.

Crosby, J. A., Janick, J., Pecknold, P. C., Korban, S. S., O'Connor, P. A., Ries, S. M., Goffreda, J. and Voordeckers, A. 1992. Breeding apples for scab resistance: 1945–1990. *Fruit Varieties Journal*, 46, 145–66.

Cummins, J. N. and Aldwinckle, H. S. 1983. Breeding apple rootstocks. *Plant Breeding Reviews*, 1, 294–394.

Cummins, J. N., Forsline, P. L. and Mackenzie, J. D. 1981. Woolly apple aphid colonization on *Malus* cultivars. *Journal of the American Society of Horticultural Science*, 106, 26–30.

Daccord, N., Celton, J.-M., Linsmith, G., Becker, C., Choisne, N., Schijlen, E., Van De Geest, H., Bianco, L., Micheletti, D., Velasco, R., Di Pierro, E. A., Gouzy, J., Rees, D. J. G., Guérif, P., Muranty, H., Durel, C.-E., Laurens, F., Lespinasse, Y., Gaillard, S., Aubourg, S., Quesneville, H., Weigel, D., van de Weg, E., Troggio, M. and Bucher, E. 2017. High-quality *de novo* assembly of the apple genome and methylome dynamics of early fruit development. *Nature Genetics*, 49, 1099–106.

Dangl, J. L. and Jones, J. D. G. 2001. Plant pathogens and integrated defence responses to infection. *Nature*, 411, 826–33.

Davé, B. A., Kaplan, H. J. and Petrie, J. F. 1980. The isolation of *Penicillium digitatum* sacc. strains tolerant to 2-AB, SOPP, TBZ and Benomyl. *Proceedings of the Florida State Horticultural Society*, 93, 344–7.

den Boer, J. H. and Green, T. L. 1995. Disease and aesthetic ratings. *Malus*, 9, 8–58.

den Boer, E., Pelgrom, K. T. B., Zhang, N. W. W., Visser, R. G. F., Niks, R. E. and Jeuken, M. J. W. 2014. Effects of stacked quantitative resistances to downy mildew in lettuce do not simply add up. *Theoretical and Applied Genetics*, 127, 1805–16.

Deng, C. H., Plummer, K. M., Jones, D. A. B., Mesarich, C. H., Shiller, J., Taranto, A. P., Robinson, A. J., Kastner, P., Hall, N. E., Templeton, M. D. and Bowen, J. K. 2017. Comparative analysis of the predicted secretomes of Rosaceae scab pathogens *Venturia inaequalis* and *V. pirina* reveals expanded effector families and putative determinants of host range. *BMC Genomics*, 18, 339.

Didelot, F., Brun, L. and Parisi, L. 2007. Effects of cultivar mixtures on scab control in apple orchards. *Plant Pathology*, 56, 1014–22.

Didelot, F., Caffier, V., Orain, G., Lemarquand, A. and Parisi, L. 2016. Sustainable management of scab control through the integration of apple resistant cultivars in a low-fungicide input system. *Agriculture, Ecosystems & Environment*, 217, 41–8.

Dodds, P. N., Lawrence, G. J. and Ellis, J. G. 2001. Six amino acid changes confined to the leucine-rich repeat B-strand/B-turn motif determine the difference between the *P* and *P2* rust resistance specificities in flax. *Plant Cell*, 13, 163–78.

Doehlemann, G., van der Linde, K., Aßmann, D., Schwammbach, D., Hof, A., Mohanty, A., Jackson, D. and Kahmann, R. 2009. Pep1, a secreted effector protein of *Ustilago maydis*, is required for successful invasion of plant cells. *PLoS Pathogens*, 5, e1000290.

Dondini, L., Gaiotti, F., Pierantoni, L. and Bazzi, C. 2006. QTLs related to fire blight resistance in pear. *Acta Horticulturae*, 704, 567–70.

Dondini, L., De Franceschi, P., Ancarani, V., Civolani, S., Fano, E. A. and Musacchi, S. 2015. Identification of a QTL for psylla resistance in pear via genome scanning approach. *Scientia Horticulturae*, 197, 568–72.

Dougherty, L., Singh, R., Brown, S., Dardick, C. and Xu, K. N. 2018. Exploring DNA variant segregation types in pooled genome sequencing enables effective mapping of weeping trait in *Malus*. *Journal of Experimental Botany*, 69, 1499–516.

Dunemann, F. and Egerer, J. 2010. A major resistance gene from Russian apple 'Antonovka' conferring field immunity against apple scab is closely linked to the *Vf* locus. *Tree Genetics & Genomes*, 6, 627–33.

Durel, C.-E., van de Weg, E., Venisse, J. S. and Parisi, L. 2000. Localisation of a major gene for scab resistance on the European genetic map of the Prima × Fiesta cross. *IOBC/WPRS Bulletin*, 23(12), 245–8.

Edae, E. A., Pumphrey, M. O. and Rouse, M. N. 2018. A genome-wide association study of field and seedling response to individual stem rust pathogen races reveals combinations of race-specific genes in North American spring wheat. *Frontiers in Plant Science*, 9, 52.

Edge-Garza, D. A., Luby, J. J. and Peace, C. 2015. Decision support for cost-efficient and logistically feasible marker-assisted seedling selection in fruit breeding. *Molecular Breeding*, 35, 223.

Ellis, J. G., Lawrence, G. J., Luck, J. E. and Dodds, P. N. 1999. Identification of regions in alleles of the flax rust resistance gene *L* that determine differences in gene-for-gene specificity. *Plant Cell*, 11, 495–506.

Evans, K. M. and James, C. M. 2003. Identification of SCAR markers linked to *Pl-w* mildew resistance in apple. *Theoretical and Applied Genetics*, 106, 1178–83.

Fahrentrapp, J., Broggini, G. A. L., Kellerhals, M., Peil, A., Richter, K., Zini, E. and Gessler, C. 2013. A candidate gene for fire blight resistance in *Malus* x *robusta* 5 is coding for a CC-NBS-LRR. *Tree Genetics and Genomes*, 9, 237–51.

Fischer, M. and Dunemann, F. 2000. Search for polygenic scab and mildew resistance in apple varieties cultivated at the Fruit Genebank Dresden-Pillnitz. *Acta Horticulturae*, 538, 71–7.

Flachowsky, H., Hanke, M. V., Peil, A., Strauss, S. H. and Fladung, M. 2009. A review on transgenic approaches to accelerate breeding of woody plants. *Plant Breeding*, 128, 217–26.

Flachowsky, H., Le Roux, P. M., Peil, A., Patocchi, A., Richter, K. and Hanke, M. V. 2011. Application of a high-speed breeding technology to apple (*Malus* × *domestica*) based on transgenic early flowering plants and marker-assisted selection. *New Phytologist*, 192, 364–77.

Flor, H. H. 1956. The complementary genic systems in flax and flax rust. *Advances in Genetics*, 8, 29–54.

Frantzeskakis, L., Von Dahlen, J. K., Panstruga, R. and Rose, L. E. 2018. Rapid evolution in the tug-of-war between microbes and plants. *New Phytologist*, 219, 12–14.

Frederick, Z. A., Villani, S. M., Cooley, D. R., Biggs, A. R., Raes, J. J. and Cox, K. D. 2014. Prevalence and stability of qualitative QoI resistance in populations of *Venturia inaequalis* in the northeastern United States. *Plant Disease*, 98, 1122–30.

Galli, P., Broggini, G. A. L., Gessler, C. and Patocchi, A. 2010. Phenotypic characterization of the *Rvi15* (*Vr2*) apple scab resistance. *Journal of Plant Pathology*, 92, 219–26.

Gao, C. X. 2018. The future of CRISPR technologies in agriculture. *Nature Reviews Molecular Cell Biology*, 19, 275–6.

Gardiner, S. E., Bus, V. G. M., Rusholme, R. L., Chagné, D. and Rikkerink, E. H. A. 2007. Apple. *In:* Kole, C. (Ed.), *Genome Mapping and Molecular Breeding in Plants: Fruits and Nuts.* Berlin and Heidelberg: Springer-Verlag.

Gates, J. E. and Millikan, D. F. 1972. Seasonal fluctuations in susceptibility of the inner bark tissues of apple to colonization by the collar rot fungus, *Phytophthora cactorum. Phytoprotection*, 53, 76–81.

Gessler, C. and Patocchi, A. 2007. Recombinant DNA technology in apple. *Green Gene Technology: Research in an Area of Social Conflict*, 107, 113–32.

Giliomee, J. H., Strydom, D. K., van Zyl, H. J. 1968. Northern Spy, Merton and Malling-Merton rootstocks susceptible to woolly aphid, *Eriosoma lanigerum*, in the Western Cape. *South African Journal of Agricultural Science*, 11, 183–6.

Gradziel, T. M. and Wang, D. 1993. Evaluation of brown rot resistance and its relation to enzymatic browning in clingstone peach germplasm. *Journal of the American Society for Horticultural Science*, 118, 675–9.

Guan, Y. Z., Peace, C., Rudell, D., Verma, S. and Evans, K. 2015. QTLs detected for individual sugars and soluble solids content in apple. *Molecular Breeding*, 35, 135.

Gygax, M., Gianfranceschi, L., Liebhard, R., Kellerhals, M., Gessler, C. and Patocchi, A. 2004. Molecular markers linked to the apple scab resistance gene *Vbj* derived from *Malus baccata jackii. Theoretical and Applied Genetics*, 109, 1702–9.

Hamdoun, S., Gao, M., Gill, M., Kwon, A., Norelli, J. L. and Lu, H. 2018. Signalling requirements for *Erwinia amylovora*-induced disease resistance, callose deposition and cell growth in the non-host *Arabidopsis thaliana. Molecular Plant Pathology*, 19, 1090–103.

Heath, M. C. 1998. Apoptosis, programmed cell death and the hypersensitive response. *European Journal of Plant Pathology*, 104, 117–24.

Hinrichs-Berger, J. and Brüstle, S. 2015. Die Marssonina-Blattfallkrankheit des Apfels. Ein Update. *Obstbau*, 72–4.

Hrazdina, G., Borejsza-Wysocki, W. and Lester, C. 1997. Phytoalexin production in an apple cultivar resistant to *Venturia inaequalis. Phytopathology*, 87, 868–76.

Hulbert, S. and Pumphrey, M. 2014. A time for more booms and fewer busts? Unraveling cereal-rust interactions. *Molecular Plant-Microbe Interactions*, 27, 207–14.

James, C. M., Clarke, J. B. and Evans, K. M. 2005. Identification of molecular markers linked to the mildew resistance gene *Pl-d* in apple. *Theoretical and Applied Genetics*, 110, 175–81.

Jänsch, M., Frey, J. E., Hilber-Bodmer, M., Broggini, G. A. L., Weger, J., Schnabel, G. and Patocchi, A. 2012. SSR marker analysis of *Monilinia fructicola* from Swiss apricots suggests introduction of the pathogen from neighbouring countries and the United States. *Plant Pathology*, 61, 247–54.

Jänsch, M., Broggini, G. A. L., Weger, J., Bus, V. G. M., Gardiner, S. E., Bassett, H. and Patocchi, A. 2015. Identification of SNPs linked to eight apple disease resistance loci. *Molecular Breeding*, 35, 1–21.

Jeffers, S. N., Aldwinckle, H. S., Burr, T. J. and Arneson, P. A. 1981. Excised twig assay for the study of apple tree crown rot pathogens *in vitro. Plant Disease*, 65, 823–5.

Jeuken, M. J. W., Pelgrom, K., Stam, P. and Lindhout, P. 2008. Efficient QTL detection for nonhost resistance in wild lettuce: Backcross inbred lines versus F-2 population. *Theoretical and Applied Genetics*, 116, 845–57.

Johnson, R. 1978. Practical breeding for durable resistance to rust diseases in self-pollinating cereals. *Euphytica*, 27, 529–40.

Johnson, R. 2000. Classical plant breeding for durable resistance to diseases. *Journal of Plant Pathology*, 82, 3–7.
Juliana, P., Singh, R. P., Singh, P. K., Crossa, J., Huerta-Espino, J., Lan, C., Bhavani, S., Rutkoski, J. E., Poland, J. A., Bergstrom, G. C. and Sorrells, M. E. 2017. Genomic and pedigree-based prediction for leaf, stem, and stripe rust resistance in wheat. *Theoretical and Applied Genetics*, 130, 1415–30.
Kalisz, S., Oszmianski, J. and Wojdylo, A. 2015. Increased content of phenolic compounds in pear leaves after infection by the pear rust pathogen. *Physiological and Molecular Plant Pathology*, 90, 113–19.
Kellerhals, M., Baumgartner, I. O., Leumann, L., Frey, J. E. and Patocchi, A. 2013. Progress in pyramiding disease resistances in apple breeding. *Acta Horticulturae*, 976, 487–91.
Khan, M. A., Han, Y., Zhao, Y. F. and Korban, S. S. 2012. A high-throughput apple SNP genotyping platform using the GoldenGate™ assay. *Gene*, 494, 196–201.
Kirkham, D. S. 1957. The significance of polyphenolic metabolites of apple and pear in the host relations of *Venturia inaequalis* and *Venturia pirina*. *Journal of General Microbiology*, 17, 491–504.
Kost, T. D., Gessler, C., Jänsch, M., Flachowsky, H., Patocchi, A. and Broggini, G. A. L. 2015. Development of the first cisgenic apple with increased resistance to fire blight. *PLoS One*, 10, e0143980.
Kourelis, J. and van der Hoorn, R. A. L. 2018. Defended to the nines: 25 years of resistance gene cloning identifies nine mechanisms for R protein function. *Plant Cell*, 30, 285–99.
Krattinger, S. G., Lagudah, E. S., Spielmeyer, W., Singh, R. P., Huerta-Espino, J., McFadden, H., Bossolini, E., Selter, L. L. and Keller, B. 2009. A putative ABC transporter confers durable resistance to multiple fungal pathogens in wheat. *Science*, 323, 1360–3.
Krens, F. A., Schaart, J. G., van der Burgh, A. M., Tinnenbroek-Capel, I. E. M., Groenwold, R., Kodde, L. P., Broggini, G. A. L., Gessler, C. and Schouten, H. J. 2015. Cisgenic apple trees: Development, characterization and performance. *Frontiers in Plant Science*, 6, 286.
Kriticos, D. J., Kean, J. M., Phillips, C. B., Senay, S. D., Acosta, H. and Haye, T. 2017. The potential global distribution of the brown marmorated stink bug, *Halyomorpha halys*, a critical threat to plant biosecurity. *Journal of Pest Science*, 90, 1033–43.
Kumar, S., Bink, M., Volz, R. and Chagné, D. 2012a. Towards genomic selection in apple (*Malus × domestica* Borkh.) breeding programmes: Prospects, challenges and strategies. *Tree Genetics and Genomes*, 8, 1–14.
Kumar, S., Chagné, D., Bink, M. C. A. M., Volz, R. K., Whitworth, C. and Carlisle, C. 2012b. Genomic selection for fruit quality traits in apple (*Malus × domestica* Borkh.). *PLoS One*, 7, e36674.
Kumar, S., Kirk, C., Deng, C., Wiedow, C., Knaebel, M. and Brewer, L. 2017. Genotyping-by-sequencing of pear (*Pyrus* spp.) accessions unravels novel patterns of genetic diversity and selection footprints. *Horticulture Research*, 4, 17015–15.
Lai, Y. and Eulgem, T. 2018. Transcript-level expression control of plant NLR genes. *Molecular Plant Pathology*, 19, 1267–81.
Laloi, G., Vergne, E., Durel, C. E., Le Cam, B. and Caffier, V. 2017. Efficiency of pyramiding of three quantitative resistance loci to apple scab. *Plant Pathology*, 66, 412–22.
Lam, E., Kato, N. and Lawton, M. 2001. Programmed cell death, mitochondria and the plant hypersensitive response. *Nature*, 411, 848–53.

Lambert, P., Campoy, J. A., Pacheco, I., Mauroux, J.-B., Da Silva Linge, C., Micheletti, D., Bassi, D., Rossini, L., Dirlewanger, E., Pascal, T., Troggio, M., Aranzana, M. J., Patocchi, A. and Arús, P. 2016. Identifying SNP markers tightly associated with six major genes in peach [*Prunus persica* (L.) Batsch] using a high-density SNP array with an objective of marker-assisted selection (MAS). *Tree Genetics and Genomes*, 12, 121.

Lasserre-Zuber, P., Caffier, V., Stievenard, R., Lemarquand, A., Le Cam, B. and Durel, C.-E. 2018. Pyramiding quantitative resistance with a major resistance gene in apple: From ephemeral to enduring effectiveness in controlling scab. *Plant Disease*, 102, 2220–3.

Lateur, M. and Populer, C. 1994. Screening fruit tree genetic resources in Belgium for disease resistance and other desirable characters. *In:* Schmidt, H. and Kellerhals, M. (Eds), *Progress in Temperate Fruit Breeding*. Dordrecht, NL: Kluwer Academic Publishers, pp. 425–31.

Lateur, M., Lefrancq, B. and Parisi, L. 2002. First evidence of the breakdown in Belgium of the scab resistance conferred by the *Vf* gene. *Mededelingen Faculteit Landbouwwetenschappen Universiteit van Gent*, 67/2, 135–43.

Laurens, F. 1999. Review of the current apple breeding programmes in the world: Objectives for scion cultivar improvement. *Acta Horticulturae*, 484, 163–70.

Laurens, F., Chevalier, M., Dolega, E., Gennari, F., Goerre, M., Fischer, C., Kellerhals, M., Lateur, M., Lefrancq, B., Parisi, L., Schouten, H. J. and Tartarini, S. 2004. Local European cultivars as sources of durable scab resistance in apple. *Acta Horticulturae*, 663, 115–21.

Laurens, F., Aranzana, M. J., Arus, P., Bassi, D., Bink, M., Bonany, J., Caprera, A., Corelli-Grappadelli, L., Costes, E., Durel, C.-E., Mauroux, J.-B., Muranty, H., Nazzicari, N., Pascal, T., Patocchi, A., Peil, A., Quilot-Turion, B., Rossini, L., Stella, A., Troggio, M., Velasco, R. and van de Weg, E. 2018. An integrated approach for increasing breeding efficiency in apple and peach in Europe. *Horticulture Research*, 5, 11.

Le Roux, P. M., Flachowsky, H., Hanke, M. V., Gessler, C. and Patocchi, A. 2012. Use of a transgenic early flowering approach in apple (*Malus* × *domestica* Borkh.) to introgress fire blight resistance from cultivar Evereste. *Molecular Breeding*, 30, 857–74.

Lê Van, A., Caffier, V., Lasserre-Zuber, P., Chauveau, A., Brunel, D., Le Cam, B. and Durel, C. E. 2013. Differential selection pressures exerted by host resistance quantitative trait loci on a pathogen population: A case study in an apple *Venturia inaequalis* pathosystem. *New Phytologist*, 197, 899–908.

Leach, J. E., Vera Cruz, C. M., Bai, J. and Leung, H. 2001. Pathogen fitness penalty as a predictor of durability of disease resistance genes. *Annual Review of Phytopathology*, 39, 187–224.

Lee, H. A., Lee, H. Y., Seo, E., Lee, J., Kim, S. B., Oh, S., Choi, E., Choi, E., Lee, S. E. and Choi, D. 2017. Current understandings of plant nonhost resistance. *Molecular Plant-Microbe Interactions*, 30, 5–15.

Lefrancq, B. and Lateur, M. 2009. Monitoring and occurrence of new races of *Venturia inaequalis* on apple in Belgium. *Communications in Agricultural and Applied Biological Sciences*, 74, 623–31.

Lemaire, C., De Gracia, M., Leroy, T., Michalecka, M., Lindhard-Pedersen, H., Guerin, F., Gladieux, P. and Le Cam, B. 2016. Emergence of new virulent populations of apple scab from nonagricultural disease reservoirs. *New Phytologist*, 209, 1220–9.

Leonard, K. J. 2007. Persistent virulence associations in sexual populations of *Puccinia coronata*. *Plant Pathology*, 56, 35–45.

Lespinasse, Y. 1983. Amélioration du pommier pour la résistance a l'oidium (*Podosphaera leucotricha*). Premiers résultats concernant la virulence du champignon. *IOBC/WPRS Bulletin*, 6(4), 96–110.

Lespinasse, Y. 1989. Breeding pome fruits with stable resistance to diseases. 3. Genes, resistance mechanisms, present work and prospects. *IOBC/WPRS Bulletin*, 12(6), 100–15.

Liebhard, R., Gianfranceschi, L., Koller, B., Ryder, C. D., Tarchini, R., van de Weg, W. E. and Gessler, C. 2002. Development and characterisation of 140 new microsatellites in apple (*Malus × domestica* Borkh.). *Molecular Breeding*, 10, 217–41.

Luby, J. J., Alspach, P. A., Bus, V. G. M. and Oraguzie, N. C. 2002. Field resistance to fire blight in a diverse apple (*Malus* sp.) germplasm collection. *Journal of the American Society of Horticultural Science*, 127, 245–53.

Malnoy, M., Viola, R., Jung, M. H., Koo, O. J., Kim, S., Kim, J. S., Velasco, R. and Kanchiswamy, C. N. 2016. DNA-free genetically edited grapevine and apple protoplast using CRISPR/Cas9 ribonucleoproteins. *Frontiers in Plant Science*, 7, 1904.

Markussen, T., Kruger, J., Schmidt, H. and Dunemann, F. 1995. Identification of PCR-based markers linked to the powdery-mildew-resistance gene *Pl1* from *Malus robusta* in cultivated apple. *Plant Breeding*, 114, 530–4.

Martinez-Bilbao, A., Ortiz-Barredo, A., Montesinos, E. and Murillo, J. 2012. *Venturia inaequalis* resistance in local Spanish cider apple germplasm under controlled and field conditions. *Euphytica*, 188, 273–83.

Martínez-García, P. J., Parfitt, D. E., Bostock, R. M., Fresnedo-Ramírez, J., Vazquez-Lobo, A., Ogundiwin, E. A., Gradziel, T. M. and Crisosto, C. H. 2013. Application of genomic and quantitative genetic tools to identify candidate resistance genes for brown rot resistance in peach. *PLoS One*, 8, e78634.

Masny, S. 2017. Occurrence of *Venturia inaequalis* races in Poland able to overcome specific apple scab resistance genes. *European Journal of Plant Pathology*, 147, 313–23.

Mayerhofer, G., Schwaiger-Nemirova, I., Kuhn, T., Girsch, L. and Allerberger, F. 2009. Detecting streptomycin in apples from orchards treated for fire blight. *Journal of Antimicrobial Chemotherapy*, 63, 1076–7.

Mayr, U., Michalek, S., Treutter, D. and Feucht, W. 1997. Phenolic compounds of apple and their relationship to scab resistance. *Journal of Phytopathology*, 145, 69–75.

McDonald, B. A. and Linde, C. 2002. The population genetics of plant pathogens and breeding strategies for durable resistance. *Euphytica*, 124, 163–80.

McGarry, R. C., Klocko, A. L., Pang, M. X., Strauss, S. H. and Ayre, B. G. 2017. Virus-induced flowering: An application of reproductive biology to benefit plant research and breeding. *Plant Physiology*, 173, 47–55.

Meng, J. W., He, D. C., Zhu, W., Yang, L. N., Wu, E. J., Xie, J. H., Shang, L. P. and Zhan, J. S. 2018. Human-mediated gene flow contributes to metapopulation genetic structure of the pathogenic fungus *Alternaria alternata* from potato. *Frontiers in Plant Science*, 9, 198.

Merwin, I. A., Brown, S. K., Rosenberger, D. A., Cooley, D. R. and Berkett, L. P. 1994. Scab-resistant apples for the northeastern United States: New prospects and old problems. *Plant Disease*, 78(1), 4–10.

Meuwissen, T. H. E., Hayes, B. J. and Goddard, M. E. 2001. Prediction of total genetic value using genome-wide dense marker maps. *Genetics*, 157, 1819-29.

Mihatsch, H. and Mildenberger, G. 1966. Beiträge zur Vererbung der Mehltau-resistenz bei *Malus* I. Freilandbeobachtungen an erwachsenen Sämlingen aus Kultursortenkreuzungen und freier Abblüte. *Der Züchter*, 36, 1-11.

Mine, A., Seyfferth, C., Kracher, B., Berens, M. L., Becker, D. and Tsuda, K. 2018. The defense phytohormone signaling network enables rapid, high-amplitude transcriptional reprogramming during effector-triggered immunity. *Plant Cell*, 30, 1199-219.

Montanari, S. 2015. *Identification and mapping of genomic regions controlling fire blight and psylla resistance and hybrid necrosis in pear*. PhD thesis, L'Université Nantes Angers Le Mans.

Montanari, S., Guérif, P., Ravon, E., Denancé, C., Muranty, H., Velasco, R., Chagné, D., Bus, V. M., Robert, P., Perchepied, L. and Durel, C.-E. 2015. Genetic mapping of *Cacopsylla pyri* resistance in an interspecific pear (*Pyrus* spp.) population. *Tree Genetics & Genomes*, 11, 1-14.

Mundt, C. C. 2018. Pyramiding for resistance durability: Theory and practice. *Phytopathology*, 108, 792-802.

Mundt, C. C., Cowger, C. and Garrett, K. A. 2002. Relevance of integrated disease management to resistance durability. *Euphytica*, 124, 245-52.

Muranty, H., Troggio, M., Sadok, I. B., Rifaï, M. A., Auwerkerken, A., Banchi, E., Velasco, R., Stevanato, P., van de Weg, W. E., di Guardo, M., Kumar, S., Laurens, F. and Bink, M. C. A. M. 2015. Accuracy and responses of genomic selection on key traits in apple breeding. *Horticulture Research*, 2, 15060.

Mysore, K. S. and Ryu, C.-M. 2004. Nonhost resistance: How much do we know? *Trends in Plant Science*, 9, 97-104.

Naerstad, R., Hermansen, A. and Bjor, T. 2007. Exploiting host resistance to reduce the use of fungicides to control potato late blight. *Plant Pathology*, 56, 156-66.

Nazareno, E. S., Li, F., Smith, M., Park, R. F., Kianian, S. F. and Figueroa, M. 2018. *Puccinia coronata* f. sp. *avenae*: A threat to global oat production. *Molecular Plant Pathology*, 19, 1047-60.

Niks, R. E. and Rubiales, D. 2002. Potentially durable resistance mechanisms in plants to specialised fungal pathogens. *Euphytica*, 124, 201-16.

Nishitani, C., Hirai, N., Komori, S., Wada, M., Okada, K., Osakabe, K., Yamamoto, T. and Osakabe, Y. 2016. Efficient genome editing in apple using a CRISPR/Cas9 system. *Scientific Reports*, 6, 31481.

Norelli, J. L., Aldwinckle, H. S. and Beer, S. V. 1986. Differential susceptibility of *Malus* spp. cultivars Robusta 5, Novole, and Ottawa 523 to *Erwinia amylovora*. *Plant Disease*, 70, 1017-19.

Norelli, J. L., Wisniewski, M. E., Flachowsky, H. and Hanke, M.-V. 2015. Using early flowering transgenic apple to accelerate the breeding of donor parents with multiple loci for disease resistance (*Malus × domestica*). *Proceedings of Plant and Animal Genome Conference XXIII, San Diego, CA, USA, P0897*.

Oliveira Lino, L., Pacheco, I., Mercier, V., Faoro, F., Bassi, D., Bornard, I. and Quilot-Turion, B. 2016. Brown rot strikes *Prunus* fruit: An ancient fight almost always lost. *Journal of Agricultural and Food Chemistry*, 64, 4029-47.

Oraguzie, N. C. and Wilcox, P. L. 2007. An overview of association mapping. *In:* Oraguzie, N. C., Rikkerink, E. H. A., Gardiner, S. E. and de Silva, H. N. (Eds), *Association Mapping in Plants*. New York, NY: Springer.

Padder, B. A., Sofi, T. A., Ahmad, M., Shah, M. U. D., Hamid, A., Saleem, S. and Ahanger, F. A. 2013. Virulence and molecular diversity of *Venturia inaequalis* in commercial apple growing regions in Kashmir. *Journal of Phytopathology*, 161, 271-9.

Padmarasu, S., Sargent, D. J., Jaensch, M., Kellerhals, M., Tartarini, S., Velasco, R., Troggio, M. and Patocchi, A. 2014. Fine-mapping of the apple scab resistance locus *Rvi12* (*Vb*) derived from 'Hansen's baccata #2'. *Molecular Breeding*, 34, 2119-29.

Pagliarani, G., Dapena, E., Minarro, M., Denancé, C., Lespinasse, Y., Rat-Morris, E., Troggio, M., Durel, C. E. and Tartarini, S. 2016. Fine mapping of the rosy apple aphid resistance locus *Dp-fl* on linkage group 8 of the apple cultivar 'Florina'. *Tree Genetics & Genomes*, 12, 56.

Palloix, A., Ayme, V. and Moury, B. 2009. Durability of plant major resistance genes to pathogens depends on the genetic background, experimental evidence and consequences for breeding strategies. *New Phytologist*, 183, 190-9.

Parisi, L., Lespinasse, Y., Guillaumes, J. and Krüger, J. 1993. A new race of *Venturia inaequalis* virulent to apples with resistance due to *Vf* gene. *Phytopathology*, 83, 533-7.

Parisi, L., Durel, C.-E. and Laurens, F. 2000. First report on the presence of *Venturia inaequalis* race 7 in French apple orchards. *IOBC/WPRS Bulletin*, 23(12), 99-104.

Parisi, L., Fouillet, V., Schouten, H. J., Groenwold, R., Laurens, F., Didelot, F., Evans, K., Fischer, C., Gennari, F., Kemp, H., Lateur, M., Patocchi, A., Thissen, J. and Tsipouridis, C. 2004. Variability of the pathogenicity of *Venturia inaequalis* in Europe. *Acta Horticulturae*, 663, 107-13.

Parisi, L., Gros, C., Combe, F., Parveaud, C. E., Gomez, C. and Brun, L. 2013. Impact of a cultivar mixture on scab, powdery mildew and rosy aphid in an organic apple orchard. *Crop Protection*, 43, 207-12.

Parlevliet, J. E. 2002. Durability of resistance against fungal, bacterial and viral pathogens; present situation. *Euphytica*, 124, 147-56.

Passey, T. A. J., Shaw, M. W. and Xu, X. M. 2016. Differentiation in populations of the apple scab fungus *Venturia inaequalis* on cultivars in a mixed orchard remain over time. *Plant Pathology*, 65, 1133-41.

Patocchi, A., Walser, M., Tartarini, S., Broggini, G. A. L., Gennari, F., Sansavini, S. and Gessler, C. 2005. Identification by genome scanning approach (GSA) of a microsatellite tightly associated with the apple scab resistance gene *Vm*. *Genome*, 48, 630-6.

Patocchi, A., Frei, A., Frey, J. E. and Kellerhals, M. 2009. Towards improvement of marker assisted selection of apple scab resistant cultivars: *Venturia inaequalis* virulence surveys and standardization of molecular marker alleles associated with resistance genes. *Molecular Breeding*, 24, 337-47.

Peil, A., Bus, V. G. M., Geider, K., Richter, K., Flachowsky, H. and Hanke, M.-V. 2009. Improvement of fire blight resistance in apple and pear. *International Journal of Plant Breeding*, 3, 1-27.

Peil, A., Patocchi, A., Hanke, M.-V. and Bus, V. G. M. 2018. Apple cultivar Regia possessing both *Rvi2* and *Rvi4* resistance genes is the source of a new race of *Venturia inaequalis*. *European Journal of Plant Pathology*, 151, 533-9.

Peng, Y. J., van Wersch, R. and Zhang, Y. L. 2018. Convergent and divergent signaling in PAMP-triggered immunity and effector-triggered immunity. *Molecular Plant-Microbe Interactions*, 31, 403-9.

Periyannan, S. 2018. Sustaining global agriculture through rapid detection and deployment of genetic resistance to deadly crop diseases. *New Phytologist*, 219, 45–51.

Pessina, S., Angeli, D., Martens, S., Visser, R. G. F., Bai, Y. L., Salamini, F., Velasco, R., Schouten, H. J. and Malnoy, M. 2016. The knock-down of the expression of MdMLO19 reduces susceptibility to powdery mildew (*Podosphaera leucotricha*) in apple (*Malus domestica*). *Plant Biotechnology Journal*, 14, 2033–44.

Pink, D. A. C. 2002. Strategies using genes for non-durable resistance. *Euphytica*, 124, 227–36.

Pitera, E. and Bogdanowicz, J. 1992. Evaluation of scab resistant apple progenies for powdery mildew tolerance. *Acta Horticulturae*, 317, 81–7.

Poland, J. A., Balint-Kurti, P. J., Wisser, R. J., Pratt, R. C. and Nelson, R. J. 2009. Shades of gray: The world of quantitative resistance. *Trends in Plant Science*, 4, 21–9.

Pontais, I., Treutter, D., Paulin, J. P. and Brisset, M. N. 2008. *Erwinia amylovora* modifies phenolic profiles of susceptible and resistant apple through its type III secretion system. *Physiologia Plantarum*, 132, 262–71.

Praz, C. R., Menardo, F., Robinson, M. D., Müller, M. C., Wicker, T., Bourras, S. and Keller, B. 2018. Non-parent of origin expression of numerous effector genes indicates a role of gene regulation in host adaption of the hybrid triticale powdery mildew pathogen. *Frontiers in Plant Science*, 9, 49.

Proffer, T. J. 2014. Valsa canker. *In:* Sutton, T. B., Aldwinckle, H. S., Agnello, A. M. and Walgenbach, J. F. (Eds), *Compendium of Apple and Pear Disease and Pests*. 2nd ed. St. Paul, MN: APS Press, p. 56.

Quintana-Rodriguez, E., Duran-Flores, D., Heil, M. and Camacho-Coronel, X. 2018. Damage-associated molecular patterns (DAMPs) as future plant vaccines that protect crops from pests. *Scientia Horticulturae*, 237, 207–20.

Ramirez-Prado, J. S., Piquerez, S. J. M., Bendahmane, A., Hirt, H., Raynaud, C. and Benhamed, M. 2018. Modify the histone to win the battle: Chromatin dynamics in plant-pathogen interactions. *Frontiers in Plant Science*, 9, 355.

Roberts, A. L. and Crute, I. R. 1994. Apple scab resistance from *Malus floribunda* 821 (*Vf*) is rendered ineffective by isolates of *Venturia inaequalis* from *Malus floribunda*. *Norwegian Journal of Agricultural Sciences*, 17, 404–6.

Rock, G. C. and Zeiger, D. C. 1974. Woolly apple aphid infests Malling and Malling-Merton rootstocks in propagation beds in North Carolina. *Journal of Economic Entomology*, 67, 137–8.

Rodriguez-Moreno, L., Song, Y. and Thomma, B. P. H. J. 2017. Transfer and engineering of immune receptors to improve recognition capacities in crops. *Current Opinion in Plant Biology*, 38, 42–9.

Roux, P. M. F. L., Khan, M. A., Broggini, G. A. L., Duffy, B., Gessler, C. and Patocchi, A. 2010. Mapping of quantitative trait loci for fire blight resistance in the apple cultivars 'Florina' and 'Nova Easygro'. *Genome*, 53, 710–22.

Rutkoski, J., Singh, R. P., Huerta-Espino, J., Bhavani, S., Poland, J., Jannink, J. L. and Sorrells, M. E. 2015. Genetic gain from phenotypic and genomic selection for quantitative resistance to stem rust of wheat. *The Plant Genome*, 8. doi:10.3835/plantgenome2014.10.0074.

Sandanayaka, W. R. M., Bus, V. G. M., Connolly, P. and Newcomb, R. 2003. Characteristics associated with woolly apple aphid, *Eriosoma lanigerum*, resistance of three apple rootstocks. *Entomologia Experimentalis et Applicata*, 109, 63–72.

Sandanayaka, W. R. M., Bus, V. G. M. and Connolly, P. 2005. Mechanisms of woolly aphid [*Eriosoma lanigerum* (Hausm.)] resistance in apple. *Journal of Applied Entomology*, 129, 534–41.

Sandskär, B. and Liljeroth, E. 2005. Incidence of races of the apple scab pathogen (*Venturia inaequalis*) in apple growing districts in Sweden. *Acta Agriculturae Scandinavica Section B-Soil and Plant Science*, 55, 143–50.

Schaart, J. G., Krens, F. A., Pelgrom, K. T., Mendes, O. and Rouwendal, G. J. 2004. Effective production of marker-free transgenic strawberry plants using inducible site-specific recombination and a bifunctional selectable marker gene. *Plant Biotechnology Journal*, 2, 233–40.

Schlathölter, I., Jänsch, M., Flachowsky, H., Broggini, G. A. L., Hanke, M.-V. and Patocchi, A. 2018. Generation of advanced fire blight-resistant apple (*Malus* × *domestica*) selections of the fifth generation within 7 years of applying the early flowering approach. *Planta*, 247, 1475–88.

Schmidt, M. 1938. *Venturia inaequalis* (Cooke) Aderhold. VIII. Weitere Untersuchungen zur Züchtung schorfwiderstandsfähiger Apfelsorten. *Der Zuchter*, 10, 280–91.

Schouten, H. J., Krens, F. A. and Jacobsen, E. 2006. Cisgenic plants are similar to traditionally bred plants: International regulations for genetically modified organisms should be altered to exempt cisgenesis. *EMBO Reports*, 7, 750–3.

Schouten, H., Brinkhuis, J., Burgh, A., Schaart, J., Groenwold, R., Broggini, G. L. and Gessler, C. 2014. Cloning and functional characterization of the *Rvi15* (*Vr2*) gene for apple scab resistance. *Tree Genetics & Genomes*, 10, 251–60.

Schulze-Lefert, P. and Panstruga, R. 2011. A molecular evolutionary concept connecting nonhost resistance, pathogen host range, and pathogen speciation. *Trends in Plant Science*, 16, 117–25.

Sen Gupta, G. C. and Miles, P. W. 1975. Studies on the susceptibility of varieties of apple to the feeding of two strains of woolly aphids (Homoptera) in relation to the chemical content of the tissues of the host. *Australian Journal of Agricultural Research*, 26, 157–68.

Shay, J. R. and Hough, L. F. 1952. Evaluation of apple scab resistance in selections of *Malus*. *American Journal of Botany*, 39, 288–97.

Shay, J. R. and Williams, E. B. 1956. Identification of three physiologic races of *Venturia inaequalis*. *Phytopathology*, 46, 190–3.

Shiller, J., Van De Wouw, A. P., Taranto, A. P., Bowen, J. K., Dubois, D., Robinson, A., Deng, C. H. and Plummer, K. M. 2015. A large family of *AvrLm6*-like genes in the apple and pear scab pathogens, *Venturia inaequalis* and *Venturia pirina*. *Frontiers in Plant Science*, 6, 980.

Sierotzki, H. and Gessler, C. 1998. Inheritance of virulence of *Venturia inaequalis* toward *Malus* × *domestica* cultivars. *Journal of Phytopathology*, 146, 509–14.

Singh, S., Sidhu, J. S., Huang, N., Vikal, Y., Li, Z., Brar, D. S., Dhaliwal, H. S. and Khush, G. S. 2001. Pyramiding three bacterial blight genes (*xa5*, *xa13* and *Xa21*) using marker-assisted selection into indica rice cultivar PR106. *Theoretical and Applied Genetics*, 102, 1011–15.

Sitther, V., Haro, P. A. G., Molineros, J. E., Garzon, C. D. and Jimenez-Gasco, M. M. 2018. Genetic diversity of apple- and crabapple-infecting isolates of *Venturia inaequalis* in Pennsylvania, the United States, determined by microsatellite markers. *Forest Pathology*, 48, 12405.

Skłodowska, M., Mikicinski, A., Wielanek, M., Kuzniak, E. and Sobiczewski, P. 2018. Phenolic profiles in apple leaves and the efficacy of selected phenols against fire blight (*Erwinia amylovora*). *European Journal of Plant Pathology*, 151, 213–28.

Slatnar, A., Petkovsek, M. M., Halbwirth, H., Stampar, F., Stich, K. and Veberic, R. 2012. Polyphenol metabolism of developing apple skin of a scab resistant and a susceptible apple cultivar. *Trees: Structure and Function*, 26, 109–19.

Soriano, J. M., Joshi, S. G., Van Kaauwen, M., Noordijk, Y., Groenwold, R., Henken, B., van de Weg, W. E. and Schouten, H. J. 2009. Identification and mapping of the novel apple scab resistance gene *Vd3*. *Tree Genetics & Genomes*, 5, 475–82.

Spiers, A. G., Brewster, D. T., Bus, V. G. and Hopcroft, D. H. 1998. Seasonal variation in susceptibility of xylem tissue of *Malus*, *Pyrus*, *Prunus*, and *Salix* species to *Chondrostereum purpureum* in New Zealand. *Mycological Research*, 102, 881–90.

Stam, R. and McDonald, B. A. 2018. When resistance gene pyramids are not durable – the role of pathogen diversity. *Molecular Plant Pathology*, 19, 521–4.

Stuthman, D. D. 2002. Contribution of durable disease resistance to sustainable agriculture. *Euphytica*, 124, 253–8.

Sutton, T. B., Aldwinckle, H. S., Agnello, A. M. and Walgenbach, J. F. 2014. *Compendium of Apple and Pear Diseases and Pests*. St. Paul, MN: APS Press.

Takahashi, S., Sawamura, K. and Sato, Y. 2014. Marssonina blotch. *In:* Sutton, T. B., Aldwinckle, H. S., Agnello, A. M. and Walgenbach, J. F. (Eds), *Compendium of Apple and Pear Diseases and Pests*. 2nd ed. St. Paul, MN: APS Press, pp. 46–7.

Tan, K.-C. and Oliver, R. P. 2017. Regulation of proteinaceous effector expression in phytopathogenic fungi. *PLoS Pathogens*, 13, e1006241.

Tancos, K. A., Villani, S., Kuehne, S., Borejsza-Wysocka, E., Breth, D., Carol, J., Aldwinckle, H. S. and Cox, K. D. 2016. Prevalence of streptomycin-resistant *Erwinia amylovora* in New York apple orchards. *Plant Disease*, 100, 802–9.

Thakur, K., Chawla, V., Bhatti, S., Swarnkar, M. K., Kaur, J., Shankar, R. and Jha, G. 2013. *De novo* transcriptome sequencing and analysis for *Venturia inaequalis*, the devastating apple scab pathogen. *PLoS One*, 8, e53937.

Upson, J. L., Zess, E. K., Białas, A., Wu, C.-H. and Kamoun, S. 2018. The coming of age of Evompmi: Evolutionary molecular plant-microbe interactions across multiple timescales. *Current Opinion in Plant Biology*, 44, 108–16.

Urbanietz, A. and Dunemann, F. 2005. Isolation, identification and molecular characterization of physiological races of apple powdery mildew (*Podosphaera leucotricha*). *Plant Pathology*, 54, 125–33.

USDA. 2011. APHIS confirmation of the regulatory status of null segregant (NS) lines derived from genetically engineered (GE) plants in an accelerated tobacco breeding program. https://www.aphis.usda.gov/biotechnology/downloads/reg_loi/Dr%20R am sey%20S%20Lew is%20NCS C%20Fin al.pdf (Accessed 15 August 2016).

USDA. 2014. Request for confirmation that potato is not a regulated article. https://www.aphis .usda.gov/biotechnology/downloads/reg_loi/aphisresponse_cellectis_pota to.pdf (Accessed 15 August 2016).

Usenik, V., Mikulic-Petkovsek, M., Solar, A. and Stampar, F. 2004. Flavonols of leaves in relation to apple scab resistance. *Journal of Plant Diseases and Protection*, 111(2), 137–44.

Utkhede, R. S. 1986. *In vitro* screening of the world apple germplasm collection for resistance to *Phytophthora cactorum* crown rot. *Scientia Horticulturae*, 29, 205–10.

Utkhede, R. S. and Quamme, H. A. 1988. Use of the excised shoot assay to evaluate resistance to *Phytophthora cactorum* of apple rootstock cultivars. *Canadian Journal of Plant Science*, 68, 851–7.

Valentin, R. E., Nielsen, A. L., Wiman, N. G., Lee, D.-H. and Fonseca, D. M. 2017. Global invasion network of the brown marmorated stink bug, *Halyomorpha halys. Scientific Reports*, 7, 9866.

Vallejo, R. L., Leeds, T. D., Gao, G., Parsons, J. E., Martin, K. E., Evenhuis, J. P., Fragomeni, B. O., Wiens, G. D. and Palti, Y. 2017. Genomic selection models double the accuracy of predicted breeding values for bacterial cold water disease resistance compared to a traditional pedigree-based model in rainbow trout aquaculture. *Genetics Selection Evolution*, 49, 17.

van de Weg, E., Di Guardo, M., Jänsch, M., Socquet-Juglard, D., Costa, F., Baumgartner, I., Broggini, G. A. L., Kellerhals, M., Troggio, M., Laurens, F., Durel, C.-E. and Patocchi, A. 2017. Epistatic fire blight resistance QTL alleles in the apple cultivar 'Enterprise' and selection X-6398 discovered and characterized through pedigree-informed analysis. *Molecular Breeding*, 38, 5.

van de Weg, E., di Guardo, M., Jänsch, M., Socquet-Juglard, D., Costa, F., Baumgartner, I., Broggini, G. A. L., Kellerhals, M., Troggio, M., Laurens, F., Durel, C.-E. and Patocchi, A. 2018. Epistatic fire blight resistance QTL alleles in the apple cultivar 'Enterprise' and selection X-6398 discovered and characterized through pedigree-informed analysis. *Molecular Breeding*, 38, 5.

van den Bosch, F., Lopez-Ruiz, F., Oliver, R., Paveley, N., Helps, J. and van den Berg, F. 2018. Identifying when it is financially beneficial to increase or decrease fungicide dose as resistance develops. *Plant Pathology*, 67, 549–60.

van der Hoorn, R. A. L. and Kamoun, S. 2008. From Guard to Decoy: A new model for perception of plant pathogen effectors. *Plant Cell*, 20, 2009–17.

van der Zwet, T. and Keil, H. L. 1979. *Fire Blight: A Bacterial Disease of Rosaceous Plants*. Washington DC: USDA.

Vanblaere, T., Szankowski, I., Schaart, J., Schouten, H., Flachowsky, H., Broggini, G. A. L. and Gessler, C. 2011. The development of a cisgenic apple plant. *Journal of Biotechnology*, 154, 304–11.

Velasco, R., Zharkikh, A., Affourtit, J., Dhingra, A., Cestaro, A., Kalyanaraman, A., Fontana, P., Bhatnagar, S. K., Troggio, M., Pruss, D., Salvi, S., Pindo, M., Baldi, P., Castelletti, S., Cavaiuolo, M., Coppola, G., Costa, F., Cova, V., Dal Ri, A., Goremykin, V., Komjanc, M., Longhi, S., Magnago, P., Malacarne, G., Malnoy, M., Micheletti, D., Moretto, M., Perazzolli, M., Si-Ammour, A., Vezzulli, S., Zini, E., Eldredge, G., Fitzgerald, L. M., Gutin, N., Lanchbury, J., Macalma, T., Mitchell, J. T., Reid, J., Wardell, B., Kodira, C., Chen, Z., Desany, B., Niazi, F., Palmer, M., Koepke, T., Jiwan, D., Schaeffer, S., Krishnan, V., Wu, C., Chu, V. T., King, S. T., Vick, J., Tao, Q., Mraz, A., Stormo, A., Stormo, K., Bogden, R., Ederle, D., Stella, A., Vecchietti, A., Kater, M. M., Masiero, S., Lasserre, P., Lespinasse, Y., Allan, A. C., Bus, V., Chagné, D., Crowhurst, R. N., Gleave, A. P., Lavezzo, E., Fawcett, J. A., Proost, S., Rouze, P., Sterck, L., Toppo, S., Lazzari, B., Hellens, R. P., Durel, C.-E., Gutin, A., Bumgarner, R. E., Gardiner, S. E., Skolnick, M., Egholm, M., Van De Peer, Y., Salamini, F. and Viola, R. 2010. The genome of the domesticated apple (*Malus × domestica* Borkh.). *Nature Genetics*, 42, 833–9.

Villarino, M., Larena, I., Martinez, F., Melgarejo, P. and De Cal, A. 2012. Analysis of genetic diversity in *Monilinia fructicola* from the Ebro Valley in Spain using ISSR and RAPD markers. *European Journal of Plant Pathology*, 132, 511–24.

Vleeshouwers, V. G. A. A. and Oliver, R. P. 2014. Effectors as tools in disease resistance breeding against biotrophic, hemibiotrophic, and necrotrophic plant pathogens. *Molecular Plant-Microbe Interactions*, 27, 196–206.

Vogt, I., Wohner, T., Richter, K., Flachowsky, H., Sundin, G. W., Wensing, A., Savory, E. A., Geider, K., Day, B., Hanke, M. V. and Peil, A. 2013. Gene-for-gene relationship in the host-pathogen system *Malus* × *robusta* 5 – *Erwinia amylovora*. *New Phytologist*, 197, 1262–75.

Volz, R. K., Rikkerink, E., Austin, P., Lawrence, T. and Bus, V. G. M. 2009. 'Fast-breeding' in apple: A strategy to accelerate introgression of new traits into elite germplasm. *Acta Horticulturae*, 814, 163–8.

Vrancken, K., Holtappels, M., Schoofs, H., Deckers, T., Treutter, D. and Valcke, R. 2013. *Erwinia amylovora* affects the phenylpropanoid-flavonoid pathway in mature leaves of *Pyrus communis* cv. conference. *Plant Physiology and Biochemistry*, 72, 134–44.

Walker, J. T. S., Suckling, D. M. and Wearing, C. H. 2017. Past, present, and future of integrated control of apple pests: The New Zealand experience. *Annual Review of Entomology*, 62, 231–48.

Waltz, E. 2015. Nonbrowning GM apple cleared for market. *Nature Biotechnology*, 33, 326.

Waltz, E. 2018. With a free pass, CRISPR-edited plants reach market in record time. *Nature Biotechnology*, 36, 6.

War, A. R., Sharma, H. C., Paulraj, M. G., War, M. Y. and Ignacimuthu, S. 2011. Herbivore induced plant volatiles: Their role in plant defense for pest management. *Plant Signaling & Behavior*, 6, 1973–8.

White, A. G. and Bus, V. G. 1999. Breeding commercial apple cultivars in New Zealand with resistances to pests and diseases. *Acta Horticulturae*, 484, 157–61.

Wicks, T. 1977. Tolerance to benzimidazole fungicides in blue mold *Penicillium expansum* on pears. *Plant Disease Reporter*, 61, 447–9.

Williams, E. B. and Brown, A. G. 1968. A new physiologic race of *Venturia inaequalis*, incitant of apple scab. *Plant Disease Reporter*, 52, 799–801.

Wolfe, M. S. and Gessler, C. 1992. The use of resistance genes in breeding: Epidemiological considerations. *In:* Boller, T. and Meins, F. (Eds), *Plant Gene Research 8: Genes Involved in Plant Disease*. New York, NY: Springer-Verlag.

Won, K., Bastiaanse, H., Kim, Y., Song, J., Kang, S., Lee, H., Cho, K., Brewer, L., Singla, G., Gardiner, S., Chagné, D. and Bus, V. M. 2014. Genetic mapping of polygenic scab (*Venturia pirina*) resistance in an interspecific pear family. *Molecular Breeding*, 34, 2179–89.

Woo, J. W., Kim, J., Kwon, S. I., Corvalán, C., Cho, S. W., Kim, H., Kim, S.-G., Kim, S.-T., Choe, S. and Kim, J.-S. 2015. DNA-free genome editing in plants with preassembled CRISPR-Cas9 ribonucleoproteins. *Nature Biotechnology*, 33, 1162–4.

Würdig, J., Flachowsky, H., Sass, A., Peil, A. and Hanke, M. V. 2015. Improving resistance of different apple cultivars using the *Rvi6* scab resistance gene in a cisgenic approach based on the Flp/*FRT* recombinase system. *Molecular Breeding*, 35, 95.

Xu, X.-M. 2012. Super-races are not likely to dominate a fungal population within a life time of a perennial crop plantation of cultivar mixtures: A simulation study. *BMC Ecology*, 12, 16.

Xu, M. and Korban, S. S. 2002. A cluster of four receptor-like genes resides in the *Vf* locus that confers resistance to apple scab disease. *Genetics*, 162, 1995–2006.

Young, E., Rock, G. C., Zeiger, D. C. and Cummins, J. N. 1982. Infestation of some *Malus* cultivars by the North Carolina woolly apple aphid biotype. *HortScience*, 17, 787–8.

Zhang, M. X. and Coaker, G. 2017. Harnessing effector-triggered immunity for durable disease resistance. *Phytopathology*, 107, 912–19.

Zhang, S. C., Wen, Z. X., Difonzo, C., Song, Q. J. and Wang, D. C. 2018. Pyramiding different aphid-resistance genes in elite soybean germplasm to combat dynamic aphid populations. *Molecular Breeding*, 38, 29.

Zhou, H. X., Zhang, R. M., Tan, X. M., Tao, Y. L., Wan, F. H., Wu, Q. and Chu, D. 2015. Invasion genetics of woolly apple aphid (Hemiptera: Aphididae) in China. *Journal of Economic Entomology*, 108, 1040–6.

Zhu, F., Zhou, Y. K., Ji, Z. L. and Chen, X. R. 2018. The plant ribosome-inactivating proteins play important roles in defense against pathogens and insect pest attacks. *Frontiers in Plant Science*, 9, 146.

Chapter 11

Improving plant propagation methods for fruit disease control

Ioannis Tzanetakis, University of Arkansas, USA; and Robert Martin, USDA-ARS, USA

1 Introduction

In the era of agricultural globalization, there is significant movement of plant propagation material across country lines and continents. This allows for the exchange of high-quality planting material, providing producers with superior germplasm, enhancing farm sustainability and profits. Nevertheless, this practice has hidden dangers, including the potential of moving infectious agents to areas where they are not endemic. Agents unknown to science, entities that are difficult to identify given our unawareness of their existence, can pose a significant threat to agricultural industries. During the latter part of the twentieth century, propagation material passed through rigorous tests, including bioindicators, aiming to identify those agents unknown to science, and laboratory protocols such as ELISA and PCR that provide sensitive detection of, primarily, agents that are well-characterized. These technologies have proven invaluable in safeguarding the agricultural industries and the ecosystem-at-large from invasive pathogens, yet they have several disadvantages. In the case of bioindicators, the process is slow and expensive. From inoculation till plants show symptoms it may be four years; during that time the inoculated plants need constant care. In addition, there are no guarantees that all agents present in the assessed material cause symptoms on a particular indicator. The laboratory techniques are fast and sensitive but they often do not capture the genetic diversity of the pathogen; an isolate may be undetectable using current knowledge and tools, leading to false negative results.

http://dx.doi.org/10.19103/AS.2019.0046.13

The new technologies that have emerged in the last ten years allow for the rapid and accurate characterization of the pathogens that are present in a plant. Known as next generation or high throughput sequencing (NGS or HTS, respectively) those technologies come in different platforms and capacities, with all having a common denominator: they provide massive amounts of sequence data which, when appropriately processed through designated bioinformatics pipelines, can determine the biome of the material tested.

Yet, there are impediments that need to be addressed before HTS becomes the preferred methodology for validation of the health status of plant propagation materials across the globe. Whereas there are studies that verify the superiority of HTS when compared to the 'traditional' methods (indicators, ELISA, PCR etc.), there is a necessity for additional high-quality side-to-side comparisons on different hosts and pathogen combinations that pass the peer-review process and can thus be accepted by regulatory agencies around the globe.

HTS is largely unbiased but cannot determine if a new agent is a pathogen or a benign endophyte. Our ability to identify the new agents is largely based on previous knowledge – the bioinformatics pipelines compare the data to known sequences and protein motifs. If a new agent is different enough from everything that has been characterized and deposited in the databases the bioinformatics pipelines may not recognize it as a potential pathogen and rather assign it to 'sequence unknown', a term used to describe the 'noise' of the technology. In this chapter we will take a systems-based approach on detection, elimination and safeguarding against systemic pathogens that affect clonally propagated fruit crops.

2 Steps taken to improve propagation material

In breeding programmes, a small number of advanced selections are being evaluated at any given time with some of those to be released as named cultivars. Advanced selections at this stage in the programme provide an excellent checkpoint where plants can be tested using traditional technologies (bioindicators and laboratory tests) in side-by-side comparisons with HTS to determine the better approach to evaluate the health status of plants. Testing advanced selections at this stage (2 years prior to release) allows time for virus testing and for virus elimination if needed, without delaying release.

Neither approach is 100% accurate. There are and will be cases where HTS will identify new agents that do not cause symptoms on bioindicators, as has been our experience with many viruses in berry crops that remain asymptomatic when grafted on standard indicators (Susaimuthu et al., 2008; Quito-Avila et al., 2011; Martin and Tzanetakis, *unpublished*). On the other hand there may be agents without any recognizable similarity to known pathogens escaping *in silico* detection yet they could cause symptoms on indicators. Risk assessment of both approaches, based on peer-reviewed, published case studies, is a prerequisite before the process moves forward and one methodology supersedes the other. There is always the possibility of using both approaches if funds and time are not limiting factors. Given the situation at the global scale, and given the experience in our laboratories, we expect that HTS will replace traditional methods (Al Rwahnih et al., 2015; Rott et al., 2017; Villamor et al., 2018). With more reliable testing that requires less time, improved germplasm (new cultivars) will enter nursery systems in a timely manner allowing producers to capture the advantages provided by growing better material.

Yet, before all parts of the proposed scheme are optimized there are simple steps that can assist in the timely release of pathogen-tested material to producers. An easy and effective approach is to test breeding parents for seed- and pollen-borne viruses so as to reduce the number of virus-infected seedlings in a population and increase the chance of selections being free of targeted pathogens, and thus possibly avoiding the sometimes lengthy virus elimination process. If plants are infected by pathogenic agents (as opposed to cryptic ones that are not known to cause symptoms in single or mixed infections) then there are several options used today to obtain 'clean' plants.

Thermo-, chemo- and cryotherapy, each employed alone or in combination with the other two have proven to be invaluable tools for the elimination of most systemic pathogens. The most commonly used method is thermotherapy. In the case of phytoplasmas like Bois Noir phytoplasma or other systemic bacteria like *Candidatus* Liberibacter asiaticus, heat treatment at higher temperatures (40–55°C) for a variable period of time (10 min – 48 h) may eliminate them from plants (Chalak et al., 2013; Caudwell et al., 1997; Hoffman et al., 2013). The temperature and time of hot water treatment need to be carefully monitored to balance between effective treatment and plant survival. In the case of viroids such as hop stunt viroid, low temperature treatment (5–8°C) for a prolonged period of time (~6 months) followed by excision of the apical meristem allows for their elimination (Paduch-Cichal and Kryczyński, 1987; El-Dougdoug et al., 2010).

In the cases of viruses, even those that are invasive and can be transmitted by pollen and/or seed (e.g. ilarviruses), plants go through periods of heat treatment at ~37°C, or a regime of cycling the temperature between 30°C and 37°C every 4 h (Nyland and Goheen, 1969; Varveri et al., 2015; Gella and Errea, 1998). A variation on this theme where heat is applied to the above ground parts of the plant with the root system maintained at room temperature has also been successful and can increase plant survival rate during the treatment. The process continues from 4 to 6 weeks depending on the plant species and its heat tolerance. High temperatures upregulate the plant response to virus infection through the RNA silencing pathway, plants grow slowly but virus replication may cease (Szittya et al., 2003; Mochizuki and Ohki, 2004). Taking this advantage into consideration, even though the plants grow very slowly, the undifferentiated apical meristem may become virus free. For phloem-limited viruses, excision of meristems without heat treatment can be effective, since there is no differentiated phloem in meristematic dome. After excision and placement of the meristematic dome onto appropriate growth media, plant regeneration, rooting and growing in a greenhouse, the plants are retested for the viruses detected during the original screening. If a plant free of targeted pathogens is identified it becomes the type plant for the clone. In the event that none of the regenerated plants are free of targeted pathogens, the therapy process is repeated. The premise of chemotherapy is similar to that of thermotherapy in that the application of virus replication inhibitors (e.g. ribavirin) may not allow for virus replication and movement to the apical meristem, and it is often that the two techniques are used in tandem (Hu et al., 2012, 2015). Cryotherapy is a relatively new technology and is based on the fact that meristematic cells lack a vacuole and can tolerate freezing (Wang et al., 2009). Tissues are conditioned before being immersed into liquid nitrogen and then reconditioned to allow growth in media post-treatment. Only cells that could sustain the treatment regenerate. Since there is no differentiated vascular tissue in the meristem the recovered tissue after cryo-treatment is likely to be virus free as proven with difficult-to-eliminate viruses such as raspberry bushy dwarf virus (RBDV; Wang et al., 2008; Li et al., 2016; Bettoni et al., 2018). Viruses that move between cells very efficiently are able to move into the meristematic dome and are challenging to eliminate with any of the three methods and that is where combinations of the methods can be most beneficial.

In the United States, once a breeding accession is tested free of all targeted pathogens and becomes a named cultivar it enters the propagation system. Large-scale propagation is usually carried out as part of a certification scheme that aims to secure the health status by providing safeguards in the form of rules and best management practices (BMPs) that are based on the biology of the pathogens and their vectors (Martin and Tzanetakis, 2015). BMPs address the weak links in the propagation pipelines in terms of pathogen entry, and provide measures and solutions to ensure that plants are not re-infected during propagation. Whereas guidelines may change based on the locality of the nursery, there are rules that apply universally. Such a rule is the testing of the material during the propagation stages or generations (G level). This initiates with the single 'mother' plant (G1) that will generate all plants of the cultivar, which as noted above, has been tested free for all known systemic pathogens of concern. Each propagation step becomes a new generation (generation is used here in terms of stage in the plant increase system rather than a seedling) (Martin and Tzanetakis, 2014). In most cases the scheme terminates with generation

4 (G4), certified plants, which is the material sold to producers. Whereas the BMPs aim to minimize the risk of infection there may be catastrophic failures, environmental or anthropogenic, and for this reason modern guidelines include pathogen testing at each propagation step (G2–G4). Unlike the G1 material that has been tested for all pathogens of concern and gone through therapy to eliminate target pathogens, such a process is unfeasible for the subsequent generations given the testing costs. For G2–G4 plants, guidelines aim to ensure the health status taking into account the financial burden to the nursery. G2–G4 are tested based on current knowledge of the pathogens that may circulate in the area where a nursery is located ('canary' viruses) (Tzanetakis and Martin, 2017; Martin and Tzanetakis, 2013, 2018). As an example G2–G4 blueberry plants grown in Oregon or Washington State are tested for blueberry shock and blueberry scorch viruses, known to be endemic in the area. By contrast, a plant grown under a certification scheme in Michigan needs to be tested for blueberry scorch, blueberry shoestring, blueberry leaf mottle, blueberry virus A and blueberry red ringspot viruses, found widespread in production areas in the state. Also, if the G2 or G3 plants are grown in soil they need to be tested for the nematode/soil-borne-transmitted viruses (tomato ringspot, tobacco ringspot, peach rosette mosaic and blueberry mosaic). The tests used for each pathogen and crop have undergone rigorous evaluations and are based on the population structure of the pathogen where available (Thekke-Veetil and Tzanetakis, 2017; Hassan et al., 2017). Tests are designed to detect a wide array of variants of a pathogen, obtained after studying isolates from different hosts and geographic areas and therefore minimizing the possibility of false negative because of isolate diversity (Poudel et al., 2013, 2014; Thekke-Veetil et al., 2016).

The number of plants changes drastically between generations and for this reason testing is based on a hypergeometric scheme that provides confidence in the certification scheme. For example, the proposed certification guidelines for berry crops in the United States (the latest versions are found at ncpnberries.org) have a testing of all G2 material for the canary viruses, whereas G4 plants are tested at levels that would ensure detection of 5% infection with 95% confidence. A 5% infection in a nursery operation is a significant number, but the testing is not designed to guarantee freedom from targeted pathogens, but rather identify a serious breach in the plant propagation process (Tzanetakis and Martin, 2017). This is done using a balance of BMPs combined with testing to provide plants of high health status at a reasonable cost. Even if every plant were tested, one could not guarantee that 100% of the plants were free of targeted pathogens since no test is 100% reliable. The goal of certification programmes is to safeguard the agricultural industries from dissemination of infected material that could trigger an epidemic once planted in commercial fields.

Even though rules are improving, there are still major regulatory issues that blockade the free movement of propagation material across state and country lines. For this reason the harmonization of rules that has been initiated for certain crops in the United States should become the yardstick to approach the problem (Gergerich et al., 2015). It is obvious that there cannot be a universally applicable certification scheme because of the different needs and special situations based on the nursery location. Having that in mind, and because of the revolutionary opportunities provided with the advancements of technology (e.g. HTS) there is a clear opportunity for the harmonization of rules and regulations based on evidence-based BMPs. Not all BMPs would apply to all operations but if the regulatory agencies accept them then the next step would be the ease on the movement of propagation material that originated in operations that followed the guidelines. The protocols could reduce quarantine times in countries such as South Africa, New Zealand, Chile and Australia where the introduction of new cultivars is often delayed by years due to difficulties on introduction and testing on plant material that has been accredited as 'clean' elsewhere in the world.

Producers around the world, and especially those in areas where there are no certification guidelines and high-quality nurseries, may be unable to cope with demand. Not only will individuals have access to high-quality material, but they will get access to that material in shorter time frames without the bureaucratic delays associated with moving material across borders. This practice will also benefit producers that have planted pathogen-tested material as it minimizes the risk of introducing a new pathogen in their production area.

The new standards better serve the health of agricultural industries when compared to those used in the past, and still used in some areas, which are solely based on visual observations. As examples we will provide two major disease outbreaks in strawberry in the United States that are directly connected with the dissemination of infected plants from nursery material that had undergone visual inspections (Martin and Tzanetakis, 2006, 2013). The 2002–04 epidemic cost producers an estimated US$50 M and it was due to the fact that nurseries were sending material infected with asymptomatic whitefly-transmitted viruses. The viruses cause symptoms on indicators but are symptomless in single infections in modern commercial cultivars. At the time there was no knowledge of whitefly-transmitted viruses in strawberry and even though there were significant populations of whiteflies in the nursery, there were no control measures implemented in propagation fields. Once the plants made it to production fields, additional viruses, primarily aphid-borne, infected the plants. The combination of the whitefly and aphid-transmitted viruses caused severe decline and plant death. The second case occurred in the 2012–13 season; nursery plants were almost universally infected with two viruses, strawberry mild yellow edge and strawberry mottle viruses. In the nursery setting, plants were well irrigated and fertilized and did not exhibit any obvious symptoms during the growing season prior to digging. Based on the visual inspections there was no reason for concern about the plant quality and plants were shipped to many areas in the southern United States. Once planted in production fields there was very little growth and little or no fruit production. This led to losses in excess of US$25 M for the producers. If there was even minimal testing in the G4 level in either of these cases the severe losses may have been averted (Tzanetakis and Martin, 2017).

3 Case studies

At the turn of the twenty-first century there were observations of severe decline in blackberry in the southern United States. Leaf symptoms included vein banding, oak leaf patterns, mottling and mosaic as well as ringspots (Martin et al., 2013). Symptoms were originally attributed to tobacco ringspot virus (TRSV), a nematode-transmitted virus and one of the pathogens propagators would have screened for. The widespread presence of the disease, even in areas with a known history of TRSV and its vector, could not be explained and a survey in areas with high disease pressure revealed that only few of the affected plants were infected by TRSV. In a parallel study, TRSV failed to cause symptoms when inoculated to virus-tested plants using the nematode vector (Gergerich, *unpublished*). This led to an in-depth study of the disorder, now named blackberry yellow vein disease (BYVD) using traditional approaches but also HTS for the detection and identification of known and new pathogens that may contribute to the onset of BYVD.

HTS has been an invaluable tool to identify new viruses in a timely manner in plants that may be infected with six or more viruses, information that would have taken years to obtain using traditional methodologies. Almost all major vector groups including whiteflies, eriophyid mites, mealybugs, nematodes, thrips or pollen play a role in vectoring the various viruses associated with BYVD disease (Martin et al., 2013; Hassan et al., 2017). This allowed for a better understanding of BYVD, including that virus identity is not as important as the sheer number of viruses infecting a plant. This fact allowed us to focus on the viruses that are present and circulate in a particular area to develop custom, systems-based approaches to combat BYVD. Given that none of these viruses (including TRSV) is known to cause BYVD in single infections, management strategies are based on vector presence and distribution, mode of transmission and which virus(es) are easiest to control (Martin et al., 2013). All viruses associated with the disease are now part of the certification scheme for blackberry and there is a good understanding of their distribution in all areas where the disease has become endemic. There have been numerous cases where virus-tested plants became symptomatic and tested positive for at least three viruses within a month of planting (Khadgi, 2015). In such cases, the solution is as simple as discouraging any propagation or production activity since disease is expressed prior to first harvest. If pollen-borne viruses are prevalent in an area, nurseries may be able to avoid infection by not allowing flowering or excising flowers before they open. For fruit producers this is obviously not an option, but if none of the other viruses are prevalent it may still be profitable to produce. On the other hand if there is a virus with limited host range and a vector that enters the field over a small window of time there are approaches to minimize the presence of this virus and even eliminate it from the system by a single, well-timed pesticide application (Halgren, 2006). Those approaches have yielded excellent results and even though the disease is still present in the southern states, the productive span of a blackberry field has expanded significantly from 5 to 7 years, that is the usual time frame when no control measures are applied for vector control. Also, knowledge of the viruses and their vectors combined with testing for the BYVD-associated viruses in nursery operations has led to the development of BMPs and greatly limited the spread of the viruses from nurseries to production fields (Martin et al., 2013).

Another system where the application of the new technologies and modern propagation guidelines have benefited growers is raspberry crumbly fruit and decline (RCFD). For decades the pollen-borne RBDV was thought to be the sole causal agent of the disease that primarily manifests itself with unmarketable crumbly fruit and plant decline. As in the case of BYVD, and using similar approaches, we were able to identify two additional viruses, raspberry leaf mottle virus (RLMV) and raspberry latent virus (RpLV) which act synergistically with RBDV and result in significantly more severe symptoms than RBDV alone. As an example, double- and triple-infected plants were less than half the height of healthy control plants in field trials and significantly lagged in several other attributes including fruit weight and firmness (Quito-Avila et al., 2014). Unlike RBDV, RLMV and RpLV are transmitted by the large raspberry aphid in semi-persistent and persistent propagative modes, respectively (Quito-Avila et al., 2012). Whereas RLMV caused symptoms on standard indicators, RpLV is asymptomatic in single infections and only with the advancement of HTS were we able to study it. The importance of the virus complex became apparent in the case of RCFD when molecular studies showed that mixed infections elevated the RBDV titre 400 times, which translated to severe crumbly fruit in the field (Quito-Avila and Martin, 2012). We hypothesize that the higher RBDV titre may increase the virus rate of spread. Based on data on the ecology of the large raspberry aphid in affected areas, control measures were developed where the vector of RLMV and RpLV could be largely controlled with a single insecticide application resulting in reduced disease impact (Quito-Avila et al., 2014). In the nursery system, testing for the new viruses and aphid control in the propagation field has become common practice, leading to the dissemination of clean plants with longer productive span, free of symptomology.

4 Summary

HTS has proved to be an invaluable tool in assessing the health status of high-value plants and for the rapid identification and characterization of unknown viruses. It allows for identification of known and unknown viruses without the biological bias of an indicator that may be asymptomatic when infected with a new virus or even a variant of a known one and provides the information to develop virus-specific tests (Al Rwahnih et al., 2015; Rott et al., 2017). Breeding material could be tested free of targeted pathogens and released in a shorter period of time when compared to the 2–4 years needed to complete bioassays for some crops like fruit trees if testing were based on HTS. This advantage, in addition to the implementation of a testing regime of nursery stock, can eliminate the nursery-fuelled epidemics of the past. Even when nurseries were following all the BMPs there was the potential of escapes because of new agents infecting the plants that are transmitted by new vectors on that host (e.g. whiteflies in strawberry). HTS minimizes such possibilities and allows for the testing for the full range of systemic agents infecting a crop, both known and unknown. If certification rules and BMPs are harmonized and implemented then producers will be able to access high-quality material in a fraction of the time needed today and provide the producer and the consumer with a better quality product. In addition, producers will conserve valuable environmental resources on plants that do not have the potential to produce their full potential. After all, there is no cultivation guide for sick plants, all recommendations are based on research done on healthy plants.

5 Future trends

As in all aspects of biological sciences, new discoveries have the potential to change our approaches on how to provide the healthiest material possible to producers. The implementation of genome editing through CRSPR/Cas (clustered regularly interspaced short palindromic repeats and CRISPR-associated protein) (Mojica et al., 2000) has the potential to revolutionize the production of clean plants and their ability to eliminate a pathogen upon infection. First discovered in bacteria as a mechanism to detect and destroy foreign elements, primarily viruses, CRSPR/Cas can be used to modify a eukaryotic genome (Miller et al., 2011; Zhang et al., 2011; Shan et al., 2013); altering, deleting or adding bases in a particular site. With this technology, new resistant plants could be developed and made available to the general public without the long studies required for all genetically modified organisms based on the insertion of a transgene, although CRSPR/Cas regulations have not been fully vented in the majority of the world. In the case of production of clean plants we could potentially alter a protein so it will not be conducive to virus infection, for example initiation factors that will not allow for potyvirus replication or modify a resistance gene to identify a resistance-breaking strain of a pathogen. Whereas there are studies on the implementation of the technology in plants for protection against viruses (Ali et al., 2015; Chandrasekaran et al., 2016; Macovei et al., 2018) we still need time to identify the 'susceptibilities' of all systemic pathogens of a particular crop and use the system as an umbrella approach against them, ensuring the health of plants during propagation but also in the field.

6 Where to look for further information

Several of the leaders in the National Clean Plant Network in the United States collaborated on a review article, 'Safeguarding fruit crops in the age of agricultural globalization', that covers many of the issues associated with the large-scale movement of plant material in terms of pathogen dispersal (Gergerich et al., 2015). The article describes the process of how plants are tested and if therapy needs to be applied to produce plants for the top tier in a clean plant production scheme. A comprehensive review of virus diseases and their control for major vegetatively propagated crops was the topic of Volume 91 of *Advances in Virus Research* (Lobenstein and Katis, 2015), which covered certification schemes, sweet potatoes, pome and stone fruits, cassava, citrus, grapevines, berries, banana and plantain. Valuable resources on virus elimination can be found in several review articles on the subject (Panattoni et al., 2013; Varveri et al., 2015; Wang et al., 2009). The G nomenclature was originally proposed by a working group of the North American Plant Protection Organization to facilitate accurate communication among regulators and nurserymen when discussing the plant propagation pipeline for vegetative crops (RSPM No. 35).

There are many publications demonstrating the utility of HTS for identifying unknown viruses in various plants, with much of the work focused on vegetative crops (Adams et al., 2009; Kreuze et al., 2009; Goodwin, et al., 2016; Levy and Myers, 2016; Di Bello et al., 2015; Shahid et al., 2017; Xiang et al., 2015) and several publications on the comparison of HTS with biological indicators (Al Rwahnih, 2015; Rott et al., 2017). Several groups have developed bioinformatics pipelines to facilitate the analysis of HTS data (Ho and Tzanetakis, 2014; Zheng et al., 2017). The article by Martin et al. (2016) provides a comprehensive review of how new technologies and regulatory guidelines could elevate the status of clonally propagative crops around the globe.

7 References

Adams, I. P., Glover, R. H., Monger, W. A., Mumford, R., Jackeviciene, E., Navalinskiene, M., Samuitiene, M. and Boonham, N. 2009. Next-generation sequencing and metagenomic analysis: A universal diagnostic tool in plant virology. *Molecular Plant Pathology*, 10(4), pp. 537–45.

Al Rwahnih, M., Daubert, S., Golino, D., Islas, C. and Rowhani, A. 2015. Comparison of next-generation sequencing versus biological indexing for the optimal detection of viral pathogens in grapevine. *Phytopathology*, 105(6), pp. 758–63.

Ali, Z., Abulfaraj, A., Idris, A., Ali, S., Tashkandi, M. and Mahfouz, M. M. 2015. CRISPR/Cas9-mediated viral interference in plants. *Genome Biology*, 16(1), p. 238.

Bettoni, J. C., Dalla Costa, M., Souza, J. A., Volk, G. M., Nickel, O., da Silva, F. N. and Kretzschmar, A. A. 2018. Cryotherapy by encapsulation-dehydration is effective for in vitro eradication of latent viruses from 'Marubakaido' apple rootstock. *Journal of Biotechnology*, 269, pp. 1–7.

Caudwell, A., Larrue, J., Boudon-Padieu, E. and McLean, G. D. 1997. Flavescence dorée elimination from dormant wood of grapevines by hot-water treatment. *Australian Journal of Grape and Wine Research*, 3(1), pp. 21–5.

Chalak, L., Dekwaneh, B., Elbitar, A., Mourad, N., Mortada, C. and Choueiri, E. 2013. Elimination of grapevine bois noir phytoplasma by tissue culture coupled or not with heat therapy or hot water treatment. *Advances in Crop Science and Technology*, 1, p. 107.

Chandrasekaran, J., Brumin, M., Wolf, D., Leibman, D., Klap, C., Pearlsman, M., Sherman, A., Arazi, T. and Gal-On, A. 2016. Development of broad virus resistance in non-transgenic cucumber using CRISPR/Cas9 technology. *Molecular Plant Pathology*, 17(7), pp. 1140–53.

Di Bello, P. L., Ho, T. and Tzanetakis, I. E. 2015. The evolution of emaraviruses is becoming more complex: Seven segments identified in the causal agent of Rose rosette disease. *Virus Research*, 210, pp.241–4.

El-Dougdoug, K. A., Osman, M. E., Abdelkader, H. S., Dawoud, R. A. and Elbaz, R. M. 2010. Elimination of Hop stunt viroid (HSVd) from infected peach and pear plants using cold therapy and chemotherapy. *Australian Journal of Basic and Applied Sciences*, 4(1), pp. 54–60.

Gella, R. and Errea, P. 1998. Application of in vitro therapy for ilarvirus elimination in three Prunus species. *Journal of Phytopathology*, 146(8–9), pp.445–9.

Gergerich, R. C., Welliver, R. A., Gettys, S., Osterbauer, N. K., Kamenidou, S., Martin, R. R., Golino, D. A., Eastwell, K., Fuchs, M., Vidalakis, G. and Tzanetakis, I. E. 2015. Safeguarding fruit crops in the age of agricultural globalization. *Plant Disease*, 99(2), pp. 176–87.

Goodwin, S., McPherson, J. D. and McCombie, W. R. 2016. Coming of age: Ten years of next-generation sequencing technologies. *Nature Reviews Genetics*, 17(6), p.333.

Halgren, A. B. 2006. Characterization, epidemiology, and ecology of a virus associated with black raspberry decline (Doctoral dissertation).

Hassan, M., Di Bello, P. L., Keller, K. E., Martin, R. R., Sabanadzovic, S. and Tzanetakis, I. E. 2017. A new, widespread emaravirus discovered in blackberry. *Virus Research*, 235, pp. 1–5.

Ho, T. and Tzanetakis, I. E. 2014. Development of a virus detection and discovery pipeline using next generation sequencing. *Virology*, 471, pp. 54–60.

Hoffman, M. T., Doud, M. S., Williams, L., Zhang, M. Q., Ding, F., Stover, E., Hall, D., Zhang, S., Jones, L., Gooch, M. and Fleites, L. 2013. Heat treatment eliminates '*Candidatus* Liberibacter asiaticus' from infected citrus trees under controlled conditions. *Phytopathology*, 103(1), pp. 15-22.

Hu, G. J., Hong, N., Wang, L. P., Hu, H. J. and Wang, G. P. 2012. Efficacy of virus elimination from in vitro-cultured sand pear (*Pyrus pyrifolia*) by chemotherapy combined with thermotherapy. *Crop Protection*, 37, pp. 20-5.

Hu, G., Dong, Y., Zhang, Z., Fan, X., Ren, F. and Zhou, J. 2015. Virus elimination from *in vitro* apple by thermotherapy combined with chemotherapy. *Plant Cell, Tissue and Organ Culture (PCTOC)*, 121(2), pp. 435-43.

Khadgi, A. 2015. *Blackberry Virosome: A Micro and Macro Approach*. University of Arkansas.

Kreuze, J. F., Perez, A., Untiveros, M., Quispe, D., Fuentes, S., Barker, I. and Simon, R. 2009. Complete viral genome sequence and discovery of novel viruses by deep sequencing of small RNAs: A generic method for diagnosis, discovery and sequencing of viruses. *Virology*, 388(1), pp. 1-7.

Levy, S. E. and Myers, R. M. 2016. Advancements in next-generation sequencing. *Annual Review of Genomics and Human Genetics*, 17, pp. 95-115.

Li, B. Q., Feng, C. H., Hu, L. Y., Wang, M. R. and Wang, Q. C. 2016. Shoot tip culture and cryopreservation for eradication of *Apple stem pitting virus* (ASPV) and *Apple stem grooving virus* (ASGV) from apple rootstocks 'M9'and 'M26'. *Annals of Applied Biology*, 168(1), pp. 142-50.

Loebenstein, G. and Katis, N. I. 2015. Control of plant virus diseases: Vegetatively-propagated crops. In: G. Loebenstein and N. I. Katis (Eds), *Advances in Virus Research* (vol. 91, 337p). Academic Press.

Macovei, A., Sevilla, N. R., Cantos, C., Jonson, G., Slamet-Loedin, I., Čermák, T., Voytas, D., Choi, I. R. and Chadha-Mohanty, P. 2018. Novel alleles of rice *eIF4G* generated by *CRISPR/Cas9*-targeted mutagenesis confer resistance to Rice tungro spherical virus. *Plant Biotechnology Journal*, 16, pp. 1918-27.

Martin, R. R. and Tzanetakis, I. E. 2006. Characterization and recent advances in detection of strawberry viruses. *Plant Disease*, 90(4), pp. 384-96.

Martin, R. R. and Tzanetakis, I. E. 2013. High risk strawberry viruses by region in the United States and Canada: Implications for certification, nurseries, and fruit production. *Plant Disease*, 97(10), pp. 1358-62.

Martin, R. R. and Tzanetakis, I. E. 2014. Pathogen-tested planting material. In *Encyclopedia of Agriculture and Food Systems* (vol. 4, pp. 304-12). Elsevier.

Martin, R. R. and Tzanetakis, I. E. 2015. Control of virus diseases of berry crops. In: G. Loebenstein and N. I. Katis (Eds), *Advances in Virus Research* (vol. 91, pp. 271-309). Academic Press.

Martin, R. R. and Tzanetakis, I. E. 2018. High risk blueberry viruses by region in North America: Implications for certification, nurseries, and fruit production. *Viruses*, 10, 342.

Martin, R. R., MacFarlane, S., Sabanadzovic, S., Quito, D., Poudel, B. and Tzanetakis, I. E. 2013. Viruses and virus diseases of *Rubus*. *Plant Disease*, 97(2), pp. 168-82.

Martin, R. R., Constable, F. and Tzanetakis, I. E. 2016. Quarantine regulations and the impact of modern detection methods. *Annual Review of Phytopathology*, 54, pp. 189-205.

Miller, J. C., Tan, S., Qiao, G., Barlow, K. A., Wang, J., Xia, D. F., Meng, X., Paschon, D. E., Leung, E., Hinkley, S. J. and Dulay, G. P. 2011. A TALE nuclease architecture for efficient genome editing. *Nature Biotechnology*, 29(2), p. 143.

Mochizuki, T. and Ohki, S. T. 2004. Shoot meristem tissue of tobacco inoculated with *Cucumber mosaic virus* is infected with the virus and subsequently recovers from infection by RNA silencing. *Journal of General Plant Pathology*, 70(6), pp. 363–6.

Mojica, F. J., Díez-Villaseñor, C., Soria, E. and Juez, G. 2000. Biological significance of a family of regularly spaced repeats in the genomes of *Archaea*, *Bacteria* and *Mitochondria*. *Molecular Microbiology*, 36(1), pp. 244–6.

Nyland, G. and Goheen, A. C. 1969. Heat therapy of virus diseases of perennial plants. *Annual Review of Phytopathology*, 7(1), pp. 331–54.

Paduch-Cichal, E. and Kryczyński, S. 1987. A low temperature therapy and meristem-tip culture for eliminating four viroids from infected plants. *Journal of Phytopathology*, 118(4), pp. 341–6.

Panattoni, A., Luvisi, A. and Triolo, E. 2013. Elimination of viruses in plants: Twenty years of progress. *Spanish Journal of Agricultural Research*, 11(1), pp. 173–88.

Poudel, B., Wintermantel, W. M., Cortez, A. A., Ho, T., Khadgi, A. and Tzanetakis, I. E. 2013. Epidemiology of Blackberry yellow vein associated virus. *Plant Disease*, 97(10), pp. 1352–7.

Poudel, B., Ho, T., Laney, A., Khadgi, A. and Tzanetakis, I. E. 2014. Epidemiology of *Blackberry chlorotic ringspot virus*. *Plant Disease*, 98(4), pp. 547–50.

Quito-Avila, D. F. and Martin, R. R. 2012. Real-time RT-PCR for detection of *Raspberry bushy dwarf virus*, *Raspberry leaf mottle virus* and characterizing synergistic interactions in mixed infections. *Journal of Virological Methods*, 179(1), pp. 38–44.

Quito-Avila, D. F., Jelkmann, W., Tzanetakis, I. E., Keller, K. and Martin, R. R. 2011. Complete sequence and genetic characterization of *Raspberry latent virus*, a novel member of the family *Reoviridae*. *Virus Research*, 155(2), pp. 397–405.

Quito-Avila, D. F., Lightle, D., Lee, J. and Martin, R. R. 2012. Transmission biology of *Raspberry latent virus*, the first aphid-borne reovirus. *Phytopathology*, 102(5), pp. 547–53.

Quito-Avila, D. F., Lightle, D. and Martin, R. R. 2014. Effect of *Raspberry bushy dwarf virus*, *Raspberry leaf mottle virus*, and *Raspberry latent virus* on plant growth and fruit crumbliness in 'Meeker' red raspberry. *Plant Disease*, 98(2), pp. 176–83.

Rott, M., Xiang, Y., Boyes, I., Belton, M., Saeed, H., Kesanakurti, P., Hayes, S., Lawrence, T., Birch, C., Bhagwat, B. and Rast, H. 2017. Application of next generation sequencing for diagnostic testing of tree fruit viruses and viroids. *Plant Disease*, 101(8), pp. 1489–99.

Shahid, M. S., Aboughanem-Sabanadzovic, N., Sabanadzovic, S. and Tzanetakis, I. E. 2017. Genomic characterization and population structure of a badnavirus infecting blackberry. *Plant Disease*, 101(1), pp. 110–15.

Shan, Q., Wang, Y., Li, J., Zhang, Y., Chen, K., Liang, Z., Zhang, K., Liu, J., Xi, J. J., Qiu, J. L. and Gao, C. 2013. Targeted genome modification of crop plants using a CRISPR-Cas system. *Nature Biotechnology*, 31(8), p. 686.

Susaimuthu, J., Tzanetakis, I. E., Gergerich, R. C., Kim, K. S. and Martin, R. R. 2008. Viral interactions lead to decline of blackberry plants. *Plant Disease*, 92(9), pp.1288–92.

Szittya, G., Silhavy, D., Molnár, A., Havelda, Z., Lovas, Á., Lakatos, L., Bánfalvi, Z. and Burgyán, J. 2003. Low temperature inhibits RNA silencing-mediated defence by the control of siRNA generation. *The EMBO Journal*, 22(3), pp. 633–40.

Thekke-Veetil, T. and Tzanetakis, I. E. 2017. Development of reliable detection assays for blueberry mosaic- and blackberry vein banding-associated viruses based on their population structures. *Journal of Virological Methods*, 248, pp. 191–4.

Thekke-Veetil, T., Ho, T., Moyer, C., Whitaker, V. M. and Tzanetakis, I. E. 2016. Detection of *Strawberry necrotic shock virus* using conventional and TaqMan® quantitative RT-PCR. *Journal of Virological Methods*, 235, pp. 176–81.

Tzanetakis, I. E. and Martin, R. R. 2017. A systems-based approach to manage strawberry virus diseases. *Canadian Journal of Plant Pathology*, 39(1), pp. 5–10.

Varveri, C., Maliogka, V. I. and Kapari-Isaia, T. 2015. Principles for supplying virus-tested material. In: G. Loebenstein and N. I. Katis (Eds), *Advances in Virus Research* (Vol. 91, pp. 1–32). Academic Press.

Villamor, D. E.V., Martin, R. R. and Tzanetakis, I. E. 2018. Comparison between high throughput sequencing and current protocol for virus detection in berry crops. *Phytopathology* (in press).

Wang, Q., Cuellar, W. J., Rajamäki, M. L., Hirata, Y. and Valkonen, J. P. 2008. Combined thermotherapy and cryotherapy for efficient virus eradication: Relation of virus distribution, subcellular changes, cell survival and viral RNA degradation in shoot tips. *Molecular Plant Pathology*, 9(2), pp. 237–50.

Wang, Q. C., Panis, B., Engelmann, F., Lambardi, M. and Valkonen, J. P. T. 2009. Cryotherapy of shoot tips: A technique for pathogen eradication to produce healthy planting materials and prepare healthy plant genetic resources for cryopreservation. *Annals of Applied Biology*, 154(3), pp. 351–63.

Xiang, Y., Bernardy, M., Bhagwat, B., Wiersma, P. A., DeYoung, R. and Bouthillier, M. 2015. The complete genome sequence of a new polerovirus in strawberry plants from eastern Canada showing strawberry decline symptoms. *Archives of Virology*, 160(2), pp. 553–6.

Zhang, F., Cong, L., Lodato, S., Kosuri, S., Church, G. M. and Arlotta, P. 2011. Efficient construction of sequence-specific TAL effectors for modulating mammalian transcription. *Nature Biotechnology*, 29(2), p. 149.

Zheng, Y., Gao, S., Padmanabhan, C., Li, R., Galvez, M., Gutierrez, D., Fuentes, S., Ling, K. S., Kreuze, J. and Fei, Z. 2017. VirusDetect: An automated pipeline for efficient virus discovery using deep sequencing of small RNAs. *Virology*, 500, pp. 130–8.

Chapter 12

Improving fungicide use in integrated fruit disease management

Mengjun Hu, University of Maryland, USA

1 Introduction

The use of fungicides has proved to be one of the most effective methods for disease management on a wide variety of agricultural crops. Compared to many field and vegetable crops, fungicides are in general used more frequently on fruit crops. Among various types of fungicides, fungicides with single-site mode of action are typically favoured in IPM (Integrated Pest Management) programmes, due to their enhanced effectiveness and reduced risk for workers, consumers and environment (Adaskaveg et al., 2004). Commonly used single-site fungicide groups include methyl benzimidazole carbamates (MBCs; FRAC 1), dicarboximides (FRAC 2), demethylation inhibitors (DMIs; FRAC 3), succinate dehydrogenase inhibitors (SDHIs, FRAC 7), quinone outside inhibitors (QoIs; FRAC 11) and phenylpyrroles (PPs; FRAC 12). FRAC codes are numbers and letters used to distinguish the fungicide groups.

The disadvantage of single-site fungicides is that they are prone to resistance development. Unlike multi-site fungicides which attack multiple critical enzymes of a fungus, they are active against only one site in a single metabolic pathway. This weakness, combined with the intensive use on fruit crops, has led to the development of resistance in a number of fungal pathogens to a wide range of single-site fungicides (Ma and Michailides, 2005). Moreover, simultaneous resistance to multiple chemical

http://dx.doi.org/10.19103/AS.2019.0046.14

classes of fungicides in a single isolate has been detected in some major fungal pathogens such as *Botrytis cinerea* (Fernández-Ortuño et al., 2014), *Colletotrichum gloeosporioides* (Hu et al., 2015a), *Venturia inaequalis* (Köller and Wilcox, 2001), *Monilinia fructicola* (Chen et al., 2013) and *Alternaria alternata* (Fan et al., 2015).

Current resistance management recommendations include rotation and/or mixture of different FRAC codes and the restriction of the number of applications per growing season (Wade, 1988; Brent and Hollomon, 2007). While these strategies are practically important to delay resistance development in situations where a fungal pathogen has not yet acquired resistance, some of them may fail in situations when multi-fungicide resistance in individual isolates is already observed in the population (Schnabel et al., 2015). Fitness cost is yet another important factor that should be taken into account when making fungicide deployment strategies. If no fitness cost is associated with a resistance mutation, then the resistant isolates will persist in the fungal population regardless of fungicide applications. As a result, the corresponding fungicide may no longer be effective, and using this fungicide in a mixture or alternation will not be beneficial (Mikaberidze and McDonald, 2015).

Monitoring of resistance in fungal pathogens can reveal shifts in sensitivity and determine the frequency of resistance to individual chemical classes of fungicides (Brent, 1988), but it is usually done at a larger, region-wide scale, and therefore may overlook the differences in sensitivity in local fungal populations (Schnabel et al., 2015). Monitoring methods generally include traditional bioassays and molecular methods. Compared to bioassay procedures, molecular methods are more sensitive and less laborious. However, molecular methods are specific for monogenetic resistance (resistance is controlled by a single gene) detection due to the lack of understanding of molecular basis of polygenetic resistance (resistance is controlled by multiple genes). Although traditional bioassays allow the detection of both mono- and polygenetic resistance, the selection of discriminatory concentrations and the interpretation of the test results can be challenging.

Fungicides are usually labelled for specific disease control regardless of pathogens, which in turn neglects the fact that one disease could be caused by multiple fungal pathogens. Fungal pathogens, even within the same genus, can have differential sensitivities to the same fungicide. Knowing what species cause a disease is thus important to choosing effective fungicides for disease control. Many cultural and molecular characteristics among fungal species have been identified and used to identify species. However, morphology is usually variable among isolates even within the same species. Internal transcribed spacer (ITS), a universal DNA barcode marker for fungi (Schoch et al., 2012), can also fail in distinguishing closely related species (Weir et al., 2012).

Disease models are developed to predict infection risk based on environmental variables and/or inoculum (Xu et al., 2000; Bulger et al., 1987). Using these models, weather-based decision support tools have been developed and implemented on a wide range of crops for disease control. For example, Network of Environment and Weather Applications (NEWA), developed by scientists at Cornell University, has integrated over 20 weather-based IPM forecast tools for multiple crops. Most tools are added when developed and validated, based on published research (http://newa.cornell.edu/). The adoption of forecast tools can not only save the cost of materials and labour, but also reduce resistance selection pressure on target pathogens. Moreover, it can largely eliminate unnecessary sprays being applied to the environment.

In this chapter we will review the development of fungicide resistance in fruit pathogens using *Botrytis cinerea* as an example. We will discuss the current spray strategies based on multi-year monitoring of resistance in *Botrytis* species from strawberry fields, address the importance of species identification in relation to fungicide use and introduce decision-support tools. The objective of this chapter is to explore possibilities that could lead to improvement of fungicide use on fruit crops.

2 Evolution of fungicide resistance

Despite a wide variety of single-site fungicides available for plant disease control, fungal pathogens can quickly develop resistance due to their plastic genome and high reproduction rate. The first case of resistance to MBCs was reported in the 1960s (Schroeder and Prowidenti, 1969), and now MBC resistance has been identified in more than 90 fungal pathogens (Hawkins and Fraaije, 2016). Reduced sensitivity to DMIs was first reported in field isolates of *V. inaequalis* (Stanis and Jones, 1985), and the resistance was later on found in a wide range of fruit pathogens, including *M. fructicola* (Luo and Schnabel, 2008), *Sphaerotheca pannosa* (Reuveni, 2001), *Uncinula necator* (Steva et al., 1990) and *V. nashicola* (Tomita and Ishii, 1998). QoI fungicides are well known for resistance development and the resistance was detected within 2 years of release (Erichsen, 1999). A new generation of succinate dehydrogenase inhibitors (SDHIs) was introduced in 2007, but by 2017 resistant field isolates were found in 17 pathogen species (Fisher et al., 2018).

2.1 Mechanisms of fungicide resistance

Fungicides inhibit fungal growth by attacking critical metabolism pathways. For single-site fungicides, there are specific sites or enzymes in a pathway to which the fungicides are binding. Different chemical classes of fungicides usually have unique target sites in different metabolism pathways. For instance, DMI fungicides have sterol 14α-Demethylase (*cyp 51*) enzyme as a target, interfering with sterol biosynthesis, whereas QoI fungicides attack cytochrome b (*cyt b*) of complex II in respiration pathway. So far, the main mechanisms of resistance characterized across different fungal pathogens include mutation in the target gene, overexpression of the target and increased efflux pump activities (Ma and Michailides, 2005). Other possible resistance mechanisms, such as fungicide detoxification or decreased uptake (Leroux et al., 2002), do not seem to play an important role in plant pathogenic fungi (Hahn, 2014).

Why do we need to study the underlying mechanisms of resistance? First, rapid molecular methods of resistance screening can be developed based on the molecular basis of resistance (Luo et al., 2008; Ma et al., 2005). Second, it provides an insight into differential resistance levels that impact spray decisions. Resistance conferred by point mutations in the target gene generally results in a higher level of resistance compared to other underlying mechanisms. For example, G143A mutation (glycine to alanine at position 143 of *cyt b*) often leads to a several hundred-fold increase in EC_{50} values (Leadbeater, 2012), whereas overexpression of *cyp* 51 (Cools et al., 2012) or enhanced drug efflux pump activity (Kretschmer et al., 2009) usually increases about ten-fold or less in EC_{50} values. Third, different mutations in the same target gene may confer differential sensitivity to the same chemical class of fungicides. One excellent example is SDHI resistance conferred by mutations in *SdhB* in *B. cinerea*. H272R (histidine to arginine at position 272 of *SdhB*) confers resistance to boscalid, H272Y (histidine to tyrosine at position 272) confers resistance to boscalid, fluxapyroxad and penthiopyrad, and N230I (asparagine to isoleucine at position 230) or P225F (proline to phenylalanine at position 225) confers resistance to all SDHI fungicides tested including boscalid, fluopyram, fluxapyroxad and penthiopyrad (Hu et al., 2016b).

2.2 Development of resistance to multiple chemical classes of fungicides in **Botrytis** *species*

Among all filamentous fungal pathogens, *B. cinerea* has great ability, if not the greatest, to develop resistance to multiple chemical classes of fungicides. Resistance monitoring conducted in strawberry fields of seven Southern US states showed that isolates of *B. cinerea* from both flowers and fruit were resistant to multiple chemical classes including FRAC 1 (thiophanate-methyl), FRAC 2 (iprodione), FRAC 7 (boscalid), FRAC 9 (cyprodinil and pyrimethanil), FRAC 11 (pyraclostrobin), FRAC 12 (fludioxonil) and FRAC 17 (fenhexamid) (Fernández-Ortuño et al., 2014). Resistance to seven chemical classes of fungicides in individual isolates was also detected shortly thereafter (Fernández-Ortuño et al., 2015). Multi-fungicide resistance was also found in *B. cinerea* on other small fruits such as blueberry, raspberry, redcurrant and blackberry worldwide

(Li et al., 2014; Fan et al., 2016; Weber, 2011). In addition to *B. cinerea*, two other species, *Botrytis fragariae* and *Botrytis* group S, recently found on strawberries, have developed resistance to multiple chemical classes of fungicides (Rupp et al., 2017; Hu et al., 2018).

Two mechanisms were found to be responsible for multi-fungicide resistance in *Botrytis* species. Kretschmer et al. (2009) first documented field *B. cinerea* isolates with resistance to multiple chemical classes of fungicides. Based on the resistance phenotypes, isolates were defined as MDR1 and MDR2 isolates. The multi-fungicide resistance was found to be conferred by mutations in a transcription factor (Mrr1) and rearrangement in the promoter region of the major facilitator superfamily transporter gene *mfsM2* for MDR1 and MDR2 isolates, respectively. However, a more common mechanism that usually confers much higher level of resistance to individual fungicides is target gene alterations (Amiri et al., 2012; Fernández-Ortuño et al., 2015) through a stepwise accumulation (Li et al., 2014).

3 Monitoring of fungicide resistance

As an early warning system, monitoring is of vital importance to resistance management. It can be used to ensure that management strategies are still appropriate in real time, to investigate cases of suspected loss of performance, to track the spread of resistance over time, to guide fungicide selection at a local level, to gain understanding of the behaviour of resistant forms in field populations and to validate predictions (Brent, 1992). The choice of monitoring methods is largely dependent on the pathogen, the host, the mechanism underlying fungicide resistance and available resources.

Molecular tools have been developed to facilitate resistance monitoring, allowing for a quick turnaround. Common molecular methods usually include PCR-RFLP (Luo et al., 2008), quantitative PCR (Banno et al., 2009) and lamp PCR (Duan et al., 2014). Molecular tools are developed based on molecular basis of resistance. If target genes and mutations correlated with fungicide resistance are unknown, bioassays must be used for monitoring. If multiple mutations and genes are involved in individual isolates with multi-fungicide resistance, then it would be difficult to develop molecular tools due to PCR limitations. This is because detection of multiple mutations in a single reaction requires more primers which would largely increase PCR complexity and reduce reliability. For bioassays, selection of discriminatory concentrations is of paramount importance as sensitivity to the same fungicide often varies between individual isolates. *In vitro* combined with *in vivo* assays of sensitivity on isolates with a wide range of EC_{50} values must be conducted to determine discriminatory concentrations that can accurately distinguish between sensitive and resistant isolates (Schnabel et al., 2015).

3.1 *Profiling resistance by bioassays in* B. cinerea

B. cinerea isolates have been found to be resistant to up to seven different chemical classes of fungicides, and resistance to each class of fungicides can be conferred by multiple mutations in target genes (Fernández-Ortuño et al., 2015). Moreover, resistant phenotypes of *Botrytis* isolates can be very different at two locations in close proximity. Hence, a location-specific resistance monitoring method was developed to help strawberry growers identify potential resistance issues and make spray recommendations (Schnabel et al., 2015). In this method, *Botrytis* isolates collected from strawberry flowers were tested for sensitivity to various fungicides. Flowers are highly susceptible to *Botrytis* infection, therefore, knowing what products can effectively protect flowers is key to managing the disease. Specifically, strawberry flowers with black pollen sacs or ovaries (most likely caused by frost damage) were collected during bloom. After sterilization, flowers were then transferred onto wet filter paper in 150-mm Petri dishes with lids and incubated for 3 days to induce sporulation at room temperature. Spores were then transferred to a set of 24-well plates containing seven different chemical classes of fungicides at discriminatory concentrations using a sterilized toothpick. Resistance was visually assessed based on presence or absence of mycelial growth on the plates after 4 days of incubation. Specific recommendations were then made to individual growers based on the testing results. In addition, a germ tube growth assay (Weber and Hahn, 2011) was also developed to detect multi-fungicide resistance in *Botrytis* isolates. Similar to the method described above, this method obtained conidia from plant tissues and transferred them onto medium containing fungicides at discriminatory concentrations. This was followed by the assessment of conidial germination and/or germ tube length under microscope after 12–16 h of incubation.

Both methods allow for large-scale resistance monitoring, despite some shortcomings. A single strawberry flower can harbour multiple haplotypes of *Botrytis* with multiple resistant phenotypes, which complicates fungicide sensitivity bioassays (Hu et al., 2018). For the mycelial growth assay, bulk conidia were transferred to medium directly from sporulating flowers. The major advantage of this type of assay is that all resistant phenotypes are captured. However, the assay assumes that conidia from the same flower are produced from the same isolate, unrecognizing the occasional variation in resistant phenotypes of single flower isolates. In contrast to the mycelial growth assay, the germ tube assay allows for an accurate determination of resistance frequency since individual spores are examined. However, only a sample of 20–30 conidia are usually examined for each fungicide, which may not be sufficient to capture all phenotypes.

3.2 Rethinking general resistance management strategies: observations through a location-specific resistance profiling

Alternation of chemical classes of fungicides may not be sustainable for the management of multi-fungicide resistance. It is believed that use of mixture or rotation of fungicides with different modes of action during a single growing season is the key to prevent or delay fungicide resistance (Brent and Hollomon, 2007). This strategy has been implemented for the management of resistance in all fungal pathogens ever since the introduction of single-site fungicides. However, it might not be sufficient for managing pathogens in which multi-fungicide resistance has developed.

Given the significance of grey mould on strawberries and the high variation in fungicide resistance in the pathogen, the above mentioned location-specific resistance profiling was developed to identify potential resistance issues and make appropriate spray recommendations to individual growers (Schnabel et al., 2015). The extension-led programme started in the 2011–12 season, and participating growers implemented specific fungicide spray programmes based on the monitoring results. Growers were advised to only apply a fungicide to which the frequency of resistance was ≤10% once per season, and to avoid fungicides to which the frequency of resistance was ≥10%. This encouraged growers to rotate fungicides, utilizing multiple chemical classes.

From 2011–12 to 2014–15, not only frequencies of resistance to all seven chemical classes of fungicides increased (Fig. 1), but also frequency of

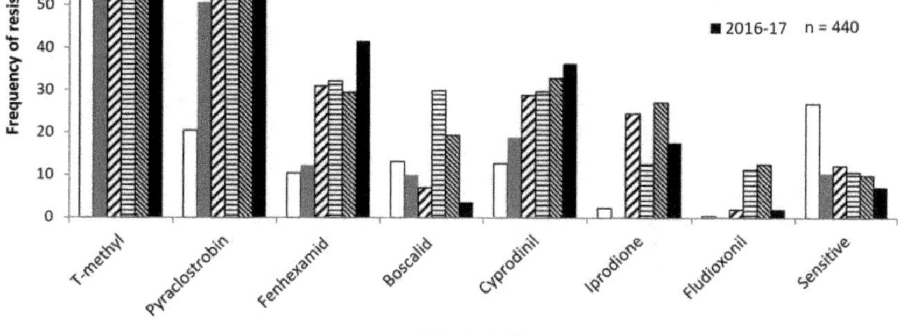

Figure 1 Frequency of fungicide resistance in *B. cinerea* from strawberry flowers from East Coast of the United States.

isolates with multi-fungicide resistance increased (Fig. 2) (Hu et al., 2016a). The majority of the isolates throughout the six seasons were found to be resistant to thiophanate-methyl, despite that the fungicide had not been used since the beginning of the monitoring due to the high frequency of resistance detected at the first season. With the use of other chemical classes of fungicides, isolates resistant to only thiophanate-methyl would be expected to be largely eliminated regardless of fitness cost. Thus, it stands to reason that resistance to thiophanate-methyl was indirectly selected. The increase in chemical class resistance within isolates was likely due to a process we termed 'selection by association', where fungicide resistance traits were often indirectly selected through genetic hitchhiking (Hu et al., 2016a). We hypothesize that the more chemical classes used in a spray programme, the faster isolates will be selected with increasing numbers of chemical class resistances if such isolates pre-exist in the population (Fig. 3a and b). The rate of selection by association is elusive. Theoretically, it is influenced by multiple factors including initial resistance ratio(s), potential fitness cost, spray intervals and chemical classes involved.

'Selection by association' would suggest that multi-fungicide resistant isolates (particularly 7CCR isolates) are selected each time a single-site fungicide is used. This would result in a rapid selection of multi-fungicide resistance. However, frequency of 5CCR, 6CCR and 7CCR isolates decreased in 2016–17 based on the monitoring results (Fig. 2). One possible explanation is that fitness cost and reduced competitive ability associated with 7CCR and other resistant phenotypes may slow down the selection process when single-site fungicides are not applied (Chen et al., 2016a). Interestingly, results of a

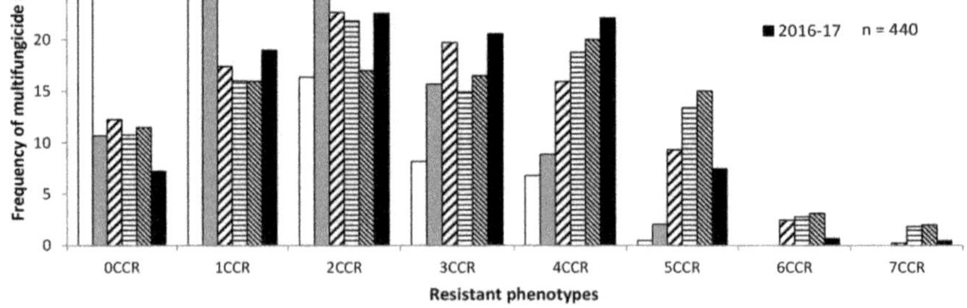

Figure 2 Frequency of *B. cinerea* isolates with resistance to multiple chemical classes of fungicides.

blackberry trial showed that artificially inoculated *Botrytis* 7CCR isolates can be recovered from both the fruit and the flowers in the absence of fungicide treatment over the course of three years, suggesting that the 7CCR isolates may not undergo fitness cost or reduced competitiveness (data not shown). Since the 2014–15 season, the growers were asked to base the disease management on multisite-fungicides (i.e. captan and thiram) and only add single-site fungicides when disease pressure was high. It is most likely that the reliance on multisite-fungicides in the last two years, at least in part, has led to a decrease of frequency of 5CCR, 6CCR and 7CCR phenotypes (Fig. 2).

In addition to being challenged by 'selection by association', the general resistance management strategy may not be suitable for incomplete cross-resistance between fungicides within the same chemical class. For example, we observed four different SDHI resistance patterns designated A, B, C or D through the monitoring. Pattern A resistant to boscalid with the mutation H272R at locus *sdhB*; pattern B resistant to boscalid and penthiopyrad with mutation H272Y or H272R; pattern C resistant to boscalid, fluxapyroxad and penthiopyrad with H272Y; and pattern D resistant to all the four SDHIs with mutation P225F or N230I (Hu et al., 2016b). This incomplete cross-resistance found in SDHI fungicides (Table 1) has raised questions as to how we should use this chemical class. Would rotation of different SDHI fungicides be an option for disease management? If so, this option would be in disagreement with the general strategy, that is, rotation of fungicides with different modes of action.

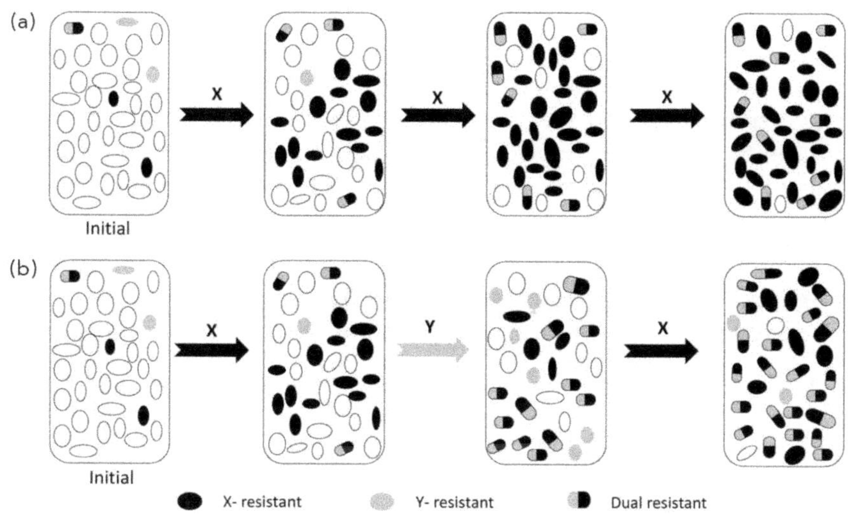

Figure 3 Selection for fungicide resistance in a pool where multiple resistant phenotypes are pre-existing (the ratio of dual-resistant isolates is assumed less than single-resistant isolates in the initial population). (a) Resistance is continuously selected by a single fungicide X; (b) resistance is selected by alternation of fungicide X and Y.

Table 1 Mutations in the SDHB subunit of *B. cinerea* and their sensitivity to five SDHI fungicides

	Active ingredients (FRAC 7)				
Genotype	Boscalid	Fluopyram	Fluxapyroxad	Penthiopyrad	Isofetamid
Wide type	S[a]	S	S	S	S
H272R	R	S	S	r	S
H272Y	R	S	R	R	S
H272L	R	R	R	R	?
N230I	R	R	R	R	S
P225F	R	R	R	R	R

[a]'S', 'R' and 'r' indicate sensitive, resistant and partial resistant respectively.
? - indicates fungicide sensitivity is unknown.

One may say that rotation of different SDHI fungicides will favour the selection of mutations conferring resistance to all SDHIs. Studies aim to compare fitness cost in different *sdhB* genotypes, and monitoring of frequencies of different resistance patterns may further guide us to use this interesting fungicide class consisting of a wide variety of chemical structures.

4 Importance of species identification

Identification of fungi to species level not only addresses basic research questions, but also has practical value as it improves the selection of fungicides. The same disease is often found to be caused by multiple fungal species, as opposed to single species. Interestingly, different species, even within the same genus, usually vary in their fungicide sensitivity, which justifies the importance of species identification. However, such sensitivity variations among species causing the same disease are often neglected. The major obstacle to an accurate species identification is the lack of understanding of genetic variations between different species with largely identical morphological features.

4.1 Differential fungicide sensitivity in fungal species responsible for same disease

Colletotrichum gloeosporioides and *Colletotrichum acutatum* are usually thought to be the causal agents of peach anthracnose fruit rot. However, both *C. gloeosporioides* and *C. acutatum* are now considered to be species complexes (Weir et al., 2012; Damm et al., 2012b). Peach anthracnose disease in the Southeast of United States can be caused by at least five different *Colletotrichum* species, including *Colletotrichum siamense*, *Colletotrichum fructicola*, *Colletotrichum fioriniae*, *Colletotrichum nymphaeae* and *Colletotrichum truncatum* (Chen et al., 2016b; Hu et al., 2015b). Among them,

C. siamense and *C. fructicola* are within *C. gloeosporioides* complex while *C. fioriniae* and *C. nymphaeae* are within *C. acutatum* complex. Dual-resistance to azoxystrobin and thiophanate-methyl was found in *C. siamense* only, which might indicate the higher risk of resistance development in this species (Hu et al., 2015a). *C. fioriniae* and *C. nymphaeae* are inherently insensitive to MBC fungicides but other species are sensitive. In sensitivity testing of several different DMI fungicides, *C. truncatum* was found to be resistant to tebuconazole, metconazole, flutriafol and fenbuconazole whereas *C. siamense* and *C. fructicola* were sensitive to all DMI fungicides tested (Chen et al., 2016b). Isolates of *C. gloeosporioides*, *C. acutatum*, *C. cereale* and *C. orbiculare* were naturally insensitive to SDHI fungicides boscalid, fluxapyroxad and fluopyram but sensitive to penthiopyrad (except for *C. orbiculare*), and benzovindiflupyr was effective against all species (Ishii et al., 2016).

The causal agents of brown rot affecting many stone fruits include *Monilinia fructicola*, *Monilinia laxa*, *Monilinia fructigena*, *Monilia mumecola* and *Monilia yunnanensis* (Hu et al., 2011). As one of the most effective fungicide classes, QoIs have been used to control brown rot on peaches since the 1990s. However, field relevant resistance to QoIs has never been reported in *Monilinia* species, although QoI fungicides are usually prone to resistance development. Sequencing analysis of the target gene *cyt b* from *M. fructicola* revealed that there is an intron right after the G143 codon, preventing G143 from being mutated into A143 to confer high resistance to QoIs. Most interestingly, the G143 intron is present in all the species except *M. fructigena*, suggesting the potential risk of QoI resistance development in *M. fructigena*.

Other fruit pathogens, such as *B. cinerea* and *B. fragariae* affecting strawberries, also differ in resistance profiles (Dowling et al., 2017). The variations in target genes and genetic background between closely related species can both impact fungicide sensitivity. These above mentioned findings highlight the importance of accurately identifying fungal isolates to species level. Traditional morphology-based identification methods are time-consuming and complicated by the existence of sibling species which are almost morphologically identical but are nonetheless isolated from one another (Filipe et al., 2008).

4.2 Identification of species with multilocus sequences

Multilocus sequence typing, also known as MLST, utilizes the variations in nucleotide sequences from multiple housekeeping genes to characterize genetic diversity. The MLST data can be used to identify fungal species and distinguish populations within species. The main feature of MLST is that all sequence data can be deposited in a database to improve the dataset for each species, and can be made readily accessible to all interested parties via the internet (Taylor and Fisher, 2003). MLST also has advantages over older methods that were sensitive to minor changes under laboratory conditions, such as restriction fragment length polymorphisms (RFLPs) and randomly amplified polymorphic DNAs (RAPDs) (Taylor and Fisher, 2003).

MLST has been widely used to study the evolution of many organisms including plant pathogenic fungi (Geiser et al., 1998; Damm et al., 2012a). Weir et al. (2012) conducted phylogenetic analyses of up to eight genes to systematically describe the *C. gloeosporioides* species complex. As a result, 22 species plus one subspecies were reported under the *C. gloeosporioides* species complex. Interestingly, all 23 species including the subspecies were morphologically similar to the broadly defined *C. gloeosporioides*, as it had been applied in the literature for the past 50 years (Weir et al., 2012). The above mentioned differential sensitivity in *Colletotrichum*, *Monilinia* or *Botrytis* species was all discovered based on species identification using MLST. As the informative sites on the loci are utilized for MLST, the loci selected for MLST largely determine the accuracy of the outcome. Unfortunately, the number of informative sites on the same loci usually varies from one species to another. It is therefore important to investigate the frequency of variations in housekeeping genes or loci commonly used in phylogeny studies between isolates of interest, if such information is unavailable.

5 Adoption of decision-support tools to improve fungicide efficacy

During crop growing seasons, growers must decide when and what to spray for managing diseases. Most of the decisions are being made largely based on their experience in consideration of weather conditions, crop developmental stages and/or presence of diseases/inoculum. However, experience-based decisions can be unreliable, and may result in delays in disease control. Real-time and data-driven decision-support tools have been developed to enable reliable and rapid decision-making. Many of those decision-support tools use environmental variables transmitted in real time from weather stations or microclimate sensors, allowing accurate monitoring of important factors affecting crop development, and disease risk.

Extension agriculture specialists strive to provide growers with up-to-date fungicide information that can be used to select effective fungicides for disease management. Such information is typically conveyed via extension publications, websites and production meetings. Smartphones are now another opportunity to distribute information that can be easily retrieved at any time. A smartphone application, namely 'MyIPM', has been developed to facilitate the communication of complicated FRAC code principles with fruit growers (Schnabel et al., 2017).

5.1 Strawberry Advisory System

Botrytis fruit rot (BFR) caused by *Botrytis* spp., and anthracnose fruit rot (AFR) caused by the *C. acutatum* species complex, are important fungal diseases on strawberries. Based on the duration of leaf wetness and the average temperature during the wet period, Bulger et al. (1987) and Wilson et al. (1990) developed models to predict AFR and BFR infection risks. Temperatures between 25°C and 30°C, and 15°C and 25°C were found to be most conducive for BFR and AFR development, respectively. Scientists at the University of Florida validated the models by carrying out multi-year field trials and then launched the web-based Strawberry Advisory System (SAS) (Pavan et al., 2011). Users can virtualize the real-time infection risks for both BFR and AFR at AgroClimate (http://agroclimate.org/tools/sas/) or through the smartphone app SAS. Growers or other users can also subscribe to email and/or text-message services for free to get timely alerts for fungicide applications when AFR or BFR infection risk reaches medium or high level (Pavan et al., 2011) (Fig. 4). The SAS was designed to better time fungicide applications to minimize unnecessary sprays. According to the trials conducted in Florida, fungicide applications were reduced, over an average of three seasons, by approximately 40% (MacKenzie and Peres, 2012a,b). With less frequent sprays, selection pressure for fungicide resistance is reduced, allowing for more sustainable fungicide use.

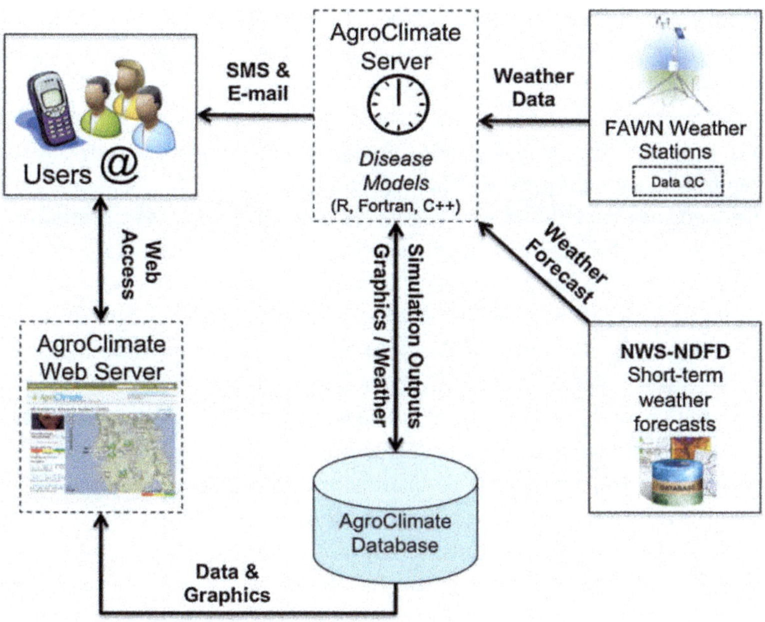

Figure 4 The Strawberry Advisory System (SAS) infrastructure. Adapted from Pavan et al. (2011).

Although such a tool has a great potential to be widely implemented, the adoption needs to take into consideration differences in production practices that may influence data accuracy. The real-time infection risk is calculated with leaf wetness and temperature data constantly obtained from weather stations. On the one hand, weather stations are not capable of monitoring the variables in an altered environment. For example, strawberry growers use straw in a matted-row system, or deploy floating row covers in a plasticulture system to protect berries from frost and accumulate degree days during autumn and winter in many regions outside of California and Florida in the United States. On the other hand, weather conditions can be dramatically different between two locations in close proximity, leading to different disease risks. Such difference could either be a result of poor maintenance on the sensors, or the difference in microclimate. In spite of the concerns described here, we believe that the development and implementation of such tools including the SAS will significantly improve fungicide efficacy by better timing and eliminating unnecessary applications.

5.2 MyIPM – a smartphone application that integrates fungicide use and resistance management

MyIPM was originally developed for strawberry and peach growers, and now it has been expanded to include a variety of fruits through collaboration among several public universities in the United States. The app is designed for growers, field advisors and specialists and it is free to download at both Google Play Store for android devices and Apple Store for iOS devices. The app has a built-in function that allows users to choose the fruit of interest to view. MyIPM uses a database that can be updated easily through a user-friendly authoring tool maintained by multiple extension specialists.

The app contains different aspects of disease and insect management in the form of text, pictures and audio (Fig. 5). The main feature of MyIPM is the complete list of colour-coded, sortable active ingredients for disease management with links to registered trade names. On the front page, there is a search bar on the top that allows searching for active ingredients or trade names registered for all the fruit crops downloaded in the app (Fig. 5a). Tapping 'Active ingredients' on the main page lists all registered active ingredients for the particular diseases. When a different disease is selected, the active ingredient list adjusts to display the correct information specific to the selected disease. The same is true for the 'Trade Names'. By tapping on the disease name or 'Select' button, users can easily move between active ingredients or registered trade names for different diseases (Fig. 5b). Active ingredients are colour-coded and sortable by FRAC code, efficacy and FRAC risk of resistance development (Fig. 5c). Rate per acre, PHI (pre-harvest

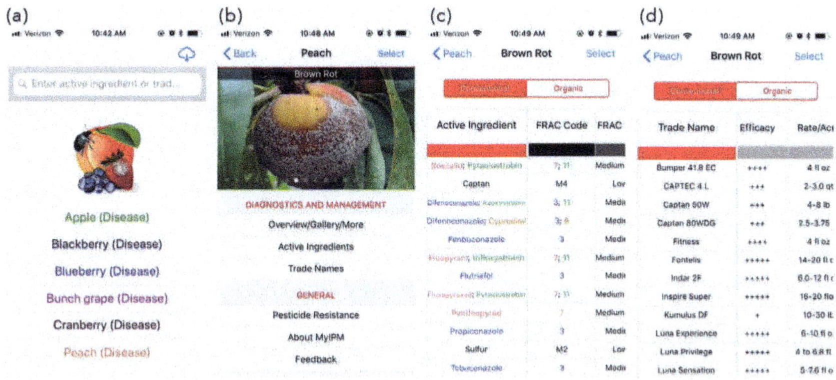

Figure 5 Screenshots of MyIPM smartphone application. (a) Main page of the MyIPM; (b) front page of a specific disease; (c) active ingredient table; (d) trade name table.

interval), REI (re-entry interval), max number of sprays and toxicity data are listed for each registered trade name (Fig. 5d). All active ingredients and registered trade names are separated by 'conventional products' and 'organic products' (Fig. 5c and d).

While the use of a smartphone application may greatly improve the timeliness of decision-making for disease management, we must be aware of potential limitations in applying the tool in a different region. For example, registered active ingredients and trade names in MyIPM would only apply to most states in the United States. Restrictions including max number of applications, PHI and REI on individual fungicides may also vary from one country or region to another. We use MyIPM just as an example here to demonstrate its usefulness as an outreach tool.

6 Future trends and conclusion

As one of the most effective tools to control diseases, fungicides will likely continue to play a significant role in IPM in the foreseeable future. However, the continuous use of synthetic fungicides is challenged by rising awareness of potential health and environmental issues and the rapid rate of emergence of pathogenic fungal isolates resistant to the limited number of fungicides (Fisher et al., 2018). The demand for novel fungicides is increasing, and resources must be leveraged to facilitate novel fungicide development. The expansion of genomic information has offered many opportunities to study many aspects of fungicides. Cools and Hammond-Kosack (2013) reviewed in detail how genomics can be exploited to identify and validate potential targets, determine the mode of action of the fungicide and study fungicide resistance. In contrast to the concerns about the use of synthetic fungicides, non-chemical control alternatives including cultural practices, physical treatments and biological materials are becoming more popular. Studies focused on identification and integration of novel non-chemical approaches for both conventional and organic productions may have a brighter future.

Resistance management strategies can be further improved by integrating more modelling approaches. For example, fitness cost has been found in many fungal isolates with resistance, resulting in less competitiveness than wild-type isolates when fungicide pressure is absent (Chen et al., 2016a). Mikaberidze and McDonald (2015) illustrated how fitness cost-associated resistance can be harnessed to improve our stewardship of extant fungicides using modelling approaches. Historically, fungicide models have not had much influence on fungicide spray strategies, presumably due to the difficulty in interpretation and inadequate connection between modelling and experimentation (van den Bosch et al., 2015). In future, more models taking into consideration a broad range of factors contributing to resistance development, such as fungicide dose, timing of application, level of overwintering inoculum and the role of full resistance or partial resistance, are expected to be developed in close collaboration with experimenters and stakeholders (Mikaberidze et al., 2017; Beckerman et al., 2015; van den Bosch et al., 2015).

Many tools have been developed in precision agriculture for better managing diseases and/or increasing crop productivity. Selective targeting of fungicide application on specialty crops has also been explored under this concept. For instance, unmanned aerial vehicles can be used to spray pesticides where the equipment and workers cannot easily reach (Mogili and Deepak, 2018). A laser-guided system was also developed to synchronize spray outputs to tree structures, with the goal of increasing pesticide usage efficiency (Shen et al., 2017). In addition, Oberti et al. (2016) tested the efficacy of using a modular agricultural robot to selectively spray for powdery mildew on grape vines under greenhouse conditions. The robot was integrated with a disease-sensing system to specifically detect powdery mildew. The results showed that 85–100% of the diseased area within the canopy was automatically detected and sprayed by the robot, resulting in 65–85% fungicide reduction compared to a conventional spray (Oberti et al., 2016). More multidisciplinary studies are expected to deliver state-of-the-art technologies with acceptable cost and high reliability for use in commercial agriculture.

We hope that the few examples above, from the development to the use of fungicides, shed light on future opportunities in fungicide research. With increasing collaboration across multiple disciplines and regions, the understanding of many aspects of fungicides will be largely improved for a sustainable agriculture system.

7 Where to look for further information

As noted above, FRAC (Fungicide Resistance Action Committee) is a Specialist Technical Group of CropLife International (http://www.frac.info/home). The main information available on FRAC includes (1) FRAC code list for all labelled fungicides; (2) documented resistance issues and molecular mechanisms; (3) known cross-resistance and resistance risk for individual active ingredients; (4) recommended procedures for use in fungicide resistance studies; (5) general guidelines and advice on the use of fungicides to reduce the risk of resistance development.

8 References

Adaskaveg, J., Förster, H., Gubler, W., Teviotdale, B. and Thompson, D. 2004. Reduced risk fungicides: A new way of managing brown rot and other fungal diseases of stone fruit crops. *Calif. Agric.*, 59, 109–14.

Amiri, A., Heath, S. M. and Peres, N. A. 2012. Phenotypic characterization of multifungicide resistance in *Botrytis cinerea* isolates from strawberry fields in Florida. *Plant Dis.*, 97, 393–401.

Banno, S., Yamashita, K., Fukumori, F., Okada, K., Uekusa, H., Takagaki, M., Kimura, M. and Fujimura, M. 2009. Characterization of QoI resistance in *Botrytis cinerea* and identification of two types of mitochondrial cytochrome b gene. *Plant Pathol.*, 58, 120–9.

Beckerman, J. L., Sundin, G. W. and Rosenberger, D. A. 2015. Do some IPM concepts contribute to the development of fungicide resistance? Lessons learned from the apple scab pathosystem in the United States. *Pest Manag. Sci.*, 71, 331–42.

Brent, K. J. 1988. Monitoring for fungicide resistance. In: Delp, C. J. (Ed.), *Fungicide Resistance in North America*. APS Press, St. Paul, MN, pp. 9–11.

Brent, K. 1992. Monitoring fungicide resistance: Purposes, procedures and progress. In: Denholm, I., Devonshire, A. L. and Hollomon, D. W. (Eds), *Resistance'91: Achievements and Developments in Combating Pesticide Resistance*. Springer, Dordrecht, The Netherlands, pp. 1–18.

Brent, K. J. and Hollomon, D. W. 2007. *Fungicide Resistance in Crop Pathogens: How Can It be Managed?* 2nd ed. Crop Life International, Belgium.

Bulger, M. A., Ellis, M. A. and Madden, L. V. 1987. Influence of temperature and wetness duration on infection of strawberry flowers by *Botrytis cinerea* and disease incidence of fruit originating from infected flowers. *Phytopathology*, 77, 1225–30.

Chen, F., Liu, X. and Schnabel, G. 2013. Field strains of *Monilinia fructicola* resistant to both MBC and DMI Fungicides isolated from stone fruit orchards in the Eastern United States. *Plant Dis.*, 97, 1063–8.

Chen, S. N., Luo, C. X., Hu, M. J. and Schnabel, G. 2016a. Fitness and competitive ability of *Botrytis cinerea* isolates with resistance to multiple chemical classes of fungicides. *Phytopathology*, 106, 997–1005.

Chen, S. N., Luo, C. X., Hu, M. J. and Schnabel, G. 2016b. Sensitivity of colletotrichum species, including *C. fioriniae* and *C. nymphaeae*, from peach to demethylation inhibitor fungicides. *Plant Dis.*, 100, 2434–41.

Cools, H. J. and Hammond-Kosack, K. E. 2013. Exploitation of genomics in fungicide research: Current status and future perspectives. *Mol. Plant Pathol.*, 14, 197–210.

Cools, H. J., Bayon, C., Atkins, S., Lucas, J. A. and Fraaije, B. A. 2012. Overexpression of the sterol 14α-demethylase gene (*MgCYP51*) in *Mycosphaerella graminicola* isolates confers a novel azole fungicide sensitivity phenotype. *Pest Manag. Sci.*, 68, 1034–40.

Damm, U., Cannon, P., Woudenberg, J., Johnston, P. and Weir, B., Tan, Y. P., Shivas, R. G. and Crous, P. W. 2012a. The *Colletotrichum boninense* species complex. *Stud. Mycol.*, 73, 1–36.

Damm, U., Cannon, P., Woudenberg, J. and Crous, P. W. 2012b. The *Colletotrichum acutatum* species complex. *Stud. Mycol.*, 73, 37–113.

Dowling, M. E., Hu, M.-J. and Schnabel, G. 2017. Fungicide resistance in *Botrytis fragariae* and species prevalence in the Mid-Atlantic United States. *Plant Dis.*, 102, 964–9.

Duan, Y., Zhang, X., GE, C., Wang, Y., Cao, J., Jia, X., Wang, J. and Zhou, M. 2014. Development and application of loop-mediated isothermal amplification for detection of the *F167Y* mutation of carbendazim-resistant isolates in *Fusarium graminearum*. *Sci. Rep.*, 4, 7094.

Erichsen, E. 1999. Problems in mildew control in northern Germany. *Getreide*, 1, 44–6.

Fan, Z., Yang, J.-H., Fan, F., Luo, C.-X. and Schnabel, G. 2015. Fitness and competitive ability of alternaria alternata field isolates with resistance to SDHI, QoI, and MBC fungicides. *Plant Dis.*, 99, 1744–50.

Fan, F., Hamada, M. S., Li, N., Li, G. Q. and Luo, C. X. 2016. Multiple fungicide resistance in *Botrytis cinerea* from greenhouse strawberries in Hubei province, China. *Plant Dis.*, 101, 601–6.

Fernández-Ortuño, D., Grabke, A., Bryson, P. K., Amiri, A., Peres, N. A. and Schnabel, G. 2014. Fungicide resistance profiles in *Botrytis cinerea* from strawberry fields of seven Southern U.S. states. *Plant Dis.*, 98, 825–33.

Fernández-Ortuño, D., Grabke, A., Li, X. and Schnabel, G. 2015. Independent emergence of resistance to seven chemical classes of fungicides in *Botrytis cinerea*. *Phytopathology*, 105, 424–32.

Filipe, P., Joao, C. and Antonio, A. 2008. Identification of Species with DNA-based technology: Current progress and challenges. *Recent Patents on DNA & Gene Sequences (Discontinued)*, 2, 187–200.

Fisher, M. C., Hawkins, N. J., Sanglard, D. and Gurr, S. J. 2018. Worldwide emergence of resistance to antifungal drugs challenges human health and food security. *Science*, 360, 739–42.

Geiser, D. M., Pitt, J. I. and Taylor, J. W. 1998. Cryptic speciation and recombination in the aflatoxin-producing fungus *Aspergillus flavus*. *Proc. Natl. Acad. Sci. USA*, 95, 388–93.

Hahn, M. 2014. The rising threat of fungicide resistance in plant pathogenic fungi: *Botrytis* as a case study. *J. Chem. Biol.*, 7, 133–41.

Hawkins, N. J. and Fraaije, B. A. 2016. Predicting resistance by mutagenesis: Lessons from 45 years of MBC resistance. *Front. Microbiol.*, 7.

Hu, M.-J., Cox, K. D., Schnabel, G. and Luo, C.-X. 2011. *Monilinia* species causing brown rot of peach in China. *PLoS ONE*, 6, e24990.

Hu, M.-J., Grabke, A., Dowling, M. E., Holstein, H. J. and Schnabel, G. 2015a. Resistance in *Colletotrichum siamense* from peach and blueberry to thiophanate-methyl and azoxystrobin. *Plant Dis.*, 99, 806–14.

Hu, M.-J., Grabke, A. and Schnabel, G. 2015b. Investigation of the *Colletotrichum gloeosporioides* species complex causing peach anthracnose in South Carolina. *Plant Dis.*, 99, 797–805.

Hu, M.-J., Cox, K. D. and Schnabel, G. 2016a. Resistance to increasing chemical classes of fungicides by virtue of 'selection by association' in *Botrytis cinerea. Phytopathology*, 106, 1513–20.

Hu, M.-J., Fernández-Ortuño, D. and Schnabel, G. 2016b. Monitoring resistance to SDHI fungicides in *Botrytis cinerea* from strawberry fields. *Plant Dis.*, 100, 959–65.

Hu, M.-J., Dowling, M. E. and Schnabel, G. 2018. Genotypic and phenotypic variations in *Botrytis* spp. isolates from single strawberry flowers. *Plant Dis.*, 102, 179–84.

Ishii, H., Zhen, F., Hu, M., Li, X. and Schnabel, G. 2016. Efficacy of SDHI fungicides, including benzovindiflupyr, against Colletotrichum species. *Pest Manag. Sci.*, 72, 1844–53.

Köller, W. and Wilcox, W. F. 2001. Evidence for the predisposition of fungicide-resistant isolates of *Venturia inaequalis* to a preferential selection for resistance to other fungicides. *Phytopathology*, 91, 776–81.

Kretschmer, M., Leroch, M., Mosbach, A., Walker, A. S., Fillinger, S., Mernke, D., Schoonbeek, H. J., Pradier, J. M., Leroux, P. and De Waard, M. A. 2009. Fungicide-driven evolution and molecular basis of multidrug resistance in field populations of the grey mould fungus *Botrytis cinerea. PLoS Pathog.*, 5, e1000696.

Leadbeater, A. 2011. Resistance risk to QoI fungicides and anti-resistance strategies. In: Thind, T. S. (Ed.), *Fungicide Resistance in Crop Protection: Risk and Management*. CABI, Wallingford, UK.

Leroux, P., Fritz, R., Debieu, D., Albertini, C., Lanen, C., Bach, J., Gredt, M. and Chapeland, F. 2002. Mechanisms of resistance to fungicides in field strains of *Botrytis cinerea. Pest Manag. Sci.*, 58, 876–88.

Li, X., Fernández-Ortuño, D., Chen, S., Grabke, A., Luo, C.-X., Bridges, W. C. and Schnabel, G. 2014. Location-specific fungicide resistance profiles and evidence for stepwise accumulation of resistance in *Botrytis cinerea. Plant Dis.*, 98, 1066–74.

Luo, C. X. and Schnabel, G. 2008. The cytochrome P450 lanosterol 14α-demethylase gene is a demethylation inhibitor fungicide resistance determinant in Monilinia fructicola field isolates from Georgia. *Appl. Environ. Microb.*, 74, 359–66.

Luo, C.-X., Cox, K., Amiri, A. and Schnabel, G. 2008. Occurrence and detection of the DMI resistance-associated genetic element 'Mona' in *Monilinia fructicola. Plant Dis.*, 92, 1099–103.

Ma, Z. and Michailides, T. J. 2005. Advances in understanding molecular mechanisms of fungicide resistance and molecular detection of resistant genotypes in phytopathogenic fungi. *Crop Prot.*, 24, 853–63.

Ma, Z. H., Yoshimura, M. A., Holtz, B. A. and Michailides, T. J. 2005. Characterization and PCR-based detection of benzimidazole-resistant isolates of *Monilinia laxa* in California. *Pest Manag. Sci.*, 61, 449–57.

MacKenzie, S. J. and Peres, N. A. 2012a. Use of leaf wetness and temperature to time fungicide applications to control anthracnose fruit rot of strawberry in Florida. *Plant Dis.*, 96, 522–8.

MacKenzie, S. J. and Peres, N. A. 2012b. Use of leaf wetness and temperature to time fungicide applications to control Botrytis fruit rot of strawberry in Florida. *Plant Dis.*, 96, 529–36.

Mikaberidze, A. and McDonald, B. A. 2015. Fitness cost of resistance: Impact on management. In: Ishii, H. and Hollomon, D. W. (Eds), *Fungicide Resistance in Plant Pathogens: Principles and a Guide to Practical Management*. Springer Japan, Tokyo.

Mikaberidze, A., Paveley, N., Bonhoeffer, S. and van den Bosch, F. 2017. Emergence of resistance to fungicides: The role of fungicide dose. *Phytopathology*, 107, 545–60.

Mogili, U. M. R. and Deepak, B. B. V. L. 2018. Review on application of drone systems in precision agriculture. *Procedia Comput. Sci.*, 133, 502–9.

Oberti, R., Marchi, M., Tirelli, P., Calcante, A., Iriti, M., Tona, E., Hočevar, M., Baur, J., Pfaff, J., Schütz, C. and Ulbrich, H. 2016. Selective spraying of grapevines for disease control using a modular agricultural robot. *Biosyst. Eng.*, 146, 203–15.

Pavan, W., Fraisse, C. W. and Peres, N. A. 2011. Development of a web-based disease forecasting system for strawberries. *Comput. Electron. Agric.*, 75, 169–75.

Reuveni, M. 2001. Improved control of powdery mildew (*Sphaerotheca pannosa*) of nectarines in Israel using strobilurin and polyoxin B fungicides; mixtures with sulfur; and early bloom applications. *Crop Prot.*, 20, 663–8.

Rupp, S., Plesken, C., Rumsey, S., Dowling, M., Schnabel, G., Weber, R. W. S. and Hahn, M. 2017. *Botrytis fragariae*, a new species causing gray mold on strawberries, shows high frequencies of specific and efflux-based fungicide resistance. *Appl. Environ. Microb.*, 83, e00269-17.

Schnabel, G., Hu, M. and Fernández-Ortuño, D. 2015. *Monitoring Resistance by Bioassay: Relating Results to Field Use Using Culturing Methods*. Springer, Japan.

Schnabel, G., Hu, M., Edison, G. and Pargas, R. 2017. *Modern Fungicides and Antifungal Compounds*. Braunschweig, Deutsche Phytomedizinische Gesellschaft.

Schoch, C. L., Seifert, K. A., Huhndorf, S., Robert, V., Spouge, J. L., Levesque, C. A. and Chen, W. 2012. Nuclear ribosomal internal transcribed spacer (ITS) region as a universal DNA barcode marker for *Fungi*. *Proc. Natl. Acad. Sci. USA*, 109, 6241–6.

Schroeder, W. T. and Prowidenti, R. 1969. Resistance to benomyl in powdery mildew of cucurbits. *Plant Dis. Rep.*, 53, 271.

Shen, Y., Zhu, H., Liu, H., Chen, Y. and Ozkan, E. 2017. Development of a laser-guided, embedded-computer-controlled, air-assisted precision sprayer. *Trans. ASABE*, 60, 1827–38.

Stanis, V. and Jones, A. 1985. Reduced sensitivity to sterol-inhibiting fungicides in field isolates of *Venturia inaequalis*. *Phytopathology*, 75, 1098–101.

Steiner, P. W. 1988. Maryblyt: A predictive model for eastern apple fire blight. *Phytopathology*, 78, 1571.

Steva, H., Cartolaro, P. and Gomes Da Silva, M. 1990. Tolerance of powdery mildew of SBI fungicides: Situation for 1989. *Phytoma*, 419, 41–4.

Taylor, J. W. and Fisher, M. C. 2003. Fungal multilocus sequence typing – it's not just for bacteria. *Curr. Opin. Microbiol.*, 6, 351–6.

Tomita, Y. and Ishii, H. 1998. Reduced sensitivity to fenarimol in Japanese field strains of *Venturia nashicola*. *Pestic. Sci.*, 54, 150–6.

van den Bosch, F., Fraaije, B., Oliver, R., Van Den Berg, F. and Paveley, N. 2015. The use of mathematical models to guide fungicide resistance management decisions. In: Ishii, H. and Hollomon, D. W. (Eds), *Fungicide Resistance in Plant Pathogens: Principles and a Guide to Practical Management*. Springer Japan, Tokyo.

Wade, M. 1988. Strategies for preventing or delaying the onset of resistance to fungicides and for managing resistance occurrences. In: Delp, C. J. (Ed.), *Fungicide Resistance in North America*. APS Press, St. Paul, MN, pp. 14–15.

Weber, R. W. S. 2011. Resistance of *Botrytis cinerea* to multiple fungicides in Northern German small-fruit production. *Plant Dis.*, 95, 1263–9.

Weber, R. W. and Hahn, M. 2011. A rapid and simple method for determining fungicide resistance in *Botrytis*. *J. Plant Dis. Prot.*, 118, 17.

Weir, B. S., Johnston, P. R. and Damm, U. 2012. The *Colletotrichum gloeosporioides* species complex. *Stud. Mycol.*, 73, 115–80.

Wilson, L. L., Madden, L. V. and Ellis, M. A. 1990. Influence of temperature and wetness duration on infection of immature and mature strawberry fruit by *Colletotrichum acutatum*. *Phytopathology*, 80, 111–16.

Xu, X., Harris, D. C. and Berrie, A. M. 2000. Modeling infection of strawberry flowers by *Botrytis cinerea* using field data. *Phytopathology*, 90, 1367–74.

Chapter 13

Use of biocontrol agents in fruit tree disease management

Jürgen Köhl, Wageningen University and Research, The Netherlands

1 Introduction

Biological control of plant diseases is the suppression of populations of plant pathogens by living organisms (Heimpel and Mills, 2017). Fungi, bacteria and viruses and members of the mesofauna can be involved in such natural suppression. The effect of naturally present beneficial consortia may be limited when broad-spectrum fungicides are applied that may affect these beneficial populations. Such undesired side effects of fungicides may result in the emergence of 'iatrogenic' diseases (Fokkema and de Nooij, 1981). 'Natural biological control' buffers cropping systems and lowers risks for disease outbreaks (Eilenberg et al., 2001; Heimpel and Mills, 2017). This natural biological control can be facilitated by 'conservation biological control' which stimulates the natural microbial buffering capacity or protects natural antagonists by avoiding unnecessary fungicide applications. Application of modern meta-genomic approaches allows a much better understanding of the beneficial functions of the microbiomes naturally present in and on crops (Berg, 2009). It is expected that this knowledge will lead to new management options for the use of natural biological control to ensure the suppression of pathogens. Amongst beneficial microorganisms, isolates can be selected which are highly effective against a targeted pathogen and can efficiently be multiplied on artificial media. Application of such selected and mass-produced antagonists in high densities once, or several times, during a growing season is called

http://dx.doi.org/10.19103/AS.2019.0046.16

'augmentative biological control' (Eilenberg et al., 2001; Heimpel and Mills, 2017; van Lenteren et al., 2018). Research on augmentative biological control leading to commercial products for disease control in fruit tree production is reviewed in this chapter.

Commercial biocontrol agents (BCA) for disease control are registered plant protection products, which are manufactured by specialized companies. Such products contain living microbial organisms such as fungi or bacteria. In some cases, metabolites produced by these microbial organisms are included in the product (Glare et al., 2012). Mycoviruses and bacteriophages can also be potential BCAs. The number of registered BCAs for disease control in fruit tree production in Australia, Brazil, Canada, European Union, Japan, New Zealand and the United States is limited: only fifteen antagonistic strains were registered belonging to four fungal genera and three bacterial genera (Table 1). These registered strains are targeted against a range of pathogens. Some strains are registered against several pathogens and, in some cases, registrations have been obtained for different pathogens in different regions. However, the majority of registered antagonists are targeted against fireblight (*Erwinia amylovora*), with seven strains, and pre- and post-harvest fruit rot diseases caused by various pathogens with eight strains.

Antagonists protect crops from damage by diseases via different modes of action. BCAs may induce resistance against infections by a pathogen in plant tissue without any direct antagonistic interaction with the pathogen. Two different pathways in the plant are involved in systemic acquired resistance (SAR) or induced systemic resistance (ISR). Also, microorganisms applied as biofertilizers or biostimulants and mycorrhizal fungi, not all of which are considered as biocontrol agents, can strengthen crops, leading to increased resilience against disease. Another indirect interaction with pathogens is in the competition for nutrients and space. Necrotrophic plant pathogens often need exogenous nutrients before being able to infect a host plant. Microorganisms with high saprophytic competitive ability can rapidly consume the nutrient sources essential for pathogen infection such as sugars, pollen and plant exudates on plant surfaces and the consequently outcompeted pathogens are not able to infect the host. An example of this efficient mode of action is the protection of wounds on fruits from pathogen invasion by fast colonizing yeasts (Spadaro and Droby, 2016). Antagonists may also interact directly with the pathogen by hyperparasitism or antibiosis. Hyperparasites are able to invade the pathogen mycelium and spores after enzymatic cell wall degradation by β-1,3-glucanases and chitinases, resulting in eradication of the invading pathogen cells. Production of antimicrobial secondary metabolites with toxic effects against pathogens is another mode of action leading to eradication of the targeted pathogen. Low amounts of *in situ* secreted secondary metabolites support antagonists to attack the pathogens, leading to a competitive

Table 1 Registered bacterial and fungal biological control agents for control of diseases in fruit tree crops[a]

Microbial strain	Targeted pathogens or fruit diseases	Countries where approved
Aureobasidium pullulans DSM 14940 and 14941	Fire blight (*Erwinia amylovora*) in pome fruit; brown rot in stone fruit	Europe, Canada, USA
Bacillus licheniformis Strain SB3086	Apple scab (*Venturia inaequalis*); rust (*Gymnosporangium* spp.)[b]	USA
Bacillus pumilus QST 2808	Fire blight (*Erwinia amylovora*) in pome fruit; scab (*Venturia* spp.), powdery mildew (*Podosphaera leuchotricha*)[c]	USA
Bacillus subtilis QST713	Fire blight (*Erwinia amylovora*) in pome fruit; scab (*Venturia* spp.) *Monilinia laxa*, *Monilinia fructicola*, *Podosphaera leucotricha*[d]	USA
Candida oleophila I-182	Wound diseases of pome fruits	USA
Candida oleophila O	Wound diseases of pome fruits	EU
Metschnikowia fructicola NRRL Y027328	Wound diseases of pome fruits	EU (pending)
Pantoea agglomerans C9-1	Fire blight (*Erwinia amylovora*) in pome fruit	CA, USA
Pantoea agglomerans E325	Fire blight (*Erwinia amylovora*) in pome fruit	CA
Pantoea agglomerans p10c	Fire blight (*Erwinia amylovora*) in pome fruit	NZ
Pseudomonas fluorescens A506	Fire blight (*Erwinia amylovora*) in pome fruit	USA
Pseudomonas rhodesiae HAI-0804	*Pseudomonas syringae* in plum[e]	Japan
Pseudomonas syringae ESC 10	Post-harvest fungal diseases of fruits of apples, pears and cherries[f]	CA, USA
Pseudomonas syringae ESC 11	Various moulds including blue mould (*P. expansum*), grey mould (*B. cinerea*) and Mucor rot on fruits of apple, pear and cherries[g]	USA
Saccharomyces cerevisiae LAS02	Fruit rots caused by *Monilinia fructigena*, *Monilinia laxa*, *Monilinia fructicola*, *Botrytis* spp., *Botrytis cinerea*, *Alternaria* spp. *Neofabraea alba* and *Penicillium* spp.[h]	EU

[a] For Australia, Brazil, Canada, European Union, Japan, New Zealand and the United States (from: van Lenteren et al., 2018).
[b] https://www3.epa.gov/pesticides/chem_search/.../decision_PC-006492_1-Feb-01.pdf
[c] https://www3.epa.gov/pesticides/chem_search/reg_actions/registration/decision_PC-006485_16-Nov-04.pdf
[d] https://www3.epa.gov/pesticides/chem_search/reg_actions/registration/decision_PC-006479_9-Aug-06.pdf
[e] http://www.abim.ch/fileadmin/abim/documents/presentations2016/3_Tetsuo_Wada_ABIM_2016.pdf
[f] http://sitem.herts.ac.uk/aeru/bpdb/Reports/1998.htm
[g] http://sitem.herts.ac.uk/aeru/bpdb/Reports/1999.htm
[h] http://sitem.herts.ac.uk/aeru/bpdb/Reports/2954.htm

advantage of the antagonist. Biocontrol agents have also been selected which secrete efficient secondary metabolites into the growth media during mass production and secondary metabolites are applied together with living cells of the antagonists in the commercial biocontrol product.

For the development of new BCA, the choice of the mode of action depends on the targeted disease, for example nutrient competition will not affect biotrophic pathogens, as well as the targeted stage of the pathogen life cycle. However, many more criteria besides the preferred modes of action have to be considered during the screening of new antagonists. Such criteria include ecological characteristics such as cold tolerance and UV resistance, and economically important characteristics such as high spore production potential to allow mass production. Also, potential risks have to be considered to exclude strains with potential pathogenicity to humans, animals or plants. Screening of antagonists for commercial use is thus a complex process that focusses not only on the efficacy of antagonists and their mode of action, but also on various other relevant criteria in a stepwise approach (Köhl et al., 2011).

2 Case study: strategies for biocontrol of apple scab – *Venturia inaequalis*

Apple scab, caused by *Venturia inaequalis*, is the most important disease worldwide affecting apple production. The life cycle of the pathogen consists of two distinct parts (MacHardy et al., 2001). During winter, the pathogen survives on apple leaf litter with a saprophytic lifestyle. Interestingly, the pathogen colonizes necrotic apple leaves but is not able to colonize other necrotic plant tissues. Survival of conidia on twigs is a second, less important, way of overwintering. The main disease source inciting new epidemics in spring comes from the ascospores produced on overwintering apple leaves. Ascospores infect the young tissues of developing leaves and young fruits. Thereafter, a polycyclic epidemic develops with many infection-sporulation cycles through conidia. During the summer part of the life cycle, the pathogen grows with a semi-biotrophic lifestyle on epidermis cells under the cuticles of leaves and fruits. The particular life cycle of *V. inaequalis* determines the options and limitations for biological control of apple scab. During the winter and summer parts of the life cycle there are differences in the lifestyle of the pathogen (saprophytic versus semi-biotrophic), niche (leaf litter on the ground versus canopy) and colonization of host tissue (internal and external colonization of dead tissue versus subcuticular colonization of living tissue), and as such, biocontrol requires different types of antagonists during these two seasons.

During the winter season, antagonists with superior saprophytic competitive colonization ability may reduce the survival of *V. inaequalis* mycelium and the formation of ascospores in fruiting bodies. Since no alternative necrotic plant tissues can be colonized by *V. inaequalis*, this application can be very well targeted at leaf litter on the orchard floor to reach potential pathogen sources. Possible ecological restrictions of this biocontrol strategy are the low temperatures during autumn and winter and the resistance of *V. inaequalis* fruiting bodies to antagonism. An epidemiological limitation of this strategy is that sanitation measures, including the removal of leaf litter, shredding of leaf litter or application of nutrients to stimulate naturally occurring microbial decomposers, are an important component in apple scab management which can delay the build-up of epidemics in spring but are not sufficient as stand-alone measures since under practical conditions, the pathogen load cannot be completely eliminated (Sutton et al., 2000; Carisse and Dewdney, 2002; Gomez et al., 2007).

The second option for biocontrol is the application of antagonists during the summer epidemic. Such antagonists may interfere with *V. inaequalis* before and during infection on the host tissue, or later in the cycle during colonization of the sub-epidermal space and subsequent pathogen sporulation. Ungerminated ascospores and conidia of *V. inaequalis* are present on the host surface during periods of low humidity, with infection processes depending on leaf wetness and temperature. Antagonists applied to prevent infections must be able to survive on host tissue surfaces during the period from application until infection. Under low humidity conditions, metabolites present in the antagonist formulation may interact with *V. aequalis*; however, during leaf wetness periods, antagonists must become active and be able to interact antagonistically with *V. aequalis*. Since *V. aequalis* spores can germinate and penetrate the cuticle within a few hours, activation of the antagonist's metabolism and interaction with the host must be sufficiently fast to achieve successful biocontrol. In cases where interaction with *V. inaequalis* occurs during colonization and sporulation in the subcuticular space, antagonists must be able to grow into this space without becoming pathogenic or causing skin disorders. A third opportunity for biocontrol during the summer epidemic may be the use of resistance-inducing antagonists (Ortega et al., 1998; Percival et al., 2009).

Biocontrol of apple scab has been an area of investigation for the past several decades (Carisse, 2000). Heye and Andrews (1983) screened fungal antagonists for their potential to reduce pseudothecium formation of *V. inaequalis* in naturally infected senesced apple leaves. The selected basiodiomycete, *Athelia bombacina,* inhibited pseudothecium formation efficiently also under orchard conditions. The potential of *A. bombacina* was confirmed by Miedtke and Kennel (1990), although a major bottleneck in this case is the required high amount of inoculum of the antagonist for orchard applications, given the difficulties in mass production of the basidiomycete (Carisse and Dewdney, 2002). Autumn application of *Microsphaeropsis ochracea*, a fungal antagonist with high potential in mass production, to the orchard floor was found to reduce the level of ascospores of the pathogen during the following spring by 70–80% (Carisse et al., 2000), allowing a reduction in the number of fungicide sprays employed during the cropping season.

The use of antagonists during the summer epidemics was first investigated by Andrews et al. (1983) who compared 50 isolates belonging to different species of bacteria and fungi. *Chaetomium globosum* had the strongest effect on *V. inaequalis* in tests under controlled conditions. However, the antagonist was not able to colonize leaves under orchard conditions and the antibiotic produced by the antagonist was not stable in the environment (Boudreau and Andrews, 1987). Burr et al. (1996) selected antagonists affecting the conidial germination and mycelial growth of *V. inaequalis*. Amongst the almost 1000 isolates tested, an isolate of *Pseudomonas syringae* was found to be superior and reduced infection by the pathogen on seedlings. Epiphytic yeasts were selected by Fiss et al. (2000) in a screening programme but no reproducible control effects were achieved under orchard conditions.

In a screening programme using bioassays with apple seedlings artificially inoculated with conidia of *V. inaequalis*, a new antagonist was discovered (Köhl, 2009; Köhl et al., 2009). *Cladosporium cladosporioides* H39, isolated from an apple leaf collected in a Dutch orchard, reduced conidial production by *V. inaequalis* when applied one to three days after pathogen inoculation. A pilot formulation of conidia of *C. cladosporioides* H39 was used in eight field trials in 2012 and 2013 in the main apple growing areas in Germany, Hungary and Poland (Köhl et al., 2015). These trials were carried out in conventional cropping systems as well as in organic systems and focussed on apple scab control during the primary or summer season. Control efficacies ranged between 42% and 98% on leaf scab and from 41% to 94% on fruit scab. In some cases the efficacies achieved were no different from those obtained by application of synthetic fungicides or copper-based fungicides, especially when applied during the summer season (Fig. 1). Interestingly, the antagonist was also effective when applied at 300–2000 degree hours after forecasted infection events, suggesting that *V. aequalis* and *C. cladosporioides* H39 can also interact after spore germination during the subcuticular colonization.

Biocontrol products for scab control are currently commercially available only in the United States, where three *Bacillus* spp. strains are registered with claimed effects on *Venturia* spp. (Table 1). Registration of new biocontrol products for scab control will have several advantages for scab management. Substitution of applications of synthetic fungicides will lower environmental risks and reduce the number and concentration of residues of synthetic fungicides on fruits. Furthermore, the use of BCA will allow a better management of strains of *V. inaequalis* with resistance to fungicides in orchard conditions (Beresford et al., 2013). In the scenario in which the number of registered chemical fungicides available for scab control decreases, and given the preference

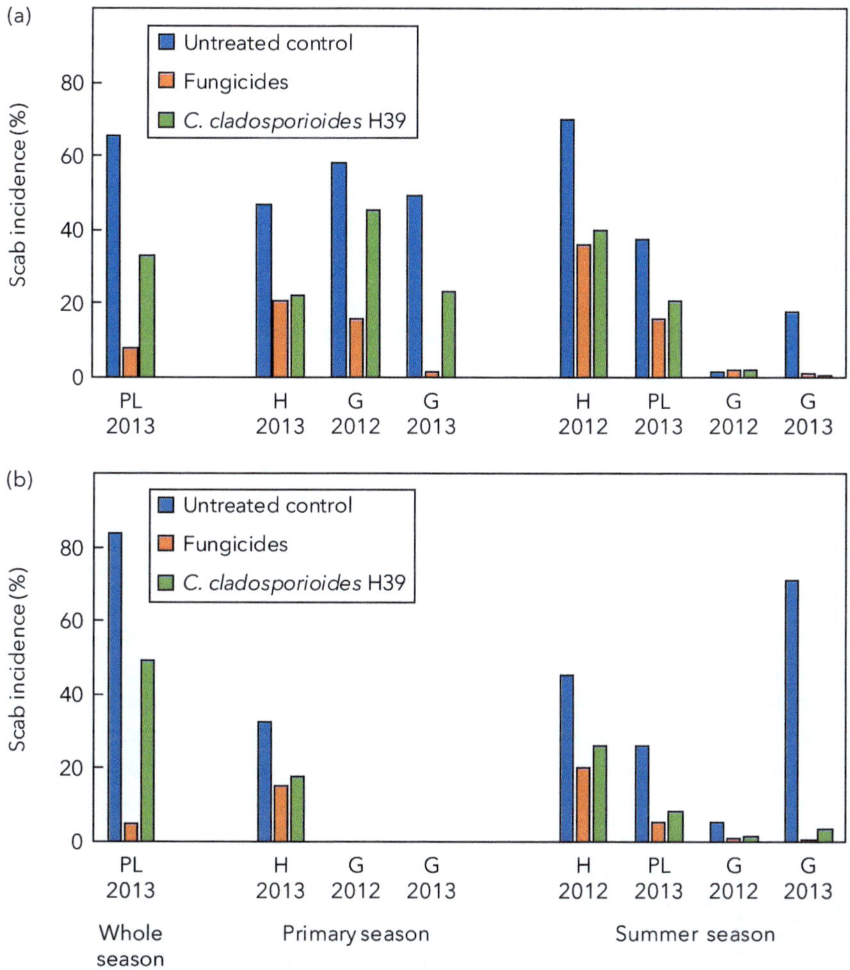

Figure 1 Effect of treatments with *Cladosporium cladosporioides* H39 in comparison to fungicides applied at regular intervals on (a) scab on leaves and (b) scab on fruit. Field trials in Poland (PL), Hungary (H) and Germany (G). Source: Adapted from Köhl et al. (2015).

for low-risk plant protection, products such as BCA will become an essential component of scab management in commercial apple production.

Scab management must fit into the management strategies applied to control other diseases and pests. The use of fungicides for the control of powdery mildew, various fruit rots or against fungi causing superficial disorders on fruits must be compatible with the use of BCA applied for scab control. There is no need to achieve full compatibility however, allowing applications with tank mixes. Established antagonist populations may be reduced by chemical fungicides applied alternately with BCA application, although antagonist populations can be re-established by subsequent application after a few days when the fungicides have lost their activity or are washed off. Alternations in the use of chemical and biocontrol products will thus offer many options for the application of biocontrol within integrated pest management systems. For the development of such alternating application schedules, knowledge of fungicide persistence and the population dynamics of the antagonist under fluctuating environmental conditions in the orchard are essential.

3 Biocontrol of fruit tree diseases: current status

3.1 Monilinia *spp. - Brown rot on stone fruit*

Monilinia fructicola, *M. laxa* and *M. fructigena* cause brown rot, the most important disease on stone fruit. The epidemiology and management of brown rot caused by *M. laxa* have recently been reviewed by Rungjindamai et al. (2014a). Infected mummies are the main substrates for overwintering and disease sources for conidium production in spring. Flowers and developing fruits can be infected by *M. fructicola* and *M. laxa*, and at harvest, the fruits are highly susceptible and infections through wounds are common. Fruit damage by brown rot occurs during the season in the orchard and during storage, where latent infections in the pre-harvest phase and wound infections caused at harvest lead to brown rot, which may further spread through contact. Infection of fruit via wounds both pre- and post-harvest can lead to the rapid rotting of the fruit. Brown rot epidemiology favours three strategies for biocontrol: microbial reduction of pathogen build-up in overwintering mummies, pre-harvest prevention of flower and fruit infection in the orchard and post-harvest prevention of wound infection before storage.

Biocontrol on harvested fruit before storage has been studied by many research groups and a broad range of bacteria and yeasts have been tested for this application. Screening of new antagonists is straightforward in this case because bioassays with artificially wounded fruit mimic practical conditions with a well-defined time window for application during a known infection period and under well-defined environmental conditions. Assessment of symptoms is also fast and directly linked to losses caused by infections. The entire fruit surface can easily be reached by antagonists via dipping and the surface area for treatment is small compared to spray applications in orchards. Progress in the use of antagonists in post-harvest control has recently been reviewed by Rungjindamai et al. (2014a). Several isolates of *Bacillus subtilis* and *B. amyloliquefaciens* have successfully been used for post-harvest brown rot control (Pusey et al., 1988; Zhou et al., 2008; Casals et al., 2010). Lipopeptides such as fengycin, produced by strains of *Bacillus* spp., are inhibitory to fungal mycelial growth and protect treated fruit wounds from invasion by *Monilinia* spp. *Bacillus subtilis* QST713 has been registered for brown rot control in the United States (Table 1). Yeasts and yeast-like fungi can also protect stone fruit from infections by *Monilinia* spp. after harvest. *Aureobasidium*, *Cryptococcus*, *Rhodotorula* and *Sporidiobolus* are common genera colonizing fruit surfaces which can quickly colonize wound tissues and may prevent pathogen infections by nutrient competition in the wounds (Janisiewicz et al., 2010). *Aureobasidium pullulans* and *Rhodotorula phylloplana* showed excellent potential for biocontrol of brown rot caused by *M. fructicola* in plums (Janisiewicz et al., 2014). Isolates of the yeasts *Debaryomyces hansenii* and *Wickerhamomyces anomalus* have a high capacity to colonize wound tissues of peach and plum fruits and protected these fruits from infection in *in vivo* assays after artificial pathogen inoculation with *M. fructigena* and *M. fructicola* (Grzegorczyk et al., 2017). In similar studies, isolates of *Pichia membranaefaciens* and *Kloeckera apiculata* were effective against infections of plums by *M. fructicola* (Zhang et al., 2017). A product containing strains of the yeast-like fungus *A. pullulans* has been registered for brown rot control in stone fruit in the European Union, Canada and the United States.

Another strategy for biocontrol of brown rot is the application of antagonists in the orchard to prevent infections of twigs, blossoms and fruits, including latent fruit infections which can lead to damage after harvest. Melgarejo et al. (1985) assessed the fungal flora of peach twigs and flowers and amongst the 19 most abundant species, isolates of *Aspergillus flavus*, *Epicoccum nigrum*, *Penicillium chrysogenum*, *P. frequentans* and *P. purpurogenum* showed *in vitro* inhibition of *M. laxa*. Isolates of *A. flavus*, *E. nigrum*, *P. frequentans* and *P. purpurogenum* were applied to peach twigs under orchard conditions (Melgarejo et al., 1986). *Aspergillus flavus*, *E. nigrum* and *P. frequentans* inhibited twig infection by *M. laxa* in the spring when applied before inoculation with the pathogen. *Penicillium frequentans* was also effective when applied after pathogen inoculation. In similar trials in the autumn, only applications of *E. nigrum* were effective. When applied at bloom and pre-harvest, *E. nigrum* moderately reduced brown rot after harvest in five out of the seven fields (Larena et al., 2005). The use of the promising antagonist *P. frequentans* Pf909 has been further investigated, including detailed studies of field application, mode of action, mass production, formulation population dynamics in the orchard and safety (for further information see Guijarro et al., 2017).

Bacillus amyloliquefaciens CPA-8 is another efficient antagonist of *M. laxa* and *M. fructicola*. Different formulations have been tested in field trials with peaches (Gotor-Vila et al., 2017a). In orchards with high pathogen pressure, four applications of the antagonist in orchards at days 30, 14, 7 and 3 before harvest had no effect on disease incidence at harvest or in the post-harvest assessment. In such cases, synthetic fungicide only managed to achieve control levels of approximately 50%. However, the orchards used for trials with such high pathogen pressure are not representative of commercial orchards. Applications of formulated *B. amyloliquefaciens* CPA-8 in an orchard with a moderate disease pressure significantly reduced brown rot at harvest by 55–64% and in the post-harvest assessment by 50%. Protocols for mass production and formulation of *B. amyloliquefaciens* CPA-8 have been developed and strain-specific markers are available for monitoring antagonist populations after field releases (Gotor-Vila et al., 2016, 2017b).

Rungjindamai et al. (2014b) investigated the effect of antagonists on sporulation of *M. laxa* in pathogen-infected mummies, which are a significant inoculum source for blossom and fruit infections in spring. A single application of *Aureobasidium pullulans* Y126 in winter reduced *M. laxa* sporulation as much as the fungicide fenbuconazole. Antagonist applications in early spring were slightly more effective, but the best results were obtained with a combination of winter and spring applications. *Aureobasidium pulluans* Y126 reduced the number of *M. laxa* conidia per plum by more than 90% from 9×10^5 to 5×10^4 conidia per mummified plum. Strains *Bacillus* sp. B91 and *B. subtilis* QST 713 (formulated as Serenade) showed some antagonistic activity. Such a biocontrol strategy during the dormant season can be an efficient component of integrated disease management with the advantage of long interaction periods between pathogen and antagonists from autumn until early spring. Screening for antagonists with high saprophytic competitive ability during colonization of necrotic host tissues is a generally neglected but promising strategy for biocontrol (Köhl and Fokkema, 1998). In mummified fruit, highly competitive saprophytic antagonists may be able to outcompete a whole range of fruit pathogens which overwinter in mummies. Besides *M. laxa*, other important fruit rot pathogens overwintering on mummified fruits, such as *Neofabraea alba* and *Cadophora luteo-olivacea* (Köhl et al., 2018), may also be controlled by such BCA.

3.2 Fruit rots caused by various pathogens

A range of fungal pathogens can cause fruit rot, leading to damage in the orchard or after harvest during storage. For several pathogens such as *Botrytis cinerea* (grey mould), *Penicillium expansum* (blue mould) and *Monilinia fructigena* (brown rot), wounds caused by insects and birds as well as by physical damage before or during harvest are important entry points (Snowdon, 1990). These pathogens typically result in a rapid decay of fruit in the pre- and post-harvest stages. Applications of BCA shortly before or after harvest are effective measures to reduce losses by wound pathogens because the timing of applications and the targeted parts of the crops are clearly defined and the environmental conditions after harvest are controlled and thus known. Antagonists can be screened for such applications in bioassays with artificially wounded fruits and incubated under commercial storage conditions. Results of bioassays can thus be directly translated to commercial application conditions.

Since the successful use of *Bacillus subtilis* for brown rot control in stone fruits in the early 1980s by Pusey and Wilson (1984), much research has been directed at selecting new antagonists for fruit rot control in fruit tree crops. In a comprehensive review of biocontrol of post-harvest diseases in fruits by microbial antagonists, Sharma et al. (2009) listed 26 antagonist species which were successfully tested against the main fruit diseases in stone and pome fruits in pre- and post-harvest applications (Table 2). The majority of the antagonist species are yeasts or yeast-like fungi or bacteria. For most of the tested isolates, competition for nutrients plays a major role, although other modes of action are also involved (Piombo et al., 2018). Sharma et al. (2009) also reviewed options for enhancing the efficacy of antagonists against post-harvest diseases through adapting the physical environment during storage using antagonist mixtures or by adding low amounts of fungicides, salts such as calcium bicarbonate or sodium carbonate, specific nutrients or plant extracts. Antagonist treatments have also been successfully combined with physical treatments such as curing or heat treatment. Progress in the research of biocontrol of fruit rot diseases resulted in the development of commercial BCAs based on the antagonists *Aureobasidium pullulans* DSM 14940 and 14941, *Candida oleophila* I-182, *C. oleophila* O, *Metschnikowia fructicola* NRRL Y027328, *Pseudomonas syringae* ESC 10, *P. syringae* ESC 11 and *Saccharomyces cerevisiae* LAS02 (Table 1). Droby et al. (2016) observed reduced and inconsistent performance of biocontrol products against post-harvest fruit pathogens because commodities vary in genetic background, physiology, post-harvest handling and susceptibility to specific pathogens. This variability makes it difficult to protect commodities with one single antagonist strain. A more promising option may be the development of synthetic microbial communities to develop more consistent and robust biocontrol solutions (Droby et al., 2016; Droby and Wisniewski, 2018).

In addition to wound infecting pathogens, another important group of pathogens infects blossoms or developing fruits during the growing season and remains quiescent without causing symptoms during the growing season and during early post-harvest storage period. Typically, symptoms of disease occur after several months in cold storage within a controlled atmosphere. Common pathogens causing such late post-harvest rot are *Neofabraea alba* (lenticel rot disease), *Neofabraea perennans* (bull's eye rot), *Cadophora malorum*, *Cadophora luteo-olivacea*, *Neonectria ditissima* (blossom-end rot), *Colletotrichum acutatum* species complex (bitter rot), *Phytophthora* spp., *Alternaria* spp., *Fusarium* spp. and *Stemphylium vesicarium* (Köhl et al., 2018). This considerable diversity of potential pathogens presents challenges to the development of pathogen-specific BCA. Furthermore, the long infection periods over which *Neonectria ditissima* and other pathogens can infect fruits

Table 2 Antagonists species successfully used for biological control of fruit rot diseases

Disease (pathogen)	Antagonist	Fruit tree crop
Grey mould (*Botrytis cinerea*)	*Acremonium brevae*	Apple
	Bacillus subtilis	Cherry
	Bacillus pumilus	Pear
	Candida guilliermondii	Nectarine, peach
	Candida sake	Apple
	Cryptococcus laurentii	Peach, pear
	Cryptococcus albidus	Apple
	Kloeckera apiculata	Cherry
	Metschnikowia pulcherrima	Apple
	Pichia guilliermondii	Apple
	Pseudomonas cepacia	Pear
	Pseudomonas fluorescens	Apple
	Pseudomonas syringae	Apple
	Rahuella aquatilis	Apple
	Rhodotorula glutinis	Apple, pear
	Trichoderma harzianum	Pear
Penicillium rot (*Penicillium digitatum, P. expansum*)	*Candida oleophila*	Apple
	Candida sake	Apple, pear
	Cryptococcus laurentii	Peach, pear
	Cryptococcus albidus	Apple
	Cryptococcus sp.	Apple
	Metschnikowia pulcherrima	Apple
	Pantoea agglomerans	Apple
	Pichia guilliermondii	Apple
	Pseudomonas cepacia	Apple, pear
	Pseudomonas syringae	Apple
	Rahuella aquatilis	Apple
	Rhodotorula glutinis	Apple, pear
Monilinia rot (*Monilinia laxa, M. fructicola*)	*Cryptococcus laurentii*	Cherry, peach
	Penicillium frequentans	Peach
	Pseudomonas cepacia	Peach
	Pseudomonas corrugata	Peach, nectarine
	Pseudomonas syringae	Peach
Rhizopus rot (*Rhizopus nigricans, R. stolonifer*)	*Candida sake*	Apple
	Cryptococcus laurentii	Peach
	Debaryomyces hansenii	Peach
	Enterobacter cloacae	Peach
	Pantoea agglomerans	Pear

Disease (pathogen)	Antagonist	Fruit tree crop
Mucor rot (*Mucor piriformis*)	*Cryptococcus laurentii*	Pear
	Cryptococcus flavus	Pear
	Cryptococcus albidus	Pear
	Pseudomonas cepacia	Apple
Bitter rot (*Glomerella cingulata*)	*Cryptococcus laurentii*	Apple
Alternaria rot (*Alternaria alternata*)	*Enterobacter aerogenes*	Cherry
Anthracnose (*Colletotrichum gloeosporioides*)	*Pestalotiopsis neglecta*	Apricot

Source: Modified from Sharma et al. (2009).

may require the continuous protection of blossoms and developing fruits by antagonists, which is not easy to achieve. There is a lack of knowledge relating to the epidemiology of this group of post-harvest pathogens so that critical infection periods cannot yet be predicted, preventing efficacious timing of antagonist application.

Generally, fruit rot pathogens have a saprophytic stage which overwinters on necrotic host tissues or necrotic plant tissues of non-hosts present in the orchard. In spring or during the entire growing season such pathogen-colonized necrotic tissues are major sources of pathogen spores which may infect blossoms, developing fruits or mature fruits. Relative importance of different pathogen sources may differ between cropping systems and, more importantly, between different pathogens. For many of the currently increasing number of storage pathogens, epidemiological knowledge regarding their saprophytic stage, such as the relative importance of different necrotic substrates in orchards and conditions for spore production and spore dispersal, is still unknown. Fruit rot pathogens are often found in mummies and fallen leaves (Köhl et al., 2018). Applications of strong saprophytic competitors selected to outcompete a broad range of fruit pathogens on these substrates may be a powerful option to reduce the general risk of fruit rot in orchards.

3.2.1 *Erwinia amylovora* – Fireblight of pome fruit

Fireblight of pome fruit is caused by *Erwinia amylovora*. The main infection sites are flowers, where the pathogen epiphytically colonizes the surfaces of nutrient-rich stigmata (Thomson, 1986; Pusey and Curry, 2004). Under favourable environmental conditions, the pathogen can multiply, move to hypanthia during rainfalls and infect the blossom if present in sufficiently high densities. This particular part of the pathogen life cycle favours the development of antagonists that block the colonization and multiplication of the pathogen on specific flower tissues. The knowledge that flowers represent a specific target, that the flowers open at a specific application time and the timing of the vulnerable pathogen stage - multiplication of the pathogen stimulated by the nutrients available on the stigmata surface - allows a very focussed selection of suitable antagonists. Main ecological characteristics of antagonists are the ability to colonize relevant flower tissues rapidly and to outcompete pathogen populations by competition for nutrients and space, possibly supported by the production of antimicrobial secondary metabolites. Changing the conditions in which bacteria grow in the niche, for example by lowering pH, can be another potential mechanism.

Although the important role of flower infections in the life cycle of *E. amylovora* favours the targeted development of BCA, disease control in the orchard can be limited because individual flowers open at different times so that multiple antagonist applications may be needed. Furthermore, infections of host tissues other than the flowers by the pathogen can occur, so that flower protection will not result in complete disease control.

In recent decades, many antagonists of *E. amylovora* have been described (Johnson and Stockwell, 2000) and a number of these have been developed into commercial biocontrol products (Venneste, 2011). Registered products include those based on bacterial isolates of *Bacillus pumilus, B. subtilis, Pantoea agglomerans*, *Pseudomonas fluorescens* and the yeast-like fungus *Aureobasidium pullulans* (Table 1). Research on the use of bacteriophages has also showed promising results (Gill et al., 2003; Born et al., 2017) and may open opportunities for future development of additional complimentary products with distinct modes of action.

The scientific successes in the biocontrol of fireblight have not adequately been translated to commercial successes (Venneste, 2011). However, the review of biological agents in the United States in 2003 by Venneste (2011) indicates that there is a substantial use of such agents in addition to the use of the antibiotics streptomycin and oxytetracycline. Venneste (2011) described variability in efficacy of control and difficulties in registration of microorganisms as the two main reasons for the limited market penetration of various biocontrol products. This situation may drastically change in the near future if the use of antibiotics and copper-based pesticides are banned or more strongly regulated.

3.2.2 Stemphylium vesicarium - Brown spot of pear

Strains of *Erwinia herbicola* and *Pseudomonas fluorescens* were effective in reducing spore germination and mycelial growth of *Stemphylium vesicarium* in assays under controlled conditions on detached pear leaves and on leaves attached to trees grown under controlled conditions with continuously moist conditions at 22°C (Montesinos et al., 1996). The sensitivity of the bacterial antagonists to low humidity conditions is expected to be the critical factor for their success under field conditions. Alternatively, antagonists were applied to fallen pear leaves on the orchard floor which are a main substrate for *S. vesicarium* for overwintering and for spore production in spring and summer. *Trichoderma* spp. decreased overwintering inoculum of the pathogen (Llorente et al., 2006). Sanitation by leaf removal had only a partial effect but leaf removal combined with application of *Trichoderma* on the orchard floor reduced brown spot incidence by more than 60% in a series of orchard trials in Spain and Italy (Llorente et al., 2010).

3.2.3 *Neonectria ditissima* - European canker in apple

There are few reports on the biocontrol of *Neonectria ditissima*. An isolate of *Bacillus subtilis* applied to defoliated apple trees under field conditions did not reduce canker development (Swinburne, 1973). Walter et al. (2017a) tested the effect of eight BCA containing different strains of *Bacillus subtilis* and *B. subtilis* var. *amyloliquefaciens* by painting canker lesions with highly concentrated products. Sporulation of the pathogen in lesions was reduced under orchard conditions by 13–84%. Canker lesion healing was observed in no case. In another set of orchard trials, freshly made picking wounds and leaf and bud scars were treated with biocontrol products and then contaminated with conidia of *N. ditissima* after 24 h (Walter et al., 2017b). None of the tested products provided wound protection and thus use of biocontrol products during harvest and leaf fall could not be recommended. Schiewe and Mendgen (1992) screened antagonists for post-harvest protection of apple fruit from wound infections by *N. ditissima*. Amongst the more than 700 isolates originating from apple trees and fruit, only four bacterial and three fungal isolates reduced lesions on fruits incubated at 20°C, although at 4°C none of the isolates were effective.

3.3 Other fruit tree diseases

For *Gymnosporangium sabinae* causing European pear rust, no scientific reports on biological control have yet been published. *Bacillus licheniformis* strain SB3086 has registration for *Gymnosporangium* spp. control in the United States (Table 1). Preliminary results on biocontrol of the *Podosphaera leucotricha* causing powdery mildew on apples have been reported by Bosshard et al. (1987). Applications of spore suspensions and culture filtrates of the phyllosphere fungi *Chaetomium* spp. reduced mildew infections on apple seedlings to some extent. It was also found that primary infections of *P. leucotricha* were severely parasitized by *Ampelomyces quisqualis*. However, the spread of the hyperparasite was delayed so that no effect on the secondary powdery mildew epidemic could be observed. *Ampelomyces quisqualis,* effective against various species of *Erysiphales*, has been commercialized under the trade name AQ10 for powdery mildew control in grapevine (Kiss et al., 2004). There are no scientific reports on the use of this product against *P. leucotricha* in apples. Products containing *Bacillus pumilus* QST 2808 and *Bacillus subtilis* QST713 have registration for use against *P. leucotricha* in the United States.

4 Use of biocontrol in orchards

Effective BCA have been selected for control of the main fruit tree diseases. Commercial companies have invested in the product development, registration and marketing of a total of 15 biocontrol products for disease control on fruit tree crops in Australia, Brazil, Canada, European Union, Japan, New Zealand and the United States (Table 1); the majority of these products are registered for control of fireblight or fruit rot. For control of other major diseases such as *Monilinia* and apple scab, antagonists with proven reliable efficacies in orchard trials are ready for commercial product development and some new potential products are already under development and in the early registration stage.

No public data are available on the commercial use of biocontrol products in fruit production. The market share of biocontrol products on the global market for crop protection products is currently around 3.5%, with an annual growth >15% (Glare et al., 2012; van Lenteren et al., 2018). This includes use of macrobial and microbial BCA in various crops. Since the commercial use of biocontrol is most developed in protected crops, it can be estimated that use in fruit tree production is far below 3.5% of the total market for crop protection products.

Although potential biocontrol solutions are available, fruit tree growers use only few biocontrol products and biocontrol manufactures may hesitate to invest in product development and the registration of new products for the fruit tree market. The main reasons for this are (1) the availability of cheap, effective and broad-spectrum chemical disease control products with existing distribution channels, (2) growers are well trained in the use of chemical control products and (3) advisers are highly experienced in the design of complex spray schedules combining various chemical control products, often supported by decision support systems which are developed for optimum timing of chemical control products.

The growth rate of the biocontrol market during the last decade indicates that there is a dynamic shift from chemical control to biocontrol options. Main drivers for this shift are increasing demands for sustainable fruit production systems with low environmental risks, low residues of crop protection products on fruit and the decreasing number of chemical pesticides. The number of registered synthetic chemical crop protection products has decreased significantly in recent years (Glare et al., 2012). The costs for developing new chemical active ingredients have increased significantly and the opportunities for finding new leads have decreased. Furthermore, pathogens continuously develop resistance against currently used fungicides so that their commercial life span is limited.

A combination of efforts is needed to implement biocontrol options in integrated tree fruit productions systems. Communication on the specific strengths and weaknesses of biocontrol solutions must be intensified between researchers, the biocontrol industry, advisers and growers. For example, efficacies of biocontrol products based on living microbial organisms depend much more on environmental conditions compared to chemical control products. Ignoring these dependencies on environmental factors, for example temperature and humidity, will often lead to failure. Thus, a better understanding of antagonist ecology and training of growers, advisers and distributors in the use of biocontrol products will allow the full exploitation of their potential. Generally, unlike most chemical pesticides, biocontrol products are pathogen specific and not broad spectrum. Recent experiences in pome fruit production show an increasing impact of 'minor' fruit rot pathogens and of fungi causing superficial cosmetic disorders (Weber and Zabel, 2011; Wenneker et al., 2017a,b). This may partly be the consequence of the disturbance of fruit microbiomes due to the extensive use of synthetic fungicides with unpredictable side effects on bacterial and fungal communities. On the other hand, side effects of pathogen-specific antagonists on beneficial naturally occurring organisms are not likely, so that natural disease suppression complements biocontrol whereas chemical control often reduces natural suppression, with the risk that new 'iatrogenic' diseases occur in fruit tree crops. Combined multi-year efforts will be needed to develop new integrated application schedules for different biocontrol products for pest and disease control, including the use of modelling (Jeger et al., 2009) and the specific adaptation of decision support systems to the use of biologicals (González-Domínguez et al., 2018).

Essential for this process of broad implementation of biocontrol into commercial fruit tree production systems will be the regulatory environment. Phasing out of groups of crop protection products such as copper-based fungicides for apple scab control and antibiotics for fireblight will have a significant impact. Changes in the current registration procedures are needed to recognize the generally lower and very different nature of the risks associated with biological solutions. Finally, societies may decide to estimate the true costs of plant protection solutions, including the risk of exposure of users and consumers to residues, reduced biodiversity, negative effects on ecosystem services and pollination, and the cleaning of drinking water (Pimentel and Burgess, 2014; Bourguet and Guillemaud, 2016; van Lenteren et al., 2018). These hidden costs of using synthetic pesticides, currently paid externally by societies and not by the manufacturers of the crop protection products, may strongly support the implementation of biocontrol solutions.

5 Future trends

The use of microbial biocontrol products in fruit production will increase, as it will for vegetable and arable crops. Main drivers for the increased use of biocontrol products include the loss of available fungicides, development and spread of fungicide resistance of pathogens, environmental considerations and growing demands by retailers and consumers for reduced use of pesticides. There is a significant challenge in moving from the use of biocontrol solutions against a single fruit crop disease to integrated biological programmes with a variety of biocontrol options against major diseases in fruit crops without the need for fungicide interventions. Such a complete set of biocontrol options, including biocontrol of pests, may strengthen the position of each single product.

It is expected that the markets for biocontrol industries will grow with a compound annual growth rate of 16% (Glare et al., 2012). Industrial expertise and infrastructure for mass production of new BCA, and their registration and marketing, will expand accordingly. The underpinning fundamental through to early development research will often be conducted under university and research organizations whereas product development will take place under public–private partnerships. For future research on the development of new BCA two trends are expected to support fruit production. Research into new BCA has often focussed on screening for efficacy; however, a more balanced approach is needed for a broad range of aspects from mass production to ecological characteristics (Glare et al., 2012; Köhl et al., 2011). For example, screening for rhizosphere and phyllosphere competence will result in sufficient persistence on the targets. Microbial inoculants beneficial for fruit crops will be more likely to be studied and exploited with BCA, biostimulants and biofertilizers. Biostimulants contain microorganisms which stimulate natural processes to enhance nutrient uptake, tolerance to stress and crop quality. They can be applied as soil amendment and foliar sprays. Biofertilizers applied to seedlings or soil contain microorganisms which improve phosphate or potassium mobilization or nitrogen fixation. These different objectives for using microbial inoculants are currently being investigated by different research groups, although their practical use is determined by different regulations. However, many of the beneficial microorganisms combine different properties. The same principles of microbial ecology, similar technologies for production, formulation and application, similar manufacturing processes and distribution channels as well as similar principles for assessment for human and environmental risks have to be considered when developing biocontrols, biostimulants and biofertilizers. Overcoming artificial separation of microbial inoculants by their intended use will lead to synergies during product development, marketing and practical use.

Current biocontrol products usually contain one selected bacterial or fungal strain with superior antagonism. Modes of action of antagonists include hyperparasitism, competition for nutrients or space, production of antimicrobial compounds or priming and ISR. Modern 'omics' approaches provide tools to better understand the modes of action of antagonists (Massart et al., 2015). For example, using genomic, transcriptomic, proteomic and metabolomic approaches showed that the metabolism of *Trichoderma* spp., one of the most studied fungal antagonists, is incredibly complex (Lorito et al., 2010). Understanding the genome of *Trichoderma* spp., its functions as a plant symbiont and its interactions with plants and environments is expected to lead to increased product development and better product deployment.

Plants do not rely on a single symbiont but support a whole range of beneficial microorganisms on and in their roots, shoots and leaves (Berg, 2009). Members of the plant microbiome families have complex interactions and usually plants benefit from these interactions. It is expected that exploiting this knowledge on the role of microbiome in biocontrol will lead to the design of synthetic microbial communities (Droby and Wisniewski, 2018; Mazzola and Freilich, 2017). Such synthetic consortia will provide better efficiency, increased robustness and resilience to environmental stress (Droby et al., 2016). Emerging strategies for the use of beneficial microorganisms in crop production will combine microbiome engineering to optimize plant microbiomes and plant breeding to optimize plants in supporting microbiomes with the application of microbial inoculants (Syed Ab Rahman et al., 2018). This will lead to robust exploitation of beneficial microorganisms supporting plant health, nutrient uptake and resilience to environmental stresses.

Knowledge of microbial ecology and the role of microbiomes in crop production are rapidly becoming available due to the implementation of 'omics' approaches. However, the possible use of synthetic consortia in biocontrol is hindered by regulations which were not developed for risk assessments of complex consortia but rather single strains. Current regulation considers members of consortia as a mix of 'active ingredients' which all have to be assessed individually and neglects basic principles of microbial ecology. The key future trends in biocontrol, combing beneficials to form synthetic consortia, needs thus a re-evaluation by the regulatory bodies before these new robust and sustainable solutions will become available for fruit growers.

6 Where to look for further information

Broad scientific information on the principles, ecology and application of biocontrol can be found in the book of Heimpel and Mills (2017), more specific information on the nature of biocontrol of plant pathogens in Cook and Baker (1983) and on the status and success of biocontrol in Nicot (2011). There are recent review articles covering specific aspects of biocontrol of soil pathogens (Berg, 2009; Lugtenberg et al., 2013), leaf pathogens (Elad, 2003; Vorholt,

2012), post-harvest pathogens (Droby et al., 2009; Liu et al., 2013; Teixido et al., 2011), mode of action (Massart et al., 2015) and the use of registered BCA (van Lenteren et al., 2018). Results of scientific research on biocontrol of plant pathogens are mainly published in three international journals specifically focussing on biocontrol (*BioControl*, *Biocontrol Science and Technology*, *Biological Control*), with an increasing portion of articles on plant pathogens. However, a significant number of results on biocontrol of plant pathogens are published in a broad range of scientific journals on plant pathology, microbial ecology, soil microbiology, crop-specific journals and journals on post-harvest technologies. The International Organisation for Biological and Integrated Control (IOBC) plays an important role in the communication of scientific results and their integrated use in cropping systems. The IOBC is organized globally with different regional sections. Within the West Palaearctic Regional Section of IOBC (IOBC-WPRS) several working groups include topics on the biocontrol of plant diseases in fruit production: 'IOBC-WPRS WG Integrated protection of fruit crops' with subgroups on 'Pome fruit diseases', 'Stone fruits' and 'Soft fruits', the IOBC-WPRS WG 'Biological and integrated control of plant pathogens', and the IOBC-WPRS WG 'Induced resistance in plants against insects and diseases'. These working groups meet on a regular basis, usually bi-yearly, with international participation of non-members as well. Contributions to the workshops are published in the *IOBC-WPRS Bulletin*. These bulletins usually also include reports on ongoing research and technical papers relating to the practical use of biocontrol in cropping systems.

Information on commercial development, registration and marketing of biocontrol products can be found on the website of the International Biocontrol Manufacturers Association (IBMA; www.ibma-global.org). IBMA organizes annually a global biocontrol industry meeting to network, discover and unveil new products, market opportunities and research areas and inform policy makers and regulators of the specific needs of the biocontrol industry. The website (www.abim.ch) includes an informative archive of presentations held during the meetings since 2006.

There is no central information system on globally available biocontrol products. Products are usually registered by single countries so that national databases on registered crop protection products contain information on biocontrol products including targeted crops and diseases (van Lenteren et al., 2018).

7 References

Andrews, J. H., Berbee, F. M. and Nordheim, E. V. (1983), 'Microbial antagonism to the imperfect stage of the apple scab pathogen, *Venturia inaequalis*', *Phytopathology*, 73, 228–34.

Beresford, R. M., Wright, P. J., Wood, P. N., Park, N. M., Larsen, N. J. and Fisher, B. M. (2013), 'Resistance of *Venturia inaequalis* to demethylation inhibitor and dodine fungicides in four New Zealand apple-growing regions', *New Zealand Plant Protection*, 66, 274–83.

Berg, G. (2009), 'Plant-microbe interactions promoting plant growth and health: Perspectives for controlled use of microorganisms in agriculture', *Applied Microbiology and Biotechnology*, 84 (1), 11–18.

Born, Y., Fieseler, L., Thöny, V., Leimer, N., Duffy, B. and Loessner, M. J. (2017), 'Engineering of bacteriophages Y2::dpoL1-C and Y2::luxAB for efficient control and rapid detection of the fire blight pathogen, *Erwinia amylovora*', *Applied and Environmental Microbiology*, 83 (12), article no. e00341-17.

Bosshard, E., Schüpp, H. and Siegfried, W. (1987), 'Concepts and methods in biological control of diseases in apple orchards', *EPPO Bulletin*, 17 (4), 655–63.

Boudreau, M. A. and Andrews, J. H. (1987), 'Factors influencing antagonism of *Chaetomium globosum* to *Venturia inaequalis*: A case study in failed biocontrol', *Phytopathology*, 77, 1470–5.

Bourguet, D. and Guillemaud, T. (2016), The hidden and external costs of pesticide use, *in* Lichtfouse, E. (Ed.), *Sustainable Agriculture Reviews*. Springer, Dordrecht, pp. 35–120.

Burr, T. J., Matteson, M. C., Smith, C. A., Corral-Garcia, M. R. and Huang, T.-C. (1996), 'Effectiveness of bacteria and yeasts from apple orchards as biological control agents of apple scab', *Biological Control*, 6, 151–7.

Carisse, O. (2000), '50 years of research on biological control', *IOBC-WPRS Bulletin*, 23, 5–10.

Carisse, O. and Dewdney, M. (2002), 'A review of non-fungicidal approaches for the control of apple scab', *Phytoprotection*, 83, 1–29.

Carisse, O., Philion, V., Rolland, D. and Bernier, J. (2000), 'Effect of fall application of fungal antagonists on spring ascospore production of apple scab pathogen, *Venturia inaequalis*', *Phytopathology*, 90, 31–7.

Casals, C., Teixido, N., Vinas, I., Silvera, E., Lamarca, N. and Usall, J. (2010), 'Combination of hot water, *Bacillus subtilis* CPA-8 and sodium bicarbonate treatments to control postharvest brown rot on peaches and nectarines', *European Journal of Plant Pathology*, 128, 51–63.

Cook, R. J. and Baker, K. F. (1983), *The Nature and Practice of Biological Control of Plant Pathogens*, American Phytopathological Society.

Droby, S. and Wisniewski, M. (2018), 'The fruit microbiome: A new frontier for postharvest biocontrol and postharvest biology', *Postharvest Biology and Technology*, 140, 107–12.

Droby, S., Wisniewski, M., Macarisin, D. and Wilson, W. (2009), 'Twenty years of postharvest biocontrol research: Is it time for a new paradigm?', *Postharvest Biology and Technology*, 52, 137–45.

Droby, S., Wisniewski, M., Teixidó, N., Spadaro, D. and Jijakli, M. H. (2016), 'The science, development, and commercialization of postharvest biocontrol products', *Postharvest Biology and Technology*, 122, 22–9.

Eilenberg, J., Hajek, A. and Lomer, C. (2001), 'Suggestions for unifying the terminology in biological control', BioControl, 46, 387–400.

Elad, Y. (2003), 'Biocontrol of foliar pathogens: Mechanisms and application', *Communications in Agricultural and Applied Biological Sciences Ghent University*, 68, 17–24.

Fiss, M., Kucheryava, N., Schönherr, J., Kollar, A., Arnold, G. and Auling, G. (2000), 'Isolation and characterization of epiphytic fungi from the phyllosphere of apple as potential biocontrol agents against apple scab (*Venturia inaequalis*)', *Zeitschrift für Pflanzenkrankheiten und Pflanzenschutz / Journal of Plant Diseases and Protection*, 107, 1-11.

Fokkema, N. J. and de Nooij, M. P. (1981), 'The effect of fungicides on the microbial balance in the phyllosphere', *EPPO Bulletin*, 11(3), 303-10.

Gill, J. J., Svircev, A. M., Smith, R. and Castle, A. J. (2003), 'Bacteriophages of *Erwinia amylovora*', *Applied and Environmental Microbiology*, 69 (4), 2133-8.

Glare, T., Caradus, J., Gelernter, W., Jackson, T., Keyhani, N., Köhl, J., Marrone, P., Morin, L. and Stewart, A. (2012), 'Have biopesticides come of age?', *Trends in Biotechnology*, 30, 250-8.

Gomez, C., Brun, L., Chauffour, D. and Vallée, D. D. L. (2007), 'Effect of leaf litter management on scab development in an organic apple orchard', *Agriculture, Ecosystems & Environment*, 118, 249-55.

González-Domínguez, E., Legler, S., Fedele, G., Si Ammour, M., Caffi, T. and Rossi, V. (2018), 'Helping farmers in timing the application of biocontrol agents in viticulture', *IOBC-WPRS Bulletin*, 133, 73-5.

Gotor-Vila, A., Teixidó, N., Usall, J., Dashevskaya, S. and Torres, R. (2016), 'Development of a SCAR marker and a strain-specific genomic marker for the detection of the biocontrol agent strain CPA-8 *Bacillus amyloliquefaciens* (formerly *B. subtilis*)', *Annals of Applied Biology*, 169 (2), 248-56.

Gotor-Vila, A., Teixidó, N., Casals, C., Torres, R., De Cal, A., Guijarro, B. and Usall, J. (2017a), 'Biological control of brown rot in stone fruit using *Bacillus amyloliquefaciens* CPA-8 under field conditions', *Crop Protection*, 102, 72-80.

Gotor-Vila, A., Usall, J., Torres, R., Abadias, M. and Teixidó, N. (2017b), 'Formulation of the biocontrol agent *Bacillus amyloliquefaciens* CPA-8 using different approaches: Liquid, freeze-drying and fluid-bed spray-drying', *BioControl*, 62 (4), 545-55.

Grzegorczyk, M., Żarowska, B., Restuccia, C. and Cirvilleri, G. (2017), 'Postharvest biocontrol ability of killer yeasts against *Monilinia fructigena* and *Monilinia fructicola* on stone fruit', *Food Microbiology*, 61, 93-101.

Guijarro, B., Larena, I., Melgarejo, P. and De Cal, A. (2017), 'Adaptive conditions and safety of the application of *Penicillium frequentans* as a biocontrol agent on stone fruit', *International Journal of Food Microbiology*, 254, 25-35.

Heimpel, G. E. and Mills, N. (2017), *Biological Control - Ecology and Applications*, Cambridge University Press.

Heye, C. C. and Andrews, J. H. (1983), 'Antagonism of *Athelia bombacina* and *Chaetomium globosum* to the apple scab pathogen *Venturia inaequalis*', *Phytopathology*, 73, 650-4.

Janisiewicz, W. J., Kurtzman, C. P. and Buyer, J. S. (2010), 'Yeasts associated with nectarines and their potential for biological control of brown rot', *Yeast*, 27, 389-98.

Janisiewicz, W. J., Jurick, W. M., Peter, K. A., Kurtzman, C. P. and Buyer, J. S. (2014), 'Yeasts associated with plums and their potential for controlling brown rot after harvest', *Yeast*, 31 (6), 207-18.

Jeger, M. J., Jeffries, P., Elad, Y. and Xu, X.-M. (2009), 'A generic theoretical model for biological control of foliar plant diseases', *Journal of Theoretical Biology*, 256 (2), 201-14.

Johnson, K. B. and Stockwell, V. O. (2000), Biological control of fire blight, *in* Vanneste, J. L. (Ed.), *Fire Blight: The Disease and Its Causative Agent, Erwinia amylovora*. CAB International, Wallingford, UK, pp. 319–37.

Kiss, L., Russell, J. C., Szentiványi, O., Xu, X. and Jeffries, P. (2004), 'Biology and biocontrol potential of Ampelomyces mycoparasites, natural antagonists of powdery mildew fungi', *Biocontrol Science and Technology*, 14, 635–51.

Köhl, J. A. (2009),'Novel micro-organisms controlling plant pathogens', *International Patent Application WO 2009/078710*.

Köhl, J. and Fokkema, N. J. (1998), 'Biological control of necrotrophic foliar fungal pathogens', *in* Boland, G. J. and Kuykendall, L. V. (Eds), *Plant-Microbe Interactions and Biological control*. Marcel Dekker, New York, pp. 49–88.

Köhl, J., Molhoek, W. M. L., Groenenboom-de Haas, B. H. and Goossen-van de Geijn, H. M. (2009), 'Selection and orchard testing of antagonists suppressing conidia production of the apple scab pathogen *Venturia inaequalis*', *European Journal of Plant Pathology*, 123, 401–14.

Köhl, J., Postma, J., Nicot, P., Ruocco, M. and Blum, B. (2011), 'Stepwise screening of microorganisms for commercial use in biological control of plant pathogenic fungi and bacteria', *Biological Control*, 57, 1–12.

Köhl, J., Scheer, C., Holb, I. J., Masny, S. and Molhoek W. M. L. (2015), 'Towards an integrated use of biological control by *Cladosporium cladosporioides* H39 in apple scab (*Venturia inaequalis*) management', *Plant Disease*, 99, 535–43.

Köhl, J., Wenneker, M., Groenenboom-de Haas, B. H., Anbergen, R., Goossen-van de Geijn, H. M., Lombaers-van der Plas, C. H., Pinto, F. A. M. F. and Kastelein, P. (2018), 'Dynamics of post-harvest pathogens *Neofabraea* spp. and *Cadophora* spp. in plant residues in Dutch apple and pear orchards', *Plant Pathology*. doi:10.1111/ppa.12854.

Larena, I., Torres, R., De Cal, A., Liñán, M., Melgarejo, P., Domenichini, P., Bellini, A., Mandrin, J. F., Lichou, J., De Eribe, X. O. and Usall, J. (2005), 'Biological control of postharvest brown rot (*Monilinia* spp.) of peaches by field applications of *Epicoccum nigrum*', *Biological Control*, 32 (2), 305–10.

Liu, J., Sui, Y., Wisniewski, M., Droby, S. and Liu, Y. (2013), 'Review: Utilization of antagonistic yeasts to manage postharvest fungal diseases of fruit', *International Journal of Food Microbiology*, 167 (2), 153–60.

Llorente, I., Vilardell, A. and Montesinos, E. (2006), 'Infection potential of *Pleospora allii* and evaluation of methods for reduction of the overwintering inoculum of brown spot of pear', *Plant Disease*, 90 (12), 1511–16.

Llorente, I., Vilardell, A., Vilardell, P., Pattori, E., Bugiani, R., Rossi, V. and Montesinos, E. (2010), 'Control of brown spot of pear by reducing the overwintering inoculum through sanitation', *European Journal of Plant Pathology*, 128 (1), 127–41.

Lorito, M., Woo, S. L., Harman, G. E. and Monte, E. (2010), 'Translational research on trichoderma: From 'omics to the field', *Annual Review of Phytopathology*, 48, 395–417.

Lugtenberg, B. J. J., Malfanova, N., Kamilova, F. and Berg, G. (2013), 'Microbial control of plant root diseases', *Molecular Microbial Ecology of the Rhizosphere*, 2, 575–86.

MacHardy, W. E., Gadoury, D. M. and Gessler, C. (2001), 'Parasitic and biological fitness of *Venturia inaequalis*: Relationship to disease management strategies', *Plant Disease*, 85, 1036–51.

Massart, S., Perazzolli, M., Höfte, M., Pertot, I. and Jijakli, M. H. (2015), 'Impact of the omic technologies for understanding the modes of action of biological control agents against plant pathogens', *BioControl*, 60 (6), 725–46.

Mazzola, M. and Freilich, S. (2017), 'Prospects for biological soilborne disease control: Application of indigenous versus synthetic microbiomes', *Phytopathology*, 107, 256–63.

Melgarejo, P., Carrillo, R. and M-Sagasta, E. (1985), 'Mycofora of peach twigs and flowers and its possible significance in biological control of *Monilinia laxa*', *Transactions of the British Mycological Society*, 85, 313–17.

Melgarejo, P., Carrillo, R. and M-Sagasta, E. (1986), 'Potential for biological control of *Monilinia laxa* in peach twigs', *Crop Protection*, 5, 422–6.

Miedtke, U. and Kennel, W. (1990), '*Athelia bombacina* and *Chaetomium globosum* as antagonists of the perfect stage of the apple scab pathogen (*Venturia inaequalis*) under field conditions', *Zeitschrift für Pflanzenkrankheiten und Pflanzenschutz / Journal of Plant Diseases and Protection*, 97, 24–32.

Montesinos, E., Bonaterra, A., Ophir, Y. and Beer, S. V. (1996), 'Antagonism of selected bacterial strains to *Stemphylium vesicarium* and biological control of brown spot of pear under controlled environment conditions', *Phytopathology*, 86 (8), 856–63.

Nicot, P. C. (2011), *Classical and Augmentative Biological Control against Diseases and Pests: Critical Status Analysis and Review of Factors Influencing Their Success*, IOBC-WPRS.

Ortega, F., Steiner, U. and Dehne, H.-W. (1998), 'Induced resistance to apple scab: Microscopic studies on the infection cycle of *Venturia inaequalis* (Cke.) Wint.', *Journal of Phytopathology*, 146, 399–405.

Percival, G. C., Noviss, K. and Haynes, I. (2009), 'Field evaluation of systemic inducing resistance chemicals at different growth stages for the control of apple (*Venturia inaequalis*) and pear (*Venturia pirina*) scab', *Crop Protection*, 28, 629–33.

Pimentel, D. and Burgess, M. (2014), 'Environmental and economic costs of the application of pesticides primarily in the United States', *in* Pimentel, D. and Peshin, R. (Eds), *Integrated Pest Management*. Springer, Dordrecht, pp. 47–71.

Piombo, E., Sela, N., Wisniewski, M., Hoffmann, M., Gullino, M. L., Allard, M. W., Levin, E., Spadaro, D. and Droby, S. (2018), 'Genome sequence, assembly and characterization of two *Metschnikowia fructicola* strains used as biocontrol agents of postharvest diseases', *Frontiers in Microbiology*, 9, art. no. 593. doi:10.3389/fmicb.2018.00593.

Pusey, P. L. and Curry, E. A. (2004), 'Temperature and pomaceous flower age related to colonization by *Erwinia amylovora* and antagonists', *Phytopathology*, 94, 901–11.

Pusey, P. L. and Wilson, C. L. (1984), 'Postharvest biological control of stone fruit brown rot by *Bacillus subtilis*', *Plant Disease*, 68, 753–6.

Pusey, P. L., Hotchkiss, M. W., Dulmage, H. T., Baumgardner, R. A., Zehr, E. I., Reilly, C. C. and Wilson, C. L. (1988), 'Pilot tests for commercial production and application of *Bacillus subtilis* (B-3) for postharvest control of peach brown rot', *Plant Disease*, 72, 622–6.

Rungjindamai, N., Jeffries, P. and Xu, X.-M. (2014a), 'Epidemiology and management of brown rot on stone fruit caused *by Monilinia laxa*', *European Journal of Plant Pathology*, 140 (1), 1–17.

Rungjindamai, N., Jeffries, P. and Xu, X.-M. (2014b), 'A novel strategy to reduce overwintering inoculum of *Monilinia laxa*', *European Journal of Plant Pathology*, 140 (3), 591–6.

Schiewe, A. and Mendgen, K. (1992), 'Identification of antagonists for biological control of the post-harvest pathogens *Pezicula malicorticis* and *Nectria galligena* on apples', *Journal of Phytopathology*, 134 (3), 229–37.

Sharma, R. R., Singh, D. and Singh, R. (2009), 'Biological control of postharvest diseases of fruits and vegetables by microbial antagonists: A review', *Biological Control*, 50 (3), 205–21.

Snowdon, A. L. (1990), 'Pome fruits', *in* Snowdon, A. L. (Ed.), *A Colour Atlas of Post-harvest Diseases and Disorders of Fruits and Vegetables. Vol. 1: General Introduction and Fruits*. Wolfe Scientific Ltd., London, pp. 170–218.

Spadaro, D. and Droby, S. (2016), 'Development of biocontrol products for postharvest diseases of fruit: The importance of elucidating the mechanisms of action of yeast antagonists', *Trends in Food Science and Technology*, 47, 39–49.

Sutton, D. K., MacHardy, W. E. and Lord, W. G. (2000), 'Effects of shredding or treating apple leaf litter with urea on ascospore dose of *Venturia inaequalis* and disease buildup', *Plant Disease*, 84, 1319–26.

Swinburne, T. R. (1973), 'Microflora of apple-leaf scars in relation to infection by *Nectria galligena*', *Transactions of the British Mycological Society*, 60, 389–403.

Syed Ab Rahman, S. F., Singh, E., Pieterse, C. M. J. and Schenk, P. M. (2018), 'Emerging microbial biocontrol strategies for plant pathogens', *Plant Science*, 267, 102–11.

Teixido, N., Torres, R., Viñas, I., Abadias, M. and Usall, J. (2011), 'Biological control of postharvest diseases in fruit and vegetables', *in Lacroix* (Ed.), *Protective Cultures, Antimicrobial Metabolites and Bacteriophages for Food and Beverage Biopreservation*. Woodhead Publishing. doi:10.1533/9780857090522.3.364.

Thomson, S. V. (1986), 'The role of the stigma in fire blight infections', *Phytopathology*, 76, 476–82.

van Lenteren, J. C., Bolckmans, K., Köhl, J., Ravensberg, W. and Urbaneja, A. (2018), 'Biological control using invertebrates and microorganisms: Plenty of new opportunities', *BioControl*, 63, 39–59. doi:10.1007/s10526-017-9801-4.

Venneste, J. L. (2011), 'Biological control agents of fire blight: Successes and challenges', *Acta Horticulturae*, 896, 409–16.

Vorholt, J. A. (2012), 'Microbial life in the phyllosphere', *Nature Reviews Microbiology*, 10 (12), 828–40.

Walter, M., Campbell, R. E., Amponsah, N. T., Turner, L., Rainham, D., Kerer, U. and Butler, R. C. (2017a), 'Can biological products control *Neonectria ditissima* picking wound and leaf scar infections in apples?', *New Zealand Plant Protection*, 70, 63–72.

Walter, M., Campbell, R. E., Amponsah, N. T., Scheper, R. W. A. and Butler, R. C. (2017b), 'Evaluation of biological and agrichemical products for control of *Neonectria ditissima* conidia production', *New Zealand Plant Protection*, 70, 87–96.

Weber, R. W. S. and Zabel, D. (2011), 'White haze and scarf skin, two little-known cosmetic defects of apples in Northern Germany', *European Journal of Horticultural Science*, 76(2), 45–50.

Wenneker, M., Pham K. T. K., Boekhoudt, L. C., de Boer, F. A., van Leeuwen, P. J., Hollinger, T. C. and Thomma, B. P. H. J. (2017a), 'First report of *Neofabraea kienholzii* causing bull's eye rot on pear (*Pyrus communis*) in the Netherlands', *Plant Disease*, 101, 634.

Wenneker, M., Pham, K. T. K., Lemmers, M. E. C., de Boer, F. A., van Leeuwen, P. J., Hollinger, T. C., van de Geijn, F. G. and Thomma, B. P. H. J. (2017b), '*Fibulorhizoctonia psychrophila* is the causal agent of lenticel spot on apple and pear fruit in the

Netherlands', *European Journal of Plant Pathology*, 148 (1), 213–17. DOI: 10.1007/s10658-016-1061-0.

Zhang, J., Xie, J., Zhou, Y., Deng, L., Yao, S. and Zeng, K. (2017), 'Inhibitory effect of *Pichia membranaefaciens* and *Kloeckera apiculata* against *Monilinia fructicola* and their biocontrol ability of brown rot in postharvest plum', *Biological Control*, 114, 51–8.

Zhou, T., Schneider, K. E. and Li, X. (2008), 'Development of biocontrol agents from food microbial isolates for controlling post-harvest peach brown rot caused by *Monilinia fructicola*', *International Journal of Food Microbiology*, 126, 180–5.

Chapter 14

New techniques for managing post-harvest diseases of fruit: physical, chemical and biological agents

Simona Marianna Sanzani and Antonio Ippolito, Università degli Studi di Bari Aldo Moro, Italy

1 Introduction

Being distributed over time and space, agricultural products are subjected to retention periods depending on their characteristics and market demands. However, being rich in water and nutrients, they are an ideal substrate for the development of pathogenic microorganisms, initiating the rotting process. Furthermore, the infection may take place in the field and the pathogen can remain latent for a long period of time, until ripening and/or obtaining a conducive environment (Sanzani et al., 2012). Therefore, serious economic damage may be incurred from apparently healthy products, in which infections had not advanced far enough to be detected during packing, storage or transportation. Modern storage technologies (controlled atmosphere, low oxygen atmosphere etc.) reduce such losses, but do not consistently ensure an acceptable level of disease control.

Currently, synthetic chemicals are the main means for controlling postharvest diseases of fruit and vegetables. However, public concern about food safety for consumers and possible harmful effects to the environment, as well as the emergence and subsequent spread of resistant pathogen strains, deregistration of old fungicides and much increased costs for registration of new ones, have recently intensified the search for efficient alternatives to control postharvest diseases of perishable fresh produce. Recent research efforts in this

http://dx.doi.org/10.19103/AS.2019.0046.17

area have aimed to assess, develop and integrate various alternative control methods to reduce reliance on synthetic fungicides (Wisniewski et al., 2016). Indeed, postharvest pathogens affect the produce quality quantitatively but also qualitatively; for example, the quantity of various free and bound amino acids and organic acids is altered and a gradual decrease in sugar and ascorbic acid content is observed with the advancement of diseases (Youssef et al., 2014). This chapter presents an updated overview of the main alternatives for controlling postharvest diseases, including physical, chemical and biological means.

2 Physical agents

Physical agents have been proposed to control postharvest diseases of fruit and vegetables. Several types of physical means are available and have been evaluated against postharvest pathogens. Their biggest potential drawback is the lack of residual effects so that spoilage can occur during subsequent processing.

2.1 Ultraviolet light treatment

Low doses of ultraviolet (UV) light, especially at 254 nm (UV-C hormesis), emerged as an alternative technology to chemical fungicides (Nigro and Ippolito, 2016). Indeed, it has been shown to induce resistance to decay in postharvest commodities and extend the shelf life by delaying pathogen development, and fruit ripening and senescence. In tomato, UV-C light induced the accumulation of phenolic compounds and the formation of a biochemical barrier containing lignin and suberin, associated with resistance against *Botrytis cinerea* (Charles et al., 2008). Very low doses of UV-C light (0.125–2 kJ/m^2) proved to be effective in reducing *B. cinerea* decay on kiwifruit, table grapes and strawberry (Nigro et al., 1998, 2000). Romaine lettuce plants exposed to UV-C were less susceptible to *B. cinerea* and *Sclerotinia minor* (Ouhibi et al., 2014). UV-C light at 34.5 kJ/m^2 combined with fluorescent light effectively controlled soft rot of potato tubers caused by *Pectobacterium carotovorum* subsp. *carotovorum*, without adversely affecting sprouting (Rocha et al., 2015). UV-C-treated tubers stored under fluorescent light showed an increased concentration of α-chaconine and α-solanine, the main glycoalkaloids produced by potato (Rocha et al., 2015). UV-C irradiation at a dose of 0.01 kJ/m^2 reduced the development of banana crown rot by 46.3% without adverse effects on respiration rate, ethylene production, weight loss, firmness, colour change, soluble solids concentration, titratable acidity and pH (Mohamed et al., 2017). The treatment also enhanced total phenolic content and antioxidant power, indicating the beneficial impact of UV-C on fruit nutritional quality.

Scanning electron micrographs confirmed that UV-C irradiation retarded the loss of cell wall compartments, thereby maintaining the cell wall integrity. The beneficial effect of UV-C on postharvest decay of mango cv. Tommy Atkins has also been reported (Terao et al., 2015): a low dose of UV-C irradiation at 2.5 kJ/m^2 is capable of reducing fruit rot severity by around 70%.

2.2 Heat treatment

Pre-storage heating has potential as a nonchemical method for controlling postharvest diseases by directly inhibiting pathogen growth, activating the natural resistance of the host and slowing down the ripening process. Heat treatments can be applied to fruit in several ways: hot water dips, rinses or brushing, vapour heat and hot air (Usall et al., 2016). The use of heat treatments in managing postharvest diseases can be both protective and curative. In the first case, it is applied before infections are established by the pathogens, thus reducing the level of inoculum load on the fruit surface and indirectly reducing the chance of infection. The effect can also be curative when treatment is applied once the infection has already been established but not yet become symptomatic, which is known as latent, quiescent or incipient infection (Fallik, 2004). Peaches exposed to hot water treatment were less prone to spoilage and showed down-regulation of cell-wall genes coding enzymes involved in ripening, such as ß-galactosidase, pectin-lyase, polygalacturonase and pectin methyl esterase, while genes coding ROS scavenging enzymes and PAL, chitinase and the heat-shock protein HSP70 showed increased expression (Spadoni et al., 2014). The effect of hot water (45°C for 10 min) was investigated recently on apple fruit against various pathogens (di Francesco et al., 2018). A significant reduction of rot incidence (69%) was observed in fruit inoculated with *Neofabraea vagabunda* at 3 h after hot water treatment as compared to the control. Moreover, crude protein extracts from hot water-treated fruit resulted in higher (83%) inhibition of conidial germination, and reduction of pathogenesis enzyme activities was also observed (di Francesco et al., 2018).

2.3 Electrolysed water

A critical point during the postharvest handling chain of fresh produce is the washing phase, in which cross-contamination by pathogens may occur. Use of sanitizers and frequent monitoring of water recirculation systems help to prevent contamination and maintain the water free of pathogens (Youssef et al., 2011). However, chlorine, one of the most common sanitizers, has some limitations concerning its risks on environmental and human health, due to the formation of chlorinated organic compounds. In addition, the use of chlorine might cause several problems in the equipment, and its efficacy has been questioned after the emergence of tolerant pathogens in the fresh-cut produce industry (Singh et al., 2002). New disinfection techniques, such as electrolysed water (EW), have been introduced. During electrolysis, a variety of oxidants can be produced depending on the reaction conditions such as electrolyte and its concentration, applied current, water flow rate, electrode material and electrolysis duration (Fallanaj et al., 2015). For example, Pinto et al. (2015) evaluated the antimicrobial activity of neutral EW (NEW), produced from solutions of potassium and sodium chloride and sodium bicarbonate (SBC), on fresh-cut vegetables under cold storage. The effect of NEW was evaluated with different combinations of free chlorine concentration and contact time. All concentrations above 100 mg/L, regardless of the salt used, were bactericidal after just 2 min, with an average 1.7-log reduction of mesophilic bacteria and *Enterobacteriaceae* loads on chicory and lettuce leaves. Furthermore, the effect of electrolysed salt solutions using thin-film diamond-coated electrodes on *Penicillium* spp. population in citrus fruit wash water and on fruit decay were evaluated by Fallanaj et al. (2013). Electrolysis conducted in the presence of SBC could inhibit spore germination and reduce *Penicillium* rot, with no deleterious effects on fruit quality. Commercial trials conducted in citrus packinghouses confirmed the ability of SBC to improve electrolysis performance. SBC reduced pathogen population by 93% within 1 h of electrolysis, whereas without SBC, 7 h were needed to achieve comparable results. In addition, rot incidence in fruit exposed to the electrolysed SBC solution was reduced by up to 100%, as compared to 70% in the absence of SBC. Recently, Khamis et al. (2018) assessed the activity of alkaline EW (alEW) and acidic EW (acEW) in the presence of various salts against grey mould caused by *B. cinerea* in table grapes. In fruit artificially inoculated with *B. cinerea*, alEW application caused a reduction of infected berries from 55% to 92%, with potassium carbonate giving the best results as electrolyte; whereas for acEW the reduction ranged from 77% to 98% and ammonium molybdate was the most effective electrolyte. With natural infection, EW generated in the presence of potassium sorbate, sodium carbonate and sodium metabisulphite was effective against grey mould development. In most cases, alEW and acEW did not affect berry quality.

3 Chemical agents

3.1 Organic and inorganic salts

A number of organic and inorganic salts are widely used in the food industry. In fact, most of them belong to the GRAS (Generally Regarded as Safe) category by the US Food and Drug Administration (FDA) (Palou et al., 2016). Salts as a postharvest treatment have been reported as safe and effective for preserving the quality and shelf life of fresh fruit (Youssef et al., 2012a, 2014; Montesinos-Herrero et al., 2016). Phosphite salts can both induce host resistance and inhibit fungal growth (Olivieri et al., 2012). Calcium chloride (CaCl2) delays ripening, and particularly softening, by altering intra- and extra-cellular processes. Its application during postharvest can reduce physiological disorders including bitter pit in apple (Conway et al., 2002; Chardonnet et al., 2003). Postharvest dipping of 'Golden Delicious' apple in $CaCl_2$ before 1-MCP application could prevent scald and reduce the bitter pit, which instead was enhanced by 1-MCP alone (Gago et al., 2016). SBC and potassium sorbate are used in the food industry for controlling pH, taste and texture, and they also exhibit broad-spectrum antifungal activity (Youssef et al., 2017). Treatments with sodium carbonate and SBC reduced the incidence of postharvest decay in citrus, preserving tissue firmness during storage (El-Mougy et al., 2008; Youssef et al., 2012b). The potential of bicarbonate salts in controlling postharvest pathogens has been demonstrated also in papaya, table grape and pear (Sivakumar et al., 2002; Nigro et al., 2006; Lai et al., 2015). SBC applied at a concentration of 2% (w/v) could control *Rhizopus*, *Alternaria* and *Fusarium* decay on 'Galia' and 'EinDor' melon fruit (Aharoni et al., 1997). Potassium sorbate, when applied as a dipping agent at 2–3% for 2–3 min at room temperature, might effectively control green and blue mould and sour rot of citrus (Smilanick et al., 2008; Montesinos-Herrero et al., 2009; D'Aquino et al., 2013). The effectiveness was further improved when applied at high temperatures (e.g. 30–60 s at 50–62°C), and influenced by host species and cultivar, maturity stage, presence of wounds and so on. It has been recently reported that dipping in potassium sorbate followed by brief exposures to high CO_2 or O_2 at curing temperature (37°C) synergistically controlled green and blue moulds (Montesinos-Herrero et al., 2016).

3.2 Chitosan

Chitosan and its derivatives, including glycolchitosan, were reported to inhibit fungal growth and to induce host-defence response in plants and harvested commodities (Romanazzi and Feliziani, 2016). Chitosan, a high-molecular-weight cationic polysaccharide, is soluble in diluted organic acids and has been used as a preservative coating material for fruit. It can form a semi-permeable film with a potential to modify the internal atmosphere as well as reduce transpiration losses in fruit, thus leading to extended storage life without causing anaerobiosis (Romanazzi and Feliziani, 2016).

When applied as a spray on strawberry plants every 5 days, chitosan reduced postharvest fruit decay by c.a. 30%, mainly grey mould and *Rhizopus* rot, without negatively affecting fruit colour and firmness (Feliziani et al., 2015).

El Guilliet al. (2016) applied chitosan against citrus green mould caused by *Penicillium digitatum*. Chitosan coating controlled green mould, and its effect might be related to its fungitoxic properties and/or the elicitation of biochemical defence responses in coated fruit. Its *in vitro* antifungal activity against *P. digitatum* increased with growing chitosan concentration; in parallel, chitinase and glucanase activities were enhanced in coated fruit. Fruit quality attributes, including fruit firmness, surface colour, juice content and total soluble solids, were not affected by chitosan during storage, whereas weight loss was less pronounced.

4 Biological agents

Droby et al. (2009) introduced a broader definition of biological control: in addition to the use of an organism, managing a plant disease could involve the use of a biological process or a product from a biological process. This represented a paradigm shift in the concept of biological control, which had a major impact on how to explore and interpret systems, towards a second generation of postharvest biocontrol products. As such, several new approaches have become available for developing biological control products and practices: (1) the classical approach (microbial antagonists), (2) natural plant resistance and (3) natural antimicrobials, which are products derived from a biological process.

4.1 Biocontrol agents

Biocontrol agents, as an alternative to the use of synthetic chemicals, have been the focus of considerable research over the last 30 years (Droby et al., 2016). In contrast to field- and soil-based biocontrol, the postharvest environment and the aetiology of infections that happen during storage are more advantageous to application of microbial antagonist to commodities because of controlled conditions. Microbial antagonists have been reported to protect a variety of harvested perishable commodities against postharvest pathogens (Mari et al., 2016).

Candida sake strain CPA-1 is a well-known antagonist. It was isolated from apple and demonstrated good efficacy against major postharvest pathogens of apple such as *P. expansum*, *B. cinerea* and *Rhizopus stolonifer* (Viñas et al., 1998). To improve its shelf life and manipulability, Carbó et al. (2017) established optimal conditions for a dry formulation of *C. sake*: 40°C for 45 min for the drying process, potato starch as a carrier, and 25°C for 10 min in phosphate buffer for recovering the dried material. The formulated products should be stored at 4°C and air-packaged. Even after 12 months of storage, the formulated products still maintained their biocontrol efficacy.

In a study by Panebianco et al. (2015), 14 *Pseudomonas* spp. and three *Trichoderma* spp. strains were tested against *P. digitatum*. They strongly inhibited pathogen growth *in vitro*, particularly when applied in combination.

Inhibitory effects were also observed with mixtures of *Pseudomonas* living cells and *Trichoderma atroviride* P1 culture filtrates. Incidence and severity of green mould on citrus fruit were consistently reduced when *Pseudomonas* (particularly applied as a mixture of six *P. syringae* strains) and *Trichoderma* strains T22, P1 and T34 were applied 72 h before inoculation with *P. digitatum*.

Progress has been made during the past two decades to understand the biocontrol agents' modes of action. Massart et al. (2015) suggested focusing on beneficial interactions within host microbiota to favour biocontrol agents. Indeed, once applied, these latter will face a complex community of microorganisms within which ecological interactions such as competition, parasitism, mutualism and commensalism can occur. Sare et al. (2018) investigated the potential of apple microbiota to improve the efficacy of *Pichia anomala* strain K against *B. cinerea*. Sampling was carried out by harvesting 15 apple varieties grown under four disease management strategies (no, light-organic, organic and conventional phytosanitary treatment). The results showed that apple skin harbours around $2 \pm 1.5 \times 10^4$ microorganisms/cm^2. Microbiota can increase until 100%, reduce or have no effect on the biocontrol efficacy of strain K, according to its composition, which is presence of strains, species or groups of taxa that are beneficial for achieving greater strain K efficacy.

4.2 Botanicals

Drawbacks of synthetic chemicals have increased interest in developing alternative control means that are environmentally sound and biodegradable. Therefore, replacement of synthetic fungicides with natural products of plant origin, which are non-toxic and often have a nutraceutical significance, has been gaining considerable attention. The plant kingdom represents an enormous reservoir of potential fungicidal compounds that could be useful alternatives to synthetic fungicides. Recently, there have been several attempts to use naturally occurring compounds for controlling postharvest decay (Palou et al., 2016). Indeed, plants also produce a variety of essential oils and volatiles that might have potential as antifungal preservatives for harvested commodities. Essential (volatile) plant oils occur in edible, medicinal and herbal plants, which minimize the risk regarding their safe use in food products. Moreover, some of their constituents (e.g. cinnamaldehyde, citral, eugenol, limonene or thymol) are safe for human consumption and have a GRAS status (de Aquino et al., 2015). The advantage of essential oils is their bioactivity in the vapour phase, which makes them act as fumigants for stored products (Lee et al., 2007). Some of the essential oils have been reported to inhibit postharvest fungi in *in vitro* conditions, for example, cinnamon oil, rich in eugenol (81%), demonstrated good fungicidal potential against *Lasiodiplodia theobromae* and *Colletotrichum gloeosporioides*, whereas pure thymol at 150 and 250 ppm inhibited the growth of *B. cinerea*, *Penicillium italicum* and *P. expansum* (Combrinck et al., 2011; Camele et al., 2012). The potential of essential oils to control postharvest decay has also been examined by spraying and dipping processes on cherries, citrus fruit, apples, peaches and cabbages (Sivakumar and Bautista-Baños, 2014). A promising development involves incorporating these antimicrobials into packaging materials, for example, *Mentha spicata* and *Lippia scaberrima* essential oils, as well as pure (d)-limonene and R-(−)-carvone, were incorporated into a commercial citrus coating, and then applied postharvest to 'Tomango' oranges with excellent disease control and maintaining fruit quality (du Plooy et al., 2009).

Phenolic compounds constitute the main class of plant secondary metabolites with more than 8000 identified structures (Ribes et al., 2018). The mode of action of phenolic compounds is related to membrane alteration (that leads to the loss of the pH gradient and electrical potential), interference with the ATP-system, inhibition of enzymes and germination, suppression of mycelium development and germ tube elongation (El-Mogy and Alsanius, 2012; da Cruz-Cabral et al., 2013). Furthermore, due to the conformation and functionality change of membrane proteins, they increase the concentration of reactive oxygen species (da Rocha Neto et al., 2015). Some researchers reported the effectiveness of different phenolic compounds as protective natural preservatives, inhibiting spoilage fungi in multiple food systems (Palou et al., 2016). A control effect against brown rot by *Monilia laxa* on apricots and nectarines, green mould by *P. digitatum* on oranges and grey mould by *B. cinerea* on table grapes was observed (Gatto et al., 2011). Furthermore, the effectiveness of esculetin, ferulic acid, quercetin, resveratrol, scopoletin, scoparone and umbelliferone against apple blue mould was evaluated by Sanzani et al. (2009). Following wound application, quercetin and umbelliferone were the most effective, either alone or combined, in reducing 'Golden Delicious' apple decay by 86–92% compared to the control samples. Dipping fruit in both quercetin and umbelliferone led to only 33% of apple with fungal decay after 8 days of treatment, compared with 63% in the control samples (Sanzani et al., 2009). Good results were also obtained against green mould of citrus (Sanzani et al., 2014). Furthermore, Quaglia et al. (2016) reported positive effects of spraying pomegranates with phenolic compounds from olive-mill wastewater. Treating pomegranate fruit with phenolic compounds at 4 and 8 mg/mL reduced incidence of fruit decay due to *Penicillium adametzioides* by 50%.

The bioactive compounds present in a pomegranate peel extract (PGE) were evaluated as a natural antifungal preparation for the control of postharvest rots (Pangallo et al., 2016). *In vitro* trials revealed a strong fungicidal activity against *B. cinerea*, *P. digitatum* and *P. expansum*. PGE at concentrations of 1.2 and 12 g/L was very effective in inhibiting decay following artificial inoculations of lemons, grapefruits and apples. Furthermore, it was also effective in reducing rots resulting from natural infections under semi-commercial conditions on sweet cherries and lemons. PGE treatment showed residual effect as it inhibited infections up to 24 h after the application of extract on fruit wounds. Additionally, PGE exhibited curative activity, reducing the incidence of rots when it was applied up to 24 h after inoculation.

5 Conclusions

This chapter presented several means as alternatives to synthetic fungicides to manage postharvest diseases. They were reported to have different degrees of efficacy; however, few of them have been actually applied in the production chain. Further steps are needed to improve their efficacy and consistency to favour their market entry. Their formulation (e.g. protection or encapsulation of natural agents into oil-in-water emulsions and nano-emulsions), incorporation into edible films and active packaging systems, as well as combination in integrated approaches with alternative/traditional means, are expected to improve their antifungal properties while reducing the doses. Furthermore, the preharvest use of alternative control means in combination with a postharvest application could reduce fruit handling, latent, quiescent and incipient infections, and pathogen inoculum in the environment, and increase host resistance. Indeed, it has been shown that integrated approaches involving applications of antagonists or other nonchemical means both before and after harvest can provide levels of a disease control comparable to or even better than those of synthetic fungicides, due to additive or synergic effects (Ippolito and Nigro, 2000; Ippolito et al., 2005).

6 Where to look for further information

For more details on specific topics, readers can refer to:

- special issue of *Postharvest Biology and Technology*, 'Alternatives to the use of synthetic fungicides for the control of postharvest decay of fruit and vegetables: current status and future perspectives', edited by P. Tonutti, S. Droby and G. Romanazzi (2016);
- special issue of *Acta Horticulturae*, 'Proceedings of III International Symposium on Postharvest Pathology: Using Science to Increase Food Availability', edited by A. Ippolito, S.M. Sanzani, M. Wisniewski and S. Droby (2016);
- the chapter 'Biological control of postharvest diseases', by G. Lima, S.M. Sanzani, De F. Curtis and A. Ippolito in the book *Advances in Postharvest Fruit and Vegetable Technology* edited by R.B.H. Wills and J.B. Golding (2015).

7 References

Aharoni, Y., Fallik, E., Copel, A., Gil, M., Grinberg, S. and Klein, J. D. (1997), Sodium bicarbonate reduces postharvest decay development on melons, *Postharvest Biol. Technol.*, 10(3), 201–6.

Camele, I., Altieri, L., de Martino, L., De Feo, V., Mancini, E. and Rana, G. L. (2012), *In vitro* control of post-harvest fruit rot fungi by some plant essential oil components, *Int. J. Mol. Sci.*, 13(2), 2290–300.

Carbó, A., Torres, R., Usall, J., Fons, E. and Teixidó, N. (2017), Dry formulations of the biocontrol agent *Candida sake* CPA-1 using fluidised bed drying to control the main postharvest diseases on fruits, *J. Sci. Food Agric.*, 97(11), 3691–8.

Chardonnet, C. O., Charron, C. S., Sams, C. E. and Conway, W. S. (2003), Chemical changes in the cortical tissue and cell walls of calcium-infiltrated 'Golden Delicious' apples during storage, *Postharvest Biol. Technol.*, 28, 97–111.

Charles, M. T., Goulet, A. and Arul, J. (2008), Physiological basis of UV-C induced resistance to *Botrytis cinerea* in tomato fruit. IV. Biochemical modification of structural barriers, *Postharvest Biol. Technol.*, 47, 41–53.

Combrinck, S., Regnier, T. and Kamatou, G. P. P. (2011), *In vitro* activity of eighteen essential oils and some major components against common postharvest fungal pathogens of fruit, *Ind. Crops Prod.*, 33(2), 344–9.

Conway, W. S., Sams, C. E. and Hickey, K. D. (2002), Pre- and postharvest calcium treatment of apple fruit and its effect on quality, *Acta Hortic.*, 594, 413–19.

D'Aquino, S., Palma, A., Angioni, A. and Schirra, M. (2013), Residue levels and efficacy of fludioxonil and thiabendazole in controlling postharvest green mold decay in citrus fruit when applied in combination with sodium bicarbonate, *J. Agric. Food Chem.*, 61(2), 296–306.

da Cruz-Cabral, L., Fernandez-Pinto, V. and Patriarca, A. (2013), Application of plant derived compounds to control fungal spoilage and mycotoxin production in foods, *Int. J. Food Microbiol.*, 166, 1–14.

da Rocha Neto, A. C., Maraschin, M. and Di Piero, R. M. (2015), Antifungal activity of salicylic acid against *Penicillium expansum* and its possible mechanisms of action, *Int. J. Food Microbiol.*, 215, 64–70.

de Aquino, A. B., Blank, A. F. and de Aquino Santana, L. C. L. (2015), Impact of edible chitosan-cassava starch coatings enriched with *Lippia gracilis* Schauer genotype mixtures on the shelf life of guavas (*Psidium guajava* L.) during storage at room temperature, *Food Chem.*, 171, 108–16.

di Francesco, A., Mari, M. and Roberti, R. (2018), Defense response against postharvest pathogens in hot water treated apples, *Sci. Hortic.*, 227, 181–6.

Droby, S., Wisniewski, M., Macarisin, D. and Wilson, C. (2009), Twenty years of postharvest biocontrol research: Is it time for a new paradigm? *Postharvest Biol. Technol.*, 52(2), 137–45.

Droby, S., Wisniewski, M., Teixidó, N., Spadaro, D. and Jijakli, M. H. (2016), The science, development, and commercialization of postharvest biocontrol products, *Postharvest Biol. Technol.*, 22, 22-9.

du Plooy, W., Regnier, T. and Combrinck, S. (2009), Essential oil amended coatings as alternatives to synthetic fungicides in citrus postharvest management, *Postharvest Biol. Technol.*, 53(3), 117–22.

El Guilli, M., Hamza, A., Clément, C., Ibriz, M. and Barka, E. A. (2016), Effectiveness of postharvest treatment with chitosan to control citrus green mold, *Agriculture*, 6(2), 1–15.

El-Mogy, M. M. and Alsanius, B. W. (2012), Cassia oil for controlling plant and human pathogens on fresh strawberries, *Food Cont.*, 28, 157–62.

El-Mougy, N. S., El-Gamal, N. G. and Abd-El-Kareem, F. (2008), Use of organic acids and salts to control postharvest diseases of lemon fruits in Egypt, *Arch. Phytopathol. Plant Prot.*, 41, 467–76.

Fallanaj, F., Sanzani, S. M., Zavanella, C. and Ippolito, A. (2013), Salt addition improves the control of citrus postharvest diseases using electrolysis with conductive diamond electrodes, *J. Plant Pathol.*, 95(2), 373–83.

Fallanaj, F., Sanzani, S. M., Youssef, K., Zavanella, C., Salerno, M. G. and Ippolito, A. (2015), A new perspective in controlling postharvest citrus rots: The use of electrolyzed water, *Acta Hortic.*, 1065, 1599–605.

Fallik, E. (2004), Prestorage hot water treatment (immersion, rinsing and brushing), *Postharvest Biol. Technol.*, 32, 125–34.

Feliziani, E., Landi, L. and Romanazzi, G. (2015), Preharvest treatments with chitosan and other alternatives to conventional fungicides to control postharvest decay of strawberry, *Carbohydr. Polym.*, 132, 111–17.

Gago, C. M., Guerreiro, A. C., Miguel, G., Panagopoulos, T., da Silva, M. M. and Antunes, M. D. (2016), Effect of Calcium chloride and 1-MCP (Smartfresh™) postharvest treatment on 'Golden Delicious' apple cold storage physiological disorders, *Sci. Hortic.*, 211, 440–8.

Gatto, M. A., Ippolito, A., Linsalata, V., Cascarano, N. A., Nigro, F., Vanadia, S. and di Venere, D. (2011), Activity of extracts from wild edible herbs against postharvest fungal diseases of fruit and vegetables, *Postharvest Biol. Technol.*, 61, 72–82.

Ippolito, A. and Nigro, F. (2000), Impact of preharvest application of biological control agents on postharvest diseases of fresh fruits and vegetables, *Crop Prot.*, 19(8–10), 715–23.

Ippolito, A., Schena, L., Pentimone, I. and Nigro, F. (2005), Control of postharvest rots of sweet cherries by pre- and postharvest applications of *Aureobasidium pullulans* in combination with calcium chloride or sodium bicarbonate, *Postharvest Biol. Technol.*, 36, 245–52.

Khamis, Y., Mustafa, Z. M. M. and Al-Essawy, A. (2018), Efficacy of alkaline and acidic electrolysed water generated by some salt solutions against graymold of table grape: Pre and postharvest applications, *J. Phytopathol. Pest Manag.*, 1(5), 1–21.

Lai, T., Bai, X., Wang, Y., Zhou, J., Shi, N. and Zhou, T. (2015), Inhibitory effect of exogenous sodium bicarbonate on development and pathogenicity of postharvest disease *Penicillium expansum*, *Sci. Hortic.*, 187, 108–14.

Lee, S. O., Choi, G. J., Jang, K. S., Lim, H. K., Cho, K. Y. and Kim, J. C. (2007), Antifungal activity of five plant essential oils as fumigant against postharvest and soilborne plant pathogenic fungi, *Plant Pathol. J.*, 23(2), 97–102.

Mari, M., Bautista-Baños, S. and Sivakumar, D. (2016), Decay control in the postharvest system: Role of microbial and plant volatile organic compounds, *Postharvest Biol. Technol.*, 22, 70–81.

Massart, S., Martinez-Medina, M. and Jijakli, M. H. (2015), Biological control in the microbiome era: Challenges and opportunities, *Biol. Control*, 89, 98–108.

Mohamed, N. T. S., Ding, P., Kadir, J. M. and Ghazali, H. (2017), Potential of UVC germicidal irradiation in suppressing crown rot disease, retaining postharvest quality and antioxidant capacity of Musa AAA 'Berangan' during fruit ripening, *Food Sci. Nutr.*, 5(5), 967–80.

Montesinos-Herrero, C., del Río, M. A., Pastor, C., Brunetti, O. and Palou, L. (2009), Evaluation of brief potassium sorbate dips to control postharvest penicillium decay on major citrus species and cultivars, *Postharvest Biol. Technol.*, 52(1), 117–25.

Montesinos-Herrero, C., Moscoso-Ramírez, P. A. and Palou, L. (2016), Evaluation of sodium benzoate and other food additives for the control of citrus postharvest green and blue molds, *Postharvest Biol. Technol.*, 115, 72–80.

Nigro, F. and Ippolito, A. (2016), UV-C light to reduce decay and improve quality of stored fruit and vegetables: A short review, *Acta Hortic.*, 1144, 293–8.

Nigro, F., Ippolito, A. and Lima, G. (1998), Use of UV-C light to reduce *Botrytis* storage rot of table grapes, *Postharvest Biol. Technol.*, 13, 171–81.

Nigro, F., Ippolito, A., Lattanzio, V., di Venere, D. and Salerno, M. (2000), Effect of ultraviolet-C light on postharvest decay of strawberry, *J. Plant Pathol.*, 82, 29–37.

Nigro, F., Schena, L., Ligorio, A., Pentimone, I., Ippolito, A. and Salerno, M. G. (2006), Control of table grape storage rots by pre-harvest applications of salts, *Postharvest Biol. Technol.*, 42(2), 142–9.

Olivieri, F., Feldman, M., Machinandiarena, M., Lobato, M., Caldiz, D., Daleo, G. and Andreu, A. (2012), Phosphite applications induce molecular modifications in potato tuber periderm and cortex that enhance resistance to pathogens, *Crop Prot.*, 32, 1–6.

Ouhibi, C., Attia, H., Nicot, P., Urban, L., Lachaâl, M. and Aarrouf, J. (2014), Effect of UV-C radiation on resistance of romaine lettuce against *Botrytis cinerea* and *Sclerotinia minor*, *J. Phytopath.*, 163, 578–82.

Palou, L., Ali, A., Fallik, E. and Romanazzi, G. (2016), GRAS, plant-and animal-derived compounds as alternatives to conventional fungicides for the control of postharvest diseases of fresh horticultural produce, *Postharvest Biol. Technol.*, 22, 41–52.

Panebianco, S., Vitale, A., Polizzi, G., Scala, F. and Cirvilleri, G. (2015), Enhanced control of postharvest citrus fruit decay by means of the combined use of compatible biocontrol agents, *Biol. Control*, 84, 19–27.

Pangallo, S., Li Destri Nicosia, M. G., Raphael, G., Levin, E., Ballistreri, G., Cacciola, S. O., Rapisarda, P., Droby, S. and Schena, L. (2016), Elicitation of resistance responses in grapefruit and lemon fruits treated with a pomegranate peel extract, *Plant Pathol.*, 66, 633–40.

Pinto, L., Ippolito, A. and Baruzzi, F. (2015), Control of spoiler *Pseudomonas* spp. on fresh cut vegetables by neutral electrolyzed water, *Food Microbiol.*, 50, 102–8.

Quaglia, M., Moretti, C., Cerri, M., Linoci, G., Cappelletti, S. U. and Taticchi, A. (2016), Effect of extracts of wastewater from olive milling in postharvest treatments of pomegranate fruit decay caused by *Penicillium adametzioides*, *Postharvest Biol. Technol.*, 118, 26–34.

Ribes, S., Fuentes, A., Talens, P. and Barat, J. M. (2018), Prevention of fungal spoilage in food products using natural compounds: A review, *Crit. Rev. Food Sci. Nutr.*, 58, 2002–16.

Rocha, A. B. O., Honório, S. L., Messias, C. L., Otón, M. and Gómez, P. A. (2015), Effect of UV-C radiation and fluorescent light to control postharvest soft rot in potato seed tubers, *Sci. Hortic.*, 181, 174–81.

Romanazzi, G. and Feliziani, E. (2016), Use of chitosan to control postharvest decay of temperate fruit: Effectiveness and mechanisms of action, *in* Bautista-Baños, S., Romanazzi, G. and Jiménez-Aparicio, A. (Eds), *Chitosan in the Preservation of Agricultural Commodities* (pp. 155–77). USA: Academic Press/Elsevier.

Sanzani, S. M., De Girolamo, A., Schena, L., Solfrizzo, M., Ippolito, A. and Visconti, A. (2009), Control of *Penicillium expansum* and patulin accumulation on apples by quercetin and umbelliferone, *Eur. Food Res. Technol.*, 228, 381–9.

Sanzani, S. M., Schena, L., de Cicco, V. and Ippolito, A. (2012), Early detection of *Botrytis cinerea* latent infections as a tool to improve postharvest quality of table grapes, *Postharvest Biol. Technol.*, 68, 64–71.

Sanzani, S. M., Schena, L. and Ippolito, A. (2014), Effectiveness of phenolic compounds against citrus green mould, *Molecules*, 19(8), 12500–8.

Sare, A. R., Ait-A, N., Jijakli, H. and Massart, S. (2018), Ecological relations inside plant microbiota can improve the efficacy of biocontrol agents: The case of *Pichia anomala* strain K against *Botrytis cinerea* on apple, *70th International Symposium on Crop Protection*, Belgium, https://orbi.uliege.be/handle/2268/223717.

Singh, N., Singh, R. K., Bhunia, A. K. and Stroshine, R. L. (2002), Effect of inoculation and washing methods on the efficacy of different sanitizers against *Escherichia coli* O147:H7 on lettuce, *Food Microbiol.*, 19, 183–93.

Sivakumar, D. and Bautista-Baños, S. (2014), A review on the use of essential oils for postharvest decay control and maintenance of fruit quality during storage, *Crop Prot.*, 64, 27–37.

Sivakumar, D., Hewarathgamagae, N. K., Wijeratnam, R. W. and Wijesundera, R. L. C. (2002), Effect of ammonium carbonate and sodium bicarbonate on anthracnose of papaya, *Phytoparasitica*, 30(5), 486–92.

Smilanick, J. L., Mansour, M. F., Gabler, F. M. and Sorenson, D. (2008), Control of citrus postharvest green mold and sour rot by potassium sorbate combined with heat and fungicides, *Postharvest Biol. Technol.*, 47(2), 226–38. doi:10.1016/j.postharvbio.2007.06.020.

Spadoni, A., Guidarelli, M., Sanzani, S. M., Ippolito, A. and Mari, M. (2014), Influence of hot water treatment on brown rot of peach and rapid fruit response to heat stress, *Postharvest Biol. Technol.*, 94, 66–73.

Terao, D., de Carvalho Campos, J. S., Benato, E. A. and Hashimoto, J. M. (2015), Alternative strategy on control of postharvest diseases of mango (*Mangifera indica* L.) by use of low dose of ultraviolet-C irradiation, *Food Eng. Rev.*, 7(2), 171–5.

Tonutti, P., Droby, S. and Romanazzi, G. (2016), Alternative approaches to synthetic fungicides to manage postharvest decay of fruit and vegetables: Needs and purposes of a special issue, *Postharvest Biol. Technol.*, 122, 1–2.

Usall, J., Ippolito, A., Sisquella, M. and Neri, F. (2016), Physical treatments to control postharvest diseases of fresh fruits and vegetables, *Postharvest Biol. Technol.*, 122, 30–40.

Viñas, I., Usall, J., Teixidó, N. and Sanchis, V. (1998), Biological control of major postharvest pathogens on apple with *Candida sake*, *Int. J. Food Microbiol.*, 40, 9–16.

Wisniewski, M., Droby, S., Norelli, J., Liu, J. and Schena, L. (2016), Alternative management technologies for postharvest disease control: The journey from simplicity to complexity, *Postharvest Biol. Technol*, 22, 3–10.

Youssef, K., Ligorio, A., Sanzani, S. M., Nigro, F. and Ippolito, A. (2011), Investigations of *Penicillium* spp. population dynamic in citrus packinghouse, *J. Plant Pathol.*, 93(4), S4.61. ISSN: 1125-4653.

Youssef, K., Ligorio, A., Nigro, F. and Ippolito, A. (2012a), Activity of salts incorporated in wax in controlling postharvest diseases of citrus fruit, *Postharvest Biol. Technol.*, 65, 39–43.

Youssef, K., Ligorio, A., Sanzani, S. M., Nigro, F., Ippolito, A. (2012b), Control of storage diseases of citrus by pre- and postharvest application of salts, *Postharvest Biol. Technol.*, 72, 57–63.

Youssef, K., Sanzani, S. M., Ligorio, A., Ippolito, A. and Terry, L. A. (2014), Sodium carbonate and bicarbonate treatments induce resistance to postharvest green mold on citrus fruit, *Postharvest Biol. Technol.*, 87, 61–9.

Youssef, K., Abd-Elsalam, K. A., Hussien, A., Sanzani, S. M. and Ippolito, A. (2017), Chapter 8: Organic and inorganic salts as postharvest alternative control means of citrus, *in* Harsimran, J. (Ed.), *Citrus Pathology* (pp. 161–77), Croatia: InTech.

Part 3

Insect pests of fruit

Chapter 15

Insect pests of fruits: aphids

Giuseppe E. Massimino Cocuzza, Università di Catania, Italy

1 Introduction

Aphids (*Homoptera*, *Sternorrhyncha*, *Aphididae*) currently include 5575 valid species (Remaudière and Remaudière, 1997; Favret, 2019). They are characterized by some peculiar morphological (e.g. siphunculi and cauda) and biological characteristics (e.g. numerous parthenogenetic generations, telescopic generations, the tendency to form colonies) that make these insects unique organisms, subject to numerous scientific studies (Kindlmann and Dixon, 1994; Blackman and Eastop, 2018). Aphids are closely related to specific host plants: about 99% are monophagous or oligophagous and only a small minority are polyphagous (Dixon, 1985; Barbagallo and Patti, 1997). This latter group includes some of the most harmful agricultural species such as *Myzus persicae*, *Aphis gossypii* and *A. spiraecola*. Although often under-reported, it is estimated that worldwide aphids cause very high economic losses each year (Dedryver et al., 2010). The importance of these insects can be indirectly measured by large sap-feeding insecticides available on the market (about 34%) and by the quantity sold every year (more than US$ 4 500 bn worldwide) (Simon-Delso et al., 2015).

In temperate climates, the development cycle of aphid populations follows precise patterns determined, above all, by the climate and the vegetative state of the host plants (Dixon and Kundu, 1994). Colonies begin to form in spring, in the presence of new vegetative growth, reaching their maximum density after

http://dx.doi.org/10.19103/AS.2019.0046.18

about one month and declining with rising temperatures and hardening of the shoots. The duration of this phase is directly correlated to temperature, relative humidity (abiotic factors), vegetative state of the plant, presence and activity of natural enemies (biotic factors), and the intrinsic characteristics of the species. In heteroecious species, throughout the alate forms, the aphids leave the primary host plants in search of the secondary host plants on which develop the summer generations. In autumn, the amphigonic forms return to the primary hosts to mate and deposit eggs, which allow many species to survive the winter. In some species or populations, living in favourable climates, there is often no amphigonic generation, as parthenogenetic generations follow each other, continuously (Hardie, 2017).

Although the vast majority of species are of little or no economic importance, some are considered harmful to agricultural crops. Direct damage caused by aphids is due to sap subtraction, vegetative deformations and honeydew production on which sooty mould fungi grow, reducing the plant's photosynthetic activity and suppressing induced plant defence (Reynolds, 1999; Schwartzberg and Tumlinson, 2014). Severe infestations, especially on young plants, can seriously alter their normal metabolism and development, since various nutrients are removed by the feeding aphids (water, mineral salts, carbohydrates and amino acids) (Goggin, 2007; Zhou et al., 2015; Nalam et al., 2018). Indirect types of damage, which have more serious consequences for plants, are caused by the ability of some species (about 190 species overall) of aphids to transmit viruses (Nault, 1997; Brault et al., 2010). Banana bunchy top virus, Tristeza citrus virus, Papaya ring spot virus and Plum pox virus are some of the most important viruses transmitted by aphids (Stevens and Lacomme, 2017).

The main aspects relating to the management of aphids, of economic interest, for citrus, apple, pear and stone fruits are briefly described below.

2 Citrus aphids

Although about 18 aphid species are reported on citrus crops worldwide (Blackman and Eastop, 2018), most of them occur sporadically and are rarely of economic importance. Only 3–4 species are considered responsible for significant direct and/or indirect damage (Barbagallo et al., 2017a). These are *Aphis gossypii* (cotton or melon aphid), *Aphis spiraecola* (green citrus aphid), *Aphis* (*Toxoptera*) *citricidus* (tropical citrus aphid) and less important as pest is *Aphis* (*Toxoptera*) *aurantii* (black citrus aphid). A recognition key for the main citrus aphids is shown in Fig. 1 and Box 1.

Aphis gossypii (Fig. 1d) is a polyphagous species recorded on more than 100 plant families and nearly 700 host plants, of which about 100 are crop species (Holman, 2009; Blackman and Eastop, 2017). *A. gossypii* is common,

Figure 1 Citrus aphids. (a) *Aphis spiraecola*; (b) *Myzus persicae*; (c) *Macrosiphum euphorbiae*; (d) *Aphis gossypii*; (e) *Aphis craccivora*; (f) *Aphis fabae*; (g) *Aphis (=Toxoptera) aurantii*; (h) *Aphis (=Toxoptera) citricidus* (courtesy of A. Hermoso de Mendoza).

Box 1 Recognition keys of principal aphids of citrus

1 Green aphids...2
- Aphids of other colours...3

2 Aphid green or yellowish-green. BL 1.2–2.2 mm. Cauda and siphunculi uniformly blackish. The colonies cause showy leaf deformation to the infested shoots. Very common on orange, mandarin, clementine and grapefruit......................................*Aphis spiraecola* (Fig. 1a)
- Pale green aphid, slightly bright. BL 1.2–2.3 mm. Siphunculi 1.9–2.5 times of the length of the cauda and depigmented, except at the extreme apex usually brownish; cauda pale. It causes modest leaf deformations to the infested shoots. Occasionally on citrus..........................*Myzus persicae* (Fig. 1b)
- Aphid green. BL 2.6–4 mm. Siphunculi 1.7–2.2 times the length of the cauda, green and blackish at the top. It does not cause leaf deformations. Rare on citrus, only infesting very young shoots..............................*Macrosiphum euphorbiae* (Fig. 1c)

3 Aphid variable in colour from greenish-ocherous to mottled greenish or dark bluish. BL 1–1.8 mm. Siphunculi blackish. Cauda clearer than cornicles. It does not deform infested shoots. Very common on all citrus fruit, rare on lemon...................*Aphis gossypii* (Fig. 1d)
- Brownish or black aphids. Siphunculi and cauda black.....................................4

4 Aphid black in colour, sometimes with small white wax spots on dorsum.......................5
- Aphids dark-brown or black, without wax on dorsum.....................................6

5 Aphid shiny black on dorsum. BL 1.2–2.3 mm. Siphunculi and cauda black. Occasionally on citrus... *Aphis craccivora* (Fig. 1e)
- Aphid matt black on dorsum, sometimes with small white wax spots. BL 1.8–2.4 mm. Occasionally on citrus................................... *Aphis fabae* (Fig. 1f)

6 Aphid brown in colour (paler in immature forms). BL 1.4–2.2 mm. Antennal segments III–VI cream coloured over 2/5 basal part and blackish at the tips giving them a ringed appearance. Front wings with black pterostigma and median vein once forked..*Toxoptera aurantii* (Fig. 1g)
- Aphid brownish-blackish in colour and quite shiny on dorsum. On apterae, antenna segment III–IV entirely pale. On alate, III entirely brownish, IV–V pale and brownish in the distal part. Front wings with pterostigma little pigmented and median vein twice furcate..*Toxoptera citricidus* (Fig. 1h)

especially on orange, to a lesser extent on clementine and mandarin, while it rarely infests lemon. It is an anholocyclic species, although cases of holocyclic populations are known (Komazaki et al., 1979; Inaizumi, 1981). One of the main characteristics of the species is the extreme variability in body colouration, probably related to the host plant upon which it is feeding. Populations vary from dark blue-green (typically on citrus) to ochre brown, greenish or pale yellow. In spring, on new vegetation, *A. gossypii* can develop large populations on citrus that rapidly colonize all shoots of the plants. Normally, infestations are well tolerated by mature plants (>5 years old), but they can delay development in young plants (2–3 years old) due to significant removal of sap and nutrients (Massimino Cocuzza and Rapisarda, 2017).

Aphis spiraecola (Fig. 1a) is a cosmopolitan and polyphagous species, reported on more than 450 host plants (Holman, 2009). On citrus the presence of the species is easily recognizable both for its colour (green or greenish-yellow with black siphunculi and cauda) and for the typical curl assumed by the young leaves as a consequence of the toxic saliva injected by the insects. As for *A. gossypii*, direct damage is a serious problem especially for young citrus plants, as large infestations cause a significant delay in development (Barbagallo et al., 2017a).

Aphis (*Toxoptera*) *citricidus* (Fig. 1h) is a native of tropical regions, where it is widely distributed and recently expanded into subtropical and temperate areas (Ilharco et al., 2005; Hermoso de Mendoza et al., 2008). The species has been found on about 70 host plants belonging mainly to the Rutaceae family (Michaud, 1998; Holman, 2009). Apterous and alate forms are shiny dark-brown in colour. Direct damage is negligible, but the species is particularly important because of high efficiency in the transmission of citrus tristeza virus (CTV) that, together with huanglongbing, is considered the most destructive disease of citrus (Bovè, 2006; Moreno et al., 2008). *Aphis* (*Toxoptera*) *aurantii* (Fig. 1g), attacks mainly lemon and is rarely of economic importance. *Aphis craccivora* (Fig. 1e), *Aphis fabae* (Fig. 1f), *Macrosiphum euphorbiae* (Fig. 1c) and *Myzus persicae* (Fig. 1b) only occasionally infest citrus, and normally their damage is rather negligible.

All the aforementioned aphid species are able to transmit CTV, including both the 'stem pitting' and 'seedling yellow' strains. However, while the efficiency of *A. citricidus* in transmitting CTV is very high, that of *A. spiraecola* and *A. aurantii* is significantly lower. Although it has a lower efficiency than *A. citricidus*, *A. gossypii* in relatively few years has transmitted CTV into large areas of citrus cultivation in Spain and Sicily (Davino et al., 2013). This has forced growers to uproot orchards grafted on sour orange and replace them with more tolerant rootstocks (Moreno et al., 2008). *A. citricidus*, *A. gossypii* and *A. spiraecola* are also vectors of citrus vein enation-wood gall virus, citrus sudden death-associated virus and citrus yellow vein clearing virus (Maharaj and da Graça, 1989; Yamamoto et al., 2011; Önelge et al., 2011).

3 Apple and pear aphids

More than a dozen aphid species can infest apple and pear trees, although only 5–6 can cause production losses (Barbagallo et al., 2017b). An identification key of the main apple and pear tree aphids is reported in Box 2.

Box 2 Recognition keys of principal aphids of apple and pear

1 Aphids yellow in colour..........2
- Aphids of other colours..........3

2 Aphid yellow lemon, body pyriform and small ~1 mm. Mainly infest the cavity of the calix of fruits. It also attacks twigs and buds..........*Aphanostigma piri* (Fig. 2a)
- Aphids without waxy secretion on body..........3
- Aphids with waxy secretion on body..........4

3 Aphid green. BL 1.3–2.3 mm. Siphunculi and cauda blackish. The colonies cause leaf deformation on infested shoots. Common on apple and less on pear.....*Aphis pomi* (Fig. 2b)
- Green or yellowish-green aphids with evident dark-green stripes on dorsum. BL 2.1–2.6 mm.......... *Rhopalosiphum oxiacanthae* (=*insertum*) (Fig. 2c)
- Aphid dark-brown with a large dark sclerotized patch on dorsal abdomen. BL 1.3–2 mm. Leaves are strongly curled and turn a yellowish colour with the progress of infestation. On pear..........*Melanaphis pyraria* (Fig. 2d)

4 Aphids with powdered waxy secretion on body..........5
- Aphids with abundant long waxy filaments on body..........6

5 Aphid brownish-purplish grey or dull pinkish. BL 2–2.6 mm. Infested leaves are strongly deformed and green in colour. On apple..........*Dysaphis plantaginea* (Fig. 2e)
- Aphid bluish-grey. BL 2–2.6 mm. Infested leaves are strongly deformed and reddish in colour. On apple.......... *Dysaphis anthrisci* (Fig. 2f)
- Aphid mauve-brown with siphunculi black. BL 1.7–3.2 mm. Infested leaves are deformed and green-yellowish colour. On pear..........*Dysaphis pyri* (Fig. 2g)

6 Aphid dark red or blackish brown. BL 1.5–2.5 mm. Siphunculi not developed. On apple..........*Eriosoma lanigerum* (Fig. 2h)
- Aphid pale brown or yellowish in colour. Siphunculi not developed. BL ~2 mm. Pear is the secondary host plant on which it infest the roots..........*Eriosoma laniginosum*

On apple, the colonies of *Disaphis plantaginea* (rosy apple aphid) (Fig. 2e) develop early in spring on the newly formed leaves, which soon become strongly deformed. The toxic saliva, once inoculated, is translocated from the

Figure 2 Apple and pear aphids. (a) *Aphanostigma piri*; (b) *Aphis pomi* (courtesy by S. Convertini); (c) *Rhopalosiphum oxyacanthae* (=*insertum*); (d) *Melanaphis pyraria*; (e) *Dysaphis plantaginea* (courtesy of S. Convertini); (f) *Dysaphis anthrisci*; (g) *Dysaphis pyri*; (h) *Eriosoma lanigerum*.

leaves to the growing fruits that also become reduced in size and deformed around the calyx. This species produces abundant honeydew on which sooty moulds develop (Blommers et al., 2004). On pear, similar damage is caused by *Dysaphis pyri* (pear-bedstraw aphid) (Barbagallo et al., 2017b) (Fig. 2g). *Eriosoma lanigerum* (woolly apple aphid) (Fig. 2h) colonizes roots, trunks, branches and shoots. Its toxic saliva causes the formation of neoplasms that, in the following years, evolve into cracking that slow down or block the growth of the affected parts. Neoplasms also impede the development of fruit buds, while on roots they are the entry routes for opportunistic pathogens (e.g. causative agent of apple canker) that cause a deterioration in tree health (Brown et al., 1991, 1995). On pear, *Eriosoma lanuginosum* (woolly pear aphid) infest only roots and the damage is analogous to those described for *E. lanigerum*. Attacks of *Aphis pomi* (green apple and pear aphid) (Fig. 2b) and the very similar *A. spiraecola* can be damaging to young trees and serious infestation can lead to a slowdown in growth (Footit et al., 2009). On pear, *Aphanostigma piri* (pear phylloxera) (Fig. 2a) and *Aphanostigma iaksuiense* (oriental pear phylloxera) infests twigs, buds and fruits. The colonies settle in the cavity of the calyx favouring the entry of opportunistic fungi that cause rotting of the injured area (Barbagallo et al., 2017b). Of marginal importance and causing scarce damage on apple are *Rhopalosiphum oxiacanthae* (=*insertum*) (apple-grass aphid) (Fig. 2c), the group of red-gall aphids (*Dysaphis devecta, D. anthrisci* (Fig. 2f), *D. brancoi, D. chaerophylli* and *D. radicola*), and on pear, *Melanaphis pyraria* (brown pear aphid) (Fig. 2d), *Anuraphis farfarae* (pear-coltsfoot aphid) and *Dysaphis reaumuri* (green waxy pear aphid) (Barbagallo et al., 2017b).

4 Stone fruit aphids

Despite the numerous species (about 40 worldwide) that can live on stone fruits (Prunoideae of the genus *Prunus* that include peach, plum, apricot, cherry and almond), only a few are of economic importance, both for the direct and/or indirect damage they cause (Barbagallo et al., 2017a). A recognition key of the main stone fruit aphids is reported in Fig. 3 and Box 3. On peach (*Prunus persicae*) and nectarines (*P. persicae* var. *nucipersica*), *Myzus persicae* (peach-potato aphid) (Fig. 2a) is the most injurious aphid species, as infestations cause leaf twisting on buds and abundant production of honeydew can contaminate foliage and fruits. *M. persicae* is also a vector of plum pox virus, the causal agent of 'sharka', probably the most serious disease of stone fruits (Gildow et al., 2004; Moreno et al., 2009).

Peach is also infested by *Brachycaudus persicae* (black peach aphid) (Fig. 2e), *B. schwartzi* (brown peach aphid) (Fig. 2g), *Hyalopterus arundiniformis* (mealy aphid of peach) (Fig. 2d), *Myzus varians* (cigar-rolling peach aphid)

Figure 3 Stone fruit aphids. (a) *Myzus persicae*; (b) *Myzus varians*; (c) *Brachycaudus helychrisi*; (d) *Hyalopterus arundiniformis*; (e) *Brachycaudus persicae*; (f) *Brachycaudus divaricatae* (courtesy of R. Rakauskas); (g) *Brachycaudus schwartzi*; (h) *Myzus cerasi*.

(Fig. 2b) and *Pterochloroides persicae* (large bark-living aphid of stone fruits). These species are easy to distinguish from one another, both morphologically and by the damage produced on plants. *B. schwartzi*, though an occasional pest, causes severe curling on leaves and can transmit plum pot virus to peach

Box 3 Recognition keys of principal aphids of stone fruits

1 Aphids infesting the woody parts of the plant. Brown with black marking. BL 2.5–4.2 mm. Siphunculi very short and truncate conical. Especially on peach and nectarine, occasionally on almond, plum, apricot and cherry.. *Pterochloroides persicae*

- Aphids infesting leaves of shoots and green or brown-green in colour............................2

2 Pale green aphid, slightly bright. BL 1.2–2.3 mm. Siphunculi 1.9–2.5 times of the length of the cauda and depigmented siphunculi, except at the extreme apex which is usually brownish; cauda pale. It causes strong leaf curling to the apical leaves of infested shoots. On peach, nectarines, almond and plum... *Myzus persicae* (Fig. 3a)

- Pale yellowish-green. Siphunculi dark pigmented from the middle to the apex. Infested leaves typically rolled-up like cigars. On peach............................*Myzus varians* (Fig. 3b)
- Yellowish-pale green in colour. BL 0.9–2 mm. Siphunculi pale, 0.8–2 times the length of the cauda. On plum and occasionally on peach and almond... *Brachycaudus helichrysi* (Fig. 3c)
- Green. BL 1.6–2. Characteristic black patch on the dorsum. Infested leaves are strongly deformed. On almond ....*Brachycaudus amygdalinus*
- Pale yellowish-green, slightly shiny. Three dark-green longitudinal strips on dorsum. BL 2–2.6 mm. It causes deformations on leaves. On plum...............*Phorodon humuli*
- Green. Body covered with mealy wax secretion. BL 1.5–2.6 mm. Siphunculi long twice the diameter of cauda, darker at the apex. The infested leaves are not deformed. On peach ....*Hyalopterus arundiniformis* (Fig. 3d)
- Morphologically very similar to the congeneric species. The infested leaves are not deformed. On almond and occasionally on apricot...................*Hyalopterus amygdali*
- Morphologically very similar to the other congeneric species. The infested leaves are not deformed. On plum ....*Hyalopterus pruni*
    - Brownish or greenish-brown aphids.......................................3

3 Brown with shiny dark dorsum. Immature brown. BL 1.5–2.2 mm. Infest twigs, buds, shoots and young fruits. Infested leaves are not deformed. On peach and sometimes on almond and plum ....*Brachycaudus persicae* (Fig. 3e)

- Brown with shiny dark dorsum. BL 2–2.2 mm. Infested leaves are rolled. On plum and *Prunus cerasifera*................................ *Brachycaudus divaricatae* (Fig. 3f)

- Shiny greenish-brown in colour, with black sclerotized bands across dorsum. Immature yellowish-pale brown. BL 1.4-2.1 mm. The infested leaves are strongly curled. On peach ....*Brachycaudus schwartzi* (Fig. 3g)
    - Black aphids..4

4 Shiny black on dorsum, siphunculi and cauda. Legs and antennae yellowish and dark at the apex of each segment. Immature dark-brown. BL 1.5-2.5 mm. The infested leaves show strong curls and deformations. On cherry...*Myzus cerasi* (Fig. 3h)

(Manachini et al., 2004). The congeneric *B. persicae* cause infestations very early in the season with dense colonies on twigs, buds and young fruits of peach and almond. *H. arundiniformis* and *H. amygdali* infest peach and almond, respectively, forming dense colonies on the lower side of the leaves that are not deformed. In very sunny areas, the abundant honeydew spreading on the tender vegetation can produce a 'lens effect' causing serious burns.

In plum and cherry, the most economically important aphids are *B. helichrysi* (leaf-curling plum aphid) (Fig. 2c) and *Myzus cerasi* (black cherry aphid) (Fig. 2h), respectively; early spring infestations are characterized by severe curls and deformation of leaves (Barbagallo et al., 2017a). On the same plants, a species of Central Asian origin, *Brachycaudus divaricatae* (Fig. 2f), is now spreading into Eastern and Central Europe (Rakauskas et al., 2015). Recently, *Myzus mumecola*, an aphid originating from East Asia and India, which is considered harmful to apricot in particular, has been detected in Italy (Panini et al., 2017).

5 Control of aphid pests

The first important step in the control of fruit crop aphids is to evaluate the real economic loss resulting from various levels of infestation. Depending on the crop, variety of plant and the aphid species, high levels of infestation do not always correspond to significant economic losses and vice versa. Where it is needed, aphid control should be based on a fundamental understanding of all components involved: crop, environment and species. Minimization of damage is achieved by adopting a precise strategy primarily to prevent initial infestations. Interactions between the various organisms present in an agroecosystem are an important key element in the management of aphids in orchards. In the medium-long term, disrupting the equilibrium between aphids and natural predators, for example, through the use of no selective pesticides, can have negative effects on both production and the environment. Pesticides should be used with caution and only if necessary, following the instructions provided by the manufacturers and only as part of a wider integrated pest management (IPM) programme.

In temperate or subtropical climates, weekly inspection of plants is advisable, starting when temperatures begin to rise and the new vegetative growth emerges in the spring. Continuous monitoring of trends in infestation allow evaluation over time of its intensity, considering that it can vary, significantly, from year to year, with climatic conditions, location and the plant variety. Normally, in spring, the period of infestation lasts about 4–8 weeks, with populations reaching a peak during the 3rd–4th week of vegetative restart (Massimino Cocuzza and Rapisarda, 2017). With the hardening of the shoots most aphid species disperse to secondary hosts. Species that remain on the plants during the summer (e.g. *A. spiraecola* in citrus, *B. schwartzi* and *P. persicae* in peach, *E. lanigerum* and *A. pomi* on apple and *A. piri* on pear) survive through small and negligible colonies. Moreover, in late spring natural antagonists drastically reduce aphid in numbers.

An important cultural method of control is management of plant nutrition, particularly nitrogen. Targeted use of fertiliser avoids excessive growth of shoots which favour an increase in aphid populations. Another, often under-estimated, cultural method of control is pruning (Barbagallo et al., 2017a). However, citrus pruning should be moderate, otherwise it can cause excessive shoot production which could promote more aphid infestation (Braham and Amor, 2018).

In apple, the availability of resistant varieties and rootstocks has greatly improved management of *E. lanigerum* (Stowke and Malan, 2016), via direct control and increasing the effectiveness of natural antagonists (Sandanayaka et al., 2005).

Aphids have numerous natural enemies that can significantly contribute to reducing their populations. Coccinellids (e.g. *Adalia bipunctata*, *Coccinella septempunctata*, *Hippodamia variegata*), syrphid flies (e.g. *Episyrphus* spp., *Pseudodorus clavatus*), aphid midge (e.g. *Aphidoletes aphidimyza*) and chrysopids (e.g. *Chrysoperla* spp., *Chrysopa* spp.) are the most common documented predators of aphids (Bellieure and Michaud, 2001; Brodeur et al., 2017). They are often active during the initial development of aphid colonies and can be very efficient. There are also numerous parasitoid hymenopterans, most commonly of the genera *Lysiphloebus* spp., *Aphidius* spp. and *Aphelinus* spp. (Kavallieratos et al., 2002; Satar et al., 2014). *L. fabarum* and *L. confusus* efficacy can be very high (accounting for up to 90–100% of aphid populations) (Starý et al., 1988). However, their main limitation is that they become abundant once aphid populations are established, meaning that the farmers are frequently forced to use pesticides, before parasitoids can be effective, in order to prevent damage to crops (Barbagallo et al., 2017). Conversely, *Aphelinus mali*, the primary parasitoid of *E. lanigerum*, can be considered as a successful example of classical biological control (Asante and Danthenaryana, 1992). The cosmopolitan wasp, of American origin, can contribute significantly to the control of *E. lanigerum*, especially together with other natural antagonists (Stokwe and Malan, 2016). The effectiveness of its control activity is rather variable as it is influenced by climatic conditions, inability to control numerically abundant colonies or insecticide treatments to which they are particularly sensitive (Mueller et al., 1992; Stiling, 1993). Other generic predators, such as some species of earwigs (Dermaptera), carabids, staphylinids (Coleoptera) and spiders are also active against aphids (Snyder and Wise, 1999; Gontijo et al., 2015). However, in general terms their control efficacy is difficult to quantify in crops. Normally, their efficiency is greater in diversified cultural systems or in organic orchards (Sunderland and Samu, 2000; Dib et al., 2016; Lefevre et al., 2017).

Ants feed on the honeydew of aphids and protect them from predators and parasitoids interfering with the activity of natural control agents and their excessive presence can compromise an IPM programme (Pekas et al., 2011; Dreistadt, 2012; Martínez-Ferrer and Campos-Rivela, 2017). Moreover, ants distribute aphids onto new shoots favouring their distribution on crop (Barbagallo and Patti, 1997). Especially for citrus and stone fruits, monitoring ant activity and abundance is necessary in some environments. Control of ants can be achieved with sticky bands fixed around the trunks of the fruit trees to hinder ants' ascent, or with poisoned baits and, finally, with insecticide applied on the ground (Dreistadt, 2012). However, it is important to restrict pesticide use to the minimum required within an overall IPM programme. The drive to reduce pesticide use is the result of growing concerns over protection of the environment, safeguarding natural enemies, avoiding the development of resistant aphid populations and, last but not least, market demand for fruits with minimal or even zero chemical residues.

For citrus, it is important to note that mature and vigorous plants can tolerate infestations of all aphid species affecting 40–50% of new shoots without the need for chemical treatment (Barbagallo et al., 2017). Organic aphicides include oils (petroleum and plant derived oils, such as neem or canola) and fatty potassium salts, products generally regarded as safe to beneficials (Dreistadt, 2012; Kolokytha and Sterk, 2017). Their effectiveness is greater at the beginning of the growing season when aphid colonies are smaller and natural enemies less active. Short persistence and ideally selective aphicides are the preferred choice as part of an IPM approach. The most effective and selective molecules against aphids are sulfoxaflor, spirotetramat, acetamiprid, pymetrozine, flupyradifurone and flonicamid. Farmers should avoid broad-spectrum products (such as pyrethroids or carbamates) that have negative side effects on the environment (including beneficials and non-target organisms) and because they favour an increase in phytophagous mite populations (Dutcher, 2007; Garcia, 2011; Antwi and Reddy, 2015). If necessary broad-spectrum insecticides must be applied, only, during early season, when natural enemies are not yet active. In areas infested with CTV, only preventive control is possible, because the use of chemicals is completely ineffective and a new citrus grove must be planted using tolerant virus-free rootstock (citrange troyer or citrange Carrizo or *Poncirus trifoliata*) (Moreno et al., 2008).

The timing of pesticide applications is key in aphid control. Especially in areas with recurring infestation of aphids. Monitoring of aphids on peach and nectarines should begin very early, just after dormancy, between the end of leaf bud swelling and the shoots reaching 20% of their total length. It is at this moment that the fundatrix emerge from the over wintering eggs and, after having completed the juvenile development stage, begin to produce offspring. Applying an aphicide at this point can be very effective in limiting the development of colonies and avoiding further aphicide applications, particularly as this is before predatory species emerge. For citrus, in new orchards (with plants 2–4 years old), it is advisable, in the spring, to monitor 10% of plants/ha and begin aphicide application once a threshold of 10% or 25% of shoots infested is reached (when these latter are between 20% and 80% of total length) for *A. spiraecola* and for *A. gossypii*, respectively (Barbagallo et al., 2017; Massimino Cocuzza and Rapisarda, 2017). For peach, the intervention threshold for *M. persicae* and *H. arundiniformis* is 3% of attached buds while, for nectarines it is 7–10% of infested shoots. On apple and pear 1–2% of infested shoots is the threshold for *D. plantaginea* and *D. pyri*, as well as for infested fruits by the pear phylloxera (Barbagallo et al., 2017). Adult plants can tolerate medium-high infestation of *A. pomi* (or *A. spiraecola*), whereas for *E. lanigerum*, the intervention threshold is 8–12% of infested shoots. However, insecticide treatments can depress the activity of natural enemies (especially *A. mali*) that often contribute to controlling the aphid (Barbagallo et al., 2017). Finally, for cherry, the intervention threshold recommended for *M. cerasi* is 5–10% of infested shoots (Barbagallo et al., 2017).

The activity of natural enemies (including ladybugs, hoverflies, lacewings and hymenopteran parasitoids) and the emigration by alate aphids at the end of spring, generally make further chemical intervention unnecessary and potentially counter-productive. Given the propensity with which *M. persicae* can develop resistance, chemical treatments must be limited to the minimum necessary (Bass et al., 2014). Other aphid species, normally, do not need to be controlled because damage is rarely economically important.

6 Future trends and conclusion

New areas of research include producing new aphid-resistant cultivars based on interactions and incompatibility at mesophyll and phloem level (Sun et al., 2018; Garzo et al., 2018). Some researchers are focussed on volatile organic compounds (VOCs) to attract natural enemies of aphids or genes that confer resistance (Di Gilio et al., 2010), whereas other studies have investigated developing varieties that are able to produce an alarm aphid pheromone, making the plants less attractive (Beagle et al., 2006; Bhatia et al., 2015; Bruce et al., 2015). Although, most of these studies have been performed on herbaceous plants (Smith and Chuang, 2014), whereas tree crop molecular research is currently active in identifying genes that confer plant resistance to aphids, both antixenosis (which prevent aphid infestation) and antibiosis (which negatively affect growth of colonies) (Pascal et al., 2002; van Emden, 2017; Farquharson, 2017). On peach, some resistance genes to *M. persicae* have been identified in cultivars (e.g. weeping flower peach, rubira) or species (*Prunus davidiana*) (Sauge et al., 2002, 2006, 2012; Lambert and Pascal, 2011; Verdugo et al., 2012). On apple, germplasm has allowed identification of resistant genes to *E. lanigerum* (Bus et al., 2008; Stoekli et al., 2008; Celton et al., 2009), *D. plantaginea* (Qubbaj et al., 2005), *D. devecta* (Cevik and King, 2002), *D. pyri* (Evans et al., 2008) and it is hypothesized that several others could be detected (Sandanayaka et al., 2005). However, resistance to date is monogenic and some populations of *E. lanigerum* can overcome this defence. The future challenge is to breed varieties with polygenic resistance capable of a longer durability (Smith and Chuang, 2014; Rousselin et al., 2017).

While it is well known that the effectiveness of predators and parasitoids in aphid control is limited by the delay in their activity on aphid colonies, some agronomic practices can be adopted to increase their efficiency. As an example, the maintenance of plant-refugees for natural enemies can ensure a loner residence in or around crops. Augmentative natural enemy releases are another area of future study. Studies on the use of semiochemicals and odorous substances to attract and maintain aphidophagous natural enemies in the crop are worthy of further development (Leroy et al., 2010; Dumont et al., 2011; Zhou et al., 2016).

Aphids remain among the most difficult insect pests to control. As Loxdale et al. (2017) state, this is linked to the extraordinary ecological/evolutionary adaptability of aphid species. These characteristics allow aphids to easily overcome control measures such as resistant plants or pesticides susceptibility. The number of aphicides approved and available on market is reducing and the risk of developing resistant populations is high. The speed with which some aphid species (i.e. *M. persicae* or *A. gossypii*) are able to develop resistance has been the subject of significant research (Bass et al., 2014, 2015; de Little et al., 2017; Wang et al., 2018). Currently, it is relatively easy to identify the genetic mutation in resistant populations, yet an increasingly restricted approval of new chemicals makes future resistance more likely. Pesticide companies are committed to developing more selective pesticides that are not deleterious to natural enemies. This risks the rapid development of resistance because the insecticide pressure rapidly selects for resistant individuals (Foster et al., 2017). Controlling aphids only with pesticides is now unthinkable and, indeed, unsustainable in the medium term, given the impact on natural enemies and the problem of resistance. The market requires an approach where the use of pesticides is as low and as selective as possible, with low environmental impact and within tight regulatory limits. However, with careful monitoring and planning control of aphids is possible by implementing IPM programmes with a focus on lowering populations below the threshold where they cause harm.

7 Where to look for further information

Aphids on the World's Plants – An online identification and information guide. www.aphidsonworldsplants.info

EPPO, European and Mediterranean Plant Protection Organization. https://www.eppo.int/

Influential Points, Aphid identification. https://influentialpoints.com/Gallery/Aphid_genera.htm

8 References

Antwi, F. B. and Reddy, G. V. 2015. Toxicological effects of pyrethroids on non-target aquatic insects. *Environmental Toxicology and Pharmacology* 40(3), 915–23. doi:10.1016/j.etap.2015.09.023.

Asante, S. K. and Danthanarayana, W. 1992. Development of *Aphelinus mali* an endoparasitoid of woolly apple aphid, *Eriosoma lanigerum* at different temperatures. *Entomologia Experimentalis et Applicata* 65(1), 31–7. doi:10.1111/j.1570-7458.1992.tb01624.x.

Barbagallo, S. and Patti, I. 1997. Linee evolutive nei rapporti afidi-piante ospiti. *ATTI Accademia Nazionale Italiana di Entomologia* XLV, 253–79.

Barbagallo, S., Massimino Cocuzza, G. E., Cravedi, P. and Komazaki, S. 2017a. IPM Case Studies: tropical and subtropical fruit trees. In: van Emden, H. F. and Harrington, R. (Eds), *Aphid as Crop Pests* (2nd edn.). CAB International, Wallingford, UK, pp. 643–54.

Barbagallo, S., Massimino Cocuzza, G. E., Cravedi, P. and Komazaki, S. 2017b. IPM Case Studies: deciduous fruit tree aphids. In: van Emden, H. F. and Harrington, R. (Eds), *Aphid as Crop Pests* (2nd edn.). CAB International, Wallingford, UK, pp. 632–42.

Bass, C., Puinean, A. M., Zimmer, C. T., Denholm, I., Field, L. M., Foster, S. P., Gutbrod, O., Nauen, R., Slater, R. and Williamson, M. S. 2014. The evolution of insecticide resistance in peach potato aphid, *Myzus persicae*. *Insect Biochemistry and Molecular Biology* 51, 41–51. doi:10.1016/j.ibmb.2014.05.003.

Bass, C., Denholm, I., Williamson, M. S. and Nauen, R. 2015. The global status of insect resistance to neonicotinoids insecticides. *Pesticide Biochemistry and Physiology* 121, 78–87. doi:10.1016/j.pestbp.2015.04.004.

Beagle, M. H., Birkett, M. A., Bruce, T. J. A., Chamberlain, K., Field, L. M., Huttly, A. K., Martin,J. L., Parker, R., Phillips, A. L., Pickett, J. A., Prosser, I. M., Shewry, P. R., Smart, L. E., Wadhams, L. J., Woodcock, C. M. and Zhang, Y. 2006. Aphid alarm pheromone produced by transgenic plants affects aphid and parasitoid behaviour. *Proceedings of the National Academy of Sciences of the United States of America* 103(27), 10509-13. doi:10.1073/pnas.0603998103.

Bellieure, B. and Michaud, J. P. 2001. Biology and behaviour of *Pseudodorus clavatus* (Diptera: Syrphidae), an important predator of citrus aphids. *Annals of the Entomological Society of America* 94(1), 91-6. doi:10.1603/0013-8746(2001)094[0091:BABOPC]2.0.CO;2.

Bhatia, V., Maisnam, J., Jain, A., Sharma, K. K. and Bhattacharya, R. 2015. Aphid-repellent pheromone E-β-farnesene is generated in transgenic *Arabidopsis thaliana* over-expressing *farnesyl diphosphate synthase2*. *Annals of Botany* 115(4), 581-91. doi:10.1093/aob/mcu250.

Blackman, R. L. and Eastop, V. F. 1994. *Aphids on the World's Trees. An Identification and Information Guide*. CAB International, Wallingford UK. Available at: www.aphidsonworldsplants.info (accessed on May 2019).

Blackman, R. L. and Eastop, V. F. 2017. Taxonomic issues. In: van Emden, H. F. and Harrington, R. (Eds), *Aphid as Crop Pests* (2nd edn.). CAB International, Wallingford, UK, pp. 1-36. Chapter 1.

Blommers, L. H. M., Helsen, H. H. M. and Vaal, F. W. N. M. 2004. Life history data of the rosy apple aphid *Dysaphis plantaginea* (Pass.) (Homopt., Aphidinae) on plantain and as migrant to apple. *Journal of Pesticide Science* 77, 155-63.

Bovè, J. 2006. Huanglongbinng: a destructive, newly-emerging, century-old disease of citrus. *Journal of Plant Pathology* 88, 7-37.

Braham, M. and Amor, N. 2018. Effect of pruning severity on the vegetative growth, density and population dynamics of the spirea aphid, *Aphis spiraecola* in citrus orchards. *Journal of Entomology and Zoology Studies* 6, 311-9.

Brault, V., Uzest, M., Monsion, B., Jacquot, E. and Blanc, S. 2010. Aphids as transport devices for plant viruses. *Comptes Rendus Biologies* 333(6-7), 524-38. doi:10.1016/j.crvi.2010.04.001.

Brodeur, J., Hajek, A. E., Heimpel, G. E., Sloggett, J. J., Mackauer, M., Pell, J. K. and Völkl, W. 2017. Predators, parasitoids and pathogens. In: In van Emden, H. F. and Harrington, R. (Eds), *Aphid as Crop Pests* (2nd edn.). CAB International, Wallingford, UK, pp. 643-54.

Brown, M. W., Glenn, D. M. and Wisniewski, M. E. 1991. Functional and anatomical disruption of apple roots by woolly apple aphid (Homoptera: Aphididae). *Journal of Economic Entomology* 84(6), 1823-6. doi:10.1093/jee/84.6.1823.

Brown, M. W., Ranger, S., Schmitt, J. J. and Hogmire, H. W. 1995. Yield reduction in apple by edafic woolly apple aphid (Homoptera, Aphididae) population. *Journal of Economic Entomology* 88, 127-33.

Bruce, T. J. A., Aradottir, G. I., Smart, L. E., Martin, J. L., Caulfield, J. C., Doherty, A., Sparks, C. A., Woodcock, C. M., Birkett, M. A., Napier, J. A., Jones, H. D. and Pickett, J. A. 2015. The first crop plant genetically engineered to release an insect pheromone for defence. *Scientific Reports* 5, 11183. doi:10.1038/srep11183.

Bus, V. G. M., Chagné, D., Bassett, H. C. M., Bowatte, D., Calenge, F., Celton, J. -M., Durel, C.-E., Malone, M. T., Patocchi, A., Ranatunga, A. C., Rikkerink, E. H. A., Tustin, D. S., Zhou, J. and Gardiner, S. E. 2008. Genome mapping of three major resistance genes to woolly apple aphid (*Eriosoma lanigerum* Hausm.). *Tree Genetics and Genomes* 4(2), 223–36. doi:10.1007/s11295-007-0103-3.

Celton, J. M., Tustin, D. S., Chagné, D. and Gardiner, S. E. 2009. Construction of a dense genetic linkage map for apple rootstocks using SSRs developed from *Malus* ESTs and *Pyrus* genomic sequences. *Tree Genetics and Genomes* 5(1), 93–107. doi:10.1007/s11295-008-0171-z.

Cevik, V. and King, J. 2002. High-resolution genetic analysis of the Sd-1 aphid resistance locus in *Malus* spp. *TAG. Theoretical and Applied Genetics. Theoretische und Angewandte Genetik* 105(2–3), 346–54. doi:10.1007/s00122-002-0904-6.

Davino, S., Willemsen, A., Panno, S., Davino, M., Catara, A., Elena, S. F. and Rubio, L. 2013. Emergence and phylodynamics of *Citrus tristeza* virus in Sicily, Italy. *PLoS ONE* 8(6), e66700. doi:10.1371/journal.pone.0066700.

Dedryver, C. A., Le Relac, A. and Fabre, F. 2010. The conflicting relationships between aphids and men: a review of aphid damage and control strategies. *Comptes Rendus Biologies* 333(6–7), 539–53. doi:10.1016/j.crvi.2010.03.009.

De Little, S. C., Edwards, O., van Rooyen, A. R., Weeks, A. and Umina, P. A. 2017. Discovery of metabolic resistance to neonicotinoids in green peach aphid (*Myzus persicae*) in Australia. *Pest Management Science* 73(8), 1611–7. doi:10.1002/ps.4495.

Dib, H., Sauphanor, B. and Capowiez, Y. 2016. Effect of management strategies on arthropod communities in the colonies of rosy apple aphid, *Dysaphis plantaginea* Passerini (Hemiptera: Aphididae) in south-eastern France. *Agriculture, Ecosystems and Environment* 216, 203–6. doi:10.1016/j.agee.2015.10.003.

Di Gilio, M. C., Corrado, G., Sasso, R., Coppola, V., Iodice, L., Pasquariello, M., Bossi, S., Maffei, M. E., Coppola, M., Pennacchio, F., Rao, R. and Guerrieri, E. 2010. Molecular and chemical mechanisms involved in aphid resistance in cultivated tomato. *The New Phytologist* 187(4), 1089–101. doi:10.1111/j.1469-8137.2010.03314.x.

Dixon, A. F. G. 1985. *Aphid Ecology: an Optimization Approach*. Blackie & Son, Glasgow, 157p.

Dixon, A. F. G. and Kundu, R. 1994. Ecology of host alternation in aphids. *European Journal of Entomology* 91, 63–70.

Dreistadt, S. 2012. *Integrated Pest Management for Citrus* (3rd edn.). UC ANR Publication 3303. University of California, Oakland, CA, 275p.

Dumont, V.-A., Trigaux, A., Moreau, A. and Hance, T. 2011. Study of two conditioning methods of parasitoids used in biological control prior to inundative releases in apple orchards. *European Journal of Environmental Sciences* 1(1), 51–6. doi:10.14712/23361964.2015.65.

Dutcher, J. D. 2007. A review of resurgence and replacement causing pest outbreaks in IPM. In: Ciancio, A. and Mukerji, K. G. (Eds), *General Concepts in Integrated Pest and Disease Management. Integrated Management of Plants Pests and Diseases* (vol. 1). Springer, Dordrecht.

Evans, K. M., Govan, C. L. and Fernandez-Fernandez, F. 2008. A new gene for resistance to *Dysaphis pyri* in pear and identification of flanking microsatellite markers. *Genome* 51(12), 1026–31. doi:10.1139/G08-093.

Farquharson, K. L. 2017. A phloem protein contributes to aphid resistance and heat stress tolerance. *The Plant Cell* 29(10), 2309–10. doi:10.1105/tpc.17.00781.

Favret, C. 2019. *Aphid Species File* (version 5.0/5.0). Available at: http://Aphid.SpeciesFile.org (accessed on July 2019).

Foottit, R. G., Lowery, D. T., Maw, H. E. L., Smirle, M. J. and Lushai, G. 2009. Identification, distribution and molecular characterization of the apple aphids *Aphis pomi* and *Aphis spiraecola* (Hemiptera: Aphididae: Aphidinae). *The Canadian Entomologist* 141(5), 478–95. doi:10.4039/n09-037.

Foster, S. P., Devine, G. and Devonshire, A. L. 2017. Insecticide resistance. In: van Emden, H. F. and Harrington, R. (Eds), *Aphid as Crop Pests* (2nd edn.). CAB International, Wallingford, UK, pp. 426–47.

Garcia, P. 2011. *Sublethal Effects of Pyrethroids on Insect Parasitoids: What We Need to Further Know, Pesticides - Formulations, Effects, Fate*. Stoytcheva, Prof. M. (Ed.). InTech. ISBN: 978-953-307-532-7. Available at: http://www.intechopen.com/books/pesticides-formulations-effects-fate/sublethal-effects-ofpyrethroids-on-insect-parasitoids-what-we-need-to-further-know.

Garzo, E., Fernàndez-Pascual, M., Morcillo, C., Fereres, A., Gómez-Guillamon, M. L. and Tjallingii, W. F. 2018. Ultrastructure of compatible and incompatible interactions in phloem sieve elements during the stylet penetration by cotton aphid in melon. *Insect Science* 25(4), 631–42. doi:10.1111/1744-7917.12447.

Gildow, F., Damsteege, V., Stone, A., Schneider, W., Luster, D. and Leuy, L. 2004. Plum pox in North America: identification of aphid vectors and a potential role for fruit in virus spread. *Phytopathology* 94(8), 868–74. doi:10.1094/PHYTO.2004.94.8.868.

Goggin, F. L. 2007. Plant-aphid interactions: molecular and ecological perspectives. *Current Opinion in Plant Biology* 10(4), 399–408. doi:10.1016/j.pbi.2007.06.004.

Gontijo, L. M., Beers, E. H. and Snyder, W. E. 2015. Complementary suppression of aphids by predators and parasitoids. *Biological Control* 90, 83–91. doi:10.1016/j.biocontrol.2015.06.002.

Hardie, J. 2017. Life cycles and polyphenism. In: van Emden, H. F. and Harrington, R. (Eds), *Aphid as Crop Pests* (2nd edn.). CAB International, Wallingford, UK, pp. 81–97.

Hermoso de Mendoza, A., Alvarez, A., Michelena, J. M., Gonzáles, P. and Cambra, M. 2008. Spread, biology and natural enemies of *Toxoptera citricidus* (Kirkaldy) (Hemiptera, Aphididae) in Spain. *Boletin de Sanidad Vegetal - Plagas* 34, 77–87.

Holman, J. 2009. *Host Plant Catalog of Aphids: Palaearctic Region*. Springer Science + Business Media, B.V., 1216p.

Ilharco, F. A., Sousa Silva, C. R. and Álvarez Álvarez, A. 2005. First report of *Toxoptera citricidus* (Kirkaldy) in continental Portugal and Spain (Homoptera, Aphidoidea). *Agronomia Lusitana* 51, 19–21.

Inaizumi, M. 1981. Life cycle of *Aphis gossypii* Glover (Homoptera, Aphididae) with special reference to biotype differentiation on various host plants. *Kontyu* 49, 219–40.

Kavallieratos, N. G., Athanassiou, C. G., Stathas, G. J. and Tomatovic, Ž 2002. Aphid parasitoids (Hymenoptera: Braconidae: Aphidinae) on citrus: seasonal abundance, association with the species of host plant, and sampling indices. *Phytoparasitica* 30(4), 365–77. doi:10.1007/BF02979684.

Kindlmann, P. and Dixon, A. F. G. 1994. Evolution of host range in aphids. *European Journal of Entomology* 91, 91–6.

Kolokytha, P. and Sterk, G. 2017. Side effects of FLiPPER/Relevant (carboxylic acid potassium salts) on beneficial agents. *IOBC-WPRS Bulletin* 125, 36–40.

Komazaki, S., Sagakami, Y. and Korenaga, R. 1979. Overwintering of aphids on citrus trees. *Japanese Journal of Applied Entomology and Zoology* 23(4), 246–50. doi:10.1303/jjaez.23.246.

Lambert, P. and Pascal, T. 2011. Mapping *Rm2* gene conferring resistance to the green peach aphid (*Myzus persicae* Sulzer) in the peach cultivar "Rubira®". *Tree Genetics and Genomes* 7(5), 1057–68. doi:10.1007/s11295-011-0394-2.

Lefevre, M., Frank, P., Olivares, J., Ricard, J.-M., Mandrin, J. F. and lavigne, C. 2017. Spider predation on rosy apple aphid in conventional, organic and insecticide-free orchards and its impact on aphid populations. *Biological Control* 104, 57–65. doi:10.1016/j.biocontrol.2016.10.009.

Leroy, P. D., Verheggen, F. J., Capella, Q., Francis, F. and Haubruge, E. 2010. An introduction device for the aphidophagous hoverfly *Episyrphus balteatus* (De Geer) (Diptera: Syrphidae). *Biological Control* 54(3), 181–8. doi:10.1016/j.biocontrol.2010.05.006.

Loxdale, H. D., Edwards, O., Tagu, D. and Vorburger, C. 2017. Population genetic issues: new insights using conventional molecular markers and genomics tools. In: van Emden, H. F. and Harrington, R. (Eds), *Aphid as Crop Pests* (2nd edn.). CAB International, Wallingford, UK, pp. 50–80.

Maharaj, S. B. and da Graça, J. V. 1989. Transmission of Citrus vein enation virus by *Toxoptera citricidus*. *Phytophylactica* 21, 81–2.

Manachini, B., Casati, P., Aliverti, I. and Cinami, L. 2004. Transmission of PPV-M to Prunus perscicae by *Brachycaudus schwartzi* and *Phorodon humuli* (Hem., Aphididae). *Journal of Applied Entomology* 128(9–10), 677–80. doi:10.1111/j.1439-0418.2004.00908.x.

Martínez-Ferrer, M. T. and Campos-Rivela, J. M. 2017 Diversity, spatial distribution, and sampling for ant management decision-making in integrated pest management programs in citrus groves. *Entomologia Experimentalis et Applicata* 162(2), 251–60. doi:10.1111/eea.12535.

Massimino Cocuzza, G. E. and Rapisarda, C. 2017. Integrated pest management in citrus. In: Rapisarda, C. and Massimino Cocuzza, G. E. (Eds), *Integrated Pest Management in Tropical Regions*. CAB International, Wallingford, UK, pp. 246–69.

Michaud, J. P. 1998. A review of the literature on *Toxoptera citricida* (Kirkaldy) (Homoptera: Aphididae). *The Florida Entomologist* 81(1), 39–61. doi:10.2307/3495995.

Moreno, P., Ambró, S. S., Albiach-Martí, M. R., Guerrí, J. And Peña, L. 2008. *Citrus tristeza virus*: a pathogen that changed the course of the citrus industry. *Molecular Plant Pathology* 9(2), 251–68. doi:10.1111/j.1364-3703.2007.00455.x.

Moreno, A., Fereres, A. and Cambra, M. 2009. Quantitative estimation of plum pox virus targets acquired and transmitted by a single *Myzus persicae*. *Archives of Virology* 154(9), 1391–9. doi:10.1007/s00705-009-0450-5.

Mueller, T. F., Blommers, L. H. M. and Mols, P. J. M. 1992. Woolly apple aphid (*Eriosoma lanigerum* Hausm., Hom., Aphidae) parasitism by *Aphelinus mali* Hal. (Hym., Aphelinidae) in relation to host stage and host colony size, shape and location. *Journal of Applied Entomology* 114(1–5), 143–54. doi:10.1111/j.1439-0418.1992.tb01109.x.

Nalam, V., Louis, J. and Shah, J. 2019. Plant defense against aphids, the pest extraordinaire. *Plant Science: an International Journal of Experimental Plant Biology* 279, 96–107. doi:10.1016/j.plantsci.2018.04.027.

Nault, L. R. 1997. Arthropod transmission of plant viruses: a new synthesis. *Annals of the Entomological Society of America* 90(5), 521–41. doi:10.1093/aesa/90.5.521.

Önelge, N., Satar, S., Elibüyük, Ö., Bozan, O. and Kamberoğlu, M. 2011. Transmission studies on Citrus yellow vein clearing virus. In: *Proceeding 18th Conference, IOCV*. IOCV, Riverside.

Panini, M., Massimino Cocuzza, G., Dradi, D., Chiesa, O. and Mazzoni, E. 2017. First report of *Myzus mumecola* (Matsumara, 1917) in Europe. *Bulletin OEPP/EPPO Bulletin* 47(1), 107–10. doi:10.1111/epp.12357.

Pascal, T., Pfeiffer, F., Kervella, J., Lacroze, J. P., Sauge, M. H. and Weber, W. 2002. Inheritance of green peach aphid resistance in the peach cultivar 'Rubira'. *Plant Breeding* 121(5), 459–61. doi:10.1046/j.1439-0523.2002.734333.x.

Pekas, A., Tena, A., Aguilar, A. and Garcia Marí, F. 2011. Spatio-temporal patterns and interactions with honeydew-producing Hemiptera of ants in a Mediterranean citrus orchard. *Agricultural and Forest Entomology* 13(1), 89–97. doi:10.1111/j.1461-9563.2010.00501.x.

Qubbaj, T., Reineke, A. and Zebitz, C. P. W. 2005. Molecular interactions between rosy apple aphids, *Dysaphis plantaginea*, and resistant and susceptible cultivars of its primary host *Malus domestica*. *Entomologia Experimentalis et Applicata* 115(1), 145–52. doi:10.1111/j.1570-7458.2005.00255.x.

Rakauskas, R., Havelka, J. and Zaremba, A. 2015. Plum (*Prunus* spp.) aphid guild structure in Lithuania (Hemiptera: Sternorrhyncha, Aphididae): any impact of an exotic aphid species? *Zemdirbystė-Agriculture* 102(1), 81–6. doi:10.13080/z-a.2015.102.010.

Remaudière, G. and Remaudière, M. 1997. *Catalogue des Aphididae du monde (Homoptera Aphidoidea) – Catalogue of the World's Aphididae*. INRA, Paris, 437p.

Reynolds, D. R. 1999. *Capnodium citri*: the sooty mold fungi comprising the taxon concept. *Mycopathologia* 148(3), 141–7. doi:10.1023/A:1007170504903.

Rousselin, A., Bevacqua, D., Sauge, M. H., Lescourret, F., Mody, K. and Jordan, M. O. 2017. Harnessing the aphid life cycle to reduce insectide reliance in apple and peach orchards. A review. *Agronomy for Sustainable Development* 37(5), 38. doi:10.1007/s13593-017-0444-8.

Sandanayaka, W. R. M., Bus, V. G. M. and Connolly, P. 2005. Mechanisms of woolly aphid [*Eriosoma lanigerum* (Hausm.)] resistance in apple. *Journal of Applied Entomology* 129(9–10), 534–41. doi:10.1111/j.1439-0418.2005.01004.x.

Satar, S., Satar, G., Karacaoglum, M., Uygun, N., Kavallieratos, N. G., Starý, P. and Athanassiou, C. G. 2014. Parasitoids and hyperparasitoids (Hymenoptera) on aphids (Hemiptera) infesting citrus in east Mediterranean region of Turkey. *Journal of Insect Science* 14(1), 1–6. doi:10.1093/jisesa/ieu040.

Sauge, M.-H., Lacroze, J.-P., Poessel, J.-L., Pascal, T. and Kervella, J. 2002. Induced resistance by *Myzus persicae* in the peach cultivar 'Rubira'. *Entomologia Experimentalis et Applicata* 102(1), 29–37. doi:10.1046/j.1570-7458.2002.00922.x.

Sauge, M.-H., Mus, F., Lacroze, J.-P., Pascal, T., Kervella, J. and Poëssel, J.-L. 2006. Genotypic variation in induced resistance and induced susceptibility in the peach-*Myzus persicae* aphid system. *Oikos* 113(2), 305–13. doi:10.1111/j.2006.0030-1299.14250.x.

Sauge, M.-H., Lambert, P. and Pascal, T. 2012. Co-localisation of host plant resistance QTLs affecting the performance and feeding behaviour of the aphid *Myzus persicae* in the peach tree. *Heredity* 108(3), 292–301. doi:10.1038/hdy.2011.74.

Schwartzberg, E. G. and Tumlinson, J. H. 2014. Aphid honeydew alters plant defence responses. *Functional Ecology* 28(2), 386–94. doi:10.1111/1365-2435.12182.

Simon-Delso, N., Amaral-Rogers, V., Belzunces, L. P., Bonmatin, J. M., Chagnon, M., Downs, C., Furlan, L., Gibbons, D. W., Giorio, C., Girolami, V., Goulson, D., Kreutzweiser, D. P., Krupke, C. H., Liess, M., Long, E., McField, M., Mineau, P. D., Mitchell, E. A., Morrissey, C. A., Noome, D. A., Pisa, L., Settele, J., Stark, J. D., Tapparo, A., Van

Dyck, H., Van Praagh, J., Van der Sluijs, J. P., Whitehorn, P. R. and Wiemers, M. 2015. Systemic insecticides (neonicotinoids and fipronil): trends, uses, mode of action and metabolites. *Environmental Science and Pollution Research* 22(1), 5–34. doi:10.1007/s11356-014-3470-y.

Smith, C. M. and Chuang, W. P. 2014. Plant resistance to aphid feeding: behavioural, physiological, genetic and molecular cues regulate aphid host selection and feeding. *Pest Management Science* 70(4), 528–40. doi:10.1002/ps.3689.

Snyder, W. E. and Wise, D. H. 1999. Predator interference and the establishment of generalist predator populations for biocontrol. *Biological Control* 15(3), 283–92. doi:10.1006/bcon.1999.0723.

Starý, P., Lyon, J. P. and Leclant, F. 1988. Biocontrol of aphids by the introduced *Lysiphlebus testaceipes* (Cress.) (Hym., Aphidiidae) in Mediterranean France. *Journal of Applied Entomology* 105(1–5), 74–87. doi:10.1111/j.1439-0418.1988.tb00163.x.

Stevens, M. and Lacomme, C. 2017. Transmission of plant viruses. In: van Emden, H. F. and Harrington, R. (Eds), *Aphid as Crop Pests* (2nd edn.). CAB International, Wallingford, UK, pp. 323–61.

Stiling, P. 1993. Why do natural enemies fall in classical biological control programs? *American Entomologist* 39, 1–37.

Stoeckli, S., Mody, K., Gessler, C., Patocchi, A., Jermini, M. and Dorn, S. 2008. QTL analysis for aphid resistance and growth traits in apple. *Tree Genetics and Genomes* 4(4), 833–47. doi:10.1007/s11295-008-0156-y.

Stokwe, N. F. and Malan, A. P. 2016. Woolly apple aphid, *Eriosoma lanigerum* (Hausmann), in South Africa: biology and management practices, with focus on the potential use of entomopathogenic nematodes and fungi. *African Entomology* 24(2), 267–78. doi:10.4001/003.024.0267.

Sun, M., Voorips, R. E., Steenhuis-Broers, G., van't Westende, W. and Vosman, B. 2018. Reduced phloem uptake of *Myzus persicae* on an aphid resistant pepper accession. *BMC Plant Biology* 18(1), 138 doi:10.1186/s12870-018-1340-3.

Sunderland, K. D. and Samu, F. 2000. Effects of agricultural diversity on the abundance, distribution, and pest control potential of spiders: a review. *Entomologia Experimentalis et Applicata* 95(1), 1–13. doi:10.1046/j.1570-7458.2000.00635.x.

van Emden, H. F. 2017. Host-plant resistance. In: van Emden, H. F. and Harrington, R. (Eds), *Aphid as Crop Pests* (2nd edn.). CAB International, Wallingford, UK, pp. 515–32.

Verdugo, J. A., Méndez, T., Ortiz-Martinez, S. A., Cumsille, R. and Ramirez, C. C. 2012. Variation in resistance mechanisms to the green peach aphid among different *Prunus persica* commercial cultivars. *Journal of Economic Entomology* 105(5), 1844–55. doi:10.1603/ec12100.

Yamamoto, P. T., Bassanezi, R. B., Wulff, N. A., Santos, M. A., Sanches, A. L., Toloy, R. S., Gimenes-Fernandes, N., Ayres, A. J., Jesus, W. C., Nagata, T., Tanaka, F. A. O., Kitajima, E. W. and Bovè, J. M. 2011. Citrus sudden death is transmitted by graft inoculation and natural transmission is prevented by individual insect-proof cages. *Plant Disease* 95(2), 104–12. doi:10.1094/PDIS-04-10-0307.

Wang, Z.-H., Gong, Y.-J., Chen, J.-C., Su, X.-C., Cao, L.-J., Hoffman, A. A. and Wei, S.-J. 2018. Laboratory selection for resistance to sulfoxaflor and fitness costs in the green peach aphid *Myzus persicae*. *Journal of Asia-Pacific Entomology* 21(1), 408–12. doi:10.1016/j.aspen.2018.01.024.

Zhou, S. Q., Lou, Y. R., Tzin, V. and Jander, G. 2015. Alteration of plant primary metabolism in response to insect herbivory. *Plant Physiology* 169(3), 1488–98. doi:10.1104/pp.15.01405.

Zhou, H., Chen, L., Liu, Y., Chen, J. and Francis, F. 2016. Use of slow-release plant infochemicals to control aphids: a first investigation in a Belgian wheat field. *Scientific Reports* 6, 31552. doi:10.1038/srep31552.

Chapter 16

Integrated management of tortricid pests of tree fruit

Alan L. Knight, USDA-ARS, USA; Gary J. R. Judd, Agriculture and Agri-food Canada, Canada; Todd Gilligan, USDA-APHIS-PPQ, USA; Eduardo Fuentes-Contreras, Universidad de Talca, Chile; and William B. Walker III, Swedish University of Agricultural Sciences, Sweden

1 Introduction

A group of relatively small moths in the family Tortricidae is commonly the most destructive insect pests of deciduous tree fruits throughout the world (van der Geest and Evenhuis 1991). Within each geographical area there is typically a defined group of tortricids. This group can include key species whose larvae feed on fruit, externally or internally, and which drives the entire seasonal pest management programme, and any number of external feeders whose larvae may cause sporadic fruit injury and require occasional management. Successful control of tortricids can have a significant impact on the global trade

http://dx.doi.org/10.19103/AS.2019.0046.19
Published by Burleigh Dodds Science Publishing Limited, 2019.

of tree fruits and various phytosanitary regulations are in place to counter their accidental introduction into new areas.

Tactics employed to manage tortricids in tree fruits have most commonly relied on the repeated applications of insecticides from a relatively large number of mode-of-action classes. However, food safety concerns (which restrict rates, timing and maximum levels of post-harvest pesticide residues) have increasingly limited the options for insecticide use. Issues with the loss of pest susceptibility due to selection for resistance, and/or loss of product and entire insecticide class registrations due to increasing environmental and human health concerns, have been major factors in the redesigning of management programmes into the twenty-first century.

Initial attempts to find alternative methods of pest management proved problematic. Integrated pest management (IPM) and the broader concept of integrated fruit production (IFP) are paradigms pioneered in the 1970s to establish sustainable agriculture based on ecological and economic principles (Kogan 1998; Avilla and Riedl 2003). However, growers throughout the world who applied these concepts during the 1990s still experienced serious problems managing tortricids with the available arsenal of broad-spectrum insecticides and insect growth regulators (Varela et al. 1993; Sauphanor and Bouvier 1995; Pree et al. 1998; Ahmad et al. 2002). IPM programmes also failed to avoid secondary pest outbreaks from phytophagous mites, aphids, true bugs and leaf miners, further increasing the chemical inputs required to grow clean fruit (Prokopy and Croft 1994; Blommers 1994).

As a result, orchard pest management in the last decade of the twentieth century was in a serious state of flux driven by the evolution of pesticide resistance in the key tortricid pests, an unpredictable status of a suite of secondary pests, and the increasing regulatory restrictions on pest management forcing lower chemical residues, greater protection for workers and their families, and stricter containment of spray drift and surface water run-off (Beers et al. 2003; Weddle et al. 2009).

From a positive perspective these ecological and societal pressures also created new opportunities for the development of more selective pesticides that can increase the role of biological control (Jones et al. 2009a; Mills et al. 2016; Beers et al. 2016a,b), sprayer technologies (Fox et al. 2008), orchard architecture (Simon et al. 2006, 2007; Dorigoni 2016), monitoring tools and implementation of action thresholds to avoid unnecessary perturbations to the ecosystem (Knight and Light 2005a,b; Knight et al. 2014), partial replacement of pesticides with sex pheromone-based mating disruption (Miller and Gut 2015; Ioriatti and Lucchi 2016) and the rapid dissemination of expert knowledge of how to best implement IPM via sophisticated computer-aided decision frameworks (Jones et al. 2009b; Damos 2015). This chapter

Published by Burleigh Dodds Science Publishing Limited, 2019.

summarizes the recent literature associated with several applied aspects of tortricid management.

2 Tortricid systematics and general biology

The Tortricidae is a single family contained within the superfamily Tortricoidea. With approximately 11 400 described species in 1170 genera (Brown 2005; Gilligan et al. 2018) it is one of the largest families of microlepidoptera, and nearly 1000 species have achieved pest status (Zhang 1994). Tortricid adults are small moths (wingspan up to 35 mm) recognized by a combination of morphological characters, such as well-developed ocelli and chaetosema above the compound eye, an unscaled proboscis and horizontal labial palpi (Horak 1991). The structure of the flat ovipositor lobes in the female is the only character that unites the entire family (Horak and Brown 1991). Recent studies have used molecular data to reconstruct robust phylogenies for the family at the tribal level (Regier et al. 2012; Fagua et al. 2016). Tortricids are distributed worldwide, with the greatest species diversity found in the Neotropics (Regier et al. 2012). Most economically important species are found in two tribes: Grapholitini (Olethreutinae) and Archipini (Tortricinae).

Tortricid biology varies substantially. Larvae feed on foliage, in galls, bore into root stems and fruits, feed on seeds and flowers, on detritus or leaf litter, and a few species are predatory (Brown et al. 2008). Pest species can be categorized as external (most Archipini) or internal fruit feeders (most Grapholitini). External feeders construct a silken nest of rolled or folded leaves and feed on leaves, buds, flowers and sometimes directly on fruit. Economic damage is caused by their indirect feeding on fruit that causes premature fruit drop during the season or leads to secondary, post-harvest microbial infections on injured fruits. Larvae of internal feeders may feed on leaves or in stems in early instars but will eventually bore directly into fruit and feed on the seeds. Larvae of most tortricids complete four to six instar stages and overwintering can occur as an egg, larva or prepupa. Larvae are solitary feeders though more than one internal feeding larva can occur in fruits at high pest densities. Following eclosion adults are reproductively mature and sexual activity is typically limited by temperature and occurs during defined circadian periods, that is dusk through scotophase. Females lay eggs individually or as multiple egg masses and the realized fecundity can range from 50 to 800 eggs. Adults can be quite dispersive, and species can have a narrow to wide larval host range of deciduous shrubs and trees. Typically, there are a number of parasitoids attacking tortricids including tachinids and several hymenopteran families. Predation of immature stages by a suite of generalist predators present in orchards including spiders, earwigs, beetles and lacewings can be important.

3 Key species, distribution and dispersal mechanisms

The importance of tortricid pests in tree fruits can be examined by considering four key species:

- codling moth (CM), *Cydia pomonella* (L.)
- oriental fruit moth (OFM), *Grapholita molesta* (Busck)
- light brown apple moth (LBAM), *Epiphyas postvittana* (Walker)
- obliquebanded leafroller (OBLR), *Choristoneura rosaceana* (Harris)

The first two species are internal fruit feeders in the tribe Grapholitini (Olethreutinae) with a worldwide distribution. OFM can also complete its development inside young shoots (Myers et al. 2006). Their management is the keystone of seasonal programmes and a driver of phytosanitary regulations and world trade of pome and stone fruits (Willett et al. 2009; Drogue and DeMaria 2012; Neven et al. 2018).

CM is apparently absent from Japan, Taiwan and Korea despite the climatic suitability of portions of the tree fruit production regions (Kumar et al. 2015). CM has been purported to be eradicated from Brazil through an organized urban host removal and pest management programme (Kovaleski and Mumford 2007). In contrast, the pest distribution of CM in China, by far the largest tree fruit producer, has increased across seven provinces (Men et al. 2013). OFM is originally from Asia (Kirk et al. 2013) but its distribution is now worldwide and in some geographical areas it has become a serious pest of pome fruits (Kovanci et al. 2004; Il'ichev et al. 2004). Geographical range expansion of OFM into new production areas such as eastern Washington State could be impacted by climate change (Neven et al. 2018).

LBAM and OBLR are in the tribe Archipini (Tortricinae) and are primarily leaf feeders with broad host ranges. LBAM originated from Australia and has been recorded from more than 500 plant species in 121 families (Brockerhoff et al. 2011). Recently, LBAM was accidentally introduced (hypothesized via importation of infested plant nursery material) into several new production areas, including California (Suckling et al. 2014). Eradication of LBAM from California with aerial applications of sex pheromone was attempted in 2007, but was terminated following public health concerns (Carey and Harder 2013). OBLR is the largest tortricid in North America and the adult dispersal into managed crops from surrounding unmanaged host plants has contributed to its importance in pome and stone fruits in both the humid eastern and arid western orchards (Chapman 1973; Evenden and Judd 1999).

Adult dispersal capacity is a variable trait with significant heritability in CM (Schumacher et al. 1997a), which can be selected under laboratory conditions and correlates with dispersal distance in the field (Keil et al. 2001a). No major differences in flight capacity between males and females were found under laboratory conditions (Schumacher et al. 1997a), but higher flight activity was found at different ages according to sex (Keil et al. 2001b). Flight activity is higher in mated than virgin females and increases during the egg-laying period; therefore, no dispersal will take place before mating and the oogenesis-flight syndrome is not present in this species (Schumacher et al. 1997b). Dispersal capacity is higher in individuals with smaller body weight associated with a significant reduction in fecundity, which suggests a trade-off between dispersal and fitness (Gu et al. 2006). Such findings indicate that CM populations are formed by a majority of sedentary individuals with low dispersal capacity and short dispersal flights, but a small proportion of dispersing individuals with smaller body sizes and lower fecundities are present and can undertake long-distance flights (Gu et al. 2006).

Spatial analyses of tortricid dispersal have mostly considered males caught in sex pheromone-baited traps (Mani and Wildbolz 1977; Sciarretta and Trematerra 2006; Basoalto et al. 2010; Margaritopoulos et al. 2012). Dispersal by females has been measured indirectly through the spread of fruit damage between generations (Knight et al. 1995), but more directly by using microsatellite markers and sibship analysis (Franck et al. 2011). The latter authors found that females mainly clustered their eggs among nearby edge trees within the same orchard (ca. 8 m away) and showed the mean dispersal distance between two oviposition sites was 30 m and the median 10 m. Geostatistical tools have been used to estimate the influence of landscape attributes on trap catch (Trematerra et al. 2004; Basoalto et al. 2010; Comas et al. 2012). A field technique which involves marking adults with low cost crude food proteins (milk, egg, wheat or soybean) and then releasing and re-capturing has been developed to study dispersal (Jones et al. 2006a,b; Basoalto et al. 2010). Overall, these studies show that CM adults can move a few 100 m between generations.

The characteristics of moth movement among host patches is important for monitoring and for optimal selection of tactics. Traps placed near orchard borders facilitate checking and are more effective indicators of pest immigration and local injury levels (Knight 2007b; Merckx et al. 2009). Dispersal from unmanaged surroundings into managed orchards is also of great relevance for insecticide resistance evolution (Fuentes-Contreras et al. 2007, 2008). Dispersal of adult CM among orchards and unmanaged hosts is affected by area and connectivity between habitat patches (Trematerra et al. 2004; Garcia-Salazar et al. 2007; Tyson et al. 2007; Ricci et al. 2009, 2011), but other factors such as patch resource quality and population density may also be relevant. For example, unmanaged host plants frequently show alternate fruit bearing, and therefore likely strongly influence seasonal dispersal patterns between habitat patches (Gu et al. 2006).

4 Insecticide programmes

Registered insecticides from 13 mode-of-action classes (Elbert et al. 2007) remain the primary approach to manage tortricids in orchards (Damos et al. 2015). Available chemistries include broad-spectrum and more selective synthetic materials, as well as microbial insecticides (Cox et al. 1995; Dhadialla et al. 1998; Thompson et al. 2000; McCann et al. 2001; Carlson et al. 2001; Charmillot et al. 2001; Lahm et al. 2005; Dripps et al. 2008; Lacey and Shapiro-Ilan 2008; Casida 2018). Sprays are primarily targeted against the neonate larvae of the internal feeders to prevent fruit injury. Unfortunately, insecticides which primarily rely on oral uptake typically have only marginal effectiveness against internal feeders, and timing is critical to contact larvae before they bore into plant tissue (Cox et al. 1995; Ballard et al. 2000a). Tortricid larvae feeding inside rolled leaves are more exposed and the spray timing is less critical (Jones et al. 2005). Some insecticide classes are also effective as ovicides (Charmillot and Pasquier 1992; Borchert et al. 2004; Gökce et al. 2009; Magalhaes and Walgenbach 2011). Activity against the adult stages can be significant though sprays are not typically timed for this life stage (Knight 2010a; Navarro-Roldán et al. 2017). Regardless, spray coverage is a key factor affecting the efficacy of insecticides and growers have adopted specialized sprayers to both minimize drift and maximize coverage (Panneton et al. 2005). Border sprays can be used to supplement programmes on uphill slopes and borders with proximity to extra-orchard populations (Trimble and Vickers 2000). The selectivity of insecticides against the natural enemies of tortricids, as well as their impact on key natural enemies of secondary pests, has allowed development of programme strategies that optimize the role of biological control during the season (Valentine et al. 1996; Beers et al. 2005; Biondi et al. 2012; Mills et al. 2016).

Insecticide-based management programmes are inherently threatened by the evolution of pest resistance (Roush and Tabashnik 1990). Levels of resistance to deltamethrin in populations of CM were found to be nearly 1000-fold in southern France (Sauphanor et al. 1997) and 10 000-fold to granulosis virus in Europe (Asser-Kaiser et al. 2007); yet, resistance levels <10-fold were associated with CM control failures in organophosphate-based pear programmes in California (Varela et al. 1993). The history of insecticide usage in tree fruits, as well as other crops, has been characterized as a 'reactive' approach to sequential failures with one class of insecticides adopted as the older class failed (Onstad 2008). Loss of pest susceptibility to sprays has tremendous practical and economic impacts which can drive major transformations in pest management, including in tree fruits (Knight and Norton 1989; Kazmierczak et al. 1993).

Baseline field surveys for resistance in tortricid populations have been conducted with several classes of insecticides across most geographical areas

Published by Burleigh Dodds Science Publishing Limited, 2019.

of these pests using a variety of bioassay approaches (Knight et al. 1994, 2001a; Sauphanor et al. 2000; Pasquier and Charmillot 2003; Knight 2010b). In some cases significant differences among bioassay methods suggested the need to standardize methodologies (Sauphanor et al. 1998a; Reyes and Sauphanor 2008; Magalhaes et al. 2012). In response, standardized surveys have been conducted across large geographical areas (Reyes et al. 2007, 2009). To date, significant resistance levels in CM have been found to all but one class of insecticides (anthranilic diamides) (detailed by Bosch et al. 2018). The susceptibility of OBLR to the newest classes of insecticides is more variable (Sial and Brunner 2010). In addition, laboratory experiments with OBLR demonstrated that resistance to chlorantraniliprole can be selected for over 12 generations, but reverted back to baseline levels after removing selection for five to six generations (Sial and Brunner 2012).

Growers have been fortunate to be able to switch to other classes of insecticides when resistance occurs, but the frequent occurrence of cross-resistance can severely limit these options (Sauphanor and Bouvier 1995; Sauphanor et al. 1998b; Dunley and Welter 2000; Smirle et al. 2002; Dunley et al. 2006; Reyes et al. 2007; Mota-Sanchez et al. 2008; Knight 2010b; Voudouris et al. 2011). Nevertheless, organized resistance management practices have been developed and implemented such that insecticides with different modes of action are rotated or limited to a single generation (Sparks and Nauen 2015). These efforts are noteworthy, but likely the further integration of management programmes with mating disruption, microbial insecticides and physical orchard manipulations will remain the most effective approach to maintain pest susceptibility to insecticides (Kienzle et al. 2003).

Numerous studies have examined the mechanisms associated with insecticide resistance in CM, including both enhanced enzymatic detoxification response (cytochrome P450 monooxygenases, glutathione transferases and esterases) (Rodriguez et al. 2011) and target site mutations (Reyes et al. 2007). A few point source mutations have also been detected including acetylcholinesterase (Cassanelli et al. 2006) and a knockdown resistance (kdr) mutation (Brun-Barale et al. 2005). Various studies have found that development of resistance has a fitness cost to adult tortricids (Carrière et al. 1994; Boivin et al. 2001, 2003, 2004; Konopka et al. 2012; Navarro-Roldán and Gemeno 2017). Changes in sex pheromone-based communication associated with insecticide resistance has also been shown (El-Sayed et al. 2001; Delisle and Vincent 2002; Trimble et al. 2004). Pear ester ((*E, Z*))-2,4-decadienoate was identified as a potent kairomone attractive to both sexes of CM and neonate larvae (Light et al. 2001; Knight and Light 2001). Sauphanor et al. (2007) suggested that adult CM's response to pear ester, in France, was greater in insecticide-resistant populations, but this was not supported in a later Spanish study (Bosch et al. 2015). A variety of sublethal effects on larvae and adults have been documented with the newer classes of insecticides which could have significant impacts on the management of tortricids (Sun and Barrett 1999; Sun et al. 2000; Borchert et al. 2004, 2005; Knight and Flexner 2007; Barrett 2008; Navarro-Roldán and Gemeno 2017). For example, levels of mating in CM field populations were found to be lower in blocks treated with the diamide, chlorantraniliprole, than in similar blocks treated with sex pheromones for mating disruption (Knight and Flexner 2007). The observed effects on calling behaviour and mating success detailed in this chapter were consistent with the primary mechanism of diamides and the depletion of intracellular calcium stores in muscle cells interfering with muscle regulation causing paralysis and ultimately death (Cordova et al. 2005).

5 Insecticide use in organic tree fruit production

The proportion of tree fruit production under certified organic regulation continues to grow around the world due to an increasing consumer demand and favourable economics (Marliac et al. 2015; Granatstein et al. 2016). This is made possible by a greater number of tools and natural products to control key tortricids. Nevertheless, tortricid management in organic orchards has a much more limited set of insecticides than is available for conventional growers (Knight 1994; Mills 2013).

Various formulations of the gram-positive bacterium *Bacillus thuringiensis* (*Bt*) (subsp. *kurstaki* and subsp. *aizawai*) contain different groups of toxic Cry proteins and are widely used against leaf rollers (Knight 1998a). Effective uptake of Bt by leaf roller larvae early in the season is strongly impacted by temperature. Sublethal exposures to Bt can shift a population's phenology and a subsequent spray's effectiveness (Knight 1997a). Sprays applied in the spring can be timed to minimize the negative effect against the various parasitoids utilizing the larval hosts (Cossentine et al. 2003). Various feeding stimulants have been used to enhance the effectiveness of Bt (Farrar and Ridgway 1995; Pszczolkowski et al. 2004). Cry proteins are also lethal when ingested by internal fruit feeders (Atanassov et al. 2002; Boncheva et al. 2006) and are used alone or in combination with CpGV in some geographical regions, that is Argentina. Resistance to Bt has been selected for in a laboratory colony of LBAM (Harris et al. 2006), but it has not been documented in field populations with any tortricid. Incorporation of Bt genes into transgenic apple has been discussed in the literature and privately demonstrated under quarantine, but has not been developed commercially (Wearing and Hokkaner 1994).

Spinosyns, neuroactive insecticides attacking nicotinic acetylcholine receptors (Biondi et al. 2012), are allowed in certified organic orchards because they are derived from fermentation of the actinomycete bacterium, *Saccharopolyspora spinosa* Mertz & Yao, and are highly effective targeting larval tortricids (Smirle et al. 2003; Arthurs et al. 2007a). However, cross-resistance to spinosad in CM (Knight 2010b) and OBLR has been noted (Ahmad et al. 2002; Smirle et al. 2003; Sial and Brunner 2010). Repeated use of this material has also been associated with outbreaks of secondary pests, such as aphids leading many organic growers to avoid its overuse (Delate et al. 2008; Biondi et al. 2012). In addition, the total seasonal use of spinosad is limited and the organic formulation has a relatively short residual effectiveness against tortricids when compared with newer formulations of this class of insecticide (Smirle et al. 2003; Depalo et al. 2016). The addition of feeding stimulants, such as monosodium glutamate or a monosodium glutamate receptor agonist can significantly improve spinosad activity against leaf-feeding tortricids; however, this approach has not been commercially developed (Pszczolkowski and Brown 2002, 2004).

Commercial formulations of the CM granulosis virus (CpGV) are widely used (estimated at 150 000 ha annually) in organic and IFP orchards (Lacey and Shapiro-Ilan 2008). Unfortunately, the effectiveness of CpGV is limited by its sensitivity to ultraviolet (280–320 nm) radiation and the need for larval ingestion (Lacey et al. 2004). Growers in Washington State can apply up to 14 applications of the virus during the season and achieve only moderately effective fruit protection within a season (Arthurs et al. 2005). The repeated use of CpGV over a number of years can be effective in reducing populations (Lacey and Shapiro-Ilan 2008).

Efforts to protect the virus from solar radiation have tried various adjuvants but improvements have been marginal (Ballard et al. 2000b; Arthurs et al. 2008). A second approach to improve the performance of CpGV has been the addition of feeding stimulants, and have included sugary baits, non-nutritive sugar substitutes, that is monosodium glutamate, and the amino acid, L-aspartate (Ballard et al. 2000a,b; Pszczolkowski and Brown 2002, 2004; Arnault et al. 2016). Results from several studies examining the potential of using pear ester in both apple and pear to enhance CpGV have been inconsistent (Arthurs et al. 2007b; Schmidt et al. 2008; Knight et al. 2015). However, field studies showed that the activity of CpGV could be enhanced by adding either the common brewer yeast *Saccharomyces cerevisiae* Meyen ex E. C. Hansen, or one of several yeasts isolated from field-collected larvae in combination with brown cane sugar (Knight and Witzgall 2013; Knight et al. 2015). Importantly, the numbers of overwintering larvae on trees was significantly reduced following a seasonal programme of CpGV plus yeasts and sugar in this latter study. Achieving reductions in fruit injury was more problematic with all adjuvants, as larvae commonly caused small 'stings' to the fruit and levels of unblemished fruit were not increased with the additives.

High levels of field resistance to CpGV have been detected in some European populations (Fritsch et al. 2005; Sauphanor et al. 2006). Unfortunately, this resistance is stable and does not confer a fitness cost (Undorf-Spahn et al. 2012). This resistance to commercial isolates of CpGV was overcome by identifying new effective virus isolates (Berling et al. 2009). CpGV isolates can also exhibit activity for OFM (Lacey et al. 2005; Zingg et al. 2012). Viruses have been isolated from other tortricids, but none have been commercialized (MacCollom and Reed 1971; Pronier et al. 2002).

Horticultural oil is widely used alone and as an adjuvant (spreader, penetrant) with various insecticide sprays targeting a range of soft-bodied arthropod pests, for example aphids, psyllids, scales and mites (Davidson et al. 1991; Hilton et al. 1992). Oil can also be effective as a topical treatment of tortricid eggs via suffocation (Riedl 1995; Fernández et al. 2005). Modern summer spray oils are highly refined with a narrow molecular weight range to minimize phytotoxic effects (Davidson et al. 1991). However, growers generally avoid their overuse and must consider the sensitivity of plant tissue based on air temperature, rate and fruit sensitivity (Willett and Westigard 1988; Hilton et al. 1992).

Various plant extracts and essential oils have been evaluated to control tortricids (Larocque et al. 1999; Landolt et al. 1999; Lowery and Smirle 2000; Machial et al. 2010; Durden et al. 2010; Pszczolkowski et al. 2011). One focus has been on the potential role of water soluble sugar metabolites on oviposition by CM (Lombarkia and Derridj 2002). Significant differences in CM oviposition across several apple cultivars was correlated with the cultivars' different sugar and sugar alcohol levels (Lombarkia and Derridj 2008). The addition of sugars with a neonicotinyl insecticide was evaluated and the observed reduction in fruit injury was associated with reduced oviposition (Arnault et al. 2016). Studies suggested that some of the activity of CpGV is due to reductions in oviposition associated with changes in leaf chemical signals (Lombarkia et al. 2013).

Pear ester sprayed on leaves or fruit causes increased larval wandering and longer arrestment times which can both extend the topical exposure of larvae to insecticides (Light and Beck 2010, 2012). Greater exposure to insecticides when combined with pear ester could allow managers to reduce rates of insecticides, boost the effectiveness of some classes of insecticides, and ameliorate the effects of poor spray coverage. Choice and no-choice bioassays demonstrated that pear ester stimulated CM oviposition (Knight and Light 2004) and increased the median distance eggs were laid from fruits which was hypothesized to contribute to the observed reduced fruit injury (Pasqualini et al. 2005a,b). A microencapsulated (MEC) formulation of pear ester (CideTrak DA-MEC, 5% A.I.) has been registered for conventional and organic production and has been carefully characterized, including capsule density, size range, emission rates and shown to have a residual attractiveness of 14 days (Light and Beck 2010). Various field trials were conducted that added pear ester to different classes of insecticides with a range of residual and oral route of exposures in both walnut, *Juglans regia* L. (Light and Knight 2011) and apple (Knight and Light 2013). In general, pear ester significantly improved the performance of synthetic insecticides with residual effectiveness and also reduced nut injury by the navel orangeworm, *Amyelois transitella* Walker, in walnut (Light and Knight 2011). Results with pear ester in both laboratory and field trials with insecticides requiring ingestion even when used with purported feeding stimulants have been more variable (Arthurs et al. 2007b; Schmidt et al. 2008; Light and Knight 2011; Knight and Witzgall 2013; Knight et al. 2015).

6 Physical crop protection

Particle films are widely used in apple production for a number of management objectives. Originally, the use of a hydrophobic kaolin formulation was developed to broadly suppress pests and diseases in tree fruits (Glenn et al. 1999). Subsequent studies showed that particle films could enhance plant growth, fruit yield and quality, in pome and stone fruits (Glenn et al. 2001; Lalancette et al. 2004). A liquid formulation was subsequently developed, and growers created specific uses for this, including early season management of pear psylla, *Cacopsylla pyricola* (Foerster), in pear orchards (Puterka et al. 2005), and mid- to late-season use in apple to reduce heat stress (Glenn et al. 2002). Studies showed that particle films could also suppress CM and OBLR populations (Unruh et al. 2000; Knight et al. 2000a; Sackett et al. 2005) and impact moth movement within and between orchards (Jones et al. 2006a,b). Unfortunately, seasonal use of kaolin films has been associated with several pest problems due to disruption of biological control of pests, such as leaf miners and phytophagous mites (Knight et al. 2000b; Sackett et al. 2007; Marko et al. 2008; Bostanian and Racette 2008).

Two additional physical methods have been considered and used on a limited basis to manage tortricids: bagging individual fruits on trees (Sharma et al. 2014; Zheng et al. 2015) and the use of overhead watering to disrupt moth flight, oviposition, and egg and larval survivorship (Knight 1998b). Bagging is obviously a highly labour-intensive method and likely has no practical use in commercial fruit production in developed countries, except very niche markets, that is special gifts in Japan. Overhead watering has been used for evaporative cooling of orchards to minimize fruit sunburn (Evans et al. 1995), and levels of fruit injury from CM could be reduced by up to 90% with daily watering. This use of water, however, had severe limitations on the availability of water to irrigate blocks concurrently, and the build-up of mineral deposits on fruits. The use of water for sunburn protection has declined as growers made greater use of netting and particle films (Gindaba and Wand 2005).

The most practical and effective physical approach to manage tortricids is likely the use of exclusion netting (Lawson et al. 1994). Originally, overhead netting of orchards provided insurance for rare hail events or was used before harvest in select orchards prone to bird predation (Whitaker and Middleton 1999). The high cost of installation compared with these risk factors limited its use. However, observations that overhead nets could also reduce fruit injury from tortricids dramatically improved the economics of installing nets (Graf et al. 1999). The occurrence of resistance to CpGV forced many growers in southern France to adopt the use of single row hail nets with side walls (Alt'carpo) to manage CM (Severac and Romet 2008). Subsequent studies have shown how netting can interfere with the sexual activity and flight of CM (Tasin et al. 2008; Sauphanor et al. 2012; Ioriatti and Tasin 2018). The occurrence of nets in eastern Washington appears to be expanding rapidly due to this effect and to protect fruits from sunburn while enhancing tree physiology (Lloyd et al. 2005; Bastias et al. 2012). Unfortunately, the use of netting may exacerbate management of other pests such as aphids (Dib et al. 2010). It is exciting to hypothesize that future tortricid management programmes could require very few chemical inputs due to the integration of netting, MD and the release of natural enemies (Chouinard et al. 2016).

The physical structure of tree canopies and entire orchards can have significant effects on the management of pests, including tortricids (Stoeckli et al. 2008; Costes et al. 2013). Canopy architecture directly impacts spray coverage and the occurrence of refugia within orchards (Xu et al. 2006; Duga et al. 2015). Tree size and canopy porosity affects the distribution of microclimates and the phenology of pests (Kührt et al. 2006). Fruit trees are complex perennial resources and canopies can be manipulated to minimize pest incidence and maximize natural enemies (Simon et al. 2006, 2007). The use of cover crops in orchards to enhance biological control of various pests has been considered (Fernández et al. 2008; Bone et al. 2009; Mullinix et al. 2010; Marko et al. 2012), including to benefit natural enemies of tortricids (Sarvary et al. 2010a,b; Mullinix et al. 2011; Song et al. 2014).

7 Biological control

The transition of tree fruit production to the use of selective insecticides has greatly increased the role of biological control to suppress tortricid populations (Beers et al. 2016a). However, the importance of biological control varies widely between the internal fruit feeders and the foliage feeders. The use of natural enemies for the key pests, CM and OFM, has not been relied on for several reasons (Mills 2005):

- the low threshold for fruit injury
- internal larval feeding which reduces the availability of susceptible life stages to parasitoids and predators

- the widespread use of pesticide sprays on tree fruits
- the paucity of hosts immediately surrounding orchards

Moderate levels of parasitism of CM occurs in unsprayed habitats from wasps that attack the egg stage, such as *Trichogramma* spp. and the egg-larval braconid, *Ascogaster quadridentata* Viereck (DeLury et al. 1999; Suckling et al. 2002). Inundative releases of *Trichogramma* spp. have had only marginal success due to the biology of the parasitoids, the quality of mass-reared insects and the significant pesticide residues in orchards (Lawson et al. 1997; McDougall and Mills 1997; Mills et al. 2000; Mills 2003; Brunner et al. 2005). One novel integrated effort demonstrated that egg parasitoid populations could be sustained after a single release in orchards via releases of sterile moths, with parasitoids developing from the sterile eggs laid in the orchard (Bloem et al. 1998). Significant classical biological control efforts importing parasitoids from Kazakhstan have been attempted and a few species have been disseminated to other fruit-producing regions (Unruh 1998; Kuhlmann and Mills 1999). The biology of the ichneumonid, *Mastrus ridibundus* (Gravenhorst), an ectoparasitoid that attacks cocooned larvae, has been intensively studied (Bezemer and Mills 2001; Hougardy et al. 2005; Devotto et al. 2010). Predation of CM eggs and diapausing larvae is likely a more important aspect of biological control (Riddick and Mills 1994; Knight et al. 1997; Monteiro et al. 2012; Boreau de Roince et al. 2012) and gut content analysis has been used to characterize the major predators, that is carabids, spiders, earwigs and harvestmen (Unruh et al. 2016).

The use of more selective insecticides has also expanded the importance of natural enemies of leaf rollers (Vakenti et al. 2001; Cossentine et al. 2004; Pluciennik and Olszak 2010; Monteiro et al. 2012). In some regions the combined activity of these guilds has likely contributed to significant declines in their pest status (Blommers 1994; Lo et al. 2018). Interestingly, parasitism of LBAM introduced into California quickly rose to levels found in Australia (Burgi and Mills 2014; Burgi et al. 2015). Predation of eggs and larvae is often important (MacLellan 1973; Miliczky and Calkins 2002). An ingenious conservation approach to establish a sustained role for biological control of leaf rollers involves the orchard planting of rose/strawberry gardens infested with the tortricid *Ancylis comptana* (Frölich), an alternate host for the eulophid parasitoid, *Colpoclypeus florus* (Walker) (Pfannenstiel et al. 2010; Unruh et al. 2012).

Parasitic nematodes have also been investigated to supplement management of tortricids (Lacey et al. 2006; Negrisoli et al. 2013). Their use requires careful handling to be effective, including selection of the best adapted nematode species for the local climate (Lacey et al. 2006). The required occurrence of suitable temperatures for nematode searching behaviours and

Published by Burleigh Dodds Science Publishing Limited, 2019.

maintenance of high humidity by wetting tree trunks are difficult factors to control during late season when larvae move into overwintering sites in the arid production areas of the world (Unruh and Lacey 2001; Lacey et al. 2006).

8 Mating disruption

Identification of tortricid sex pheromones has been a key driver in the design of effective management programmes allowing both improved decision-making for the need and timing of sprays and in direct control via disruption of mating (Witzgall et al. 2008; Ioriatti and Lucchi 2016). Grower adoption of various MD tactics, that is sprayable microencapsulated formulations, various hand-applied dispensers, and the use of aerosol emitters (Miller and Gut 2015) and has allowed significant reductions in insecticide applications (Knight 1995, 2008; Gut and Brunner 1998). Other tortricids have become more problematic following the initial adoption of MD for CM and associated reductions in broad-spectrum insecticides (Knight 1995; Walker and Welter 2001). The level of female mating under MD appears to be high for CM (Knight 2006, 2007a) and OBLR (Knight et al. 2017), and much lower with OFM (Cichon et al. 2013; Mujica et al. 2018). The effectiveness of MD may also be achieved through causing a delay in mating (Knight 1997b; Vickers 1997; Fraser and Trimble 2001; Foster and Howard 1999; Jones et al. 2008a). Sex pheromone dispensers for MD are a sizeable upfront cost for growers, but dual dispensers for CM and OFM (Stelinski et al. 2007, 2009) and CM and leaf rollers (Judd and Gardiner 2006a; Lo et al. 2013; Porcel et al. 2014) have been registered to help reduce application costs.

Pear ester has also been developed to improve mating disruption of CM (Light and Knight 2005). Studies have shown that the proportion of unmated female moths is significantly higher in blocks treated with various dispensers releasing both sex pheromone and pear ester than only sex pheromone (Knight et al. 2012a,b). The repeated applications of a microencapsulated formulation of pear ester to improve insecticidal activity (Knight and Light 2012) can also increase the level of MD in treated blocks (Knight and Light 2014).

9 Precision pest management

With the rapid development of GPS-driven applications including mapping, record keeping and variable-rate spray technology, new opportunities are available to further optimize aspects of orchard management (Sonka and Coaldrake 1997). Combined with new monitoring tools and the ability to process voluminous data remotely site-specific pest management seems plausible (Fleischer et al. 1997, 1999; Zijlstra et al. 2011). Development of more effective traps and lures including female attractants has improved correlations of moth catches with local pest levels (Knight et al. 2005, 2014; Knight and Light 2005b; Padilha et al. 2017). Standardized methodologies for trap density, placement within the orchard and proximity to MD dispensers are some of the important factors affecting the reliability of traps (Gut and Brunner 1996; Knight et al. 1999; Knight 2007b; Hsu et al. 2009). Kairomone blends including acetic acid combined with, either pear ester or nonatriene for CM (Landolt et al. 2007; Knight and Light 2012); 2-phenylethanol for OBLR and *Pandemis* spp. (Judd et al. 2017a; Knight et al. 2017); phenylacetonitrile for LBAM and eye-spotted bud moth, *Spilonota ocellana* (Denis and Schiffermüller) (El-Sayed et al. 2016; Judd et al. 2017b); and terpinyl acetate for OFM (Mujica et al. 2018) have significantly increased growers' ability to monitor moths even under MD programmes. Improved phenological monitoring, including more accurate models (Knight 2007c; Joshi et al. 2016), are more effectively timed with female moth catches (Knight and Light 2005c; Barros-Parada et al. 2015).

The ability to monitor female tortricid populations creates an opportunity to redesign orchard pest management (Light et al. 2001; Hughes et al. 2003). Precision management of CM can be defined as an approach that restricts the spatial application of sprays based on site-specific pest monitoring data (triggers), such as moth catch, that is conveniently matched to the coverage provided by spray tank capacity (Knight et al. 2009). The first step is to map the orchard and create management subunits surrounding each trap. Treatments are then applied to designated areas in response to the magnitude of moth catches and consideration of other factors, such as risk preference, pest history, status of other pests and biological control agents in the orchard, and the requirements imposed by other crop production practices. This approach has been demonstrated over several years to reduce total production costs through reduced spray applications. Additional benefits from reduced spraying may be derived from conservation of biological control agents (Knight et al. 2011). The orchard managers also appreciated the programme, because treating only portions of orchards allowed them to move labour crews around the spray teams within the same location and reduced the costs of moving labour and associated equipment to alternate orchard work locations during treatment and re-entry intervals. The result was less downtime and increased efficiencies in other production tasks. Development of smart traps, especially if multiple species can be monitored together, will likely facilitate this approach (Guarnieri et al. 2011; Kim et al. 2011).

A complicating factor in improving precision in pest management is climate change. The predicted scenario of a gradual warming trend across the major fruit production areas will impact tortricid pests by increasing their generation numbers per year and disrupting diapause induction and termination (Bale and Hayward 2010; Luedeling et al. 2011; Stoeckli et al. 2012). Phenological models are widely used with tortricid pests to target the most susceptible life stages (Knight 2007c). Development of female-based models using new bisexual attractants can reduce the error associated with predicting egg hatch (Knight and Light 2005c; Barros-Parada et al. 2015). The approach of using a 'Biofix' (first sustained male catch in sex pheromone-baited traps) to shorten the time between the start of male flight and egg hatch has been widely adopted (Riedl et al. 1976). However, the widespread use of MD and a general decline in the amplitude of moth flights has made selecting a Biofix more difficult (Jones et al. 2008b). Instead, predicting the timing of egg hatch with the accumulation of degree days from 1 January has been developed (Jones et al. 2013). Unfortunately, this method is particularly sensitive to the warming trends that are occurring. In particular, variable accumulations of degree days in the winter months can disrupt the typical spring emergence patterns of CM which could dramatically increase the error in predicting egg hatch (Neven et al. 2000). The magnitude of this error could be more severe in the warmer geographical areas supporting tree fruit, that is California in North America.

10 Area-wide IPM

Tortricid adults readily disperse among both managed agricultural habitats and a variety of surrounding unmanaged host plants (Jeanneret and Charmillot 1995; Knight 2001; Thistlewood and Judd 2003a; Ellis and Hull 2013; Ferrer et al. 2014). Urban encroachment into traditional agricultural areas affects the proximity of unmanaged hosts for tortricids impacting orchard management and production, for example homeowners want their own fruit trees but not the added labour of growing their trees to mitigate risks to commercial crops.

Given that many tortricids are mobile as both juveniles and adults, and both managed and unmanaged hosts occur within any given production region, the design of effective orchard management programmes should consider pest population dynamics across large areas, including neighbouring orchards and alternate hosts. Implementation of the first area-wide management programme for CM in the western United States (CAMP) demonstrated that increased sampling and monitoring efforts greatly diminished the use of insecticides (Calkins and Faust 2003). The programme in Washington State was very successful in terms of pest management, but more importantly, it served as a vehicle to quickly disseminate new knowledge to the grower community and extension agents, particularly as it related to MD (Knight 2008). The key lesson learned from the CAMP programme was that pest management requires attention, experience and skill to be effective (Knight 1999). Several operational and organizational factors contributed to the success of the CAMP programme. First, effective and selective tactics for both the key and secondary pests backed by technical support were available. Second, the project was well-funded, coordinated and directly involved growers, researchers, industry leaders and

Published by Burleigh Dodds Science Publishing Limited, 2019.

government administrators. However, the single key factor affecting the success of each area-wide programme was the creation of an organizational structure that allowed neighbouring growers to come together to eliminate 'problem orchards or blocks' through cooperation, peer pressure and education (Knight 2008).

Sterile insect technology (SIT), MD and host-plant removal have all been used in area-wide programmes in different pome fruit producing areas with varying levels of success. The programme established along the United States–Canada border in 1995 combined the use of sterile moths and MD (Calkins et al. 2000). A similar programme to CAMP was conducted across both apple and peach growing regions in eastern USA by implementing MD for CM, OFM and OBLR (Agnello et al. 2009; McGhee et al. 2011). The area-wide management of CM and OBLR in Washington State used MD for both pest species supplemented by targeted use of selective insecticides including microbials (Knight et al. 2001b). Development of a landscape-scale programme for OFM in Australia required coordination between pome and stone fruit growers (Il'ichev et al. 2002). The effort to eradicate LBAM from California used aerial applications of pheromones (Brockerhoff et al. 2012) and the area-wide eradication programme for CM in Brazil involved removal of host plants across a large urban region outside of agricultural lands (Kovaleski and Mumford 2007).

The longest-running, most successful area-wide effort to control CM in North America, and arguably the world, is the Okanagan-Kootenay Sterile Insect Release Program, OKSIR (Dyck et al. 1993). This socially and environmentally responsible initiative serves as a unique example of the power of community-based IPM that has clearly enabled the largest organic tree fruit industry in Canada. The programme began in 1992 as an effort to eradicate CM from 8500 ha of pome fruit and surrounding urban properties in the Okanagan, Kootenay and Similkameen valleys located in the southern interior of British Columbia (Dyck et al. 1993). Eradication of CM promised producers and local homeowners the opportunity to grow pome fruits without the use of insecticides during the fruit development period. Elimination of OPs was the key environmental target of this programme, which, ironically, was eventually achieved in totality in 2012 through legislative powers, not pest management. The lofty goal of eradication was abandoned in 1998, but the SIT effort continues as part of a more sustainable programme supplemented by targeted use of MD and selective insecticides (Thistlewood and Judd 2003b; Judd and Gardiner 2005). The programme currently services ca. 3550 ha of apples and pears, and reports that throughout the past decade at least 90% of the orchards it services have suffered less than 0.2% CM fruit damage at harvest (www.oksir.org).

Several aspects of the Canadian SIT programme are significant to its success. Firstly, special legislative powers were granted by senior government to local municipal governments to allow the taxation of all homeowners and pome fruit growers residing within the designated service area. This taxation provides an annual operating budget (Dyck and Gardiner 1992) that now exceeds $3 million CAD annually (www.oksir.org). Programme implementation also needed senior levels of government to provide the capital funding ($7.7 million CDN in 1992) to build a large CM rearing facility, and another $1 million CDN in 1994 to pay growers to apply insecticides to reduce CM populations. Often overlooked is the fact that during the first 10 years of the programme nearly 80% of the existing apple orchards were replanted into high-density dwarfing systems under a government-assisted industry-wide replant programme. Removal of old, derelict orchards, often harbouring CM populations, and with higher overwintering survivorship than new orchards, helped to reduce existing wild populations. Urban property owners were encouraged to remove unmanaged pome trees often enticed by free tree- and fruit-replacement programmes.

Weaknesses of this ongoing SIT programme include continuing concern about the fitness of the sterilized moths to compete with wild moths under field conditions (Judd and Gardiner 2006b; Judd et al. 2011). Currently moths are sterilized with ionizing gamma radiation from a Cobalt^{60} source (Bloem and Bloem 2000) but advances in alternative sterilization methods using safer, and less debilitating X-ray methods (Mastrangelo et al. 2010; Yamada et al. 2014) may improve moth quality. Poor performance of sterile CM, especially in cool springs (Judd et al. 2004), is also related to the mass-rearing system and cold storage before release (Judd et al. 2006a,b, 2011). Studies on thermal acclimation of CM offer hope that use of appropriate thermal pre-treatments may improve field performance of sterile CM under adverse thermal conditions (Chidawanyika and Terblanche 2011).

Currently the OKSIR programme mostly releases a uniform predetermined number of moths (2000 mixed-sex moths per ha) once each week to each orchard across an established grid using ground-based delivery from all-terrain vehicles over a 4-month period (May–August), regardless of local pest pressure (Bloem and Bloem 2000). The programme is unable to adjust moth production quickly in the event of an early spring or during peaks of moth flight as needed. The programme is unable to target problem sites that emerge during the season with additional releases of moths due to the lack of adequate resources to assess its effectiveness. Moth monitoring efforts are carried out weekly and automated trapping systems (Kim et al. 2011) that could provide daily information are not yet able to distinguish wild and sterile moths. This means the programme often reacts to information about inadequate or even excessive sterile to wild ratios long after the problem has occurred. Targeted moth releases based on daily trapping would be advantageous because moths could be shunted from where they are not needed to where they are. The potential for fast targeted release might be easier if human drivers with fixed daily schedules could be taken out of the equation. The potential for using drone technology to deliver moths in this way is being investigated by a number of research groups most notably in New Zealand (Horner at al. 2016). Researchers are partnering with commercial drone companies like M3 Consulting Group (www.m3cg.us) that are looking to expand the range of services they can provide to the agricultural industry.

One of the challenges frequently faced by new SIT initiatives is the capital cost of building an insect rearing facility (Addison 2005; Botto and Glaz 2010; Blomefield et al. 2011; Horner et al. 2016). Researchers in South Africa (Blomefield et al. 2011) and New Zealand (Horner at al. 2016) have overcome this capital investment challenge by importing sterile Canadian moths for local release. Sales of OKSIR moths into southern hemisphere countries, especially during the Canadian winter, may provide additional funding which could be used to refine or subsidize the Canadian programme. If moths from the Canadian programme can be produced, shipped and released into other production regions for costs that are competitive with local methods of CM pest control, then SIT can potentially become another useful management tactic that could be integrated into existing programmes needing help, such as organic production (Judd and Gardiner 2005).

11 Post-harvest management

The need to manage tortricids continues after harvest to avoid international trade disruptions (Vail et al. 1993). Irradiation and heating fruit has been investigated as a substitute to methyl bromide treatments (Follett and Snook 2012; Wang et al. 2006). A systems approach has been conceptualized that includes both within-season and post-harvest practices to minimize the probability (probit 9 efficacy = 99.9968329% mortality) of the introduction of quarantine tortricid pests (Follett and Neven 2006). Effective culling of fruits on packing lines is influenced by both fruit characteristics, that is colour, size of insect injury, the co-occurrence of other visual defects and operational factors, such as the numbers of fruits sorted per worker-second (Knight and Moffitt 1991). Progress in developing automated defect inspection processes based on machine vision and learning is progressing rapidly (Sofu et al. 2016; Gupta et al. 2018) and the possibility of detecting internal pests in apples in packinghouses has been explored but does not seem practical due to the handling times required (Li et al. 2018).

Wooden bins used to harvest fruit are placed in orchards at variable times before harvest and can become infested with CM larvae that build cocoons in the bins (Higbee et al. 2001). Stacking of bins during the off-season changes the temperatures that larvae are exposed to in different areas of the stack and interferes with moth emergence times during the season. This is problematic for timing effective sprays and also causes unexpected hot spots to be created in orchards near bin stacks. Sharing bins among infested and uninfested orchards contributes to the regional spread of CM. Bin sterilization with acetic acid to eliminate insects and disease pathogens has been investigated (Randall et al. 2011); some packinghouses have developed routine bin sterilization programmes using hot water emersion (Hansen et al. 2006). The concept of disinfesting wooden bins through treatment with nematodes has also been studied (Lacey and Chauvin 1999) and the transition to picking fruit into plastic instead of wooden bins may reduce the incidence of overwintering larvae (Waelti 1992).

12 Molecular tools

Advances in molecular biology, genetics and genomics research have provided various tools and technologies that have and will likely contribute more towards the management of tortricids. Several molecular genetic tools have been developed that facilitate identification of these internal feeding fruit pests relevant to quarantine, importation and management situations. DNA barcoding via polymerase chain reaction (PCR), quantitative real-time PCR (qRT-PCR) and PCR restriction fragment length polymorphism (PCR-RFLP) amplification of mitochondrial cytochrome oxidase I (mCOI) have been utilized for this purpose (Barcenas et al. 2009; Chen and Dorn 2009; Kwon et al. 2018). qRT-PCR assay of ribosomal RNA has also been used for diagnostic identification of LBAM in North America (Barr et al. 2011). Simple sequence repeat microsatellites are rapidly evolving genetic loci that are selectively neutral and vary in number of repeated units. These features make microsatellites very useful for studying population genetics and gene flow (Franck et al. 2005, 2011; Fuentes-Contreras et al. 2008; Chen and Dorn 2010; Kirk et al. 2013). A better understanding of the population structures of tortricid pests at the genetic level informs IPM through identification of source populations for invasive species, identification of host races and characterization of dispersal rates and migratory patterns (reviewed in Chen and Dorn 2010; Kirk et al. 2013). Next-generation sequencing methods may also be used for characterization of genetic structure of populations and study of gene flow within and across populations. Genotyping-by-sequencing combines high-throughput sequencing of biological samples with broad-scale single nucleotide polymorphism analysis. With this methodology it has recently been shown for OFM moth in Brazil that geographic distance is the primary factor that affects genetic structure and gene flow (Silva-Brandao et al. 2015). Insecticide resistance mechanisms in European populations of CM have been characterized at the genetic level with individual genes identified that correlate with resistance (Reyes et al. 2007). Genetic studies of the genes and genotypes that mediate resistance facilitate assessments of the prevalence of resistance within populations (Franck et al. 2007; Franck and Timm 2010).

Genetic engineering of apple trees has proven to enhance apple varieties with various traits including food quality (Murata et al. 2001), fire blight (Norelli et al. 2003) and resistance to pest insects (Markwick et al. 2003; Maheswaran et al. 2007). Genetically modified non-browning apples have recently been approved for market in the United States (Waltz 2015). Transgenic cv. Royal Gala apples that express biotin-binding proteins avidin and streptavidin in shoot foliage displayed increased resistance to LBAM damage; larvae feeding on transgenic plants displayed increased mortality and decreased fitness compared to those feeding on controls (Markwick et al. 2003). LBAM fitness parameters were also impacted in larvae feeding on Royal Gala leaves in which a *Nicotiana alata,* Link & Otto, proteinase inhibitor was transgenically expressed (Maheswaran et al. 2007).

CRISPR/CAS9 technology has recently revolutionized biological research by providing methodologies for efficient, site-specific genome editing (Jinek et al. 2012). To date, utility of this technology has been applied to several plant species (Xu 2013). Given that creating transgenic apple is well established, and CRISPR/CAS9 utilizes similar entry point methods, CRISPR/CAS9 editing of the apple genome is anticipated. This has implications for pest control vis-à-vis generation of edited cultivars that display greater resistance to pests. This may be carried out through the introduction of foreign genes (transgenesis) or knockout of native genes, the latter of which may be beneficial in avoiding regulatory hurdles that come with the introduction of foreign genetic materials. Moreover, CRISPR/CAS9 has been demonstrated, proof of principle, for CM (Garczynski et al. 2017). Development of this technology in CM, and its application in other tortricid pest species may enhance multiple fronts of pest control. Utilization of molecular genetic resources to improve the effectiveness and efficiency of SIT/IS with CM has been proposed (Vreysen et al. 2007). Specifically, CRISPR-mediated transgenic introduction of dominant conditional lethal mutations that specifically target females would facilitate production of genetic sexing strains, which has been proven to dramatically improve the efficiency and economics of SIT with the medfly, *Ceratitis capitata* (Wiedemann). Additionally, utilization of repressible dominant lethal genes that are present in males, but active only in females, would circumvent the need for sterilizing radiation, which can reduce fitness of released insects (Alphey and Bonsall 2018; Thomas et al. 2000). Recent biotechnology efforts have been aimed at bioengineering transgenic plants or fungi that produce insect pheromones. Transgenic *Nicotiana benthamiana* Domin engineered to produce the pheromones of two small ermine moths were shown to be effective in trapping male moths of the respective species (Ding et al. 2014) and reconstitution of moth pheromone biosynthetic pathways in yeast demonstrate an additional means of pheromone production (Ding et al. 2016; Hagstrom et al. 2013). Application of these technological systems to the production of tortricid moth pheromones may directly contribute to IPM of target species by providing alternative, efficient and cost-effective pheromone production and delivery systems. A final point with respect to basic research: *in situ* genetic manipulation of key tortricid pest species has the potential to add to our basic understanding of the biology and ecology of these organisms.

13 Future trends and conclusion

Tortricids utilizing tree fruits are lovely little moths (an entomological researcher's perspective) with a worldwide importance as a serious threat to the production and shipping of valuable, pristine commodities. These are exciting times in the realm of tortricid management in tree fruits. The tools and knowledge of how to manage tortricids is available to all growers, and the development and marketing of new more valuable cultivars has provided the funds for many growers to implement effective management programmes. Overall, it seems that the future heralds a reduced pest status for tortricids within orchards. Instead, growers will have to continually refocus on effectively managing the introduction of other pests, and be forced to rebalance their IFP programmes in order to maintain low pesticide residues and the important ecological services provided from biological control. Perhaps this is too optimistic and too ethnocentric, ignoring the dogma that 'Nature always finds a way!' We will see, and growers can always worry and remain on guard each growing season armed with an arsenal of tactics for these seemingly gentle moths dispersing into their orchards to wreak havoc!

The lessons learned from the success of area-wide management will continue to shine attention on the value of substituting out prophylactic sprays with increased knowledge about the pest status of tortricids. The most effective programmes will be where unmanaged hosts are not located near orchards, where all growers use high-level management programmes, where monitoring and thresholds are used to minimize the use of disruptive tactics and to maximize their effectiveness. Protection of orchards with netting will likely increase. It is uncertain if new classes of insecticides will be developed so custodianship of the active materials will continue to be critical through their judicious use. Targeting localized problem areas with mass trapping, releases of sterile moths, and removal and replanting of orchards will become more important.

14 Where to look for further information

The first place to look when researching information on tortricids is the amazing book edited by Geest and Evenhuis. A second source would be to scan the *Annual Review of Entomology* for reviews of tortricid biology and management. Also, the key journals listed in the literature review are likely to continue to be the most important sources of new information on tortricids. Several IOBC working groups in Europe, including the 'Integrated protection of fruit crops', and 'Pheromones and other semio-chemicals in integrated production' hold meetings every few years and consistently bring together the experts in various management fields related to tortricids. In addition, the annual meetings of the Entomological Society of America and the Orchard Pest Management Disease Conference held in the second week of January in Portland, Oregon, are always extremely educating and excellent opportunities to hear the latest scientific developments related to tortricids and other insect groups impacting tree fruit.

15 References

Addison, M. F. (2005), Suppression of codling moth *Cydia pomonella* L. (Lepidoptera: Tortricidae) populations in South African apple and pear orchards using sterile insect release, *in* Theron, K. I. (Ed.), *Proceedings of the Ninth International Pear Symposium*, Stellenbosch, South Africa, 1-6 February 2004, International Society for Horticultural Science, Lueven, Belgium.

Agnello, A. M., Atanassov, A., Bergh, J. C., Biddinger, D. J., Gut, L. J., et al. (2009), Reduced-risk pest management programs for eastern U.S. apple and peach orchards: A 4-year project, *Am. Entomol.* 55(3), 184-97.

Ahmad, M., Hollingworth, R. M. and Wise, J. C. (2002), Broad-spectrum insecticide resistance in obliquebanded leafroller *Choristoneura rosaceana* (Lepidoptera: Tortricidae) from Michigan, *Pest Manag. Sci.* 58, 834-8.

Alphey, N. and Bonsall, M. B. (2018), Genetics-based methods for agricultural insect pest management, *Agric. For. Entomol.* 20, 131-40.

Arnault, I., Lombarkia, N., Joy-Ondet, S., Romet, L., Brahim, I., et al. (2016), Foliar application of microdoses of sucrose to reduce codling moth *Cydia pomonella* L. (Lepidoptera: Tortricidae) damage to apple trees, *Pest Manag. Sci.* 72, 1901-9.

Arthurs, S. P., Lacey, L. A. and Fritts, R. (2005), Optimizing use of codling moth granulovirus: Effects of application rate and spraying frequency on control of codling moth larvae in Pacific Northwest apple orchards, *J. Econ. Entomol.* 98(5), 1459-68.

Arthurs, S. P., Lacey, L. A. and Miliczky, E. R. (2007a), Evaluation of the codling moth granulovirus and spinosad for codling moth control and impact on non-target species in pear orchards, *Biolog. Control* 41, 99-109.

Arthurs, S. P., Hilton, R., Knight, A. L. and Lacey, L. A. (2007b), Evaluation of the pear ester kairomone as a formulation additive for the granulosis virus of codling moth (Lepidoptera: Tortricidae) in pome fruit, *J. Econ. Entomol.* 100(3), 702-9.

Arthurs, S. P., Lacey, L. A. and Behle, R. W. (2008), Evaluation of lignins and particle films as solar protectants for the granulovirus of the codling moth, *Cydia pomonella*, *Biocontrol Sci. Technol.* 18(8), 829-39.

Asser-Kaiser, S., Fritsch, E., Undorf-Spahn, K., Kienzle, J., Eberle, K. E., et al. (2007), Rapid emergence of baculovirus resistance in codling moth due to dominant, sex-linked inheritance, *Science* 317(5846), 1916-18.

Atanassov, A., Shearer, P. W., Hamilton, G. and Polk, D. (2002), Development and implementation of a reduced risk peach arthropod management program in New Jersey, *J. Econ. Entomol.* 95(4), 803–12.

Avilla, J. and Riedl, H. (2003), Integrated fruit production for apples – principles and guidelines, *in* Ferree, D. C. and Warrington, I. J. (Eds), *Apples: Botany, Production and Uses*, CABI Publishing, Wallingford, UK, pp. 539–49.

Bale, J. S. and Hayward, A. L. (2010), Insect overwintering in a changing climate, *J. Exp. Biol.* 213, 980–94.

Ballard, J., Ellis, D. J. and Payne, C. C. (2000a), Uptake of granulovirus from the surface of apples and leaves by first instar larvae of the codling moth *Cydia pomonella* L. (Lepidoptera: Olethreutidae), *Biocontrol Sci. Technol.* 10, 617–25.

Ballard, J., Ellis, D. J. and Payne, C. C. (2000b), The role of formulation additives in increasing the potency of *Cydia pomonella* granulovirus for codling moth larvae, in laboratory and field experiments, *Biocontrol Sci. Technol.* 10, 627–40.

Barcenas, N. M., Unruh, T. R. and Neven, L. G. (2009), DNA diagnostics to identify internal feeders (Lepidoptera: Tortricidae) of pome fruits of quarantine importance, *J. Econ. Entomol.* 98(2), 299–306.

Barr, N. B., Ledezma, L. A., Farris, R. E., Epstein, M. E. and Gilligan, T. M. (2011), A multiplex real-time polymerase chain reaction assay to diagnose *Epiphyas postvittana* (Lepidoptera: Tortricidae), *J. Econ. Entomol.* 104(5), 1706–19.

Barrett, B. A. (2008), Assessment of methoxyfenozide exposure on the sexual attractiveness and responsiveness of adult codling moth, *Cydia pomonella* L., in small orchard blocks, *Pest Manag. Sci.* 64, 916–22.

Barros-Parada, W., Knight, A. L. and Fuentes-Contreras, E. (2015), Modeling codling moth (Lepidoptera: Tortricidae) phenology and predicting egg hatch in apple orchards of the Maule region, Chile, *Chilean J. Agric. Res.* 75(1), 57–62.

Basoalto, E., Miranda, M., Knight, A. L. and Fuentes-Contreras, E. (2010), Landscape analysis of adult codling moth (Lepidoptera: Tortricidae) distribution and dispersal within typical agroecosystems dominated by apple production in central Chile, *Environ. Entomol.* 39, 1399–408.

Bastias, R. M., Manfrini, L. and Grappadelli, L. C. (2012), Exploring the potential use of photo-selective nets for fruit growth regulation in apple, *Chilean J. Agric. Res.* 72(2), 224–31.

Beers, E. H., Suckling, D. M., Prokopy, R. J. and Avilla, J. (2003), Ecology and management of apple arthropod pests, *in* Ferree, D. C. and Warrington, I. J. (Eds), *Apples: Botany, Production and Uses*, CABI Publishing, Wallingford, UK, pp. 489–519.

Beers, E. H., Brunner, J. F., Dunley, J. E., Doerr, M. and Granger, K. (2005), Role of neonicotinyl insecticides in Washington apple integrated pest management. Part II. Nontarget effects on integrated mite control, http://www.insectscience.org/ (accessed 16 May 2005).

Beers, E. H., Horton, D. R. and Miliczky, E. (2016a), Pesticides used against *Cydia pomonella* disrupt biological control of secondary pests of apple, *Biol. Control* 102, 35–43.

Beers, E. H., Mills, N. J., Shearer, P. W., Horton, D. R., Milickzy, E. R., et al. (2016b), Nontarget effects of orchard practices on natural enemies: Lessons from the field and laboratory, *Biol. Control* 102, 44–52.

Berling, M., Blachere-Lopez, C., Soubabere, O., Lery, X., Bonhomme, A., et al. (2009), *Cydia pomonella granulovirus* genotypes overcome virus resistance in the codling moth and improve virus efficiency by selection against resistant hosts, *Appl. Environ. Microbiol.* 75(4), 925–30.

Bezemer, T. M. and Mills, N. J. (2001), Host density responses of *Mastrus ridibundus*, a parasitoid of the codling moth, *Cydia pomonella*, *Biol. Control* 22, 169–75.

Biondi, A., Mommaerts, V., Smagghe, G., Vinuela, E., Zappala, L., et al. (2012), The non-target impact of spinosyns on beneficial arthropods, *Pest Manag. Sci.* 68, 1523–36.

Bloem, K. A. and Bloem, S. (2000), SIT for codling moth eradication in British Columbia, Canada, *in* Tan, K.-H. (Ed.), *Area-Wide Control of Fruit Flies and Other Insect Pests. Joint Proceedings of the International Conference on Area-wide Control of Insect Pests, and the Fifth International Symposium on Fruit Flies of Economic Importance*, May–June 1998, Penerbit Universiti Sains Malaysia, Pulau Penang, pp. 207–14.

Bloem, S., Bloem, K. A. and Knight, A. L. (1998), Oviposition by sterile codling moths, *Cydia pomonella* (Lepidoptera: Tortricidae) and control of wild populations with combined releases of sterile moths and egg parasitoids, *J. Entomol. Soc. Brit. Columbia* 95, 99–109.

Blomefield, T., Carpenter, J. E. and Vreysen, M. J. B. (2011), Quality of mass-reared codling moth (Lepidoptera: Tortricidae) after long distance transportation: 1. Logistics of shipping procedures and quality parameters as measured in the laboratory, *J. Econ. Entomol.* 104, 814–22.

Blommers, L. H. M. (1994), Integrated pest-management in European apple orchards, *Annu. Rev. Entomol.* 39, 213–41.

Boivin, T., d'Hières, C. C., Bouvier, J. C., Beslay, D. and Sauphanor, B. (2001), Pleiotropy of insecticide resistance in the codling moth, *Cydia pomonella*, *Entomol. Exp. Appl.* 99, 381–6.

Boivin, T., Bouvier, J. C., Chadoeuf, J., Beslay, D. and Sauphanor, B. (2003), Constraints on adaptive mutations in the codling moth *Cydia pomonella* (L.): Measuring fitness trade-offs and natural selection, *Heredity* 90, 107–13.

Boivin, T., Bouvier, J. C., Beslay, D. and Sauphanor, B. (2004), Variability in diapause propensity within populations of a temperate insect species: Interactions between insecticide resistance genes and photoperiodism, *Biol. J. Linnean Soc.* 83, 341–51.

Boncheva, R., Dukiandjiev, S., Minkov, I., de Maagd, R. A. and Naimov, S. (2006), Activity of *Bacillus thuringiensis* gamma-endotoxins against codling moth (*Cydia pomonella* L.) larvae, *J. Invert. Path.* 92, 96–9.

Bone, N. J., Thomson, L. J., Ridland, P. M., Cole, P. and Hoffman, A. A. (2009), Cover crops in Victorian apple orchards: Effects on production, natural enemies and pests across a season, *Crop Prot.* 28, 675–83.

Borchert, D. M., Walgenbach, J. F., Kennedy, G. G. and Long, J. W. (2004), Toxicity and residual activity of methoxyfenozide and tebufenozide to codling moth (Lepidoptera: Tortricidae) and oriental fruit moth (Lepidoptera: Tortricidae), *J. Econ. Entomol.* 97(4), 1342–52.

Borchert, D. M., Walgenbach, J. F. and Kennedy, G. G. (2005), Assessment of sublethal effects of methoxyfenozide on oriental fruit moth (Lepidoptera: Tortricidae), *J. Econ. Entomol.* 98(3), 765–71.

Boreau de Roince, C., Lavigne, C., Ricard, J.-M., Franck, P., Bouvier, J.-C., et al. (2012), Predation by generalist predators on the codling moth versus a closely-related emerging pest the oriental fruit moth: A molecular analysis, *Agric. Forest Entomol.* 14, 260–9.

Bosch, D., Rodriguez, M. A. and Avilla, J. (2015), Captures of MFO-resistant *Cydia pomonella* adults as affected by lure, crop management system and flight, *Bull. Entomol. Res.* 106(1), 54–62.

Bosch, D., Rodriguez, M. A., Depalo, L. and Avilla, J. (2018), Determination of the baseline susceptibility of European populations of *Cydia pomonella* (Lepidoptera: Tortricidae) to chlorantraniliprole and the role of Cytochrome P450 monooxygenases, *J. Econ. Entomol.* 111(2), 844–52.

Bostanian, N. J. and Racette, G. (2008), Particle films for managing arthropod pests of apple, *J. Econ. Entomol.* 101(1), 145–50.

Botto, E. and Glaz, P. (2010), Potential for controlling codling moth *Cydia pomonella* (L.) (Lepidoptera: Tortricidae) in Argentina using the sterile insect technique and egg parasitoids, *J. Appl. Entomol.* 134, 251–60.

Brockerhoff, E. G., Suckling, D. M., Ecroyd, C. E., Wagstaff, S. J., Raabe, M. C., et al. (2011), Worldwide host plants of the highly polyphagous, invasive *Epiphyas postvittana* (Lepidoptera: Tortricidae), *J. Econ. Entomol.* 104 (5), 1514–24.

Brockerhoff, E. G., Suckling, D. M., Kimberly, M., Richardson, B., Coker, G., et al. (2012), Aerial application of pheromones for mating disruption of an invasive moth as a potential eradication tool, *PLoS ONE* 7(8), 1–8.

Brown, J. W. (2005), *World Catalogue of Insects, Volume 5: Tortricidae (Lepidoptera)*, Apollo Books, Stenstrup, Denmark, 741pp.

Brown, J. W., Robinson, G. and Powell, J. A. (2008), *Food plant database of the leafrollers of the world (Lepidoptera: Tortricidae) (Version 1.0.0)*, available at: http://www.tortricidae.com/foodplants.asp (accessed 13 April 2018).

Brun-Barale, A., Bouvier, J.-C., Pauron, D., Bergé, J. B. and Sauphanor, B. (2005), Involvement of a sodium channel mutation in pyrethroid resistance in *Cydia pomonella* L, and development of a diagnostic test, *Pest Manag. Sci.* 61(6), 549–54.

Brunner, J. F., Dunley, J. E. and Doerr, M. (2005), Effect of pesticides on *Colpoclypeus florus* (Hymenoptera: Eulophidae) and *Trichogramma platneri* (Hymenoptera: Trichogrammatidae), parasitoids of leafrollers in Washington, *J. Econ. Entomol.* 94, 1075–84.

Burgi, L. P. and Mills, N. J. (2014), Lack of enemy release for an invasive leafroller in California: Temporal patterns and influence of host plant origin, *Biol. Invasions* 16, 1021–34.

Burgi, L. P., Roltsch, W. J. and Mills, N. J. (2015), Allee effects and population regulation: A test for biotic resistance against an invasive leafroller by resident parasitoids, *Popul. Ecol.* 57, 215–25.

Calkins, C. O. and Faust, R. J. (2003), Overview of areawide programs and the program for suppression of codling moth in the western USA directed by the United States Department of Agriculture – Agricultural Research Service, *Pest Manag. Sci.* 59, 601–4.

Calkins, C. O., Knight, A. L., Richardson, G. and Bloem, K. A. (2000), Area-wide suppression of codling moth. Area-wide control of fruit flies and other insect pests, *Joint Proceedings of the International Conference on Area-wide Control of Insect Pests*, 28 May – 2 June 1998 and the Fifth International Symposium on Fruit Flies of Economic Importance, Penang, Malaysia, 1–5 June 1998.

Carey, J. R. and Harder, D. (2013), Clear, present, significant, & imminent danger: Questions for the California light brown apple moth (*Epiphyas postvittana*) technical working group, *Am. Entomol.* 59(4), 240–7.

Carlson, G. R., Dhadialla, T. S., Hunter, R., Jansson, R. K., Jany, C. S., et al. (2001), The chemical and biological properties of methoxyfenozide a new insecticidal ecdysteroid agonist, *Pest Manag. Sci.* 57, 115–19.

Carrière, Y., Deland, J. P., Roff, D. A. and Vincent, C. (1994), Life-history costs associated with the evolution of insecticide resistance, *Proc. R Soc. Lond. B Biol. Sci.* 258, 35–40.

Casida, J. E. (2018), Neonicotinoids and other insect nicotinic receptor competitive modulators: Progress and prospects, *Annu. Rev. Entomol.* 63, 125–44.

Cassanelli, S., Reyes, M., Rault, M., Manicardi, G. C. and Sauphanor, B. (2006), Acetylcholinesterase mutation in an insecticide-resistant population of the codling moth *Cydia pomonella* (L.), *Insect Biochem. Mol. Biol.* 36(8), 642–53.

Chapman, P. J. (1973), Bionomics of the apple-feeding Tortricidae, *Annu. Rev. Entomol.* 18, 73–96.

Charmillot, P. J. and Pasquier, D. (1992), Modification de la fertilite du carpocapse *Cydia pomonella* a la suite du contact des adultes avec un régulateur ou un inhibiteur de croissance d'insectes, *Entomol. Exp. Appl.* 63, 87–93.

Charmillot, P. J., Gourmelon, A. L. and Pasquier, D. (2001), Ovicidal and larvicidal effectiveness of several insect growth inhibitors and régulators on the codling moth *Cydia pomonella* L. (Lep., Tortricidae), *J. Appl. Ent.* 125, 147–53.

Chen, M. H. and Dorn, S. (2009), Reliable and efficient discrimination of four internal fruit-feeding *Cydia* and *Grapholita* species (Lepidoptera: Tortricidae) by polymerase chain reaction-restriction fragment length polymorphism, *J. Econ. Entomol.* 102(6), 2209–16.

Chen, M. H. and Dorn, S. (2010), Microsatellites reveal genetic differentiation among populations in an insect species with high genetic variability in dispersal, the codling moth, *Cydia pomonella* (L.) (Lepidoptera: Tortricidae), *Bull. Entomol. Res.* 100, 75–85.

Chidawanyika, F. and Terblanche, J. S. (2011), Costs and benefits of thermal acclimation for codling moth, *Cydia pomonella* (Lepidoptera: Tortricidae): Implications for pest control and the sterile insect release programme, *Evol. Appl.* 4, 534–44.

Chouinard, G., Firle J. A. and Cormier, D. (2016), Going beyond sprays and killing agents: Exclusion, sterilization and disruption for insect pest control in pome and stone fruit orchards, *Sci. Hortic.* 208, 13–27.

Cichon, L., Fuentes-Contreras, E., Garrido, S., Lago, J., Barros-Parada, W., et al. (2013), Monitoring oriental fruit moth (Lepidoptera: Tortricidae) with sticky traps baited with terpinyl acetate and sex pheromone, *J. Appl. Entomol.* 137, 275–81.

Comas, C., Avilla, J., Sarasúa, M. J., Albajes, R. and Ribes-Dasi, M. (2012), Lack of anisotropic effects in the spatial distribution of *Cydia pomonella* pheromone trap catches in Catalonia, NE Spain, *Crop Prot.* 34, 88–95.

Cordova, D., Benner, E. A., Sacher M. D., Rauh, J. J., Sopa, J. S., et al. (2005), Anthranilic diamides: A new class of insecticides with a novel mode of action, ryanodine receptor activation, *Pest. Biochem. Physiol.* 84, 196–214.

Cossentine, J., Jensen, L. B. and Deglow, E. K. (2003), Strategy for orchard use of *Bacillus thuringiensis* while minimizing impact on *Choristoneura rosaceana* parasitoids, *Entomol. Exp. Appl.* 109(3), 205–10.

Cossentine, J., Jensen, L., Dedglow, E., Bennett, A., Goulet, H., et al. (2004), The parasitoid complex affecting *Choristoneura rosaceana* and *Pandemis limitata* in organically managed apple orchards, *Biocontrol* 49, 359–72.

Costes, E., Lauri, P. E. and Andrieu, B. (2013), Plant architecture, its diversity and manipulation in agronomic conditions, in relation with pest and pathogen attacks, *Eur. J. Plant Pathol.* 135, 455–70.

Cox, D. L., Knight, A. L., Biddinger, D. J., Lasota, J. A., Pikounis, B., et al. (1995), Toxicity and field efficacy of avermectins against codling moth (Lepidoptera: Tortricidae) on apples, *J. Econ. Entomol.* 88(3), 708–15.

Damos, P. (2015), Modular structure of web- based decision support systems for integrated pest management: A review, *Agron. Sustain. Dev.* 35, 1347–72.

Damos, P., Escudero-Colomar, L. A. and Ioratti, C. (2015), Integrated fruit production and pest management in Europe: The apple case study and how far we are from the original concept? *Insects* 6, 626–57.

Davidson, N. A., Dibble, J. E., Flint, M. L., Marer, P. J. and Guye, A. (1991), Managing insects and mites with spray oils, *IPM Edu. Pub. Statewide Integr. Pest Manag. Project University California* pub. 3347.

Delisle, J. and Vincent, C. (2002), Modified pheromone communication associated with insecticidal resistance in the obliquebanded leafroller, *Choristoneura rosaceana* (Lepidoptera: Tortricidae), *Chemoecol.* 12, 47–51.

Delate, K., McKern, A., Turnbull, R., Walker, J., Volz, R., et al. (2008), Organic apple production in two humid regions: Comparing progress in pest management strategies in Iowa and New Zealand, *HortSci.* 43(1), 12–20.

DeLury, N. C., Gries, R., Gries, G., Judd, G. J. R. and Khaskin, G. (1999), Moth scale-derived kairomones used by egg-larval parasitoid *Ascogaster quadridentate* to locate eggs of its host, *Cydia pomonella, J. Chem. Ecol.* 25, 2419–30.

Depalo, L., Masettia, A., Avilla, J., Bosch, D. and Pasqualini, E. (2016), Toxicity and residual activity of spinetoram to neonate larvae of *Grapholita molesta* (Busck) and *Cydia pomonella* (L.) (Lepidoptera: Tortricidae): Semi-field and laboratory trials, *Crop Prot.* 89, 32–7.

Devotto, L., del Valle, C., Ceballos, R. and Gerding, M. (2010), Biology of *Mastrus ridibundus* (Gravenhorst), a potential biological control agent for area-wide management of *Cydia pomonella* (Linneaus) (Lepidoptera: Tortricidae), *J. Appl. Entomol.* 134, 243–50.

Dhadialla, T. S., Carlson, G. R. and Le, D. P. (1998), New insecticides with ecdysteroidal and juvenile hormone activity, *Annu. Rev. Entomol.* 43, 545–69.

Dib, H., Sauphanor, B. and Capowiez, Y. (2010), Effect of codling moth exclusion nets on the rosy apple aphid, *Dysaphis plantaginea*, and its control by natural enemies, *Crop Prot.* 29, 1502–13.

Ding, B. J., Hofvander, P., Wang, H. L., Durrett, T. P., Stymne, S. and Lofstedt, C. (2014), A plant factory for moth pheromone production, *Nat. Commun.* 5, 3353.

Ding, B. J., Lager, I., Bansal, S., Durrett, T. P., Stymne, S. and Lofstedt, C. (2016), The yeast ATF1 acetyltransferase efficiently acetylates insect pheromone alcohols: Implications for the biological production of moth pheromones, *Lipids* 51(4), 469–75.

Dorigoni, A. (2016), Innovative fruit tree architecture as a nexus to improve sustainability in orchards, *Acta Hortic.* 1137, 1–10.

Dripps, J., Olson, B., Sparks, T. and Crouse, G. (2008), Spinetoram: How artificial intelligence combined natural fermentation with synthetic chemistry to produce a new spinosyn insecticide, *Plant Health Prog.* doi:10.1094/PHP-2008-0822-01-PS.

Drogue, S. and DeMaria, F. (2012), Pesticide residues and trade, the apple of discord? *Food Policy* 37, 641–9.

Duga, A. T., Ruysen, K., Dekeyser, D., Nuyttens, D. Bylemans, D., et al. (2015), Spray deposition profiles in pome fruit trees: Effects of sprayer design, training system and tree canopy characteristics, *Crop Prot.* 67, 200–13.

Dunley, J. E. and Welter, S. C. (2000), Correlated insecticide cross-resistance in azinphos-methyl resistant codling moth (Lepidoptera: Tortricidae), *J. Econ. Entomol.* 93, 955–62.

Dunley, J. E., Brunner, J. F., Doerr, M. D. and Beers, E. H. (2006), Resistance and cross resistance in populations of the leafrollers, *Choristoneura rosaceana* and *Pandemis pyrusana*, in Washington apples, *J. Insect Sci.* 6, 14. Available online: insectscience.org/6.14

Durden, K., Sellars, S., Cowell, B. Brown, J. J. and Pszczolkowski, M. A. (2010), *Artemisia annua* extracts, artemisinin and 1,8-cineole prevent fruit infestation by major cosmopolitan pest of apples, *Pharma. Biol.* 49, 563–8. doi:10.3109/13880209.2010.528433.

Dyck, V. A. and Gardiner, G. T. (1992), Sterile-insect release program to control the codling moth *Cydia pomonella* (L.) (Lepidoptera: Olethreutidae) in British Columbia, Canada, *Acta Phytopath. Entomol. Hung.* 27(1–4), 219–22.

Dyck, V. A., Graham, S. H. and Bloem, K. A. (1993), Implementation of the sterile insect release programme to eradicate the codling moth, *Cydia pomonella* (L.) (Lepidoptera: Olethreutidae), in British Columbia, Canada, *in Proceedings of the FAO/IAEA International Symposium on Management of Insect Pests: Nuclear and Related Molecular and Genetic Techniques*, 19–23 October 1992, Vienna, IAEA-SM-327/29, International Atomic Energy Agency, Vienna, Austria, pp. 285–97.

Elbert, A., Nauen, R. and McCaffery, A. (2007), IRAC, insecticide resistance and mode of action classification of insecticides, *in* Kramer, W. and Schirmer U. (Eds), *Modern Crop Protection Compounds*, Wiley, New York, NY, 600pp.

Ellis, N. H. and Hull, L. A. (2013), Factors influencing adult male *Grapholita molesta* dispersal in commercial *Malus* and *Prunus* host crops, *Entomol. Exp. Appl.* 146, 232–41.

El-Sayed, A. M., Fraser, H. M. and Trimble, R. M. (2001), Modification of the sex-pheromone communication system associated with organophosphorus-insecticide resistance in the obliquebanded leafroller (Lepidoptera: Tortricidae), *Can. Entomol.*, 133, 867–81.

El-Sayed, A. M., Knight, A. L., Byers, J. A., Judd, G. J. R. and Suckling, D. M. (2016), Caterpillar-induced plant volatiles attract conspecific adults in nature, *Sci. Rep.* 6, 37555. doi:10.1038/srep37555.

Evans, R. G., Kroeger, M. W. and Mahan, M. O. (1995), Evaporative cooling of apples by overtree sprinkling, *Appl. Eng. Agric.* 11(1), 93–9.

Evenden, M. L. and Judd, G. J. R. (1999), Adult eclosion, flight and oviposition of *Choristoneura rosaceana* (Lepidoptera: Tortricidae), in British Columbia apple orchards, *J. Entomol. Soc. Brit. Columbia* 96, 77–88.

Fagua, G., Condamine, F. L., Horak, M., Zwick, A. and Sperling, F. A. H. (2016), Diversification shifts in leafroller moths linked to continental colonization and the rise of angiosperms, *Cladistics* 33, 449–66.

Farrar, R. R. and Ridgway, R. L. (1995), Enhancement of activity of *Bacillus thuringiensis* Berliner against four lepidopterous pests by nutrient-based phagostimulants, *J. Entomol. Sci.* 30(1), 29–42.

Fernández, D. E., Beers, E. H., Brunner, J. F, Doerr, M. D. and Dunley, J. E. (2005), Effects of seasonal mineral oil applications on the pest and natural enemy complexes of apple, *J. Econ. Entomol.* 98, 1630–40.

Fernández, D. E., Cichon, L. I., Sánchez, E. E., Garrido, S. A. and Gittins, C. (2008), Effect of different cover crops on the presence of arthropods in an organic apple (*Malus domestica* Borkh.) orchard, *J. Sustain. Agric.* 32(2), 197–211.

Ferrer, A., Mazzi, D. and Dorn, S. (2014), Stay cool, travel far: Cold-acclimated oriental fruit moth females have enhanced flight performance but lay fewer eggs, *Entomol. Exp. Appl.* 151, 11–18.

Fleischer, S. J., Weisz, R., Smilowitz, Z. and Midgarden, D. (1997), Spatial variation in insect populations and site-specific integrated pest management, *Site-Spefic. Manag. Agric. Syst.*, pp. 101–30.

Fleischer, S. J., Blom, P. E. and Weisz, R. (1999), Sampling in Precision IPM: When the object is a map, *PhytoPath.* 89(11), 1112–18.

Follett, P. A. and Neven, L. G. (2006), Current trends in quarantine entomology, *Annu. Rev. Entomol.* 51, 359–85.

Follett, P. A. and Snook, K. (2012), Irradiation for quarantine control of the invasive light brown apple moth (Lepidoptera: Tortricidae) and a generic dose for tortricid eggs and larvae, *J. Econ. Entomol.* 105(6), 1971–8.

Foster, S. P. and Howard, A. J. (1999), The effects of mating, age at mating, and plant stimuli, on the lifetime fecundity and fertility of the generalist herbivore *Epiphyas postvittana, Entomol. Exp. Appl.* 91, 287–95.

Fox, R. D., Derksen, R. C., Zhu, H., Brazee, R. D. and Svensson, S. A. (2008), A history of air-blast sprayer development and future prospects, *Trans. ASABE* 51(2), 405–10.

Franck, P. and Timm, A. E. (2010), Population genetic structure of *Cydia pomonella*: A review and case study comparing spatiotemporal variation, *J. Appl. Entomol.* 134, 191–200.

Franck, P., Guerin, F., Loiseau, A. and Sauphanor, B. (2005), Isolation and characterization of microsatellite loci in the codling moth *Cydia pomonella* L. (Lepidoptera: Tortricidae), *Mol. Ecol.* 5(1), 99–102.

Franck, P., Reyes, M., Olivares, J. and Sauphanor, B. (2007), Genetic architecture in codling moth populations: Comparison between microsatellite and insecticide resistance markers, *Mol. Ecol.* 16, 3554–64.

Franck, P., Ricci, B., Klein, E. K., Olivares, J., Simon, S., et al. (2011), Genetic inferences about the population dynamics of codling moth females at a local scale, *Genetica* 139, 949–60.

Fraser, H. W. and Trimble, R. M. (2001), Effect of delayed mating on reproductive biology of the oriental fruit moth (Lepidoptera: Tortricidae), *Can. Entomol.* 133(2), 219–27.

Fritsch, E., Undorf-Spahn, K., Kienzle, J., Zebitz, C. P. W. and Huber, J. (2005), Apfelwicklel Granulovirus: Erste Hinweise auf Unterschiede in der Empfindlichkeit lokaler Apfelwickler-Populationen, *Nach. Deut. Pflanzen.* 57, 29–34.

Fuentes-Contreras, E., Reyes, M., Barros, W. and Sauphanor, B. (2007), Evaluation of azinphos-methyl resistance and activity of detoxifying enzymes in codling moth (Lepidoptera: Tortricidae) from central Chile, *J. Econ. Entomol.* 100, 551–6.

Fuentes-Contreras, E., Espinoza, J. L., Lavandero, B. and Ramirez, C. C. (2008), Population genetic structure of codling moth (Lepidoptera: Tortricidae) from apple orchards in central Chile, *J. Econ. Entomol.* 101, 190–8.

Garcia-Salazar, C., Gut, L. J. and Whalon, M. E. (2007), Hedgerow barriers and other reduced-risk controls for managing oriental fruit moth, *Grapholitha molesta* (Busck) (Lepidoptera: Tortricidae) in apples, *Renew. Agric. Food Syst.* 22, 181–8.

Garczynski, S. F., Martin, J. A., Griset, M., Willett, L. S., Cooper, W. R., et al. (2017), CRISPR/Cas9 editing of the codling moth (Lepidoptera: Tortricidae) *CpomOR1* gene affects egg production and viability, *J. Econ. Entomol.* 110(4), 1847–55.

Gilligan, T. M., Baixeras, J. and Brown, J. W. (2018), *T@RTS: Online World Catalogue of the Tortricidae (Ver. 4.0)*, available at: http://www.tortricid.net/catalogue.asp (accessed 28 December 2018).

Gindaba, J. and Wand, J. E. (2005), Comparative effects of evaporative cooling, kaolin particle film, and shade net on sunburn and fruit quality in apples, *HortSci.* 40(3), 592–6.

Glenn, D. M., Puterka, G. J., Vanderzwet, T., Byers, R. E. and Feldhake, C. (1999), Hydrophobic particle films: A new paradigm for suppression of arthropod pests and plant diseases, *J. Econ. Entomol.* 92(4), 759–71.

Glenn, D. M., Puterka, G. J., Drake, S. R., Unruh, T. R. and Knight, A. L. (2001), Particle film application influences apple leaf physiology, fruit yield, and fruit quality, *J. Am. Soc. Hortic. Sci.* 126(2), 175–81.

Glenn, D. M., Prado, E., Erez, A., McFerson, J. and Puterka, G. J. (2002), A reflective, processed-kaolin particle film affects fruit temperature, radiation reflection, and solar injury in apple, *J. Am. Soc. Hortic. Sci.* 127(2), 188–93.

Gökce, A., Kim, S. S.-H., Wise, J. C. and Whalon, M. E. (2009), Reduced egg viability in codling moth *Cydia pomonella* (L.) (Lepidoptera: Tortricidae) following adult exposure to novaluron, *Pest Manag. Sci.* 65, 283–7.

Graf, B., Hopli, H., Rauscher, S. and Hohn, H. (1999), Hail nets influence the migratory behavior of codling moth and leaf roller, *Obst- und Weinbau* 135, 289–92.

Granatstein, D. Kirby, E., Ostenson, H. and Willer, H. (2016), Global situation for organic tree fruits, *Sci. Hortic.* 208, 3–12.

Gu, H. N., Hughes, J. and Dorn, S. (2006), Trade-off between mobility and fitness in *Cydia pomonella* L. (Lepidoptera: Tortricidae), *Ecol. Entomol.* 31, 68–74.

Guarnieri, A., Maini, S. Molaro, G. and Rondelli, V. (2011), Automatic trap for moth detection in integrated pest management, *Bull Insect.* 64(2), 247–51.

Gupta, O., Das, A. J., Hellerstein, J. and Raskar, R. (2018), Machine learning approaches for large scale classification of produce, *Sci. Rep.* 8, 5226. doi: 10.1038/s41598-018-23394-3.

Gut, L. J. and Brunner, J. F. (1996), Implementing codling moth mating disruption in Washington pome fruit orchards, Washington State Tree Fruit Research Extension Center Bulletin 1, 12pp.

Gut, L. J. and Brunner, J. F. (1998), Pheromone-based management of the codling moth (Lepidoptera: Tortricidae) in Washington apple orchards, *J. Agric. Entomol.* 15, 387–405.

Hagstrom, A., Wang, H. L., Lienard, M. A., Lassance, J. M., Johansson, T. and Lofstedt, C. (2013), A moth pheromone brewery: Production of (Z)-11-hexadecenol by heterologous co-expression of two biosynthetic genes from a noctuid moth in a yeast cell factory, *Microb. Cell Fact.* 12, 125.

Hansen, J. D., Heidt, M. L. and Anderson, P. A. (2006), Bin sterilization to prevent reintroduction of codling moth, *J. Agric. Urban Entomol.* 23(1), 17–26.

Harris, M. O., Markwick, N. and Sandanayake, M. (2006), Is resistance to *Bacillus thuringiensis* endotoxin Cry1Ac associated with a change in the behavior of light brown apple moth larvae (Lepidoptera: Tortricidae)? *J. Econ. Entomol.* 99(2), 508–18.

Higbee, B. S., Calkins, C. O. and Temple, C. A. (2001), Overwintering of codling moth (Lepidoptera: Tortricidae) larvae in apple harvest bins and subsequent moth emergence, *J. Econ. Entomol.* 94, 1511–17.

Hilton, R. J., Riedl, H. and Westigard, P. H. (1992), Phytotoxicity response of pear to applications of abamectin-oil combinations, *HortSci.* 27 (12), 1280–2.

Horak, M. (1991), Morphology, *in* van der Geest, L. P. S. and Evenhuis, H. H. (Eds), *Tortricid Pests: Their Biology, Natural Enemies and Control*, Elsevier, Amsterdam, The Netherlands, pp. 1–22.

Horak, M. and Brown, R. L. (1991), Taxonomy and phylogeny, *in* van der Geest, L. P. S. and Evenhuis, H. H. (Eds), *Tortricid Pests: Their Biology, Natural Enemies and Control*, Elsevier, Amsterdam, The Netherlands, pp. 23–50.

Horner, R. M., Walker, J. T. S., Rogers, D. J., Lo, P. L. and Suckling, D. M. (2016), Use of the sterile insect technique in New Zealand: Benefits and constraints, *N. Z. Plant Prot.* 69, 296–304.

Hougardy, E., Bezemer, T. M. and Mills, N. J. (2005), Influence of host deprivation and egg expenditure on the reproduction capacity of *Mastrus ridibundus*, an introduced parasitoid for the biological control of codling moth in California, *Biol. Control* 33, 96–106.

Hsu, C. L., Agnello, A. M., and Reissig, W. H (2009), Edge effects in the directionally biased distribution of *Choristoneura rosaceana* (Lepidoptera: Tortricidae) in apple orchards, *Environ. Entomol.* 38, 433–41.

Hughes, W. O., Gailey, D. and Knapp, J. J. (2003), Host location by adult and larval codling moth and the potential for its disruption by the application of kariomones, *Entomol. Exp. Appl.* 106, 147–53.

Il'ichev, A. L., Gut, L. J., Williams, M. S. and Jerie, P. H. (2002), Area-wide approach for improved control of oriental fruit moth *Grapholita molesta* (Busck) (Lepidoptera: Tortricidae) by mating disruption, *Gen. Appl. Ent.* 31, 1–16.

Il'ichev, A. L., Williams, D. G. and Milner, A. D. (2004), Mating disruption barriers in pome fruit for improved control of oriental fruit moth *Grapholita molesta* Busck (Lepidoptera: Tortricidae) in stone fruit under mating disruption, *J. Appl. Entomol.* 128(2), 126–32.

Ioriatti, C. and Lucchi, A. (2016), Semiochemical strategies for tortricid moth control in apple orchards and vineyards in Italy, *J. Chem. Ecol.* 42(7), 571–83.

Ioriatti, C. and Tasin, M. (2018), Hail nets enhance disruption of sexual communication by synthetic pheromone in codling moth, *Entomol. Gen.* 37(1), 7–18.

Jeanneret, P. and Charmillot, P. J. (1995), Movements of tortricid moths (Lep., Tortricidae) between apple orchards and adjacent ecosystems, *Agric. Ecosyst. Environ.* 55, 37–49.

Jinek, M., Chylinski, K., Fonfara, I., Hauer, M., Doudna, J. A. and Charpentier, E. (2012), A programmable dual-RNA-guided DNA endonuclease in adaptive bacterial immunity, *Science* 337(6096), 816–21.

Jones, V. P., Eastburn, C. C., Wilburn, T. D. and Brunner J. F. (2005), Instar-specific phenology of *Pandemis pyrusana* and *Choristoneura rosaceana* (Lepidoptera: Tortricidae) in Washington apple orchards, *J. Econ. Entomol.* 98(3), 875–83.

Jones, V. P., Hagler, J. R., Brunner, J. F., Baker, C. C. and Wilburn, T. D. (2006a), An inexpensive immunomarking technique for studying movement patterns of naturally occurring insect populations, *Environ. Entomol.* 35, 827–36.

Jones, V. P., Wilburn, T. D. and Baker, C. C. (2006b), Movement of codling moth between abandoned and commercial orchards, *Proc. Wash. State Hortic. Soc.* 100, 180–4.

Jones, V. P., Wiman, N. G. and Brunner, J. F. (2008a), Comparison of delayed female mating on reproductive biology of codling moth and obliquebanded leafroller, *Environ. Entomol.* 37(3), 679–85.

Jones, V. P., Doerr, M. and Brunner, J. F. (2008b), Is biofix necessary for predicting codling moth (Lepidoptera: Tortricidae) emergence in Washington state apple orchards? *J. Econ. Entomol.* 101(5), 1651–7.

Jones, V. P., Unruh, T. R., Horton, D. R., Mills, N. J., Brunner, J. F., et al. (2009a), Tree fruit IPM programs in the western United States: The challenge of enhancing biological control through intensive management, *Pest Manag. Sci.* 65, 1305–10.

Jones, V. P., Brunner, J. F., Grove, Petit, B., Tangren, G. V. and Jones, W. E. (2009b), A web-based decision support system to enhance IPM programs in Washington tree fruit, *Pest Manag. Sci.* 66, 587–95.

Jones, V. P., Hilton, R., Brunner, J. F., Bentley, W. J., Alston, D. G., et al. (2013), Predicting the emergence of the codling moth, *Cydia pomonella* (Lepidoptera: Tortricidae), on a degree-day scale in North America, *Pest Manag. Sci.* doi:10.1002/ps.3519.

Joshi, N. K., Rajotte, E. G., Naithani, K. J., Krawczyk, G. and Hull, L. A. (2016), Population dynamics and flight phenology model of codling moth differ between commercial and abandoned apple orchard ecosystems, *Front. Physiol.* 7, 408, 1–13.

Judd, G. J. R. and Gardiner, M. G. T. (2005), Towards eradication of codling moth in British Columbia by complimentary actions of mating disruption, tree banding and sterile insect technique: Five-year study in organic orchards, *Crop Prot.* 24, 718–33.

Judd, G. J. R. and Gardiner, M. G. T. (2006a), Simultaneous disruption of pheromone communication and mating in *Cydia pomonella*, *Choristoneura rosaceana*, and *Pandemis limitata* (Lepidoptera: Tortricidae) using Isomate-CM/LR in apple orchards, *J. Entomol. Soc. Brit. Columbia* 101, 3–13.

Judd, G. J. R. and Gardiner, M. G. T. (2006b), Temperature, irradiation and delivery as factors affecting spring time flight activity and recapture of mass reared male codling moths released by the Okanagan-Kootenay sterile insect programme, *J. Entomol. Soc. Brit. Columbia* 103, 19–32.

Judd, G. J. R., Knight, A. L. and El-Sayed, A. M. (2017a), Trapping *Pandemis limitata* (Lepidoptera: Tortricidae) moths with mixtures of acetic acid, caterpillar-induced apple-leaf volatiles, and sex pheromone, *Can. Entomol.* 149, 813–22.

Judd, G. J. R., Knight, A. L. and El-Sayed, A. M. (2017b), Development of kairomone-based lures and traps targeting *Spilonota ocellana* (Lepidoptera: Tortricidae) in apple orchards treated with sex pheromones, *Can. Entomol.* 149, 662–76.

Judd, G. J. R., Gardiner, M. G. T. and Thistlewood, H. M. A. (2004), Seasonal variation in recapture of mass-reared sterile codling moth, *Cydia pomonella* (L.) (Lepidoptera: Tortricidae): Implications for control by sterile insect technique in British Columbia, *J. Entomol. Soc. Brit. Columbia* 101, 29–43.

Judd, G. J. R., Thistlewood, H. M. A., Gardiner, M. G. T. and Lannard, B. L. (2006a), Is lack of mating competitiveness in spring linked to mating asynchrony between wild and mass-reared male codling moth from an operational sterile insect programme? *Entomol. Exp. Appl.* 120, 113–24.

Judd, G. J. R., Cockburn, S., Eby, C., Gardiner, M. G. T. and Wood, S. (2006b), Diapause improves spring-time mating competitiveness of male codling moth mass-reared for a sterile insect programme, *Entomol. Exp. Appl.* 120, 161-6.

Judd, G. J. R., Arthur, S., Deglow, K. and Gardiner, M. G. T. (2011), Operational mark-release-recapture field tests comparing competitiveness of wild and differentially mass-reared codling moths from the Okanagan-Kootenay sterile insect program, *Can. Entomol.* 143, 300–16.

Kazmierczak, R. F., Norton, G. W., Knight, A. L., Rajotte, E. G. and Hull, L. A. (1993), Economic effects of resistance and withdrawal of organophosphate pesticides on an apple production system, *J. Econ. Entomol.* 86(3), 684–96.

Keil, S., Gu, H. N. and Dorn, S. (2001a), Response of *Cydia pomonella* to selection on mobility: Laboratory evaluation and field verification, *Ecol. Entomol.* 26, 495–501.

Keil, S., Gu, H. N. and Dorn, S. (2001b), Diel patterns of locomotor activity in *Cydia pomonella*: Age and sex related differences and effects of insect hormone mimics, *Physiol. Entomol.* 26, 306–14.

Kienzle, J., Schultz, C., Zebitz, C. P. W. and Huber, J. (2003), Codling moth granulovirus as a tool for resistance management and area-wide population control, *IOBC/WPRS Bull.* 26(11),69–74.

Kim, Y., Yung, S., Kim, Y. and Lee, Y. (2011), Real-time monitoring of oriental fruit moth, *Grapholita molesta* populations using a remote sensing pheromone trap in apple orchards, *J. Asia Pac. Entomol.* 14, 259–62

Kirk, H., Dorn, S. and Mazzi, D. (2013), Worldwide population genetic structure of the oriental fruit moth (*Grapholita molesta)*, a globally invasive pest, *BMC Ecol.* 13, 12.

Knight, A. L. (1994), Insect pest and natural enemy populations in paired organic and conventional apple orchards in the Yakima Valley, Washington, *J. Entomol. Soc. Brit. Columbia* 91, 27-36.

Knight, A. L. (1995), The impact of codling moth (Lepidoptera: Tortricidae) mating disruption on apple pest management in Yakima Valley, Washington, *J. Entomol. Soc. Brit. Columbia* 92, 29-38.

Knight, A. L. (1997a), Optimizing the use of Bts for leafroller control, *Good Fruit Grower* 48, 47–9. 1997.

Knight, A. L. (1997b), Delay of mating of codling moth in pheromone-treated orchards, *IOBC/WPRS Bull.* 20, 203–6.

Knight, A. L. (1998a), Activity of Cry1 endotoxin of *Bacillus thuringiensis* for four tree fruit leafroller pest species (Lepidoptera: Tortricidae), *J. Agric. Entomol.* 15(20), 93–103.

Knight, A. L. (1998b), Management of codling moth (Lepidoptera: Tortricidae) in apple with overhead watering, *J. Econ. Entomol.* 91(1), 209–16.

Knight, A. L. (1999), Spatial versus sociological-based pest management: Biological and economic constraints, *Proc. Wash. Hortic. Assoc.* 95, 167–8.

Knight, A. L. (2001), Monitoring the seasonal population density of *Pandemis pyrusana* (Lepidoptera: Tortricidae) within a diverse fruit crop production area in the Yakima Valley, WA, *J. Entomol. Soc. Brit. Columbia* 217-25.

Knight, A. L. (2006), Assessing the mating status of female codling moth (Lepidoptera: Tortricidae) in orchards treated with sex pheromone using traps baited with ethyl (*E, Z*)-2,4-decadienoate, *Environ. Entomol.* 35(4), 894-900.

Knight, A. L. (2007a), Multiple mating of male and female codling moth (Lepidoptera, Tortricidae) in apple orchards treated with sex pheromone, *Environ. Entomol.* 36, 157–64.

Knight, A. L. (2007b), Influence of within-orchard trap placement on catch of codling moth (Lepidoptera: Tortricidae) in sex pheromone-treated orchards, *Environ. Entomol.* 36(2), 425–32.

Knight, A. L. (2007c), Adjusting the phenology model of codling moth (Lepidoptera: Tortricidae) in Washington State apple orchards, *Environ. Entomol.* 36(6), 1485–93.

Knight, A. L. (2008), Areawide pest management, theory and implementation, *in* Koul, O., Cuperus, G. W. and Elliot, N. (Eds), *Codling Moth Areawide Integrated Pest Management*, CAB International, Wallingford, UK, pp. 159–90.

Knight, A. L. (2010a), Targeting *Cydia pomonella* (l.) (Lepidoptera: Tortricidae) adults with low-volume applications of insecticides alone and in combination with sex pheromone, *Pest Manage. Sci.* 66, 709–17.

Knight, A. L. (2010b), Cross-resistance between azinphos-methyl and acetamiprid in populations of codling moth, *Cydia pomonella* (L.) (Lepidoptera; Tortricidae), from Washington State, *Pest Manag. Sci.* 66 (8), 865–74.

Knight, A. L. and Flexner, L. (2007), Disruption of mating in codling moth (Lepidoptera: Tortricidae) by chlorantranilipole, an anthranilic diamide insecticide, *Pest Manag. Sci.* 63, 180–9.

Knight, A. L. and Light, D. M. (2001), Attractants from Bartlett pear for codling moth, *Cydia pomonella* (L.) larvae, *Naturwissenschaften* 88(8), 339–42.

Knight, A. L. and Light, D. M. (2004), Use of ethyl (*E,Z*)-2,4-decadienoate in codling moth management, stimulation of oviposition, *J. Entomol. Soc. British Columbia* 101, 53–60.

Knight, A. L. and Light, D. M. (2005a), Dose-response of codling moth (Lepidoptera: Tortricidae) to ethyl (*E,Z*)-2,4-Decadienoate in apple orchards treated with sex pheromone dispensers, *Environ. Entomol.* 34(3), 604–9.

Knight, A. L. and Light, D. M. (2005b), Developing action thresholds for codling moth (Lepidoptera: Tortricidae) with pear ester- and codlemone-baited traps in apple orchards treated with sex pheromone mating disruption, *Can. Entomol.* 137, 739–47.

Knight, A. L. and Light, D. M. (2005c), Timing of egg hatch by early-season codling moth (Lepidoptera: Tortricidae) predicted by moth catch in pear ester- and codlemone-baited traps, *Can. Entomol.* 137, 728–38.

Knight, A. L. and Light, D. M. (2012), Monitoring codling moth (Lepidoptera: Tortricidae) in sex pheromone-treated orchards with (*E*)-4,8-dimethyl-1,3,7-nonatriene or pear ester in combination with codlemone and acetic acid, *Environ. Entomol.* 41, 407–14.

Knight, A. L. and Light, D. M. (2013), Adding microencapsulated pear ester to insecticides for control of *Cydia pomonella* (Lepidoptera, Tortricidae) in apple, *Pest Manag. Sci.* 69, 66–74.

Knight, A. L. and Light, D. M. (2014), Combined approaches using sex pheromone and pear ester for behavioural disruption of codling moth (Lepidoptera: Tortricidae), *J. Appl. Entomol.* 138, 96–108.

Knight, A. L. and Moffitt, H. R. (1991), Evaluation of packinghouse removal of codling moth-injured apples, *Tree Fruit Postharvest J.* 2(1), 14–15.

Knight, A. L. and Norton, G. W. (1989), Economics of agricultural pesticide resistance in arthropods, *Annu. Rev. Entomol.* 34, 293–313.

Knight, A. L. and Witzgall, P. (2013), Combining mutualistic yeast and pathogenic virus – a novel method for codling moth control, *J. Chem. Ecol.* 39, 1019–26.

Knight, A. L., Brunner, J. F. and Alston, D. (1994), Survey of azinphosmethyl resistance in codling moth (Lepidoptera: Tortricidae) in Washington and Utah, *J. Econ. Entomol.* 87(2), 285–92.

Knight, A. L., Howell, J. F., McDonough and Weiss, M. (1995), Mating disruption of codling moth (Lepidoptera: Tortricidae) with polyethylene tube dispensers: Determining emission rates and the distribution of fruit injuries, *J. Agric. Entomol.* 12(2), 85–100.

Knight, A. L., Turner, J. E. and Brachula, B. (1997), Predation on eggs of codling moth (Lepidoptera: Tortricidae) in mating disrupted and conventional orchards in Washington, *J. Entomol. Soc. Brit. Columbia* 94, 67–74.

Knight, A. L., Croft, B. A. and Bloem, K. A. (1999), Effect of mating disruption dispenser placement on trap performance for monitoring codling moth (Lepidoptera: Tortricidae), *J. Entomol. Brit. Columbia* 96, 95–102.

Knight, A. L., Unruh, T. R., Christianson, B. A., Puterka, G. J. and Glenn, D. M. (2000a), Effects of a kaolin-based particle film on obliquebanded leafroller (Lepidoptera: Tortricidae), *J. Econ. Entomol.* 93(3), 744–9.

Knight, A. L., Unruh, T. R., Christianson, B. A., Puterka, G. J. and Glenn, D. M. (2000b), Impacts of seasonal kaolin particle films on apple pest management, *Can. Entomol.* 133, 413–28.

Knight, A. L., Dunley, J. E. and Jansson, R. K. (2001a), Baseline monitoring of codling moth (Lepidoptera: Tortricidae) larval response to benzoylhydrazine insecticides, *J. Econ. Entomol.* 94(1), 264–70.

Knight, A. L., Christianson, B., Cockfield, S. and Dunley, J. (2001b), Areawide management of codling moth and leafrollers, *Good Fruit Grower* 52, 31–3.

Knight, A. L., Hilton, R. and Light, D. M. (2005), Monitoring codling moth (Lepidoptera: Tortricidae) in apple with blends of ethyl (*E,Z*)-2,4-Decadienoate and codlemone, *Environ. Entomol.* 34(3), 598–603.

Knight, A., Hawkins, L.; McNamara, K. and Hilton, R. (2009), Monitoring, managing codling moth clearly and precisely, *Good Fruit Grower* 60(4), 26–7.

Knight, A., Hawkins, L., McNamara, K., Borman, M. and Hilton, R. (2011), Managing codling moth clearly and precisely with semiochemicals, *IOBC/WPRS Bull.* 54, 415–18.

Knight, A. L., Stelinski, L. L., Hebert, V., Gut, L. Light, D., et al. (2012a), Evaluation of novel semiochemical dispensers simultaneously releasing pear ester and sex pheromone for mating disruption of codling moth (Lepidoptera: Tortricidae), *J. Appl. Entomol.* 136, 79–86.

Knight, A., Light, D. and Chebny, V. (2012b), Evaluating dispensers loaded with codlemone and pearester for disruption of codling moth (Lepidoptera: Tortricidae), *Environ. Entomol.* 41, 399–406.

Knight, A. L., Hilton, R. Basoalto, E. and Stelinski, L. L. (2014), Use of acetic acid to enhance bisexual monitoring of tortricid pests with kairomone lures in pome fruits, *Environ. Entomol.* 43 (6), 1628–40.

Knight, A. L., Basoalto, E. and Witzgall, P. (2015), Improving the performance of the granulosis virus of codling moth (Lepidoptera, Tortricidae) by adding the yeast *Saccharomyces cerevisiae* with sugar, *Environ. Entomol.* 44, 252–9.

Knight, A. L., El-Sayed, A. M., Judd, G. J. R. and Basoalto, E. (2017), Development of 2-phenylethanol plus acetic acid lures to monitor obliquebanded leafroller (Lepidoptera: Tortricidae) under mating disruption, *J. Appl. Entomol.* 141, 729–39.

Kogan, M. (1998), Integrated pest management: historical perspectives and contemporary developments, *Annu. Rev. Entomol.* 43, 243–70.

Konopka, J. K., Scott, I. M. and McNeil, J. N. (2012), Costs of insecticide resistance in *Cydia pomonella* (Lepidoptera: Tortricidae), *J. Econ. Entomol.* 105 (3), 872–7.

Kovaleski, A. and Mumford, J. (2007), Pulling out the evil by the root: The codling moth *Cydia pomonella* eradication programme in Brazil, *in* Vreysen, M. J. B., Robinson, A. S. and Hendrichs, J. (Eds), *Area-wide Control of Insect Pests, From Research to Field Implementation*, Springer, Dordrecht, pp. 581–90.

Kovanci, O. B., Walgenbach, J. F. and Kennedy, G. G. (2004), Evaluation of extended-season mating disruption of the oriental fruit moth *Grapholita molesta* (Busck) (Lepidopeta: Tortricidae) in apples, *J. Appl. Entomol.* 128(9), 664–9.

Kuhlmann, U. and Mills, N. J. (1999), Exploring the diversity of central Asia to assess specialized parasitoids for biological control of apple pests in Europe and North America, *IOBC/WPRS Bull.* 22, 1–6.

Kührt, U., Samietz, J. and Dorn, S. (2006), Effect of plant architecture and hail nets on temperature of codling moth habitats in apple orchards, *Entomol. Exp. Appl.* 118, 245–59.

Kumar, S., Neven, L. G., Zhu, H. and Zhang, R. (2015), Assessing the global risk of establishment of *Cydia pomonella* (Lepidoptera: Tortricidae) using CLIMEX and MaxEnt niche models, *J. Econ. Entomol.* 108, 1708–19.

Kwon, D. H., Kwon, H. K., Kim, D. H. and Yang, C. Y. (2018), Larval species composition and genetic structures of *Carposina sasakii*, *Grapholita dimorpha*, and *Grapholita molesta* from Korea, *Bull. Entomol. Res.* 108, 241–52.

Lacey, L. A. and Chauvin, R. L. (1999), Entomopathogenic nematodes for control of diapausing codling moth (Lepidoptera: Tortricidae) in fruit bins, *J. Econ. Entomol.* 92(1), 104–9.

Lacey, L. A. and Shapiro-Ilan, D. I. (2008), Microbial control of insect pests in temperate orchard systems: Potential for incorporation into IPM, *Annu. Rev. Entomol.* 53, 121–44.

Lacey, L. A., Arthurs, S., Knight, A. L. Becker, K. and Headrich, H. (2004), Efficacy of codling moth granulovirus: Effect of adjuvants on persistence of activity and comparison with other larvicides in a Pacific Northwest apple orchard, *J. Entomol. Sci.* 39(4), 500–13.

Lacey, L. A., Arthurs, S. P. and Headrick, H. (2005), Comparative activity of the codling moth granulovirus against *Grapholita molesta* and *Cydia pomonella* (Lepidoptera: Tortricidae), *J. Entomol. Soc. Brit. Columbia* 102, 79–80.

Lacey, L. A., Arthurs, S. P., Unruh, T. R., Headricks, H. and Fritts, R. (2006), Entomopathogenic nematodes for control of codling moth (lepidoptera: Tortricidae) in apple and pear orchards: Effect of nematode species and seasonal temperatures, adjuvants, application equipment, and post-application irrigation, *Biol. Control* 37, 214–23.

Lahm, G. P., Selby, T. P., Freudenberger, J. H., Stevenson, T. M., Myers, B. J., et al. (2005), Insecticidal anthranilic diamides: A new class of potent ryanodine receptor activators, *Bioorg. Med. Chem. Lett.* 15, 4898–906.

Lalancette, N., Belding, R. D., Shearer, P. W., Frecon, J. L. and Tietjen, W. H. (2004), Evaluation of hydrophobic and hydrophilic kaolin particle films for peach crop, arthropod and disease management, *Pest Manage. Sci.* 61(1), 25–39.

Landolt, P. J., Hofstetter, R. W. and Biddick, L. L. (1999), Plant essential oils as arrestants and repellents for neonate larvae of the codling moth (Lepidoptera: Tortricidae), *Environ. Entomol.* 28(6), 954–60.

Landolt, P. J., Suckling, D. M. and Judd, G. (2007), Positive interaction of a feeding attractant and host kairomone for trapping the codling moth (Lepidoptera: Tortricidae), *J. Chem. Ecol.* 33, 2236–44.

Larocque, N., Vincent, C., Belanger A. and Bourassa, J. P. (1999), Effects of tansy essential oil from *Tanacetum vulgare* on biology of obliquebanded leafroller, *Choristoneura rosaceana, J. Chem. Ecol.* 25(6), 1319–30.

Lawson, D. S., Reissig, W. H., Nyrop, J. P. and Brown, S. K. (1994), Management of arthropods on columnar apple trees using exclusionary cages, *Crop Prot.* 13(5), 346–56.

Lawson, D. S., Nyrop, J. P. and Reissig, W. H. (1997), Assays with commercially produced *Trichogramma* (Hymenoptera: Trichogrammatidae) to determine suitability for obliquebanded leafroller (Lepidoptera: Tortricidae) control, *Environ. Entomol.* 26(3), 684–93.

Li, M., Ekramirad, N. Radym A. and Adedeji, A. (2018), Application of acoustic emission and machine learning to detect moth infested apples, *Trans. ASABE* 37, 1–15.

Light, D. M. and Beck, J. J. (2010), Characterization of microencapsulated pear ester, (2*E*,4*Z*)-ethyl-2,4-decadienoate, a kairomonal spray adjuvant against neonate codling moth larvae, *J. Agric. Food Chem.* 58, 7836–45.

Light, D. M. and Beck, J. J. (2012), Behavior of codling moth (Lepidoptera, Tortricidae) neonate larvae on surfaces treated with microencapsulated pear ester, *Environ. Entomol.* 41(3), 603–11.

Light, D. M. and Knight, A. L. (2005), Kairomone-augmented mating disruption control for codling moth in Californian walnuts and apples, *IOBC/WPRS Bull.* 28, 300–3.

Light, D. M. and Knight, A. L. (2011), Microencapsulated pear ester enhances insecticide efficacy in walnuts for codling moth (Lepidoptera, Tortricidae) and navel orangeworm (Lepidoptera, Pyralidae), *J. Econ. Entomol.* 104(4), 1309–15.

Light. D. M., Knight, A. L., Henrick, C. A., Rajapaska, D., Lingren, B., et al. (2001), A pear-derived kairomone with pheromonal potency that attracts male and female codling moth, *Cydia pomonella* (L.), *Naturwissenschaften* 88(8), 333–8.

Lloyd, A., Hamacek, E, George, A., Nissen, R. and Waite, G. (2005), Evaluation of exclusion netting for insect pest control and fruit quality enhancement in tree crops, *Acta Hortic.* 694, 253–8.

Lo, P. L., Walker, J. T. S., Horner, R. M. and Hedderley, D. I. (2013), Development of multiple species mating disruption to control codling moth and leafrollers (Lepidoptera: Tortricidae), in New Zealand, *Plant Prot.* 66, 264–9.

Lo, P. L., Walker, J. T. S., Hedderley, D. I. and Cole, L. M. (2018), Reduction in leafroller (Lepidoptera; Tortricidae) abundance in orchards and vineyards 1976–2016 in Hawke's bay, New Zealand, *Agric. For. Entomol.* doi:10.1111/afe.12283.

Lombarkia, N. and Derridj, S. (2002), Incidence of apple fruit and leaf surface metabolites on *Cydia pomonella* oviposition, *Entomol. Exp. Appl.* 104(1), 79–97.

Lombarkia, N. and Derridj, S. (2008), Resistance of apple trees to *Cydia pomonella* egg-laying due to leaf surface metabolites, *Entomol. Exp. Appl.* 128(1), 57–65.

Lombarkia, N., Derridj, S., Ioriatti, C. and Bourget, E. (2013), Effect of a granulovirus larvicide, Madex, on egg-laying of *Cydia pomonella* L. (Lepidoptera: Tortricidae) due to changes in chemical signalization on the apple leaf surface, *Afr. Entomol.* 21(2), 196–208.

Lowery, D. T. and Smirle, M. J. (2000), Toxicity of insecticides to obliquebanded leafroller, *Choristoneura rosaceana*, larvae and adults exposed previously to neem seed oil, *Entomol. Exp. Appl.* 95, 201–7.

Luedeling, E., Steinmann, K. P., Zhang, M., Brown, P. H., Grant, J., et al. (2011), Climate change effects on walnut pests in California, *Global Change Biol.* 17, 228–38.

MacCollom, G. B. and Reed, E. M. (1971), A nuclear polyhedrosis virus of the light brown apple moth, *Epiphyas postvittana*, *J. Invert. Pathol.* 18(3), 337–43.

Machial, C. M., Ikkei, S., Smirle, M., Bradbury, R. and Isman, M. B. (2010), Evaluation of the toxicity of 17 essential oils against *Choristoneura rosaceana* (Lepidoptera: Tortricidae) and *Trichoplusia ni* (Lepidoptera: Noctuidae), *Pest Manag. Sci.*, 66, 1116–21.

MacLellan, C. R. (1973), Natural enemies of the light brown apple moth, *Epiphyas postvittana* in the Australian Capital Territory, *Can. Entomol.* 105(3), 681–700.

Magalhaes, L. C. and Walgenbach, J. F. (2011), Life stage toxicity and residual activity of insecticides to codling moth and oriental fruit moth (Lepidoptera: Tortricidae), *J. Econ. Entomol.* 104(6), 1950–9.

Magalhaes, L. C., Van Kretschmar, J. B., Barlow, V. M., Roe, R. M. and Walgenbach, J. F. (2012), Development of a rapid resistance monitoring bioassay for codling moth larvae, *Pest Manag. Sci.* 68, 883–8.

Maheswaran, G., Pridmore, L., Franz, P. and Anderson, M. A. (2007), A proteinase inhibitor from *Nicotiana alata* inhibits the normal development of light-brown apple moth, *Epiphyas postvittana* in transgenic apple plants, *Plant Cell Rep.* 26(6), 773–82.

Mani, E. and Wildbolz, T. (1977), The dispersal of male codling moths (*Laspeyresia pomonella* L.) in the upper Rhine valley, *J. Appl. Entomol.* 83(1), 161–8.

Margaritopoulos, J. T., Voudouris, C. C., Olivares, J., Sauphanor, B., Mamuris, Z., et al. (2012), Dispersal ability in codling moth: Mark-release-recapture experiments and kinship analysis, *Agric. Forest Entomol.* 14, 399–407.

Marko, V., Blommers, L. H. M., Bogya, S. and Helsen, H. (2008), Kaolin particle films suppress many apple pests, disrupt natural enemies and promote woolly apple aphid, *J. Appl. Entomol.* 132, 26–35.

Marko, V., Jenser, G., Muhalyi, K. Hegyi, T. and Balazs, K. (2012), Flowers for better pest control? Effects of apple orchard groundcover management on mites (Acari), leafminers (Lepidoptera: Scitellidae), and fruit pests, *Biocontrol Sci. Technol.* 22(1), 39–60.

Markwick, N. P., Docherty, L. C., Phung, M. M., Lester, M. T., Murray, C., et al. (2003), Transgenic tobacco and apple plants expressing biotin-binding proteins are resistant to two cosmopolitan insect pests, potato tuber moth and light brown apple moth, respectively, *Transgenic Res.* 12, 671–81.

Marliac, G., Penvern, S., Barbier, J.-M., Lescourret, F. and Capowiez, Y. (2015), Impact of crop protection strategies on natural enemies in organic apple production, *Agron. Sustain. Dev.* 35, 803–13.

Mastrangelo, T., Parker, A. G., Jessup, A., Pereira, R., Orozco-Dávila, D., et al. (2010), A new generation of X ray irradiators for insect sterilization, *J. Econ. Entomol.* 103, 85–94.

McCann, S. F., Annis, G. D., Shapiro, R., Piotrowski, D. W., Lahm, G. P., et al. (2001), The discovery of indoxacarb: Oxadiazines as a new class of pyrazoline-type insecticides, *Pest Manag. Sci.* 57, 153–64.

McDougall, S. J. and Mills, N. J. (1997), The influence of hosts, temperature and food sources on the longevity of *Trichogramma platneri*, *Entomol. Exp. Appl.* 83, 195–203.

McGhee, P. S., Epstein, D. L. and Gut, L. J. (2011), Quantifying the benefits of areawide pheromone mating disruption programs that target codling moth (Lepidoptera: Tortricidae), *Am. Entomol.* 57(2), 93–100.

Men, Q.-L., Chen, M.-H., Zhang, Y.-L. and Feng, L.-L. (2013), Genetic structure and diversity of a newly invasive species, the codling moth, *Cydia pomonella* (L.) (Lepidoptera: Tortricidae) in China, *Biol. Invasions* 15, 447–58.

Merckx, T., Feber, R. E., Dulieu, R. L., Townsend, M. C. and Parsons, M. S. (2009), Effect of field margins on moths depends on species mobility: Field-based evidence for landscape-scale conservation, *Agric. Ecosyst. Environ.* 129, 302–9.

Miliczky, E. R. and Calkins, C. O. (2002), Spiders (*Araneae*) as potential predators of leafroller larvae and egg masses (Lepidoptera: Tortricidae) in Central Washington apple and pear orchards, *Pan-Pacif. Entomol.* 78(2), 140–50.

Miller, J. R. and Gut, L. J. (2015), Mating disruption for the 21st century: Matching technology with mechanism, *Environ. Entomol.* 44(3), 427–53.

Mills, N. (2003), Augmentation in orchards: Improving the efficacy of *Trichogramma* inundation, *in* van Driesche, R. (Ed.), *1st International Symposium on Biological Control of Arthropods*, USDA Forest Service, pp. 130–5.

Mills, N. (2005), Selecting effective parasitoids for biological control introductions: Codling moth as a case study, *Biol. Control* 34, 274–82.

Mills, N. (2013), Biological pest control in organic tree crops in the western U.S.: An overview. *Acta Hortic.* 1001, 105–10.

Mills, N. J., Pickel, C., Mansfield S., McDougall, S., Buchner, R., et al. (2000), *Trichogramma* inundation: Integrating parasitism into management of codling moth, *Calif. Agric.* 54(6), 22–5.

Mills, N. J., Beers, E. H., Shearer, P. W., Unruh, T. R. and Amarasekare, K. G. (2016), Comparative analysis of pesticide effects on natural enemies in western orchards: A synthesis of laboratory bioassay data, *Biol. Control* 102, 17–25.

Monteiro, L. B., Lavigne, C., Ricci, B., Franck, P., Toubon, J.-F., et al. (2012), Predation of codling moth eggs is affected by pest management practices at orchard and landscape levels, *Agric. Ecosyst. Environ.* doi:10.1016/j.agee.2011.10.012.

Mota-Sanchez, D., Wise, J. C., VanderPoppen, R., Gut, L. J. and Hollingworth, R. M. (2008), Resistance of codling moth, *Cydia pomonella* (L.) (Lepidoptera: Tortricidae), larvae in Michigan to insecticides with different modes of action and the impact on field residual activity, *Pest Manag. Sci.* 64, 881–90.

Mujica, V., Preti, M., Basoalto, E., Cichon, L., Fuentes-Contreras, E., et al. (2018), Improved monitoring of oriental fruit moth (Lepidoptera: Tortricidae) with terpinyl acetate plus acetic acid membrane lures, *J. Appl. Entomol.* 142, 731–44.

Mullinix, K., Isman, M. B. and Brunner, J. F. (2010), Key and secondary arthropod pest population trends in apple cultivated over four seasons with no insecticides and a legume cover, *J. Sustain. Agric.* 34, 584–94.

Mullinix, K., Brunner, J. F. and Isman, M. B. (2011), Apple leafroller (Lepidoptera: Tortricidae) populations and parasitism in an orchard managed with either a grass or alfalfa cover and without insecticides over four growing seasons, *Int. J. Fruit Sci.* 11(2), 99–110.

Murata, M., Nishimura, M., Murai, N., Haruta, M., Homma, S., et al. (2001), A transgenic apple callus showing reduced polyphenol oxidase activity and lower browning potential, *Biosci. Biotechnol. Biochem.* 65(2), 383–8.

Myers, C. T., Hull, L. A. and Krawczyk, G. (2006), Comparative survival rates of oriental fruit moth (Lepidoptera: Tortricidae) larvae on shoots and fruit of apple and peach, *J. Econ. Entomol.* 99(4), 1299–309.

Navarro-Roldan, M. A. and Gemeno, C. (2017), Sublethal effects of neonicotinoid insecticide on calling behavior and pheromone production of tortricid moths, *J. Chem. Ecol.* 43, 881–90.

Navarro-Roldan, M. A., Avilla, J., Bosch, D., Valls, J. and Gemeno, C. (2017), Comparative effect of three neurotoxic insecticides with different modes of action on adult males and females of three tortricid moth pests, *J. Econ. Entomol.* 110(4), 1740–9.

Negrisoli, C. R., Negrisoli, A. S. Garcia, M. S., Dolinski, C. and Bernardi, D. (2013), Control of *Grapholita molesta* (Busck 1916) (Lepidoptera: Tortricidae) with entomopathogenic nematodes (Rhabditida: Heterorhabditidae, Steinernematidae) in peach orchards, *Exp. Parasit.* 135, 466–70.

Neven, L., Ferguson, H. J. and Knight A. (2000), Sub-zero cooling synchronizes post-diapause development of codling moth, *Cydia pomonella, Cryo Letters* 21(4), 203–14.

Neven, L. G., Kumar, S., Yee, W. L. and Wakie, W. (2018), Current and future potential risk of establishment of *Grapholita molesta* (Lepidoptera: Tortricidae) in Washington State, *Environ. Entomol.* 47 (2), 448–56.

Norelli, J. L., Jones, A. L. and Aldwinckle, H. S. (2003), Fire blight management in the twenty-first century - Using new technologies that enhance host resistance in apple. *Plant Disease*. 87(7):756–65.

Onstad, D. W. (2008), *Insect Resistance Management, Biology, Economics and Prediction*, Elsevier, Amsterdam, 305pp.

Padilha, A. C., Arioli, C. J., Boff, M. I. C., Rosa, J. M. and Botton, M. (2017), Traps and baits for luring *Grapholita molesta* (Busck) adults in mating disruption-treated apple orchards, *Neotrop. Entomol.* doi:10.1007/s13744-017-0517-z.

Panneton, B., Lacasse, B. and Theriault, R. (2005), Penetration of spray in apple trees as a function of airspeed, airflow, and power for tower sprayers, *Can. Biosys. Eng.* 47, 213–20.

Pasqualini, E., Villa, M., Civolani, S., Espinha, I., Ioriatti, C., et al. (2005a), The pear ester ethyl (*E,Z*)-2,4-decadienoate as a potential tool for control of *Cydia pomonella* larvae, preliminary investigation, *Bull. Insect.* 58, 65–9.

Pasqualini, E., Schmidt, S., Espinha, I., Civolani, S., Ioriatti, C., et al. (2005b), Effects of the kairomone ethyl (2*E*, 4*Z*)-2,4-decadienoate (DA 2313) on the oviposition behaviour of *Cydia pomonella*, preliminary investigations, *Bull. Insect.* 58, 119–24.

Pasquier, D. and Charmillot, P. J. (2003), Effectiveness of twelve insecticides applied topically to diapausing larvae of the codling moth, *Cydia pomonella* L., *Pest Manag. Sci.* 60, 305–8.

Pfannestiel, R. S., Unruh, T. R. and Brunner, J. F. (2010), Overwintering hosts for the exotic leafroller parasitoid, *Colpoclypeus florus*: Implications for habitat manipulation to augment biological control of leafrollers in pome fruits, *J. Insect Sci.* 10, a75.

Pluciennik, Z. and Olszak, R. W. (2010), The role of parasitoids in limiting the harmfulness of leafrollers in apple orchards, *J. Plant Protec. Res.* 50(1), doi:10.2478/v10045-010-0001-9 (accessed September 2018).

Porcel, M., Sjoberg, P., Swiergiel, W., Dinwiddle, R., Ramert, B. and Tasin, M. (2014), Mating disruption of *Spilonota ocellana* and other apple orchard tortricids using a multispecies reservoir, *Pest Manag. Sci.* 71(4), 562–70.

Pree, D. J., Whitty, K. J. and Van Driel, L. (1998), Resistance to insecticides in oriental fruit moth populations (*Grapholita molesta*) from the Niagara peninsula of Ontario, *Can. Entomol.* 230, 245–56.

Prokopy, R. J. and Croft, B. A. (1994), Apple insect pest management, *in* Metcalf, R. L. and Luckman, W. H. (Eds), *Introduction to Insect Pest Management*. Wiley, New York, NY, pp. 543–85.
Pronier, I., Pare, J., Wissocq, J.-C. and Vincent, C. (2002), Nucleopolyhedrovirus infection in obliquebanded leafroller (Lepidoptera: Tortricidae), *Can. Entomol.* 134, 303–9.
Pszczolkowski, M. A. and Brown, J. J. (2002), Prospects of monosodium glutamate use for enhancement of spinosad toxicity against codling moth neonates, *Phytoparasitica* 31, 283–91.
Pszczolkowski, M. A. and Brown, J. J. (2004), Enhancement of spinosad toxicity to *Cydia pomonella* neonates by monosodium glutamate receptor agonist, *Phytoparas.* 32(4), 342–50.
Pszczolkowski, M. A., Brunner, J. F., Doerr, M. D. and Brown, J. J. (2004), Enhancement of *Bacillus thuringiensis* with monosodium glutamate against larvae of obliquebanded leafroller (Lep.: Tortricidae), *J. Appl. Entomol.* 128(7), 474–7.
Pszczolkowski, M. A., Durden, K., Sellars, S., Cowell, B. and Brown, J. J. (2011), Effects of *Ginkgo biloba* constituents on fruit-infesting behavior of codling moth (*Cydia pomonella*) in apples, *J. Agric. Food Chem.* 59, 10879–86.
Puterka, G. J., Glenn, D. M. and Pluta, R. C. (2005), Action of particle films on the biology and behavior or pear psylla (Homoptera: Psyllidae), *J. Econ. Entomol.* 98(6), 2079–88.
Randall, P., Sholberg, P., Judd, G. and Cossentine, J. (2011), Acetic acid fumigation of fruit storage bins to control diapausing codling moth larvae, *HortSci.* 46(12), 1634–9.
Regier, J. C., Brown, J. W., Mitter, C., Baixeras, J., Cho, S., et al. (2012), A molecular phylogeny for the leaf-roller moths (Lepidoptera: Tortricidae) and its implications for classification and life history evolution, *PLoS ONE* 7, e35574. doi:10.1371/journal.pone.0035574.
Reyes, M. and Sauphanor, B. (2008), Resistance monitoring in codling moth: A need for standardization, *Pest Manag. Sci.* 64,945–53.
Reyes, M., Franck, P., Charmillot, P. J., Ioriatti, C. and Olivares, J. (2007), Diversity of insecticide resistance mechanisms and spectrum in European populations of the codling moth, *Cydia pomonella*, *Pest Manag. Sci.* 63, 890–902.
Reyes, M., Franck, P., Olivares, J., Margaritopoulos, J., Knight, A. L., et al. (2009), Worldwide variability of insecticide resistance mechanisms in the codling moth, *Cydia pomonella* L. (Lepidoptera: Tortricidae), *Bull. Entomol. Res.* 99, 359–69.
Ricci, B., Franck, P., Toubon, J. F., Bouvier, J. C. and Sauphanor, B. (2009), The influence of landscape on insect pest dynamics: A case study in southeastern France, *Landsc. Ecol.* 24, 337–49.
Ricci, B. R., Franck, B. P., Bouvier, J. C., Casado, D. and Lavigne, C. (2011), Effects of hedgerow characteristics on intra-orchard distribution of larval codling moth, *Agric. Ecosyst. Environ.* 140, 395–400.
Riddick, E. W. and Mills, N. J. (1994), Potential of adult carabids (Coleoptera: Carabidae) as predators of fifth-instar codling moth (Lepidoptera; Tortricidae) in apple orchards in California, *Environ. Entomol.* 23(5), 1338–45.
Riedl, H. (1995), Laboratory evaluation of mineral oils for control of codling moth (Lepidoptera: Tortricidae), *J. Econ. Entomol.* 88(1), 140–7.
Riedl, H., Croft, B. A. and Howitt, A. J. (1976), Forecasting codling moth phenology based on pheromone trap catches and physiological time models, *Can. Entomol.* 108, 449–60.

Rodriguez, M. A., Bosch, D. and Avilla, J. (2011), Resistance of Spanish codling moth (*Cydia pomonella*) populations to insecticides and activity of detoxifying enzymatic systems, *Entomol. Exp. Appl.* 138(3), 184–92.

Roush, R. T. and Tabashnik, B. E. (Eds) (1990), *Pesticide Resistance in Arthropod*, Chapman and Hall, New York, NY, 303pp.

Sackett, T. E., Buddle, C. M. and Vincent, C. (2005), Effects of kaolin on fitness and behavior of *Choristoneura rosaceana* (Lepidoptera: Tortricidae) larvae, *J. Econ. Entomol.* 98(5), 1648–53.

Sackett, T. E., Buddle, C. M. and Vincent, C. (2007), Effects of kaolin on the composition of generalist predator assemblages and parasitism of *Choristoneura rosaceana* (Lep., Tortricidae) in apple orchards, *J. Appl. Entomol.* 131(7), 478–85.

Sarvary, M. A., Nyrop, J. and Reissing, H. (2010a), Effects of natural enemies and host plants in wild and orchards habitats on obliquebanded leafroller (Lepidoptera: Tortricidae) larval survival, *Biolog. Control* 55, 110–17.

Sarvary, M. A., Nyrop, J., Reissing, H. and Gifford, K. M. (2010b), Potential for conservation biological control of the obliquebanded leafroller (OBLR) *Choristoneura rosaceana* (Harris) in orchards systems managed with reduced-risk insecticides, *Biolog. Control* 40, 37–47.

Sauphanor, B. and Bouvier, J. C. (1995), Cross-resistance between benzoylureas and benzoylhydrazines in the codling moth, *Cydia pomonella* L. *Pest Manag. Sci.* 45, 369–75.

Sauphanor, B., Cuany, A., Bouvier, J. C., Brosse, V., Amichot, M., et al. (1997), Mechanism of resistance to deltamethrin in *Cydia pomonella* (L.) (Lepidoptera: Tortricidae), *Pest Biochem. Physiol.* 58, 109–17.

Sauphanor, B., Brosse, V., Monier, C. and Bouvier, J. C. (1998a), Differential ovicidal and larvicidal resistance to benzoylureas in the codling moth, *Cydia pomonella*, *Entomol. Exp. Appl.* 88, 247–53.

Sauphanor, B., Bouvier, J. C. and Brosse, V. (1998b), Spectrum of insecticide resistance in *Cydia pomonella* (Lepidoptera: Tortricidae) in southeastern France, *J. Econ. Entomol.* 91, 1225–31.

Sauphanor, B., Brosse, V., Bouvier, J. C., Speich, P., Micoud, A., et al. (2000), Monitoring resistance to diflubenzuron and deltamethrin in French codling moth populations (*Cydia pomonella*), *Pest Manage. Sci.* 56, 74–82.

Sauphanor, B., Berling, M., Toubon, J. F., Reyes, M., Delnatte, J., et al. (2006), Carpocapse des pommes, Cas de résistance au virus de la granulose en vergers biologiques, *Phytoma* 590, 24–7.

Sauphanor, B., Franck, P., Lasnier, T., Toubon, J. F. and Beslay, D., et al. (2007), Insecticide resistance may enhance the response to host-plant volatile kairomone for the codling moth, *Cydia pomonella*, *Naturwissen.* 94(6), 449–58.

Sauphanor, B., Severac, G., Maugin, S., Toubon, J. F. and Capowiez, Y. (2012), Exclusion netting may alter reproduction of the codling moth (Cydia pomonella) and prevent associated fruit damage to apple orchards, *Entomol. Experi. Appl.* 145(2), 134–42.

Schmidt, S., Tomasi, C., Pasqualini, E. and Ioriatti, C. (2008), The biological efficacy of pear ester on the activity of granulosis virus for codling moth, *J. Pest Sci.* 81, 29–34.

Schumacher, P., Weber, D. C., Hagger, C. and Dorn, S. (1997a), Heritability of flight distance for *Cydia pomonella*, *Entomol. Exp. Appl.* 85, 169–75.

Schumacher, P., Weyeneth, A., Weber, D. C. and Dorn, S. (1997b), Long flights in *Cydia pomonella* L. (Lepidoptera: Tortricidae) measured by a flight mill: Influence of sex, mated status and age, *Physiol. Entomol.* 22, 149–60.

Sciarretta, A. and Trematerra, P. (2006), Geostatistical characterization of the spatial distribution of *Grapholita molesta* and *Anarsia lineatella* males in an agricultural landscape, *J. Appl. Entomol.* 130, 73–83.

Severac, G. and Romet, L. (2008), Alt'Carpo, une alternative efficace, *Phytoma* 612, 16–20.

Sharma, R. R., Reddy, S. V. R. and Jhalegar, M. J. (2014), Pre-harvest fruit bagging: A useful approach for plant protection and improved post-harvest fruit quality – a review, *J. Hortic. Sci. Biotechnol.* 89(2), 101–13.

Sial, A. Q. and Brunner, J. F. (2010), Toxicity and residual efficacy of chlorantraniliprole, spinoteram, and emamectin benzoate to obliquebanded leafroller (Lepidoptera: Tortricidae), *J. Econ. Entomol.* 103(4), 1277–85.

Sial, A. A. and Brunner, J. F. (2012), Selection for resistance, reversion towards susceptibility and synergism of chlorantraniliprole and spinetoram in obliquebanded leafroller, *Choristoneura rosaceana* (Lepidoptera: Tortricidae), *Pest Manag. Sci.* 68, 462–8.

Silva-Brandao, K. L., Silva, O. A. B. N. E., Brandao, M. M., Omoto, C. and Sperling, F. A. H. (2015), Genotyping-by-sequencing approach indicates geographic distance as the main factor affecting genetic structure and gene flow in Brazilian populations of *Grapholita molesta* (Lepidoptera, Tortricidae), *Evol. Appl.* 8(5), 476–85.

Simon, S., Lauri, P. E., DeFrance, H. and Sauphanor, B. (2006), Does manipulation of fruit-tree architecture affect the development of pests and pathogens? A case study in an organic apple orchard, *J. Hortic. Sci. Biotechnol.* 81 (4), 765–73.

Simon, S., Sauphanor, B. and Lauri, P.-E. (2007), Control of fruit tree pests through manipulation of tree architecture, *Pest Technol.* 191, 33–7.

Smirle, M. J., Lowery, T. and Zurowski, C. L. (2002), Resistance and cross-resistance to four insecticides in populations of obliquebanded leafroller (Lepidoptera: Tortricidae), *J. Econ. Entomol.* 95(4), 820–5.

Smirle, M. J., Lowery, D. T. and Zurowski, C. L. (2003), Susceptibility of leafrollers (Lepidoptera: Tortricidae) from organic and conventional orchards to azinphos-methyl, spinosad, and *Bacillus Thuringiensis*, *J. Econ. Entomol.* 96, 879–84.

Sofu, M. M., Er, O., Kayacan, M. C. and Cetisli, B. (2016), Design of an automatic apple sorting system using machine vision, *Comput. Electron. Agric.* 127, 395–405.

Song, B., Jiao, H., Tang, G. and Yao, Y. (2014), Combining repellent and attractive aromatic plants to enhance biological control of three tortricid species (Lepidoptera: Tortricidae) in an apple orchard, *Florida Entomol.* 97 (4), 1679–89.

Sonka, S. T. and Coaldrake, K. F. (1997), Cyberfarm: What does it look like? what does it mean? *Am. J. Agric. Econ.* 78, 1263–8.

Sparks, T. C. and Nauen, R. (2015), IRAC: Mode of action classification and insecticide resistance management, *Pest. Biochem. Physiol.* 121, 122–8.

Stelinski, L. L., Gut, L. J., Haas, M., McGhee, P. and Epstein, D. (2007), Evaluation of aerosol devices for simultaneous disruption of sex pheromone communication in *Cydia pomonella* and *Grapholita molesta* (Lepidoptera: Tortricidae), *J. Pest Sci.* 80, 225–33.

Stelinski, L. L., Il'Ichev, A. L. and Gut, L. J. (2009), Efficacy and release rate of reservoir pheromone dispensers for simultaneous mating disruption of codling moth and oriental fruit moth (Lepidoptera: Tortricidae), *J. Econ. Entomol.* 102(1), 315–23.

Stoeckli, S., Mody, K. and Dorn, S. (2008), Influence of canopy aspect and height on codling moth (Lepidoptera: Tortricidae) larval infestation in apple and relationship between infestation and fruit size, *J. Econ. Entomol.* 101(1), 81-9.

Stoeckli, S., Hirschi, M., Spirig, C., Calanca, P., Rotach, M. W., et al. (2012), Impact of climate change on voltinism and prospective diapause induction of a global pest insect *Cydia pomonella* (L.), *PLoS ONE*, 7, e35723.

Suckling, D. M., Gibb, A. R., Burnip, G. M. and Delury, N. C. (2002), Can parasitoid sex pheromones help in insect biocontrol? A case study of codling moth (Lepidoptera: Tortricidae) and its parasitoid *Ascogaster quadridentata* (Hymenoptera: Braconidae), *Environ. Entomol.* 31(6), 947-52.

Suckling, D. M., Stringer, L. D., Baird, D. B., Butler R. C., Sullivan, T. E. S., et al. (2014), Light brown apple moth (*Epiphyas postvittana*) (Lepidoptera; Tortricidae) colonization of California, *Biol. Invasions* 16, 1851-63.

Sun, X. and Barrett, B. A. (1999), Fecundity and fertility changes in adult codling moth (Lepidoptera: Tortricidae) exposed to surfaces treated with tebufenozide and methoxyfenozide, *J. Econ. Entomol.* 92, 1039-44.

Sun, X., Barrett, B. A. and Biddinger, D. J. (2000), Fecundity and fertility reductions in adult leafroller exposed to surfaces treated with the ecdysteroid agonists tebufenozide and methoxyfenozide, *Entomol. Exp. Appl.* 94, 75-83.

Tasin, M., Demaria, D., Ryne, C. Cesano, A., Galliano, A, et al. (2008), Effect of anti-hail nets on *Cydia pomonella* behavior in apple orchards, *Entomol. Exp. Appl.* 129, 32-6.

Thistlewood, H. and Judd, G. J. R. (2003a), Movement and dispersal of codling moth, *Cydia pomonella*, in rural and orchard areas, *IOBC/WPRS Bull.* 26, 111-16.

Thistlewood, H. and Judd, G. J. R. (2003b), Area-wide management of codling moth, *Cydia pomonella*, at very low densities, *IOBC/WPRS Bull.* 26(11), 103-9.

Thomas, D. D., Donnelly, C. A., Wood, R. J. and Alphey, L. S. (2000), Insect population control using a dominant, repressible, lethal genetic system, *Science* 287(5462), 2474-6.

Thompson, G. D., Dutton, R., and Sparks, T. C. (2000), Spinosad – a case study: An example from a natural products discovery programme, *Pest Manag. Sci.* 56, 696-702.

Trematerra, P., Gentile, P. and Sciarretta, A. (2004), Spatial analysis of pheromone trap catches of codling moth (*Cydia pomonella*) in two heterogeneous agro-ecosystems, using geostatistical techniques, *Phytoparasitica* 32, 325-41.

Trimble, R. M. and Vickers, P. M. (2000), Evaluation of border sprays for managing the codling moth (Tortricidae: Lepidoptera) and the apple maggot (Tephritidae: Diptera) in Ontario apple orchards, *J. Econ. Entomol.* 93(3), 777-87.

Trimble, R. M., El-Sayed, A. M. and Pree, D. J. (2004), Impact of sub-lethal residues of azinphos-methyl on the pheromone-communication systems of insecticide-susceptible and insecticide-resistant obliquebanded leafrollers *Choristoneura rosaceana* (Lepidoptera: Tortricidae), *Pest Manag. Sci.* 60, 660-8.

Tyson, R., Thistlewood, H. and Judd, G. J. R. (2007), Modelling dispersal of sterile male codling moths, *Cydia pomonella*, across orchard boundaries, *Ecol. Mod.* 205, 1-12.

Undorf-Spahn, K., Fritsch, E., Huber, J., Kienzle, J., Zebitz, P. W., et al. (2012), High stability and no fitness costs of the resistance of codling moth to *Cydia pomonella* granulovirus (CpGV-M), *J. Invert. Path.* 111, 136-42.

Unruh, T. R. (1998), From Russia with love: New predators and parasite for control of tree fruit insect pests, *Proc. Wash. Hortic. Assoc.* 94, 42-9.

Unruh, T. R. and Lacey, L. A. (2001), Control of codling moth, *Cydia pomonella* (Lepidoptera: Tortricidae), with *Steinernema carpocapsae*: Effects of supplemental wetting and pupation site on infection rate, *Biol. Control* 20, 48–56.

Unruh, T. R., Knight, A. L., Upton, J., Glenn, D. M. and Puterka, P. J. (2000), Particle films for suppression of the codling moth (Lepidoptera; Tortricidae) in apple and pear orchards, *J. Econ. Entomol.* 93, 737–43.

Unruh, T. R., Pfannenstiel, R. S., Peters, C., Brunner, J. F. and Jones, V. P. (2012), Parasitism of leafrollers in Washington fruit orchards is enhanced by perimeter plantings of rose and strawberry, *Biol. Control* 62, 162–72.

Unruh, T. R., Miliczky, E. R., Horton, D. R., Thomsen-Archer, K., Rehfield-Ray, L., et al. (2016), Gut content analysis of arthropod predators of codling moth in Washington apple orchards, *Biol. Control* 102, 85–92.

Vail, P. V., Tebbets, J. S., Mackey, B. E. and Curtis, C. E. (1993), Quarantine treatments: A biological approach to decision-making for selected hosts of codling moth (Lepidoptera; Tortricidae), *J. Econ. Entomol.* 86(1), 70–5.

Vakenti, J. M., Cossentine, J. E., Cooper, B. E., Sharkey, M. J., Yoshimoto, C. M., et al. (2001), Host-plant range and parasitoids of obliquebanded and three-lined leafrollers (Lepidoptera: Tortricidae), *Can. Entomol.* 133, 139–46.

Valentine, B. J., Gurr, G. M. and Thwaite, W. G. (1996), Efficacy of the insect growth regulators tebufenozide and fenoxycarb for lepidopteran pest control in apples, and their compatibility with biological control for integrated pest management, *Aust. J. Exp. Agric.* 36(4), 501-6.

Varela, L. G., Welter, S. C., Jones, V. P., Brunner, J. F. and Riedl, H. (1993), Monitoring and characterization of insecticide resistance in codling moth (Lepidoptera: Tortricidae) in four western states, *J. Econ. Entomol.* 86(1), 1-10.

Vickers, R. A. (1997), Effect of delayed mating on oviposition pattern, fecundity and fertility in codling moth, *Cydia pomonella* (L.) (Lepidoptera: Tortricidae), *Aust. J. Entomol.* 36, 179–82.

Voudouris, C. C., Sauphanor, B., Franck, P., Reyes, M., Mamuris, Z., et al. (2011), Insecticide resistance status of the codling moth *Cydia pomonella* (Lepidoptera: Tortricidae) from Greece, *Pest. Biochem. Physiol.* 100(3), 229–328.

Vreysen, M. J. B., Robinson, A. S. and Hendrichs, J. (2007), *Area-wide Control of Insect Pests: From Research to Field Implementation*, Springer, Dordrecht, The Netherlands.

van der Geest, L. P. S. and Evenhuis. H. H. (Eds), (1991), *Tortricid Pests: Their Biology, Natural Enemies and Control*, Elsevier, Amsterdam, The Netherlands, 808pp.

Waelti, H. (1992), Should we use plastic bins? *Tree Fruit Postharvest J.* 3, 14–17.

Walker, K. R. and Welter, S. C. (2001), Potential for outbreaks of leafrollers (Lepidoptera: Tortricidae) in California apple orchards using mating disruption for codling moth suppression, *J. Econ. Entomol.* 94(2), 373–80.

Waltz, E. (2015), Nonbrowning GM apple cleared for market, *Nat. Biotechnol.* 33(4), 326–7.

Wang, S., Birla, S. L., Tang, J. and Hansen, J. D. (2006), Postharvest treatment to control codling moth in fresh apples using water assisted radio frequency heating, *Postharvest Biol. Technol.* 40, 89–96.

Wearing, C. H. and Hokkaner, H. M. T. (1994), Pest resistance to *Bacillus thuringiensis*: Case studies of ecological crop assessment for Bt gene incorporation and strategies of management, *Biocontr. Sci. Technol.* 4, 573–90.

Weddle, P. W., Welter, S. C. and Thomson, D. (2009), History of IPM in California pears – 50 years of pesticide use and the transition to biologically intensive IPM, *Pest Manag. Sci.* 65, 1287–92.

Whitaker, K. and Middleton, S. (1999), The profitability of hail netting in apple orchards, *Aust. Agric. Res. Econ.* 43, 1-15.

Willett, M. and Westigard, P. H. (1988), Using horticultural spray oils to control orchard pests, *Pacific Northwest Coop. Ext. Bull.* 328, Corvalis, OR.

Willett, M. J., Neven, L. and Miller, C. E. (2009), The occurrence of codling moth in low latitude countries: Validation of pest distribution reports, *HortTech.* 19(3), 633-7.

Witzgall, P., Stelinski, L. L., Gut, L. J. and Thomson, D. (2008), Codling moth management and chemical ecology, *Annu. Rev. Entomol.* 53, 503-22.

Xu, K. (2013), The next generation biotechnology for apple improvement and beyond: The CRISPR/CAS9 story, *N. Y. Fruit Quart.* 21(4), 19-22.

Xu, X., Wu, P., Thorbek, P. and Hyder, K. (2006), Variability in initial spray deposit in apple trees in space and time, *Pest Manag. Sci.* 62, 947-56.

Yamada, H., Parker, A. G., Oliva, C. F., Balestrino, F. and Gilles, J. R. L. (2014), X-ray-induced sterility in *Aedes albopictus* (Diptera: Culicidae) and male longevity following irradiation, *J. Med. Entomol.* 51, 811-16.

Zhang, B.-C. (1994), *Index of Economically Important Lepidoptera*, CAB International, Wallingford, UK, 599pp.

Zheng, Y., Qiao, X., Wang, K., Dorn, S. and Chen, M. (2015), Population genetics affected by pest management using fruit-bagging: A case study with *Grapholita molesta* in China, *Netherlands Entomol. Soc. Entomol. Experi.* 156, 117-27.

Zijlstra, C., Lund, I., Justesen, A. F., Nicolaisen, M., Jensen, P. K., et al. (2011), Combining novel monitoring tools and precision application technologies for integrated high-tech crop protection in the future, *Pest Manag. Sci.* 67, 616-25.

Zingg, D., Kraaz, I., Wandeler, H. and Zuger, M. (2012), Madex twin, a new *Cydia pomonella* granulovirus isolate for the control of both codling moth *Cydia pomonella* and oriental fruit moth *Grapholita molesta*, *Proceedings of the 15th Ecofruit International Conference on Organic Fruit Growing*, Hohenheim, Germany, 20-22 February 2012, pp. 187-93.

Chapter 17

Integrated management of mite pests of tree fruit

Rebecca A. Schmidt-Jeffris, Clemson University, USA; Elizabeth H. Beers, Washington State University, USA; and Carlo Duso, University of Padua, Italy

1 Introduction

Mites are serious and chronic pests of many tree fruit crops worldwide. This chapter will focus on mite pests of pome (apple, pear) and stone (cherry, peach, apricot etc.) fruits. Mites that cause damage to these crops include the tetranychids (spider mites) and eriophyids. Depending on their life history and damage, various species of eriophyids on tree fruits are referred to as rust, blister or bud mites. We will discuss the identification, biology and management of tetranychids and eriophyids in tree fruit crops, with management including cultural, biological and chemical control.

1.1 Spider mites

Spider mites pierce individual cells of leaves and consume the contents, decreasing photosynthetic activity, reducing plant vigour and, consequently, crop yield (Huffaker and McMurtry, 1969; Hall and Ferree, 1975; Sances et al., 1981; Lakso et al., 1996; Park and Lee, 2005). Plant response can vary considerably in severity; leaf bronzing (a brownish discolouration) is a common outcome of mite feeding, but in more sensitive crops (almonds, certain pear cultivars), leaf necrosis or abscission results from even moderate levels of damage (Huffaker and McMurtry, 1969; Rice and Jones, 1978; Hoyt,

http://dx.doi.org/10.19103/AS.2019.0046.20

1991; Beers and Hoyt, 1993). Spider mite outbreaks occur worldwide on many different crops. These outbreaks are primarily attributed to the harmful non-target effects of agricultural chemicals on spider mite natural enemies, especially predatory mites in the family Phytoseiidae (see Section 3.2) (Huffaker et al., 1970; McMurtry et al., 1970; DeBach, 1974; Gerson and Cohen, 1989). Spider mites are 'induced pests', which without perturbation, remain at low densities due to the action of natural enemies. While pesticide resistance has been implicated in outbreaks, this is a consequence of outbreaks rather than the cause.

Spider mites have five life stages: egg, larva, protonymph, deutonymph and adult (Fig. 1). Eggs are spherical in shape and vary in colour depending on species. *Tetranychus* spp. eggs (~0.14 mm diameter) are fairly translucent, *Panonychus* and *Bryobia* eggs (~0.15 mm diameter) are red or brownish. *Panonychus* eggs also have a thicker waxy coating and a long filament, called a stipe, that projects from the top of the egg. Tetranychid larvae are roughly the size of the eggs and have three pairs of legs. Protonymphs and deutonymphs are progressively larger and have four pairs of legs. Females (~0.35–0.45 mm) are larger than males (~0.30–0.40 mm), and males have pointed abdomens. Many spider mites are haplodiploid; unmated females lay unfertilized eggs that produce male offspring, whereas fertilized eggs result in female offspring. Post-mating, there is a short pre-oviposition period before the females begin

Figure 1 The life cycle of a typical spider mite, *T. urticae*: (a) egg, (b) larva, (c) protonymph, (d) deutonymph, (e) adult female, (f) male (bottom) and female (top) mating.

laying eggs. In some genera (e.g. most species of *Bryobia*), reproduction is entirely thelyotokous; no males are produced (Norton et al., 1993; Weeks and Breeuwer, 2001). The overwintering stage varies by species. In *Panonychus ulmi* (Koch), the egg is the overwintering stage, whereas it is the adult females for *Tetranychus urticae* Koch. In some species, like *T. urticae*, the diapausing females undergo a colour change, changing from a pale green/white colour to reddish orange (Veerman, 1974).

Common pest spider mites of tree fruits vary worldwide, but some of the most important are members of the genera *Tetranychus*, *Panonychus*, *Amphitetranychus*, *Eotetranychus* and *Bryobia*. *Tetranychus urticae* is notorious for its wide host range, with over 1000 recorded host plants (Grbic et al., 2011). It can reach pest status in all tree fruits. In the United States, it is more likely to be a pest of pears, while *P. ulmi* is more frequently found in apples and stone fruits. However, in Europe *T. urticae* can become problematic in cherries grown under protection (e.g. hail netting). Until the 1990s, *T. mcdanieli* was considered the main pest of apples in the northwestern U.S., but is rarely found now. Instead, mixed infestations of *P. ulmi* and *T. urticae* or *P. ulmi* infestations alone are more typical. *Panonychus ulmi* (Fig. 2) can be found on all tree fruits and is a major pest of grapes in Europe (Duso et al., 2012). *P. citri* (McGregor), in addition to attacking citrus, can be problematic on pears in some areas (Hoy, 2011e). *P. mori* Yokoyama infests peach and pear trees in Japan. *Amphitetranychus viennensis* (Zacher) is a pest of many fruit trees, including cherry, apricot, peach, pear and apple. Its current distribution includes Europe and Asia, and it is a quarantine pest in North America. *Eotetranychus pruni* (Oudemans) is an important pest of apples and cherries in many parts of the world, including Europe, China and Iran. *Bryobia rubrioculus* (Scheuten) is a

Figure 2 *Panonychus ulmi*, a worldwide pest of tree fruit; adult female and eggs.

cosmopolitan species found on pome and stone fruits and can reach pest status on peaches and almonds. In other fruits, it can be found on abandoned trees, but almost never causes economic damage. However, as apple management programmes in the northwestern U.S. continue to decrease their pesticide use, *B. rubrioculus* is becoming increasingly common and may reach pest status.

1.2 Eriophyids

Like spider mites, eriophyids feed on leaf cellular contents and the appearance of leaf damage is similar, although eriophyid leaf russeting is more finely textured. In some crops (e.g. peach), feeding damage causes the leaves to take on a silver hue. Unlike tetranychids, some rust mite species feed directly on the skin of fruits, which is readily apparent on light-skinned varieties and can reduce marketability. Other species of eriophyids cause formation of galls or small blisters on leaf and fruit surfaces. Like russeting, this damage reduces the photosynthetic capability of the plant. Damage to the fruit can cause deformation, russeting and reduced fruit size, decreasing marketability. Many eriophyids are host specific.

Eriophyids found on fruit trees are very small mites (~150 μm) with four legs positioned at the front of their bodies. A typical life cycle is egg, larva, nymph and adult. In some species, the adult female has two forms that are morphologically distinct, protogynes and deutogynes. Protogyne females are found in the summer, whereas the deutogyne is the overwintering stage.

Apple eriophyids include the very common *Aculus schlechtendali* (Nalepa) (apple rust mite) (Fig. 3) and the relatively rare *Eriophyes mali* (Burts) (apple leaf blister mite). In the United States, control of apple rust mite is usually confined to fruit russet prevention on light-skinned cultivars (e.g. cv. Golden

Figure 3 *Aculus schlechtendali*, a common eriophyid pest of apples.

Delicious). In Europe, apple rust mites are controlled to prevent russeting and reduced fruit size. Pears are attacked by *Epitrimerus pyri* (Nalepa) (pear rust mite) and *Eriophyes pyri* Pagenstecher (pear leaf blister mite) (Fig. 4). Pear rust mite can cause serious damage early in the season by russeting fruit, and is considered a chronic threat to fruit quality. Pear leaf blister mite is rare in orchards managed with conventional insecticides, and control occurs as a by-product of those applications. However, pear leaf blister mite damage is becoming more common in orchards in the United Kingdom, possibly due to reduced pesticide programmes. *Aculus cornutus* (Banks) (peach silver mite) is synonymous with the species formerly called plum nursery mite, plum rust mite and cherry rust mite *Aculus* (*Phyllocoptes*) *fockeui* (Nalepa and Trouessart). This species may more accurately be called *Prunus* rust mite, as it can reach pest status on cherry, nectarine and peach, as well as European, myrobalan and Damson plums (but not American or Japanese plum varieties)

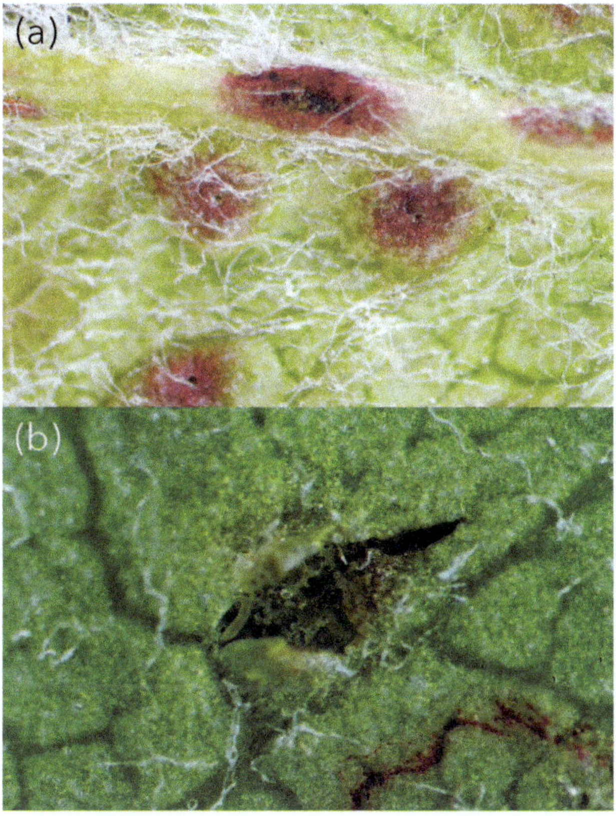

Figure 4 (a) *Eriophyes pyri*, pear leaf blister mite inside blisters and (b) pear leaf blister mite inside a dissected blister.

(Beers and Oldfield, 1993). Damage from this species can usually be tolerated without chemical intervention. Miticide resistance has not been an issue with rust mites; applications of sulphur or specific miticides are typically highly effective (Oldfield et al., 1993).

Because blister mites spend most or all of their life cycle inside the blisters they form, they are not well controlled by natural enemies. Conversely, rust mites have numerous predators that can adequately control populations below an economic level; the exception is pear rust mite causing fruit damage. Some species of eriophyids may serve as an alternative food source for predatory mites in periods of spider mite scarcity, improving biological control (Hoy, 2011b; Wearing et al., 2014). Because spider mites tend to be both more damaging and more difficult to control than eriophyids, low populations of rust mites and their damage may be tolerated, and provide a net benefit to tree fruit production due to their role in biological control (Hoyt, 1969).

2 Cultural control

2.1 Tolerant varieties

Tree fruit varieties have differing levels of susceptibility to mite damage, although none are considered resistant. Selection of varieties or even fruit species that are more tolerant to damage is a method that reduces management issues, especially in areas where pressure is known to be perennially challenging (Beers and Hull, 1987). For apples, varieties that are darker-skinned will not show russeting symptoms (Beers et al., 1993). Pears in general are highly susceptible to spider mite damage, but varieties like Anjou are known to be extremely sensitive (Oldfield et al., 1993). Detailed studies on the impact of rust mites to leaf physiology have been described by Spieser et al. (1998; 1999). In these investigations cv. Jonagold was more susceptible than cv. Golden Delicious. However, studies conducted in Italy revealed that rust mites can also produce serious damage on the latter cultivar depending on environmental conditions (Simoni et al., 2018). Additional studies confirm the role of cultivar and rootstock in mediating mite-plant relationships (Ioriatti et al., 1996; Duso et al., 2003).

2.2 Weed management

Weed management can have significant impacts on spider mite outbreaks. Many weeds serve as non-crop hosts for generalist species of spider mites, especially *T. urticae*. Spider mites can build up their populations in orchard floor weeds and then migrate into trees after weeds senesce, due to hot, dry conditions, herbicide applications or mowing. It is recommended that weed management occur on a regular basis to prevent weed growth and spider mite establishment. When weeds that have gone unmanaged are mowed or sprayed while the trees are fruiting, the spider mites migrate to the trees during the most critical part of the season (Flexner et al., 1991). However, newer research is indicating that relaxed orchard weed management (or habitat modification through beneficial plantings) may improve biological control (Section 3.2).

2.3 Fertilization

Over-fertilization, especially with nitrogen, can result in pest outbreaks. Studies on a variety of crops have shown increased *T. urticae* abundance at higher nitrogen rates (Garman and Kennedy, 1949; Hamstead and Gould, 1957; van de Vrie et al., 1972; Wermelinger et al., 1991; Chow et al., 2009). Increased foliar nitrogen content has been associated with increased *T. urticae* fecundity, and reduced development time and pre-oviposition period (Wermelinger et al., 1985, 1991). Less has been published on the role of fertilization and eriophyid infestations, although some earlier research suggests that nitrogen levels do not affect eriophyid abundance (Sternlicht et al., 1975; Leite et al., 1999). Following research-based fertilization recommendations is important for growers seeking to reduce pest outbreaks. Continued investigation of nutrient-input rates that balance yield increases with reduced risk of damaging arthropod populations is needed.

2.4 Dust control and irrigation

Dusty conditions (dust residue on leaves) have been implicated in causing mite outbreaks, through harmful effects on some beneficial insects (Bartlett, 1951; Fleschner, 1958; Flaherty et al., 1969; Fukushima and Stalford, 1969; Kinn et al., 1972). This effect has been attributed to the water absorbent properties of dust particles, resulting in dehydration of arthropods coated in dust (Bartlett, 1951; Fleschner, 1958). Conversely, more recent research has not validated the role of dust in mite outbreaks (Oi and Barnes, 1989; Pringle et al., 2014). Dust is a sign of dry conditions, which can stress both plants and predatory mites (Kinn et al., 1972; Oi, 1987; Pringle et al., 2014). This results in a crop that is more susceptible to herbivory and at the same time lacking natural enemies that could reduce herbivory. Conditions that tend to be dust-free (adequate humidity, rainfall or irrigation) also promote ground cover, which may also improve biological control of mites. Spider mites may also be less tolerant of wet conditions (Duso et al., 2004a). Therefore, current research suggests that dry conditions, not dust in and of itself, should be prevented where possible to decrease mite outbreaks.

3 Biological control

3.1 Augmentative control

Augmentative biological control involves mass rearing and releasing natural enemies where the natural enemy is not expected to establish permanently (DeBach, 1964; Wilson, 1966). It has been divided into two types: inoculative and inundative. Inoculative releases use small numbers of natural enemies, and biological control is provided by the successive generations of the released natural enemies, not the released enemies themselves (DeBach, 1964; Wilson, 1966). Conversely, inundative releases require large numbers of natural enemies, with the generation released providing control (DeBach, 1964; Wilson, 1966). Inundative releases of natural enemies have been referred to as 'biological pesticides' due to the immediacy of the effect desired (Stehr, 1982; Parrella et al., 1992; Hoy, 2011d). The distinction between these two types may not be clear in specific situations (Parrella et al., 1992).

A variation on inoculative biological control is that of introducing a pesticide-resistant strain of predatory mite. This was tested extensively in orchards in the United States and Australia in the 1970s–90s as a means of spreading various species of pesticide-resistant predatory mites (see Section 3.2) into new locations. Releases of *Galendromus occidentalis* (Nesbitt) (Readshaw, 1975a; Weiss and Julien, 1975; Field, 1978; Hoy et al., 1983), *Neoseiulus fallacis* (Garman) (Meyer, 1975; Whalon et al., 1982) and *Typhlodromus pyri* Scheuten (Croft, 1976) were typically deemed successful in terms of predator establishment and management. In some cases, resistant strains of predators were deliberately created through repeated pesticide applications and mass rearing (Hoy, 1982; Hoy et al., 1982). However, not all releases were successful (Prokopy and Christie, 1992). In some cases, these releases could also be termed classical biological control, because they involved the introduction of a species to a different part of the world. For instance, pyrethroid-resistant *G. occidentalis* was introduced from North America to Australia (Readshaw, 1975a; Field, 1978).

Inundative biological control remains uncommon in tree fruit crops and is primarily used in greenhouses. Many studies have emphasized the number of predators released, the number of prey present or the time of year released as the most likely cause of augmentation failure. All of these aspects address the same issue: the ratio of predators to prey (prey numbers tend to be lower earlier in the season). Many experimental releases of phytoseiids have concluded that successful control requires a high ratio of predator to prey, and that releases are best done when pest numbers are low (Smith and Papacek, 1991; Hoddle et al., 1999; Faber et al., 2000; Stavrinides, 2010; Schmidt et al., 2013). There have also been failures when pesticide applications have had non-target effects on released phytoseiids (Mansour et al., 1993; Ahmad et al., 2013; Schmidt et al., 2013). However, augmentative releases of *Phytoseiulus persimilis* Athias-Henriot, especially to treat spider mite 'hot spots' is showing some promise for cherry grown under protection in Europe.

Most studies of inundative control do not examine the cost-effectiveness of release. While cost-effectiveness has been established for strawberries (Trumble and Morse, 1993), it has been questioned for tree crops (Prokopy and Christie, 1992; Hoddle et al., 2000; Shrewsbury and Hardin, 2003; Schmidt et al., 2013). Indeed, past successful augmentation programmes in tree crops, including tree fruits, required extremely high release rates (60–2000 mites/tree) (Croft and McMurtry, 1972; Hoddle et al., 2000; Fadamiro et al., 2013). This is prohibitively expensive in most instances, especially when compared to the cost of a pesticide application. While augmentation may be a suitable strategy for small canopy or protected crops, a more permanent solution, like conservation biological control, may be required for large perennial agroecosystems.

3.2 Conservation biological control

Integrated mite management in tree fruit includes many of the best-known conservation biological control systems worldwide. These systems are primarily driven by predatory mites in the family Phytoseiidae, but also include stigmaeids (another predatory mite family) and coccinellids (*Stethorus* spp.). *Zetzellia mali* (Ewing) (Fig. 5) is a common stigmaeid that thrives in both unsprayed and commercial orchards, where it can supplement the predatory activity of phytoseiids (Woolhouse and Harmsen, 1985; Strickler et al., 1987; Thistlewood, 1991; Schmidt-Jeffris and Beers, 2016). It consumes eggs of spider mites and motile stages of eriophyids. *Z. mali* is also a known intraguild

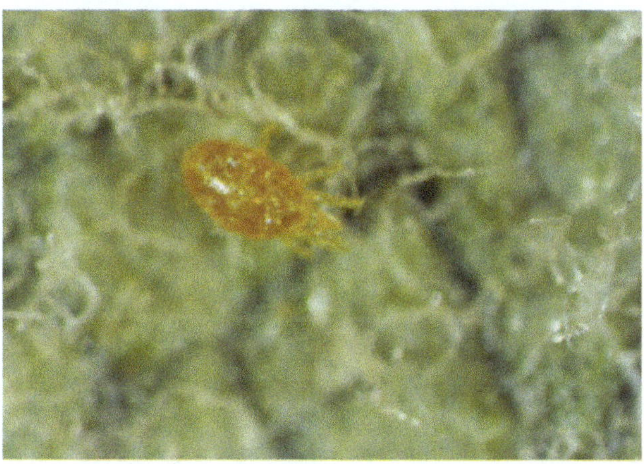

Figure 5 *Zetzellia mali*, a stigmaeid predator of spider mites.

Figure 6 *Stethorus picipes* adult, a coccinellid predator of spider mites.

predator of phytoseiid eggs (Clements and Harmsen, 1990), which may limit its benefit to biocontrol. *Stethorus* spp. (Fig. 6) are mite specialist predators found in a variety of agroecosystems that consume all life stages of spider mites. While these mites are voracious tetranychid predators, they are not sufficiently abundant in orchards until spider mite populations have reached or surpassed economic thresholds (McMurtry et al., 1970). Other organisms, such as lacewings, generalist ladybeetles, predatory thrips and predatory heteropterans like anthocorids (especially *Orius* spp.) and *Campylomma* spp. (Miridae) can also contribute to the natural control of spider mites (Collyer, 1953; Huffaker and McMurtry, 1969; Readshaw, 1975b).

Phytoseiids (Fig. 7) are the best-known and arguably the most important group of predatory mites because of their ability to reduce or maintain

Figure 7 Phytoseiid consuming an adult *T. urticae*.

populations of spider mites below economic thresholds (Gerson et al., 2003c). They have a near-worldwide distribution and are found in a wide range of climates (Hoy, 2011c). Like spider mites, phytoseiids have a rapid life cycle that consists of egg, larva, protonymph, deutonymph and adult stages. Their prey preferences vary widely, with some species consuming prey only within certain genera, while others are less specialized (McMurtry and Croft, 1997). In tree fruit systems, integrated mite management programmes involving phytoseiids (Table 1) are primarily based on selective pesticide use. New programmes developing in other parts of the world, such as Japan (*Neoseiulus womersleyi* (Shicha), *Typhlodromus vulgaris* Ehara, *Amblyseius tsugawai* Ehara), are employing similar strategies to maintain populations of their predators (Funayama et al., 2015).

Selective use of pesticides was one of the first conservation biocontrol techniques developed, and has been the main emphasis of this field (Wilson, 1966; Hull and Beers, 1985; Landis et al., 2000). Predators tend to develop resistance more slowly or less often than pests, perhaps because selection pressure is stronger on the pest than on the predator; the pest is more closely associated with the area where pesticides are applied (Georghiou, 1972; Croft and Brown, 1975). The predator relies on the presence of the pest (and will die or emigrate when the pest is eliminated), so the pest will develop resistance first (Georghiou, 1972; Croft, 1990b). Many generalist pests are also pre-adapted for pesticide resistance due to their ability to detoxify a variety of secondary plant compounds (Grbic et al., 2011). After the pest becomes resistant, the pesticide used is switched, giving the predator insufficient time to gain resistance (Georghiou, 1972; Stehr, 1982). Some phytoseiids have managed to develop resistance despite these factors and appear to be some of the few exceptions to the rule; these were the first known cases of a natural enemy becoming resistant to pesticides (Croft and Brown, 1975). This could be due to

Table 1 Phytoseiids that are key components of integrated mite management systems in tree fruit production

Species	Area where commonly used	References
Amblyseius andersoni (Chant)	Northern hemisphere, especially important in Europe	Blommers (1994), McMurtry and Croft (1997)
Euseius victoriensis (Womersley)	Australia	James (1989), James et al. (2001)
Galendromus occidentalis	Arid northwestern North America	Hoyt (1969), Gerson et al. (2003c)
Neoseiulus fallacis	Eastern U.S.	McMurtry (1982), Gerson et al. (2003c)
Typhlodromus pyri	Humid areas in North America, New Zealand, Europe	Huffaker et al. (1970), Collyer (1976), McMurtry (1982)

their close association with spider mites (increasing selection pressure), while still having the ability to persist without them on other food sources (allowing them to continue to be exposed to pesticides after the pest is eliminated) (Croft and Brown, 1975). Like the spider mites, they also have short generation times (McMurtry et al., 1970), allowing them to evolve quickly (Croft and Brown, 1975). Their limited ability to disperse also increases the selection pressure for resistance (Georghiou, 1972; Croft and Brown, 1975; Stehr, 1982).

Specific cases of resistance in phytoseiids allowed for the development of integrated mite management (IMM) programmes, where selected pesticides were used for control of key pests while allowing survival of a phytoseiid resistant to those pesticides (Madsen, 1964; Hoyt, 1969; McMurtry, 1982; Croft, 1990a; Gerson et al., 2003c; Hoy, 2011c). A natural extension of this work was to introduce resistant strains (selected *in situ* or in the lab) of phytoseiids to new areas through inoculative releases. The success of IMM has prompted continual screening of new pesticides as they are brought to the market for testing of non-target effects on phytoseiids (Villanueva and Walgenbach, 2005; van Driesche et al., 2006; Duso et al., 2008; Bostanian et al., 2009, 2010; Lefebvre et al., 2011, 2012; Beers and Schmidt, 2014; Fountain and Medd, 2015). While screening efforts in the past have emphasized acute mortality (Croft, 1990d), there is currently an increased interest in sublethal effects, which may have significant population-level impacts (Croft, 1990d; Blumel et al., 1993; Desneux et al., 2007; Beers and Schmidt, 2014; Beers and Schmidt-Jeffris, 2015). Neonicotinoids and pyrethroids have been implicated in mite outbreaks due to non-target effects on predators (Croft, 1990c; Beers et al., 2005; Beers and Schmidt, 2014; Mills et al., 2016; Schmidt-Jeffris and Beers, 2018), as have novaluron and sulphur (Martinez-Rocha et al., 2008; Beers et al., 2009; Lefebvre et al., 2011, 2012; Beers and Schmidt, 2014; Duso et al., 2014; Mills et al., 2016; Schmidt-Jeffris and Beers, 2018). Some non-selective miticides are also harmful to phytoseiids and can worsen issues with spider mite control, especially if the spider mite population has developed resistance. This is still a developing field, and as new pesticides are registered, research regarding their effects on natural enemies will need to continue if biological control efforts are to succeed.

Conservation biocontrol also involves the provision of multiple resources, including shelter. For phytoseiids, shelter can be provided in the form of increased ground cover, either through relaxed weed control, or deliberate plantings (Alston, 1994; Mailloux et al., 2010). Dead plant material (Morris et al., 1996) and hedgerows (Boller et al., 1988; Tixier et al., 1998, 2002; Duso et al., 2004b) may also serve as shelter. Ground cover provides a microclimate with lower temperatures and higher humidity than the canopy (Croft and McGroarty, 1977; Huang et al., 1983; Liang and Huang, 1994), conditions preferred by phytoseiids (Croft et al., 1993; Schausberger, 1998; Gerson et al., 2003a). Hedgerows and ground cover can also provide refuge sites from pesticide applications (Boller et al., 1988; Liang and Huang, 1994). Finally, in addition to the top-down controls provided by maintaining weedy ground cover, there may be bottom-up effects. Some weeds act as reservoirs for *T. urticae*; they will disperse into orchard trees if these weeds are removed (Flexner et al., 1991), but may remain in the ground cover if weeds are maintained (Aguilar-Fenollosa et al., 2011).

The same structures that provide shelter may also provide sources of food other than the target prey species. Hedgerows (Altieri and Letourneau, 1982; Duso et al., 2004b) and cover crops can increase natural enemy densities by providing floral resources. Generalist phytoseiids benefit from the pollen as an added nutrient source (Kennett et al., 1979; McMurtry, 1982; Onzo et al., 2005; González-Fernández et al., 2009). The pollen provided by hedgerows and ground cover has been found to increase phytoseiid populations, possibly by providing higher quality or quantity of pollen than the crop, and over a longer period of time (Boller et al., 1988; Grout and Richards, 1992; Tixier et al., 1998; Duso et al., 2004b; Mailloux et al., 2010). In addition to pollen, non-crop plants can also provide a source of spider mites or alternative prey (Boller et al., 1988; Flexner et al., 1991). Phytoseiids use these resources in periods of prey scarcity in the crop, keeping their numbers high and preventing the occurrence of pest outbreaks (Waite, 1988; Takahashi et al., 1998). Leaving weedy row middles or herbicide strips minimally managed may increase phytoseiid populations and biological control services (Schmidt-Jeffris et al., 2015).

4 Chemical control

4.1 Scouting and thresholds

Thresholds for indirect pests can be difficult to estimate, especially in perennial crops where damage to the plant can accumulate between seasons (Beers and Hull, 1990, 1995). However, general guidelines have been developed for many tree fruit crops. For instance, 0.5–2 spider mites per leaf is a commonly recommended threshold in Washington pear (Beers and Hoyt, 1993). Relationships between mite population size and yield and/or fruit quality can be affected by a number of factors such as soil features, water availability, plant vigour, cultivar susceptibility, cultural practices and so on. The time of attack is also crucial for evaluating the impact of both *P. ulmi* and *A. schlechtendali* (Hoyt et al., 1979; Beers and Hull, 1990; Easterbrook, 1996). It is not surprising that action thresholds proposed in different countries/areas reflect the diversity and variability of factors affecting the interaction between host plants and herbivore mites (van de Vrie, 1985; Easterbrook, 1996; Duso et al., 2010). When available, growers should use thresholds that have been calculated for their climate, crop phenology and horticultural conditions. Cultivars may be differentially susceptible (Downing and Moilliet, 1967; Beers and Hull, 1987; Hogmire and Miller, 2005), and pest populations can develop more quickly in warmer environments (Tanigoshi et al., 1975; Herbert, 1981a,b; Roy et al., 2003).

Scouting for mites typically involves the inspection of individual leaves with a hand lens (magnification 10× or greater). This magnification allows spider mites to be counted and roughly identified to species, while allowing for the detection of rust mites and a general sense of their abundance. For more accurate counts of both groups, a brushing machine (Fig. 8) and microscope counts are recommended. The machine has brushes that spin in opposite directions; leaves are inserted between the brushes, which knock all small arthropods off of the leaf onto a revolving soap-coated glass plate positioned beneath the brushes. After brushing, the glass plate is placed on a grid (Fig. 8) and the contents of the plate can be counted with reasonable accuracy under a microscope. A shake-and-wash technique for monitoring *P. ulmi* and *A. schlechtendali* in commercial apple orchards proved to be more effective than directly counting the mites on plant material under a dissection microscope. Leaves, shoots and spurs were agitated in alcohol and mites were removed using a separating funnel (Zacharda et al., 1988). Hossain (1992) confirmed that the washing technique was efficient for estimating spider mite abundance but reported that the brushing technique was more efficient for rust mites.

P. ulmi's spatial distribution is aggregated (Herbert and Butler, 1975) and the negative binomial model is appropriate for its description (Baumgaertner and Zahner, 1984). Therefore observations on a relatively high number of plant leaves or shoots are needed for estimating mite densities and to decide if control measures should be applied. A monitoring protocol based on tripartite sampling plans (sample again in one week, sample again in two weeks or intervene) that use binomial counts rather than complete enumeration has been proposed (Nyrop et al., 1994). This method involves counting the numbers of infested and uninfested leaves in the field and using species-specific models

Figure 8 Leaf brushing machine (centre), counting grid (left) and glass plates (right).

to estimate density, with higher levels of confidence in the estimate for larger sample sizes. Sequential sampling may be used in conjunction with this technique to enhance accuracy (Wilson et al., 1984). Such models are available for a number of pest and predatory mite species (Nyrop et al., 1989; Jones, 1990a,b; Nyrop and Binns, 1992). Recently, forecasting tools have been used to optimize monitoring, management and control of pests in fruit orchards. Phenology models were developed based on solar radiation and air and soil temperatures. The effect of temperature on the development of individual stages of key pests of apple (including *P. ulmi*) has been investigated and model predictions were compared with field observations allowing consultants to receive precise information (Samietz et al., 2015).

4.2 Available products

There are many available pesticides for controlling mite outbreaks. Here, we list the most common active ingredients (AI) registered for use in the United States (Table 2); trade names, AI, formulations and concentrations will vary by region and country. In previous decades, miticidal and insecticidal activity was often present in the same AI, but newer products tend to be more selective to pest mites. These products may be relatively selective in terms of non-target impacts on beneficial organisms, including predatory mites, and therefore tend to be a better fit for integrated biological control programmes. However, few of the miticides are effective on eriophyids. Important considerations when making a miticide application include (1) density and stages of pest mites present; (2) densities or ratios of predators to pest mites; (3) efficacy of the product against various life stages; (4) non-target effects on natural enemies; (5) mode of action (MOA) group for rotation to slow resistance development; (6) re-entry interval and its effect on orchard operations; (7) pre-harvest interval and maximum residue levels.

4.3 Resistance issues

While there are no known resistance issues with eriophyids in tree fruits, spider mites are notorious for rapid resistance development. *Tetranychus urticae* in particular is considered the 'most resistant pest', with documented resistance to 96 different AIs (Arthropod Pesticide Resistance Database, 2018). Resistance to a number of AIs has also been reported for other tree fruit mites, including *P. ulmi* (47), *P. citri* (28), *T. mcdanieli* (13), *A. viennensis* (7) and *T. pacificus* McGregor (5) (Arthropod Pesticide Resistance Database, 2018). In *T. urticae*, resistance is commonly reported within a few years of the adoption of new AIs (Dermauw et al., 2012). In some cases, particularly with the pyrethroids, resistance can be so high that there is a 2000-fold difference in susceptibility

Table 2 Miticides used in U.S. tree fruit production

Active ingredient	IRAC group number	Mode of action	Labelled selective?	U.S. trade name
Abamectin	6	Glutamate-gated chloride channel allosteric modulators	No	Agri-Mek
Clofentezine	10A		Yes	Apollo
Hexythiazox		Mite growth inhibitors	No	Onager/Savey
Etoxazole	10B		No	Zeal
Fenbutatin oxide	12B	Inhibitors of mitochondrial ATP synthase	No	Vendex
Bifenazate	20D	Mitochondrial complex III electron transport inhibitors	Yes	Acramite/Vigilant
Acequinocyl	20B		Yes	Kanemite
Fenpyroximate	21A	Mitochondrial complex I electron transport inhibitors	No	FujiMite/Portal
Pyridaben			No	Nexter
Spirodiclofen	23	Inhibitors of acetyl CoA carboxylase	No	Envidor
Cyflumetofen	25A	Mitochondrial complex II electron transport inhibitors	Yes	Nealta

Insecticide Resistance Action Committee (IRAC) group numbers from IRAC Mode of Action Classification Scheme v. 8.4. Selectivity information taken from pesticide labels. These classifications are correct at the time of writing but subject to change.

between resistant and susceptible populations (van Leeuwen et al., 2009, 2010). Pesticide applicators should carefully follow label directions to ensure maximum efficacy of pesticides. Rotating between MOA groups prevents sequential generations from being exposed to the same selective pressure and can slow the development of resistance. Indeed, the labels of many newer miticides allow only a single application per growing season specifically for resistance management purposes. Using thresholds for spray application decisions should decrease the frequency of the applications, thus lowering selection pressure. However, use of integrated biological control ultimately may be the single most important component in a resistance management programme, primarily in that it minimizes miticide use. Secondarily, it reduces the number of resistant survivors, because predators do not discriminate between resistant and susceptible prey. In crops that have a low damage threshold (e.g. pears and almonds), and thus a high frequency of miticide applications, resistance issues can become critical and make integrated pest management difficult.

5 Conclusion

There have been many successes with IMM in tree fruit systems worldwide. Key components of these systems have involved tolerating some level of foliar damage from mites and favouring the use of pesticides that are not disruptive to local predator communities, especially phytoseiids. Augmentative releases of predatory mites require further investigation, especially as orchard acreage grown under protection increases. Economic analysis will be required alongside rigorous testing for control efficacy. The use of habitat improvement to enhance spider mite control is a relatively new field and will likely play a role in the future of mite management. Future research will need to determine what cover plants are ideal and how to best manage understoreys. As new pesticides are developed for use in tree fruits, research will need to be conducted to determine their compatibility with biological control. Due to their sensitivity to mite damage, pears in particular lag behind other tree fruit crops in their ability to manage mites without substantial chemical intervention.

6 Where to look for further information

A recent book on Agricultural Acarology contains chapters on spider mites, eriophyids, mite biological control and case studies of IMM in tree fruits (Hoy, 2011a). Another text on the use of mites for control of pests is also available (Gerson et al., 2003b). Some biological control companies host websites with pesticide non-target effects databases (e.g. https://www.koppert.com/side-effects/). Several U.S. land-grant institutions have extension websites where information on mite pests in tree fruits can be found, including Washington State

University (treefruit.wsu.edu/crop-protection/opm/), Penn State (extension.psu.edu), University of California (ipm.ucanr.edu/), Cornell University (blogs.cornell.edu/treefruit/ipm/) and Michigan State (msue.anr.msu.edu/). Many of these states also produce annual tree fruit production guides, which contain information on mite management.

7 References

Aguilar-Fenollosa, E., Ibáñez-Gual, M. V., Pascual-Ruiz, S., Hurtado-Ruiz, M. and Jacas, J. A. 2011. Effect of ground-cover management on spider mites and their phytoseiid natural enemies in clementine mandarin orchards (I): Tetranychid mite dynamics. *Biol. Control*, 59, 171–9.

Ahmad, S., Pozzebon, A. and Duso, C. 2013. Augmentative releases of the predatory mite *Kampimodromus aberrans* in organic and conventional apple orchards. *Crop Prot.*, 52, 47–56.

Alston, D. G. 1994. Effect of apple orchard floor vegetation on density and dispersal of phytophagous and predaceous mites in Utah. *Agric. Ecosyst. Environ.*, 50, 73–84.

Altieri, M. A. and Letourneau, D. K. 1982. Vegetation management and biological control in agroecosystems. *Crop Prot.*, 1, 405–30.

Arthropod Pesticide Resistance Database. 2018. Available online: https://www.pesticideresistance.org/

Bartlett, B. R. 1951. The action of certain 'inert' dust materials on parasitic hymenoptera. *J. Econ. Entomol.*, 44, 891–6.

Baumgaertner, J. and Zahner, P. 1984. Simulation experiments with stochastic population models as a tool to explore pest management strategies in an apple-tree mite system (*Panonychus ulmi* Koch, *Tetranychus urticae* Koch). *In:* Cavalloro, R. (Ed.), *Statistical and Mathematical Models in Population Dynamics*. Rotterdam and Boston: A.A. Balkema.

Beers, E. H. and Hoyt, S. C. 1993. European red mite. *In:* Beers, E. H., Brunner, J. F., Willett, M. J. and Warner, G. M. (Eds), *Orchard Pest Management: A Resource Book for the Pacific Northwest*. Yakima, WA: Good Fruit Grower.

Beers, E. H. and Hull, L. A. 1987. Effect of European red mite (Acari: Tetranychidae) injury on vegetative growth and flowering of four cultivars of apples. *Environ. Entomol.*, 16, 569–74.

Beers, E. H. and Hull, L. A. 1990. Timing of mite injury affects the bloom and fruit development of apple. *J. Econ. Entomol.*, 83, 547–51.

Beers, E. H. and Hull, L. A. 1995. Dry matter partitioning of nonbearing apple trees following injury by European red mite (Acari: Tetranychidae). *J. Econ. Entomol.*, 88, 998–1003.

Beers, E. H. and Oldfield, G. N. 1993. Prunus rust mites. *In:* Beers, E. H., Brunner, J. F., Willett, M. J. and Warner, G. M. (Eds), *Orchard Pest Management: A Resource Book for the Pacific Northwest*. Yakima, WA: Good Fruit Grower.

Beers, E. H. and Schmidt, R. A. 2014. Impacts of orchard pesticides on *Galendromus occidentalis*: Lethal and sublethal effects. *Crop Prot.*, 56, 16–24.

Beers, E. H. and Schmidt-Jeffris, R. A. 2015. Effects of orchard pesticides on *Galendromus occidentalis* (Acari: Phytoseiidae): Repellency and irritancy. *J. Econ. Entomol.*, 108, 259–65.

Beers, E. H., Oldfield, G. N. and Westigard, P. H. 1993. Apple rust mite. *In:* Beers, E. H., Brunner, J. F., Willett, M. J. and Warner, G. M. (Eds), *Orchard Pest Management: A Resource Book for the Pacific Northwest*. Yakima, WA: Good Fruit Grower.

Beers, E. H., Brunner, J. F., Dunley, J. E., Doerr, M. and Granger, K. 2005. Role of neonicotinyl insecticides in Washington apple integrated pest management. Part II. Nontarget effects on integrated mite control. *J. Insect Sci.*, 5(16) (online).

Beers, E. H., Martinez-Rocha, L., Talley, R. R. and Dunley, J. E. 2009. Lethal, sublethal, and behavioral effects of sulfur-containing products in bioassays of three species of orchard mites. *J. Econ. Entomol.*, 102, 324–35.

Blommers, L. H. M. 1994. Integrated pest management in European apple orchards. *Annu. Rev. Entomol.*, 39, 213–41.

Blumel, S., Bakker, F. and Grove, A. 1993. Evaluation of different methods to assess the side-effects of pesticides on *Phytoseiulus persimilis* A.-H. *Exp. Appl. Acarol.*, 17, 161–9.

Boller, E. F., Remund, U. and Candofi, M. P. 1988. Hedges as potential sources of *Typhlodromus pyri*, the most important predatory mite in vineyards of northern Switzerland. *Entomophaga*, 33, 249–55.

Bostanian, N. J., Thistlewood, H. A., Hardman, J. M., Laurin, M.-C. and Racette, G. 2009. Effect of seven new orchard pesticides on *Galendromus occidentalis* in laboratory studies. *Pest Manag. Sci.*, 65, 635–9.

Bostanian, N. J., Hardman, J. M., Thistlewood, H. A. and Racette, G. 2010. Effects of six selected orchard insecticides on *Neoseiulus fallacis*. *Pest Manag. Sci.*, 66, 1263–7.

Chow, A., Chau, A. and Heinz, K. M. 2009. Reducing fertilization for cut roses: Effect on crop productivity and twospotted spider mite abundance, distribution, and management. *J. Econ. Entomol.*, 102, 1896–907.

Clements, D. R. and Harmsen, R. 1990. Predatory behavior and prey-stage preferences of stigmaeid and phytoseiid mites and their potential compatibility in biological control. *Can. Entomol.*, 122, 321–8.

Collyer, E. 1953. Biology of some predatory insects and mites associated with the fruit tree red spider mite *Metatetranychus ulmi* (Koch) in South-eastern England. *Journal of Horticultural Science*, 28, 85–112.

Collyer, E. 1976. Integrated control of apple pests in New Zealand 6. Incidence of European red mite, *Panonychus ulmi* (Koch), and its predators. *N.Z. J. Zool.*, 3, 39–50.

Croft, B. A. 1976. Establishing insecticide-resistant phytoseiid mite predators in deciduous tree fruit orchards. *Entomophaga*, 21, 383–99.

Croft, B. A. 1990a. Endemic species. *Arthropod Biological Control Agents and Pesticides*. New York, NY: John Wiley & Sons.

Croft, B. A. 1990b. Factors affecting resistance. *Arthropod Biological Control Agents and Pesticides*. New York, NY: John Wiley & Sons.

Croft, B. A. 1990c. Pesticide selectivity: Pyrethroids. *Arthropod Biological Control Agents and Pesticides*. New York, NY: John Wiley & Sons.

Croft, B. A. 1990d. Sublethal influences. *Arthropod Biological Control Agents and Pesticides*. New York, NY: John Wiley & Sons.

Croft, B. A. and Brown, A. W. A. 1975. Responses of arthropod natural enemies to insecticides. *Annu. Rev. Entomol.*, 20, 285–335.

Croft, B. A. and McGroarty, D. L. 1977. The role of *Amblyseius fallacis* (Acarina: Phytoseiidae) in Michigan apple orchards. *Farm Science*, 333, 2–22.

Croft, B. A. and McMurtry, J. A. 1972. Minimum release of *Typhlodromus occidentalis* to control *Tetranychus mcdanieli* on apple. *J. Econ. Entomol.*, 65, 188–91.

Croft, B. A., Messing, R. H., Dunley, J. E. and Strong, W. B. 1993. Effects of humidity on eggs and immatures of *Neoseiulus fallacis, Amblysieus andersoni, Metaseiulus occidentalis* and *Typhlodromus pyri* (Phytoseiidae): Implications for biological control on apple, caneberry, strawberry and hop. *Exp. Appl. Acarol.*, 17, 451–9.

DeBach, P. (Ed.). 1964. *Biological Control of Insect Pests and Weeds*. London, UK: Chapman and Hall.

DeBach, P. 1974. Fostering pests through misuse of chemicals. *Biological Control by Natural Enemies*. London, UK: Cambridge University Press.

Dermauw, W., Wybouw, N., Rombauts, S., Menten, B., Vontas, J., Grbić, M., et al. 2012. A link between host plant adaptation and pesticide resistance in the polyphagous spider mite *Tetranychus urticae*. *Proc. Natl. Acad. Sci. U.S.A.*, 110, E113–22.

Desneux, N., Decourtye, A. and Delpuech, J.-M. 2007. The sublethal effects of pesticides on beneficial arthropods. *Annu. Rev. Entomol.*, 52, 81–106.

Downing, R. S. and Moilliet, T. K. 1967. Relative densities of predaceous and phytophagous mites on three varieties of apple trees. *Can. Entomol.*, 99, 738–41.

Duso, C., Pasini, M. and Pellegrini, M. 2003. Distribution of the predatory mite *Typhlodromus pyri* (Acari: Phytoseiidae) on different apple cultivars. *Bicontrol Sci. Techn.*, 13, 671–81.

Duso, C., Chiarini, F., Conte, L., Bonora, V., Dalla Montá, L. and Otto, S. 2004a. Fogging can control *Tetranychus urticae* on greenhouse cucumbers. *J. Pest Sci.*, 77, 105–11.

Duso, C., Malagnini, V., Paganelli, A., Aldegheri, L., Bottini, M. and Otto, S. 2004b. Pollen availability and abundance of predatory phytoseiid mites on natural and secondary hedgerows. *BioControl*, 49, 397–415.

Duso, C., Malagnini, V., Pozzebon, A., Castagnolia, M., Liguori, M. and Simoni, S. 2008. Comparative toxicity of botanical and reduced-risk insecticides to Mediterranean populations of *Tetranychus urticae* and *Phytoseiulus persimilis* (Acari Tetranychidae, Phytoseiidae). *Biol. Control*, 47, 16–21.

Duso, C., Castagnoli, M., Simoni, S. and Angeli, G. 2010. The impact of eriophyoids on crops: Recent issues on *Aculus schlechtendali, Calepitrimerus vitis* and *Aculops lycopersici*. *Exp. Appl. Acarol.*, 51, 151–68.

Duso, C., Pozzebon, A., Kreiter, S., Tixier, M.-S. and Candofi, M. P. 2012. Management of phytophagous mites in European vineyards. *In:* Bostanian, N. J., Vincent, C. and Isaacs, R. (Eds), *Arthropod Management in Vineyards: Pests, Approaches, and Future Directions*. Dordrecht, Netherlands: Springer.

Duso, C., Ahmad, S., Tirello, P., Pozzebon, A., Klaric, V., Baldessari, M., Malagnini, V. and Angeli, G. 2014. The impact of insecticides applied in apple orchards on the predatory mite *Kampimodromus aberrans* (Acari: Phytoseiidae). *Exp. Appl. Acarol.*, 62, 391–414.

Easterbrook, M. A. 1996. Damage and control of eriophyoid mites in apple and pear. *World Crop Pests*, 6, 527–41.

Faber, B., Yee, W. L., Phillips, P. and Rakha, M. 2000. Effect of predaceous mite release timing on persea mite (Acari: Tetranychidae) populations in avocado orchards. *California Avocado Society Yearbook*, 84, 111–25.

Fadamiro, H. Y., Akotsen-Mensah, C., Xiao, Y. and Anikwe, J. 2013. Field evaluation of predacious mites (Acari: Phytoseiidae) for biological control of citrus red mite, *Panonychus citri* (Trombidiformes: Tetranychidae). *Fla. Entomol.*, 96, 80–91.

Field, R. P. 1978. Control of the two-spotted spider mite in a Victorian peach orchard with an introduced insecticide-resistant strain of the predatory mite *Typhlodromus occidentalis* Nesbitt (Acarina: Phytoseiidae). *Aust. J. Zool.*, 26, 519–27.

Flaherty, D., Lynn, C. D., Jensen, F. L. and Luvisi, D. A. 1969. Ecology and integrated control of spider mites in San Joaquin vineyards. *Calif. Agric.*, 23, 11.

Fleschner, C. A. 1958. The effect of orchard dust on the biological control of avocado pests. *California Avocado Society Yearbook 1958*, 42, 94–8.

Flexner, J. L., Westigard, P. H., Gonzalves, P. and Hilton, R. 1991. The effect of groundcover and herbicide treatment on twospotted spider mite density and dispersal in southern Oregon pear orchards. *Entomol. Exp. Appl.*, 60, 111–23.

Fountain, M. T. and Medd, N. 2015. Integrated pesticides and predatory mites in soft fruit crops. *Phytoparasitica*, 43, 657–67.

Fukushima, C. and Stalford, E. M. 1969. Effects of road dust on spider mites. *Calif. Agric.*, 23, 10.

Funayama, K., Komatus, M., Sonoda, S., Takahashi, I. and Hara, K. 2015. Management of apple orchards to conserve generalist phytoseiid mites suppresses two-spotted spider mite, *Tetranychus urticae* (Acari: Tetranychidae). *Exp. Appl. Acarol.*, 65, 43–54.

Garman, P. and Kennedy, B. H. 1949. Effect of soil fertilization on the rate of reproduction of the two-spotted spider mite. *J. Econ. Entomol.*, 42, 157–8.

Georghiou, G. P. 1972. The evolution of resistance to pesticides. *Annu. Rev. Entomol.*, 3, 133–68.

Gerson, U. and Cohen, E. 1989. Resurgences of spider mites (Acari: Tetranychidae) induced by synthetic pyrethroids. *Exp. Appl. Acarol.*, 6, 29–46.

Gerson, U., Smiley, R. L. and Ochoa, R. 2003a. The effect of host plants and the ground cover on acarine biocontrol agents. *Mites (Acari) for Pest Control*. Oxford, UK: Blackwell Science.

Gerson, U., Smiley, R. L. and Ochoa, R. 2003b. *Mites (Acari) for Pest Control*. Oxford, UK: Blackwell Science.

Gerson, U., Smiley, R. L. and Ochoa, R. 2003c. The Phytoseiidae. *Mites (Acari) for Pest Control*. Oxford, UK: Blackwell Science.

González-Fernández, J. J., de la Peña, F., Hormaza, J. I., Boyero, J. R., Vela, J. M., Wong, E., Trigo, M. M. and Montserrat, M. 2009. Alternative food improves the combined effect of an omnivore and a predator on biological pest control. A case study in avocado orchards. *Bull. Entomol. Res.*, 99, 433–44.

Grbic, M., van Leeuwen, T., Clark, R. M., Rombauts, S., Rouze, P., Grbic, V., Osborne, E. J., Dermauw, W., Ngoc, P. C., Ortego, F., et al. 2011. The genome of *Tetranychus urticae* reveals herbivorous pest adaptations. *Nature*, 479, 487–92.

Grout, T. G. and Richards, R. I. 1992. The dietary effect of windbreak pollens on longevity and fecundity of a predacious mite *Euseius addoensis addoensis* (Acari: Phytoseiidae) found in citrus orchards in South Africa. *Bull. Entomol. Res.*, 82, 317–20.

Hall, F. R. and Ferree, D. C. 1975. Influence of twospotted spider mite populations on photosynthesis of apple leaves. *J. Econ. Entomol.*, 68, 517–20.

Hamstead, E. O. and Gould, E. 1957. Relation of mite populations to seasonal leaf nitrogen levels in apple orchards. *J. Econ. Entomol.*, 50, 109–10.

Herbert, H. J. 1981a. Biology, life tables, and innate capacity for increase of the twospotted spider mite, *Tetranychus urticae* (Acarina: Tetranychidae). *Can. Entomol.*, 113, 371–8.

Herbert, H. J. 1981b. Biology, life tables, and intrinsic rate of increase of the European red mite, *Panonychus ulmi* (Acarina: Tetranychidae). *Can. Entomol.*, 113, 65–71.

Herbert, H. J. and Butler, K. P. 1975. Sampling systems for European red mite, *Panonychus ulmi* (Acarine: Tetranychidae): II. Effect of spur cluster size on mite distribution. *Can. Entomol.*, 107, 547–50.

Hoddle, M. S., Aponte, O., Kerguelen, V. and Heraty, J. 1999. Biological control of *Oligonychus perseae* (Acari: Tetranychidae) on avocado: I. Evaluating release timings, recovery and efficacy of six commercially available phytoseiids. *Int. J. Acarol.*, 25, 211–19.

Hoddle, M. S., Robinson, L. and Virzi, J. 2000. Biological control of *Oligonychus perseae* (Acari: Tetranychidae) on avocado: III. Evaluating the efficacy of varying release rates and release frequency of *Neoseiulus californicus* (Acari: Phytoseiidae). *Int. J. Acarol.*, 26, 203–14.

Hogmire, H. W. and Miller, S. S. 2005. Relative susceptibility of new apple cultivars to arthropods. *HortScience*, 40, 2071–5.

Hossain, S. M. 1992. Comparison of sampling techniques for the European red mite, *Panonychus ulmi* (Koch) (Acari: Tetranychidae) and the apple rust mite, *Aculus schlechtendali* (Nalepa) (Acari: Eriophyidae). *Acta Agric Scand. B-Soil Plant Sci.*, 42, 128–32.

Hoy, M. A. 1982. Genetic improvement of *Metaseiulus occidentalis* implementing pesticide resistant strains – progress and problems. *In:* Cameron, P. J., Wearing, C. H. and Kain, W. M. (Eds), *Australasian Workshop on Development and Implementation of IPM*, 20–22 July 1982, Aukland, New Zealand. Entomology Division, Department of Scientific and Industrial Research, pp. 18–28.

Hoy, M. A. 2011a. *Agricultural Acarology: Introduction to Integrated Mite Management.* Boca Raton, FL: Taylor and Francis Group, LLC.

Hoy, M. A. 2011b. The Eriophyoidea: The good, the bad, and the unknown. *Agricultural Acarology: Introduction to Integrated Mite Management.* Boca Raton, FL: Taylor and Francis Group, LLC.

Hoy, M. A. 2011c. The Phytoseiidae: Effective natural enemies. *Agricultural Acarology: Introduction to Integrated Mite Management.* Boca Raton, FL: Taylor and Francis Group, LLC.

Hoy, M. A. 2011d. The strategy of integrated mite management. *Agricultural Acarology: Introduction to Integrated Mite Management.* Boca Raton, FL: Taylor and Francis Group, LLC.

Hoy, M. A. 2011e. Tetranychidae: Premier plant pests. *Agricultural Acarology: Introduction to Integrated Mite Management.* Boca Raton, FL: Taylor and Francis Group, LLC.

Hoy, M. A., Castro, D. and Cahn, D. 1982. Two methods for large scale production of pesticide-resistant strains of the spider mite predator *Metaseiulus occidentalis* (Nesbitt) (Acarina, Phytoseiidae). *Z. Angew. Entomol.*, 94, 1–9.

Hoy, M. A., Westigard, P. H. and Hoyt, S. C. 1983. Release and evaluation of a laboratory-selected pyrethroid-resistant strain of the predaceous mite *Metaseiulus occidentalis* (Acari: Phytoseiidae) in southern Oregon pear orchards and a Washington apple orchard. *J. Econ. Entomol.*, 76, 383–8.

Hoyt, S. C. 1969. Integrated chemical control of insects and biological control of mites on apple in Washington. *J. Econ. Entomol.*, 62, 74–86.

Hoyt, S. C. 1991. Biology, ecology and control of mites in Washington orchards. *In:* Williams, K. (Ed.), *New Directions in Tree Fruit Pest Management*. Yakima, WA: Good Fruit Grower.

Hoyt, S. C., Tanigoshi, L. K. and Browne, R. W. 1979. Economic injury level studies in relation to mites on apple. *Recent Advances in Acarology*, 1, 3–12.

Huang, M.-D., Mai, S.-W., Li, S.-X. and Situ, J. 1983. Biological control of citrus red mite, *Panonychus citri* (McG.) in Guangdong Province. *In:* Matsumoto, K., ed. 4th International Citrus Congress, 9–12 November 1981, Tokyo, Japan. International Society of Citriculture, pp. 643–6.

Huffaker, C. B. and McMurtry, J. A. 1969. The ecology of tetranychid mites and their natural control. *Annu. Rev. Entomol.*, 14, 125–74.

Huffaker, C. B., van de Vrie, M. and McMurtry, J. A. 1970. Ecology of tetranychid mites and their natural enemies: A review: II. Tetranychid populations and their possible control by predators: An evaluation. *Hilgardia*, 40, 391–458.

Hull, L. A. and Beers, E. H. 1985. Ecological selectivity: Modifying chemical control practices to preserve natural enemies. *In:* Hoy, M. A. and Herzog, D. C. (Eds), *Biological Control in Agricultural IPM Systems*. New York, NY: Academic Press.

Ioriatti, C., Forti, D., Angeli, G. and Molignoni, R. 1996. L'eriofide del melo (*Aculus schlechtendali*): Morfologia, biologia e danno. *Informatore Fitopatologico*, 12, 9–14.

James, D. G. 1989. Effect of pesticides on survival of *Amblyseius victoriensis* (Womersley), an important predatory mite in southern New South Wales peach orchards. *Plant Prot. Q.*, 4, 141–3.

James, D. G., Price, T., Wright, L. C., Coyle, J. and Perez, J. 2001. Mite abundance and phenology on commercial and escaped hops in Washington State, USA. *Int. J. Acarol.*, 27, 151–6.

Jones, V. P. 1990a. Developing sampling plans for spider mites (Acari: Tetranychidae): Those who don't remember the past may have to repeat it. *J. Econ. Entomol.*, 83, 1656–64.

Jones, V. P. 1990b. Sampling and dispersion of the twospotted spider mite (Acari: Tetranychidae) and the western orchard predatory mite (Acari: Phytoseiidae) on tart cherry. *J. Econ. Entomol.*, 83, 1376–80.

Kennett, C. C., Flaherty, D. L. and Hoffmann, R. W. 1979. Effect of wind-borne pollens in the population dynamics of *Amblyseius hibisci*. *Entomophaga*, 24, 83–98.

Kinn, D. N., Joos, J. L. and Doutt, R. L. 1972. Influence of overhead sprinkler systems on spider mite populations in north coast vineyards of California. *Environ. Entomol.*, 1, 795–6.

Lakso, A. N., Mattii, G. B., Nyrop, J. P. and Denning, S. S. 1996. Influence of European red mite on leaf and whole-canopy carbon dioxide exchange, yield, fruit size, quality, and return cropping in 'Starkrimson Delicious' apple trees. *Journal of the American Society for Horticultural Science*, 121, 954–8.

Landis, D. A., Wratten, S. D. and Gurr, G. M. 2000. Habitat management to conserve natural enemies of arthropod pests in agriculture. *Annu. Rev. Entomol.*, 45, 175–201.

Lefebvre, M., Bostanian, N. J., Thistlewood, H. M. A., Mauffette, Y. and Racette, G. 2011. A laboratory assessment of the toxic attributes of six 'reduced risk insecticides' on *Galendromus occidentalis* (Acari: Phytoseiidae). *Chemosphere*, 84, 25–30.

Lefebvre, M., Bostanian, N. J., Mauffette, Y., Racette, G., Thistlewood, H. A. and Hardman, J. M. 2012. Laboratory-based toxicological assessments of new insecticides on mortality and fecundity of *Neoseiulus fallacis* (Acari: Phytoseiidae). *J. Econ. Entomol.*, 105, 866–71.

Leite, G. L. D., Picanço, M., Guedes, R. N. C. and Zanuncio, J. C. 1999. Influence of canopy height and fertilization levels on the resistance of *Lycopersicon hirsutum* to *Aculops lycopersici* (Acari: Eriophyidae). *Exp. Appl. Acarol.*, 23, 633–42.

Liang, W. and Huang, M. 1994. Influence of citrus orchard ground cover plants on arthropod communities in China: A review. *Agric. Ecosyst. Environ.*, 50, 29–37.

Madsen, H. F. 1964. Integrated control of phytophagous mites on apple and pear. *In:* 60th Annual Meeting of the Washington State Horticultural Association, 7–9 December 1964. Wenatchee, WA: Washington State Horticultural Association, pp. 75–8.

Mailloux, J., Le Bellec, F., Kreiter, S., Tixier, M.-S. and Dubois, P. 2010. Influence of ground cover management on diversity and density of phytoseiid mites (Acari: Phytoseiidae) in Guadeloupean citrus orchards. *Exp. Appl. Acarol.*, 52, 275–90.

Mansour, F., Cohen, H. and Shain, Z. 1993. Integrated mite management in apples in Israel: Augmentation of a beneficial mite and sensitivity of tetranychid and phytoseiid mites to pesticides. *Phytoparasitica*, 21, 39–51.

Martinez-Rocha, L., Beers, E. H. and Dunley, J. E. 2008. Effect of pesticides on integrated mite management in Washington State. *J. Entomol. Soc. B.C.*, 105, 1–12.

McMurtry, J. A. 1982. The use of phytoseiids for biological control: Progress and future prospects. *In:* Hoy, M. A. (Ed.), *Conference of the Acarology Society of America*, 29 November – 3 December 1981, San Diego, CA. Oakland, CA. University of California Press, pp. 23–48.

McMurtry, J. A. and Croft, B. A. 1997. Life-styles of phytoseiid mites and their roles in biological control. *Annu. Rev. Entomol.*, 42, 291–321.

McMurtry, J. A., Huffaker, C. B. and van de Vrie, M. 1970. Ecology of tetranychid mites and their natural enemies: A review: I. Tetranychid enemies: Their biological characters and the impact of spray practices. *Hilgardia*, 40, 331–90.

Meyer, R. H. 1975. Release of carbaryl-resistant predatory mites in apple orchards. *Environ. Entomol.*, 4, 48–51.

Mills, N. J., Beers, E. H., Shearer, P. W., Unruh, T. R. and Amarasekare, K. G. 2016. Comparative analysis of pesticide effects on natural enemies in western orchards: A synthesis of laboratory bioassay data. *Biol. Control*, 102, 17–25.

Morris, M. A., Croft, B. A. and Berry, R. E. 1996. Overwintering and effects of autumn habitat manipulation and carbofuran on *Neoseiulus fallacis* and *Tetranychus urticae* in peppermint. *Exp. Appl. Acarol.*, 20, 249–58.

Norton, R. A., Kethley, J. B., Johnston, D. E. and O'Connor, B. M. 1993. Phylogenetic perspectives on genetic systems and reproductive modes of mites. *In:* Wrensch, D. L. and Ebbert, M. A. (Eds), *Evolution and Diversity of Sex Ratio in Insects and Mites*. New York: Chapman and Hall.

Nyrop, J. P. and Binns, M. R. 1992. Algorithms for computing operating characteristic and average sample number functions for sequential sampling plans based on binomial count models and revised plans for European red mite (Acari: Tetranychidae) on apple. *J. Econ. Entomol.*, 85, 1253–73.

Nyrop, J. P., Agnello, A. M., Kovach, J. and Reissig, W. H. 1989. Binomial sequential classification sampling plans for European red mite (Acari: Tetranychidae) with special reference to performance criteria. *J. Econ. Entomol.*, 82, 482–90.

Nyrop, J. P., Binns, M. R., van der Werf, W. and Kovach, J. 1994. Cascading tripartite binomial classification plans to monitor European red mite (Acari, Tetranychidae) through a season; development and evaluation of a new methodology for pest monitoring. *Exp. Appl. Acarol.*, 18, 123–53.

Oi, D. H. 1987. *Factors related to Pacific spider mite outbreaks in roadside almond trees*. Ph.D. dissertation, University of California.

Oi, D. H. and Barnes, M. M. 1989. Predation by the western predatory mite (Acari: Phytoseiidae) on the Pacific spider mite (Acari: Tetranychidae) in the presence of road dust. *Environ. Entomol.*, 18, 892–6.

Oldfield, G. N., Westigard, P. H., Smirle, M. and Dunley, J. E. 1993. Pear rust mite. *In:* Beers, E. H., Brunner, J. F., Willett, M. J. and Warner, G. M. (Eds), *Orchard Pest Management: A Resource Book for the Pacific Northwest*. Yakima, WA: Good Fruit Grower.

Onzo, A., Hanna, R., Negloh, K., Toko, M. and Sabelis, M. W. 2005. Biological control of cassava green mite with exotic and indigenous phytoseiid predators - effects of intraguild predation and supplementary food. *Biol. Control*, 33, 143–52.

Park, Y.-L. and Lee, J.-H. 2005. Impact of twospotted spider mite (Acari: Tetranychidae) on growth and productivity of glasshouse cucumbers. *J. Econ. Entomol.*, 98, 457–63.

Parrella, M. P., Heinz, K. M. and Nunney, L. 1992. Biological control through augmentative releases of natural enemies: A strategy whose time has come. *Am. Entomol.*, 38, 172–9.

Pringle, K. L., Heunis, J. M. and de Villiers, M. 2014. Does dust result in mite outbreaks in apple orchards? *J. Appl. Entomol.*, 138, 307–14.

Prokopy, R. J. and Christie, M. 1992. Studies on releases of mass-reared organophosphate resistant *Amblyseius fallacis* (Garm.) predatory mites in Massachusetts commercial apple orchards. *J. Appl. Entomol.*, 112, 131–7.

Readshaw, J. L. 1975a. Biological control of orchard mites in Australia with an insecticide-resistant predator. *J. Aust. Inst. Agric. Sci.*, 41, 213–14.

Readshaw, J. L. 1975b. The ecology of tetranychid mites in Australian orchards. *J. Appl. Ecol.*, 12, 473–95.

Rice, R. E. and Jones, R. A. 1978. Mites in almonds and stone fruits. *Calif. Agric.*, 32, 20–1.

Roy, M., Brodeur, J. and Cloutier, C. 2003. Effect of temperature on intrinsic rates of natural increase (r_m) of a coccinellid and its spider mite prey. *BioControl*, 48, 57–72.

Samietz, J., Höhn, H., Razavi, E., Schaub, L. and Graf, B. 2015. Decision support for sustainable orchard pest management with the Swiss forecasting system SOPRA. *Acta Hort.*, 1099, 383–90.

Sances, F. V., Wyman, J. A., Ting, I. P., van Steenwyk, R. A. and Oatman, E. R. 1981. Spider mite interactions with photosynthesis, transpiration and productivity of strawberry. *Environ. Entomol.*, 10, 442–8.

Schausberger, P. 1998. The influence of relative humidity on egg hatch in *Euseius finlandicus*, *Typhlodromus pyri* and *Kampimodromus aberrans* (Acari, Phytoseiidae). *J. Appl. Entomol.*, 122, 497–500.

Schmidt, R. A., Beers, E. H., Unruh, T. R. and Horton, D. R. 2013. Releases of insectary-reared *Galendromus occidentalis* (Acari: Phytoseiidae) in commercial apple orchards. *J. Econ. Entomol.*, 106, 1996–2005.

Schmidt-Jeffris, R. A. and Beers, E. H. 2016. Phenology and structure of a phytoseiid community in an insecticide-free apple orchard. *Exp. Appl. Acarol.*, 68, 173–82.

Schmidt-Jeffris, R. A. and Beers, E. H. 2018. Potential impacts of orchard pesticides on *Tetranychus urticae*: A predator-prey perspective. *Crop Prot.*, 103, 56–64.

Schmidt-Jeffris, R. A., Beers, E. H. and Crowder, D. W. 2015. Phytoseiids in Washington commercial apple orchards: Biodiversity and factors affecting abundance. *Exp. Appl. Acarol.*, 67, 21–34.

Shrewsbury, P. M. and Hardin, M. R. 2003. Evaluation of predatory mite (Acari: Phytoseiidae) releases to suppress spruce spider mites, *Oligonychus ununguis* (Acari: Tetranychidae), on juniper. *J. Econ. Entomol.*, 96, 1675–84.

Simoni, S., Angeli, G., Baldessari, M. and Duso, C. 2018. Effects of *Aculus schlechtendali* (Acari: Eriophyidae) population densities on Golden Delicious apple production. *Acarol.*, 58(Suppl), 134–44.

Smith, D. and Papacek, D. F. 1991. Studies of the predatory mite *Amblyseius victoriensis* (Acarina: Phytoseiidae) in citrus orchards in south-east Queensland: Control of *Tegolophus australis* and *Phyllocoptruta oleivora* (Acarina: Eriophyidae), effect of pesticides, alternative host plants and augmentative release. *Exp. Appl. Acarol.*, 12, 195–217.

Spieser, F., Graf, B., Walther, P. and Noesberger, J. 1998. Impact of apple rust mite (Acari: Eriophyiidae) feeding on apple leaf gas exchange and leaf color associated with changes in leaf tissue. *Environ. Entomol.*, 27, 1149–56.

Spieser, F., Graf, B., Höhn, H. and Höpli, H.-U. 1999. Effects of high apple rust mite population densities on gas exchange, yield, fruit quality, tree growth and flower formation. *IOBC/WPRS Bull.*, 22, 77–85.

Stavrinides, M. C. 2010. The effects of timing and rate of release on population growth of *Phytoseiulus persimilis* reared on *Tetranychus urticae*. *Phytoparasitica*, 38, 349–54.

Stehr, F. W. 1982. Parasitoids and predators in pest management. *In:* Metcalf, R. L. and Luckmann, W. H. (Eds), *Introduction to Insect Pest Management*. New York, NY: Wiley-Interscience.

Sternlicht, M., Regev, S. and Goldenberg, S. 1975. Effect of chemical element deficiencies in nutrient solutions on the reproduction of *Aceria sheldoni* (Ewing) (Acarina, Eriophyidae). *Bull. Entomol. Res.*, 65, 433–42.

Strickler, K., Cushing, N., Whalon, M. and Croft, B. A. 1987. Mite (Acari) species composition in Michigan apple orchards. *Environ. Entomol.*, 16, 30–6.

Takahashi, M., Inoue, M. and Takafuji, A. 1998. Management of the spider-mite population in a vinylhouse vinery by releasing *Phytoseiulus persimilis* Athias-Henriot onto the ground cover. *Jap. J. Appl. Entomol. Zool.*, 42, 71–6.

Tanigoshi, L. K., Hoyt, S. C., Browne, R. W. and Logan, J. A. 1975. Influence of temperature on population increase of *Tetranychus mcdanieli* (Acarina: Tetranychidae). *Ann. Entomol. Soc. Am.*, 68, 972–8.

Thistlewood, H. 1991. A survey of predatory mites in Ontario apple orchards with diverse pesticide programmes. *Can. Entomol.*, 123, 1163–74.

Tixier, M.-S., Kreiter, S., Auger, P. and Weber, M. 1998. Colonization of Languedoc vineyards by phytoseiid mites (Acari: Phytoseiidae): Influence of wind and crop environment. *Exp. Appl. Acarol.*, 22, 523–42.

Tixier, M.-S., Kreiter, S., Croft, B. A. and Auger, P. 2002. Colonization of vineyards by *Kampimodromus aberrans* (Oudemans) (Acari: Phytoseiidae): Dispersal from surrounding plants as indicated by random amplified polymorphism DNA typing. *Agric. Forest Entomol.*, 4, 255–64.

Trumble, J. T. and Morse, J. P. 1993. Economics of integrating the predaceous mite *Phytoseiulus persimilis* (Acari: Phytoseiidae) with pesticides in strawberries. *J. Econ. Entomol.*, 86, 879–85.

van de Vrie, M. 1985. Apple. *In:* Helle, W. and Sabelis, M. W. (Eds), *Spider Mites: Their Biology, Natural Enemies and Control. Vol. 1B*. Oxford and New York: Tokyo.

van de Vrie, M., McMurtry, J. A. and Huffaker, C. B. 1972. Ecology of tetranychid mites and their natural enemies: A review. III. Biology, ecology, and pest status, and host-plant relations of tetranychids. *Hilgardia*, 41, 343–432.

van Driesche, R. G., Lyon, S. and Nunn, C. 2006. Compatibility of spinosad with predacious mites (Acari: Phytoseiidae) used to control western flower thrips (Thysanoptera: Thripidae) in greenhouse crops. *Fla. Entomol.*, 89, 396–401.

van Leeuwen, T., Vontas, J., Tsagarakou, A. and Tirry, L. 2009. Mechanisms of acaricide resistance in the two-spotted spider mite *Tetranychus urticae*. *In:* Ishaaya, I. and Horowitz, A. (Eds), *Biorational Control of Arthropod Pests*. Dordrecht, Netherlands: Springer.

van Leeuwen, T., Vontas, J., Tsagarakou, A., Dermauw, W. and Tirry, L. 2010. Acaricide resistance mechanisms in the two-spotted spider mite *Tetranychus urticae* and other important Acari: A review. *Insect Biochem. Mol. Biol.*, 40, 563–72.

Veerman, A. 1974. Carotenoid metabolism in *Tetranychus urticae* Koch (Acari: Tetranychidae). *Comp. Biochem. Physiol. B, Biochem. Mol. Biol.*, 47, 101–16.

Villanueva, R. and Walgenbach, J. F. 2005. Development, oviposition, and mortality of *Neoseiulus fallacis* (Acari: Phytoseiidae) in response to reduced-risk insecticides. *J. Econ. Entomol.*, 98, 2114–20.

Waite, G. K. 1988. Integrated control of *Tetranychus urticae* in strawberries in south-east Queensland. *Exp. Appl. Acarol.*, 5, 23–32.

Wearing, C. H., Marshall, R. R., Colhoun, C. and Attfield, B. A. 2014. Phytophagous mites and their predators during the establishment of apple orchards under biological and integrated fruit production in Central Otago, New Zealand. *N.Z. J. Crop Hort. Sci.*, 42, 127–44.

Weeks, A. R. and Breeuwer, J. A. J. 2001. *Wolbachia* - induced parthenogenesis in a genus of phytophagous mites. *Proc. R. Soc. B.*, 268, 2245–51.

Weiss, P. W. and Julien, M. H. 1975. A comparison of two species of spiny emex (*Emex australis* and *E. spinosa*) in northwestern Victoria. *J. Aust. Inst. Agric. Sci.*, 41, 211–13.

Wermelinger, B., Oertli, J. J. and Delucchi, V. 1985. Effect of host plant nitrogen fertilization on the biology of the two-spotted spider mite, *Tetranychus urticae*. *Entomol. Exp. Appl.*, 38, 23–8.

Wermelinger, B., Oertli, J. J. and Baumgärtner, J. 1991. Environmental factors affecting the life-tables of *Tetranychus urticae* (Acari: Tetranychidae) III. Host-plant nutrition. *Exp. Appl. Acarol.*, 12, 259–74.

Whalon, M. E., Croft, B. A. and Mowry, T. M. 1982. Introduction and survival of susceptible and pyrethroid-resistant strains of *Amblyseius fallacis* (Acari: Phytoseiidae) in a Michigan apple orchard. *Environ. Entomol.*, 11, 1096–9.

Wilson, F. 1966. The conservation and augmentation of natural enemies. *Pest Articles and News Summaries. Section A. Insect Control*, 12, 40–6.

Wilson, L. T., Hoy, M. A., Zalom, F. G. and Smilanick, J. M. 1984. Sampling mites in almonds: I. Within-tree distribution and clumping pattern of mites with comments on predator-prey interactions. *Hilgardia*, 52, 1–13.

Woolhouse, M. E. J. and Harmsen, R. 1985. The mite complex on the foliage of a pesticide-free apple orchard: Population dynamics and habitat associations. *In:* Ellis, C. R., Kevan, P. G., Pree, D. J. and Trimble, R. M. (Eds), *Proceedings of the Entomological Society of Ontario*, 115, pp. 1–11.

Zacharda, M., Pultar, O. and Muška, J. 1988. Washing technique for monitoring mites in apple orchards. *Exp. Appl. Acarol.*, 5, 181–3.

Chapter 18

Integrated management of tree fruit insect pests: *Drosophila suzukii* (Spotted Wing Drosophila)

Neil Audsley, Fera Science Ltd, UK; and Lorenzo Tonina and Nicola Mori, University of Padova, Italy

1 Introduction

The vinegar fly *Drosophila suzukii* Matsumura (Diptera: Drosophilidae), commonly known as spotted winged drosophila (Fig. 1a), is a highly polyphagous invasive pest, native to Asia, which causes significant damage to a wide variety of berry and stone fruit crops. Unlike other *Drosophila* species, which lay eggs in overripe or decaying fruits, *D. suzukii* lays eggs, not only in ripe and ripening fruits, but also in unripe fruits which have not yet been harvested (Keesey et al., 2015; Shawer et al., 2018). A serrated oviscapt (Fig. 1b) allows the female to slice the skin and insert her ovipositor and deposit eggs inside the fruit. Here the larvae feed and develop causing fruit to soften and rapidly collapse (Kanzawa, 1936, 1939; Walsh et al., 2011). The resulting reduction in crop yields and significant financial losses are often compounded by secondary infection due to pathogens (e.g. bacteria, fungi) which enter via the ovipositor wound (Louise et al., 1996; Cini et al., 2012), causing various fruit rots to occur.

Summer-morphotype adult *D. suzukii* are 2–3 mm long and have a wingspan of 6–8 mm. Usually, females are, on average, larger than males. The thorax is light yellow-brown in colour. The abdomen is characterized by a light-yellow

http://dx.doi.org/10.19103/AS.2019.0046.22

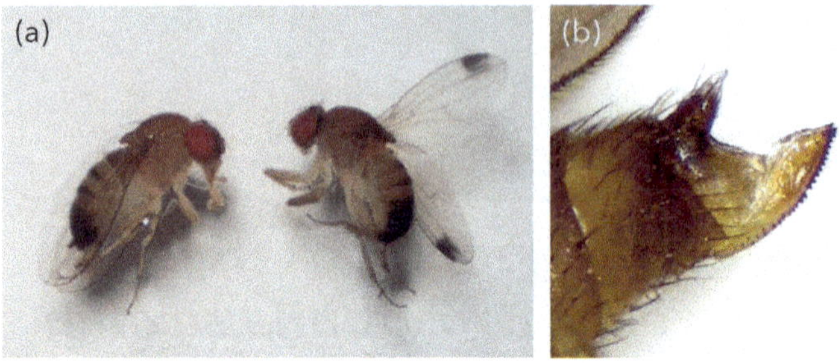

Figure 1 (a) Female (left) and male (right) adult, summer-morphotype, *Drosophila suzukii* and (b) female serrated oviscapt.

colour and presents unbroken bands at the posterior ends of the tergites (Kanzawa, 1939; Kawase and Uchino, 2005). The male has wide black spots at the tip of the wing on the second vein and two dark coloured combs (sex combs) on the first and second segments of the front legs (foretarsi). Females have a large sclerotized, hard, shiny and dark oviscapt with strong serrations of teeth, with saw-like edges when fully exposed (Fig. 1b, Vlach, 2010; Walsh et al., 2011).

Compared to summer morphotypes, winter morphotypes have a darker pigmentation, longer wing length and larger body size. These features are advantageous in colder environments by improving thermoregulation, capability to store sugars and fats and the ability to disperse (Stephens et al., 2015; Shearer et al., 2016).

Drosophila suzukii originates in the temperate and subtropical regions of Asia (Kanzawa, 1939; Lemeunier et al., 1986). It was first described in Japan in 1916, but it is uncertain whether it is native to Japan, or introduced there (Kanzawa, 1936). *Drosophila suzukii* has been reported in China, North and South Korea, Pakistan, India, Myanmar, Thailand, Taiwan and the Russian Far East (reviewed by Calabria et al., 2012; Cini et al., 2012).

This invasive pest was first reported in both Europe and North America in 2008, since which it has rapidly spread throughout both continents most likely as a result of the movement of infested plant material and fruit, in addition to establishment and dispersal (EPPO, 2010; Lee et al., 2011a; Asplen et al., 2015).

In North America, *D. suzukii* was first reported in 2008 in California, USA, on berry crops, and by 2009 had rapidly spread throughout California and the Pacific Coast region of the United States to British Columbia, Canada, and on the Atlantic coast, in Florida (Goodue et al., 2011; Hauser, 2011; Lee et al., 2011a). It is now widespread throughout the United States and Canada (EPPO, 2018).

The first report of *D. suzukii* in South America was in Southern Brazil in 2013, and it has since been reported in Uruguay, Argentina and Chile (Deprá et al., 2014; EPPO, 2018).

In Europe, *D. suzukii* was first identified in samples collected in Spain and Italy in 2008 (Calabria et al., 2012; Cini et al., 2012), but is now present in virtually every European country becoming a major pest on a wide variety of berry and stone fruits (Asplen et al., 2015; EPPO, 2018).

The routes of invasion of *D. suzukii* have been determined using genetic markers (Fraimout et al., 2017). Japan was the most likely source of *D. suzukii* populations in Hawaii, which in turn, together with China, are the likely source of *D. suzukii* in western United States. European populations are most probably derived from Northeast China, with evidence of limited relationship with populations in eastern United States (Fraimout et al., 2017).

Drosophila suzukii is considered a minor pest in its area of origin, hence there were few studies on its biology and ecology and as a consequence little information on how to control this pest when it invaded North America and Europe. Significant progress has since been made in understanding the biology of the fly which will be reviewed in this chapter, providing key information required for the development of sustainable and integrated control methods.

2 Impact on crops

Drosophila suzukii is a particularly onerous pest as it prefers undamaged ripening fruits from a wide range of plant species (Table 1). Eggs are laid within the fruit and direct damage is caused by the developing larvae feeding within the fruit causing it to become soft and rot rapidly, thereby reducing its marketable value (Fig. 2).

The holes caused by the female's serrated oviscapt are also a route for secondary infection by pathogens (bacteria and yeast). The deteriorating and rotting fruit can attract and allow the entry of other *Drosophila* spp. (Louise et al., 1996; Walsh et al., 2011; Cini et al., 2012; Ioriatti et al., 2015; Hamby et al., 2016). In addition to lower production, the economic losses are related to the increased cost for harvesting (fruit sorting time and increased picking frequency), insect management (plant protection products, traps, insect exclusion mesh etc.), removal, treatment and disposal of waste fruit and for the value of the production (Grassi et al., 2012).

Targeting larvae within fruit with chemical insecticides is a major challenge, especially as larvae are most destructive during the pre-harvest interval when pesticides should not be applied. As a result, there are few effective options available to growers. The presence of *D. suzukii* has also disrupted established and reliable integrated pest management (IPM) systems in fruit growing areas (Mazzi et al., 2017), hence growers have had to change their pest management practices and production methods to deal with the challenges of this pest. The EPPO PRA report concluded that the potential economic consequences of *D.*

Table 1 List of *Drosophila suzukii* host plant species

Family	Species	Family	Species
Actinidiaceae	*Actinidia chinensis*	Rosaceae	*Amelanchier lamarckii*
	Actinidia arguta		*Amelanchier ovalis*
Adoxaceae	*Sambucus ebulus*		*Cotoneaster franchetii*
	Sambucus nigra		*Cotoneaster lacteus*
	Sambucus nigra spp. *cerulea*		*Cotoneaster rehderi*
	Sambucus racemosa		*Crataegus chrysocarpa*
	Sambucus racemose var. *melanocarpa*		*Crataegus monogyna*
			Duchesnea indica
	Viburnum dilatatum		*Eriobotrya japonica*
	Viburnum lantana		*Fragaria vesca*
	Viburnum rhytidophyllum		*Malus baccata*
Araceae	*Arum italicum*		*Malus pumila*
	Arum maculatum		*Photinia beauverdiana*
Araliaceae	*Hedera helix*		*Photinia villosa*
Berberidaceae	*Mahonia aquifolium*		*Photinia prunifolia*
	Mahonia × media		*Prunus armeniaca*
			Prunus avium
Buxaceae	*Sarcococca confusa*		*Prunus buergeriana*
Caprifoliaceae	*Lonicera alpigena*		*Prunus cerasifera*
	Lonicera × bella		*Prunus cerasus*
	Lonicera caerulea		*Prunus domestica*
	Lonicera caprifolium		*Prunus donarium*
	Lonicera ferdinandii		*Prunus japonica*
	Lonicera morrowii		*Prunus laurocerasus*
	Lonicera nigra		*Prunus lusitanica*
	Lonicera nitida		*Prunus mahaleb*
			Prunus nipponica
	Lonicera tatarica		*Prunus padus*
	Lonicera xylosteum		*Prunus persica*
	Symphoricarpos albus		*Prunus salicina*
Cornaceae	*Alangium platanifoluium*		*Prunus sargentii*
	Cornus alba		*Prunus serotina*
	Cornus amomum		*Prunus spinosa*
	Cornus controversa		*Prunus virginiana*
	Cornus foemina		*Prunus yedoensis*
	Cornus kousa		*Pyracantha sp.*
	Cornus mas		*Pyrus calleryana*
	Cornus racemose		*Pyrus sinensis*
	Cornus sanguinea		*Ribes rubrum*
	Cornus sericea		*Ribes sanguineum*

Family	Species	Family	Species
Cucurbitaceae	*Bryonia cretica*		*Ribes uva-crispa*
Dioscoreaceae	*Tamus communis*		*Rosa acicularis*
Ebenaceae	*Diospyros kaki*		*Rosa canina*
Ericaceae	*Arbutus unedo*		*Rosa glauca*
	Gaultheria adenothrix		*Rosa pimpinellifolia*
	Gaultheria shallon		*Rosa rugosa*
	Gaultheria × wisleyensis		*Rubus allegheniensis*
	Vaccinium macrocarpon		*Rubus aremeniacus*
	Vaccinium myrtilloides		*Rubus caesius*
	Vaccinium myrtillus		*Rubus crataegifolius*
	Vaccinium oldhamii		*Rubus fruticosus*
	Vaccinium ovatum		*Rubus idaeus*
	Vaccinium praestans		*Rubus mirophyllus*
	Vaccinium vitis-idea		*Rubus parvifolius*
	Vaccinium vitis-idaea		*Rubus phoenicolasius*
Elaeagnaceae	*Elaeagnus × ebbingei*		*Rubus saxatilis*
	Elaeagnus multiflora		*Rubus spectabilis*
	Eleaegnus umbellate		*Rubus ulmifolius*
	Hippophae rhamnoides		*Sorbus aria*
Garryaceae	*Aucuba japonica*		*Sorbus aucuparia*
Lauraceae	*Lindera benzoin*		*Sorbus sitchensis*
Liliaceae	*Polygonatum multiflorum*	Rhamnaceae	*Frangula alnus*
Melanthiaceae	*Paris quadrifolia*		*Rhamnus cathartica*
Moraceae	*Ficus carica*	Rutaceae	*Murraya paniculata*
	Morus alba		*Citrus X sinensis*
	Morus alba × rubra	Santalaceae	*Viscum album*
	Morus australis	Solanaceae	*Atropa belladonna*
	Morus nigra		*Lycium barbarum*
			Physalis alkekengi
	Morus rubra		*Solanum chenopodioides*
Myricaceae	*Morella rubra*		*Solanum dulcamara*
Myrtaceae	*Eugenia uniflora*		*Solanum lycopersicum*
Phytolaccaceae	*Phytolacca americana*		*Solanum nigrum*
	Phytolacca esculenta		*Solanum villosum*
Rhamnaceae	*Frangula purshiana*	Taxaceae	*Taxus baccata*
			Torreya nucifera
		Thymelaeaceae	*Daphne mezereum*
		Vitaceae	*Parthenocissus quinquefolia*
			Vitis vinifera

Source: adapted from Poyet et al. (2015), Lee et al. (2015) and Kenis et al. (2016).

Figure 2 (a) *Drosophila suzukii* damage on cherry and (b) puncture holes showing eggs and filaments (spiracles) identified by arrows.

suzukii were high, crop production costs would increase and harvested fruit may be downgraded due to damage (EPPO, 2010).

In Japan, studies carried out during the 1930s reported damage up to 100% on cherries and 80% on grapes on different orchards within a few years (Kanzawa, 1939). There are few studies of the actual economic impact on the fruit industry due to damage caused by *D. suzukii* in Europe and North America (Asplen et al., 2015; Steffen et al., 2015).

In 2008, yield losses on five cultivated crops (strawberries, blueberries, raspberries, blackberries and cherries) in the Pacific Coast states of California, Oregon and Washington were estimated to be >US$400 million, which is approximately 16% of the total farmgate value (Bolda et al., 2010). A similar study by Walsh et al. (2011) estimated losses of >US$500 million, assuming a 20% yield loss in the same crops in the same states. A more in-depth study by Goodhue et al. (2011) based on the 2008 production of untreated strawberries and raspberries in California estimated yield reductions of 50% and 20%, respectively, and revenue reductions of 35% in raspberries and 20% in strawberries.

In a small soft fruit growing area in the Trento region of Italy estimated losses of €0.5 million and €3 million in 2010 and 2011, respectively, were reported (Ioriatti et al., 2011 reported in Cini et al., 2012). In the same region, an estimate of the economic impact of *D. suzukii* on the most affected fruit crops (strawberry, raspberry, blueberry, blackberry and cherry) in 2011 also indicated revenue losses of more than €3 million.

3 Biology and ecology

3.1 Life cycle and environmental requirements

3.1.1 Life cycle

Drosophila suzukii requires 24–48 h to reach the sexual maturity after emergence from the pupal cocoon (Dreves et al., 2009). Males and females mate soon after emergence and females can lay between 7 and 16 eggs each day, and up to 350–400 eggs during her life. On each fruit, one to three eggs are laid on average (Kanzawa, 1939; Grassi et al., 2012) resulting in a high proportion of fruit being damaged and the emergence of many larvae (Smyth and Saverimuttu, 2010). The eggs are white, 0.62 × 0.18 mm, with two respiratory processes that protrude from the epicarp of the fruit (Fig 2b). Through these structures, called spiracles, the supply of oxygen to the embryo is achieved (Kanzawa, 1939; Walsh et al., 2011). After 12–72 h from egg hatch, the larva begins to feed on the fruit pulp rapidly.

The larva is legless, cylindrical and a white-cream colour. The length is 0.5–4 mm, depending on larval stage (three instars). The mouthparts are formed of black mandibular teeth. Spiracles are also present in the larval stages, serving to breathe in a low oxygen environment (Kanzawa, 1939; Walsh et al., 2011).

The pupa of *D. suzukii* is 3–3.3 mm long and 1 mm wide. It has a yellow-golden colour, which gradually turns to opaque red and finally to dark red. As the pupa matures, some adult structures become visible, including eyes and wings. Two breathing processes are present (spiracles, 0.3 mm (Fig. 3)). These pupal spiracles project outward from the puparium and terminate in tube-like extensions, called spiracular papillae. Pupation can take place in the fruit or externally in the immediate vicinity (Kanzawa, 1936; Walsh et al., 2011).

The adult fly is present all year round and has a cold-resistant adult morphotype that is in reproductive diapause over the winter (Mitsui et al., 2010; Walsh et al., 2011; Dalton et al., 2011; Ometto et al., 2013; Zerulla et al., 2015; Rossi-Stacconi et al., 2016; Shearer et al., 2016). The overwintering phase begins when the average environmental temperature is lower than 5°C. Sexually mature adults enter this physiological state in the autumn and resume their reproductive activities, in spring, as the temperature begins to increase (Kanzawa, 1939; Mitsui et al., 2006; Walsh et al., 2011). During the winter, this

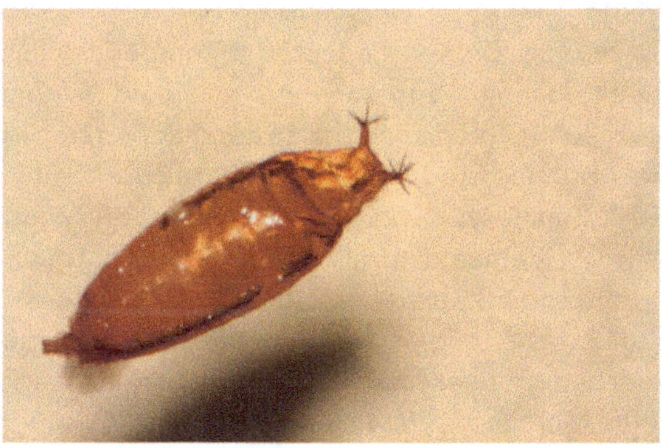

Figure 3 *Drosophila suzukii* pupa showing breathing processes and finger-like projections.

winter form can live on average 200 days, up to a maximum of 301 days (Dreves et al., 2009). Several places are suitable for overwintering, leaf litter in forests, natural vegetation and anthropic sites (Zerulla et al., 2015). High levels of snow cover during the winter may provide protection that enhances winter survival (Pelton et al., 2016) in regions where it regularly freezes, such as the Alps (Zerulla et al., 2015). During early spring, cherry or other stone fruit blossoms may provide a food source for *D. suzukii* adults; this carbohydrate-protein rich diet enhances their survival (Tochen et al., 2016a).

3.1.2 Environmental requirements

Drosophila suzukii is resilient to different climatic conditions. Temperature is the most influential factor on the development of local populations of *D. suzukii* and related to altitude, latitude and seasonal climate trends (Kimura, 1988). The life cycle, from egg laying to adult emergence, ranges from 10 to 50 days at 30°C and 10°C, respectively (Kanzawa, 1939; Asplen et al., 2015). The rapid life cycle in optimum conditions (20–25°C) allows *D. suzukii* to complete up to 13 generations in a fruit harvesting period in some regions, permitting an exponential population growth (Kanzawa, 1939; Dreves et al., 2009; Walsh et al., 2011; Grassi et al., 2012).

The larval development was also possible as low as 10°C on cherry under laboratory conditions (79.4 days), although oviposition was not observed at this temperature (Tochen et al., 2014). Asplen et al. (2015) estimated the lower threshold of development at 6°C. Under natural fluctuating conditions, the minimum threshold of average daily temperature for the egg-to-adult development of *D. suzukii* corresponds to 11.6°C (Tonina et al., 2016), a value lower than the lower threshold (13.4°C) for intrinsic rate of population increase assessed by Tochen et al. (2014) using a nonlinear estimation. This indicates that *D. suzukii* is particularly adapted to exploit the naturally fluctuating temperatures typical of mountain weather, for both development rate and yield (Tonina et al., 2016).

Drosophila suzukii is a pest well adapted to temperate climatic conditions characterized by alternating warm summers and cold winters (Mitsui et al., 2010; Ometto et al., 2013). Tolerance to harsh conditions may be physiologically dependent or may be mediated by adaptations such as altitudinal migration (Mitsui et al., 2010; Tonina et al., 2016), acclimatization (Walsh et al., 2011), phenotypic plasticity (Clemente et al., 2018; Stockton et al., 2018) and/or overwintering in anthropic habitats or other protected sites (Kimura, 2004; Zerulla et al., 2015). The migration along elevation gradients has been explained as a mechanism to escape summer heat and to track available host fruits (Kimura, 2004; Kinjo et al., 2014; Tonina et al., 2016).

Despite the presence of *D. suzukii* in sites where the winter average temperatures are between −12°C and −4°C (island of Hokkaido in Japan;

Walsh et al., 2011), this insect is a relatively cold-intolerant and chill-susceptible species as 50% of adult flies are killed following a 24 h exposure to temperatures just below 0°C (Kimura, 2004). Adults can also be killed by long exposure to 0°C or by short exposure to chilling temperatures (Jakobs et al., 2015). Stockton et al. (2018) found the lower lethal limit of adult *D. suzukii* was −1.1°C in winter morphotype (WM) adults compared to 1.7°C in non-acclimated summer morphotype (SM) adults, following static acclimation. However, dynamic acclimation reduced the lower limit to −5°C in SM flies and 50% of WM flies survived 72 h at −7.5°C (Stockton et al., 2018).

In contrast, *D. suzukii* is quite sensitive to warm temperature. Above 29–30°C, the adult male becomes sterile and ceases reproducing (Walsh et al., 2011). When adults mated at 31°C for 4 days, none of the eggs hatched. Female oviposition and egg hatching rate are also reduced as temperature increased during the oviposition period. When *D. suzukii* larvae developed above 31°C, pupation and adult eclosion were abolished. According to field observations, adult *D. suzukii* ceased to appear when the average temperature exceeded 28°C or when the temperature within a day exceeded 33°C for 8 h or more (Kinjo et al., 2014).

Temperature and humidity likely affect *D. suzukii* in an interactive fashion (Tochen et al., 2016b). The fly is quite sensitive to drought. In the absence of water, individuals will die within 24 h. In addition, humid climates reduce dehydration. Tochen et al. (2016b) recorded the highest reproductive status of females at 82% and 94% RH, compared to 71% RH. Moreover, short survival times and lack of reproduction was recorded at 20% RH, indicating that this humidity level is below the conditions under which the pest can reproduce.

In open field conditions, the presence of small rivers also promotes the development (Walsh et al., 2011; Grassi et al., 2017; Eben et al., 2018).

3.2 Spatial ecology

3.2.1 Influence of semi-natural habitat

Landscape composition has the potential to affect the population dynamics of mobile insect species (Tscharntke et al., 2005; Gardiner et al., 2009). The amount of natural habitat in the landscape can be an important driver if it provides resources such as food, nesting or overwintering sites (Rusch et al., 2011, 2013). As a generalist pest, for *D. suzukii*, landscape complexity can provide more suitable conditions compared to landscapes dominated by a single crop (Tscharntke et al., 2016).

The high polyphagy of *D. suzukii* allows the pest to feed on cultivated, wild and ornamental species usually abundant in woody habitats such as forests or hedgerows (Poyet et al., 2014, 2015; Lee et al., 2015; Kenis et al., 2016; Tonina et al., 2016). Therefore, forests and hedgerows may affect pest population dynamics by providing alternative host resources that promote the build-up of *D. suzukii* populations during fruit ripening, overwintering habitats in forest leaf litter or refuge areas during cold, warm or dry periods or when crops are sprayed with insecticides (Lee et al., 2015; Zerulla et al., 2015; Kenis et al., 2016; Pelton et al., 2016; Tochen et al., 2016a; Bal et al., 2017; Santoiemma et al., 2018). For these reasons, *D. suzukii* adults can benefit from forest-dominated landscapes (Pelton et al., 2016). Therefore, a higher presence of semi-natural habitats in the landscape can facilitate the winter survival of the pest (Pelton et al., 2016). Spring wild hosts and woodland microclimates may provide additional resources and protection to build up reservoirs of local populations early in the season which can then move into susceptible fruit crops. In autumn, wild hosts may provide late-season resources which build up populations before the winter and suitable microclimates may increase survival rates (Pelton et al., 2016). In forest habitats, relative humidity maintains higher levels compared to open habitats, thereby creating suitable conditions for improved *D. suzukii* survival and fitness (Calabria et al., 2012; Tochen et al., 2016b). The digestive tract of *D. suzukii,* sampled from woodlands in the winter, contained microbiota associated with cold, wet climatic conditions and vegetation and decomposition (Fountain et al., 2018). The authors proposed that because the fly is so moisture dependent, this provides evidence that WM flies may be exploiting food sources associated with moisture on decomposing vegetation during this time.

In contrast, crop-dominated landscapes do not provide the same amount of shelter. Santoiemma et al. (2018) found a reduced density of *D. suzukii* in landscapes characterized by a high proportion of cherry orchards and linked this to insecticide applications which occur repeatedly throughout the entire harvest season (Bruck et al., 2011; Van Timmeren and Isaacs, 2013; Gautam et al., 2016; Shawer et al., 2018). In landscapes with large expanses of orchards, the almost synchronous pesticide applications over large areas probably reduce population refuge areas from which the pest could recolonize the crop fields, which has a negative effect on pest density but not on pest damage (Santoiemma et al., 2018).

If the protection provided by woodland habitats improves overwintering survival (Zerulla et al., 2015), proximity of the orchards to semi-natural and other overwintering habitats can contribute to *D. suzukii* colonization at the beginning of the growing season (Pelton et al., 2016) and enhance pest density and/or damage during the ripening period (Basoalto et al., 2010; Asplen et al., 2015; Klick et al., 2016; Pelton et al., 2016; Santoiemma et al., 2018).

Klick et al. (2016) found higher *D. suzukii* populations within the cultivated crops where the field margin contains Himalayan blackberry (*Rubus armeniacus*, Focke 1874) and other host plants due to higher activity levels in this habitat;

in comparison to where no host plants were present in the field margin, zero to low numbers of flies were captured.

Pelton et al. (2016) found earlier activity of *D. suzukii* in the season in landscapes characterized by high amounts of woodland area (1 week earlier in farms surrounded by 60% of semi-natural area within 1.5 km instead of 10% woodland) and higher presence throughout the summer and autumn in resource-rich woodlands bordering crop fields. Wild hosts are used by adults for food in the autumn rather than oviposition sites. To meet overwintering energy needs, females may need to build up sufficient fat reserves through feeding or, if lacking, be required to reabsorb eggs (Zerulla et al., 2015; Pelton et al., 2016). Wild hosts may provide much needed food resources for females, such as yeast, to overwinter successfully in cold climates; higher fat reserves, in turn, could lead to a higher proportion of females with mature eggs in the spring (Pelton et al., 2016).

In warmer geographic regions, wild hosts in woodland landscapes may be more important to *D. suzukii* populations, and therefore crop risk, by providing temporal continuity of resources. If fly populations are not reduced by high winter mortality, wild hosts can create temporal resource continuity (Pelton et al., 2016). Santoiemma et al. (2018) found that *D. suzukii* responded positively to increasing cover of forest in the landscape. Cherry orchards characterized by a greater amount of semi-natural habitat in the close surrounding area (250–500 m) appeared to be more susceptible to pest attack. However, in this study, Santoiemma et al. (2018) found that, even in orchards located in crop-dominated landscapes with no forest, the pest density and damage were substantial. Even a reduction in harvest loss may have considerable economic effects (Mazzi et al., 2017).

Pelton et al. (2016) found an opposite relationship in the woods, with fewer flies trapped in wood traps where there was a high proportion of woodlands in the landscape. One possible explanation for this pattern is a resource dilution effect. That is, at farms with low amounts of woodland adjacent to the fruit crop field, the woodlands present may serve as the only refuge for overwintering or source of wild hosts. When there are high amounts of woodland in the landscape, adults emerging from the crop can disperse across a larger area on the farm scale. Adults may be able to select from a much larger area to find the ideal microclimatic habitat for overwintering (Zerulla et al., 2015). Baited traps in the woods may also be relatively more attractive at farms with minimal woodland due to the reduced availability of wild host fruits (Pelton et al., 2016).

In native areas, early studies on vertical flight in semi-natural and natural habitats found that *D. suzukii* mainly flies above tree tops while few individuals are usually found close to the ground (Beppu, 1980; Toda, 1987, 1992). *Drosophila suzukii* shows a strong and constant preference for the canopy layer depending on the vertical foliage complexity of the trees (Tanabe, 2002).

Tonina et al. (2018a) showed that *D. suzukii* adult populations in sweet cherry orchards generally decreased with increasing distance from the forest margin. Overall, a gradient of decreasing density with increasing height above the ground was also present. These patterns varied over phenological stages of the cherry indicating that the pest used multiple habitats during the seasons. When the cultivated host plant was not suitable for reproduction, *D. suzukii* preferred to fly near the forest margin and close to ground level. However, when the cherries were suitable for egg development, *D. suzukii* adults colonized the orchards both horizontally and vertically exploring the canopy volume in greater depth (Tonina et al., 2018a). The large availability of ripe cherries likely attracted individuals from the forests into the orchards (Asplen et al., 2015). This was confirmed by the fact that, after harvest, a higher activity density was observed in forests than in orchards, indicating a return of *D. suzukii* to the refuge areas, rich in host plants (Kenis et al., 2016) and favourable microclimatic conditions (Tanabe, 2002; Hamby et al., 2016; Tochen et al., 2016b). The capture difference between low (1–2 m) and high (5–6 m) traps was more evident up to cherry flowering and then decreased when fruits were present, indicating an increased vertical exploration of the tree canopy. Trap catch increased, again, after cherry picking, indicating a higher insect activity density at lower heights (Tonina et al., 2018a) where conditions are more humid and cooler (shading by foliage and grass; Hamby et al., 2016; Tochen et al., 2016b).

Within the forest, *D. suzukii* populations were more localized at greater heights, with the exception in the early spring where the individuals were concentrated nearer to the ground (Tonina et al., 2018a), probably due to their overwintering in the forest litter (Zerulla et al., 2015). Studying the flying height inside the forest is very important to better understand the potential of *D. suzukii* to infest wild fruits present at the top of native trees (e.g. *Viscum album*; Poyet et al., 2015). In addition, it is useful to understand the canopy colonization dynamics in taller cultivated plants and the height from which *D. suzukii* can invade neighbouring cultivated orchards.

3.2.2 Movement among habitat

It is known that diverse arthropod species move seasonally from one area to another (Coyne et al., 1982; Riley et al., 1995; Ahmed et al., 2009; Stefanescu et al., 2013). In *Drosophila*, some temperate species migrate from low to high elevations in early summer and probably return to low elevations in the autumn (Kimura et al., 1977; Kimura and Beppu, 1993; Kimura, 2004; Mitsui et al., 2010). Mitsui et al. (2010) described *D. suzukii* as a fruit specialist species that occurs at specific times of the year in mountainous regions of Japan. For this reason, it may be reasonable to assume that *D. suzukii* is able to seasonally migrate between low and high elevations (Tait et al., 2018).

The spatial dynamics of *D. suzukii* have been investigated using red traps with a liquid food bait (apple cider vinegar, red wine and sugar). Flies are attracted by fermenting volatiles, but the efficacy of traps is dependent on climatic conditions that affect fly activity (e.g. wind, humidity and temperature), elevation and competition from other wild and crop food sources (Grassi et al., 2015).

According to the number of trap catches throughout the season and across elevational gradients, Mitsui et al. (2010) and Rossi-Stacconi et al. (2016) hypothesized that *D. suzukii* migrates progressively towards higher elevations during summer and returns to progressively lower elevations as temperatures decrease. It was speculated by Mitsui et al. (2010) that this behaviour would enable *D. suzukii* to minimize exposure to suboptimal high temperatures and resource-limited conditions at low elevations during summer. Consequently, this behaviour would also allow such populations, generation after generation, to exploit suitable food resources, that is ripening fruits, as they progressively become susceptible for colonization at higher elevations as the season progresses (Tonina et al., 2016).

Tait et al. (2018) documented that *D. suzukii* can disperse, in less than a month, over distances of up to 9 km and further, suggesting that during spring and summer, flies move from low to higher elevations. Temperature (Tochen et al., 2014; Ryan et al., 2016; Wiman et al., 2014; Stephens et al., 2015; Tonina et al., 2016; Grassi et al., 2017) and searching for additional oviposition sites and food resources (Mitsui et al., 2010; Tochen et al., 2014, 2016a; Wiman et al., 2014; Evans et al., 2017) may play a role in this dispersion. Tait et al. (2018) report that *D. suzukii* movement was most likely facilitated by updrafts during spring (Beerwinkle et al., 1994; Brattström et al., 2008; Giovannini et al., 2013), while during autumn, air movement from high to low elevations likely promotes the dispersal of flies towards lower elevations (Giovannini et al., 2013).

3.2.3 Implication for IPM

These biological and ecological factors make *D. suzukii* control particularly challenging. Understanding ecological aspects, landscape complexity, pest behaviour and activity are important for designing effective pest and landscape management programmes that enhance pest control and increase populations of beneficial arthropods (Schneider, 1989; Dyer and Landis, 1997; Holland and Fahrig, 2000; Tscharntke et al., 2005; Blackshaw and Vernon, 2006; Carriere et al., 2006; Rand et al., 2006; Gladbach et al., 2011).

The effect of forest cover on both adult density and crop damage suggests that a reduction of forest cover is not a viable option for controlling *D. suzukii* (Santoiemma et al., 2018). Therefore, any removal or management of semi-natural habitats should be carefully considered as these habitats deliver multiple fundamental ecosystem services to several crops and to other pest species (Taki et al., 2007; Rusch et al., 2011, 2013; Chaplin-Kramer and Kremen, 2012; Kennedy et al., 2013; Veres et al., 2013; Nayak et al., 2015) and increase farmland biodiversity (e.g. Dainese et al., 2015). However, the high availability of natural habitats in the landscape can complicate the current IPM options (Santoiemma et al., 2018). Even when whole crops are sprayed with conventional products, *D. suzukii* can continue to re-invade neighbouring orchards (Mitsui et al., 2010; Santoiemma et al., 2018) and high dispersal ability (Tait et al., 2018) means even natural habitats at long distances from the crop can act as a source (Santoiemma et al., 2018).

Current *D. suzukii* control relies heavily on the use of conventional insecticides with several negative environmental impacts (Walsh et al., 2011). To be effective, pesticide applications should be particularly timely in orchards close to forests due to the higher and quicker colonization of this pest.

Focussing on field margins, management tools proven for other pests such as bait sprays (Vargas et al., 2001; Prokopy et al., 2003), mass trapping (Cohen and Yuval, 2000) and non-crop host removal (i.e. ecological management; Prokopy, 2003; Suckling and Brockerhoff, 2010) may significantly reduce *D. suzukii* populations, especially if implemented in area-wide programmes (Klick et al., 2016). However, these practices are subject to regional and local pesticide and habitat legislation and are not achievable in all areas.

Less susceptible or early ripening fruit varieties could be planted close to the source of infestation (Tonina et al., 2018a). Planting crops in rows parallel to the source of infestation to intercept the individuals dispersing from the forest onto the first rows of plants to protect the inner part of the orchard should be investigated (Tonina et al., 2018a). Targeted sprays at the margins of crops should be considered, respecting the buffer zones according to the insecticide label. It is also important to achieve good spray coverage of the upper parts of the crop canopy for effective control (Tonina et al., 2018a).

To obtain maximum efficacy with mass trapping (Hampton et al., 2014), traps should be placed at the forest margin before flowering when adult *D. suzukii* are still in the forest and before they colonize the crop (Tonina et al., 2018a).

Netting at borders close to forest margins may reduce the numbers of *D. suzukii* entering an orchard, but because they can fly at heights greater than 6 metres, for complete exclusion it is necessary to cover the entire orchard (Chouinard et al., 2016; Tonina et al., 2018a).

Furthermore, field margins could provide a refuge to reduce *D. suzukii* insecticide resistance development as wild-type alleles may dilute selective pressure to develop insecticide resistance (Tabashnik and Croft, 1985).

4 Host range and host susceptibility

Lee and Sial (2016) created an ongoing host list (last updated, October 2016) which is available at: http://eorganic.info/spottedwingorganic/resources. Host fruits are also listed on the CABI Invasive Species Compendium (https://www.cabi.org/isc/datasheet/109283/aqb).

Drosophila suzukii tend to infest and damage ripening fruit, when their skins soften, resistance to skin puncture is reduced (penetration force) and sugar levels in fruits are elevated (Burrack et al., 2013; Kinjo et al., 2013; Ioriatti et al., 2015; Lee et al., 2015). The fruit penetration force could potentially be used to assess host susceptibility (Burrack et al., 2013).

The host range of *D. suzukii* has been assessed through field observations of damaged fruits, assessment of infestation in fruit collected from the field and oviposition in fruits under laboratory conditions.

Infestations of *D. suzukii* were first reported on cherry in Japan (Kanzawa, 1936). Since its invasion into North America and Europe, the main commercial crops affected are varieties of *Rubus* (e.g. blackberry, raspberry), *Vaccinium* (e.g. blueberry), *Prunus* (e.g. cherry), Fragaria (*strawberry*) and *Vitis* (winegrape) (Bellamy et al., 2013; Ioriatti et al., 2015; Lee et al., 2011b, 2015).

In field tests, the infestation of raspberries was greater than most blackberry varieties, with strawberries being affected the least. Similar results were obtained in laboratory choice experiments where oviposition was greatest in raspberries > blackberries > strawberries > blueberries (Burrack et al., 2013). All four fruits were more susceptible than grape cultivars (Lee et al., 2011b). There are also notable differences in susceptibility between varieties of the same species. Susceptibility of fruit (and larval success) is dependent on a number of factors, including fruit structure and changes associated with fruit ripening such as fruit firmness, pH and sugar content (Lee et al., 2011b, 2015; Burrack et al., 2013; Poyet et al., 2015). For raspberry and blackberry, susceptibility also differs between floricane- and primocane-fruiting varieties, the latter being generally less susceptible, and fruit grown outdoors is also infested at higher rates than that grown under tunnels (Burrack et al., 2013).

Studies on wild and ornamental fruits have been undertaken in both North America by Lee et al. (2015) and in Europe by Poyet et al. (2015) and Kenis et al. (2016) contributing to an extensive list of plants that are susceptible to *D. suzukii*.

Using survey fruits in the field in Michigan and/or Oregon for natural infestation of *D. suzukii* and the susceptibility of fruits in laboratory no-choice and choice studies, Lee et al. (2015) identified over 20 new hosts and previously reported hosts of this invasive fly. Fleshy fruits collected from plants in forests, hedgerows, grasslands, coastal areas, gardens and urban areas of the Picardy region of France were exposed to adult *D. suzukii* and assessed for egg laying, larval and adult development. Out of 67 plant species tested, adult flies emerged from 33 species belonging to 17 different families. Some plant species contained eggs or larvae that did not develop further (Poyet et al., 2015). From 2-year field surveys in Italy, the Netherlands and Switzerland, Kenis et al. (2016) reported that *D. suzukii* were found in 84 plant species primarily from the genera *Cornus*, *Prunus*, *Rubus*, *Sambucus* and *Vaccinium*, out of 165 potential wild and ornamental host plants collected.

Table 1 gives the plants that have been shown as hosts for *D. suzukii*. A total of 159 species from 30 families can act as hosts for this pest.

4.1 Implications for IPM

The identification of wild and alternative hosts has important implications in the development of management strategies. Non-crop hosts may be a source of *D. suzukii* infestation of neighbouring fruit crops; they also provide an early reproductive and food source before commercial fruits become available. They allow *D. suzukii* survival and population growth outside the cropping area (Lee and Sial, 2016). The effect of neighbouring non-crop hosts could be a negative - source of inoculum for commercial crops, or positive - distraction from commercial crops, but more research is needed to quantify this. In addition, non-cropping areas harbour natural enemies for *D. suzukii*, including parasitoids and generalist predators (Rossi-Stacconia, 2017; Wolf et al., 2018). Furthermore, some plants attract female *D. suzukii* and stimulate oviposition, but larvae do not develop fully or eggs do not hatch. These are known as dead-end hosts and could be used as part of an IPM control strategy (Poyet et al., 2015).

Given the information outlined above, it may be prudent to focus on non-pesticide approaches in wild areas to control *D. suzukii* (Kenis et al., 2016; Haye et al., 2016).

5 Detection

The need to recognize invasion of *D. suzukii* into new areas, time spray applications, monitor changes in the landscape and monitor control success has promoted the development of traps and attractant baits (Grassi et al., 2012; Lee et al., 2011a; Cini et al., 2012; Ioriatti et al., 2015; Tonina et al., 2018b). Effective tools for monitoring constitute the essential prerequisite to better predicting pest outbreaks and implementing IPM strategies in order to reduce insecticide applications (Cini et al., 2012; Wiman et al., 2014; Hamby et al., 2016; Hamby and Becher, 2016; Tonina et al., 2018b). To develop an effective monitoring method, a trap and lure is needed which is sensitive to low populations, species selective, easy to use, low cost and with low environmental impacts (Tonina et al., 2018b). High selectivity minimizes detrimental effects on the biodiversity of non-target species and facilitates the screening operations during trap sorting (Burrack et al., 2015).

The majority of baits are based blends of fruit juices reliant upon the odour of fermenting fruits to attract *D. suzukii* or on synthetic semiochemicals (Cha et al., 2014). These traps are not species selective and, hence, catch a range of other insects including similar looking *Drosophila* species. They measure activity density rather than pure population density and are not competitive with surrounding fruiting crops (Keesey et al., 2015; Tonina et al., 2018b).

The most widely used are fermenting substances (e.g. vinegar) either alone or in a mixture with wine or in combination with peptides (Landolt et al., 2012; Lee et al., 2012, 2013; Basoalto et al., 2013; Cha et al., 2013, 2014, 2015a,b; De los Santo Ramos et al., 2014; Iglesias et al., 2014; Kleiber et al., 2014; Burrack et al., 2015; Grassi et al., 2015; Lasa and Tadeo, 2017; Tonina et al., 2018b).

Fermenting substances are highly effective at attracting this species (Cha et al., 2013; Grassi et al., 2015; Keesey et al., 2015). Yeast is important as a food resource for *D. suzukii* and a potential indicator of habitat quality for drosophilid flies (Hamby and Becher, 2016). Wine-based mixtures with apple cider vinegar (ACV) (Landolt et al., 2012) (e.g. Droskidrink 75% of ACV, 25% red wine and 20 g/L of cane sugar) are an effective non-specific attractant (Grassi et al., 2015; Tonina et al., 2018b). These baits are readily available, cheap and easily prepared by the user, but are labour intensive to change regularly and can form into a solid in the summer months (Tonina et al., 2018b).

Yeast and microorganisms play an important role in determining the volatile profile of attractants; yeasts can be incorporated into baits with sugar to improve attraction (Hamby et al., 2014; Iglesias et al., 2014; Scheidler et al., 2015; Huang et al., 2017). More *D. suzukii* were captured in traps baited with comboS (supernatant from a yeast) than those baited with the wine or yeast alone (Huang et al., 2017). *Drosophila suzukii* is also associated with the yeast *Hanseniaspora uvarum* which is found in the midgut of the fly (Hamby et al., 2016). There is the potential to develop distinctive yeast volatiles into selective baits in the future (Huang et al., 2017).

Long-lasting baits are useful for their longer interval substitution, particularly in mass trapping or attract-and-kill programmes; furthermore, long-lasting baits involve a lower use of material, decreasing costs and supplying activities (Tonina et al., 2018b).

Several physical features for trapping are effective including colour, size, shape, entry holes and surface area of the bait-air interface (Grassi et al., 2012; Lee et al., 2012; Basoalto et al., 2013; Hampton et al., 2014; Renkema et al., 2014; Vaccari et al., 2014). Red-coloured traps with lateral entry holes (e.g. Fig. 4), and sometimes with a black stripe for visual stimulus, gave the best efficacy compared with clear or white plastic cups or bottles (Grassi et al., 2012; Basoalto et al., 2013; Hampton et al., 2014; Renkema et al., 2014; Tonina et al., 2018b). Moreover, the use of a clear dome lid favoured fly retention and reduced the probability of flies escaping once they had entered the trap (Lasa et al., 2017).

Figure 4 Drosotrap® on a cherry tree.

Factors related to the local climate, crop type, the low selectivity of food baits and the competition with ripe fruits can lead to an underestimation of pest population density and subsequently erroneous management of the pest (Iglesias et al., 2014; Hamby and Becher, 2016; Huang et al., 2017; Tonina et al., 2018b). Pelton et al. (2016) outlined the high value of fruit sampling to check egg and/or larvae presence, rather than relying on adult trap catches, to assess field risk. Although larval infestation and concurrent adult abundance were positively correlated, this was highly variable and a weak predictor of larval infestation (Pelton et al., 2016; Hamby et al., 2014; Burrack et al., 2015).

6 Chemical, physical and cultural management and control strategies

The management of *D. suzukii* requires a holistic approach, involving the many actors along the food production chain to ensure adequate post-harvest transport, storage and distribution.

The research of *D. suzukii* population containment tools and control is extremely complicated, because of the high biological performance of the species (Griffo et al., 2012). Removal of the pest from areas where it is already present is impossible (Eppo, 2010), hence strategies for continuous suppression of pest populations are required in order to reduce the subsequent infestation (Grassi et al., 2013).

6.1 Monitoring

To achieve effective control of the pest, it is essential to monitor the presence of adults and oviposition on the fruits (Ioriatti et al., 2015).

In the field, monitoring traps should be placed at the height of the crop, inside the vegetation and in the most shady and cool areas (Tonina et al., 2018b). Traps should also be placed in the neighbouring wild areas in order to provide information of *D. suzukii* populations in areas surrounding the crop.

The content of each trap needs to be replaced or filtered and refilled each week during warm weather and twice a month during cold periods and the numbers of male and female *D. suzukii* recorded.

However, data on catches from traps is problematic due to the low attractiveness and selectivity of lures and baits with no relationship between the captures and crop damage, attraction competition between fruits and lures (Burrack et al., 2013; Hamby et al., 2014; Ioriatti et al., 2015; Pelton et al., 2016).

Hence, the following information (Table 2) needs to be taken into account to help interpret trap catches:

- Trends in catches from previous weeks;
- Climatic parameters;
- Comparison with the catches in semi-natural habitat (could be higher for disturbing factors in the orchards, abundance of host-plant and refuge areas in wild conditions);
- Careful inspection at the stereomicroscope, avoiding mere observation in the field;
- State of ovarian maturity of captured females;
- Fruit competition.

Trap information must be modulated for each situation, according to different climate and culture sensitivity.

Table 2 Interpretation of trap data

	Low risk	High risk
Abundance of catches	Low catches	High catches
Trends in catches from previous weeks	Decreasing in combination with spray programme and crop hygiene measures	Increasing
Climatic parameters: current and planned for the following days	Average temperature >28°C or temperature >33°C for 8 h/day (Kinjo et al., 2014) Average temperatures <11°C (Tonina et al., 2016)	Temperature between 11°C and 28°C with high relative humidity
Comparison with the catches in wild condition	Lower than catches in semi-natural habitat	Similar or higher than catches in semi-natural habitat
Maturity of captured females	<10% fertile females	>30% fertile females
Fruit competition	Fruit green	Fruit in ripening or ripe

6.2 Chemical control

The *D. suzukii* polyphagia, the infestation during ripening stage close to harvest, the length and scale of the harvesting period and presence of different varieties greatly complicate chemical control (Ioriatti et al., 2011).

An adulticide-ovicide and residual approach is needed because effective control of larvae is not possible. In Europe and the United States, broad-spectrum insecticides (e.g. organophosphates, carbamates, pyrethroids, spynosins and diamides) have been shown to be effective (Beers et al., 2011; Bruck et al., 2011; Angeli et al., 2012; Profaizer et al., 2012, 2015; Van Timmeren and Isaacs, 2013; Shawer et al., 2018).

To maintain a residual protection throughout the harvest period, two to four applications are needed (Haviland and Beers, 2012; Shawer et al., 2018) but the number of insecticide applications can increase to five to eight depending on pest abundance, crop susceptibility, longevity of cropping season and other environmental factors (Van Timmeren and Isaacs, 2013).

The intensive use of insecticides poses serious concerns that residues on fruits will exceed maximum residue limits (Haviland and Beers, 2012), the development of insecticide resistance in *D. suzukii* (Asplen et al., 2015; Wiman et al., 2016) and negative impacts on the environment beyond beneficials.

Moreover, the management of *D. suzukii* in cold habitats, which are very suitable for the cultivation of high-quality berries in mountain farming systems, has to take into consideration that at low temperature the performance of neonicotinoids and organophosphates is lower (Saeed et al., 2018).

Chemical control is further limited by the reduced availability of active substances registered for *D. suzukii* (Ioriatti et al., 2015).

6.3 Mass trapping

The aim of mass trapping is to contribute to the containment of the *D. suzukii* damage by reducing the adult population density using numerous traps. Kanzawa (1939) reported a reduction of damage in some regions of Japan by using bottle traps baited with diluted molasses, grape wine and rice vinegar, whereas in other regions this technique was ineffective. Investigation in North Italy, using Drosotraps® (Biobest) baited with Droskidrink® (Prantil), showed that this method is suitable and cost-effective only for cultivations where the pest pressure is considerably low. This is where crops are of medium to large size and regular shape, not surrounded by forest or hedgerow, in a geographically flat area with a hot and dry microclimate (Sancassani et al., 2016) or for small area fruit cultivation under exclusion mesh cover (Tonina et al., 2017). However, the effectiveness of this method is limited due to the lack of a highly specific and attractive lures that are able to compete with ripe fruit (Ioriatti et al., 2015; Tonina et al., 2018b) and the deterrent effect that yeast-associated smells may have at high concentration (Cowles, 2013). To protect crops, traps for mass trapping should be placed around the perimeter outside fruit fields (at least 5 m from the fruit trees) while insecticides need to be applied to the surface of traps or on nearby fruit to function as an attract-and-kill strategy (Hampton et al., 2014). Potentially a mesh screen could be incorporated between the traps and the crop to intercept *D. suzukii* that are not instantly captured in the traps.

6.4 Insect-proof nets

Among the sustainable protection techniques for the control of *D. suzukii*, the use of insect-proof nets has proved effective, reducing or completely replacing the use of insecticides in some instances, and providing high levels of exclusion of *D. suzukii* from the crop (Grassi et al., 2013). Mesh size should not exceed 1 mm^2. Different net application methods were evaluated, 'single plant' and 'single row' netting were most effective, while 'whole orchard' netting was not as efficient due to the continuous lifting of the nets to allow the movement of farmers and machinery (Grassi et al., 2013). The netting needs to be installed after flowering, in the case of cherry to allow pollination, but prior to fruits changing colour and should be kept closed until the end of the harvest (Grassi et al., 2013). Growers use a variety of systems, but it is generally agreed by practitioners, based on experience, that the most effective are perimeter netting where the farm machinery and fruit pickers can enter and leave through one door so that the integrity of the barrier around the crop is not constantly compromised. These netting systems can be compromised by workers forgetting to close doors on exiting the crop or damage to the mesh that is not repaired in a timely fashion (Ghelfi et al., 2016; Caruso et al., 2017).

Although the initial investment is extremely expensive, the high economic value of fruit and negating potential losses due to *D. suzukii* and adverse weather events, the long-term returns can be favourable (Ghelfi et al., 2016).

6.5 Cultural practices

Cultural control can be integrated with insecticide applications in *D. suzukii* management (Haye et al., 2016) and have proved effective in blueberry and raspberry (Van Timmeren and Isaacs, 2013; Lee et al., 2015; Tochen et al., 2016b).

During the ripening season, sanitary measures such as removal of dropped, infested and overripe fruits have been suggested (Lee et al., 2011b; Shi, 2015). Infested fruits left on plants and/or those fall to the ground provide a high source of *D. suzukii* (Dreves et al., 2009; Walsh et al., 2011). The collection and treatment of infested fruit through sun exposure, disposal in closed containers, crushing, low temperature treatments, bagging and burying, to destroy *D. suzukii* eggs and larvae, are essential IPM procedures to limit the infestation of healthy fruit (Cini et al., 2012; Grassi et al., 2012; Noble et al., 2017).

Moreover, considering that the viability of *D. suzukii* eggs is lower under dry, warm conditions (Tochen et al., 2016b), cool humid microhabitats should be avoided by pruning to open up the canopy and by using wider tree spacing to increase airflow to the canopy and reduce shading. Open canopies also help increase the penetration of the insecticide applications and to concentrate the ripening period of the whole plant and speed up the harvest. In addition, the use of mulches to reduce standing water can further contribute to the reduction of humidity in fruit orchards (Hoashi-Erhardt and Bixby-Brosi, 2014). Precision irrigation should also be incorporated to reduce pooling of water on the ground.

7 Biological control strategies

The wide-host range and natural habitats that *D. suzukii* inhabits require an area-wide strategy to reduce populations (Kenis et al., 2016; Haye et al., 2016). Biocontrol using natural enemies could play an important role in managing *D. suzukii* as part of an IPM strategy.

The development of a suitable biological control agent against *D. suzukii* has received considerable attention, and involved searching for natural enemies in both its region of origin and regions of invasion. Studies have included laboratory and field testing parasitoids and generalist predators. In Drosophilidae, larval parasitoids have a major impact on mortality (Carton et al., 1986; Fleury et al., 2009) and it is preferable to use natural enemies that are specialized to a specific pest species (i.e. *D. suzukii*) rather than generalists that will also affect non-target species.

7.1 Parasitoids of **Drosophila suzukii**

In the native region (Asia), a number of hymenopteran parasitoids of *D. suzukii* have been reported. In contrast, mainly generalist natural enemies have been found to attack *D. suzukii* in its regions of invasion such as Europe and America (Chabert et al., 2012; Kacsoh and Schlenke, 2012), including earwigs, ground beetles, mirid bugs, spiders (Wolf et al., 2018; Schmidt et al., 2019) and parasitoids (Chabert et al., 2012; Rossi-Stacconi et al., 2013). Invasive species often do not arrive with their specific diseases or parasitoids, or the latter fail to establish in the new environment further enabling the success of non-native species establishment (Williamson, 2006).

7.1.1 Parasitoids from the region of origin (Asia)

The first report on parasitism of *D. suzukii* was from a study on its biology where, in Japan, *D. suzukii* larvae were parasitized by a hymenopteran from the genus *Phaenopria* (Kanzawa, 1939). More recent studies of parasitoids from the region of origin of *D. suzukii* have identified species of the genera *Ganaspis* and *Leptopilina* (Hymenoptera: Figitidae), and *Trichopria* (Hymenoptera: Diapriidae) which have been found to either emerge from field-collected *D. suzukii* or have attacked *D. suzukii* in laboratory trials.

Ganaspis sp. was the main larval parasitoid from wild cherries collected from the field in Tokyo, Japan (Kasuya et al., 2013a,b). This parasitoid did not attack other *Drosophila* species associated with fruit.

Generalist parasitoids (*Trichopria* spp., *Leptopilina japonica* Novković & Kimura and *Asobara* spp.) were also identified from *D. suzukii* in wild cherry. Matsuura et al. (2018) confirmed that the *D. suzukii*-associated *Ganaspis* spp. (*Ganaspis brasiliensis* Ihering) was also the most abundant parasitoid to emerge from *D. suzukii* in wild cherries and *Asobara* sp. TK1 and *Leptopilina japonica* also emerged from field-collected wild cherries in Tokyo.

Nomano et al. (2015) trapped eight *Asobara* spp. from field collections across Japan and tested their ability to parasitize *D. suzukii* larvae in the laboratory. Of the five species that oviposited in *D. suzukii* larvae, only two, *A. japonica* and *Asobara* sp. TK1, survived to adulthood. *Asobara japonica* Belokobylskij is a generalist parasitoid which attacks other drosophilids. It is widely distributed in Japan and found in >25 drosophilid species, but only rarely associated with *D. suzukii* in the field (Ideo et al., 2008; Mitsui and Kimura, 2010; Novković et al., 2012).

In contrast, *Asobara* sp. TK1 only attacked *D. suzukii* and, therefore, may be a more suitable candidate for biological control. However, the parasitism rate of *Asobara* sp. TK1 in *D. suzukii* appears to be low (13.3%) and over 80% of *D. suzukii* larvae died before these parasitoids emerged (Namano et al., 2015). Hence, *D. suzukii* may not sustain continuous populations of this parasitoid so continuous releases of *Asobara* sp. TK1 would most likely be required to reduce *D. suzukii* populations in a biological control programme. Field surveys of parasitoids in South Korea, using uninfested fruit-baited traps, and from collecting commercial or wild host fruits, identified six species of larval parasitoids *A. japonica, A. leveri, A. brevicauda, Ganaspis brasiliensis, Leptopilina japonica, L. j. formosana* Novković & Kimura, one pupal parasitoid, *Trichopria drosophilae* Perkins from *D. suzukii*. Laboratory assays confirmed that *A. japonica, G. brasiliensis, L. j. japonica* and *T. drosophilae* were able to develop from *D. suzukii*. The pupal parasitoid *Pachycrepoideus vindemiae* Rondani (Hymenoptera: Pteromalidae) was identified from field-collected *D. melanogaster*, and was also found to attack *D. suzukii* under laboratory conditions (Daane et al., 2016).

The ability of Asian hymenopteran parasitoids (*Leptopilina japonica, Ganaspis* sp. and *Asobara japonica*) to develop in *D. suzukii* larvae were compared to European generalist species *Leptopilina heterotoma* (Girod et al., 2018a,b). All three Asian parasitoids successfully developed in larvae in fresh fruit and it appears that *Ganapsis* sp. may have a high degree of specialization on *D. suzukii*, and use host-finding cues associated with fresh fruit. *Leptopilina japonica* and *A. japonica* have a wider Drosophilidae host range and parasitize larvae on ripe and rotten fruits, mushrooms and decaying leaves, and hence can detect larvae in different substrates (Girod et al., 2018a). The European *L. heterotoma* was less successful at parasitizing *D. suzukii* as most of the parasitoid eggs were encapsulated (defensive reaction from the immune system of *D. suzukii*). This species also has a broad *Drosophila* host range and is attracted to fermentation volatiles (ethanol, ethyl acetate and acetaldehyde) associated with decaying fruit and plant materials (Carton et al., 1986; Girod et al., 2018a). Hence, it is unlikely that volatiles from fresh blueberries would induce host-seeking behaviours (Girod et al., 2018a).

Girod et al. (2018b) evaluated the host specificity of Asian parasitoids (*A. japonica, L. japonica* and *Ganaspis cf. brasiliensis*) on larvae of *D. suzukii* and five non-target *Drosophila* spp. (*D. busckii, D. hydei, D. immigrans, D. melanogaster* and *D. subobscura*) including the Tephritidae, *Ceratitis capitate* under laboratory conditions. *Asobara japonica* developed in all *Drosophila* spp., whereas *Leptopilina japonica* only successfully parasitized *D. melanogaster* and *D. subobscura*. *Ganapsis cf. brasiliensis,* from Japan, was highly specific, attacking only *D. suzukii*. The same species from China parasitized *D. suzukii* and *D. melanogaster* and occasionally *D. subobscura*.

The introduction of Asian parasitoids, that have co-evolved with *D. suzukii*, has the potential to be used in classical biological control strategies to provide a sustainable solution for the area-wide management of this pest, and contribute to reducing populations below economic threshold levels. A parasitoid specialized to *D. suzukii* is preferable to a generalist, which may affect species of other non-target species (Girod et al., 2018a,b). The surveys in Asia have shown that *G. cf. brasiliensis* derived from Japan would be the preferred candidate for introduction in a classical biological control programme (Girod et al., 2018a,b).

7.1.2 Parasitoids in region of invasion

A number of surveys have been carried out to assess the presence of indigenous parasitoids attacking *D. suzukii* in its region of invasion (North America and Europe). There are numerous (>40) different species of hymenopteran parasitoids that have been reported to attack Drosophilidae (Carton et al., 1986). Laboratory trials have also been undertaken to assess the ability of some of these parasitoids to attack and develop in *D. suzukii* larvae and pupae.

Field sentinel traps baited with *D. suzukii* larvae and pupae in Northern Italy and in the Pacific Northwest United States have been used in combination with laboratory studies to determine which native parasitoids would utilize *D. suzukii* as a host. The generalist pupal parasitoid, *P. vindemiae*, was identified from *D. suzukii* pupae in traps in both Italy and the United States (also found in Korea) and was confirmed to parasitize *D. suzukii* pupae under laboratory conditions (Chabert et al., 2012; Rossi-Stacconi et al., 2013). *Pachycrepoideus vindemiae* has a wide-host range (>60 dipterans), including *D. melanogaster* (Rossi-Stacconi et al., 2013) but has been shown to parasitize significant numbers of sentinel pupae, in field conditions, in the United Kingdom (Fountain, pers. comm.). Similarly, the pupal parasitoid, *Trichopria cf drosophilae*, was confirmed to parasitize and develop on *D. suzukii* in the laboratory (Chabert et al., 2012; Rossi-Stacconi et al., 2015; Mazzetto et al., 2016). Laboratory testing showed high parasitism success rates in *D. suzukii* pupae for both species and they were both found to significantly reduce adult emergence of adult *D. suzukii* from infested fruit (Gabarra et al., 2015).

In laboratory choice experiments, where both *D. suzukii* and *D. melanogaster* pupae were offered, both *Drosophila* spp. were favourable hosts of *T* cf. *drosophilae* (Wang et al., 2016).

In both Northern Italy and in Oregon, USA, the generalist parasitoids, *L. heterotoma* and *P. Vindemiae*, emerged from *D. suzukii* or *D. melanogaster* pupae collected from sentinel traps (Miller et al., 2015). In Italy, *T.* cf. *drosophilae* was also collected. Under laboratory conditions, Rossi-Stacconi et al. (2015) reported that both Italian and US strains of *P. vindemiae* were shown to parasitize *D. suzukii*. The Italian *L. heterotoma* strain parasitized *D. suzukii*, whereas the US *L. heterotoma* strain did not.

Studies, in both North America and Europe, demonstrate that generalist larval parasitoids (*L. heterotoma* and *L. boulardi*) attack *D. suzukii* (under laboratory conditions), but fail to develop, primarily due to the strong immune response of *D. suzukii* larvae (Chabert et al., 2012; Poyet et al., 2013; Mazzetto et al., 2016). In contrast, the pupal parasitoids *T.* cf. *drosophilae* and *P. vindemiae* attack and develop inside *D. suzukii* (Chabert et al., 2012; Gabarra et al., 2015; Rossi-Stacconi et al., 2015; Miller et al., 2015). Chabert et al. (2012) concluded that their results were consistent with the natural enemy release hypothesis; natural enemies specialized on native species (i.e. larval parasitoids) will be less effective against invasive species, whereas pupal parasitoids, which are much more generalist, are able to adapt to a new host.

Research is ongoing to identify effective native and exotic parasitoids that could potentially be used in augmentative or classical biological control programmes, respectively. The evidence to date suggests that the introduction of an Asian parasitoid (e.g. *Ganaspis* sp.) that is specialized to *D. suzukii* is the most favourable option (Girod et al., 2018a,b).

A preliminary field study, on the dispersal ability and host location of *T. drosophilae* and the impact on adult *D. suzukii* emergence in eight crops in Italy, reported that parasitoids were found in sentinel traps ≤40 m from the release point. The emergence of *D. suzukii* adults was significantly reduced in traps 10 m from the *T. drosophilae* release point (Rossi-Stacconi et al., 2018). Only 500 male and 500 female parasitoids were released each week (over a 5-week period) during the trials (Rossi-Stacconi et al., 2018). Potentially optimizing the frequency and numbers of parasitoids could increase the frequency of *D. suzukii* parasitism as part of an IPM control strategy.

Recently, Pfab et al. (2018) have provided a model for optimal timing of release of the pupal *T. drosophilae*, based on data gathered in the Trento region of Northern Italy. The success of the biological control strategy was strongly dependent on the timing of parasitoid release. The estimated optimal time of release would be between late spring and early summer; but late enough that *D. suzukii* pupae would be available in the wild.

7.2 Natural predators

Commercially available native natural predators have been evaluated for their potential use in augmentative releases. Generalist predators such as canopy active *Anthocoridae* (Hemiptera) and ground feeding Staphylinidae (Coleoptera) have the potential to feed on *D. suzukii* larvae and pupae in infested hanging berries or on those that drop to the ground. *D. suzukii* larvae pupate either within the fruit, or emerge and pupate in the soil (Gabarra et al., 2015; Woltz and Lee, 2017).

Several laboratory trials have assessed predation on different *D. suzukii* life stages in confined arenas. When starved, the earwig *Labidura riparia* (Dermaptera: Labiduridae) effectively killed *D. suzukii* larvae and pupae in raspberries confined in petri dishes (Gabarra et al., 2015). In similar conditions, European generalist predators (Cuthbertson et al., 2014a) and two US predators have been tested (Woltz et al., 2015).

Atheta coriaria (a Staphylinid beetle) and *Hypoaspis miles* (a predatory mite) had no major impact on *D. suzukii* larvae or pupae. *Orius majusculus*, *O. laevigatus* and *Anthocoris nemoralis* (Anthocoridae, Hemiptera) all preyed upon adult *D. suzukii* (Cuthbertson et al., 2014a). *Orius laevigatus* fed on *D. suzukii* eggs in strawberry fruits, but not on larvae (Gabarra et al., 2015).

O. insidiosus reduced *D. suzukii* larvae survival in blueberries but had little impact on *D. suzukii* survival when tested on outdoor-infested potted blueberries, or on infested blueberries on outdoor plants when isolated on bagged branches (Woltz et al., 2015). The soil-dwelling staphylinid, *Dalotia coriaria,* had no effect on *D. suzukii* larval survival in blueberries in the laboratory (Woltz et al., 2015).

Woltz and Lee (2017) used predator exclusion experiments in the field to determine whether natural enemies could reduce *D. suzukii* larvae in strawberry and blueberry fruits, and pupae in the soil below strawberry, blueberry and blackberry. Predators significantly reduced the number of *D. suzukii* larvae in fruits and pupae in the soil. Ants and spiders were commonly observed in a variety of sampling methods, and ants were observed on video to remove pupae from the soil.

To date, there is little evidence to support augmentative release of current commercially available biocontrol predators. However, naturally occurring predators within the ecosystem may contribute to population suppression (Cuthbertson et al., 2014a) as identified by Wolf et al. (2018) who found *D. suzukii* DNA in the digestive tracts of earwigs, pentatomids and spiders. Furthermore, *Orius* spp. have been observed feeding on *D. suzukii* larvae in raspberries (Walsh et al., 2011) and among infested blueberries (Woltz et al., 2015) in the field. In Spain, four hemipteran predators (*Orius laevigatus, Cardiastethus nazarenus, C. fasciventris* and *Dicyphus tamaninii*) were found in *D. suzukii*-infested strawberries and raspberries (Arnó et al., 2012).

7.2.1 Implications for IPM

Limiting fruit on the ground is already considered good sanitation practice and critical for *D. suzukii* management. Additional management of pupae in the ground, which have dropped from hanging fruit, is also an important consideration, and where natural predators may have a role. Natural populations of predators would most likely need supplementing with augmentative releases to have any impact in an IPM strategy.

Further work on the effectiveness of native parasitoids and generalist predators in Europe and the United States, in the field, is required; however, the augmentative release of pupal parasitoids and conservation biocontrol of generalist predators, potentially, could contribute to the integrated management of *D. suzukii* populations, especially in natural habitats close to commercial crops. The effects of current management practices (e.g. chemicals, pruning) may need to be modified and other approaches introduced to provide more favourable conditions for natural enemies (Rossi-Stacconi et al., 2018; Wolf et al., 2018).

7.3 Entomopathogens

Commercially available entomopathogens have been tested under laboratory and/or field conditions.

7.3.1 Entomopathogenic fungi

Cuthbertson et al. (2014b) tested the impact of *Beauveria bassiana* on infested blueberries and by direct application to adults under laboratory conditions. Although *B. bassiana* caused significant mortality in adults, it did not prevent females laying eggs in blueberries and subsequent adult emergence, so did not suppress *D. suzukii* populations in these trials. When infested berries were treated, *B. bassiana* had little impact on adult *D. suzukii* emergence (Cuthbertson et al., 2014b).

Similar tests were undertaken with *Metarhizium anisopliae* and *Isaria fumosorosea* (Cuthbertson and Audsley, 2016). By applying to infested blueberries, both fungal species reduced *D. suzukii* populations developing from these berries; *I. fumosorosea* had the greatest impact. Exposing adults to uninfested treated berries resulted in mortality, but fungi had little impact on the number of flies that subsequently developed from these berries (Cuthbertson and Audsley, 2016).

Cossentine et al. (2016) also demonstrated mortality of adult *D. suzukii* through contact with *Metarhizium brunneum*, *B. bassiana*, *I. fumosorosea* and *Lecanicillium lecanii* conidia under laboratory conditions. Furthermore, *M. brunneum*-reduced fly fecundity and conidia were transferred between sexes.

Naranjo-Lázaro et al. (2014) assessed the susceptibility of *D. suzukii* to four EPF fungal strains (*I. fumosorosea* Pf21, Pf17, Pf15 and *M. anisopliae* Ma59), using *in vitro* assays. Fly mortality varied between strains from 12% (*M. anisopliae*) to 85% (*I. fumosorosea* Pf21).

Further testing is required to assess the impact of EPF on suppressing *D. suzukii* populations in the field and whether EPF may contribute to an overall IPM programme.

7.3.2 Entomopathogenic nematodes

Three entomopathogenic nematodes (*Steinernema carpocapsae*, *S. feltiae*, *S. kraussei*) were tested against larvae in infested blueberries under laboratory conditions and shown to have no impact on the number of adult *D. suzukii* emerging (Cuthbertson et al., 2014b). Similar results were reported by Woltz et al. (2015) who tested *Heterorhabditis bacteriophora*, *S. feltiae* and *S. carpocapsae* against larvae and pupae in infested blueberries or in diet medium, leading them to conclude that there was little evidence supporting the use of nematodes against *D. suzukii*.

The above nematode species were also tested as soil drenches (on sand in petri dishes) against *D. suzukii* larvae and pupae. All nematodes had an impact on survival of both *D. suzukii* pupae and larvae, and on subsequent adult emergence. *Steinernema kraussei* caused the highest pupal mortality, whereas *H. bacteriophora* was the most effective against larvae. Mortality in larvae (>90%) was greater than that in pupae (*c.* 55%).

It would appear that nematodes do not readily infect *D. suzukii* larvae or pupae when applied to berries, under laboratory conditions, but would appear to be much more effective when applied as a soil drench. However, this needs to be tested under field conditions to assess whether nematodes could be used in this manner as a component of IPM.

7.3.3 Multiple natural enemies

There is little evidence to support augmentative release of a single native predator; however, combinations of predators (*D. coriaria* and *O. insidiosus*) and/or an entomopathogenic nematode (*H. bacteriophora*) have been tested on strawberries and blueberries in the field (Renkema and Cuthbertson, 2018). Only a combination of *O. insidiosus* and *H. bacteriophora* had any impact on survival of *D. suzukii*, compared to individual treatments, whereas other combinations (both predators, *D. coriaria* + *H. bacteriophora*) did not significantly increase *D. suzukii* mortality (Renkema and Cuthbertson, 2018). Wolf et al. (2018) have also shown that from multiple predators collected from the field, earwigs, spiders and predatory bugs were regular feeders on *D. suzukii* and could potentially be used in a (bio) control strategy.

These data show that combinations of biological control agents may provide enhanced suppression of *D. suzukii* populations. However, further research in the field is required to substantiate these data and to determine whether other combinations of natural enemies may also impact on *D. suzukii* populations.

8 Conclusion

The arrival of *D. suzukii* in Europe and America from Asia caused extensive damage to cherry and soft fruit strongly affecting their cultivation. Currently, to control the pest, there is an increase in the number of insecticide applications that are sprayed from fruit changing colour until the harvest period. This increase in the use of chemicals has resulted in higher production costs, greater risk to consumers and the environment, compromise of current IPM strategies for other pests and may hasten the development of pesticide resistance (Boselli et al., 2012; Grassi et al., 2012; Mori et al., 2015). It is clear that an integrated approach to managing *D. suzukii* is required, as no one control option is sufficiently effective. Knowledge on the biology and ecology of any pest is a prerequisite to developing and/or improving an IPM strategy and consequently reducing the use of insecticides, and these have been extensively studied for *D. suzukii* since its invasion of North America and Europe.

The numbers of *D. suzukii* trapped and the damage caused on different crops are highly variable between areas and location of orchards and the surrounding environment. Temperature, landscape composition and host plant abundance all play an important role on pest abundance and population dynamics. Studies show that catches and damage on fruit are higher at the margins of the orchards, close to hedges and forest, due to the presence of numerous host plants that provide food and refuge areas (Walsh et al., 2011; Cini et al., 2012; Grassi et al., 2013; Klick et al., 2016; Pelton et al., 2016; Tonina et al., 2018a). Semi-natural areas have an important role providing alternative host resources, overwintering habitats or refuge areas when crops are sprayed with insecticides (Zerulla et al., 2015; Kenis et al., 2016; Pelton et al., 2016; Santoiemma et al., 2018), with implication on distribution and population dynamics inside the orchards. When the host plant was not suitable for reproduction, *D. suzukii* preferred to fly closer to the forest margin and near the grass. Differently, when the host plant was suitable, *D. suzukii* dispersed further into the orchards both horizontally and vertically. These gradients must be considered in the design of new orchards and in insecticide applications (Tonina et al., 2018a). In addition, catches and damage of *D. suzukii* are not uniform but vary widely between years. Years characterized by mild and humid winters, springs and summers are favourable for the development of pest, in contrast with cold, dry winters and hot summers (Sancassani et al., 2016). These considerations indicate the need for a management at agroecosystem level and not just limited to the orchard.

All cultural and agronomical practices, in order to disadvantage the development of the pest, should be adopted in the orchard as pruning, mowing, borders management and ripe abandoned fruit destruction. Problems related with residues and side effects on the environment require careful management of chemical treatments, to be limited to the most sensitive fruit stages. It is of crucial importance that a rapid and timely harvesting of all fruits and proper management of discarded fruits is carried out (Dreves et al., 2009; Walsh et al., 2011; Cini et al., 2012; Grassi et al., 2012; Noble et al., 2017).

The wide-host range and natural habitats that harbour *D. suzukii* has led to the conclusion that an area-wide strategy is required to reduce populations of *D. suzukii* (Kenis et al., 2016; Haye et al., 2016). Hence, biocontrol using natural enemies may play an important role in managing *D. suzukii* as part of an IPM strategy.

An area-wide strategy which can suppress *D. suzukii* populations in the natural environment, where chemical pesticides cannot be applied, is required as part of an IPM programme. One such option would be biological control using natural enemies, and an Asian parasitoid (*Ganapsis* sp), which specifically attacks *D. suzukii* pupae, has the potential to fulfil this role. However, the release of non-native species in Europe is tightly regulated, and a risk assessment, including non-target impacts, must be undertaken as part of obtaining a licence for import and release of such parasitoids. Until such time, it may be worth pursuing the augmentative release of native natural enemies, and implementing attract and infect strategies using entomopathogenic fungi as tools in IPM strategies.

Other targeted strategies to reduce *D. suzukii* populations are currently being investigated. Area-wide approaches such as sterile insect techniques (using irradiation) and incompatible insect technique (utilizing *Wolbachia* which reduces fertility; Hamby et al., 2016) have been reviewed by Nikolouli et al. (2018). Self-limiting technology is a novel alternative to SIT (Alphey et al., 2013), where (male) insects carrying self-limiting genes induce mortality in progeny at an immature life-stage thereby reducing pest populations, is currently being developed for *D. suzukii* (Oxitech, 2018). Downregulation of expression of essential genes through RNA interference (RNAi) using double-stranded RNA or targeting hormones and receptors to disrupt physiology or behaviour for pesticide discovery are other approaches under development (Audsley and Down, 2015; Taning et al., 2016).

9 References

Ahmed, S., Compton, S. G., Butlin, R. K. and Gilmartin, P. M. 2009. Wind-borne insects mediate directional pollen transfer between desert trees 160 kilometers apart. *Proc. Natl. Acad. Sci. U.S.A.* 106(48), 20342–7. doi:10.1073/pnas.0902213106.

Alphey, L., McKemey, A., Nimmo, D., Neira Oviedo, M., Lacroix, R., Matzen, K. and Beech, C. 2013. Genetic control of *Aedes* mosquitoes. *Pathog. Glob. Health* 107(4), 170–9. doi:10.1179/2047773213Y.0000000095.

Angeli, G., Profaizer, D., Chiesa, S., Sofia, M., Trainotti, D. and Zadra, E. 2012. I prodotti efficaci contro *Drosophila suzukii*. *L'informatore agrario* 68(26), 64–7.

Arno, J., Riudavets, J. and Gabarra, R. 2012. Survey of host plants and natural enemies of *Drosophila suzukii* in an area of strawberry production in Catalonia (northeast Spain). *IOBC/WPRS Bull.* 80, 29–34.

Asplen, M. K., Anfora, G., Biondi, A., Choi, D. S., Chu, D., Daane, K. M., Gibert, P., Gutierrez, A. P., Hoelmer, K. A., Hutchison, W. D., et al. 2015. Invasion biology of spotted-wing Drosophila (*Drosophila suzukii*): a global perspective and future priorities. *J. Pest. Sci.* 88(3), 469–94. doi:10.1007/s10340-015-0681-z.

Audsley, N. and Down, R. E. 2015. G protein coupled receptors as targets for next generation pesticides. *Insect Biochem. Mol. Biol.* 67, 27-37. doi:10.1016/j.ibmb.2015.07.014.

Bal, H. K., Adams, C. and Grieshop, M. 2017. Evaluation of off-season potential breeding sources for spotted wing Drosophila (*Drosophila suzukii* Matsumura) in Michigan. *J. Econ. Entomol.* 110(6), 2466–70. doi:10.1093/jee/tox252.

Basoalto, E., Miranda, M., Knight, A. L. and Fuentes-Contreras, E. 2010. Landscape analysis of adult codling moth (Lepidoptera: Tortricidae) distribution and dispersal within typical agroecosystems dominated by apple production in central Chile. *Environ. Entomol.* 39(5), 1399–408. doi:10.1603/EN09371.

Basoalto, E., Hilton, R. and Knight, A. L. 2013. Factors affecting the efficacy of a vinegar trap for *Drosophila suzukii* (Diptera: Drosophilidae). *J. Appl. Entomol.* 137(8), 561–70. doi:10.1111/jen.12053.

Beers, E. H., Van Steenwyk, R. A., Shearer, P. W., Coates, W. W. and Grant, J. A. 2011. Developing *Drosophila suzukii* management programs for sweet cherry in the western United States. *Pest Manag. Sci.* 67(11), 1386–95. doi:10.1002/ps.2279.

Beerwinkle, K. R., Lopez, J. D., Witz, J. A., Schleider, P. G., Eyster, R. S. and Lingren, P. D. 1994. Seasonal radar and meteorological observations associated with nocturnal insect flight at altitudes to 900 meters. *Environ. Entomol.* 23(3), 676–83. doi:10.1093/ee/23.3.676.

Bellamy, D. E., Sisterson, M. S. and Walse, S. S. 2013. Quantifying host potentials: indexing postharvest fresh fruits for spotted wing drosophila, *Drosophila suzukii*. *PLoS ONE* 8(4), e61227. doi:10.1371/journal.pone.0061227.

Beppu, K. 1980. Vertical microdistribution of the Drosophilidae (Diptera) within various forests in Hokkaido: II. Streamside in natural broad leaved forests. *Kontyû* 48(4), 549–57.

Blackshaw, R. P. and Vernon, R. S. 2006. Spatiotemporal stability of two beetle populations in non-farmed habitats in an agricultural landscape. *J. Appl. Ecol.* 43(4), 680–9. doi:10.1111/j.1365-2664.2006.01167.x.

Bolda, M. P., Goodhue, R. E. and Zalom, F. G. 2010. Spotted wing drosophila: potential economic impact of a newly established pest. *Agric. Res. Econ. Update* 13(3), 5–8.

Boselli, M., Tiso, R., Nannini, R., Bortolotti, P., Caruso, S. and Dradi, D. 2012. Monitoraggio di *Drosophila suzukii* in Emilia - Romagna. *Atti Giornate Fitopatologiche* 1, 429–32.

Brattström, O., Kjellen, N., Alerstam, T. and Akesson, S. 2008. Effects of wind and weather on red admiral, *Vanessa atalanta*, migration at a coastal site in southern Sweden. *Anim. Behav.* 76(2), 335–44.

Bruck, D. J., Bolda, M., Tanigoshi, L., Klick, J., Kleiber, J., Defrancesco, J., Gerdeman, B. and Spitler, H. 2011. Laboratory and field comparisons of insecticides to reduce infestation of *Drosophila suzukii* in berry crops. *Pest Manag. Sci.* 67(11), 1375–85. doi:10.1002/ps.2242.

Burrack, H. J., Fernandez, G. E., Spivey, T. and Kraus, D. A. 2013. Variation in selection and utilization of host crops in the field and laboratory by *Drosophila suzukii* Matsumara (Diptera: Drosophilidae), an invasive frugivore. *Pest Manag. Sci.* 69(10), 1173–80. doi:10.1002/ps.3489.

Burrack, H. J., Asplen, M., Bahder, L., Collins, J., Drummond, F. A., Guédot, C., Isaacs, R., Johnson, D., Blanton, A., Lee, J. C., et al. 2015. Multistate comparison of attractants for monitoring *Drosophila suzukii* (Diptera: Drosophilidae) in blueberries and caneberries. *Environ. Entomol.* 44(3), 704–12. doi:10.1093/ee/nvv022.

Calabria, G., Máca, J., Bächli, G., Serra, L. and Pascual, M. 2012. First records of the potential pest species *Drosophila suzukii* (Diptera: Drosophilidae) in Europe. *J. Appl. Entomol.* 136(1–2), 139–47. doi:10.1111/j.1439-0418.2010.01583.x.

Carriere, Y., Ellsworth, P. C., Dutilleul, P., Ellers-Kirk, C., Barkley, V. and Antilla, L. 2006. A GIS-based approach for area- wide pest management: the scales of *Lygus hesperus* movements to cotton from alfalfa, weeds, and cotton. *Entomol. Exp. Appl.* 118(3), 203–10. doi:10.1111/j.1570-7458.2006.00384.x.

Carton, Y., Boule´treau, B., Van Alphen, J. J. M. and Van Lenteren, J. C. 1986. The *Drosophila* parasitic wasps. In: Ashburner, M., Carson, H. L. and Thompson, J. N. (Eds), *The Genetics and Biology of Drosophila*. Academic Press, London, pp. 347–94.

Caruso, S., Vaccari, G., Lugli, S., Bellelli, S., Francati, S., Maistrello, L. and Cassanelli, S. 2017. *Drosophila*, efficaci le reti ma servono soluzioni più convenienti. *Rivista di Frutticoltura e di Ortofloricoltura* 4, 14–9.

Cha, D. H., Hesler, S. P., Cowles, R. S., Vogt, H., Loeb, G. M. and Landolt, P. J. 2013. Comparison of a synthetic chemical lure and standard fermented baits for trapping *Drosophila suzukii* (Diptera: Drosophilidae). *Environ. Entomol.* 42(5), 1052–60. doi:10.1603/EN13154.

Cha, D. H., Adams, T., Werle, C. T., Sampson, B. J., Adamczyk, J. J., Rogg, H. and Landolt, P. J. 2014. A four-component synthetic attractant for *Drosophila suzukii* (Diptera: Drosophilidae) isolated from fermented bait headspace. *Pest Manag. Sci.* 70(2), 324–31. doi:10.1002/ps.3568.

Cha, D. H., Gill, M. A., Epsky, N. D., Werle, C. T., Adamczyk, J. J. and Landolt, P. J. 2015a. From a non-target to a target: identification of a fermentation volatile blend attractive to *Zaprionus indianus*. *J. Appl. Entomol.* 139(1-2), 114–22. doi:10.1111/jen.12164.

Cha, D. H., Hesler, S. P., Park, S., Adams, T. B., Zack, R. S., Rogg, H., Loeb, G. M. and Landolt, P. J. 2015b. Simpler is better: fewer non-target insects trapped with a four-component chemical lure vs. a chemically more complex food-type bait for *Drosophila suzukii*. *Entomol. Exp. Appl.* 154(3), 251–60. doi:10.1111/eea.12276.

Chabert, S., Allemand, R., Poyet, M., Eslin, P. and Gilbert, P. 2012. Ability of European parasitoids (Hymenoptera) to control a new invasive Asiatic pest, *Drosophila suzukii*. *Biol. Control* 63(1), 40–7. doi:10.1016/j.biocontrol.2012.05.005.

Chaplin-Kramer, R. and Kremen, C. 2012. Pest control experiments show benefits of complexity at landscape and local scales. *Ecol. Appl.* 22(7), 1936–48. doi:10.1890/11-1844.1.

Chouinard, G., Firlej, A. and Cormier, D. 2016. Going beyond sprays and killing agents: exclusion, sterilization and disruption for insect pest control in pome and stone fruit orchards. *Sci. Hortic.* 208, 13–27. doi:10.1016/j.scienta.2016.03.014.

Cini, A., Ioratti, C. and Anfora, G. 2012. A review of the invasion of *Drosophila suzukii* in Europe and a draft research agenda for integrated pest management. *Bull. Insectology* 65, 149–60.

Clemente, M., Fusco, G., Tonina, L. and Giomi, F. 2018. Temperature-induced phenotypic plasticity in the ovipositor of the invasive species *Drosophila suzukii*. *J. Therm. Biol.* 75, 62–8. doi:10.1016/j.jtherbio.2018.05.001.

Cohen, H. and Yuval, B. 2000. Perimeter trapping strategy to reduce Mediterranean fruit fly (Diptera: Tephritidae) damage on different host species in Israel. *J. Econ. Entomol.* 93(3), 721–5. doi:10.1603/0022-0493-93.3.721.

Cossentine, J., Robertson, M. and Xu, D. 2016. Biological activity of *Bacillus thuringiensis* in *Drosophila suzukii* (Diptera: Drosophilidae). *J. Econ. Entomol.* 109(3), 1071–8. doi:10.1093/jee/tow062.

Cowles, R. S. 2013. Can we mass-trap SWD (spotted wing drosophila)? Available at: http://www.stopswd.org/stopswd/assets/File/SWDWG/SWD-Mass-trapping-report-2013.pdf (accessed July 2018).

Coyne, J. A., Boussy, I. A., Prout, T., Bryant, S. H., Jones, J. S. and Moore, J. A. 1982. Long-distance migration of *Drosophila*. *Am. Nat.* 119(4), 589–95. doi:10.1086/283936.

Cuthbertson, A. G. S. and Audsley, N. 2016. Further screening of entomopathogenic fungi and nematodes as control agents for *Drosophila suzukii*. *Insects* 7(2), 24–33. doi:10.3390/insects7020024.

Cuthbertson, A. G. S., Blackburn, L. F. and Audsley, N. 2014a. Efficacy of commercially available invertebrate predators against *Drosophila suzukii*. *Insects* 5(4), 952–60. doi:10.3390/insects5040952.

Cuthbertson, A. G. S., Collins, D. A., Blackburn, L. F., Audsley, N. and Bell, H. A. 2014b. Preliminary screening of potential control products against *Drosophila suzukii*. *Insects* 5(2), 488–98. doi:10.3390/insects5020488.

Daane, K. M., Wang, X.-G., Biondi, A., Miller, B., Miller, J. C., Riedl, H., Shearer, P. W., Guerrieri, E., Giorgini, M., Buffington, M., et al. 2016. First exploration of parasitoids of *Drosophila suzukii* in South Korea as potential classical biological agents. *J. Pest. Sci.* 89(3), 823–35. doi:10.1007/s10340-016-0740-0.

Dainese, M., Luna, D. I., Sitzia, T. and Marini, L. 2015. Testing scale-dependent effects of semi-natural habitats on farmland biodiversity. *Ecol. Appl.* 25(6), 1681–90. doi:10.1890/14-1321.1.

Dalton, D. T., Walton, V. M., Shearer, P. W., Walsh, D. B., Caprile, J. and Isaacs, R. 2011. Laboratory survival of *Drosophila suzukii* under simulated winter conditions of the Pacific Northwest and seasonal field trapping in five primary regions of small and stone fruit production in the United States. *Pest Manag. Sci.* 67(11), 1368–74. doi:10.1002/ps.2280.

De los Santos Ramos, M., Bello, A., Gómez, R. J., Hernández, R. and Iribe, T. J. 2014. Monitoring of spotted wing drosophila (*Drosophila suzukii* Mats.) and assessment of the new attractant SuzukiiTrap® in Tijuana, Baja California, Mexico. *J. Food Agric. Environ.* 12, 349–55.

Deprá, M., Poppe, J. L., Schmitz, H. J., De Toni, D. C. and Valente, V. L. S. 2014. The first records of the invasive pest *Drosophila suzukii* in the South American continent. *J. Pest. Sci.* 87(3), 379–83. doi:10.1007/s10340-014-0591-5.

Dreves, A. J., Walton, V. and Fisher, G. 2009. A new pest attacking healthy ripening fruit in Oregon: spotted wing drosophila, *Drosophila suzukii* (Matsumura). Oregon State University Extension Publication EM 8991. Available at: https://ir.library.oregonstate.edu/downloads/7h149q24d

Dyer, L. E. and Landis, D. A. 1997. Influence of noncrop habitats on the distribution of *Eriborus terebrans* (Hymenoptera: Ichneumonidae) in cornfields. *Environ. Entomol.* 26(4), 924–32.

Eben, A., Reifenrath, M., Briem, F., Pink, S. and Vogt, H. 2018. Response of *Drosophila suzukii* (Diptera: Drosophilidae) to extreme heat and dryness. *Agric. Forest Entomol.* 20(1), 113–21. doi:10.1111/afe.12235.

EPPO. 2010. Pest risk analysis: *Drosophila suzukii*. Available at: https://gd.eppo.int/taxon/DROSSU/documents (accessed June 2018).

EPPO. 2018. EPPO global database. Available at: https://gd.eppo.int/taxon/DROSSU/distribution (accessed August 2018).

Evans, R. K., Toews, M. D. and Sial, A. A. 2017. Diel periodicity of *Drosophila suzukii* (Diptera: Drosophilidae) under field conditions. *PLoS ONE* 12(2), e0171718. doi:10.1371/journal.pone.0171718.

Fleury, F., Gibert, P., Ris, N. and Allemand, R. 2009. Ecology and life history evolution of frugivorous *Drosophila* parasitoids. *Adv. Parasitol.* 70, 3–44. doi:10.1016/S0065-308X(09)70001-6.

Fountain, M. T., Bennett, J., Cobo-Medina, M., Conde Ruiz, R., Deakin, G., Delgado, A. and Harrison, N. 2018. Alimentary microbes of winter-form *Drosophila suzukii*. *Insect Molec. Biol.* 27(3), 383–92.

Fraimout, A., Debat, V., Fellous, S., Hufbauer, R. A., Foucaud, J., Pudlo, P. and Chen, X. 2017. Deciphering the routes of invasion of *Drosophila suzukii* by means of ABC random forest. *Mol. Biol. Evol.* 34, 980–96.

Gabarra, R., Riudavets, J., Rodríguez, G. A., Pujade-Villar, J. and Arnó, J. 2015. Prospects for the biological control of *Drosophila suzukii*. *BioControl* 60(3), 331–9. doi:10.1007/s10526-014-9646-z.

Gardiner, M. M., Landis, D. A., Gratton, C., Schmidt, N., O'Neal, M., Mueller, E., Chacon, J., Heimpel, G. E. and DiFonzo, C. D. 2009. Landscape composition influences patterns of native and exotic lady beetle abundance. *Divers. Distrib.* 15(4), 554–64. doi:10.1111/j.1472-4642.2009.00563.x.

Gautam, B. K., Little, B. A., Taylor, M. D., Jacobs, J. L., Lovett, W. E., Holland, R. M. and Sial, A. A. 2016. Effect of simulated rainfall on the effectiveness of insecticides against spotted wing drosophila in blueberries. *Crop Prot.* 81, 122–8. doi:10.1016/j.cropro.2015.12.017.

Ghelfi, R., Palmieri, A., Francati, S., Dindo, M. L. and Muzzi, E. 2016. Analisi tecniche ed economiche sulle coperture multifunzionali. *Rivista di frutticoltura e di ortofloricoltura* 78(4), 42–9.

Giovannini, L., Antonacci, G., Zardi, D., Laiti, L. and Panziera, L. 2013. Sensitivity of simulated wind speed to spatial resolution over complex terrain. *Energy Proc.* 59C, 323–9.

Girod, P., Lierhmann, O., Urvois, T., Turlings, T. C. J., Kenis, M. and Haye, T. 2018a. Host specificity of Asian parasitoids for potential classical biological control of *Drosophila suzukii*. *J. Pest. Sci.* 91(4), 1241–50. doi:10.1007/s10340-018-1003-z.

Girod, P., Rossignaud, L., Haye, T., Turlings, T. C. J. and Kenis, M. 2018b. Development of Asian parasitoids in larvae of *Drosophila suzukii* feeding on blueberry and artificial diet. *J. Appl. Entomol.* 142(5), 483–94. doi:10.1111/jen.12496.

Gladbach, D. J., Holzschuh, A., Scherber, C., Thies, C., Dormann, C. F. and Tscharntke, T. 2011. Crop-noncrop spillover: arable fields affect trophic interactions on wild plants in surrounding habitats. *Oecologia* 166(2), 433–41. doi:10.1007/s00442-010-1870-3.

Goodhue, R. E., Bolda, M., Farnsworth, D., Williams, J. C. and Zalom, F. G. 2011. Spotted wing drosophila infestation of California strawberries and raspberries: economic analysis of potential revenue losses and control costs. *Pest Manag. Sci.* 67(11), 1396–402. doi:10.1002/ps.2259.

Grassi, A., Palmieri, L. and Giongo, L. 2012. *Drosophila* (Sophophora) *suzukii* (Matsumura), new pest of soft fruits in Trentino (North-Italy) and in Europe. *IOBC/WPRS Bull.* 70, 121–8.

Grassi, A., Maistri, S., Eccher, F. and Pezzè, M. 2013. Biologia dell'insetto e strategie di contenimento da applicarsi sui piccoli frutti e ciliegio. Giornata tecnica *Drosophila suzukii* Fondazione Mach 13 febbraio 2013.

Grassi, A., Anfora, G., Maistri, S., Maddalena, G., De Cristofaro, A., Savini, G. and Ioriatti, C. 2015. Development and efficacy of Droskidrink, a food bait for trapping *Drosophila suzukii*. *IOBC/WPRS Bull.* 109, 197–204.

Grassi, A., Gottardello, A., Dalton, D. T., Tait, G., Rendon, D., Ioriatti, C., Gibeaut, D., Rossi-Stacconi, M. V. and Walton, M. V. 2017. Seasonal reproductive biology of *Drosophila suzukii* (Diptera: Drosophilidae) in temperate climates. *Environ Entomol.* 47, 166–74.

Griffo, R., Frontuto, A., Cesaroni, C. and Desantis, M. 2012. L'insetto *Drosophila suzukii* sempre più presente in Italia. *L'Informatore Agrario* 68(9), 56–60.

Hamby, K. A. and Becher, P. G. 2016. Current knowledge of interactions between *Drosophila suzukii* and microbes, and their potential utility for pest management. *J. Pest. Sci.* 89(3), 621–30. doi:10.1007/s10340-016-0768-1.

Hamby, K. A., Bolda, M. P., Sheehan, M. E. and Zalom, F. G. 2014. Seasonal monitoring for *Drosophila suzukii* (Diptera: Drosophilidae) in California commercial raspberries. *Environ. Entomol.* 43(4), 1008–18. doi:10.1603/EN13245.

Hamby, K. A., Bellamy, D. E., Chiu, J. C., Lee, J. C., Walton, V. M., Wiman, N. G., York, R. M. and Biondi, A. 2016. Biotic and abiotic factors impacting development, behavior, phenology, and reproductive biology of *Drosophila suzukii*. *J. Pest. Sci.* 89(3), 605–19. doi:10.1007/s10340-016-0756-5.

Hampton, E., Koski, C., Barsoian, O., Faubert, H., Cowles, R. S. and Alm, S. R. 2014. Use of early ripening cultivars to avoid infestation and mass trapping to manage *Drosophila suzukii* (Diptera: Drosophilidae) in *Vaccinium corymbosum* (Ericales: Ericaceae). *J. Econ. Entomol.* 107(5), 1849–57. doi:10.1603/EC14232.

Hauser, M. 2011. A historic account of the invasion of *Drosophila suzukii* (Matsumura) (Diptera:drosophilae) in the continental United States, with remarks on their identification. *Pest Manag. Sci.* 67(11), 1352–7. doi:10.1002/ps.2265.

Haviland, D. R. and Beers, E. H. 2012. Chemical control programs for *Drosophila suzukii* that comply with international limitations on pesticide residues for exported sweet cherries. *J. Integr. Pest Manag.* 3(F1), 1–6. doi:10.1603/IPM11034.

Haye, T., Girod, P., Cuthbertson, A. G. S., Wang, X. G., Daane, K. M., Hoelmer, K. A., Baroffio, C., Zhang, J. P. and Desneux, N. 2016. Current SWD IPM tactics and their practical implementation in fruit crops across different regions around the world. *J. Pest. Sci.* 89(3), 643–51. doi:10.1007/s10340-016-0737-8.

Hoashi-Erhardt, W. and Bixby-Brosi, A. 2014. Spotted wing drosophila (SWD) management for organic producers: organic options for invasive pest of northwest fruits. *Tilth Prod. Quart. J.* (24), 2. Available at: http://tilthproducers.org/farmer-resources/wiki/spotted-wing-drosophila-swd-management-for-organic-producers-organic-options-for-invasive-pest-of-northwest-fruits/ (accessed June 2018).

Holland, J. and Fahrig, L. 2000. Effect of woody borders on insect density and diversity in crop fields: a landscape- scale analysis. *Agric. Ecosyst. Environ.* 78(2), 115–22. doi:10.1016/S0167-8809(99)00123-1.

Huang, J., Gut, L. and Grieshop, M. 2017. Evaluation of food-based attractants for *Drosophila suzukii* (Diptera: Drosophilidae). *Environ. Entomol.* 46(4), 878–84. doi:10.1093/ee/nvx097.

Ideo, S., Watada, M., Mitsui, H. and Kimura, M. T. 2008. Host range of *Asobara japonica* (Hymenoptera: Braconidae), a larval parasitoid of drosophilid flies. *Entomol. Sci.* 11(1), 1–6. doi:10.1111/j.1479-8298.2007.00244.x.

Iglesias, L. E., Nyoike, T. W. and Liburd, O. E. 2014. Effect of trap design, bait type, and age on captures of *Drosophila suzukii* (Diptera: Drosophilidae) in berry crops. *J. Econ. Entomol.* 107(4), 1508–18. doi:10.1603/EC13538.

Ioriatti, C., Grassi, A. and Agnello, A. M. 2011. Due insetti minacciano la frutticoltura integrata. *L'Informatore Agrario* 67(4), 67–70.

Ioriatti, C., Boselli, M., Caruso, S., Galassi, T., Gottardello, A., Grassi, A., Tonina, L., Vaccari, G. and Mori, N. 2015. Approccio integrato per la difesa dalla *Drosophila suzukii*. *Frutticoltura* 4, 32–6.

Jakobs, R., Gariepy, T. D. and Sinclair, B. J. 2015. Adult plasticity of cold tolerance in a continental-temperate population of *Drosophila suzukii*. *J. Ins. Physiol.* 79, 1–9. doi:10.1016/j.jinsphys.2015.05.003.

Kacsoh, B. Z. and Schlenke, T. A. 2012. High hemocyte load is associated with increased resistance against parasitoids in *Drosophila suzukii*, a relative of *D. melanogaster*. *PLoS ONE* 7(4), e34721. doi:10.1371/journal.pone.0034721.

Kanzawa, T. 1936. Studies on *Drosophila suzukii* Mats. *J. Plant Prot. (Tokyo)* 23, 66–70, 127–32, 183–91. Abstract in *Review of Applied Entomology* 24, 315.

Kanzawa, T. 1939. Studies on *Drosophila suzukii* Mats. Yamanashi Agricultural Experiment Station, Kofu, 49p. Abstracts in *Review of Applied Entomology* 29, 622.

Kasuya, N., Mitsui, H., Aotsuka, T. and Kimura, M. T. 2013a. Diversity and host association of parasitoids attacking mycophagous drosophilids (Diptera: Drosophilidae) in northern and central Japan. *Entomol. Sci.* 16(2), 227–34. doi:10.1111/j.1479-8298.2012.00544.x.

Kasuya, N., Mitsui, H., Ideo, S., Watada, M. and Kimura, M. T. 2013b. Ecological, morphological and molecular studies on Ganaspis individuals (Hymenoptera: Figitidae) attacking *Drosophila suzukii* (Diptera: Drosophilidae). *Appl. Entomol. Zool.* 48(1), 87–92. doi:10.1007/s13355-012-0156-0.

Kawase, S. and Uchino, K. 2005. Influence of maturity and wounding on blueberry fruit injury by *Drosophila suzukii*. *J. Jpn. Soc. Hortic. Sci.* 74 (Suppl. 1), 281.

Keesey, I. W., Knaden, M. and Hansson, B. S. 2015. Olfactory specialization in *Drosophila suzukii* supports an ecological shift in host preference from rotten to fresh fruit. *J. Chem. Ecol.* 41(2), 121–8. doi:10.1007/s10886-015-0544-3.

Kenis, M., Tonina, L., Eschen, R., van der Sluis, B., Sancassani, M., Mori, N., Haye, T. and Helsen, H. 2016. Non-crop plants used as hosts by *Drosophila suzukii* in Europe. *J. Pest. Sci.* 89(3), 735–48. doi:10.1007/s10340-016-0755-6.

Kennedy, C. M., Lonsdorf, E., Neel, M. C., Williams, N. M., Ricketts, T. H., Winfree, R., Bommarco, R., Brittain, C., Burley, A. L., Cariveau, D., et al. 2013. A global quantitative synthesis of local and landscape effects on wild bee pollinators in agroecosystems. *Ecol. Lett.* 16(5), 584–99. doi:10.1111/ele.12082.

Kimura, M. T. 1988. Adaptation to temperate climates and evolution of over-wintering strategies in the *Drosophila melanogaster* species group. *Evolution* 42(6), 1288–97. doi:10.1111/j.1558-5646.1988.tb04188.x.

Kimura, M. T. 2004. Cold and heat tolerance of drosophilid flies with reference to their latitudinal distributions. *Oecologia* 140(3), 442–9. doi:10.1007/s00442-004-1605-4.

Kimura, M. T. and Beppu, K. 1993. Climatic adaptations in the Drosophila immigrans species group-seasonal migration and thermal tolerance. *Ecol. Entomol.* 18(2), 141–9. doi:10.1111/j.1365-2311.1993.tb01195.x.

Kimura, M. T., Toda, M. J., Beppu, A. and Watabe, H. 1977. Breeding sites of drosophilidies in and near Sapporo, northern Japan, with supplementary notes on adult feeding habits. *Kontyu* 45, 571–82.

Kinjo, H., Kunimi, Y., Ban, T. and Nakai, M. 2013. Oviposition efficacy of *Drosophila suzukii* (Diptera: Drosophilidae) on different cultivars of blueberry. *J. Econ. Entomol.* 106(4), 1767–71.

Kinjo, H., Kunimi, Y. and Nakai, M. 2014. Effects of temperature on the reproduction and development of *Drosophila suzukii* (Diptera: Drosophilidae). *Appl. Entomol. Zool.* 49(2), 297–304. doi:10.1007/s13355-014-0249-z.

Kleiber, J. R., Unelius, C. R., Lee, J. C., Suckling, D. M., Qian, M. C. and Bruck, D. J. 2014. Attractiveness of fermentation and related products to spotted wing drosophila (Diptera: Drosophilidae). *Environ. Entomol.* 43(2), 439–47. doi:10.1603/EN13224.

Klick, J., Yang, W. Q., Walton, V. M., Dalton, D. T., Hagler, J. R., Dreves, A. J., Lee, J. C. and Bruck, D. J. 2016. Distribution and activity of *Drosophila suzukii* in cultivated raspberry and surrounding vegetation. *J. Appl. Entomol.* 140(1-2), 37–46. doi:10.1111/jen.12234.

Landolt, P. J., Adams, T. and Rogg, H. 2012. Trapping spotted wing drosophila, *Drosophila suzukii* (Matsumura) (Diptera: Drosophilidae), with combinations of vinegar and wine, and acetic acid and ethanol. *J. Appl. Entomol.* 136(1–2), 148–54. doi:10.1111/j.1439-0418.2011.01646.x.

Lasa, R., Tadeo, E., Toledo-Hérnandez, R. A., Carmona, L., Lima, I. and Williams, T. 2017. Improved capture of *Drosophila suzukii* by a trap baited with two attractants in the same device. *PLoS ONE* 12(11), e0188350. doi: 10.1371/journal.pone.0188350.

Lee, J. C. and Sial, A. 2016. Reference list of fruits with *Drosophila suzukii*. Available at: http://eorganic.info/sites/eorganic.info/files/u461/SWD-hostlist-quick.pdf (accessed March 2018).

Lee, J. C., Bruck, D. J., Dreves, A. J., Ioriatti, C., Vogt, H. and Baufeld, P. 2011a. In focus: spotted wing drosophila, *Drosophila suzukii*, across perspectives. *Pest Manag. Sci.* 67(11), 1349–51. doi:10.1002/ps.2271.

Lee, J. C., Bruck, D. J., Curry, H., Edwards, D., Haviland, D. R., Van Steenwyk, R. A. and Yorgey, B. M. 2011b. The susceptibility of small fruits and cherries to the spotted-wing drosophila, *Drosophila suzukii*. *Pest Manag. Sci.* 67(11), 1358–67. doi:10.1002/ps.2225.

Lee, J. C., Burrack, H. J., Barrantes, L. D., Beers, E. H., Dreves, A. J., Hamby, K. A., Haviland, D. R., Isaacs, R., Richardson, T. A., Shearer, P. W., et al. 2012. Evaluation of monitoring traps for *Drosophila suzukii* (Diptera: Drosophilidae) in North America. *J. Econ. Entomol.* 105(4), 1350–7.

Lee, J. C., Shearer, P. W., Barrantes, L. D., Beers, E. H., Burrack, H. J., Dalton, D. T., Dreves, A. J., Gut, L. J., Hamby, K. A., Haviland, D. R., et al. 2013. Trap designs for monitoring *Drosophila suzukii* (Diptera: Drosophilidae). *Environ. Entomol.* 42(6), 1348–55. doi:10.1603/EN13148.

Lee, J. C., Dreves, A. J., Cave, A. M., Kawai, S., Isaacs, R., Miller, J. C., van Timmeren, S. and Bruck, D. J. 2015. Infestation of wild and ornamental non-crop fruits by *Drosophila suzukii* (Diptera: Drosophilidae). *Ann. Entomol. Soc. Am.* 108(2), 117–29. doi:10.1093/aesa/sau014.

Lemeunier, F., David, J. R., Tsacas, L. and Ashburner, M. 1986. The *melanogaster* species group. In: Ashburner, M., Carson, H. L. and Thompson, J. N. (Eds), *The Genetics and Biology of Drosophila*. Academic Press, New York, pp.148–239.

Louise, C., Girard, M., Kuhl, G. and Lopez-Ferber, M. 1996. Persistence of *Botrytis cinerea* in its vector *Drosophila melanogaster*. *Phytopathology* 86(9), 934–9. doi:10.1094/Phyto-86-934.

Matsuura, A., Mitsui, H. and Kimura, M. T. 2018. A preliminary study on distributions and oviposition sites of *Drosophila suzukii* (Diptera: Drosophilidae) and its parasitoids on wild cherry tree in Tokyo, central Japan. *Appl. Entomol. Zool.* 53(1), 47–53.

Mazzetto, F., Marchetti, E., Amiresmaeili, N., Sacco, D., Francati, S., Jucker, C., Dindo, M. L., Lupi, D. and Tavella, L. 2016. Drosophila parasitoids in northern Italy and their potential to attack the exotic pest *Drosophila suzukii*. *J. Pest. Sci.* 89(3), 837–50. doi:10.1007/s10340-016-0746-7.

Mazzi, D., Bravin, E., Meraner, M., Finger, R. and Kuske, S. 2017. Economic impact of the introduction and establishment of *Drosophila suzukii* on sweet cherry production in Switzerland. *Insects* 8(1), 1–13. doi:10.3390/insects8010018.

Miller, B., Anfora, G., Buffington, M., Daane, K. M., Dalton, D. T., Hoelmer, K. M., Rossi-Stacconi, M. V., Grassi, A., Ioriatti, C., Miller, J., et al. 2015. Seasonal occurrence of resident parasitoids associated with *Drosophila suzukii* in two small fruit production regions of Italy and the USA. *Bull. Insectology* 68, 255–63.

Mitsui, H. and Kimura, M. T. 2010. Distribution, abundance and host association of two parasitoid species attacking frugivorous drosophilid larvae in central Japan. *Eur. J. Entomol.* 107(4), 535–40. doi:10.14411/eje.2010.061.

Mitsui, H., Takahashi, K. H. and Kimura, M. T. 2006. Spatial distribution and clutch sizes of Drosophila species ovipositing on cherry fruits of different stages. *Pop. Ecol.* 48(3), 233–7. doi:10.1007/s10144-006-0260-5.

Mitsui, H., Beppu, K. and Kimura, M. T. 2010. Seasonal life cycles and resource uses of flower- and fruit-feeding drosophilid flies (Diptera: Drosophilidae) in central Japan. *Entomol. Sci.* 13(1), 60–7. doi:10.1111/j.1479-8298.2010.00372.x.

Mori, N., Carli, C., Caruso, S., Ceredi, G., Galassi, T., Grassi, A., Lupi, D., Maistrello, L., Maini, S., Marchesini, E., et al. 2015. Drosophila suzukii, un problema per tutto il nord Italia. I lavori del tavolo tecnico interregionale. Convegno nazionale del ciliegio 2.0 25.

Naranjo-Lázaro, J. M., Mellín-Rosas, M. A., González-Padilla, V. D., Sánchez-González, J. A., Moreno-Carrillo, G. and Arredondo-Bernal, H. C. 2014. Susceptibility of *Drosophila suzukii* Matsumura (Diptera: Drosophilidae) to entomophatogenic fungi. *Southwest. Entomol.* 39(1), 201–3. doi:10.3958/059.039.0119.

Nayak, G. K., Roberts, S. P. M., Garratt, M., Breeze, T. D., Tscheulin, T., Harrison-Cripps, J., Vogiatzakis, I. N., Stirpe, M. T. and Potts, S. G. 2015. Interactive effect of floral abundance and semi-natural habitats on pollinators in field beans (*Vicia faba*). *Agric. Ecosyst. Environ.* 199, 58–66. doi:10.1016/j.agee.2014.08.016.

Nikolouli, K., Colinet, H., Renault, D., Enriquez, T., Mouton, L., Gibert, P., Sassu, F., Cáceres, C., Stauffer, C., Pereira, R. and Bourtzis, K. 2018. Sterile insect technique and Wolbachia symbiosis as potential tools for the control of the invasive species *Drosophila suzukii*. *J. Pest. Sci.* 91(2), 489–503.

Noble, R., Dobrovin-Pennington, A., Shaw, B., Buss, D. S., Cross, J. V. and Fountain, M. T. 2017. Fermentation for disinfesting fruit waste from *Drosophila* species (Diptera: Drosophilidae). *Environ. Entomol.* 46(4), 939–45. doi:10.1093/ee/nvx107.

Nomano, F. Y., Mitsui, H. and Kimura, M. T. 2015. Capacity of Japanese *Asobara* species (Hymenoptera; Braconidae) to parasitize a fruit pest *Drosophila suzukii* (Diptera; Drosophilidae). *J. Appl. Entomol.* 139(1–2), 105–13. doi:10.1111/jen.12141.

Novković, B., Oikawa, A., Murata, Y., Mitsui, H. and Kimura, M. T. 2012. Abundance and host association of parasitoids attacking frugivorous drosophilids in Iriomote-jima, a subtropical island of Japan. *Eur. J. Entomol.* 109(4), 517–26. doi:10.14411/eje.2012.065.

Ometto, L., Cestaro, A., Ramasamy, S., Grassi, A., Revadi, S., Siozios, S., Moretto, M., Fontana, P., Varotto, C., Pisani, D., et al. 2013. Linking genomics and ecology to investigate the complex evolution of an invasive Drosophila pest. *Genome Biol. Evol.* 5(4), 745–57. doi:10.1093/gbe/evt034.

Oxitech. 2018. Precision agriculture pest suppression. Available at: https://www.oxitec.com/crop-protection/ (accessed September 2018).

Pelton, E., Gratton, C., Isaacs, R., van Timmeren, S., Blanton, A. and Guédot, C. 2016. Earlier activity of *Drosophila suzukii* in high woodland landscapes but relative abundance is unaffected. *J. Pest. Sci.* 89(3), 725–33. doi:10.1007/s10340-016-0733-z.

Pfab, F., Stacconi, M. V. R., Anfora, G., Grassi, A., Walton, V. and Pugliese, A. 2018. Optimized timing of parasitoid release: a mathematical model for biological control of *Drosophila suzukii*. *Theor. Ecol.* 11(4), 489–501. doi:10.1007/s12080-018-0382-3.

Poyet, M., Havard, S., Prevost, G., Chabrerie, O., Doury, G., Gibert, P. and Eslin, P. 2013. Resistance of *Drosophila suzukii* to the larval parasitoids *Leptopilina heterotoma* and *Asobara japonica* is related to haemocyte load. *Physiol. Entomol.* 38(1), 45–53.

Poyet, M., Eslin, P., Héraude, M., Le Roux, V., Prévost, G., Gibert, P. and Chabrerie, O. 2014. Invasive host for invasive pest: when the Asiatic cherry fly (*Drosophila suzukii*) meets the American black cherry (*Prunus serotina*) in Europe. *Agric. Forest Entomol.* 16(3), 251–9. doi:10.1111/afe.12052.

Poyet, M., Le Roux, V., Gibert, P., Meirland, A., Pre´vost, G., Eslin, P. and Chabrerie, O. 2015. The wide potential trophic niche of the Asiatic fruit fly *Drosophila suzukii*: the key of its invasion success in temperate Europe? *PLoS ONE* 10(11), e0142785. doi:10.1371/journal.pone.0142785.

Profaizer, D., Angeli, G., Trainotti, D., Marchel, L., Zadra, E., Sofia, M. and Ioriatti, C. 2012. *Drosophila suzukii*: valutazione di agrofarmaci e analisi sul corretto posizionamento in campo. *ATTI Giornate Fitopatologiche* 1, 229–35.

Profaizer, D., Angeli, G., Sofia, M. and Zadra, E. 2015. Chemical control of *Drosophila suzukii* infesting cherries. *Inf. Agr.* 71, 51–5.

Prokopy, R. J., Miller, N. W., Piñero, J. C., Barry, J. D., Tran, L. C., Oride, L. and Vargas, R. I. 2003. Effectiveness of GF-120 fruit fly bait spray applied to border area plants for control of melon flies (Diptera: Tephritidae). *J. Econ. Entomol.* 96(5), 1485–93. doi:10.1093/jee/96.5.1485.

Rand, T. A., Tylianakis, J. M. and Tscharntke, T. 2006. Spillover edge effects: the dispersal of agriculturally subsidized insect natural enemies into adjacent natural habitats. *Ecol. Lett.* 9(5), 603–14. doi:10.1111/j.1461-0248.2006.00911.x.

Renkema, J. M. and Cuthbertson, A. G. 2018. Impact of multiple natural enemies on immature *Drosophila suzukii* in strawberries and blueberries. *BioControl* 63(5), 719–28.

Renkema, J. M., Buitenhuis, R. and Hallett, R. H. 2014. Optimizing TRAP design and trapping protocols for *Drosophila suzukii* (Diptera: Drosophilidae). *J. Econ. Entomol.* 107(6), 2107–18: doi:10.1603/EC14254.

Riley, J. R., Reynolds, D. R., Mukhopadhyay, S. and Ghosh, M. R. 1995. Long distance migration of aphids and other small insects in northeast India. *Eur. J., Entomol.* 92, 639–53.

Rossi-Stacconi, M. V., Grassi, A., Dalton, D. T., Miller, B., Ouantar, M., Loni, A., Ioriatti, C., Walton, V. M. and Anfora, G. 2013. First field records of *Pachycrepoideus vindemiae* as a parasitoid of *Drosophila suzukii* in European and Oregon small fruit production areas. *Entomologia* 1, 1–15.

Rossi-Stacconi, M. V., Buffington, M., Daane, K. M., Dalton, D. T., Grassi, A., Kaçar, G., Miller, B., Miller, J. C., Baser, N., Ioriatti, C., et al. 2015. Host stage preference, efficacy and fecundity of parasitoids attacking *Drosophila suzukii* in newly invaded areas. *Biol. Control* 84, 28–35. doi:10.1016/j.biocontrol.2015.02.003.

Rossi-Stacconi, M. V., Kaur, R., Mazzoni, V., Ometto, L., Grassi, A., Gottardello, A., Rota-Stabelli, O. and Anfora, G. 2016. Multiple lines of evidence for reproductive winter diapause in the invasive pest *Drosophila suzukii*: useful clues for control strategies. *J. Pest. Sci.* 89(3), 689–700. doi:10.1007/s10340-016-0753-8.

Rossi-Stacconi, M. V., Panel, A., Basey, N., Ioriatti, C., Pantezzi, T. and Anfora, G. 2017. Comparative life history traits of indigenous Italian parasitoids of *Drosophila suzukii* and their effectiveness at different temperatures. *Biol. Control* 112, 20–7. doi:10.1016/j.biocontrol.2017.06.003.

Rossi-Stacconi, M. V., Amiresmaeili, N., Biondi, A., Carli, C., Caruso, S., Dindo, M. L., Francati, S., Gottardello, A., Grassi, A., Lupi, D., et al. 2018. Host location and dispersal ability of the cosmopolitan parasitoid *Trichopria drosophilae* released to control the invasive spotted wing Drosophila. *Biol. Control* 117, 188–96. doi:10.1016/j.biocontrol.2017.11.013.

Rusch, A., Valantin-Morison, M., Sarthou, J. P. and Roger-Estrade, J. 2011. Multi-scale effects of landscape complexity and crop management on pollen beetle parasitism rate. *Landsc. Ecol.* 26(4), 473–86. doi:10.1007/s10980-011-9573-7.

Rusch, A., Bommarco, R., Jonsson, M., Smith, H. G. and Ekbom, B. 2013. Flow and stability of natural pest control services depend on complexity and crop rotation at the landscape scale. *J. Appl. Ecol.* 50(2), 345–54. doi:10.1111/1365-2664.12055.

Ryan, G. D., Emiljanowicz, L., Wilkinson, F., Kornya, M. and Newman, J. A. 2016. Thermal tolerances of the spotted-wing drosophila *Drosophila suzukii* (Diptera: Drosophilidae). *J. Econ. Entomol.* 109(2), 746–52. doi:10.1093/jee/tow006.

Saeed, N., Tonina, L., Battisti, A. and Mori, N. 2018. Temperature alters the response to insecticides in *Drosophila suzukii* (Diptera: Drosophilidae). *J. Econ. Entomol.* 111(3), 1306–12. doi:10.1093/jee/toy080.

Sancassani, M., Tonina, L., Tirello, P., Giomi, F., Marchesini, E., Zanini, G. and Mori, N. 2016. *Drosophila suzukii* su ciliegio, esperienze di lotta integrata. *L'Informatore Agrario* 15, 59–63.

Santoiemma, G., Mori, N., Tonina, L. and Marini, L. 2018. Semi-natural habitats boost *Drosophila suzukii* populations and crop damage in sweet cherry. *Agric. Ecosys. Environ.* 257, 152–8.

Scheidler, N. H., Liu, C., Hamby, K. A., Zalom, F. G. and Syed, Z. 2015. Volatile codes: correlation of olfactory signals and reception in Drosophila-yeast chemical communication. *Sci. Rep.* 5, 14059. doi:10.1038/srep14059.

Schmidt, J. M., Whitehouse, T. S., Green, K., Krehenwinkel, H., Schmidt-Jeffris, R. and Sial, A. A. 2019. Local and landscape-scale heterogeneity shape spotted wing drosophila (*Drosophila suzukii*) activity and natural enemy abundance: implications for trophic interactions. *Agric. Ecosys. Environ.* 272, 86–94.

Schneider, J. C. 1989. Role of movement in evaluation of area wide insect pest management tactics. *Environ. Entomol.* 18(5), 868–74. doi:10.1093/ee/18.5.868.

Shawer, R., Tonina, L., Tirello, P., Duso, C. and Mori, N. 2018. Laboratory and field trials to identify effective chemical control strategies for integrated management of *Drosophila suzukii* in European cherry orchards. *Crop Protect.* 103, 73–80. doi:10.1016/j.cropro.2017.09.010.

Shearer, P. W., West, J. D., Walton, V. M., Brown, P. H., Svetec, N. and Chiu, J. C. 2016. Seasonal cues induce phenotypic plasticity of *Drosophila suzukii* to enhance winter survival. *BMC Ecol.* 16(1), 11. doi:10.1186/s12898-016-0070-3.

Shi, J. C. 2015. Occurrence characteristics and integrated control technology of fruit fly on cherry in Tianshui (Gansu). *Pomol. Hosp.* 5, 35–6.

Smyth, M. J. and Saverimuttu, N. 2010. Verification visit to the USA. *Drosophila suzukii*. Trip Report. Biosecurity Services Group, Department of Agriculture, Fisheries and Forestry.

Stephens, A. R., Asplen, M. K., Hutchison, W. D. and Venette, R. C. 2015. Cold hardiness of winter-acclimated *Drosophila suzukii* (Diptera: Drosophilidae) adults. *Environ. Entomol.* 44(6), 1619–26. doi:10.1093/ee/nvv134.

Stockton, D., Wallingford, A. and Loeb, G. 2018. Phenotypic plasticity promotes overwintering survival in a globally invasive crop pest, *Drosophila suzukii*. *Insects* 9(3), 105. doi:10.3390/insects9030105.

Stefanescu, C., Páramo, F., Åkesson, S., Alarcón, M., Ávila, A., Brereton, T., Carnicer, J., Cassar, L. F., Fox, R., Heliölä, J., et al. 2013. Multi-generational long-distance migration of insects: studying the painted lady butterfly in the western Palaearctic. *Ecography* 36(4), 474–86. doi:10.1111/j.1600-0587.2012.07738.x.

Steffen, K., Grousset, F., Petter, F., Suffert, M. and Schrader, G. 2015. EU-project DROPSA: first achievements regarding pathway analyses for fruit pests. *EPPO Bull.* 45(1), 148–52. doi:10.1111/epp.12179.

Suckling, D. M. and Brockerhoff, E. G. 2010. Invasion biology, ecology, and management of the light brown apple moth (Tortricidae). *Annu. Rev. Entomol.* 55, 285–306. doi:10.1146/annurev-ento-112408-085311.

Tabashnik, B. E. and Croft, B. A. 1985. Evolution of pesticide resistance in apple pests and their natural enemies. *Entomophaga* 30(1), 37–49. doi:10.1007/BF02372281.

Tait, G., Grassi, A., Pfab, F., Crava, C. M., Dalton, D. T., Magarey, R., Ometto, L., Vezzulli, S., Rossi-Stacconi, M. V., Gottardello, A., et al. 2018. Large-scale spatial dynamics of *Drosophila suzukii* in Trentino, Italy. *J. Pest. Sci.* 91(4), 1213–24. doi:10.1007/s10340-018-0985-x.

Taki, H., Kevan, P. G. and Ascher, J. S. 2007. Landscape effects of forest loss in a pollination system. *Landsc. Ecol.* 22(10), 1575–87. doi:10.1007/s10980-007-9153-z.

Tanabe, S. I. 2002. Between-forest variation in vertical stratification of drosophilid populations. *Ecol. Entomol.* 27(6), 720–31. doi:10.1046/j.1365-2311.2002.00469.x.

Taning, C. N. T., Christiaens, O., Berkvens, N., Casteels, H., Maes, M. and Smagghe, G. 2016. Oral RNAi to control *Drosophila suzukii*: laboratory testing against larval and adult stages. *J. Pest. Sci.* 89(3), 803–14. doi:10.1007/s10340-016-0736-9.

Tochen, S., Dalton, D. T., Wiman, N. G., Hamm, C., Shearer, P. W. and Walton, V. M. 2014. Temperature-related development and population parameters for *Drosophila suzukii* (Diptera: Drosophilidae) on cherry and blueberry. *Environ. Entomol.* 43(2), 501–10. doi:10.1603/EN13200.

Tochen, S., Walton, V. M. and Lee, J. C. 2016a. Impact of floral feeding on adult *Drosophila suzukii* survival and nutrient status. *J. Pest. Sci.* 89(3), 793–802. doi:10.1007/s10340-016-0762-7.

Tochen, S., Woltz, J. M., Dalton, D. T., Lee, J. C., Wiman, N. G. and Walton, V. M. 2016b. Humidity affects populations of *Drosophila suzukii* (Diptera: Drosophilidae) in blueberry. *J. Appl. Entomol.* 140(1–2), 47–57. doi:10.1111/jen.12247.

Toda, M. J. 1987. Vertical microdistribution of Drosophilidae (Diptera) within various forests in Hokkaido: III. The Tomakomai experiment forest, Hokkaido University. *Res. Bull. Coll. Exper. For.* 44(2), 611–32.

Toda, M. J. 1992. Three-dimensional dispersion of drosophilid flies in a cool temperate forest of northern Japan. *Ecol. Res.* 7(3), 283–95. doi:10.1007/BF02347097.

Tonina, L., Mori, N., Giomi, F. and Battisti, A. 2016. Development of *Drosophila suzukii* at low temperatures in mountain areas. *J. Pest. Sci.* 89(3), 667–78. doi:10.1007/s10340-016-0730-2.

Tonina, L., Gottardello, A., Rossi-Stacconi, M. V., Santoiemma, G. and Vaccari, G. 2017. Ecologia e gestione integrata di Drosophila suzukii al Nord. *L'Informatore Agrario* 12, 40–4.

Tonina, L., Grassi, A., Caruso, S., Mori, N., Gottardello, A., Anfora, G., Giomi, F., Vaccari, G. and Ioriatti, C. 2018a. Comparison of attractants for monitoring *Drosophila suzukii* in sweet cherry orchards in Italy. *J. Appl. Entom.* 142(1-2), 18–25. doi:10.1111/jen.12416.

Tonina, L., Mori, N., Sancassani, M., Dall'Ara, P. and Marini, L. 2018b. Spillover of *Drosophila suzukii* between noncrop and crop areas: implications for pest management. *Agric. Forest Entomol.* 20(4), 575–81. doi:10.1111/afe.12290.

Tscharntke, T., Rand, T. A. and Bianchi, F. J. 2005. The landscape context of trophic interactions: insect spillover across the crop-noncrop interface. *Annales Zoologici Fennici* 42, 421–32.

Tscharntke, T., Karp, D. S., Chaplin-Kramer, R., Batáry, P., DeClerck, F., Gratton, C., Hunt, L., Ives, A., Jonsson, M., Larsen, A., et al. 2016. When natural habitat fails to enhance biological pest control – five hypotheses. *Biol. Conserv.* 204, 449–58. doi:10.1016/j.biocon.2016.10.001.

Vaccari, G., Caruso, S., Nouhaud, C. and Maistrello, L. 2014. Comparison between different trap types for *Drosophila suzukii* in cherry orchards. In: *8th Meeting IOBC Workshop on Integrated Soft Fruit Production – Pergine Valsugana (Italy)*, 25–28 May 2014.

Van Timmeren, S. and Isaacs, R. 2013. Control of spotted wing drosophila, *Drosophila suzukii*, by specific insecticides and by conventional and organic crop protection programs. *Crop Protect.* 54, 126–33. doi:10.1016/j.cropro.2013.08.003.

Vargas, R. I., Peck, S. L., McQuate, G. T., Jackson, C. G., Stark, J. D. and Armstrong, J. W. 2001. Potential for areawide integrated management of Mediterranean fruit fly (Diptera: Tephritidae) with a braconid parasitoid and a novel bait spray. *J. Econ. Entomol.* 94(4), 817–25. doi:10.1603/0022-0493-94.4.817.

Veres, A., Petit, S., Conord, C. and Lavigne, C. 2013. Does landscape composition affect pest abundance and their control by natural enemies? A review. *Agric. Ecosyst. Environ.* 166, 110–7. doi:10.1016/j.agee.2011.05.027.

Vlach, J. 2010. *Identifying Drosophila suzukii*. Oregon Department of Agriculture. Available at: http://www.oregon.gov/ODA/PLANT/docs/pdf/ippm_d_suzukii_id_guide 10.pdf?ga=t (accessed April 2018).

Walsh, D. B., Bolda, M. P., Goodhue, R. E., Dreves, A. J., Lee, J., Bruck, D. J., Walton, V. M., O'Neal, S. D. and Zalom, F. G. 2011. *Drosophila suzukii* (Diptera: Drosophilidae): invasive pest of ripening soft fruit expanding its geographic range and damage potential. *J. Integ. Pest Manag.* 2(1), G1–7. doi:10.1603/IPM10010.

Wang, X. G., Kaçar, G., Biondi, A. and Daane, K. M. 2016. Life-history and host preference of *Trichopria drosophilae*, a pupal parasitoid of spotted wing drosophila. *BioControl* 61(4), 387–97. doi:10.1007/s10526-016-9720-9.

Williamson, M. 2006. Explaining and predicting the success of invading species at different stages of invasion. *Biol. Invas.* 8(7), 1561–8. doi:10.1007/s10530-005-5849-7.

Wiman, N. G., Walton, V. M., Dalton, D. T., Anfora, G., Burrack, H. J., Chiu, J. C., Daane, K. M., Grassi, A., Miller, B., Tochen, S., et al. 2014. Integrating temperature-dependent life table data into a matrix projection model for *Drosophila suzukii* population estimation. *PLoS ONE* 9(9), e106909. doi:10.1371/journal.pone.0106909.

Wiman, N. G., Dalton, D. T., Anfora, G., Biondi, A., Chiu, J. C., Daane, K. M., Gerdeman, B., Gottardello, A., Hamby, K. A., Isaacs, I., Grassi, A., Ioriatti, C., Lee, J. C., Miller, B., Valerio Rossi-Stacconi, M., Shearer, P. W., Tanigoshi, L., Wang, X. and Walton, V. M. 2016. *Drosophila suzukii* population response to environment and management strategies. *J. Pest Sci.* 89(3), 653–65.

Wolf, S., Zeisler, C., Sint, D., Romeis, J., Traugot, M. and Collatz, J. 2018. A simple and cost-effective molecular method to track predation on *Drosophila suzukii* in the field. *J. Pest. Sci.* 91(2), 927–35. doi:10.1007/s10340-017-0948-7.

Woltz, J. M. and Lee, J. C. 2017. Pupation behavior and larval and pupal biocontrol of *Drosophila suzukii* in the field. *Bio. Control* 110, 62–9. doi:10.1016/j.biocontrol.2017.04.007.

Woltz, J. M., Donahue, K. M., Bruck, D. J. and Lee, J. C. 2015. Efficacy of commercially available predators, nematodes and fungal entomopathogens for augmentative control of *Drosophila suzukii*. *J. Appl. Entomol.* 139(10), 759–70. doi:10.1111/jen.12200.

Zerulla, F. N., Schmidt, S., Streitberger, M., Zebitz, C. P. W. and Zelger, R. 2015. On the overwintering ability of *Drosophila suzukii* in South Tyrol. *J. Berry Res.* 5(1), 41–8. doi:10.3233/JBR-150089.

Part 4

Integrated management of fruit insect pests

Chapter 19

Cultural control of arthropod pests in temperate tree fruit

Matthew J. Grieshop, Michigan State University, USA

1 Introduction

Cultural management of insect pests takes myriad forms because it exists at the intersection of horticultural and pest management practices. Norris et al. (2003) define cultural management as follows: 'The cultural management of pests involves changes to the way a crop is grown in order to make the crop less suitable for the pest, to make it more suitable for natural enemies or to enhance the ability of the crop to withstand pest attack'. Across cropping systems varietal selection, sanitation, crop rotation, planting timing, crop spacing, trap crops, intercropping, fertility management and harvest timing are all commonly considered to belong under the cultural management umbrella.

Cultural management tactics are often considered to be the cornerstone of both integrated pest management (IPM) and especially organic pest

http://dx.doi.org/10.19103/AS.2019.0046.24

management (OPM) programmes because they are typically preventative/prophylactic in nature and do not involve the use of pesticides. This is evident in IPM pyramids, which place cultural management at the base of the pyramid and pesticides at the top (Naranjo, 2001) (Fig. 1), and is legally mandated for OPM in the United States. At the time of writing, compliance with the United States Department of Agriculture (USDA) National Organic Program (NOP) requires that growers develop a farming plan based on the 'Organic Hierarchy of Pest Management' as outlined in §205.206 of the NOP standard. This hierarchy states that before growers may graduate to the use of a NOP compliant insecticide, they must first exhaust cultural, biological and physical pest management tactics, in that order.

Perennial systems, such as temperate tree fruit, present interesting challenges and opportunities for the practice of cultural management tactics. The first of these is that fruit crops typically have a high value and equally high cosmetic standard, making the tolerance for damage very low. A second challenge is that the year-round presence of trees, trellis and irrigation systems limits the grower's ability to utilize tillage and cultivation, relative to annual cropping systems. Crop rotation, perhaps the most obvious cultural control used in annual systems, is exceptionally challenging in orchard systems. For example, given the current 20–30-year expected life of an apple orchard an orchardist would be hard pressed to 'rotate' crops more than two to three times in their farming lifetime! Conversely the perennial nature of tree fruit

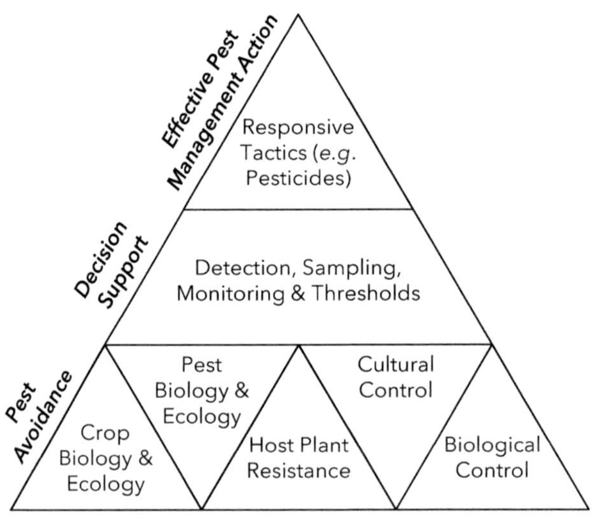

Figure 1 Conceptual model of an IPM program emphasizing a graduation from preventative pest management tactics to increasingly responsive tactics. Source: redrawn from Naranjo (2001).

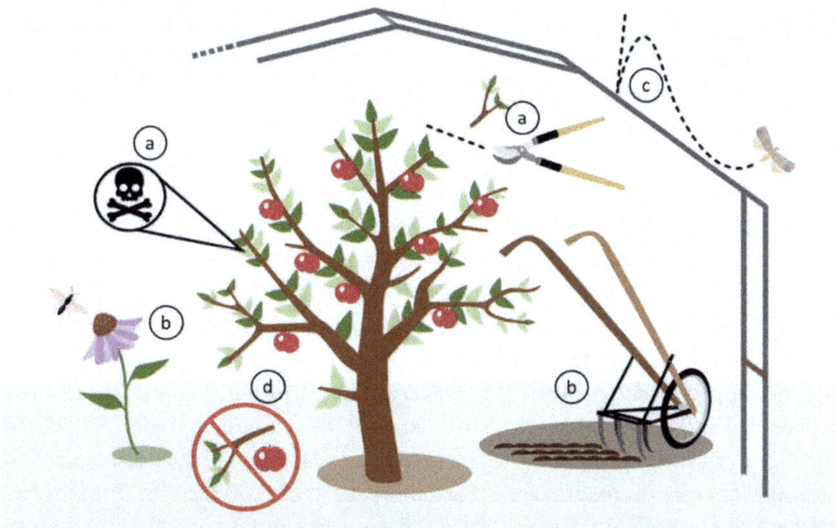

Figure 2 Conceptual diagram of the major modes of cultural pest management in temperate tree fruit: (a) modification of trees; (b) modification of the orchard floor; (c) protected culture and physical barriers; and (d) postharvest sanitation. Source: image by Holly Hooper.

crops results in increased ecosystem stability within and between production years, providing unique opportunities, such as the incorporation of perennial cover crops in orchard systems. In this chapter, cultural management of insect pests in tree fruit is delimited to the following subtypes: modification of trees; modification of the orchard floor; protected culture and physical barriers; and postharvest sanitation (Fig. 2).

2 Modification of trees

Cultural control of insect pests through fruit tree modification has been accomplished in three major pathways: the selection of insect-resistant cultivars, nutrient management and through training and pruning. For the most part, these tactics have largely been deployed to manage indirect pests (e.g. aphids, leaf-feeding insects) but have also seen some application towards direct pests (e.g. fruit-feeding insects).

3 Host-plant resistance

Most insect-resistant tree fruit cultivars provide resistance to indirect pests such as aphids, mites and leaf-feeding insects. Breeding insect pest resistance into tree fruit cultivars has been a challenging proposition. The development and release of new tree fruit varieties is typically a multi-decade endeavour and there is a long list of other breeding priorities that come into play before insect pest resistance including precociousness, consumer preference, tendency for alternate bearing, cold hardiness and disease resistance (Brown, 2012). Woolly apple aphid, *Eriosoma lanigerum* (Hausmann), resistance is a classic example of tree fruit cultivar insect pest resistance/susceptibility.

Woolly apple aphids establish gall-forming colonies on both the foliage and roots of apple trees with root colonies leading to a decline in fruit quality and in severe cases, particularly on younger or nursery trees, tree death. Because root-feeding colonies typically escape both predation and insecticides, host-plant resistance has historically been the most important management tactic. Resistant apple cultivars were first described in the early twentieth century (Staniland, 1923, 1924; Underhill and Cox, 1938) with three major resistance genes identified: *Er1, Er2 and Er3*, isolated from Northern Spy, Robusta 5 and Aotea apple varieties, respectively (Sandanayaka et al., 2003, 2005). These three cultivars have been used extensively in the development of resistant rootstocks including some of the Malling Merton (e.g. MM.106 and MM.111) and Geneva varieties in Europe and the Americas and the Aotea variety in New Zealand. The fundamental mechanisms underlying this resistance is not well understood but likely involves physical or chemical traits that interfere with aphid feeding behaviour resulting in a failure to establish colonies (Sandanayaka et al., 2005). Resistance has been reportedly overcome for the *Er1* and *Er3,* but not the *Er2* genes (Giliomee et al., 1968; Sandanayaka et al., 2003).

Evaluation of tree fruit susceptibility/resistance to other aphids has been the subject of research in both pome and stone fruit. Resistance to the rosy apple aphid, *Dysaphis plantaginea* (Passarini), the leaf-curling aphid, *Dysaphis cf. devecta* (Walker), and the green apple aphid, *Aphis pomi* (de Geer), has been documented in some apple cultivars. Wagener and Cox's Orange Pippin, Florina Querina and Golden Orange all have lower susceptibility to rosy apple and leaf curling aphids (Angeli and Simoni, 2006; Stoeckli et al., 2008a). Apple cultivars with reduced susceptibility or increased resistance to the green apple aphid include Baldwin, Bogatyr, Brat, Florina, La Nationale, Landgridge 1, McIntosh, Northern Spy, Pobedy, Pobeda, Resi, Ralls Janet, Red Delicious, Resi, Suvorovets and Vered (Briggs and Alston, 1969; Caesar et al., 1934; Habekuss et al., 2000; Stoeckli et al., 2008a). Evans et al. (2008) have identified genes conferring resistance to *Dysaphis pyri* (Boyer de Fonscolombe) in pear but resistant varieties have yet to be developed. In peaches, the variety Rubira has reduced susceptibility to the green peach aphid, *Myzus persicae* (Sulzer) and demonstrates potential induced resistance (Sauge et al., 1998a,b, 2002, 2006). Staudt et al. (2010) determined that resistant peach cultivars produced higher levels of volatile organic compounds that might be implicated in the attraction of natural enemies. Arnaudov and Kolev (2009) reported reduced susceptibility of Katalin, Summit, Hudson, Sunburst and Regina in the sweet cherry cultivars to the black cherry aphid, *Myzus cerasi* (Fab.).

Pear psylla, *Cacopsylla pyri* (Foerster), is a critical pest of pear in both the old and new world, damaging the crop through contamination of the fruit with honeydew, leading to sooty moulds or by vectoring diseases through feeding. Resistance has been characterized in a wide range of pear cultivars, many from wild *Pyrus* spp. (as reviewed in Nin et al., 2012). Mechanisms of resistance are varied ranging from feeding inhibition, delayed development and increased juvenile mortality and ovipositional non-preference. Both Bartlett and Anjou pear varieties are highly susceptible to pear psylla. Commercial pear cultivars with pear psylla resistance include Honeysweet, Doyenné de Poitiers, Erabasma, Haydeaa, Magness, Sierra, Spadona and William Precoce Morettini (Bell and Puterka, 2004; Nin et al., 2012).

Host-plant resistance to mites has been explored for apples and other tree fruit with most attention paid to two: spotted spider mite, *Tetranychus urticae* (Koch), and European red mite, *Panonychus ulmi* (Koch). Mite resistance has often been attributed to the amount of leaf pubescence possessed by varieties with heavier pubescence proposed to be less susceptible to mites (Goonewardene et al., 1976; Hogmire and Miller, 2005). Other studies, however, have not established a correlation between the amount of pubescence and mite populations (Goonewardene et al., 1980). It has been suggested that 'resistance' in pubescent varieties is actually mediated by mite predators which can show variable responses to pubescence (Duso et al., 2003; Roda et al., 2001; Schmidt, 2014; Schmidt-Jeffris and Beers, 2017). Rust mite (Acari: Eriophyidae) resistance in apple has also been explored, most recently using QTL markers and is present in Cox's Orange Pippin, Florina, Golden Delicious, Jonagold, McIntosh and Red Delicious cultivars (Downing and Moilliet, 1967; Easterbrook and Palmer, 1996; Herbert, 1974; Spieser et al., 1998; Stoeckli et al., 2009b).

Both chemical and physical modes of resistance have been suggested for tree fruit resistance to other leaf-feeding arthropods. Golden Delicious, Rome Beauty and Yorking apples are resistant/tolerant to tufted apple bud moth, *Platynota idaeusalis* (Walker) with the causal mechanism suggested to be leaf toughness or water or nitrogen content (Hunter et al., 1994). Furthermore, Hogmire and Miller (2005) reported that low-level injury (5%) by tufted apple bud moth and redbanded leafroller, *Argyrotaenia velutinana* Walker, for Honeycrisp, Sunrise and Yataka variety apples with high levels of injury (>20%) recorded for Gala Supreme and Goldrush, but the authors did not evaluate causal mechanisms. The *Cab1* gene has provided resistance to the brownheaded leafroller, *Ctenopseustis obliquana* (Walker), in New Zealand with the larva unable to complete development on the apple variety Prima

(Wearing et al., 2003). Excised leaf bioassays have demonstrated that the apple varieties Lady, Viking and Northern Spy show high levels of resistance to oblique banded leafroller, *Choristoneura rosaceana* (Harris), whereas Granny Smith was a highly susceptible variety (Schwarz and Brunner, 2017).

Resistance to internally feeding fruit pests has been recorded for pests in the orders Lepidoptera, Diptera, Coleoptera and Hymenoptera, but typically at lower levels than for the indirect pests discussed up to this point. Apple resistance to codling moth, *Cydia pomonella* (L.), is present in both traditional and modern apple varieties. Traditional varieties including Starkcrimson, Batul and Pónyik are 'moderately resistant to codling moth' (Bálint et al., 2013). Likewise, the modern apple varieties, Sunrise, Sansa, Pristine, Ginger Gold and Arlet are less susceptible to both codling moth and the oriental fruit moth, *Grapholita molesta* (Busck) (Hogmire and Miller, 2005). The resistance mechanisms for codling moth are not well understood; however, Lombarkia and Derridj (2008) demonstrated that in some cultivars resistance is mediated by reduced ovipositional preference due to changes in leaf surface metabolites. Furthermore, Stoeckli et al. (2009a) identified a quantitative trait locus (QTL) associated with codling moth susceptibility on the Discovery linkage group 10, which may assist, in the future, marker-assisted breeding programmes. Host-plant resistance/susceptibility has also been evaluated for both the plum curculio, *Conotrachelus nenuphar* (Herbst), and the apple maggot, *Rhagoletis pomonella* (Walsh), in crab apples and in early selections (Myers et al., 2007, 2008; Pree, 1977). The basis for resistance for both these pests may be, largely, due to the fruit developmental rate. Messina and Jones (1990) suggested that apparent host resistance in apple orchards in Utah, United States, were due to phenological incompatibility with later maturing varieties experiencing less damage. Burgart et al. (2016) evaluated 11 apple cultivars for susceptibility to damage by the European apple sawfly, *Hoplocampa testudinea* (Klug), and found that Zestar and Pinova were particularly susceptible varieties, while Ambrosia, Silken and Jubilee Fugi were somewhat less susceptible.

4 Tree nutrient status

Like host-plant resistance, nutrient status of crops can have measurable effects on a wide variety of arthropod plant pests. In orchard crops increased indirect pest populations have been positively correlated with the presence of young, nitrogen-rich tissues (Breukel and Post, 1959; McMullen and Jong, 1972; Rodriguez, 1958). In temperate orchard systems almost all research, in this area, has focused on hemipteran and mite (Acari) pests. For example, Haltrich et al. (2000) found that green and rosy apple aphid infestations were reduced on Jonathan but not Idared apple cultivars under urea versus ammonium fertilization and that aphid populations were higher on trees of both cultivars receiving low and high ammonium fertilization rates (0 and 200 kg/ha, respectively) compared to an intermediate rate (50 kg/ha). A sister study evaluating the same treatments for European red mite concluded that there was a positive relationship between increased nitrogen fertilization, leaf nitrogen and resultant mite populations (Papp et al., 2001). Mite populations, however, never exceeded the economic injury level regardless of nitrogen application. These findings are largely consistent with other work showing that phytophagous mite populations on apple respond positively to increased leaf nitrogen content (Breukel and Post, 1959; Rodriguez, 1958; Wermelinger et al., 1985). Nutrient source has also been evaluated for impacts on apple pests. Brown and Tortoski (2004) concluded that both spotted tentiform leafminer, *Phyllonorycter blancardella* (F.), and woolly apple aphid nymphs were reduced in apple plantings receiving compost compared with those receiving chemical fertilizers for up to 2 years following compost application. A similar study demonstrated that aphid predators were upregulated in orchards receiving compost and that this difference was most likely due to biological rather than physical or chemical aspects of the compost application (Bell et al., 2008).

Tree nutrient quality can also affect pear psylla population development. Both pear leaf nitrogen and protein content have been positively associated with increased psylla reproduction and size (Pfeiffer and Burts, 1983, 1984). Interestingly, pear psylla protein content was highest for psylla collected from pear trees with intermediate levels of fertilization (0.5 g/tree/week) compared to unfertilized and heavily fertilized trees (2 g/tree/week). Pear psylla population development rate responded differently; it was higher under either unfertilized or 2 g urea/tree/week compared to 0.5 g urea/tree/week. A subsequent study by Daugherty et al. (2007) demonstrated an increase in psylla populations under high nitrogen levels but that top-down pear psylla population regulation by *Anthocoris nemoralis* (Fab.), an important psyllid predator in Western United States orchards, was not affected by increased leaf nitrogen. These findings suggest that pear resistance/tolerance to psylla can be compromised by nitrogen fertilization regime, but biological control is less susceptible to disruption.

5 Tree architecture

Tree training and pruning can affect arthropod pests directly through changes in microclimate, changes in nutrient status, and habitat destruction or indirectly by modifying the efficacy of chemical or biological management tactics (Simon et al., 2007; Stoeckii et al., 2008b). Pruning and training that increases porosity are hypothesized to increase the temperature within canopies and have been suggested to increase the developmental rate of codling moth (Kührt et al., 2006). Nutrient-related impacts on pests have been less consistent. Holb et al.

(2001) reported that heavy pruning in apples contributed to European red mite management compared to light pruning, while Grechi et al. (2008) reported that heavy pruning of peach trees stimulated aphid populations. In both cases the authors suggest that the vigorous new growth resulting from pruning was responsible for the observed pest population response. Heavy winter pruning in apples has also been implicated in reductions of European red mite and rosy apple aphids by removing overwintering habitat and by reducing the tendency of ants to tend and protect aphids from natural enemies (Simon, 2007). Pruning can also interact with chemical control strategies by facilitating improved penetration, coverage and deposition of insecticides. For example, chemical control of *Leidosaphes* sp. (Homoptera: Diaspididae) in citrus orchards was improved on more heavily pruned tress (Juste et al., 1990) and coverage and deposition of pesticides on apple trees is significantly affected by apple tree size and architecture (Cross et al., 2001; Duga et al., 2015).

6 Orchard floor management

The majority of floor management in industrialized temperate fruit orchards divides the orchard floor into two zones: (1) a strip that extends to the drip line of the trees and is largely devoid of vegetation and (2) an alleyway consisting of a mix of grasses and forbs. Management of floor canopy within the tree row is especially important during the establishment phase of orchards with calliper, and scaffold growth increased when competition from floor canopy plants is eliminated within the tree row (Merwin, 2003). Management of the alleyway is also important because the maintenance of healthy turf in the alleyway enables the transit of heavy equipment (tractors, sprayers, platforms) through the orchard. Maintenance of these two orchard floor zones is typically done using herbicides, mowing, cultivation/tillage, mulches and/or cover crops. All these practices can impact pest arthropod populations either directly or mediated through natural enemies.

7 Cultivation

The maintenance of an 'herbicide strip' in the tree row is probably the most common approach to the management of orchard floor canopies (Merwin, 2003). Natural enemy populations are typically expected to be adversely affected by this management practice. For example, Altieri and Schmidt (1985) reported a decrease in natural enemies and increase in aphids and leafhoppers in California orchards with cover crops versus orchards that were clean cultivated. Herbicide strips, however, can also contribute to the management of both arthropod and vertebrate pests by reducing the quality of overwintering habitat (Bobb, 1949; Lafleur et al., 1987; Smith and Flessel, 1968). A survey study by Killian and Meyer (1984) showed that herbicide management of orchard floors was correlated with reduced 'cat-facing' damage to peach fruits by lygus bugs in North Carolina, United States, peach orchards.

Organic and conventional farmers wishing to avoid herbicide use can maintain a mowed and/or cultivated strip around the trees. In these systems, the floor canopy is allowed to rebound between cultivation and/or mowing events resulting in increased amounts of plant debris, compared to herbicide-based management. Cultivation events may have variable impacts on specific pests depending on the behaviour and biology of the pest as mortality from cultivation is most likely due to burial, rather than impacts with the cultivator. Baughman et al. (2015) reported that running a Lilliston-style cultivator over codling moth pupa and plum curculio-infested apples resulted in minimal direct mortality, while burial of moth pupa under even 3 cm of soil resulted in near complete reductions in adult emergence. Burying apples containing pre-pupal plum curculio up to 15 cm deep, increased adult emergence compared to unburied apples. Earlier literature, however, reports that frequent disc cultivation of peach orchards reduces plum curculio emergence, although in these studies it is suggested that exposure to the elements and/or predators and diseases were the likely sources of mortality, not burial or mechanical damage (Quaintance and Jenne, 1912; Snapp, 1930). Tree row cultivation has also been evaluated as a control method for the dogwood borer, *Synanthedon scitula* Harris, an important pest of young apple plantings. In this case, cultivation equipment was used to hill soil above the graft union of young trees to hide it from oviposition by female moths, yielding up to 99% control of larval densities (Gut et al., 2005). Growers utilizing this technique must uncover the graft union following the oviposition period to avoid scion-rooting which can lead to unacceptable tree vigour.

Rotational grazing of livestock and the 'Swiss Sandwich' approach are two specific systems that incorporate intentional orchard floor disturbance as a cultural practice. Hogs and other animals have a long history of grazing in traditional, low-density orchards and were recognized as providing weed and insect control as early as the eighteenth century (Walsh and Riley, 1868). More recently, rotational grazing of feeder hogs (<50 kg) in a Michigan apple orchard provided reductions in both codling moth and plum curculio damage in the years following grazing (see Section 12). In a sister study, fully grown hogs were grazed postharvest in a succession of certified organic tart cherry, apple and pear orchards, and provided significant reductions in damage by plum curculio in cherries and apples and codling moth and oriental fruit moth in pears and apples, in the year following grazing (Buehrer and Grieshop, 2014). The 'Swiss Sandwich' approach involves repeated shallow cultivation up to the trees, resulting in a narrow strip of uncultivated orchard floor in line with the trees (Schmid et al., 2004; Stefanelli et al., 2009; Weibel, 2002; Zoppolo et al., 2011).

The result is management of weed competition in much of the tree rooting zone with the maintenance of a small amount of plant material which can serve as a potential refuge for beneficial arthropods. A trial conducted in an apple orchard reported increases in predator taxa in 'sandwich' managed plots as compared to 'herbicide strip' plots, including ants (Formicidae), ground beetles (Carabidae) and Lampyridae but did not increase the removal of sentinel prey (Baughman, 2014).

8 Mowing

The timing and frequency of mowing of alleyways has also been evaluated for its impact on pest management. The timing of mowing in Australian pecans affected the density of the pentatomid pest *Nezara viridula* (L.) by limiting its primary reproductive hosts: a diverse range of annual plants on the orchard floor (Coombs, 2000). Phytoseiid mite populations, important predators of phytophagous mites, were enhanced under reduced mowing regimes in Japan (Funayama et al., 2015). In the Pacific Northwest, decreasing mowing frequency to once per month was correlated with an increase in both generalist and some specialist predators/parasitoids including their phytophagous prey/hosts (Horton et al., 2003).

9 Orchard cover crops

Cover crops planted in the tree or alleyway is an additional weed management tactic that can impact orchard arthropod populations. Cover crops modify the environment by adding structural complexity and changing soil moisture and nutrient dynamics. For example, dense stands of centipede grass, *Eremochloa ophiuroides* (Munro), reduced second generation plum curculio adult emergence from the soil under peach trees in the Southern United States by interfering with the insect's ability to enter the soil to pupate (Akotsen-Mensah et al., 2012). However, earlier studies evaluating bare ground or mowed grass versus unmanaged floor canopies report reduced plum curculio overwintering survival under bare ground or mowed grass conditions (Bobb, 1949; Lafleur et al., 1987; Smith and Flessel, 1968). Vegetative cover within the tree row can also promote tree-damaging rodent populations by providing protection from predators and overwintering habitat (Wiman et al., 2009).

Drive row plant composition and management is increasingly being recognized as influencing orchard arthropod diversity and associated ecosystem services, including biological control. For example, a study evaluating arthropod diversity in apple orchards in Northern Patagonia, Argentina, demonstrated that ratios of herbivorous to predaceous arthropods were significantly impacted across five different floor management programmes. In this case, floor mixes based on either an alfalfa, *Medicago sativa* (L.), fescue, *Festuca* spp., mix and strawberry clover, *Trifolium fragiferum* (L.), had higher ratios of predators and higher overall arthropod diversity compared to orchard floors composed of native vegetation (Fernandez et al., 2008). In Hungary, woolly apple aphids were increased in flowering alleyway plantings in apple orchards compared to bare ground or mowed grass but green apple aphid populations did not differ among bare, mowed grass and flowering ground covers (Markó et al., 2013). Fye (1983) concluded that including grain crops in pear orchards in Washington State, United States, had a slight tendency to increase pear psylla predators, most likely by providing alternative prey, and in Georgia pecan orchards, cover crops of hairy vetch, *Vicia villosa* Roth, led to increased biological control of the yellow pecan aphid, *Monelliopsis pecanis* (Bissell), by the convergent lady beetle, *Hippodamia convergens* (Guérin-Méneville) (Tedders, 1983). Similar increases in generalist predators in response to diversified orchard floor plant diversity have also been observed in Western United States orchards (Bugg and Dutcher, 1989; Bugg et al., 1991) as well as in Oklahoma, United States (Smith et al., 1996). Many of these differences however did not translate to the pecan canopy, that is largely floor-level predator communities were affected.

10 Protected culture and physical exclusion

Protected culture is becoming increasingly common in tree fruit production and is typically used to either shorten or extend the growing season and protect fruit from rainfall or hail events (Alaphilippe et al., 2016; Lang, 2009). Enclosures made of either transparent plastic (in the case of high tunnels or greenhouses) or netting (in the case of hail nets) can provide the additional benefit of excluding pest insects but may require modification to do so. In particular, these systems typically have open ends allowing insects to move in and out of the protected area. This can be remedied by adding exclusion netting to structure ends, an approach first widely adopted in French and Italian orchards (Alaphilippe et al., 2016). Exclusion netting may be especially useful for managing direct fruit pests such as the apple maggot, codling moth, spotted wing Drosophila (*Drosophila suzukii* (Matsumura)) and true bugs (Heteroptera) (Chouinard et al., 2017). For example, exclusion netting, on its own, reduced codling moth damage of apples by 90% in Croatia (Pajač Živković et al., 2016). In addition to preventing immigration of codling moth, netting has also been implicated in altering the reproductive behaviour of males foraging for females, reducing subsequent generations (Sauphanor et al., 2012). Codling moth has apparently overcome this behavioural mechanism of action in France, where increasing incidence of codling moth damage has been observed under nets (Siegwart et al., 2013). Wrapping single cherry trees or entire rows in insect proof netting has been evaluated for the management of the cherry fly, *Rhagoletis cerasi* (L.), with good success, but requires that nets be put in place as soon as or just before flies are detected (Ughini et al., 2010). The addition of insect exclusion nets to raspberries grown in high tunnels not only reduces fruit damage by spotted wing Drosophila, but also delays first detection of maggots and flies in screened tunnels, relative to berries grown in unscreened tunnels (Leach et al., 2016). Physical exclusion by bagging individual fruit with expandable nylon mesh bags has been evaluated for both peaches and apples (Grasswitz and Fimbres, 2013). This approach showed no effect on apple maggot but did reduce codling moth damage by up to 25% in apples and significantly reduced damage on peaches by twig borer, *Anarsia lineatella* (Zeller), birds and stink bugs (Pentatomidae). Due to the high labour demand for this approach it is unlikely to be economical for growers producing at the greater than 0.5 ha scale.

11 Sanitation

Harvest and postharvest sanitation practices are important cultural arthropod control tactics and can consist of postharvest treatment/storage of fruits, disposal/sanitation of waste fruit or fruit remnants and the cleaning and/or sterilization of orchard equipment. Postharvest refrigeration of soft fruits including blueberries and raspberries can kill both eggs and larvae of spotted wing Drosophila with 72 h at 1.1°C providing 100% control (Aly et al., 2017). Harvest frequency can also impact spotted wing Drosophila populations. For example, infestation of late season raspberries is reduced when growers harvest more frequently with significant reductions reported when berries are harvested every 1–2 days versus 2–3 times/week (Leach et al., 2018). The same study also evaluated bagging and solarization of culled or dropped raspberries and demonstrated that a 32 h solarization period reduced larval survival by 99%. Fermentation of postharvest fruit wastes is an alternative approach for managing spotted wing Drosophila with hypoxia suggested as the mechanism of action. Spotted wing Drosophila were eliminated from fruit wastes after a fermentation period of 3 days, with 4 day providing control of *D. melanogaster* (Meigen) (Noble et al., 2017).

Pruning wastes left in the orchard can serve as an important refuge for arthropod pests. Sanitation of dead trees and large prunings provides a first line of defence against shothole borers and ambrosia beetles (Curculionidae: Scolininae) across temperate tree fruit production systems. Doerr et al. (2008) found that *Scolytus rugulosus* Müller and *Xyleborinus saxeseni* Ratzburg are the primary wood-boring Curculionidae found in Washington, United States, cherry orchards with the latter accounting for the majority of damage to young plantings. Both species build up populations on cut wood or prunings in or near the orchards and disposal of these resources (burning, chipping or physical removal by more than 100 m) greatly reduced the incidence of these pests. Chipping and solarization of cut wood were also evaluated in Southern California for the protection of hardwood trees from wood-boring beetles. Chipping wood to sizes <5 cm and solarization when temperatures were >34°C led to significant reductions in colonization of adjacent trees and logs (Eatough Jones and Paine, 2015). Solarization at cooler temperatures did not reduce populations, with beetles persisting for more than 2 months under fall and winter conditions.

Orchard equipment, such as fruit bins or boxes have been implicated as overwintering refuges for some insects. Codling moth will overwinter on bins left in apple orchards and can even be transported from orchards if larvae spin hibernacula on the fruit bins (Higbee et al., 2001). Submersion of bins in high-temperature water has been evaluated as a method of disinfecting bins with 30 s of submersion in 80°C water eliminating >99% of larvae and pupae (Hansen et al., 2006). Likewise, fumigation of bins with acetic acid can eliminate up to 100% of codling moth larva residing in plastic bins but is ineffective when used on wooden bins (Randall et al., 2011). In contrast, fumigation with 13% carbon dioxide for 20 days only provided 88% mortality (Cossentine et al., 2004).

12 Case study: integration of rotational hog grazing and apple production

Management of the over two dozen potential insect pests common to Eastern United States apple orchards presents a major challenge to apple production in both conventional and organic systems. Two insects of special importance are codling moth, *C. pomonella*, and plum curculio, *C. nenuphar*, both of which feed internally on developing apples. Plum curculio is especially hard to manage in organic systems because there are no effective organically approved insecticides. Rotational grazing of livestock has a long history of use in orchard and silvicultural systems, providing weed and sometimes arthropod pest management benefits (Walsh and Riley, 1868). The objectives of this case study were to assess the impact of rotationally grazed hogs on plum curculio, codling moth and orchard floor canopy in a mature, semi-dwarf apple orchard.

The experiment took place in six 0.2 ha orchard plots within a 2.5 ha commercial Michigan organic apple orchard in 2007 and 2008. Trees were free-standing, semi-dwarf McIntosh on M26 rootstock at >20th leaf planted at a density of 716 trees/ha (4.6 m row spacing × 3 m tree spacing). Experimental plots were separated by at least 9 m and half were selected to receive grazing and half remained ungrazed (n = 3). Twenty-five to thirty, 6- to 8-week-old, 30–45 kg Berkshire hogs, *Sus domesticus* (L.), were rotationally grazed for repeated, 1-week intervals, such that each plot was grazed every 14 days between late May and late July (Fig. 3). Hogs were farrowed at the study farm from existing Berkshire stock and were contained within the grazed paddocks with a three-wire electric fence with wires positioned 15, 30 and 45 cm above the ground and a 2000 J livestock fence energizer (Zareba, Lititz Pennsylvania, United States). Hogs were provided access to water and supplementary feed (ground soya beans, spelt and corn) throughout the study period. Grazing period was selected to intercept the late larval and pupal life stages of plum curculio located within dropped apples and the soil.

Impacts on plum curculio were evaluated through damage evaluations conducted during late June and at harvest in 2007 and 2008. Four hundred fruits were selected from 20 trees (20 fruit/tree) per plot at each sampling period. Plum curculio oviposition scars were recorded during the June evaluations and adult feeding during the harvest evaluation. In 2008 codling moth damage evaluations were conducted during the harvest damage evaluation. Weekly assessments of the impact of hog grazing on orchard floor canopy were also

Figure 3 (Above) 30-45 kg Berkshire hogs grazing in an apple orchard. (Below) Level of soil disturbance typical after a 1-week grazing period at a stocking rate of 100 hogs/ha: note that most soil disturbance occurs under the tree row. Source: image by David Epstein.

done in 2008. These were estimated using 3, 30 m transects per plot with a Daubenmire quadrat read at 1 m intervals (Daubenmire, 1959). Transects were placed within the dripline of the trees (tree row) and the percentage bare ground, and plant-covered ground estimated for each quadrat. Data were analysed using two sample *t*-tests assuming unequal variance. The *t*-tests were performed to compare percentage plum curculio-damaged fruit in late June (oviposition scars) and at harvest (adult feeding injury) in 2007 and 2008 and codling moth damaged fruit (fruit with entries and/or exits) at harvest during 2008. Percentage data were arcsine transformed to improve conformation to a normal distribution. Percentage floor cover data were averaged across all frames within a plot, and the resulting proportion data visualized over time. All statistics were conducted in the R computing environment (R Core Team, 2016).

Rotational grazing of hogs had a marked impact on insect pest damage and floor canopy coverage in the tree row. In 2007 there was a slightly higher rate of plum curculio oviposition scars in the grazed plots compared to the ungrazed plots (Fig. 4); however, these measurements were taken prior to rotational grazing. Plum curculio adult feeding damage was reduced nearly fourfold, in 2007, suggesting that grazing interrupted the developing generation of plum curculio. In June 2008, grazed plots had a 50% reduction of fruit oviposition scars and a significant reduction in adult feeding damage to fruits at harvest (Fig. 4). Also, in 2008, significantly fewer apples were damaged by codling moth in the grazed plots, with 7% and 2.5% damage recorded in the rotationally grazed and ungrazed plots, respectively (Fig. 5). Rotational grazing also provided significant disruption of the floor canopy within the tree row,

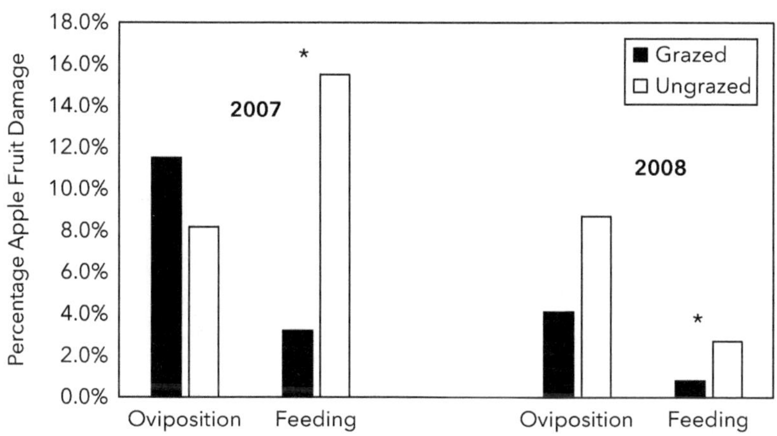

Figure 4 Mean percentage plum curculio oviposition scars and feeding damage to apples from 0.2 ha grazed or ungrazed orchard plots in 2007 and 2008. * indicates a significant difference between grazed and ungrazed plots (Student's *t*-test alpha = 0.05). Source: redrawn from Epstein and Grieshop (unpublished).

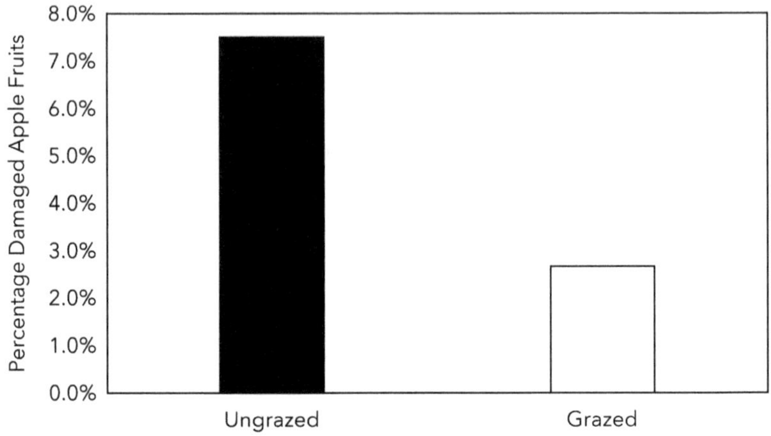

Figure 5 Mean percentage codling moth larvae apple entries from 0.2 ha grazed or ungrazed orchard plots in 2008. A significant difference was detected (t = 3.83, d.f. = 4, P = 0.019). Source: redrawn from Epstein and Grieshop (unpublished).

increasing the percentage of bare ground by 50% or more compared to the ungrazed plots in 2008 (Fig. 6). The increased percentage of bare ground in the rotationally grazed plots persisted months after the cessation of rotational grazing (ca. 30 July 2008).

Rotational grazing of young hogs during late spring and early summer can provide cultural pest management of key insect pests and weeds within the tree row at a density of 100 hogs/ha. This study did not evaluate the rooting depth of hogs, but this was not viewed as a concern for the collaborating farmer. It

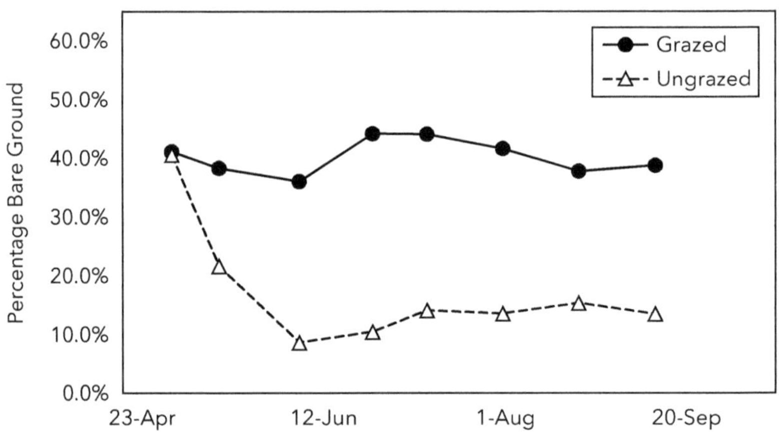

Figure 6 Mean percentage bare ground within the tree row of 0.2 ha grazed or ungrazed apple orchard plots in 2008. Note the consistently higher percentage of bare ground throughout the season. Source: redrawn from Epstein and Grieshop (unpublished).

is important to note that the age/size of hogs likely had a marked impact on observed results. The <45 kg hogs were unable to damage the semi-dwarf trees in this orchard or damage the orchard alleyways with rooting behaviour. Instead the young hogs preferred to root in the somewhat less dense vegetation under the tree row, disturbing plant cover. The shade provided by trees likely also kept hogs in this area.

Future proof-of-concept studies are needed to evaluate whether small hogs are capable of damaging modern, high-density orchards (e.g. >2000 trees/ha), given the smaller stature of trees in these systems. From an economic perspective, the scalability of this approach to larger plantings is not known, although this case study required the addition of at least one full-time employee to manage the hogs. Finally, there are a variety of other potential impacts of rotational grazing on the orchard system ranging from how hogs interact with the full arthropod pest and natural enemy complex, vertebrate pests (e.g. rodents) and beneficial species (e.g. amphibians, birds, bats), soil tilth and nutrient status, and the potential spread of dangerous microbes associated with hog faeces (e.g. *E. coli* and *Salmonella*). This case study further supports the hypothesis that fruit removal and ground disturbance may contribute to the management of these two serious orchard pests.

13 Future trends and conclusion

The preponderance of evidence suggests that, while cultural arthropod management tactics are unlikely to provide economic control of direct orchard pests, they can contribute to and (in some cases) provide economic control of indirect pests. Furthermore, the incorporation of cultural pest management tactics may help reduce the development of resistance to insecticides/acaricides by introducing additional bottom-up and/or top-down selection pressures to pest populations. For example, manipulation of tree vigour that results in reduced nitrogen availability to arthropod pests may make them more susceptible to chemical control and in some cases top-down pressures from natural enemies. Likewise, reductions in overwintering pest insects through floor management practices that disturb habitat may not only reduce next year's 'inoculum' but do so using a mode of action entirely different than either toxic or pathogenic insecticides. The major challenge to applying cultural tactics, however, is that, unlike chemical pest management, their efficacy is context dependent. Practices that function well in one tree fruit system in a specific region may not function in others depending on the pest and beneficial complex present, microclimatic factors, the amount of labour required for practices, and finally, the market forces determining the grower's tolerance for damage. The two areas of future research that may have the most promise for the development of cultural control methods in temperate tree fruit are the application of genetic engineering to develop resistant varieties and research that evaluates how combinations of cultural tactics can be optimized to negatively impact broad complexes of insect pests without compromising tree health/production and/or damaging beneficial arthropod communities.

Host-plant resistance in temperate tree fruit has traditionally been of limited economic utility due to the breeding priority placed on horticultural and market demands of tree fruit (Brown, 2012). Technological developments in molecular genetics provide new opportunities to introduce resistance factors into cultivars with no history of resistance. The first step of this process, mapping of resistance genes, is already under way in some crops (Stoeckli et al., 2009a,b). *Cis*-genetic manipulation, the artificial movement of genes within a recognized species, has been recently used to develop apple cultivars with resistance to both apple scab and fire blight; however, this work has yet to result in released cultivars (Gessler et al., 2014; Kost et al., 2015; Krens et al., 2015; Vanblaere et al., 2014). Okanagan Specialty Fruits (Summerland, BC, Canada), however, has recently released two Arctic® apple cultivars, which have used gene silencing to effect non-browning apples (Waltz, 2015). While molecular editing of existing tree fruit genomes may provide some outlets for host-plant resistance to direct pests, the real potential for this likely lies in 'traditional', transgenic genetic manipulation; the artificial movement of genes from across recognized species. An obvious first approach is the insertion of genes coding for *Bacillus thuringiensis* crys toxins, which have already had massive impacts on corn, soybean, alfalfa and eggplant breeding programmes (Sanchis, 2011). At present, these approaches have not been applied to tree fruit breeding programmes, largely due to consumer concerns around genetic engineering (Costa-Font et al., 2008; Lucht and Lucht, 2015). Given the now almost worldwide acceptance of transgenic annual crops, it is quite likely that transgenic tree fruit will be created within the next few decades.

Due to the inherent complexity of research in perennial crops, very few studies have been completed that evaluate how combinations of cultural practices affect tree fruit pest complexes, the trees themselves and beneficial arthropods. Furthermore, the inferential space of these studies is typically limited to a narrow geographic range due to the high level of abiotic and biotic factors in different fruit-growing regions. The late Ron Prokopy (2003) conducted a 20-year study evaluating combined arthropod and disease management in a small orchard Conway, Massachusetts USA which combined habitat management, resistant cultivars, floor sanitation, mass trapping and a few insecticide applications per year, primarily directed at direct fruit pests. This unique demonstration project showed that pest arthropods could be maintained below economic levels over the course of two decades. Plant pathogens, however built up during the last 5 years and fungicide applications became necessary to maintain adequate fruit quality.

An alternative approach towards this goal is to evaluate broad patterns of how single or combined cultural practices affect arthropod communities. For example, research on how specific pestilent and beneficial arthropod species and community assemblages respond to broad disturbance such as herbicide strips, cultivation or mowing might be more translatable if the organisms are viewed by ecological function, that is trophic positioning or feeding guilds. An important question from this perspective might be the following: How do different feeding guilds of herbivores respond to different patterns of floor disturbance compared to predator or parasitoid guilds? In the case of protected culture, understanding how plastics or mesh affect abiotic factors such as microclimate, light quality and intensity affect arthropod developmental times, arthropod sensory systems or pesticide decay rates may provide information that can be translated across many crops and regions. Such broad inferences might then be combined to develop composite cultural control strategies that could be evaluated in regional demonstrations.

14 Where to look for further information

A standard introduction to cultural controls in Integrated Pest Management (IPM) programmes can be found in a variety of textbooks such as *Concepts in Integrated Pest Management* (see Norris et al., 2003).

University and governmental centres with further information on cultural control applications in perennial fruit crops include:

- Cornell Fruit Resources: Tree Fruit (https://blogs.cornell.edu/treefruit/)
- Michigan State University Integrated Pest Management Program (https://www.canr.msu.edu/ipm/agriculture/fruit/)
- PennState Extension Pest Management and Education Program (https://extension.psu.edu/pests-and-diseases/pest-management-and-education)
- The University of California Statewide Integrated Pest Management Program (http://ipm.ucanr.edu/)
- Washington State University Comprehensive Tree Fruit Site (http://treefruit.wsu.edu/)

15 References

Akotsen-Mensah, C., Boozer, R. T. and Fadamiro, H. Y. 2012. Influence of orchard weed management practices on soil dwelling stages of plum curculio, *Conotrachelus nenuphar* (Coleoptera: Curculionidae). *Florida Entomologist* 95(4), 882–9. doi:10.1653/024.095.0410.

Alaphilippe, A., Capowiez, Y., Severac, G., Simon, S., Saudreau, M., Siegwart, M., Stefano, C. and Stefano, V. 2016. Codling moth exclusion netting: an overview of French and Italian experiences. *IOBC-WPRS Bulletin* 112, 31–5.

Altieri, M. A. and Schmidt, L. L. 1985. Cover crop manipulation in northern California orchards and vineyards: effects on arthropod communities. *Biological Agriculture and Horticulture* 3(1), 1–24. doi:10.1080/01448765.1985.9754453.

Aly, M. F. K., Kraus, D. A. and Burrack, H. J. 2017. Effects of postharvest cold storage on the development and survival of immature *Drosophila suzukii* (Diptera: Drosophilidae) in artificial diet and fruit. *Journal of Economic Entomology* 110(1), 87–93. doi:10.1093/jee/tow289.

Angeli, G. and Simoni, S. 2006. Apple cultivars acceptance by *Dysaphis plantaginea* Passerini (Homoptera: Aphididae). *Journal of Pest Science* 79(3), 175–9. doi:10.1007/s10340-006-0129-6.

Arnaudov, V. A. and Kolev, K. K. 2009. Susceptibility of some introduced sweet cherry cultivars to the attacks of black cherry aphids *Myzus cerasi* fab. (Homoptera: Aphididae). *Acta Horticulturae* 825(825), 401–6. doi:10.17660/ActaHortic.2009.825.63.

Bálint, J., Thiesz, R., Nyárádi, I.-I. and Szabó, K.-A. 2013. Field evaluation of traditional apple cultivars to induced diseases and pests. *Notulae Botanicae* 41(1), 238–43.

Baughman, W. B. 2014. Impacts of strip cultivation in apple and grape systems. M.S. thesis. Michigan State University. Available at: https://search.proquest.com/docview/1532214864/abstract/31B748FE9A7E4991PQ/1.

Baughman, W. B., Nelson, P. N. and Grieshop, M. J. 2015. Impact of cultivation and subsequent burial on *Cydia pomonella* (Lepidoptera: Tortricidae) and *Conotrachelus nenuphar* (Coleoptera: Curculionidae). *Journal of Economic Entomology* 108(3), 1215–20. doi:10.1093/jee/tov071.

Bell, R. L. and Puterka, G. L. 2004. Modes of host plant resistance to pear psylla: a review. *Acta Horticulturae* 663(663), 183–8. doi:10.17660/ActaHortic.2004.663.26.

Bell, J. R., Traugott, M., Sunderland, K. D., Skirvin, D. J., Mead, A., Kravar-Garde, L., Reynolds, K., Fenlon, J. S. and Symondson, W. O. C. 2008. Beneficial links for the control of aphids: the effects of compost application on predators and prey. *Journal of Applied Ecology* 45(4), 1266–73. doi:10.1111/j.1365-2664.2008.01479.x.

Bobb, M. L. 1949. Hibernation of the plum curculio in Virginia. *Journal of Economic Entomology* 42(1), 19–22. doi:10.1093/jee/42.1.19.

Breukel, L. M. and Post, A. 1959. The influence of the manurial treatment of orchards on the population density of *Metatetranychus ulmi* (Koch) (Acari, Tetranychidae). *Entomologia Experimentalis et Applicata* 2(1), 38–47. doi:10.1111/j.1570-7458.1959.tb02095.x.

Briggs, J. B. and Alston, F. H. 1969. *Sources of Pest Resistance in Apple Cultivars*. East Malling Research Station Annual Report.

Brown, S. 2012. Apple. In: Badenes, M. L. and Byrne, D. H. (Eds), *Fruit Breeding (Handbook of Plant Breeding)*. Springer US, Boston, MA, pp. 329–67. doi:10.1007/978-1-4419-0763-9_10.

Brown, M. W. and Tworkoski, T. 2004. Pest management benefits of compost mulch in apple orchards. *Agriculture, Ecosystems and Environment* 103(3), 465–72. doi:10.1016/j.agee.2003.11.006.

Buehrer, K. A. and Grieshop, M. J. 2014. Postharvest grazing of hogs in organic fruit orchards for weed, fruit, and insect pest management. *Organic Agriculture* 4(3), 223–32. doi:10.1007/s13165-014-0076-0.

Bugg, R. L. and Dutcher, J. D. 1989. Warm-season cover crops for pecan orchards: horticultural and entomological implications. *Biological Agriculture and Horticulture* 6(2), 123–48. doi:10.1080/01448765.1989.9754511.

Bugg, R. L., Dutcher, J. D. and McNeill, P. J. 1991. Cool-season cover crops in the pecan orchard understory: effects on Coccinellidae (Coleoptera) and pecan aphids (Homoptera: Aphididae). *Biological Control* 1(1), 8–15. doi:10.1016/1049-9644(91)90095-H.

Burgart, C., Hillier, N. K. and Blatt, S. 2016. Apple cultivar preferences by *Hoplocampa testudinea* (Hymenoptera: Tenthredinidae) in the Annapolis Valley of Nova Scotia, Canada. *The Canadian Entomologist* 148(6), 724–35. doi:10.4039/tce.2016.16.

Caesar, L., Hall, J. A. and Kelsall, A. 1934. The influence of cultural practices on orchard insects. Report of the Entomological Society of Ontario 65, 40–2.

Chouinard, G., Veilleux, J., Pelletier, F., Larose, M., Philion, V. and Cormier, D. 2017. Impact of exclusion netting row covers on arthropod presence and crop damage to 'Honeycrisp' apple trees in North America: a five-year study. *Crop Protection* 98, 248–54. doi:10.1016/j.cropro.2017.04.008.

Coombs, M. T. 2000. Seasonal phenology, parasitism, and evaluation of mowing as a control measure for *Nezara viridula* (Hemiptera: Pentatomidae) in Australian pecans. *Environmental Entomology* 29(5), 1027–33. doi:10.1603/0046-225X-29.5.1027.

Cossentine, J. E., Sholberg, P. L., Jensen, L. B. J., Bedford, K. E. and Shephard, T. C. 2004. Fumigation of empty fruit bins with carbon dioxide to control diapausing codling moth larvae and *Penicillium expansum* Link. ex Thom Spores. *HortScience* 39(2), 429–32.

Costa-Font, M., Gil, J. M. and Traill, W. B. 2008. Consumer acceptance, valuation of and attitudes towards genetically modified food: review and implications for food policy. *Food Policy* 33(2), 99–111. doi:10.1016/j.foodpol.2007.07.002.

Cross, J. V., Walklate, P. J., Murray, R. A. and Richardson, G. M. 2001. Spray deposits and losses in different sized apple trees from an axial fan orchard sprayer: 1. Effects of spray liquid flow rate. *Crop Protection* 20(1), 13–30. doi:10.1016/S0261-2194(00)00046-6.

Daubenmire, R. 1959. Canopy coverage method of vegetation analysis. *Northwest Science* 33, 39–64.

Daugherty, M. P., Briggs, C. J. and Welter, S. C. 2007. Bottom-up and top-down control of pear psylla (*Cacopsylla pyricola*): fertilization, plant quality, and the efficacy of the predator *Anthocoris nemoralis*. *Biological Control* 43(3), 257–64. doi:10.1016/j.biocontrol.2007.09.001.

Doerr, M. D., Brunner, J. F. and Smith, T. J. 2008. Biology and management of bark beetles (Coleoptera: Curculionidae) in Washington cherry orchards. *Journal of the Entomological Society of British Columbia* 105, 69–81.

Downing, R. S. and Moilliet, T. K. 1967. Relative densities of predacious and phytophagous mites on three varieties of apple trees. *The Canadian Entomologist* 99(7), 738–41. doi:10.4039/Ent99738-7.

Duga, A. T., Ruysen, K., Dekeyser, D., Nuyttens, D., Bylemans, D., Nicolai, B. M. and Verboven, P. 2015. Spray deposition profiles in pome fruit trees: effects of sprayer design, training system and tree canopy characteristics. *Crop Protection* 67, 200–13. doi:10.1016/j.cropro.2014.10.016.

Duso, C., Pasini, M. and Pellegrini, M. 2003. Distribution of the predatory mite *Typhlodromus pyri* (Acari: Phytoseiidae) on different apple cultivars. *Biocontrol Science and Technology* 13(7), 671–81. doi:10.1080/09583150310001606264.

Easterbrook, M. A. and Palmer, J. W. 1996. The relationship between early-season leaf feeding by apple rust mite, *Aculus schlechtendali* (Nal.), and fruit set and photosynthesis of apple. *Journal of Horticultural Science* 71(6), 939–44. doi:10.1080/14620316.1996.11515478.

Eatough Jones, M. and Paine, T. D. 2015. Effect of chipping and solarization on emergence and boring activity of a recently introduced ambrosia beetle (*Euwallacea* sp., Coleoptera: Curculionidae: Scolytinae) in Southern California. *Journal of Economic Entomology* 108(4), 1852–9. doi:10.1093/jee/tov169.

Evans, K. M., Govan, C. L. and Fernández-Fernández, F. 2008. A new gene for resistance to *Dysaphis pyri* in pear and identification of flanking microsatellite markers. *Genome* 51(12), 1026–31. doi:10.1139/G08-093.

Fernández, D. E., Cichón, L. I., Sánchez, E. E., Garrido, S. A. and Gittins, C. 2008. Effect of different cover crops on the presence of arthropods in an organic apple (*Malus domestica* Borkh) Orchard. *Journal of Sustainable Agriculture* 32(2), 197–211. doi:10.1080/10440040802170624.

Funayama, K., Komatus, M., Sonoda, S., Takahashi, I. and Hara, K. 2015. Management of apple orchards to conserve generalist phytoseiid mites suppresses two-spotted spider mite, *Tetranychus urticae* (Acari: Tetranychidae). *Experimental and Applied Acarology* 65(1), 43–54. doi:10.1007/s10493-014-9850-7.

Fye, R. E. 1983. Cover crop manipulation for building pear psylla (Homoptera: Psyllidae) predator populations in pear orchards. *Journal of Economic Entomology* 76(2), 306–10. doi:10.1093/jee/76.2.306.

Gessler, C., Vanblaere, T., Parravicini, G. and Broggini, G. A. L. 2014. Cisgenic 'Gala' containing the scab resistance gene from *Malus floribunda* 821 and the fire blight resistance genes from m. 'Evereste'. *Acta Horticulturae* 1048(1048), 43–9. doi:10.17660/ActaHortic.2014.1048.4.

Giliomee, J. H., Strydom, D. K. and Van Zyl, H. J. 1968. Northern Spy, Merton and Malling-Merton rootstocks susceptible to woolly aphid, *Eriosoma lanigerum*, in the Western Cape. *South African Journal of Agricultural Science* 11, 183–6.

Goonewardene, H. F., Williams, E. B., Kwolek, W. F. and McCabe, L. D. 1976. Resistance to European red mite, *Panonychus ulmi* (Koch), in apple. *Journal of the American Society for Horticultural Science* 101(5), 532–7.

Goonewardene, H. F., Kwolek, W. F., Dayton, D. F. and Hayden, R. A. 1980. Preference of the European red mite for strains of 'delicious' apple with differences in leaf pubescence. *Journal of Economic Entomology* 73(1), 101–3. doi:10.1093/jee/73.1.101.

Grasswitz, T. R. and Fimbres, O. 2013. Efficacy of a physical method for control of direct pests of apples and peaches. *Journal of Applied Entomology* 137(10), 790–800. doi:10.1111/jen.12064.

Grechi, I., Sauge, M., Sauphanor, B., Hilgert, N., Senoussi, R. and Lescourret, F. 2008. How does winter pruning affect peach tree–*Myzus persicae* interactions? *Entomologia Experimentalis et Applicata* 128(3), 369–79. doi:10.1111/j.1570-7458.2008.00720.x.

Gut, L. J., McGhee, P. H. and Perry, R. 2005. Soil mounding as a control for dogwood borer in apple. *HortScience* 40(7), 2066–70. doi:10.21273/HORTSCI.40.7.2066.

Habekuss, A., Proesler, G., Schliephake, E. and Fischer, C. 2000. Resistance of apple to spider mites and aphids. *Acta Horticulturae* 538(538), 271–6. doi:10.17660/ActaHortic.2000.538.47.

Haltrich, A., Papp, J., Fail, J. and Kis, L. 2000. Effect of nitrogen-fertilizers and apple cultivars on aphids under IPM treatment conditions. *Acta Horticulturae* 525(525), 209–16. doi:10.17660/ActaHortic.2000.525.25.

Hansen, J. D., Heidt, M. L. and Anderson, P. A. 2006. Bin sterilization to prevent reintroduction of codling moth. *Journal of Agricultural and Urban Entomology* 23(1), 17–26.

Herbert, H. J. 1974. Notes on the biology of the apple rust mite, *Aculus schlechtendali* (Prostigmata: Eriophyoidae), and its density on several cultivars of apple in Nova Scotia. *The Canadian Entomologist* 106(10), 1035–8. doi:10.4039/Ent1061035-10.

Higbee, B. S., Calkins, C. O. and Temple, C. A. 2001. Overwintering of codling moth (Lepidoptera: Tortricidae) larvae in apple harvest bins and subsequent moth emergence. *Journal of Economic Entomology* 94(6), 1511–7. doi:10.1603/0022-0493-94.6.1511.

Hogmire, H. W. and Miller, S. S. 2005. Relative susceptibility of new apple cultivars to arthropod pests. *HortScience* 40(7), 2071–5. doi:10.21273/HORTSCI.40.7.2071.

Holb, I. J., Gonda, I. and Bitskey, K. 2001. Pruning and incidences of diseases and pests in environmentally oriented apple growing systems: some aspects. *International Journal of Horticultural Science* 7(1), 24–9.

Horton, D. R., Broers, D. A., Lewis, R. R., Granatstein, D., Zack, R. S., Unruh, T. R., Moldenke, A. R. and Brown, J. J. 2003. Effects of mowing frequency on densities of natural enemies in three Pacific Northwest pear orchards. *Entomologia Experimentalis et Applicata* 106(2), 135–45. doi:10.1046/j.1570-7458.2003.00018.x.

Hunter, M. D., Hull, L. A. and Schultz, J. C. 1994. Evaluation of resistance to tufted apple bud moth (Lepidoptera: Tortricidae) within and among apple cultivars. *Environmental Entomology* 23(2), 282–91. doi:10.1093/ee/23.2.282.

Juste, F., Sánchez, S., Ibañez, R., Val, L. and García, C. 1990. Measurement of spray deposition and efficiency of pesticide application in citrus orchards. *Journal of Agricultural Engineering Research* 46, 187–96.

Killian, J. C. and Meyer, J. R. 1984. Effect of orchard weed management on catfacing damage to peaches in North Carolina. *Journal of Economic Entomology* 77(6), 1596–600. doi:10.1093/jee/77.6.1596.

Kost, T. D., Gessler, C., Jänsch, M., Flachowsky, H., Patocchi, A. and Broggini, G. A. 2015. Development of the first cisgenic apple with increased resistance to fire blight. *PLOS ONE* 10(12), e0143980. doi:10.1371/journal.pone.0143980.

Krens, F. A., Schaart, J. G., van der Burgh, A. M., Tinnenbroek-Capel, I. E., Groenwold, R., Kodde, L. P., Broggini, G. A., Gessler, C. and Schouten, H. J. 2015. Cisgenic apple trees; development, characterization, and performance. *Frontiers in Plant Science* 6, 286. doi:10.3389/fpls.2015.00286.

Kührt, U., Samietz, J. and Dorn, S. 2006. Effect of plant architecture and hail nets on temperature of codling moth habitats in apple orchards. *Entomologia Experimentalis et Applicata* 118(3), 245–59. doi:10.1111/j.1570-7458.2006.00385.x.

Lafleur, G., Hill, S. B. and Vincent, C. 1987. Fall migration, hibernation site selection, and associated winter mortality of plum curculio (Coleoptera: Curculionidae) in a Quebec apple orchard. *Journal of Economic Entomology* 80(6), 1152–72. doi:10.1093/jee/80.6.1152.

Lang, G. A. 2009. High tunnel tree fruit production: the final frontier? *HortTechnology* 19(1), 50–5. doi:10.21273/HORTSCI.19.1.50.

Leach, H., Van Timmeren, S. and Isaacs, R. 2016. Exclusion netting delays and reduces *Drosophila suzukii* (Diptera: Drosophilidae) infestation in raspberries. *Journal of Economic Entomology* 109(5), 2151–8. doi:10.1093/jee/tow157.

Leach, H., Moses, J., Hanson, E., Fanning, P. and Isaacs, R. 2018. Rapid harvest schedules and fruit removal as non-chemical approaches for managing spotted wing Drosophila. *Journal of Pest Science* 91(1), 219–26. doi:10.1007/s10340-017-0873-9.

Lombarkia, N. and Derridj, S. 2008. Resistance of apple trees to *Cydia pomonella* egg-laying due to leaf surface metabolites. *Entomologia Experimentalis et Applicata* 128, 57–65.

Lucht, J. M. 2015. Public acceptance of plant biotechnology and GM crops. *Viruses* 7(8), 4254–81. doi:10.3390/v7082819.

Markó, V., Jenser, G., Kondorosy, E., Ábrahám, L. and Balázs, K. 2013. Flowers for better pest control? The effects of apple orchard ground cover management on green apple aphids (*Aphis* spp.) (Hemiptera: Aphididae), their predators and the canopy insect community. *Biocontrol Science and Technology* 23(2), 126–45. doi:10.1080/09583157.2012.743972.

McMullen, R. D. and Jong, C. 1972. Influence of temperature and host vigor on fecundity of the pear psylla (Homoptera: Psyllidae). *The Canadian Entomologist* 104(8), 1209–12. doi:10.4039/Ent1041209-8.

Merwin, I. 2003. Orchard-floor management systems. In: Ferree, D. and Warrington, I. (Eds), *Apple: Botany, Production and Uses*. CABI Publishing, Wallingford – Oxford, UK.

Messina, F. J. and Jones, V. P. 1990. Relationship between fruit phenology and infestation by the apple maggot (Diptera: Tephritidae) in Utah. *Annals of the Entomological Society of America* 83(4), 742–52. doi:10.1093/aesa/83.4.742.

Myers, C. T., Leskey, T. C. and Forsline, P. L. 2007. Susceptibility of fruit from diverse apple and crabapple germplasm to attack by plum curculio (Coleoptera: Curculionidae). *Journal of Economic Entomology* 100(5), 1663–71. doi:10.1093/jee/100.5.1663.

Myers, C. T., Harvey Reissig, W. H. and Forsline, P. L. 2008. Susceptibility of fruit from diverse apple and crabapple germplasm to attack from apple maggot (Diptera: Tephritidae). *Journal of Economic Entomology* 101(1), 206–15. doi:10.1093/jee/101.1.206.

Naranjo, S. E. 2001. Conservation and evaluation of natural enemies in IPM systems for *Bemisia tabaci*. *Crop Protection* 20(9), 835–52. doi:10.1016/S0261-2194(01)00115-6.

Nin, S., Ferri, A., Sacchetti, P. and Giordani, E. 2012. Pear resistance to Psilla (*Cacopsylla pyri* L.). A review. *Advances in Horticultural Science* 26(2), 59–74.

Noble, R., Dobrovin-Pennington, A., Shaw, B., Buss, D. S., Cross, J. V. and Fountain, M. T. 2017. Fermentation for disinfesting fruit waste from Drosophila species (Diptera: Drosophilidae). *Environmental Entomology* 46(4), 939–45. doi:10.1093/ee/nvx107.

Norris, R. F., Chen, C. and Kogan, M. 2003. Concepts in Integrated Pest Management.

Pajač Živković, I., Jemrić, T., Fruk, M., Buhin, J. and Barić, B. 2016. Influence of different netting structures on codling moth and apple fruit damages in northwest Croatia. *Agriculturae Conspectus Scientificus* 81(2), 99–102.

Papp, J., Jenser, G. and Haltrich, A. 2001. Effect of nitrogen supply on the population of European red spider mite and green apple aphid in an IPM apple orchard. *Acta Horticulturae* 564(564), 407–12. doi:10.17660/ActaHortic.2001.564.50.

Pfeiffer, D. G. and Burts, E. C. 1983. Effect of tree fertilization on numbers and development of pear psylla (Homoptera: Psyllidae) and on fruit damage. *Environmental Entomology* 12(3), 895–901. doi:10.1093/ee/12.3.895.

Pfeiffer, D. G. and Burts, E. C. 1984. Effect of tree fertilization on protein and free amino acid content and feeding rate of pear psylla (Homoptera: Psyllidae). *Environmental Entomology* 13(6), 1487–90. doi:10.1093/ee/13.6.1487.

Pree, D. J. 1977. Resistance to development of larvae of the apple maggot in crab apples. *Journal of Economic Entomology* 70(5), 611–4. doi:10.1093/jee/70.5.611.

Prokopy, R. J. 2003. Two decades of bottom-up, ecologically based pest management in a small commercial apple orchard in Massachusetts. *Agriculture, Ecosystems and Environment* 94(3), 299–309. doi:10.1016/S0167-8809(02)00036-1.

Quaintance, A. L. and Jenne, E. L. 1912. The Plum Curculio. United States Department of Agriculture, Bureau of Entomology.

R Core Team. 2016. *R: A Language and Environment for Statistical Computing*. R Foundation for Statistical Computing, Vienna.

Randall, P., Sholberg, P., Judd, G. and Cossentine, J. 2011. Acetic acid fumigation of fruit storage bins to control diapausing codling moth larvae. *HortScience* 46(12), 1634–9. doi:10.21273/HORTSCI.46.12.1634.

Roda, A., Nyrop, J., English-Loeb, G. and Dicke, M. 2001. Leaf pubescence and two-spotted spider mite webbing influence phytoseiid behavior and population density. *Oecologia* 129, 551–60.

Rodriguez, J. G. 1958. The Yarative NPK Y of *Panonychus ulmi* (Koch) and *Tetranychus telarius* (L.) on apple trees. *Journal of Economic Entomology* 51(3), 369–73. doi:10.1093/jee/51.3.369.

Sanchis, V. 2011. From microbial sprays to insect-resistant transgenic plants: history of the biospesticide *Bacillus thuringiensis*. A review. *Agronomy for Sustainable Development* 31(1), 217–31. doi:10.1051/agro/2010027.

Sandanayaka, W. R. M., Bus, V. G. M., Connolly, P. and Newcomb, R. 2003. Characteristics associated with Woolly Apple Aphid *Eriosoma lanigerum*, resistance of three apple rootstocks. *Entomologia Experimentalis et Applicata* 109(1), 63–72. doi:10.1046/j.1570-7458.2003.00095.x.

Sandanayaka, W. R. M., Bus, V. G. M. and Connolly, P. 2005. Mechanisms of woolly aphid [*Eriosoma lanigerum* (Hausm.)] resistance in apple. *Journal of Applied Entomology* 129(9–10), 534–41. doi:10.1111/j.1439-0418.2005.01004.x.

Sauge, M.-H., Kervella, J. and Pascal, T. 1998a. Settling behaviour and reproductive potential of the green peach aphid *Myzus persicae* on peach varieties and a related wild *Prunus*. *Entomologia Experimentalis et Applicata* 89(3), 233–42. doi:10.1046/j.1570-7458.1998.00404.x.

Sauge, M.-H., Kervella, J. and Rahbé, Y. 1998b. Probing behaviour of the green peach aphid *Myzus persicae* on resistant *Prunus* genotypes. *Entomologia Experimentalis et Applicata* 89(3), 223–32. doi:10.1046/j.1570-7458.1998.00403.x.

Sauge, M.-H., Lacroze, J., Poessel, J., Pascal, T. and Kervella, J. 2002. Induced resistance by *Myzus persicae* in the peach cultivar 'Rubira'. *Entomologia Experimentalis et Applicata* 102(1), 29–37. doi:10.1046/j.1570-7458.2002.00922.x.

Sauge, M.-H., Mus, F., Lacroze, J., Pascal, T., Kervella, J. and Poëssel, J. 2006. Genotypic variation in induced resistance and induced susceptibility in the peach-*Myzus persicae* aphid system. *Oikos* 113(2), 305–13. doi:10.1111/j.2006.0030-1299.14250.x.

Sauphanor, B., Severac, G., Maugin, S., Toubon, J. F. and Capowiez, Y. 2012. Exclusion netting may alter reproduction of the codling moth (*Cydia pomonella*) and prevent associated fruit damage to apple orchards. *Entomologia Experimentalis et Applicata* 145(2), 134–42. doi:10.1111/j.1570-7458.2012.01320.x.

Schmid, A., Weibel, F., Allemannn, P. and Santini, D. 2004. Sandwich-system. *Erwerbs-Obstbau* 46(3), 87–94. doi:10.1007/s10341-004-0034-1.

Schmidt, R. A. 2014. Leaf structures affect predatory mites (Acari: Phytoseiidae) and biological control: a review. *Experimental and Applied Acarology* 62(1), 1–17. doi:10.1007/s10493-013-9730-6.

Schmidt-Jeffris, R. A. and Beers, E. H. 2017. Effects of foliar apple trichomes on *Galendromus occidentalis* (Nesbitt) (Acari: Phytoseiidae): preferences, fecundity and prey consumption. *Agricultural and Forest Entomology* 19(1), 93–100. doi:10.1111/afe.12184.

Schwarz, J. J. and Brunner, J. F. 2017. Excised leaf bioassay: phenotypic expression of natural resistance in apple against the obliquebanded leafroller. *Advances in Entomology* 05(1), 1–25. doi:10.4236/ae.2017.51001.

Siegwart, M., Pierrot, M., Toubon, J.-F., Maugin, S. and Lavigne, C. 2013. Adaptation to exclusion netting of the codling moth (*Cydia pomonella* L.) in apple orchards. *IOBC-WPRS Bulletin* 91, 127–31.

Simon, S., Sauphanor, B. and Lauri, P.-E. 2007. Control of fruit tree pests through manipulation of tree architecture. *Pest Technology* 1(1), 33–7.

Simon, S., Bouvier, J., Debras, J. and Sauphanor, B. 2010. Biodiversity and pest management in orchard systems. A review. *Agronomy for Sustainable Development* 30(1), 139–52. doi:10.1051/agro/2009013.

Simon, S., Morel, K., Durand, E., Brevalle, G., Girard, T. and Lauri, P. 2012a. Aphids at crossroads: when branch architecture alters aphid infestation patterns in the apple tree. *Trees* 26(1), 273–82. doi:10.1007/s00468-011-0629-8.

Simon, S., Morel, K., Defrance, H., Hemptinne, J. L., Regnard, J. L. and Lauri, P. E. 2012b. Development of the rosy apple aphid within its habitat: some structural and physiological aspects in apple trees. *Acta Horticulturae* 933(933), 491–8. doi:10.17660/ActaHortic.2012.933.64.

Smith, E. H. and Flessel, J. K. 1968. Hibernation of the plum curculio and its spring migration to host trees. *Journal of Economic Entomology* 61(1), 193–203. doi:10.1093/jee/61.1.193.

Smith, M. W., Arnold, D. C., Eikenbary, R. D., Rice, N. R., Shiferaw, A., Cheary, B. S. and Carroll, B. L. 1996. Influence of ground cover on beneficial arthropods in pecan. *Biological Control* 6(2), 164–76. doi:10.1006/bcon.1996.0021.

Snapp, O. I. 1930. Life history and habits of the plum curculio in the Georgia peach belt. *USDA Technical Bulletin* 188.

Spieser, F., Graf, B., Walther, P. and Noesberger, J. 1998. Impact of apple rust mite (Acari: Eriophyiidae) feeding on apple leaf gas exchange and leaf color associated with changes in leaf tissue. *Environmental Entomology* 27(5), 1149–56. doi:10.1093/ee/27.5.1149.

Staniland, L. N. 1923. The immunity of apple stocks from attacks of woolly aphis. Part I. the relative resistance of different root stocks. *Journal of Pomology and Horticultural Science* 3(2), 85–95. doi:10.1080/03683621.1923.11513253.

Staniland, L. N. 1924. The immunity of apple stocks from attacks of woolly aphis (*Eriosoma lanigerum*, Hausmann). Part II. The causes of the relative resistance of the stocks. *Bulletin of Entomological Research* 15(2), 157–70.

Staudt, M., Jackson, B., El-Aouni, H., Buatois, B., Lacroze, J. P., Poëssel, J. L. and Sauge, M. H. 2010. Volatile organic compound emissions induced by the aphid *Myzus persicae* differ among resistant and susceptible peach cultivars and a wild relative. *Tree Physiology* 30(10), 1320–34. doi:10.1093/treephys/tpq072.

Stefanelli, D., Zoppolo, R. J., Perry, R. L. and Weibel, F. 2009. Organic orchard floor management systems for apple effect on rootstock performance in the Midwestern United States. *HortScience* 44(2), 263–7. doi:10.21273/HORTSCI.44.2.263.

Stoeckli, S., Mody, K. and Dorn, S. 2008a. *Aphis pomi* (Hemiptera: Aphididae) population development, shoot characteristics, and antibiosis resistance in different apple genotypes. *Journal of Economic Entomology* 101(4), 1341–8. doi:10.1093/jee/101.4.1341.

Stoeckli, S., Mody, K. and Dorn, S. 2008b. Influence of canopy aspect and height on codling moth (Lepidoptera: Tortricidae) larval infestation in apple, and relationship between infestation and fruit size. *Journal of Economic Entomology* 101(1), 81–9. doi:10.1093/jee/101.1.81.

Stoeckli, S., Mody, K., Gessler, C., Christen, D. and Dorn, S. 2009a. Quantitative trait locus mapping of resistance in apple to *Cydia pomonella* and *Lyonetia clerkella* and of two selected fruit traits. *Annals of Applied Biology* 154(3), 377–87. doi:10.1111/j.1744-7348.2008.00307.x.

Stoeckli, S., Mody, K., Patocchi, A., Kellerhals, M. and Dorn, S. 2009b. Rust mite resistance in apple assessed by quantitative trait loci analysis. *Tree Genetics and Genomes* 5(1), 257–67. doi:10.1007/s11295-008-0186-5.

Tedders, W. L. 1983. Insect management in deciduous orchard ecosystems: habitat manipulation. *Environmental Management* 7(1), 29–34. doi:10.1007/BF01867038.

Ughini, V., Malvicini, G. L., Pisaroni, F., Plessi, C. and Caruso, S. 2010. Trials on the use of nets in the Vignola cherry district against cherry fruit fly (*Rhagoletis cerasi* L.). *Acta Horticulturae* 873(873), 337–42. doi:10.17660/ActaHortic.2010.873.40.

Underhill, G. W. and Cox, J. A. 1938. Studies on the resistance of apple to the woolly apple aphid, *Eriosoma lanigerum* (Hausm.). *Journal of Economic Entomology* 31(5), 622–5. doi:10.1093/jee/31.5.622.

Vanblaere, T., Flachowsky, H., Gessler, C. and Broggini, G. A. 2014. Molecular characterization of cisgenic lines of apple 'Gala' carrying the *Rvi6* resistance gene. *Plant Biotechnology Journal* 12(1), 2–9. doi:10.1111/pbi.12110.

Walsh, B. and Riley, C. 1868. *The American Entomologist: an Illustrated Magazine of Popular and Practical Entomology*. R. P. Studley & Company, St. Louis, MO.

Waltz, E. 2015. Nonbrowning GM apple cleared for market. *Nature Biotechnology* 33(4), 326–7. doi:10.1038/nbt0415-326c.

Wearing, C. H., Colhoun, K., McLaren, G. F., Attfield, B. and Bus, V. G. M. 2003. Evidence for single gene resistance in apple to brownheaded leafroller, *Ctenopseustis obliquana*, and implications for resistance to other New Zealand leafrollers. *Entomologia Experimentalis et Applicata* 108(1), 1–10. doi:10.1046/j.1570-7458.2003.00062.x.

Weibel, F. 2002. Soil management and in-row weed control in organic apple production. *Compact Fruit Tree* 35(4), 118–21.

Wermelinger, B., Oertli, J. J. and Delucchi, V. 1985. Effect of host plant nitrogen fertilization on the biology of the two-spotted spider mite, *Tetranychus urticae*. *Entomologia Experimentalis et Applicata* 38(1), 23–8. doi:10.1111/j.1570-7458.1985.tb03493.x.

Wiman, M. R., Kirby, E. M., Granatstein, D. M. and Sullivan, T. P. 2009. Cover crops influence meadow vole presence in organic orchards. *HortTechnology* 19(3), 558–62. doi:10.21273/HORTSCI.19.3.558.

Zoppolo, R. J., Stefanelli, D., Bird, G. W. and Perry, R. L. 2011. Soil properties under different orchard floor management systems for organic apple production. *Organic Agriculture* 1(4), 231–46. doi:10.1007/s13165-011-0018-z.

Chapter 20

Improving monitoring and forecasting in integrated management of fruit arthropod pests

Tim Belien, Proefcentrum Fruitteelt VZW, Belgium; Slawomir Lux, inSilico-IPM, Poland; Bart De Ketelaere, Katholieke Universiteit Leuven, Belgium; and Dany Bylemans, Proefcentrum Fruitteelt VZW, Belgium

1 Introduction

The efficient and accurate monitoring of insects and mites is a cornerstone of integrated pest management (IPM) in orchards (Baggiolini et al., 1971, 2001; Binns and Nyrop, 1992; Nyrop et al., 1999). Having a good idea of the status of pest populations (and, ideally, also of their natural enemies) throughout the season forms the necessary basis for extension people (agronomists) and fruit growers to choose between management options. In modern IPM the process of decision-making often relies on three components: (1) monitoring the presence of relevant insects and mites, (2) linking monitoring data to actual pest densities and economic thresholds and (3) model-based forecasting. Since the start of the new millennium the latter component has increasingly grown in importance (Binns et al., 2000; Damos et al., 2015). Although accurate pest forecasting and population dynamics predictions by phenology models can depend on monitoring data, for example for determining starting dates or biofix (indicator of pest developmental event) to synchronize the model and field populations (Knutson and Muegge, 2010), models can also be helpful to determine the appropriate time and localization to execute field observations and collect monitoring data (Binns et al., 2000; Lux et al., 2016). Furthermore,

http://dx.doi.org/10.19103/AS.2019.0046.25

due to ever-changing weather conditions and integration of information from multiple sources together with complexities associated with the decision process in IPM, intelligent model-based systems are also starting to play a role in optimizing management options and, hence, represent a fourth element in the decision-making process (Padma et al., 2017).

In the first part of this chapter key issues and challenges regarding the first two basic components (monitoring methods and economic thresholds) are described. A key tool for insect pest monitoring in tree fruit is monitoring traps baited with sex or aggregation pheromones (Witzgall et al., 2010; Gross and Gündermann, 2016). However, the widely applied use of mating disruption has complicated monitoring using these classic sex pheromone-baited traps. This is mainly being addressed by the development of alternative (bisexual/food-based) baits (Knight, 2010; Padilha et al., 2018). Another challenge involves the labour-intensive time taken to process trap samples and monitoring devices and the requirement of specific entomological expertise to identify pest species (Potamitis et al., 2015). New trends and advances in automated monitoring technologies may help resolve these hurdles (Potamitis et al., 2017a,b). In the second part of this chapter current developments in predictive models for pest forecasting are discussed.

2 Manual monitoring techniques and economic thresholds

Numerous techniques have been developed for monitoring pests in orchard crops (Beers et al., 1994). The use of monitoring traps is probably the most known and convenient method for the detection of the presence (or absence) and abundance of insects pest species (Witzgall et al., 2010). However, monitoring traps are only effective if (1) the target (life stage of the) pest is sufficiently mobile in order to move to the trap and (2) an effective and selective attractant is known and available. If one of both requirements is not fulfilled, other, less passive, monitoring methods are used (Gross and Gündermann, 2016).

2.1 Trap-independent monitoring techniques

Currently several techniques exist for monitoring relatively immobile (life stages of) insects and mites in fruit orchards. However, little or no improvements in terms of efficiency and accuracy have been achieved over the past decades with these rather labour-intensive trap-independent methods. Standard, regular, visual inspections in different parts of the trees (trunk, branches, buds, flowers, shoots, leaves, fruits and even roots) remain essential to follow up the presence and development of a variety of pests. In many cases, the first symptom is not the detection of a pest itself but, often, early damage (Alford, 2007). For visual inspections to identify many small arthropods or their eggs or larvae, a hand lens (in the range from 10× to 30× magnification) is an essential tool (Augustin et al., 2012; Dreistadt and Flint, 1995). Specific sampling methods are required for assessing particular life stages of pests. It is recommended to execute inspections on regular timings (using a 1–3-week interval) during the period in which target pests are expected to become active (or to start migrating into orchards). Such a monitoring event of a common orchard of 1–5 ha involves the checking of 100–600 randomly chosen plant parts spread over the orchard. The exact plant part (leaf, flower, bud etc.) and number of checks depends on the target pest and the economic damage threshold value. For low threshold values (≤1% damage level) it is recommended to check 500–600 plant parts, for intermediate threshold values (1–5%) 300 plant parts and for high threshold values (≥5%) checking of 100 plant parts should be sufficient (Baggiolini et al., 2001). The advised spreading of sampling/checking spots within the orchard also depends on the nature of the target pest. For sedentary pests (spending their whole lifetime in the orchard, not migrating) one should follow a predetermined sampling route (e.g. zigzag or diagonally W-shaped route) leading to monitoring spots randomly distributed over the whole orchard. Doing this, one should be aware of (less mobile) pests that typically start in 'hot spots' from which they increase their populations and colonize the rest of the orchard (e.g. spider mites and scales). For non-sedentary migrating pests increased attention should be focussed on the borders and edges of orchards (where in-flights/immigration occurs, e.g. stink bugs) (Beers et al., 1994; Binns et al., 2000; Baggiolini et al., 2001).

One of the most prevalently used trap-independent active monitoring techniques is the beating tray method (Hutchins, 1994). Limb beating is a conventional sampling method and has several variations but at its most basic the idea is to beat a branch with a stick over a collection device (e.g. tray, cloth, polyethylene sheet) to collect arthropods that drop from the tree. This is a simple and efficient method, but limited to the monitoring of relatively immobile species or stages of arthropods, for example Lepidoptera larvae (caterpillars), larvae and adults of Coleoptera, such as blossom weevils, and Hemiptera, such as stink bugs, myrid bugs, adult psyllids (Beers et al., 1994; Hutchins, 1994). As a general rule, one beating tray monitoring event involves beatings of 100 different branches spread over the orchard (Baggiolini et al., 2001). It is recommended to take beating samples in the morning, as later on with increasing temperatures and insect activity, most of the adults quickly fly or jump away (Deronzier and Atger, 1980; Horton, 1994). Over the years several studies have compared beating tray data with other monitoring methods for correlation of actual pest densities. For instance, for pear psyllids (*Cacopsylla pyri* and *C. pyricola*), good correlations were obtained between beating tray data and absolute density estimation methods (Horton, 1994; Jenser et al.,

2010). In the latter study, a wash-down method was used in which the number of psylla larvae in samples was exactly quantified by extracting them from flowers or shoot samples in a plastic bag containing a 1% detergent-water solution, and subsequently rinsing the solution through filter paper and counting the larvae (Jenser et al., 2010). This method allows exact quantification and, unlike the beating tray method, the outcomes are not affected by weather conditions. However, the latter method is too time-consuming for use by growers in practice. In a comparative study with sticky traps pear psylla the beating methods was comparable for predicting pest damage, and thus pest-management decisions (Horton and Lewis, 1997), although for more precise quantification of absolute densities of adult psylla the authors suggested that beat trays were preferred. Beating numbers of adult pear psylla were also found to correlate with observed numbers of psylla nymphs on shoots, confirming the usefulness of the beating method for population density estimations of this pest (Sanchez and Ortin-Angulo, 2011). Conversely, for monitoring Western flower thrips in nectarine orchards beating of branches and infield visual (bud) inspections were insufficiently accurate. Using sticky cards gave an indication of population trends, but were less efficient than counting thrips in nectarine buds (Pearsall and Myers, 2000).

Other active trap-independent monitoring techniques involve the use of sweep nets or suction devices (Hutchins, 1994); using these methods more mobile insects can be sampled. Despite efforts to standardize these techniques for specific pest monitoring, outcomes remain rather variable due to the non-focussed way of sampling and quite large differences from one user to the next (Pedigo and Buntin, 1994). As a consequence they are not commonly used for IPM decision-making in tree fruit, with the exception of mobile pests that also frequently occur in ground cover such as *Lygus* and stink bugs (Binns et al., 2000; Leskey and Hogmire, 2005; Ohlendorf, 1999; Pedigo and Buntin, 1994).

An effective way to monitor small arthropods in hidden cryptic habitats (like soil, leaf litter, bark of flower cluster samples etc.) is the Berlese funnel (Agnello et al., 2003). A Berlese funnel consists of a funnel with a wire mesh insert that supports the sample material (plant or soil sample) above a collection reservoir containing ethanol or another preservative fluid. Usually, a light bulb (providing light and heat) is used to encourage the arthropods move downwards and eventually through the funnel into the collection container. Subsequently they can be identified and counted, if needed with the aid of a magnifying glass or microscope. Examples of arthropods that can be monitored with the Berlese funnel include mites (pests and predatory mites) and thrips (Agnello et al., 2003; Kumral et al., 2010).

With very small arthropods inhabiting leaves (e.g. mites) a leaf-brushing machine is also an effective tool to enforce accuracy and standardizing of monitoring (Beers et al., 1994; Macmillan and Costello, 2015). This small machine consists of two revolving spiral brushes that dislodge the arthropods from the leaves onto a (often sticky) collection plate (Hutchins, 1994). However, in practice, this procedure involving manual counts is generally still considered as too laborious and time-consuming for commercial IPM programmes. Recent progress in automatic counts using image analysis (Miranda et al., 2014) is a promising new area of development (see Section 3). Besides a leaf-brushing machine, several washing techniques are also described for the efficient isolation of small arthropods from leaves, shoots or buds, for example for mites (Zacharda et al., 1988; Schreiner et al., 2014) or pear psylla larvae (Jenser et al., 2010). Ethanol extraction consistently yielded the highest numbers of mites and was the most rapid method for recovering the phytophagous mite, *Tetranychus urticae*, from apple leaf samples compared to leaf brushing (Harris et al., 2017).

2.2 Trap-dependent monitoring techniques

The most convenient and widely used monitoring method is based on the use of insect traps (Gross and Gündermann, 2016; Suckling, 2000). The traps are designed in such a way that insects are attracted to enter the trap and once inside they are unable to leave. Non-selective traps, like black light traps, are not used in practice, because identifying the target pest in the complex mixture of multiple captured insects is very time-consuming (Beers et al., 1994). In order to be useful for monitoring in fruit crops a trap should display at least some (and preferably a high) degree of selectivity for the target pest. Various trap features can confer selectivity to a monitoring trap; like positioning and timing of the trap placement, the colour, shape, design, mode of capture and, of course, the bait or lure used. For instance, sticky tape traps placed around branches can be used to predict when scale crawlers are migrating (Coccoidea, Diaspididae) (Dreistadt, 2012). Coloured sticky plates represent quite effective monitoring traps for thrips (Thysanoptera) in orchards, but with limited selectivity (Broughton and Harrison, 2012). Well-known examples of widely used traps in orchards that are somewhat selective due to colour/reflectance and timing of placement are yellow sticky traps for monitoring of cherry fruit flies (*Rhagoletis* sp., Tephritidae) (orienting on ripening (yellowing) cherry fruits) (Daniel and Grunder, 2012), and white sticky traps for monitoring of sawflies (*Hoplocampa* sp., Tenthredinidae) (orienting on white flowers) in pome and plum fruit orchards (Owens and Prokopy, 1978). These traps are typically cross-shaped indicating the importance of the shape making them visible from all directions (Daniel et al., 2014). The shape of the trap can also enhance the visual stimulus by mimicking the form of the host fruits, for example spherical traps for monitoring of various maggot flies (Tephritidae) (Prokopy and Hauschild, 1979; Liburd et al., 1998; Mayer et al., 2000). The influence of the mode of capture on trap efficacy and selectivity can be illustrated by the impact of the size and shape of the entrance holes, for example *Drosophila suzukii* traps (Lee et al., 2013; Renkema et al., 2014). However, usually chemical cues originating from baits/lures that are combined with the trap are the most decisive parameter for the selectivity/specificity of a monitoring trap.

Significant advances in the use of insect traps for monitoring were reached following the identification and commercialization of pheromone lures (Witzgall et al., 2010). To date, numerous pheromone compounds have been identified for use in pest monitoring, mostly in Lepidoptera, but also for Coleoptera and Diptera (El-Sayed, 2018). Sex pheromones are the most common semiochemicals employed, but other semiochemicals increasingly find their way to application, for instance aggregation pheromones in the monitoring of noxious beetles (Hanks and Millar, 2016), thrips (Hamilton et al., 2005) and stink bugs (Pentatomidae) (Acebes-Doria et al., 2018). Pheromone traps are highly selective with sex pheromones often exclusively attracting a single species, but commonly several related species (Bell, 2014). However, most sex pheromone-based monitoring exclusively attracts male pest insects and the presence of females in the orchard remains undetected. This is important because the females are often the damaging sex due to their oviposition activity (Landolt, 1997). Moreover, the wide adoption of mating disruption techniques in orchards has complicated the use of sex pheromone traps for monitoring.

2.2.1 Effective monitoring traps in orchards under mating disruption

The mating disruption technique consists of releasing a high amount of synthetic sex pheromone into an orchard or landscape, so males are not able to orient towards the natural sex pheromone plume being released by the females, preventing them finding a mate and subsequent egg laying and larvae feeding damage (Witzgall et al., 2010). However, this also makes sex pheromone-baited monitoring traps difficult to locate by males (Knight, 2007).

In the past decades, several solutions have been developed to address the challenge of accurate and reliable monitoring in sex pheromone-treated orchards. The first solution is the use of monitoring traps baited with a high-load sex pheromone lure (Charmillot, 1990). Further progress was made by improvements in trap design (longer lasting) emission rates and isomeric purity of pheromone lures (Knight, 2002; Knight et al., 2002) including better trap placement strategies (Knight, 2007). Substantial advances in monitoring pests in mating disrupted orchards were achieved with traps and lures that are effective in also attracting females. In the case of codling moth (*Cydia pomonella*) combinational lures were developed consisting of sex pheromone (codlemone) and kairomones (interspecific cues acting between species) for example ethyl (*E,Z*)-2,4-decadienoate (pear ester) or (*E*)-4,8-dimethyl-1,3,7-nonatriene

(nonatriene) (Light et al., 2001; Light and Knight, 2005; Knight and Light, 2012). In recent years monitoring lures were further improved by the addition of acetic acid and host plant volatiles to further enhance bisexual monitoring of tortricid pests in mating-disrupted orchards (El-Sayed et al., 2013; Knight et al., 2014, 2015).

2.2.2 Monitoring of invasive species

The recent intensified global trade in fruit and fruit plants in addition to climate change has led to increasingly serious problems with alien invasive species threatening local agricultural crops (Chapman et al., 2017; Barzman et al., 2015). The monitoring of these pests is often a challenge because of a general lack of knowledge about their behaviour and ecology in the invaded range (Lockwood et al., 2013). The spotted wing Drosophila (*Drosophila suzukii*) (Asplen et al., 2015) and the brown marmorated stink bug (*Halyomorpha halys*) (Haye et al., 2015), both originating from Asia and rapidly spreading over different continents in the world, are the most notorious examples in fruit orchards. As a consequence, many studies have been executed by the monitoring of both fruit pests. Nowadays, monitoring traps play an important role in the interception and gauging the dispersal of both invasive pests.

Many trap designs and baits have been developed and tested for the invasive fruit fly *D. suzukii* (Lee et al., 2012, 2013; Renkema et al., 2014; Iglesias et al., 2014). Despite variability in these studies due, in part, to crop type, trap position in the crop, climatological and environmental conditions, in general red-coloured traps were found to be consistently the most effective ones (Kirkpatrick et al., 2016; Rice et al., 2016, 2017). The use of home-made apple cider vinegar and yeast-based baited traps has been superseded by commercial proprietary liquid baits and now, dry, synthetic lures (Burrack et al., 2015; Cha et al., 2018; Cloonan et al., 2018; Jaffe et al., 2018). Furthermore, important insights were generated on the physiology and occurrence of two morph phenotypes (winter and summer morphs) (Shearer et al., 2016; Fountain et al., 2018) and the impact on behavioural responses of *D. suzukii* to attractive odours, and hence also on monitoring traps (Belien et al., 2016; Kirkpatrick et al., 2018b; Wong et al., 2018). However, to date (November 2018), *D. suzukii*-monitoring traps and their baits and lures still need improvement as they are not sufficiently selective, capturing too many non-target organisms and compromising the uptake of monitoring (Kirkpatrick et al., 2018b; Cloonan et al., 2018).

Numerous recent research efforts on *H. halys* have led to significant progress on the monitoring of this invasive stink bug. An important milestone was the identification of the male-produced two-component aggregation pheromone with the components (3S,6S,7R,10S)-10,11-epoxy-1-bisabolen-3 -ol and (3R,6S,7R,10S)-10,11-epoxy-1-bisabolen 3-ol (Khrimian et al., 2014).

Currently, this attractive aggregation pheromone mixture and pheromone synergist methyl (2E,4E,6Z)-decatrienoate (Weber et al., 2014, 2017) are used in most monitoring efforts. Combined with visually attractive black coroplast pyramid traps (Leskey et al., 2012), these olfactory stimuli enable the season-long detection and monitoring of *H. halys* adults and nymphs (Weber et al., 2014; Leskey et al., 2015a; Morrison et al., 2015). The phenology of *H. halys* in a wide range of crops and wild host habitats has been determined (Leskey et al., 2012, 2015a,b; Morrison et al., 2015, 2017; Akotsen-Mensah et al., 2018). Recently, the design, lure deployment location (placing lures on the outside of collection jars on pyramid traps) and retention mechanism (using insecticide-impregnated netting) of these monitoring traps were further improved (Rice et al., 2018b). Furthermore ethyl (*2E*,*4E*,*6Z*)-decatrienoate was found to enhance the attraction of the monitoring traps (Rice et al., 2018a). Interestingly, the citizen science initiative 'BugMap' was conceived to complement and assist researchers in monitoring the distribution and seasonality of *H. halys* in Italy (Malek et al., 2018). Via a user-friendly, freely available mobile application lots of volunteer-collected data were coupled to traditional monitoring efforts of researchers, unravelling the invasion characteristics of *H. halys*. Future research should focus on interpreting catches in traps for management decisions, which may be clarified through studies on placement of traps in fruit orchards and comparisons on the behavioural response of *H. halys* in different fruit crops (Akotsen-Mensah et al., 2018).

2.3 Linking monitoring data to absolute pest densities and economic thresholds

The purpose of monitoring in IPM is to determine if, and when, the economic threshold is exceeded, that is the pest density at which management action should be taken to prevent an increasing pest population from reaching the economic injury level (i.e., the smallest number of insects/mites that will cause yield losses equal to the crop protection management costs). Economic thresholds have been established for a number of pests, but they generally rely upon comparative analyses of historical monitoring data and evaluations of fruit damage at the end of the season with no estimates of pest population numbers (e.g. Brunner et al., 2002; Knight and Light, 2005; Perez-Rodriguez et al., 2017; Robertson et al., 2005; Thomson et al., 2001; Mudavanhu et al., 2011). For instance, a well-documented overview of sampling methods for pests and their derived economic thresholds in agricultural crops, including apple and citrus, was published at the end of the twentieth century when IPM became the standard in many crops (Beers et al., 1994; Hare, 1994).

Despite the considerable progress in the development of lure-baited monitoring traps, there is still a knowledge gap between the numbers captured (monitoring data) and their translation into absolute pest-density estimates, which is the key variable for setting economic injury levels (Adams et al., 2017). Accurate estimations of absolute pest density from a catch number recorded in a lure-baited trap requires knowledge of (1) the pest's distribution in space and time; (2) the pest's movement patterns before and after encountering the lure; (3) the farthest distance from which the trap can be reached, both unassisted and assisted by the lure; and (4) the average probability of capture for all individuals in the trapping area of known size (Miller et al., 2015b). It also requires knowledge on the vulnerability of the crop at the time of capture and the numbers and efficacy of natural enemies in the crop. In this respect one has to realize that estimating absolute pest density by monitoring traps only becomes feasible if insects orientating to the lure (but not yet encountering the trap) move randomly, that is execute 'random walks' (Berg, 1993) that ironically cause the overall population to exhibit highly reproducible properties like those well quantified in physics and chemistry, for example in phenomena such as diffusion or heat transfer (Miller et al., 2015b).

An excellent case study illustrating the derivation of the absolute pest density from capture numbers in attractive traps was recently reported for codling moth (*C. pomonella*) in apple orchards using standard codlemone-baited traps (Adams et al., 2017). A remarkable outcome of this study was the consistency recorded for the pheromone plume reach (<5 m), the trapping radius and trapping area (240–275 m and 18–24 ha, respectively), as well as the average probability of capture (= proportion caught: 0.01–0.025), despite the different trial conditions across three growing seasons at eight different experimental sites. This consistency and pattern of trapping outcomes provides strong evidence that codling moth is a 'random walker'. Once the trapping area and the average probability of capture (=proportion caught) are known and shown to be reproducible, the absolute pest density, given in units of pests per trapping area, can then simply be obtained by dividing the capture number by the proportion caught (Miller et al., 2015b). Hence, in the cited study a codling moth male catch of 1, 3, 10 and 30 equates to (divided by 0.02) 50, 150, 500 and 1500 males per 21 ha, or 2.4, 7, 24 and 71 males per ha, respectively (Adams et al., 2017). Given that the sex ratio of codling moth is 1:1, this estimate holds for females.

Combined with life table data on the pest's fecundity and mortality, alongside data on crop yield per trapping area, this fundamental knowledge of how to interpret catch numbers enables pest managers to make considerably more precise projections of damage and, therefore, more precise and reliable decisions on whether crop protection management applications are justified.

For the cited case study (Adams et al., 2017) this was done as follows. Based on the mean number of eggs per female (50) and the mean fraction of eggs producing larvae that successfully attack apples (0.33), the injury frequency expected from each codling moth female could be estimated at 17 apples (50 × 0.33). Taking into account the mean total production of apples/ha and the fact that the threshold for rejection of apples from the fresh market followed by diversion to apple juice is 0.5% infestation, Adams et al. elegantly demonstrated that the catch threshold indicative of severe economic loss is approached at 30 male codling moth per trap per growing season, and exceeded at a catch of 45 (at which the economic threshold of 0.5% infestation level is reached). The current action threshold of a cumulative catch is approximately 12 codling moth per season. This was derived by trial and error based on historical field experiences and comparative analyses of monitoring data and fruit injury evaluations. Hence the estimate was reasonable, though strongly biased towards safe limits. Based on the intelligent scientific approach and case study of Adams et al. (2017), the currently used capture threshold for codling moth can be raised, improving IPM in apple orchards and elevating grower profits by reducing pest control costs. Moreover, the principles and methods established could be broadly applicable to any pest or beneficial arthropod that forages locally by random walks.

Measures of pheromone plume reach and trapping radius are critical for understanding pest detectability and should be a research priority for any invasive species and new attractant/lure. The fundamental knowledge of how to interpret catch numbers (with knowledge of the trapping area and the average probability of capture), combined with life table data on pest fecundity and mortality, and data on crop yield per trapping area will enable pest managers to make considerably more precise projections of possible economic damage levels. Hence, the principles and methods established in this case study pave the way to the setting of a new standard for IPM decisions based on trapping. Indeed, recently, the same approach was followed to reveal the link between monitoring data, pest density and the economic threshold value for the invasive *Drosophila suzukii* fruit fly in cherries (Kirkpatrick et al., 2018a).

3 Automatic monitoring of pests

Currently, traps are monitored manually in fruit orchards. This is a labour-intensive practice and therefore costly, due to the need of weekly visual inspections of spatially distributed traps. It is also error-prone, as it relies on visual judgement of the operator (Potamitis et al., 2014). Because of these limitations, the manually collected data are typically heterogeneous and sparse in terms of geographic and temporal coverage, making them unsuitable for accurate mapping of pest population dynamics (Potamitis et al., 2017a). However, electronic devices that can automatically detect/identify pests have the potential to systematically and simultaneously monitor a large number of sites on a daily basis without requiring human presence. Moreover, such automated intensive monitoring would permit continuous follow up of insect activity throughout the diurnal and nocturnal cycles. Combined with the near-continuous data, already recorded by modern automated weather stations, this would greatly increase the precision of phenological modelling (Tobin et al., 2009). Therefore, more attention has been paid to the development of such electronic devices and their incorporation into automatic monitoring networks (Lopez et al., 2012; Miller et al., 2015a).

Generally such an automatic monitoring network consists of (1) a series of battery or solar panel powered monitoring devices (usually traps), (2) a wireless communication system, (3) a base station and (4) a cloud-based server with a web-based user interface where collected information can be viewed, stored and analysed (see a schematic overview in Fig. 1). In the development of the automated monitoring systems, two approaches can be distinguished: (1) imaging of trapped insects (= camera-based automatic monitoring) and (2) non-imaging sensors that capture an insect-specific signal such as wingbeat characteristics (= camera-independent automatic monitoring). Table 1 displays an overview of the currently existing automatic monitoring systems for IPM in fruit production sites and their main characteristics.

3.1 Camera-based automatic monitoring

In the camera-based approach, digital images of trapped target pest(s) are automatically captured and sent to a server (Lopez et al., 2012). The pests are then identified and counted, either manually by human experts looking at them on the screen, or automatically by image-recognition software, which can gradually improve by machine-learning methods (so-called artificial intelligence or deep-learning methods). Examples for the first approach are described for a number of pests in fruit orchards, for instance plum curculio (Selby et al., 2014), fruit flies (Doitsidis et al., 2017) and several lepidopterous insects including codling moth (Gonzalez et al., 2014; Guarnieri et al., 2011). Considering the automatic monitoring, the most basic approach involves the automatic counting of the number of all pests trapped on a sticky surface without identifying them. This practice has been reported in several studies during the last decade (Tirelli et al., 2011; Wang et al., 2013). At present, commercial applications of such systems are already on the market, such as the e-GLEEK® system (Advansee, Nantes, France) (e-GLEEK, 2018), which is able to automatically count trapped insects on sticky plates and classify these insects into different groups according to their size.

Automatic image-recognition technology has advanced dramatically in recent years, and an increasing number of studies are published demonstrating successes in machine image-based classification and even identification of pest species (Solis-Sánchez et al., 2009; Wen and Guyer, 2012; Doitsidis

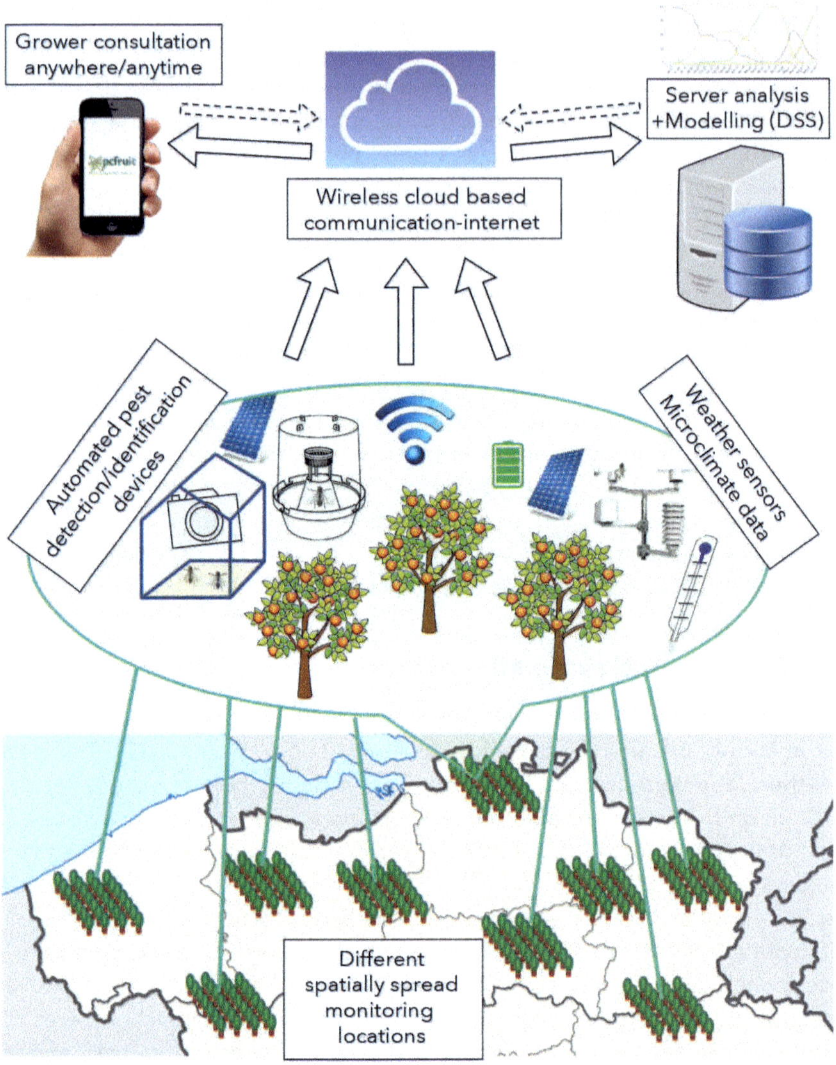

Figure 1 Schematic strategy of automatic monitoring network. In various spatially spread monitoring locations (orchards), a series of automatic monitoring devices (usually traps) (battery or solar panel powered) equipped with a wireless communication system is installed as well as basic weather stations to record all relevant microclimate parameters. All collected data (pest and meteorology) are sent to a cloud-based server where the data are stored, analysed and whenever possible evaluated in the frame of running pest phenology models and decision support systems (DSS). The collected and analysed data and generated decision support advice can be consulted anytime and anywhere by growers or advisers through the cloud-based communication system using a web-based user-friendly interface.

Table 1 Overview of automatic monitoring systems for integrated pest management in fruit production sites and their main characteristics

Pest detection system	Automatic or manual handling?	Automatic pest species identification?	Major advantage(s)	Major disadvantage(s)	References (examples)
Camera-based visible light-imaging systems					
Digital camera built into trap and/or directed on sticky trapping plate in the field	Automatic[a]	No	Automatic sampling considerably reducing labour costs	Relatively high cost per camera-trap device, still time-consuming manual pest identification process	Doitsidis et al. (2017), e-GLEEK (2018), Gonzalez et al. (2014), Guarnieri et al. (2011), Selby et al. (2014)
		Yes, if identification algorithm is available	Automatic sampling and pest identification considerably reducing labour costs	Relatively high cost per camera-trap device	Maric et al. (2016), Semios (2018), Spensa (2018), TRAPVIEW (2018)
Image-capturing device with controlled predefined settings ensuring photographing in a uniform way under well-defined conditions	Manual	No	Standardized images facilitate reliable and consistent pest counting Relatively low cost as there is only one image-capturing device for analysis of multiple monitoring locations in the field	Time-consuming manual sampling (collecting sticky plates) and manual pest identification still required	Qiao et al. (2008), Veres et al. (2017)
		Yes, if identification algorithm is available	Standardized images facilitate reliable and consistent pest identification and counting Relatively low cost as there is only one image-capturing device for analysis of multiple monitoring spots	Time-consuming manual sampling (collecting sticky plates) still required	Scoutbox (2018), Solis-Sánchez et al. (2011)

(*Continued*)

Table 1 (*Continued*)

Pest detection system	Automatic or manual handling?	Automatic pest species identification?	Major advantage(s)	Major disadvantage(s)	References (examples)
Camera-independent systems					
Infrared/piezoelectric light-interruption/ vibration counter built into trap	Automatic	No	Simple robust, low cost	Manual pest identity confirmation is still required	Beerwinkle (2001), Tobin et al. (2009), Holguin et al. (2010), Bromenshenk et al. (2015)
Opto-electronic wingbeat detection sensor built into trap	Automatic	Yes	High signal-to-noise ratios, automatic sampling/pest detection and pest identification considerably reducing labour costs	Only for winged insects	Moore and Miller (2002), Potamitis et al. (2014), Potamitis et al. (2015)
Bioimpedance sensor built into trap	Automatic[a]	Yes	Automatic sampling/pest detection and pest identification considerably reducing labour costs	Relative high cost per trapping device	Krawczyk et al. (2016), Spensa (2018)
Acoustic sensor	Automatic or manual	Yes	Possible to detect pests inside plant tissue	Low signal-to-noise ratios due to interference with background noise	Martin et al. (2013), Pinhas et al. (2008)

[a] Note: Automatic sampling (without need for trap visits/manual handling) is only possible if a mechanism for automatic replacement of sticky surfaces is available in the trapping device.

et al., 2017; Zhong et al., 2018). Generally speaking, this requires insects to be mapped (fixated) in two-dimensional (2D) spaces with uniform backgrounds, such as sticky trapping boards where the distances from the target insects to the camera are constrained. Recognition of pests in a free three-dimensional space, such as on a plant, is substantially more complex and has not been well studied (Liu et al., 2017). For the automatic identification of photographed pests on 2D sticky trapping plates from the field, two main challenges exist. The first challenge is low image quality, due to constraints such as the cost of the camera, power consumption and the speed by which images can be transmitted (Ding and Taylor, 2016). The second challenge originates from inconsistencies which are driven by many factors, including variation in illumination, movement of the trap, movement of the target insect, decay or damage to the target insect, appearance of other objects (such as non-target insects or leaves) and so forth (Ding and Taylor, 2016). An approach to circumvent the latter difficulties is to ensure the standardization of the image-capturing settings and the context of the acquisition (Martineau et al., 2017). In this case, images are not produced in an autonomously functioning trap by an inbuilt camera, but by, manually, placing handled samples or sticky sampling plates in an image-capturing device with controlled predefined settings ensuring photographing in a uniform way under fixed constrained conditions. Examples of this approach are the development of a digital mite recognition and counting system to monitor spider mites (Veres et al., 2017) and a scanner device to determine the density of whiteflies on sticky plates (Qiao et al., 2008). An example of a commercially available device of this approach is the Scoutbox (SoilCares, Wageningen, The Netherlands) (Scoutbox, 2018). This methodology still requires field visits and sample handling and therefore is not fully automatic in contrast to an autonomously functioning trap equipped with an inbuilt camera. In the ideal situation, the sticky surface can automatically be replaced by an inbuilt mechanism in the trap so that regular visits for the replacement of the sticky plate are not required.

Several commercial pheromone-based camera traps for automatic monitoring of pests in orchards are currently available on the market. Probably, the best-known examples are the TRAPVIEW system (EFOS d.o.o., Hruševje, Slovenia) (Maric et al., 2016; TRAPVIEW, 2018) (see example in Fig. 2), the Sentinel Trap (Spensa technologies, West Lafayette, Indiana, United States) (Spensa, 2018) and the Semios system (SemiosBio Technologies Inc, Vancouver, Canada) (Semios, 2018).

The most common approach to image-based automated identification is to create a set of reference images of the target, and to use these training images to extract a set of machine-interpretable features that are, in turn, used to evaluate a captured image of a monitoring trap to identify and count the target pest. However, as the current camera-based automatic monitoring applications are based on 2D images, automatic identification algorithms are

Figure 2 An example of a camera-based automatic monitoring trap (TRAPVIEW®) equipped with solar panel and a basic weather station installed in an apple orchard.

often focussing on the (relatively flat) insect wings (Li and Cao, 2015; Martineau et al., 2017; Wen et al., 2009; Feng et al., 2016), while other important anatomical features such as genitalia cannot be taken into account. This makes species identification by image-recognition software a challenge for several target pests. Accordingly, the camera-based automatic monitoring traps are currently of limited effectiveness for small insects. Indeed, most of the image-based traps were designed for relatively large (winged) insects such as the red-palm weevil, *Rhynchophorus ferrugineus* (Lopez et al., 2012), and the rice bug, *Leptocorisa chinensis* (Fukatsu et al., 2012). Nevertheless, more recently, good results are reported for the machine identification of small insects such as aphids, thrips and whiteflies (Cho et al., 2007; Solis-Sánchez et al., 2011; Xia et al., 2012; Feng et al., 2016; Liu et al., 2016). Driven by advances in computer vision algorithms, even a procedure for image processing based on low-resolution images (and hence low computational costs) suitable for identification of small size insects was recently reported, illustrating the fast progress in this technology field (Xia et al., 2015). Advances in deep-learning technologies such as the use of convolutional neural networks further spur the applicability of camera-based automatic monitoring approaches for a variety of insects (Ding and Taylor, 2016).

3.2 Camera-independent automatic monitoring

In the camera-independent automatic monitoring field, different types of sensors can be used to detect and/or identify pest insects. In its most basic form, a sensor (infrared or piezoelectric) is placed inside a lure trap, measuring a light interruption or vibration to count the number of times a target insect enters or falls into the trap (e.g. red-palm weevil traps, pitfall traps, funnel traps, beehives) (Beerwinkle, 2001; Tobin et al., 2009; Holguin et al., 2010; Bromenshenk et al., 2015). Unlike the imaging systems, the insects are not identified in this approach. Hence, it is crucial that the lure is exclusively specific for the target pest to avoid erroneous counts caused by non-target species. Therefore, in recent years a more advanced sensor-based system was developed capable of identifying counted insects based on measuring their wingbeat characteristics (Potamitis et al., 2014; van Roy et al., 2014; Batista et al., 2011; Welsh et al., 2017). In this approach, an optoelectronic sensor is used that records the wingbeat of an entering insect. Subsequently, the insect can be identified to species level based on the optoacoustic spectrum analysis of its wingbeat (Potamitis et al., 2015, 2017a,b; Chen et al., 2014). This sensor system can be integrated in a so-called electronic McPhail trap. A sketch summarizing the concept of this autonomous monitoring trap is in Fig. 3 (Potamitis et al., 2015). Since most insects beat their wings with frequencies below 1 kHz and the operating frequency of the inbuilt photodiodes can be designed to receive light pulses at 450 kHz - thus far higher than any biological organism can reach

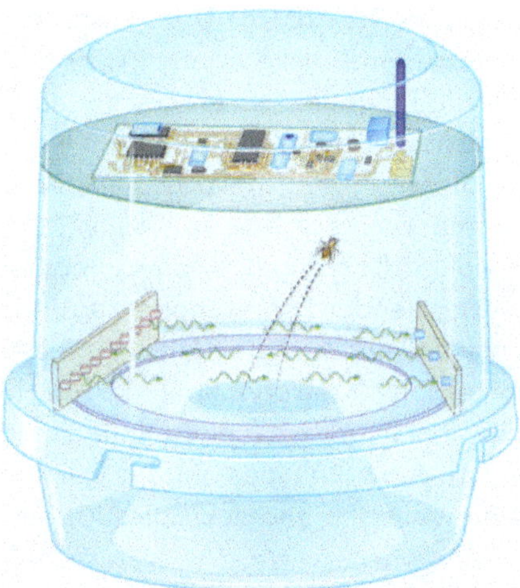

Figure 3 A sketch summarizing the concept of the 'electronic McPhail trap'. In the proposed sensor design, the entered insect occludes the path of light from emitter to receiver with its body and flapping wings. The electronics of the trap then record and analyse the light fluctuation of the receiver. Light-intensity fluctuations constitute a 'biometric signature' directly related to insect's wingbeat frequency, size and shape of its wings. The signature is compared to pre-embedded patterns from the target pest. Finally, counts of the target pest, temperature, humidity and GPS coordinates are transmitted through the wireless communication network from the field to a base station and can further be consulted via a web-based interface. Source: adapted from Potamitis et al. (2015).

with its wings – potentially any insect starting from low-frequency Lepidoptera to high-frequency midges can be detected. Moreover, using frequency modulation the background interference from environmental light fluctuations (e.g. artificial light sources attached to an AC source) is thus avoided (Potamitis and Rigakis, 2016). Another important advantage of this optoelectronic sensor is its relatively low cost. According to Potamitis, the cost of an automatic monitoring trap equipped with this type of sensor (= the proposed 'electronic McPhail Trap') can be less than €100 (Potamitis et al., 2017b, 2018), making it more affordable than camera-based automatic monitoring devices.

The wingbeat event can also be detected by measuring the sound that is produced by the beating using an acoustic sensor (Raman et al., 2007). Moreover, other wavelengths such as communication sounds or vibrations created by adult insects or their larvae can be detected (Liu et al., 2017; Martin et al., 2013; Pinhas et al., 2008). Hence, acoustic sensing has the advantage of being able to detect non-flying pests in the soil or inside crops (plant tissue or even inside the wood). However, acoustic signals generated by insects are often hard to interpret because of background noise and wind interference, which raises practical difficulties for infield applications (Liu et al., 2017).

Another sensor-based automatic pest detection system is the technology with which the Z-trap® is equipped (Spensa Technologies, West Lafayette, Indiana, United States) (Spensa, 2018). The insects that enter these traps, attracted by a pheromone lure, make contact with a bioimpedance sensor to stimulate a shock ('zap') impulse, which creates an electrical current that is evaluated by the device's software. Based on the measurement of the electrical impedance of the insect, the pest can be identified. In a study aiming to determine the accuracy and reliability of these Z-traps for the automatic monitoring of codling moth, *Cydia pomonella*, and Oriental fruit moth, *Grapholita molesta*, standard pheromone traps were found to generally capture higher cumulative numbers of moths than the Z-traps throughout different followed-up seasons (Krawczyk et al., 2016). However, the trends in observed flight activity characteristics (i.e. the main flight/mating activity and peaks of egg laying) between both types of traps were consistent, making the Z-traps useful for moth monitoring and decision support.

4 Modelling of pest populations to support pest management

Recent advances in global climate research, Earth imagery and positioning and our ability to emulate complex biological processes – together – open up new avenues for data integration and prognostic modelling of pest survival, establishment, distribution and expansion in different spatiotemporal scales. Such approaches significantly extend the scope of direct (trap-based) pest monitoring and enable the development of innovative applications supporting decision-making processes in pest management at various operational levels. Several conceptual modelling frameworks have been proposed, and the choice of an approach depends largely on the objective of the model and the availability of data (Tonnang et al., 2017). An extensive discussion of this rapidly growing and conceptually heterogeneous domain exceeds the boundaries and scope of this chapter, which is why only selected examples are outlined to illustrate the main approaches, from global or macroregional to local farm levels.

4.1 Projecting macroregional and global pest distributions, invasion risk and range expansions

The scale and intensity of contemporary human travel and the transport of plant products together with the ongoing climate change pose an unprecedented risk of pest transmission. The latter raises serious biosecurity, socioeconomic and biodiversity concerns and generates demand for tools and methods for assessing and forecasting such emerging threats on a macroregional and global scale. In the absence of feasible empirical approaches, a number of complex models have been developed, such as CLIMEX (Sutherst et al., 2000, 2007), ILCYM (Tonnang et al., 2009; Sporleder et al., 2013), MaxEnt and GARP (Philips et al., 2006) (the latter two only require occurrence data), to simulate the development of pests in various temperature regimes and map pest-vulnerability areas that are potentially suitable for pest establishment (see examples in Fig. 4). The availability of satellite-based climate and land surface imagery allows high-resolution modelling of pest establishment, distribution and expansion on large spatial scales. Both CLIMEX and ILCYM are climate-driven (not weather!), and thus provide generic predictions about the potential geographical distribution of pests. The projections are based on a detailed analyses of the current or anticipated climate and comparisons between the locations of the known and presumed presence of pests. These models use a combination of parameters required for population growth and stress conditions that barely allow survival at an unfavourable time. These parameters are then combined to describe the responses of pests in the full range of climatic conditions that occur in different seasons and places in the world, and respectively, allocate an 'ecoclimatic index' describing the climatic suitability of any site for the pest. Such maps and predictions derived from such models may be used to warn about the potential expansion of the pest range, focus pest surveillance operations in the areas at risk and optimize in time and space the operation of 'on-the-ground' pest detection systems, usually consisting of a network of traps.

Users of such maps should, however, be aware of frequent discrepancies in the quality and resolution of the information used to guide these models in

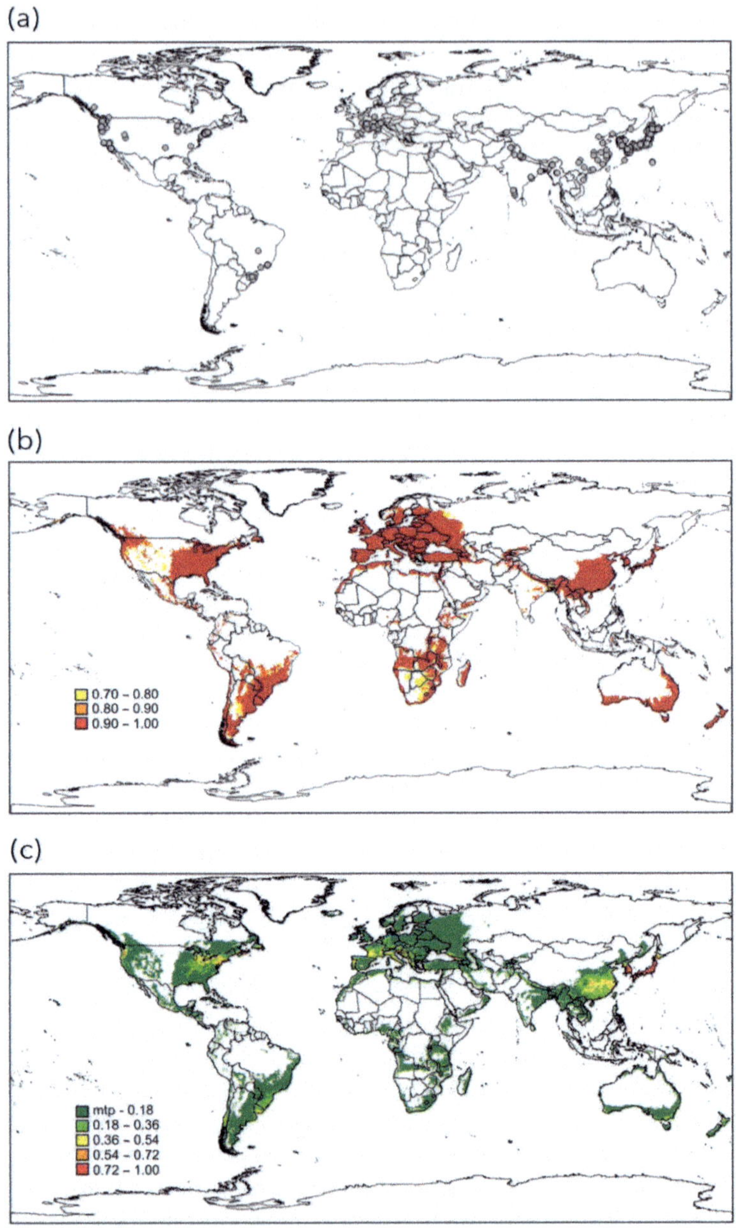

Figure 4 Maps of model-predicted distributions of *D. suzukii*. (a) Known existing sites of occurrence of *D. suzukii* used to generate predictive models. (b) Predictive model of the geographic distribution of *D. suzukii* generated by the GARP algorithm. (c) Predictive model of the geographic distribution of *D. suzukii* generated by the MaxEnt algorithm. The legend indicates low (0) and high (1) environmental suitability for *D. suzukii*. Source: adapted from dos Santos et al. (2017).

relation to the climate and geo-imagery compared to pest biology. The latter depends on the state of our knowledge, and therefore is often limited to more or less rudimentary temperature-related pest development parameters. In extreme cases, where even basic life-cycle data are not available, maps can be inferred from the already confirmed observations of geographical presence and seasonal abundance of pests and extrapolation to areas with similar climatic conditions. It has to be acknowledged that even such a simplistic 'climate-matching' approach provides useful predictions about the potential distribution of pests in cases that would otherwise be impossible due to lack of adequate data for building population models (Sutherst et al., 2000). When more comprehensive demographic information about pests is available, detailed maps can be generated (Ni et al., 2012; de Villiers et al., 2013, 2016, 2017; Kriticos et al., 2017; Yonow et al., 2017; Wakie et al., 2018), allowing even the distinction of the potential distribution for closely related sibling species, such as *Ceratitis rosa* and *C. quilici* (Tanga et al., 2018). However, even then, the reliability of such predictions varies and depends strictly on the relevance of the input data of the pest biology (Elith et al., 2010). For example, the frequent source of uncertainty or error results from the premise adopted in the absence of empirical data that the demographic parameters established for the pest under optimal conditions will remain representative for all pest cohorts, including those living on the edges of the pest range. Indeed, often the pest demographic parameters used to drive the model are determined on the basis of laboratory insect populations tested at optimal and different constant temperatures. Such data only to a certain extent resemble the parameters of pest development in natural conditions with daily and seasonal fluctuations in temperature, and even less of its various local strains adapted to different local conditions within the pest range. Growing evidence of significant variability in resistance to climate stress among pest subpopulations (Terblanche et al., 2006; Weldon et al., 2018) suggests that such a simplified approach may lead to substantial miscalculations of the potential ranges of pest occurrence.

In addition, despite triviality, it is worth re-emphasizing that potential suitability of the climate should be interpreted only as 'the possibility of pest existence' and should not be confused with the threat of 'pest invasion risk'. The impressive maps generated by these models only visualize the potential distribution of pests, predicted according to climate suitability, and more specifically, the adequacy of local temperature regimes for the development of pests. Typically, the climate-matching models assume no limiting factors other than climate (Sutherst et al., 2007). But the actual pest invasion risk depends on many other factors that are not included in the models, which affect the actual transmission of pests and determine the establishment of invasive propagules at the place of their arrival. Most of the events related to the transmission of pests are of anthropogenic origin, caused by the trade in plant products and human travel. Seasonal fluctuations in the direction and intensity of trade and travel modulate the risk of transporting pests and cause it to fluctuate accordingly in time and space. These relations, although known to play a role, are rarely analysed in a comprehensive way and few examples are available of their large-scale quantification (Szyniszewska and Tatem, 2014). Furthermore, anthropogenic transmission implies that most of the invasive propagules arrive within urbanized locations (ports of entry), which differ significantly in the traits and complexity of their landscapes. The chances of establishing a pest depend on the availability of suitable host plants at the time and place of propagule arrival. In the case of long-lived multivoltine and polyphagous pests, this also depends on the completeness of the host chain necessary to maintain the annual pest development cycle. If it is known that the required host plants are generally present in the affected area, propagule establishment is usually 'taken for granted' and therefore this process is treated superficially or neglected. However, usually the size of invasive propagules is small, limited to just a few individuals (depending on the frequency of propagation), and the process of pest establishment depends on their individual behaviour and fate. Even if essential hosts are indeed present, their role is modified by the spatial and phenological distribution. The distance in space and time between successive hosts and their relationship to the pest's lifespan - and individual mobility - play a role. Recent attempts to elucidate this process by modelling lifetime behaviours of individual insects - members of invasive propagules of *C. capitata* - have shown that the complexity and traits of the local landscape play a central role in determining the fate and detectability of such incipient populations, and that the presence of necessary hosts in the area does not always guarantee pest establishment (Lux, 2017).

In conclusion, models generating regional and global maps that visualize the potential climatic suitability for pest development constitute a major advance in our ability to understand the dynamics and scale of threats caused by pests, human activity and climate change. Although their interpretation requires a certain degree of caution and experience, their potential for improving pest management is unquestioned and difficult to overestimate. However, in order to fully assess the risk of pest invasion, all key elements of the 'pest-risk triad' should be considered: (1) climate suitability, (2) threats related to the anthropogenic pest transmission and (3) the suitability of local landscape at the 'port of entry' for establishment of invasive propagules. Unfortunately, each of these elements is still largely treated as a separate 'academic' domain, and basic models and methods are largely developed as disparate and loosely related 'academic' tools and have not yet been integrated and transformed into truly functional comprehensive decision support systems (DSS).

4.2 Forecasting pest phenology and seasonal abundance at regional scales

Typically, an IPM decision to undertake pest control is made after receiving information that (1) the pest has just emerged and/or (2) has reached (or is likely to reach) the density that justifies the control action. To obtain such 'decision-triggering information', farmers are advised to monitor pests 'on the farm' using bait traps or otherwise. Although this approach is justified from the point of view of the IPM concept, its practical implementation often turns out to be problematic. The main reason for the difficulties stems directly from the primary purpose of IPM: preventing the growth of pest populations. This means that when you need to make IPM decisions, your target pests still exist at low densities. In addition, many important pests such as Tephritid fruit flies, for example *Ceratitis capitata* or *Rhagoletis cerasi*, occur naturally at low-to-moderate densities even at the peaks of unmanaged populations (several hundred or several thousand per ha), unlike mites, thrips, aphids and pollen beetles, reaching hundreds of thousands or a dozen million per ha. Thus, in the early stages of the appearance of pests, the number of individuals caught can be very low and very random, which, with a small number of traps usually used to monitor pests on the farm, often provides uncertain information difficult to interpret and use. In addition, farmers usually have to deal with many pests occurring on different crops in each season and gathering reliable information about the arrival date and density of each pest may become unfeasible. This is one of the main reasons for abandoning IPM and switching to calendar-based preventive use of pesticides, which although suboptimal, is convenient and effective in most cases.

To promote and facilitate the widespread implementation of IPM, a variety of services have been established, public or private, to free the farmer from the burden of collecting the locally relevant pest-related information. In the past, such information has been obtained empirically through establishing local stocks of overwintering pests kept in 'experimental gardens' and monitoring their emergence in the spring, local pest monitoring using traps, scouting plant infestation and so forth. Over the past decades, development of numerous population models devoted to various pests has led to a significant improvement in such services and a reduction in their operating costs. Such models are 'weather-driven', and their predictions are usually based on the well-known cumulative degree-day Uvarov model (Uvarov, 1931) or its improved versions (Ikemoto and Takai, 2000), developed for each pest stage, and sometimes also gender. Most of these models use experimentally determined pest demographic parameters and combine them with local weather forecasts to predict and map the spring appearance and abundance of pest populations at medium spatial scales, thus supporting many locations in a given country or region (Damos and Karabatakis, 2013).

Based on this approach, several advanced regional IPM decision support platforms have been developed, such as SIMLEP (Jörg et al., 2007) or SOPRA (Samietz et al., 2007, 2011). SIMLEP is temperature-driven DSS for managing the Colorado potato beetle in Germany, Austria and Poland. The system consists of two main parts: (1) regional forecasting of the time of spring emergence of adult beetles and the beginning of egg laying and (2) field-specific model forecasting the occurrence of different larval stages. It determines the time of application of control measures according to the occurrence of pesticide-sensitive larval stages. The system is efficient and reliable; during the field assessment in 2006–09 only minor deviations between the SIMLEP predictions and empirical egg laying dates and occurrence of young larvae were observed, thus the forecasts allowed optimal application of conventional or biological pesticides (Wójtowicz, 2010). SOPRA was developed by Agroscope Changins-Wädenswil to optimize the timing of management operations for a number of pests of economic importance in fruit orchards in Switzerland and southern Germany, such as *Dysaphis plantaginea*, *Hoplocampa testudinea*, *Cydia pomonella*, *Grapholita lobarzewskii*, *Cacopsylla pyri*, *Rhagoletis cerasi*, *Anthonomus pomorum* and *Adoxophyes orana*. To generate predictions about pest phenology, the system uses a time-varying distributed delay approach to pest phenology models driven by detailed data on solar radiation, air temperature and soil, accessed with hourly temporal resolution from 14 meteorological stations located in the main regions of Switzerland. This information is used for biophysical modelling of habitat conditions in places of interest. Initial parameters for the temperature and stage-specific pest development rates were established based on insects reared under strictly controlled laboratory conditions. Then, the system was validated by means of multi-seasonal comparisons of the model-generated predictions and empirical field observations, which were then used to fine-tune algorithms that emulate pest development. The pest-dedicated models, driven by local weather data, generate the age structure of the pest populations and, based on it, issue IPM advice. The advice is provided to agricultural extension and advisory personnel and fruit producers through online services (www.sopra.info).

Such 'weather-driven' population dynamics models were developed not only for pest insects, but recently also for beneficial arthropods in orchards. One example is the 'Earwig Management Tool': a software tool that integrates an earwig phenology model with management recommendations (Belien et al., 2012). The programme is based on a day-degree model for earwigs which is fed by local temperature data. In addition, a pesticide database with known side effects of a wide range of products on the different life stages of earwigs is integrated in the system. The output gives the current status of the earwig population and management recommendations for activities critical for their survival (Belien et al., 2013). Hence, by consultation of this user-friendly software fruit growers can predict the earwig development in the field at any time and organize the timing of orchard management actions taking into account the presence of (sensitive) life stages of this important predator. The earwig model was also integrated into a temperature-driven population dynamics model for pear psyllids (*C. pyri*) (Belien et al., 2017). Hence, by online consultation of this model (at https://www.pcfruit.be/en) growers can simultaneously obtain information on the development of the pest and the natural enemies, allowing optimization of their spraying schedules in order to achieve an optimal balance between natural suppression by predators and highly effective correction sprays whenever needed (Belien et al., 2017).

In conclusion, comprehensive regional pest forecast systems play a major role in improving plant protection standards by optimizing the time and rationalizing the frequency of local pest control efforts and thus reducing pesticide use. Importantly, they also remove the main obstacle hindering the extensive implementation of a genuine IPM based on pest biology that uses pesticide-free or pesticide-restricted management approaches. The conceptual and methodological foundations for building such systems are robust, well-developed, accessible and time-tested. The rapidly expanding global and European Earth observation infrastructure, such as Copernicus (http://www.copernicus.eu) which is freely and openly accessible to European users, ensures the availability of weather data and land imagery with continuous improvement of quality, which in turn will provide constant improvement in the spatial resolution and reliability of such pest prediction systems. In practice, therefore, the development of similar systems for other pests of economic concern is hampered mainly by our limited knowledge of the biology of these pests, in particular the lack of reliable and locally relevant demographic parameters for these pest species and their local strains.

4.3 Modelling local (on-farm) pest development in mosaic landscapes

In the 'classic' IPM, the information obtained from the traps serves as the basis for inferring the density of the pest population and to extrapolate it further to the economic injury and control-triggering thresholds. Usually, the economic injury thresholds are based on the direct correlation of multiannual trap catches with the final crop infestation. They are fairly reliable for broadly distributed pests operating in spatially homogenous terrain, but their generic use outside the area where such 'primary' correlations have been established is prone to error.

Locally relevant injury and IPM-guiding thresholds can, in principle, be derived from the analytical translation of the local trap catches into absolute pest density, and in turn to crop damage (Adams et al., 2017). However, it should be remembered that insect monitoring (trapping) only *de facto* reveals the cumulative patterns of their seasonal response to the bait used in traps. Despite this, in IPM practice, such bait response patterns are commonly interpreted as an accurate representation of the seasonal density of pest population. This interpretation implies that (1) the effectiveness of the bait is considered constant in time and space, (2) the insect's response to the bait is also constant throughout its lifetime regardless of environmental conditions and (3) exploration and dispersion of individual insects is random. Such assumptions are rarely consciously pronounced and are rarely correct. Baits used in traps mimic, rarely perfectly, various insect-relevant objects or resources, such as pheromone-emitting conspecifics or food sources. At the same time, the 'original' sources of attraction also exist on the farm and compete with the baits for the 'attention' of the exploring insects. Such competitive 'modulation' of bait performance varies in time and space and is difficult to remedy. In addition, pest responsiveness seldom remains constant, in particular for long-lived insects with an adult phase lasting several months. For example, the behaviour of fruit flies, *Ceratitis capitata* or *Rhagoletis cerasi,* varies according to age, sex, reproductive and nutritional status. All these factors not only modulate their response to baits, but also determine the patterns and extent of spatial exploration and dispersion (both within and outside tree canopy), which in heterogeneous landscapes are far from random (Manrakhan and Lux, 2009; Lux et al., 2016, 2017; Lux, 2017; Pimentel et al., 2017). Finally, crop damage depends not only on the density of the pests, but also on the presence or absence of alternative hosts, their attractiveness, phenology and spatial distribution. Practically, the IPM action thresholds derived from trap catches tend to become problematic with the increase in the complexity of the terrain, so in fragmented landscapes their use is too often reminiscent of a simplistic 'rule-of-thumb' approach.

The recognition of this limitation stimulates the development of alternative approaches towards guiding IPM, such as modelling spatial and temporal development of pest populations. Currently, modelling the role of landscape in insect dispersion and development remains largely limited to large-scale ecological research, with occasional applications in area-wide pest control or eradication (Barclay et al., 2014; Pagabeleguem et al., 2016). However, generic population development and dispersion algorithms commonly used in such models tend to be less applicable to highly localized processes involving very small insect populations (e.g. invasive propagules) or pest development at smaller spatial scales (e.g. at farm level). It is recognized that, in such cases, 'individual-based' models that emulate behaviour of individual insects provide more realistic representation of local insect-driven processes (Lux, 2014; Manoukis and Hoffman, 2014). Such approaches are innovative, used in current ecological research (Grimm and Railsback, 2005), but they are still rare in horticulture. For example, in the extensively studied and technically advanced field of frugivorous Tephritid fruit flies, to the best of our knowledge, so far only few applications of this approach exist (Lux, 2014, 2017; Lux et al., 2017; Manoukis et al., 2014, 2015; Manoukis and Hoffman, 2014). One of them, the Manoukis model, was developed to estimate the chance of detection (Manoukis and Hoffman, 2014; Manoukis et al., 2015) or eradication (Manoukis et al., 2014) of isolated incipient pest propagules, or assessment of quarantine length (Collier and Manoukis, 2017). The model simulates the development of pests using temperature-related demographic traits and high-resolution (hourly) weather inputs. Other aspects of the local system, such as representation of terrain structure and IPM, are less detailed. Similarly, the exploration and translocation of individual insects is simulated using generic, age-independent 'random-dispersion' algorithms. Such simplifications accelerate simulation runs, thus improving the model's performance, while ensuring sufficient emulation of short (several months) episodes of insect development in fairly homogeneous landscapes, under uniform area-wide pest management (e.g. sterile insect technique).

Another model (PESTonFARM) was developed for guiding seasonal IPM operations at farm level in the spatiotemporally complex agro-landscapes (Lux, 2014, 2017; Lux et al., 2016, 2017). The model enables holistic emulation of the local 'pest–terrain–climate–IPM' system. It contains a detailed 'virtual' representation of the spatial structure and pest-relevant traits of the local terrain, such as fruit distribution, phenology, attractiveness and suitability for pest development, canopy size and structure, time and location of IPM treatments. For each sector of such a 'virtual farm', lifetime behaviour, dispersion and fate (from egg to adult) are simulated for each individual 'virtual' insect – members of multiple age cohorts and overlapping generations of pest present on the farm throughout the season. The model emulates the combined effects of (1) the complexity of the terrain with mosaics of scattered crops, (2) age dependence of behavioural patterns in long-lived insects, (3) unsynchronized local IPM operations conducted on individual crops and plots and (4) climate/weather effects on pest development, behaviour and mortality. Projections of pest density, fruit infestation and cost/benefits of the applied IPM are derived from the behaviour of the modelled 'virtual' insects. Individual behaviours and events are randomized and determined only in the 'stochastic' sense. Therefore, each simulation run emulates a single replica of the on-farm experiment, and alike, it generates a unique, stochastically equivalent set of results. The model can be applied for *in silico* emulation and pre-assessment of various IPM scenarios before their implementation in the field. Although such an approach will not completely eliminate field experiments, it can radically shorten the usual trajectory of IPM development by substituting a large part of long-term and expensive on-farm experiments with their 'virtual' emulations and focussing real empiric efforts on the most promising scenarios. The model was on-farm validated for *R. cerasi* and postulated for *in silico* designing of locally optimized IPM strategies in sweet cherry orchards (Lux et al., 2016). Potentially, application of such *in silico* approach opens new prospects for local, farm-level IPM. It allows the design of 'pest-resilient production systems' through purposeful exploitation of structural elements of landscape heterogeneity, such as pest-detrimental spatial arrangement of cultivars with diverse phenology, diversification in the canopy structure, creation of buffering and/or pest interception zones (Lux et al., 2016; Papadopoulos et al., 2017). In perennial horticulture, the development of such locally optimized pest-resilient farm designs through classical experimentation is unfeasible due to exorbitant costs and the multiannual timescale of experimental cycles. A more expanded version of the model, capable of simulating multivoltine fruit flies, was recently proposed for assessment of the impacts of local landscape structure on the distribution, fate and detectability of the ultra-low-density incipient medfly populations (Lux, 2017).

In conclusion, individual-based models offer unique insights into the mechanisms underlying local insect-driven processes and reveal them in empirically unattainable detail and resolution. The technique of deriving locally adjusted precision IPM strategies and schemes from the emulation of behaviour and development of on-farm present insects constitutes substantial, yet unexploited potential. However, this approach also has limitations. The development of such models requires extensive, detailed and reliable knowledge about the biology and behaviour of pests. So far, such quantitative information is available only for several groups of pests. Ideally, the modelled area with host and non-host plant mosaics should constitute a kind of 'autonomous ecological unit' capable of sustaining a full annual cycle of the local pest population. The size of the area strongly depends on the behaviour and mobility of the modelled pests, for example, in the case of Tephritid fruit flies, the effectively modelled area typically ranges from 10 to 100 ha, with the size of each farm sector ranging from 10 x 10 to 25 x 25 m. Characterizing each sector of such an area, although it is a 'one-time' exercise, is tedious, even if it can be facilitated by the analysis of satellite imagery or photographs taken by drones. Simulating the behaviour of individual insects and events in each farm sector entails the execution of a huge number of operations, which even for modern computers is time-consuming and usually lasts up to several hours. Increasing the modelled area and density of pest population exponentially extends working time. So, in practice, this approach is most suitable for local IPM scenarios in complex agro-landscapes, and pests that in normal IPM practices do not reach extremely high densities.

5 Conclusion and future trends

In a rapidly-changing and more integrated world, increased global trade has allowed pests to invade new regions. Therefore, future research will need to focus on the development of effective monitoring and forecasting methodologies in order to support IPM strategies for these new threats. A key aspect is the translation of monitoring data to pest population density estimates and determination of accurate economic threshold values. Measures of plume reach of (new) attractants used in monitoring traps and trapping radius are critical for understanding pest detectability and should be a research priority for any invasive species, let alone current pest trapping systems. These and other fundamental knowledge gaps, for example how to interpret monitored numbers, will remain an important research subject over the next decade.

Recent advances in real-time data transfer, automated image analysis, electronic sensors, data-interpretation software and decision support algorithms will continue to fuel the development of automated pest surveillance and decision-making systems. Hence, future research will increasingly deal with comprehensive systems integrating monitoring, forecasting and decision support in IPM, like for example the 'Location Awareness Pest Management System' (http://fruitflynet.aua.gr) (Pontikakos et al., 2012; Papadopoulos et al., 2017). This system is based on (1) automated trap for adult *Rhagoletis cerasi* combined with soil and air temperature, precipitation, wind speed and humidity monitoring sensors; (2) a wireless network to collect and transfer data in a real-time mode to cloud storage, accessible to designated users; (3) a spatial DSS that estimates the local adult emergence and development; and (4) software that analyses the acquired information and generates local pest management decisions.

6 Where to look for further information

For introductions as well as comprehensive treatises on the subject of this chapter, the authors strongly recommend the following excellent books and review articles:

- Binns, M. R., Nyrop, J. P. and Van Der Werf, W. (2000). *Sampling and Monitoring in Crop Protection. The Theoretical Basis for Developing Practical Decision Guides*. CAB International, London.
- Liu, H., Lee, S.-H. and Chahl, J. S. (2017). A review of recent sensing technologies to detect invertebrates on crops. *Precision Agriculture* 18, 635–66.
- Miller, J. R., Adams, C. G., Weston, P. A. and Schenker, J. H. 2015. *Trapping of Small Organisms Moving Randomly – Principles and Applications to Pest Monitoring and Management*. Springer International Publishing, New York.

- Pedigo, L. A. and Buntin, G. D. 1994. *Handbook of Sampling Methods for Arthropods in Agriculture*. CRC Press, Boca Raton, FL.
- Tonnang, H. E. Z., Hervé, B. D. B., Biber-Freudenberger, L., Salifu, D., Subramanian, S., Ngowi, V. B., Guimapi, R. Y. A., Anani, B., Kakmeni, F. M. M., Affognon, H., Niassy, S., Landmann, T., Ndjomatchoua, F. T., Pedro, S. A., Johansson, T., Tanga, C. M., Nana, P., Fiaboe, K. M., Mohamed, S. F., Maniania, N. K., Nedorezov, L. V., Ekesi, S. & Borgemeister, C. 2017. Advances in crop insect modelling methods–Towards a whole system approach. *Ecological Modelling* 354, 88–103.

Important Research Centres for the subject of this chapter are:

- Department of Entomology, Michigan State University, MI, United States: group of Prof. L. J. Gut and J. R. Miller (https://www.canr.msu.edu/ent/research/): pioneers in research concerning the principles and methods to link monitoring data to absolute pest densities and economic threshold values
- Technological Educational Institute of Crete (https://www.teicrete.gr/en) (Departments Music Technology & Acoustics and Electronics) and Technical University of Crete (https://www.ece.tuc.gr) (Department of Electrical and Computer Engineering), Greece: pioneers in developments of sensor-based automatic monitoring systems
- Catholic University of Leuven, Belgium. Division of Mechatronics, Biostatistics and Sensors (MeBioS) (https://www.biw.kuleuven.be/biosyst/mebios): research on sensor-based automatic monitoring systems
- Research Station for Fruit Growing, pcfruit, Belgium (https://www.pcfruit.be/en): applied research on monitoring, modelling and decision support systems in fruit growing in practice
- NIAB EMR, Pest & Pathogen Ecology Department, UK. (http://www.emr.ac.uk/horticultural-research-programmes/pest-pathogen-ecology/): applied research on monitoring, modelling and decision support systems in fruit growing in practice
- ARS, Temperate Tree Fruit and Vegetable Research: Wapato, WA, United States.(https://www.ars.usda.gov/pacific-west-area/wapato-wa/temperate-tree-fruit-and-vegetable-research/): research on pheromone/bait traps for monitoring
- AGROSCOPE, the Swiss centre of excellence for agricultural research: Switzerland.(https://www.agroscope.admin.ch/agroscope/en/home/topics/plant-production/plant-protection/zoology/pest-forecasting-monitoring-fruits-vegetables.html): research and development of pest forecasting models

The International Organization for Biological and Integrated Control (IOBC) (https://www.iobc-wprs.org/) is the key organization concerning meetings and activities to keep up to date with trends on the chapter's subject.

7 References

Acebes-Doria, A. L., Morrison, W. R., Short, B. D., Rice, K. B., Bush, H. G., Kuhar, T. P., Duthie, C. and Leskey, T. C. 2018. Monitoring and biosurveillance tools for the brown marmorated stink bug, *Halyomorpha halys* (Stål) (Hemiptera: Pentatomidae). *Insects* 9(3), 82. doi:10.3390/insects9030082.

Adams, C. G., Schenker, J. H., McGhee, P. S., Gut, L. J., Brunner, J. F. and Miller, J. R. 2017. Maximizing information yield from pheromone-baited monitoring traps: estimating plume reach, trapping radius, and absolute density of *Cydia pomonella* (Lepidoptera: Tortricidae) in Michigan apple. *Journal of Economic Entomology* 110(2), 305–18. doi:10.1093/jee/tow258.

Agnello, A. M., Reissig, W. H., Kovach, J. and Nyrop, J. P. 2003. Integrated apple pest management in New York State using predatory mites and selective pesticides. *Agriculture, Ecosystems and Environment* 94(2), 183–95. doi:10.1016/S0167-8809(02)00026-9.

Akotsen-Mensah, C., Kaser, J. M., Leskey, T. C. and Nielsen, A. L. 2018. *Halyomorpha halys* (Hemiptera: Pentatomidae) responses to traps baited with pheromones in peach and apple orchards. *Journal of Economic Entomology* 111(5), 2153–62. doi:10.1093/jee/toy200.

Alford, D. V. 2007. *Pests of Fruit Crops, A Colour Handbook*. Manson Publishing Ltd, London.

Asplen, M. K., Anfora, G., Biondi, A., Choi, D. S., Chu, D., Daane, K. M., Gibert, P., Gutierrez, A. P., Hoelmer, K. A., Hutchison, W. D., et al. 2015. Invasion biology of spotted wing Drosophila (*Drosophila suzukii*): a global perspective and future priorities. *Journal of Pest Science* 88(3), 469–94. doi:10.1007/s10340-015-0681-z.

Augustin, S., Boonham, N., De Kogel, W. J., Donner, P., Faccoli, M., Lees, D. C., Marini, L., Mori, N., Petrucco Toffolo, E., Quilici, S., et al. 2012. A review of pest surveillance techniques for detecting quarantine pests in Europe. *EPPO Bulletin* 42(3), 515–51. doi:10.1111/epp.2600.

Baggiolini, M., Keller, E., Milaire, H. G. and Steiner, H. 1971. *Guide de détermination pour les controles periodiques en vergers de pommiers 'Contrôle visuelle'. Brochure no. 2; 1ère edition*.

Baggiolini, M., Keller, E., Milaire, H. G. and Steiner, H. 2001. Guide de détermination pour les controles periodiques en vergers de pommiers 'Contrôle visuelle' (edited and extended from P. Galli & H. Höhn). *Translated and reworked by Ch. de Schaetzen). Visuele Controles in de Appelboomgaarden*.

Barclay, H. J., McInnis, D. and Hendrichs, J. 2014. Modeling the area-wide integration of male annihilation and the simultaneous release of methyl eugenol-exposed *Bactrocera* spp. sterile males. *Annals of the Entomological Society of America* 107(1), 97–112. doi:10.1603/AN13010.

Barzman, M., Lamichhane, J. R., Booij, K., Boonekamp, P., Desneux, N., Huber, L., Kudsk, P., Langrell, S. R. H., Ratnadass, A., Ricci, P., et al. 2015. *Research and Development Priorities in the Face of Climate Change and Rapidly Evolving Pests*. Springer, New York.

Batista, G. E. A. P. A., Hao, Y., Keogh, E. and Mafra-Neto, A. 2011. Towards automatic classification on flying insects using inexpensive sensors. In: *10th International Conference on Machine Learning and Applications and Workshops*, Honolulu, Hawaii, 18–21 December 2011, pp. 364–9.

Beers, E. H., Hull, L. A. and Jones, V. P. 1994. Sampling pest and beneficial arthropods of apple. In: Pedigo, L. P. and Buntin, G. D. (Eds), *Handbook of Sampling Methods for Arthropods in Agriculture*. CRC Press, Boca Raton, pp. 384–416.

Beerwinkle, K. R. 2001. An automatic capture-detection, time-logging instrumentation system for boll weevil pheromone traps. *Applied Engineering in Agriculture* 17, 893–8.

Belien, T., Moerkens, R., Leirs, H. and Peusens, G. 2012. Earwig management tool: an IPM decision aid system for augmentation of European earwig populations (Dermaptera: Forficulidae) in pip fruit orchards. *Communications in Agricultural and Applied Biological Sciences* 77(4), 657–62.

Belien, T., Moerkens, R., Leirs, H., Peusens, G. and Bylemans, D. 2013. 'Earwig management tool': transferring knowledge of population dynamics and side effects on earwigs (*Forficula auricularia* L.) into practical sustainable plant protection strategies in pip fruit growing. *IOBC-WPRS Bulletin* 91, 411–8.

Belien, T., Bangels, E., Goffin, J., Berkvens, N., De Ro, M., Casteels, H. and Bylemans, D. 2016. Factors influencing attractant preferences of *Drosophila suzukii*: implications for monitoring population dynamics and IPM measures. In: *9th International Conference on Integrated Fruit Production*, 4–8 September 2016. IOBC, Thessaloniki, Greece.

Belien, T., Bangels, E., Brenard, N., Reijniers, J., Leirs, H. and Bylemans, D. 2017. Optimized timing of IPM treatments against pear Psylla (*Cacopsylla pyri*) based on a temperature driven population dynamics model. *IOBC-WPRS Bulletin* 123, 96–100.

Bell, C. H. 2014. Chapter 29 – Pest management. In: Motarjemi, Y. and Lelieveld, H. (Eds), *Food Safety Management*. Academic Press, San Diego.

Berg, H. C. 1993. *Random Walks in Biology*. Princeton University Press, Princeton, NJ.

Binns, M. R. and Nyrop, J. P. 1992. Sampling insect populations for the purpose of IPM decision making. *Annual Review of Entomology* 37(1), 427–53. doi:10.1146/annurev.en.37.010192.002235.

Binns, M. R., Nyrop, J. P. and Van Der Werf, W. 2000. *Sampling and Monitoring in Crop Protection. The Theoretical Basis for Developing Practical Decision Guides*. CAB International, London.

Bromenshenk, J. J., Henderson, C. B., Seccomb, R. A., Welch, P. M., Debnam, S. E. and Firth, D. R. 2015. Bees as biosensors: chemosensory ability, honey bee monitoring systems, and emergent sensor technologies derived from the pollinator syndrome. *Biosensors* 5(4), 678–711. doi:10.3390/bios5040678.

Broughton, S. and Harrison, J. 2012. Evaluation of monitoring methods for thrips and the effect of trap colour and semiochemicals on sticky trap capture of thrips (Thysanoptera) and beneficial insects (Syrphidae, Hemerobiidae) in deciduous fruit trees in Western Australia. *Crop Protection* 42, 156–63. doi:10.1016/j.cropro.2012.05.004.

Brunner, J., Welter, S., Calkins, C., Hilton, R., Beer, E., Dunley, J. and Van Buskirk, P. 2002. Mating disruption of codling moth: a perspective from the Western United States. *IOBC WPRS Bulletin* 25, 1–11.

Burrack, H. J., Asplen, M., Bahder, L., Collins, J., Drummond, F. A., Guedot, C., Isaacs, R., Johnson, D., Blanton, A., Lee, J. C., et al. 2015. Multistate comparison of attractants for monitoring *Drosophila suzukii* (Diptera: Drosophilidae) in blueberries and caneberries. *Environmental Entomology* 44(3), 704–12. doi:10.1093/ee/nvv022.

Cha, D. H., Hesler, S. P., Wallingford, A. K., Zaman, F., Jentsch, P., Nyrop, J. and Loeb, G. M. 2018. Comparison of commercial lures and food baits for early detection of fruit infestation risk by *Drosophila suzukii* (Diptera: Drosophilidae). *Journal of Economic Entomology* 111(2), 645–52. doi:10.1093/jee/tox369.

Chapman, D., Purse, B. V., Roy, H. E. and Bullock, J. M. 2017. Global trade networks determine the distribution of invasive non-native species. *Global Ecology and Biogeography* 26(8), 907–17. doi:10.1111/geb.12599.

Charmillot, P. J. 1990. Mating disruption technique to control codling moth in Western Switzerland. In: Ridgway, R. L., Silverstein, R. M. and Inscoe, M. N. (Eds.), *Behavior-Modifying Chemicals for Insect Pest Management*. Marcel Dekker, New York, pp. 165–82.

Chen, Y., Why, A., Batista, G., Mafra-Neto, A. and Keogh, E. 2014. Flying insect detection and classification with inexpensive sensors. *Journal of Visualized Experiments* (92), e52111. doi:10.3791/52111.

Cho, J., Choi, J., Qiao, M., Ji, C. W., Kim, H. Y., Uhm, K. B. and Chon, T. S. 2007. Automatic identification of whiteflies, aphids and thrips in greenhouse based on image analysis. *International Journal of Mathematics and Computers in Simulation* 1, 46–53.

Cloonan, K. R., Abraham, J., Angeli, S., Syed, Z. and Rodriguez-Saona, C. 2018. Advances in the chemical ecology of the spotted wing Drosophila (*Drosophila suzukii*) and its applications. *Journal of Chemical Ecology* 44(10), 922–39. doi:10.1007/s10886-018-1000-y.

Collier, T. and Manoukis, N. 2017. Evaluation of predicted Medfly (*Ceratitis capitata*) quarantine length in the United States utilizing degree-day and agent-based models. *F1000Research* 6, 1863. doi:10.12688/f1000research.12817.2.

Damos, P. and Karabatakis, S. 2013. Real time pest modeling through the World Wide Web: decision making from theory to praxis. *IOBC-WPRS Bulletin* 91, 253–8.

Damos, P., Colomar, L. A. and Ioriatti, C. 2015. Integrated fruit production and pest management in Europe: the apple case study and how far we are from the original concept? *Insects* 6(3), 626–57. doi:10.3390/insects6030626.

Daniel, C. and Grunder, J. 2012. Integrated management of European cherry fruit fly *Rhagoletis cerasi* (L.): situation in Switzerland and Europe. *Insects* 3(4), 956–88. doi:10.3390/insects3040956.

Daniel, C., Mathis, S. and Feichtinger, G. 2014. A new visual trap for *Rhagoletis cerasi* (L.) (Diptera: Tephritidae). *Insects* 5(3), 564–76. doi:10.3390/insects5030564.

De Villiers, M., Hattingh, V. and Kriticos, D. J. 2013. Combining field phenological observations with distribution data to model the potential distribution of the fruit fly *Ceratitis rosa* Karsch (Diptera: Tephritidae). *Bulletin of Entomological Research* 103(1), 60–73. doi:10.1017/S0007485312000454.

De Villiers, M., Hattingh, V., Kriticos, D. J., Brunel, S., Vayssieres, J. -F., Sinzogan, A., Billah, M. K., Mohamed, S. A., Mwatawala, M., Abdelgader, H., et al. 2016. The potential distribution of *Bactrocera dorsalis*: considering phenology and irrigation patterns. *Bulletin of Entomological Research* 106(1), 19–33. doi:10.1017/S0007485315000693.

De Villiers, M., Kriticos, D. J. and Veldtman, R. 2017. Including irrigation in niche modelling of the invasive wasp *Vespula germanica* (Fabricius) improves model fit to predict potential for further spread. *PLoS ONE* 12(7), e0181397. doi:10.1371/journal.pone.0181397.

Deronzier, S. and Atger, P. 1980. Éléments d'étude de la dynamique des populations de Psylla pyri L. dans la Basse Vallée du Rhône: période hivernale et printanière. *Acta Oecologica* 1, 247–58.

Ding, W. and Taylor, G. 2016. Automatic moth detection from trap images for pest management. *Computers and Electronics in Agriculture* 123, 17–28. doi:10.1016/j.compag.2016.02.003.

Doitsidis, L., Fouskitakis, G. N., Varikou, K. N., Rigakis, I. I., Chatzichristofis, S. A., Papafilippaki, A. K., Birouraki, A. E. and Birouraki, A. E. 2017. Remote monitoring of the *Bactrocera oleae* (Gmelin) (Diptera: Tephritidae) population using an automated McPhail trap. *Computers and Electronics in Agriculture* 137, 69–78. doi:10.1016/j.compag.2017.03.014.

Dos Santos, L. A., Mendes, M. F., Krüger, A. P., Blauth, M. L., Gottschalk, M. S. and Garcia, F. R. M. 2017. Global potential distribution of *Drosophila suzukii* (Diptera, Drosophilidae). *PLoS ONE* 12(3), e0174318. doi:10.1371/journal.pone.0174318.

Dreistadt, S. H. 2012. *Integrated Pest Management for Citrus* (3rd ed.). University of California – Agriculture and Natural Resources, Richmond, CA.

Dreistadt, S. H. and Flint, M. 1995. Landscape pest monitoring methods and training managers to use them. *Journal of Arboriculture* 21, 1–6.

E-Gleek. 2018. *The Automatic Trapping of Flying Pests* (Online). Available at: http://www.e-gleek.com (accessed on 27 June 2018).

El-Sayed, A. M. 2018. *The Pherobase: Database of Pheromones and Semiochemicals* (Online). Available at: http://www.pherobase.com (accessed on 27 March 2018).

El-Sayed, A. M., Cole, L., Revell, J., Manning, L. A., Twidle, A., Knight, A. L., Bus, V. G. and Suckling, D. M. 2013. Apple volatiles synergize the response of codling moth to pear ester. *Journal of Chemical Ecology* 39(5), 643–52. doi:10.1007/s10886-013-0277-0.

Elith, J., Kearney, M. and Phillips, S. 2010. The art of modelling range-shifting species. *Methods in Ecology and Evolution* 1(4), 330–42. doi:10.1111/j.2041-210X.2010.00036.x.

Feng, L., Bhanu, B. and Heraty, J. 2016. A software system for automated identification and retrieval of moth images based on wing attributes. *Pattern Recognition* 51, 225–41. doi:10.1016/j.patcog.2015.09.012.

Fountain, M. T., Bennett, J., Cobo-Medina, M., Ruiz, R. C., Deakin, G., Delgado, A., Harrison, R. and Harrison, N. 2018. Alimentary microbes of winter-form *Drosophila suzukii*. *Insect Molecular Biology* 27(3), 383–92. doi:10.1111/imb.12377.

Fukatsu, T., Watanabe, T., Hu, H., Yoichi, H. and Hirafuji, M. 2012. Field monitoring support system for the occurrence of *Leptocorisa chinensis* Dallas (Hemiptera: Alydidae) using synthetic attractants, field servers, and image analysis. *Computers and Electronics in Agriculture* 80, 8–16. doi:10.1016/j.compag.2011.10.005.

Gonzalez, M., Schandy, J., Wainstein, N., Barboni, L., Gomez, A. and Croce, C. 2014. Wireless image-sensor network application for population monitoring of lepidopterous insects pest (moths) in fruit crops. In: *2014 IEEE International Instrumentation and Measurement Technology Conference (I2MTC) Proceedings*, Montevideo, Uruguay, 12–15 May 2014, pp. 1394–8.

Grimm, V. and Railsback, S. F. 2005. *Individual-Based Modeling and Ecology*. Princeton University Press, Princeton, NJ.

Gross, J. and Gündermann, G. 2016. Principles of IPM in cultivated crops and implementation of innovative strategies for sustainable plant protection. In: Horowitz,

A. R. and Ishaaya, I. (Eds), *Advances in Insect Control and Resistance Management*. Springer International Publishing, Switzerland, pp. 9–26.

Guarnieri, A., Maini, S., Molari, G. and Rondelli, V. 2011. Automatic trap for moth detection in integrated pest management. *Bulletin of Insectology* 64, 247–51.

Hamilton, J. G. C., Hall, D. R. and Kirk, W. D. J. 2005. Identification of a male-produced aggregation pheromone in the Western flower thrips *Frankliniella occidentalis*. *Journal of Chemical Ecology* 31(6), 1369–79. doi:10.1007/s10886-005-1351-z.

Hanks, L. M. and Millar, J. G. 2016. Sex and aggregation-sex pheromones of cerambycid beetles: basic science and practical applications. *Journal of Chemical Ecology* 42(7), 631–54. doi:10.1007/s10886-016-0733-8.

Hare, J. D. 1994. Sampling arthropod pests in citrus. In: Pedigo, L. P. and Buntin, G. D. (Eds) *Handbook of Sampling Methods for Arthropods in Agriculture*. CRC Press, Boca Raton, pp. 73–98.

Harris, A. L., Ullah, R. and Fountain, M. T. 2017. The evaluation of extraction techniques for *Tetranychus urticae* (Trombidiformes: Tetranychidae) from apple (*Malus domestica*) and cherry (*Prunus avium* L.) leaves. *Experimental and Applied Acarology* 72(4), 367–77. doi:10.1007/s10493-017-0154-6.

Haye, T., Gariepy, T., Hoelmer, K., Rossi, J. P., Streito, J. C., Tassus, X. and Desneux, N. 2015. Range expansion of the invasive brown marmorated stinkbug, *Halyomorpha halys*: an increasing threat to field, fruit and vegetable crops worldwide. *Journal of Pest Science* 88(4), 665–73. doi:10.1007/s10340-015-0670-2.

Holguin, G. A., Lehman, B. L., Hull, L. A., Jones, V. P., Jones, V. P. and Park, J. 2010. Electronic traps for automated monitoring of insect populations. *IFAC Proceedings Volumes* 43(26), 49–54. doi:10.3182/20101206-3-JP-3009.00008.

Horton, D. R. 1994. Relationship among sampling methods in density estimates of pear psylla (Homoptera: Psyllidae): implications of sex, reproductive maturity, and sampling location. *Annals of the Entomological Society of America* 87(5), 583–91. doi:10.1093/aesa/87.5.583.

Horton, D. R. and Lewis, T. M. 1997. Quantitative relationship Between sticky trap catch and beat tray counts of pear psylla (Homoptera: Psyllidae): seasonal, sex, and morphotypic effects. *Journal of Economic Entomology* 90(1), 170–7. doi:10.1093/jee/90.1.170.

Hutchins, S. H. 1994. Techniques for sampling arthropods in integrated pest management. In: Pedigo, L. P. and Buntin, G. D. (Eds), *Handbook of Sampling Methods for Arthropods in Agriculture*. CRC Press, Boca Raton, pp. 73–98.

Iglesias, L. E., Nyoike, T. W. and Liburd, O. E. 2014. Effect of trap design, bait type, and age on captures of *Drosophila suzukii* (Diptera: Drosophilidae) in berry crops. *Journal of Economic Entomology* 107(4), 1508–18. doi:10.1603/EC13538.

Ikemoto, T. and Takai, K. 2000. A new linearized formula for the law of total effective temperature and the evaluation of line-fitting methods with both variables subject to error. *Environmental Entomology* 29(4), 671–82. doi:10.1603/0046-225X-29.4.671.

Jaffe, B. D., Avanesyan, A., Bal, H. K., Feng, Y., Grant, J., Grieshop, M. J., Lee, J. C., Liburd, O. E., Rhodes, E., Rodriguez-Saona, C., et al. 2018. Multistate comparison of attractants and the impact of fruit development stage on trapping *Drosophila suzukii* (Diptera: Drosophilidae) in raspberry and blueberry. *Environmental Entomology* 47(4), 935–45. doi:10.1093/ee/nvy052.

Jenser, G., Szita, É. and Bálint, J. 2010. Measuring pear psylla population density (*Cacopsylla pyri* L. and *C. pyricola* Förster): review of previous methods and evaluation of a new technique. *Northwest. Journal of Zoology* 6, 54–62.

Jörg, E., Racca, P., Preiß, U., Butturini, A., Schmiedl, J. and Wójtowicz, A. 2007. Control of Colorado potato beetle with the SIMLEP decision support system. *EPPO Bulletin* 37(2), 353–8. doi:10.1111/j.1365-2338.2007.01137.x.

Khrimian, A., Zhang, A. J., Weber, D. C., Ho, H. Y., Aldrich, J. R., Vermillion, K. E., Siegler, M. A., Shirali, S., Guzman, F. and Leskey, T. C. 2014. Discovery of the aggregation pheromone of the brown marmorated stink bug (*Halyomorpha halys*) through the creation of stereoisomeric libraries of 1-Bisabolen-3-ols. *Journal of Natural Products* 77(7), 1708–17. doi:10.1021/np5003753.

Kirkpatrick, D. M., McGhee, P. S., Hermann, S. L., Gut, L. J. and Miller, J. R. 2016. Alightment of spotted wing Drosophila (Diptera: Drosophilidae) on odorless disks varying in color (Diptera: Drosophilidae). *Environmental Entomology* 45(1), 185–91. doi:10.1093/ee/nvv155.

Kirkpatrick, D. M., Gut, L. J. and Miller, J. R. 2018a. Estimating monitoring TRAP plume reach and trapping area for *Drosophila suzukii* (Diptera: Drosophilidae) in Michigan tart cherry. *Journal of Economic Entomology* 111(3), 1285–9. doi:10.1093/jee/toy062.

Kirkpatrick, D. M., Leach, H. L., Xu, P., Dong, K., Isaacs, R. and Gut, L. J. 2018b. Comparative antennal and behavioral responses of summer and winter morph *Drosophila suzukii* (Diptera: Drosophilidae) to ecologically relevant volatiles. *Environmental Entomology* 47(3), 700–6. doi:10.1093/ee/nvy046.

Knight, A. L. 2002. A comparison of gray halo-butyl elastomer and red rubber septa to monitor codling moth (Lepidoptera: Tortricidae) in sex pheromone-treated orchards. *Journal of the Entomological Society of British Columbia* 99, 123–32.

Knight, A. L. 2007. Influence of within-orchard trap placement on catch of codling moth (Lepidoptera: Tortricidae) in sex pheromone-treated orchards. *Environmental Entomology* 36(2), 425–32.

Knight, A. L. 2010. Improved monitoring of female codling moth (Lepidoptera: Tortricidae) with pear ester plus acetic acid in sex pheromone-treated orchards. *Environmental Entomology* 39(4), 1283–90. doi:10.1603/EN10034.

Knight, A. L. and Light, D. M. 2005. Developing action thresholds for codling moth (Lepidoptera: Tortricidae) with pear ester- and codlemone-baited traps in apple orchards treated with sex pheromone mating disruption. *The Canadian Entomologist* 137(6), 739–47. doi:10.4039/n05-040.

Knight, A. L. and Light, D. M. 2012. Monitoring codling moth (Lepidoptera: Tortricidae) in sex pheromone-treated orchards with (E)-4,8-dimethyl-1,3,7-nonatriene or pear ester in combination with codlemone and acetic acid. *Environmental Entomology* 41(2), 407–14. doi:10.1603/EN11310.

Knight, A. L., Larson, D. J. and Christianson, B. 2002. Flight tunnel and field evaluations of sticky traps for monitoring codling moth (Lepidoptera: Tortricidae) in sex pheromone-treated orchards. *Journal of the Entomological Society of British Columbia* 99, 107–16.

Knight, A. L., Hilton, R., Basoalto, E. and Stelinski, L. L. 2014. Use of glacial acetic acid to enhance bisexual monitoring of tortricid pests with kairomone lures in pome fruits. *Environmental Entomology* 43(6), 1628–40. doi:10.1603/EN14153.

Knight, A. L., Basoalto, E., Katalin, J. and El-Sayed, A. M. 2015. A binary host plant volatile lure combined with acetic acid to monitor codling moth (Lepidoptera: Tortricidae). *Environmental Entomology* 44(5), 1434–40. doi:10.1093/ee/nvv116.

Knutson, A. E. and Muegge, M. A. 2010. A degree-day model initiated by pheromone trap captures for managing pecan nut casebearer (Lepidoptera: Pyralidae) in pecans. *Journal of Economic Entomology* 103(3), 735–43. doi:10.1603/EC09319.

Krawczyk, G., Lehman, B., Hull, L. A. and Park, J. 2016. Z-trap technology for monitoring insect pest in orchard-field observations. In: *9th International Conference on Integrated Fruit Production*, 4–8 September 2016. IOBC, Thessaloniki, Greece, p. 130.

Kriticos, D. J., Kean, J. M., Phillips, C. B., Senay, S. D., Acosta, H. and Haye, T. 2017. The potential global distribution of the brown marmorated stink bug, *Halyomorpha halys*, a critical threat to plant biosecurity. *Journal of Pest Science* 90(4), 1033–43. doi:10.1007/s10340-017-0869-5.

Kumral, N. A., Susurluk, H. and Cobanoglu, S. 2010. Interactions among populations of predatory mites and insect and mite pests on olive trees in Turkey. *International Journal of Acarology* 36(6), 463–71. doi:10.1080/01647950903373416.

Landolt, P. J. 1997. Sex attractant and aggregation pheromones of male phytophagous insects. *American Entomologist* 43(1), 12–22. doi:10.1093/ae/43.1.12.

Lee, J. C., Burrack, H. J., Barrantes, L. D., Beers, E. H., Dreves, A. J., Hamby, K. A., Haviland, D. R., Isaacs, R., Richardson, T. A., Shearer, P. W., et al. 2012. Evaluation of monitoring traps for *Drosophila suzukii* (Diptera: Drosophilidae) in North America. *Journal of Economic Entomology* 105(4), 1350–7.

Lee, J. C., Shearer, P. W., Barrantes, L. D., Beers, E. H., Burrack, H. J., Dalton, D. T., Dreves, A. J., Gut, L. J., Hamby, K. A., Haviland, D. R., et al. 2013. Trap designs for monitoring *Drosophila suzukii* (Diptera: Drosophilidae). *Environmental Entomology* 42(6), 1348–55. doi:10.1603/EN13148.

Leskey, T. C. and Hogmire, H. W. 2005. Monitoring stink bugs (Hemiptera: Pentatomidae) in mid-Atlantic apple and peach orchards. *Journal of Economic Entomology* 98(1), 143–53. doi:10.1093/jee/98.1.143.

Leskey, T. C., Wright, S. E., Short, B. D. and Khrimian, A. 2012. Development of behaviorally-based monitoring tools for the brown marmorated stink bug (Heteroptera: Pentatomidae) in commercial tree fruit orchards. *Journal of Entomological Science* 47(1), 76–85. doi:10.18474/0749-8004-47.1.76.

Leskey, T. C., Agnello, A., Bergh, J. C., Dively, G. P., Hamilton, G. C., Jentsch, P., Khrimian, A., Krawczyk, G., Kuhar, T. P., Lee, D. H., et al. 2015a. Attraction of the invasive *Halyomorpha halys* (Hemiptera: Pentatomidae) to traps baited with semiochemical stimuli Across the United States. *Environmental Entomology* 44(3), 746–56. doi:10.1093/ee/nvv049.

Leskey, T. C., Khrimian, A., Weber, D. C., Aldrich, J. C., Short, B. D., Lee, D. H. and Morrison, W. R. 2015b. Behavioral responses of the invasive *Halyomorpha halys* (Stål) to traps baited with stereoisomeric mixtures of 10,11-Epoxy-1-bisabolen-3-OL. *Journal of Chemical Ecology* 41(4), 418–29. doi:10.1007/s10886-015-0566-x.

Li, F. and Cao, W. 2015. Automatic identification of buttery species. *Journal of Computational Information Systems* 11, 2529–38.

Liburd, O. E., Alm, S. R., Casagrande, R. A. and Polavarapu, S. 1998. Effect of trap color, bait, shape, and orientation in attraction of blueberry maggot (Diptera : Tephritidae) flies. *Journal of Economic Entomology* 91(1), 243–9. doi:10.1093/jee/91.1.243.

Light, D. M. and Knight, A. 2005. Specificity of codling moth (Lepidoptera: Tortricidae) for the host plant kairomone, ethyl (2E,4Z)-2,4-decadienoate: field bioassays with pome fruit volatiles, analogue, and isomeric compounds. *Journal of Agricultural and Food Chemistry* 53(10), 4046–53. doi:10.1021/jf040431r.

Light, D. M., Knight, A. L., Henrick, C. A., Rajapaska, D., Lingren, B., Dickens, J. C., Reynolds, K. M., Buttery, R. G., Merrill, G., Roitman, J., et al. 2001. A pear-derived kairomone with pheromonal potency that attracts male and female codling moth, *Cydia pomonella* (L.). *Die Naturwissenschaften* 88(8), 333–8. doi:10.1007/s001140100243.

Liu, T., Chen, W., Wu, W., Sun, C., Guo, W. and Zhu, X. 2016. Detection of aphids in wheat fields using a computer vision technique. *Biosystems Engineering* 141, 82–93. doi:10.1016/j.biosystemseng.2015.11.005.

Liu, H., Lee, S.-H. and Chahl, J. S. 2017. A review of recent sensing technologies to detect invertebrates on crops. *Precision Agriculture* 18(4), 635–66. doi:10.1007/s11119-016-9473-6.

Lockwood, J. L., Hoopes, M. F. and Marchetti, M. P. 2013. *Invasion Ecology* (2nd ed.). Wiley-Blackwell, West Sussex.

Lopez, O., Rach, M. M., Migallon, H., Malumbres, M. P., Bonastre, A. and Serrano, J. J. 2012. Monitoring pest insect traps by means of low-power image sensor technologies. *Sensors* 12(11), 15801–19. doi:10.3390/s121115801.

Lux, S. A. 2014. PESTonFARM – stochastic model of on-farm insect behaviour and their response to IPM interventions. *Journal of Applied Entomology* 138, 458–67.

Lux, S. A. 2017. Individual-based modeling approach to assessment of the impacts of landscape complexity and climate on dispersion, detectability and fate of incipient medfly populations. *Frontiers in Physiology* 8, 1121. doi:10.3389/fphys.2017.01121.

Lux, S. A., Wnuk, A., Vogt, H., Belien, T., Spornberger, A. and Studnicki, M. 2016. Validation of individual-based Markov-like stochastic process model of insect behavior and a 'virtual farm' concept for enhancement of site-specific IPM. *Frontiers in Physiology* 7, 363. doi:10.3389/fphys.2016.00363.

Lux, S. A., Michalska, K., Wnuk, A., Paliczuk, D., Vogt, H., Hernik, T. and Studnicki, M. 2017. Female age as a factor determining the patterns of tree canopy utilisation by the European cherry fruit fly, *Rhagoletis cerasi*. *International Journal of Pest Management* 63(4), 371–81. doi:10.1080/09670874.2016.1277802.

Macmillan, C. D. and Costello, M. J. 2015. Evaluation of a brushing machine for estimating density of spider mites on grape leaves. *Experimental and Applied Acarology* 67(4), 583–94. doi:10.1007/s10493-015-9972-6.

Malek, R., Tattoni, C., Ciolli, M., Corradini, S., Andreis, D., Ibrahim, A., Mazzoni, V., Eriksson, A. and Anfora, G. 2018. Coupling traditional monitoring and citizen science to disentangle the invasion of *Halyomorpha halys*. *ISPRS International Journal of Geo-Information* 7(5), 171. doi:10.3390/ijgi7050171.

Manoukis, N. C. and Hoffman, K. 2014. An agent-based simulation of extirpation of *Ceratitis capitata* applied to invasions in California. *Journal of Pest Science* 87, 39–51. doi:10.1007/s10340-013-0513-y.

Manoukis, N. C., Hall, B. and Geib, S. M. 2014. A computer model of insect traps in a landscape. *Scientific Reports* 4, 7015. doi:10.1038/srep07015.

Manoukis, N. C., Siderhurst, M. and Jang, E. B. 2015. Field estimates of attraction of *Ceratitis capitata* to trimedlure and *Bactrocera dorsalis* (Diptera: Tephritidae) to methyl eugenol in varying environments. *Environmental Entomology* 44(3), 695–703. doi:10.1093/ee/nvv020.

Manrakhan, A. and Lux, S. A. 2009. Diel and lifetime patterns of feeding and reproductive activities of three African fruit flies, *Ceratitis cosyra*, *C. fasciventris* and *C. capitata* (Diptera: Tephritidae), in semi-field cages of different spatial scales. *African Entomology* 17, 8–22.

Maric, M., Orovic, I. and Stankovic, S. 2016. *Compressive Sensing Based Image Processing in TrapView Pest Monitoring System*. IEEE Publications, New York.

Martin, B., Juliet, V., Sankaranarayanan, P. E., Gopal, A. and Rajkumar, I. 2013. Wireless implementation of mems accelerometer to detect red palm weevil on palms. In: *2013 International Conference on Advanced Electronic Systems (ICAES)*, Pilani, India, 21–23 September 2013, pp. 248–52.

Martineau, M., Conte, D., Raveaux, R., Arnault, I., Munier, D. and Venturini, G. 2017. A survey on image-based insect classification. *Pattern Recognition* 65, 273–84. doi:10.1016/j.patcog.2016.12.020.

Mayer, D. F., Long, L. E., Smith, T. J., Olsen, J., Riedl, H., Heath, R. R., Leskey, T. C. and Prokopy, R. J. 2000. Attraction of adult *Rhagoletis indifferens* (Diptera: Tephritidae) to unbaited and odor-baited red spheres and yellow rectangles. *Journal of Economic Entomology* 93(2), 347–51. doi:10.1603/0022-0493-93.2.347.

Miller, J. R., Adams, C. G., Weston, P. A. and Schenker, J. H. 2015a. *Automated Systems for Recording, Reporting, and Analyzing Trapping Data*. *Trapping of Small Organisms Moving Randomly - Principles and Applications to Pest Monitoring and Management* (1st ed.). Springer International Publishing, New York.

Miller, J. R., Adams, C. G., Weston, P. A. and Schenker, J. H. 2015b. *Trapping of Small Organisms Moving Randomly - Principles and Applications to Pest Monitoring and Management*. Springer International Publishing, New York.

Miranda, J. L., Gerardo, B. D. and Tanguilig III, B. T. 2014. Pest detection and extraction using image processing techniques. *International Journal of Computer and Communication Engineering* 3(3), 189–92. doi:10.7763/IJCCE.2014.V3.317.

Moore, A. and Miller, R. H. 2002. Automated identification of optically sensed aphid (Homoptera: Aphidae) wingbeat waveforms. *Annals of the Entomological Society of America* 95(1), 1–8. doi:10.1603/0013-8746(2002)095[0001:AIOOSA]2.0.CO;2.

Morrison, W. R., Cullum, J. P. and Leskey, T. C. 2015. Evaluation of trap designs and deployment strategies for capturing *Halyomorpha halys* (Hemiptera: Pentatomidae). *Journal of Economic Entomology* 108(4), 1683–92. doi:10.1093/jee/tov159.

Morrison, W. R., Milonas, P., Kapantaidaki, D. E., Cesari, M., Di Bella, E., Guidetti, R., Haye, T., Maistrello, L., Moraglio, S. T., Piemontese, L., et al. 2017. Attraction of *Halyomorpha halys* (Hemiptera: Pentatomidae) haplotypes in North America and Europe to baited traps. *Scientific Reports* 7(1), 16941. doi:10.1038/s41598-017-17233-0.

Mudavanhu, P., Addison, P. and Ken, L. P. 2011. Monitoring and action threshold determination for the obscure mealybug *Pseudococcus viburni* (Signoret) (Hemiptera: Pseudococcidae) using pheromone-baited traps. *Crop Protection* 30(7), 919–24. doi:10.1016/j.cropro.2011.02.034.

Ni, W. L., Li, Z. H., Chen, H. J., Wan, F. H., Qu, W. W., Zhang, Z. and Kriticos, D. J. 2012. Including climate change in pest risk assessment: the peach fruit fly, *Bactrocera zonata* (Diptera: Tephritidae). *Bulletin of Entomological Research* 102(2), 173–83. doi:10.1017/S0007485311000538.

Nyrop, J. P., Binns, M. R. and Van Der Werf, W. 1999. Sampling for IPM decision making: where should we invest time and resources? *Phytopathology* 89(11), 1104–11. doi:10.1094/PHYTO.1999.89.11.1104.

Ohlendorf, B. 1999. *Integrated Pest Management for Apples and Pears* (2nd ed.). Agriculture & Natural Resources, Oakland, Publication 3340.

Owens, E. D. and Prokopy, R. J. 1978. Visual monitoring trap for European apple sawfly Hymenoptera-Tenthredinidae. *Journal of Economic Entomology* 71(4), 576–8. doi:10.1093/jee/71.4.576.

Padilha, A. C., Arioli, C. J., Boff, M. I. C., Rosa, J. M. and Botton, M. 2018. Traps and baits for luring *Grapholita molesta* (Busck) adults in mating disruption-treated apple orchards. *Neotropical Entomology* 47(1), 152–9. doi:10.1007/s13744-017-0517-z.

Padma, T., Mir, S. A. and Shantharajah, S. P. 2017. Intelligent decision support system for an integrated pest management in apple orchard. In: Sangaiah, A. K., Abraham, A., Siarry, P. and Sheng, M. (Eds), *Intelligent Decision Support Systems for Sustainable Computing* (1st ed.). Springer International Publishing, Cham, pp. 225–46.

Pagabeleguem, S., Ravel, S., Dicko, A. H., Vreysen, M. J., Parker, A., Takac, P., Huber, K., Sidibe, I., Gimonneau, G. and Bouyer, J. 2016. Influence of temperature and relative humidity on survival and fecundity of three tsetse strains. *Parasites and Vectors* 9(1), 520. doi:10.1186/s13071-016-1805-x.

Papadopoulos, N. T., Lux, S. A., Köppler, K. and Beliën, T. 2017. Invertebrate and vertebrate pests: biology and management. In: Quero-García, J., Lezzoni, A., Puławska, J. and Lang, G. (Eds), *Cherries: Botany, Production and Uses*. CABI Publishing, Boston, MA, pp. 305–37.

Pearsall, I. A. and Myers, J. H. 2000. Evaluation of sampling methodology for determining the phenology, relative density, and dispersion of Western flower thrips (Thysanoptera: Thripidae) in nectarine orchards. *Journal of Economic Entomology* 93(2), 494-502. doi:10.1603/0022-0493-93.2.494.

Pedigo, L. A. and Buntin, G. D. 1994. *Handbook of Sampling Methods for Arthropods in Agriculture*. CRC Press, Boca Raton.

Perez-Rodriguez, J., Martinez-Blay, V., Soto, A., Selfa, J., Monzo, C., Urbaneja, A. and Tena, A. 2017. Aggregation patterns, sampling plan, and economic injury levels for the new citrus pest *Delottococcus aberiae* (Hemiptera: Pseudococcidae). *Journal of Economic Entomology* 110(6), 2699–706. doi:10.1093/jee/tox258.

Phillips, S. J., Anderson, R. P. and Schapire, R. E. 2006. Maximum entropy modeling of species geographic distributions. *Ecological Modelling* 190, 231–59.

Pimentel, R., Lopes, D. J. H., Mexia, A. M. M. and Mumford, J. D. 2017. Validation of a geographic weighted regression analysis as a tool for area-wide integrated pest management programs for *Ceratitis capitata* Wiedemann (Diptera: Tephritidae) on Terceira Island, Azores. *International Journal of Pest Management* 63(2), 172–84. doi :10.1080/09670874.2016.1256512.

Pinhas, J., Soroker, V., Hetzroni, A., Mizrach, A., Teicher, M. and Goldberger, J. 2008. Automatic acoustic detection of the red palm weevil. *Computers and Electronics in Agriculture* 63, 131–9.

Pontikakos, C. M., Tsiligiridis, T. A., Yialouris, C. P. and Kontodimas, D. C. 2012. Pest management control of olive fruit fly (*Bactrocera oleae*) based on a location-aware agro-environmental system. *Computers and Electronics in Agriculture* 87, 39–50.

Potamitis, I. and Rigakis, I. 2016. Large aperture optoelectronic devices to record and time-stamp insects' wingbeats. *IEEE Sensors Journal* 16, 6053–61.

Potamitis, I., Rigakis, I. and Fysarakis, K. 2014. The electronic McPhail trap. *Sensors* 14(12), 22285–99. doi:10.3390/s141222285.

Potamitis, I., Rigakis, I. and Fysarakis, K. 2015. Insect biometrics: optoacoustic signal processing and its applications to remote monitoring of McPhail type traps. *PLoS ONE* 10(11), e0140474. doi:10.1371/journal.pone.0140474.

Potamitis, I., Eliopoulos, P. and Rigakis, I. 2017a. Automated remote insect surveillance at a global scale and the internet of things. *Robotics* 6(3), 19. doi:10.3390/robotics6030019.

Potamitis, I., Rigakis, I. and Tatlas, N. A. 2017b. Automated surveillance of fruit flies. *Sensors* 17(1), 110. doi:10.3390/s17010110.

Potamitis, I., Rigakis, I., Vidakis, N., Petousis, M. and Weber, M. 2018. Affordable bimodal optical sensors to spread the use of automated insect monitoring. *Journal of Sensors* 2018, 1–25. doi:10.1155/2018/3949415.

Prokopy, R. J. and Hauschild, K. I. 1979. Comparative effectiveness of sticky red spheres and PHEROCON AM standard traps for monitoring apple maggot flies (Diptera, Tephritidae) in commercial orchards. *Environmental Entomology* 8, 696–700.

Qiao, M., Lim, J., Ji, C. W., Chung, B.-K., Kim, H.-Y., Uhm, K.-B., Myung, C. S., Cho, J. and Chon, T.-S. 2008. Density estimation of *Bemisia tabaci* (Hemiptera: Aleyrodidae) in a greenhouse using sticky traps in conjunction with an image processing system. *Journal of Asia-Pacific Entomology* 11(1), 25–9. doi:10.1016/j.aspen.2008.03.002.

Raman, D., Gerhardt, R. R. and Wilkerson, J. B. 2007. Detecting insect flight sounds in the field: implications for acoustical counting of mosquitoes. *Agricultural and Biosystems Engineering Publications* 50, 1481–1485.

Renkema, J. M., Buitenhuis, R. and Hallett, R. H. 2014. Optimizing TRAP Design and Trapping Protocols for *Drosophila suzukii* (Diptera: Drosophilidae). *Journal of Economic Entomology* 107(6), 2107–18. doi:10.1603/EC14254.

Rice, K. B., Short, B. D., Jones, S. K. and Leskey, T. C. 2016. Behavioral responses of *Drosophila suzukii* (Diptera: Drosophilidae) to visual stimuli under laboratory, semifield, and field conditions (Diptera: Drosophilidae). *Environmental Entomology* 45(6), 1480–8. doi:10.1093/ee/nvw123.

Rice, K. B., Short, B. D. and Leskey, T. C. 2017. Development of an attract-and-kill strategy for *Drosophila suzukii* (Diptera: Drosophilidae): evaluation of attracticidal spheres under laboratory and field conditions. *Journal of Economic Entomology* 110(2), 535–42. doi:10.1093/jee/tow319.

Rice, K. B., Bedoukian, R. H., Hamilton, G. C., Jentsch, P., Khrimian, A., Maclean, P., Morrison, W. R., Short, B. D., Shrewsbury, P., Weber, D. C., et al. 2018a. Enhanced response of *Halyomorpha halys* (Hemiptera: Pentatomidae) to its aggregation pheromone with ethyl decatrienoate. *Journal of Economic Entomology* 111(1), 495–9. doi:10.1093/jee/tox316.

Rice, K. B., Morrison, W. R., Short, B. D., Acebes-Doria, A., Bergh, J. C. and Leskey, T. C. 2018b. Improved trap designs and retention mechanisms for *Halyomorpha halys* (Hemiptera: Pentatomidae). *Journal of Economic Entomology* 111(5), 2136–42. doi:10.1093/jee/toy185.

Robertson, S. P., Hull, L. A. and Calvin, D. D. 2005. Tufted apple bud moth (Lepidoptera: Tortricidae) management model for processing apples based on early season pheromone trap capture. *Journal of Economic Entomology* 98(4), 1229–35. doi:10.1603/0022-0493-98.4.1229.

Samietz, J., Graf, B., Höhn, H., Schaub, L. and Höpli, H. U. 2007. Phenology modelling of major insect pests in fruit orchards from biological basics to decision support: the forecasting tool SOPRA. *EPPO Bulletin* 37(2), 255–60. doi:10.1111/j.1365-2338.2007.01121.x.

Samietz, J., Graf, B., Hohn, H., Schaub, L., Ulrich, H. and Razavi, E. 2011. Web-based decision support for sustainable pest management in fruit orchards: development of the Swiss system SOPRA. In: Jao, C. (Ed.), *Efficient Decision Support Systems-Practice and Challenges From Current to Future*. InTech, Rijeka, Croatia, pp. 373–88.

Sanchez, J. A. and Ortin-Angulo, M. C. 2011. Sampling of *Cacopsylla pyri* (Hemiptera: Psyllidae) and *Pilophorus gallicus* (Hemiptera: Miridae) in pear orchards. *Journal of Economic Entomology* 104(5), 1742–51. doi:10.1603/EC11040.

Schreiner, R. P., Skinkis, P. A. and Dreves, A. J. 2014. A rapid method to assess grape rust mites on leaves and observations from case studies in Western Oregon vineyards. *HortTechnology* 24, 38–47.

SCOUTbox. 2018. Insect scouting made easy (Online). Available at: http://www.soilcares.com/en/products/scoutbox/ (accessed on 27 June 2018).

Selby, R. D., Gage, S. H. and Whalon, M. E. 2014. Precise and low-cost monitoring of plum curculio (Coleoptera: Curculionidae) pest activity in pyramid traps with cameras. *Environmental Entomology* 43(2), 421–31. doi:10.1603/EN13136.

Semios. 2018. Automated, remotely controlled climate, insect and disease monitoring and treatment (Online). Available at: http://semios.com (accessed on 24 June 2018).

Shearer, P. W., West, J. D., Walton, V. M., Brown, P. H., Svetec, N. and Chiu, J. C. 2016. Seasonal cues induce phenotypic plasticity of *Drosophila suzukii* to enhance winter survival. *BMC Ecology* 16, 11. doi:10.1186/s12898-016-0070-3.

Solis-Sánchez, L. O., García-Escalante, J. J., Castañeda-Miranda, R., Torres-Pacheco, I. and Guevara-González, R. 2009. Machine vision algorithm for whiteflies (*Bemisia tabaci* Genn.) scouting under greenhouse environment. *Journal of Applied Entomology* 133(7), 546–52. doi:10.1111/j.1439-0418.2009.01400.x.

Solis-Sánchez, L. O., Castañeda-Miranda, R., García-Escalante, J. J., Torres-Pacheco, I., Guevara-González, R. G., Castañeda-Miranda, C. L. and Alaniz-Lumbreras, P. D. 2011. Scale invariant feature approach for insect monitoring. *Computers and Electronics in Agriculture* 75(1), 92–9. doi:10.1016/j.compag.2010.10.001.

Spensa. 2018. Automated insect monitoring and smart data-driven scouting (Online). Available at: https://spensatech.com (accessed on 24 June 2018).

Sporleder, M., Tonnang, H. E. Z., Carhuapoma, P., Gonzales, J. C., Juarez, H. and Kroschel, J. 2013. Insect life cycle modelling (ILCYM) software – a new tool for regional and global insect pest risk assessments under current and future climate change scenarios. In: Peña, J. E. (Ed.), *Potential Invasive Pests of Agricultural Crops*. CABI, Croydon.

Suckling, D. M. 2000. Issues affecting the use of pheromones and other semiochemicals in orchards. *Crop Protection* 19(8–10), 677–83. doi:10.1016/S0261-2194(00)00090-9.

Sutherst, R. W., Maywald, G. F. and Russell, B. L. 2000. Estimating vulnerability under global change: modular modelling of pests. *Agriculture, Ecosystems and Environment* 82(1–3), 303–19. doi:10.1016/S0167-8809(00)00234-6.

Sutherst, R. W., Maywald, G. F. and Kriticos, D. J. 2007. CLIMEX version 3: user's guide (Online). Available at: https://www.hearne.software/Software/CLIMEX-DYMEX/Editions (accessed on 23 July 2018).

Szyniszewska, A. M. and Tatem, A. J. 2014. Global assessment of seasonal potential distribution of Mediterranean fruit fly, *Ceratitis capitata* (Diptera: Tephritidae). *PLoS ONE* 9(11), e111582. doi:10.1371/journal.pone.0111582.

Tanga, C. M., Khamis, F. M., Tonnang, H. E. Z., Rwomushana, I., Mosomtai, G., Mohamed, S. A. and Ekesi, S. 2018. Risk assessment and spread of the potentially invasive *Ceratitis rosa* Karsch and *Ceratitis quilicii* De Meyer, Mwatawala & Virgilio sp. Nov.

using life-cycle simulation models: implications for phytosanitary measures and management. *PLoS ONE* 13(1), e0189138. doi:10.1371/journal.pone.0189138.

Terblanche, J. S., Klok, C. J., Krafsur, E. S. and Chown, S. L. 2006. Phenotypic plasticity and geographic variation in thermal tolerance and water loss of the tsetse *Glossina pallidipes* (Diptera: Glossinidae): implications for distribution modelling. *The American Journal of Tropical Medicine and Hygiene* 74(5), 786–94. doi:10.4269/ajtmh.2006.74.786.

Thomson, D., Brunner, J., Gut, L., Judd, G. and Knight, A. 2001. Ten years implementing codling moth mating disruption in the orchards of Washington and British Columbia : starting right and managing for success ! *IOBC WPRS Bulletin* 24, 23–30.

Tirelli, P., Borghese, N. A., Pedersini, F., Galassi, G. and Oberti, R. 2011. Automatic monitoring of pest insects traps by Zigbee-based wireless networking of image sensors. In: *2011 IEEE International Instrumentation and Measurement Technology Conference*, Hangzhou, China, 10–12 May 2011, pp. 1–5.

Tobin, P. C., Klein, K. T. and Leonard, D. S. 2009. Gypsy moth (Lepidoptera: Lymantriidae) flight behavior and phenology based on field-deployed automated pheromone-baited traps. *Environmental Entomology* 38(6), 1555–62. doi:10.1603/022.038.0606.

Tonnang, E. Z. H., Juarez, H., Carhuapoma, P., Gonzales, J. C., Mendoza, D., Sporleder, M., Simon, R. and Kroschel, J. 2009. ILCYM – Insect Life Cycle Modeling. A software package for developing temperature-based insect phenology models with applications for local, regional and global analysis of insect population and mapping. 15th Triennial ISTRC Symposium. International Society for Tropical Root Crops (ISTRC), Lima, Peru.

Tonnang, H. E. Z., Hervé, B. D. B., Biber-Freudenberger, L., Salifu, D., Subramanian, S., Ngowi, V. B., Guimapi, R. Y. A., Anani, B., Kakmeni, F. M. M., Affognon, H., et al. 2017. Advances in crop insect modelling methods–towards a whole system approach. *Ecological Modelling* 354, 88–103.

TRAPVIEW. 2018. Automated pest monitoring system (Online). Available at: http://www.trapview.com (accessed on 24 June 2018).

Uvarov, B. P. 1931. Insects and climate. *Transactions of the Royal Entomological Society of London* 79(1), 1–232. doi:10.1111/j.1365-2311.1931.tb00696.x.

Van Roy, J., De Baerdemaeker, J., Saeys, W. and De Ketelaere, B. 2014. Optical identification of bumblebee species: effect of morphology on wingbeat frequency. *Computers and Electronics in Agriculture* 109, 94–100. doi:10.1016/j.compag.2014.09.014.

Veres, A., Kaszó, M., Kaszó, L., Petővári, B., Vasileiadis, V. P., Pantazis, S. and Kárpinszky, A. 2017. Development of digital mite recognition system to monitor two-spotted spider mite (*Tetranychus urticae*) and its potential applicability in fruit crops. *IOBC-WPRS Bulletin* 123, 120–21.

Wakie, T. T., Yee, W. L. and Neven, L. G. 2018. Assessing the risk of establishment of *Rhagoletis cerasi* (Diptera: Tephritidae) in the United States and globally. *Journal of Economic Entomology* 111(3), 1275–84. doi:10.1093/jee/toy054.

Wang, K., Zhang, S., Wang, Z., Liu, Z. and Yang, F. 2013. Mobile smart device-based vegetable disease and insect pest recognition method. *Intelligent Automation and Soft Computing* 19, 263–73.

Weber, D. C., Leskey, T. C., Walsh, G. C. and Khrimian, A. 2014. Synergy of aggregation pheromone with methyl (E,E,Z)-2,4,6-decatrienoate in attraction of *Halyomorpha halys* (Hemiptera: Pentatomidae). *Journal of Economic Entomology* 107(3), 1061–8. doi:10.1603/EC13502.

Weber, D. C., Morrison, W. R., Khrimian, A., Rice, K. B., Leskey, T. C., Rodriguez-Saona, C., Nielsen, A. L. and Blaauw, B. R. 2017. Chemical ecology of *Halyomorpha halys*: discoveries and applications. *Journal of Pest Science* 90(4), 989–1008. doi:10.1007/s10340-017-0876-6.

Weldon, C. W., Nyamukondiwa, C., Karsten, M., Chown, S. L. and Terblanche, J. S. 2018. Geographic variation and plasticity in climate stress resistance among southern African populations of *Ceratitis capitata* (Wiedemann) (Diptera: Tephritidae). *Scientific Reports* 8(1), 9849. doi:10.1038/s41598-018-28259-3.

Welsh, T., Singh, F. S. and Suckling, D. M. 2017. Automatic detection and identification of invasive arthropod pests for pest management and biosecurity. In: *7th Asian-Australasian Conference on Precision Agriculture*, Hamilton, New Zealand.

Wen, C. and Guyer, D. 2012. Image-based orchard insect automated identification and classification method. *Computers and Electronics in Agriculture* 89, 110–5. doi:10.1016/j.compag.2012.08.008.

Wen, C., Guyer, D. E. and Li, W. 2009. Local feature-based identification and classification for orchard insects. *Biosystems Engineering* 104, 299–307.

Witzgall, P., Kirsch, P. and Cork, A. 2010. Sex pheromones and their impact on pest management. *Journal of Chemical Ecology* 36(1), 80–100. doi:10.1007/s10886-009-9737-y.

Wójtowicz, A. 2010. Field assessment of reliability of forecasting the development of Colorado potato beetle population using decision support system SimLep 3. *Progress in Plant Protection* 50, 1318–21.

Wong, J. S., Wallingford, A. K., Loeb, G. M. and Lee, J. C. 2018. Physiological status of *Drosophila suzukii* (Diptera: Drosophilidae) affects their response to attractive odours (Diptera: Drosophilidae). *Journal of Applied Entomology* 142(5), 473–82. doi:10.1111/jen.12497.

Xia, C. L., Lee, J. M., Li, Y., Chung, B. K. and Chon, T. S. 2012. In situ detection of small-size insect pests sampled on traps using multifractal analysis. *Optical Engineering* 51, 12.

Xia, C., Chon, T.-S., Ren, Z. and Lee, J.-M. 2015. Automatic identification and counting of small size pests in greenhouse conditions with low computational cost. *Ecological Informatics* 29, 139–46. doi:10.1016/j.ecoinf.2014.09.006.

Yonow, T., Kriticos, D. J. and Ota, N. 2017. The potential distribution of cassava mealybug (*Phenacoccus manihoti*), a threat to food security for the poor. *PLoS ONE* 12(3), e0173265. doi:10.1371/journal.pone.0173265.

Zacharda, M., Pultar, O. and Muška, J. 1988. Washing technique for monitoring mites in apple orchards. *Experimental and Applied Acarology* 5(1-2), 181–3. doi:10.1007/BF02053827.

Zhong, Y., Gao, J., Lei, Q. and Zhou, Y. 2018. A vision-based counting and recognition system for flying insects in intelligent agriculture. *Sensors* 18, 1489.

Chapter 21

Biological control in integrated management of deciduous fruit insect pests: the use of semiochemicals

Larry Gut, Christopher Adams and James Miller, Michigan State University, USA; and Peter McGhee and Donald Thomson, Pacific Biocontrol Corporation, USA

1 Introduction

Many insects and other organisms rely on chemical messages for reproduction and survival. Essential insect behaviours that may be stimulated or inhibited by olfactory cues include mating, host finding and oviposition. Behaviour-modifying chemicals are collectively referred to as semiochemicals. Over the past 50 years, considerable headway in understanding insect behaviour coupled with advances in analytical chemistry have led to the identification of thousands of semiochemicals (El-Sayed, 2016; Hardie and Minks, 1999; Mayer and McLaughlin, 1991).

Progress in understanding chemical communication in insects spurred the development of numerous practical applications, many of which are now integral components of pest management programmes (Cork, 2016; Evenden, 2016; Suckling, 2016; Miller and Gut, 2015; Witzgall et al., 2010; Cardé, 2007; Gut et al., 2004; Suckling and Karg, 2000). Applied scientists readily established the utility of semiochemicals for trapping insects and recognized the advantages of behaviourally active chemicals over insecticides for direct control. The deployment of semiochemicals to control insects offered advantages including

http://dx.doi.org/10.19103/AS.2019.0046.27

a limited environmental footprint and strong compatibility with biological and cultural methods of control. For the last 40 years, there has been an intense effort to develop and commercialize control tactics based on semiochemicals.

While enthusiasm for finding novel alternatives to insecticides was strong, inherent weaknesses of behaviourally based controls, as well as practical and economic considerations, slowed their development and adoption. Most are pest density dependent and consequently, behavioural controls often are more suited for management of low rather than high-density pest populations (Knipling, 1979). In contrast, a great strength of insecticides is their propensity to operate independently of pest density. Insecticides also have the advantages of greatly reducing pest density through their lethal action and only requiring a short period of activity to provide the desired outcome. Behavioural controls, on the other hand, generally do not kill the pest and must remain effective for longer durations, usually throughout a generation or the lifetime of a pest (Miller and Cowles, 1990).

Economics played a crucial role in limiting the use of semiochemical-based products. Although pheromones and other attractants are used in small quantities, high purity requirements, complex synthetic pathways and the small scale of production have kept the cost of the active ingredients relatively high in comparison to insecticides. Furthermore, to be economically viable, the use of behavioural controls must be accompanied by a reduction in insecticide use. This can be difficult to achieve when the approach is designed to control a specifically targeted pest species. Most production systems have a diverse pest complex. A strength of insecticides is that a single product often will control multiple pest species. On the other hand, the application of mating disruption and other behavioural tactics often must be accompanied by the use of insecticides or other controls to manage non-target pests.

Despite these challenges, over the past 40 years there has been considerable progress in the development and implementation of control tactics based on semiochemicals. While scientists, biopesticide companies and practitioners were instrumental in forging a path forward, progress has to a great extent been crisis driven. New regulations governing the use of pesticides, increasing environmental concerns, food and worker safety issues and the regular occurrence of resistance to chemical controls have provided strong impetus for adoption of behavioural controls. In addition, growing numbers of producers have recognized the advantages these 'soft' controls provide by conserving and enhancing populations of natural enemies.

Deciduous fruit producers have been especially proactive in adopting behaviourally based technologies. Traps baited with sex pheromones or attractant kairomones have become the principal means of detecting the presence of pests and monitoring their seasonal activity. Captures in attractant-baited traps also are used to determine the need for and to time insecticide treatments. Semiochemicals form the basis for three direct control measures that have been incorporated into fruit pest management programmes. Two strategies, mass trapping and attract-and-kill, are designed to permanently remove individuals from the population. They require a high level of attraction to the device and an efficient means of capturing or killing them. Instead of eliminating males, the most developed and commercialized direct control technique prevents them from locating females. The approach, commonly referred to as mating disruption, entails dispersing synthetic sex pheromone into the crop to interfere with the male's normal behavioural response to these compounds, thereby controlling the pest by truncating reproduction. The technique is now used on more than 1.1 million ha worldwide; about 80% of the treated acreage are fruit crops (Table 1; Miller and Gut, 2015). All three semiochemical-based tactics are used alone or in combination with other control measures.

This chapter focusses on the uses of behaviour-modifying chemicals for direct control of insect pests of deciduous tree fruits and berry crops. Over the past 40 years, our personal efforts have centred around the practical and commercial use of various semiochemical-based technologies in these perennial production systems. In keeping with the intent of this book, the aim is to review the progress that researchers and practitioners have made as they conceptualized, developed and implemented behaviourally based controls. Although the potential of mass trapping and attract-and-kill are reviewed, the core of the chapter focusses on the development and future prospects of mating disruption. We have emphasized this approach over others largely for practical

Table 1 Mating disruption formulations registered for control of deciduous fruit and nut pests and the estimated area-treated worldwide

Species	Principal crop(s)	Types of formulations[a]	Hectares treated[b]
Amyelois transitella	Pistachio, almond	LD	120 000
Cydia pomonella	Pome fruit, walnut	HD, MD, LD	243 000
Grapholita molesta	Stone and pome fruit	HD, LD	65 000
Lobesia botrana	Grape	LD, MD	400 000
Eupoecilia ambiguella	Grape	MD	60 000
Endopiza viteana	Grape	MD	1 000
Leafrollers	Apple, pear, peach, grape	MD	15 000
Synanthedon spp.	Peach, apricot, current	MD	6 000
Zeuzera pyrina	Pear	MD	3 000
Total			913 000

[a] HD = High density such as sprayable, MD = Moderate density such as solid matrix, LD = Low density such as aerosol emitter.
[b] Estimates courtesy of Pacific Biocontrol and Shin-Etsu Corporation.

reasons. A review of the literature quickly reveals that mating disruption is the most highly developed and widely commercialized semiochemical-based control tactic. Many of the successful applications are in deciduous fruit crops, as are a majority of the studies on how mating disruption operates to control insect pests (Evenden, 2016).

2 Pheromone-mediated mating disruption

Many insects, especially moth species, rely on sex pheromones to locate mates. The possibility of controlling insects by dispersing large amounts of sex attractant into a crop in an effort to interfere with mating was first demonstrated for *Trichoplusia ni* over 40 years ago (Gaston et al., 1967). This groundbreaking research, combined with advances in analytical methods that facilitated pheromone identification and synthesis for many insect species, stimulated a groundswell of research confirming that dispensing large quantities of pheromone into a crop could disrupt mate location, thereby controlling the pest by curtailing reproduction (Ridgeway et al., 1990).

Commercialization of mating disruption, however, lagged behind the research. The first registered product in the United States was for the pink bollworm (*Pectinophora gossypiella*) in 1978 (Brooks et al., 1979). Registrations for additional species came slowly for at least a decade, but then accelerated over the past 25 years (Miller and Gut, 2015). The first mating disruption products for the two key pests of deciduous fruit crops Oriental fruit moth (*Grapholita molesta*) and codling moth (*Cydia pomonella*) were granted by the US Environmental Protection Agency in 1986 and 1991, respectively. Currently mating disruptants are registered in the United States for control of more than 20 pest species.

Through the combined efforts of researchers, private industry, extension educators and others, a viable and expanding mating disruption industry has emerged. We estimate that in excess of 1 million ha are currently treated worldwide with this novel method of control (Table 1; Miller and Gut, 2015). This represents more than a doubling in the commercially treated area since 2002 (Cardé, 2007). Mating disruption products have largely been developed for control of lepidopteran species, targeting fruit, forest, cotton and vegetable pests. Deciduous fruit and nut pests have garnered a great deal of attention (Ioriatti et al., 2008; Witzgal et al., 2008; Sexton and Il'ichev, 2000; Thomson et al., 1998; Charmillot, 1990; Rice and Kirsch, 1990; Snow, 1990; Vickers, 1990) and are decidedly the most commercially successful use of mating disruption. A dozen pests of these crops account for over 800 000 ha, or 80% of the total area-treated worldwide with pheromone (Table 1). Two pests of tree fruit, *C. pomonella* and *G. molesta,* and a single grape pest, European grapevine moth (*Lobesia botrana*), account for over 700 000 of the treated ha.

The discussion that follows is a review of the developments that led to mating disruption becoming an efficacious and commercially accepted control tactic for deciduous fruit producers. Progress in understanding moth behaviour and other biological characteristics that influence the outcome of pheromone-based control programmes were instrumental. This includes an improved understanding of the principal mechanisms responsible for mating disruption. However, many crucial advances were technical or operational in nature, such as the development of devices that could deliver the optimal dose of pheromone in space and time and new tools for monitoring pest populations under a pheromone regime. The review concludes with an attempt to identify key areas of research needed to optimize mating disruption formulations, select pest species that are most amenable to the method and enhance opportunities for successful mating disruption in deciduous fruit production systems.

2.1 Categories of mating disruption

Four main mechanisms have been proposed to explain how treatment with synthetic pheromone might influence mate location (Evenden, 2016; Cardé, 1990, 2007; Bartell, 1982). (1) *Competition*: Synthetic and natural pheromone compete for the attention of mate searching males. Early researchers used the terms confusion (Shorey, 1977), false-trail following (Bartell, 1982) or false-plume following (Cardé, 1990) to describe this means of achieving disruption. (2) *Desensitization*: Impairment of the sensory system interferes with the male's ability to sense and respond normally to pheromone. (3) *Camouflage*: The spatial integrity of the female signal is compromised resulting in the inability of males to recognize and locate her because the guiding boundaries of the pheromone plume are obscured by ubiquitously present synthetic pheromone. (4) *Sensory imbalance*: The active ingredient released from dispensers distorts the normal perception of the female signal. Pheromone treatments likely invoke more than one mechanism. For example, dispensers may first compete with females by drawing males towards the synthetic source, while exposure to a high dose of pheromone once males approach close to the dispenser may result in desensitization (Baker et al., 2016; Miller et al., 2010).

Identifying the principal mechanism of mating disruption for a particular species, dispenser type, release rate and environmental conditions is crucial to furthering the development and use of this unique control method (see later discussion). However, sorting out which mechanism is responsible for disruption in the field has been challenging. In most studies, there is empirical evidence supporting more than one possible cause for disruption. To address the problem of differentiating among the mechanisms of mating disruption, Miller et al. (2006a) and Miller and Gut (2015) developed a framework that places the various mechanisms into two broad categories, competitive and non-competitive. The two principal causes of mating disruption are differentiated based on whether or not the disruptant has impaired the physiology of the male, the female or the female's signal prior to commencing sexual activity. The major mechanism operating in the absence of initial impairment entails competition between authentic and false females. Disruption is achieved by enticing males to interact with dispensers releasing synthetic pheromone, making them unavailable to females for some period of time. Competitive disruption is essentially a numbers game; it is strongest when competing sources greatly outnumber females (Miller and Gut, 2015).

Non-competitive mechanisms operate by interfering with the male's ability to respond normally to pheromone, or by suppressing female calling or mating (Fig. 1). The impact occurs before any sexual activity is initiated, thus precluding attraction to the disruptant in most cases. The remaining major modes of disruption introduced previously, desensitization, camouflage and sensory imbalance, are non-competitive mechanisms. Disruption treatments that involve impairment to the sensory system causing males to be less responsive to pheromone fall under desensitization. The effect may be due to adaptation of sensory receptors, habituation of the central nervous system or both. Pheromone exposure may not completely impair responsiveness. It is possible that exposure to a high dose of pheromone elevates the male response threshold allowing him to orient to high-releasing sources but not to the low dose of pheromone emitted by a lure or calling female (Reinke et al., 2014).

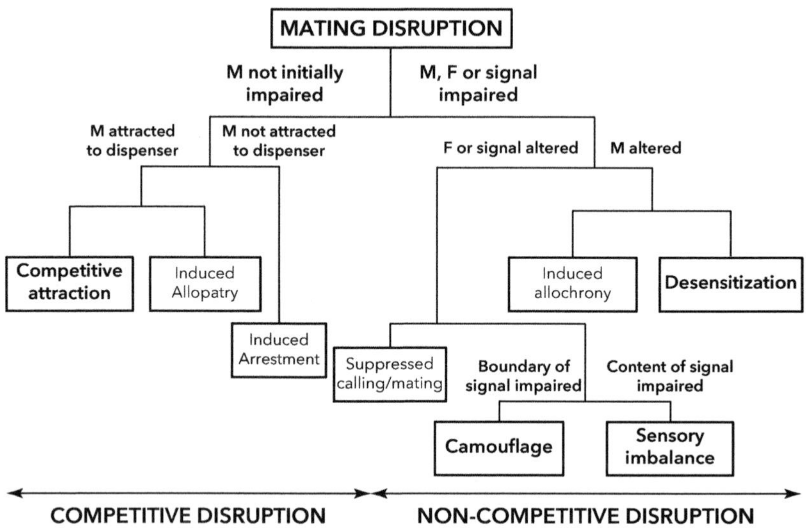

Figure 1 A framework that places the various mechanisms proposed for mating disruption into two broad categories and the criteria for differentiating among the four principal mechanisms that are shown in bold. Source: adapted from Miller and Gut (2016).

In some cases, non-competitive disruption may occur without physiological impairment to males. Rather, the disruption treatment prevents males from locating females through some behaviourally based mode of interference. Camouflage entails a compromise in the spatial integrity of the female signal; males cannot find a calling female because her pheromone plume is concealed by the background of synthetic pheromone. Sensory imbalance entails a compromise in the chemical composition of the female signal; the active ingredient released from dispensers distorts the normal perception of the signal. Any disruption resulting from exposure to off-blend stimuli is often considered as operating by sensory imbalance; however, if the exposure impairs male physiology, it more appropriately belongs in the desensitization category (Miller and Gut, 2015).

Along with impeding the ability of males to find females, a disruption treatment may suppress mating through direct effects on the female (Fig. 1). Autodetection of the female-produced pheromone has been demonstrated for several deciduous fruit pests (den Otter et al., 1978; DeLury et al., 2005; Stelinski et al., 2006b; Gökce et al., 2007). Exposure of females to pheromone-permeated air in a laboratory setting can affect the onset, duration or propensity of calling (Weissling and Knight, 1996; Gökce et al., 2007). Pheromone exposure in a wind tunnel reduced flight activity but did not affect calling behaviour of three-lined leafroller female (*Pandemis limitata*) (DeLury et al., 2005). Egg laying by redbanded leafroller (*Argyrotaenia velutina*) and obliquebanded leafroller (*Choristoneura rosaceana*) was significantly reduced in the presence of pheromone-laden air (Gökce et al., 2007). Pheromone pre-exposure of Oriental fruit moth and pandemis leafroller (*Pandemis pyrusana*) females reduced copulation, but not calling (Kuhns et al., 2012). Although these studies indicate that mating disruption treatments have the potential to impact female behaviour, few studies have documented direct effects in the field. European grapevine moth females caged in vineyards treated with mating disruption moved more, but called less frequently, than females in untreated vineyards (Harari et al., 2015). The modifying influence of field exposure of females to pheromone, and its possible contribution to the efficacy of mating disruption commands greater attention.

2.2 Differentiating among disruption mechanisms

Disentangling the principal cause of mating disruption in the field for a particular pest and specific type of disruptant has not been easy. In many cases, experimental evidence seemed to support multiple mechanisms (Evenden, 2016; Cardé, 2007). Research to uncover the cause of disruption has primarily focussed on characterizing the behavioural effects of exposing insects to a high dose of pheromone or to a disruptant-releasing synthetic pheromone

(Evenden, 2016; Cardé, 2007; Gut et al., 2004). Although these kinds of findings produced the menu of mechanisms proposed in the previous section, they fall short of allowing us to know which mechanism is responsible for disruption as it occurs in the field. For example, while electroantennogram and flight tunnel experiments have documented that males of some species, such as codling moth, can be desensitized by exposing them to very high doses of pheromone (Stelinski et al., 2003a,b, 2006a; Judd et al., 2005b), encounters with field-relevant concentrations do not affect their behaviour (Judd et al., 2005b; Stelinski et al., 2003b; Schmitz et al., 1997). Observations of males orientating to a disruptant in the field (Epstein et al., 2006; Stelinski et al., 2004b, 2005a) is most consistent with competitive attraction; however it is possible that disruption could have occurred non-competitively by desensitization and a behavioural threshold rise in the males (Reinke et al., 2014) or simply was a consequence of low-population densities making it highly unlikely that a visitation event would be observed within the sampling time allotted. Dosage-response experiments yielding disruption profiles (Miller et al., 2006a,b; Miller and Gut, 2015) are proving to be a far better approach to elucidating whether disruption is being achieved via a numbers game based on simple probabilities (competitive) or not (non-competitive).

2.2.1 Competitive and non-competitive disruption profiles

Competitive and non-competitive disruption mechanisms differ fundamentally in how they operate. Competitive disruption functions primarily by division; the attention of males is divided between a given female and nearby dispensers. On the other hand, non-competitive disruption operates mainly by subtraction; moths in pheromone-treated areas are effectively removed from the pool of potential mates. Recognizing this, Miller et al. (2006a) developed a set of comparative graphical plots based on simple mathematical simulations of the two phenomena. Profiles of simulated male-visitation rates to pheromone-baited traps deployed in pheromone-treated crops were graphed against density of pheromone dispensers using various types of axes. The distinctly different shapes of dosage-response profiles generated by the two processes led them to develop this graphical approach as a robust means of discerning whether a particular formulation operatives competitively or non-competitively (Miller et al., 2006a; Miller and Gut, 2015; Fig. 2).

For competitive disruption, a plot of male catch in traps (on the *y*-axis) against density of synthetic pheromone dispensers (on the *x*-axis) generates a concave profile (Fig. 2a). Male-visitation rate initially falls sharply with small incremental increases in dispenser density. However, the additional declines in visitation rate become progressively smaller as dispenser density continues to increase; approach to zero is asymptotic. Plotting 1 / male-visitation rate to

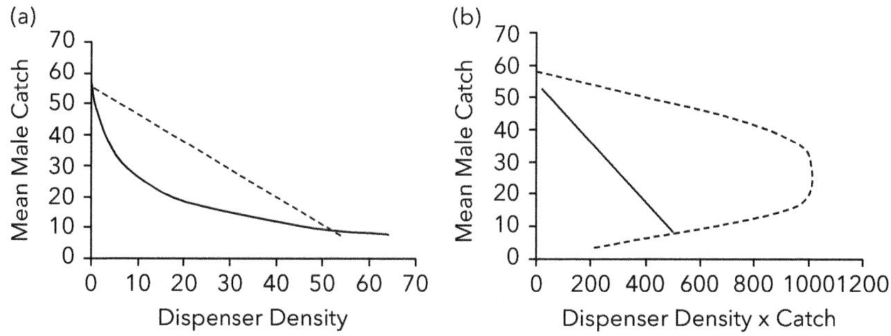

Figure 2 Untransformed dosage-response profiles (a) and Miller-deLame profiles (b) of trapping data for mating disruption mediated competitively (solid lines) or non-competitively (dashed lines). Source: adapted from Miller and Gut (2016).

a given source on the *y*-axis against dispenser density on the *x*-axis yields a straight line with positive slope. Plotting male-visitation rate to a given source on the *y*-axis against (dispenser density x visitation rate) on the *x*-axis yields a straight line with negative slope (Fig. 2b).

Non-competitive phenomena do not share this set of properties. Mating disruption by purely non-competitive mechanisms generate a straight line on non-transformed graphical axes in plots of disruptive effect vs. dispenser density (Fig. 2a) and curved lines on transformed plots (Miller et al., 2006a). The most striking feature of non-competitive profiles is a distinct recurve in a plot of catch vs. dispenser density x catch, with its inflection point occurring near the midpoint of the *y*-axis (Fig. 2b). This Miller de-Lame plot is the most stringent test for discerning between the two types of disruption.

A number of data sets have subsequently been generated from field experiments comparing male catch in pheromone-baited traps across a range of dispenser densities (Reinke et al., 2014; Miller et al., 2006b, 2010; Rodriguez-Saona et al., 2010; Teixeira et al., 2010a). Graphical analysis of results from these dosage-response studies has substantiated the theoretical suite of curves obtained from simulations, as well as their utility for determining the principal mechanism of disruption for a particular pest species and formulation (Miller et al., 2006a,b; Miller and Gut, 2015). The outcomes of 16 dosage-response experiments for 10 deciduous fruit pests are summarized in Table 2. A majority of the disruption profiles have identified competition as the principal means whereby mate finding is suppressed.

2.3 Types of formulations and how disruption is achieved in the field

In practice, mating disruption is achieved by releasing synthetic pheromone into a crop using a variety of delivery systems that protect the sensitive active ingredient from degradation while gradually releasing pheromone over an extended period of time and at a rate that interferes with normal mating behaviour. The ease of application and cost to the producer are also important considerations when designing mating disruption products. Formulation types fall into three broad categories: high-, moderate- and low-density dispensers. Along with major differences in the number of dispensers deployed per unit area, the three formulation types vary considerably in the rate of pheromone emission per source. The higher the application density, the lower the release rate per device. They also differ in the method of deployment; high-density formulations are machine-applied, moderate-density formulations are typically hand-applied and low-density formulations are also deployed by hand but require less labour. The high- and moderate-density approaches use the application process to distribute the pheromone, aiming for uniform coverage. In contrast, sparsely deployed dispensers depend on the wind to distribute pheromone, potentially leaving gaps in coverage. The substantial variability in application density and pheromone release rate among dispenser types result in differences in efficacy and the means by which disruption is attained.

2.3.1 High-density dispensers

High-density dispensers are among the first formulation types developed for mating disruption. They include microfibre and laminated flakes that are designed to release nanograms of pheromone/h from more than 10 000 dispensers/ha (Doane, 1999; Swenson and Weatherston, 1989; Brooks, 1980; Kydonius and Quisumbing, 1980). Fibres and flakes are mixed with stickers to ensure adherence to crop surfaces. The thinking behind these technologies is that sources releasing female-equivalent amounts of pheromone should be attractive to males and machine application of the disruptant by air or ground should provide uniform coverage. A fibre formulation was highly effective at disrupting codling moth males when carefully applied by hand at a density of 50/tree (Stelinski et al., 2008a). However, specialized equipment was required to apply the fibres to commercial orchards and many dispensers failed to contact or adhere to the apple trees. Poor retention of fibres that did hit the target and limited rainfastness further hindered the performance of the fibre formulation in on-farm trials (Stelinski et al., 2008a).

Paraffin-wax emulsion is another option explored as a potential sprayable and biodegradable matrix to deliver pheromones for mating disruption of orchard pests (Atterholt et al., 1998). Paraffin-wax drops (0.1 mL) emitting codling moth pheromone at a female-equivalent rate and deployed at 27 300 sources/ha provided 94–99% trap shutdown compared to the untreated control (Epstein et al., 2006). Similar-sized wax drops releasing Oriental fruit moth pheromone and applied at densities of 30 and 100 per tree (8200 and 27300/ha) provided over 99% disruption of male orientation to pheromone traps and completely eliminated mating of tethered virgin females (Stelinski et al., 2005c). Deployment of 0.04 mL wax drops from a mechanized applicator provided a high level of Oriental fruit moth mating disruption early and late in the growing season, but efficacy and longevity declined substantially in mid-summer (Stelinski et al., 2007c). The reduced efficacy was attributed to an elevated pheromone emission rate when the small, flattened wax drops were exposed to high summer temperatures. Although high-density deployment of female-equivalent formulations show promise for disrupting several tortricid pests of fruit crops, the need for reliable mechanized applicators remains a major impediment to their development and commercial use.

Pheromone also has been formulated in a novel dispensing system based on an electrostatically chargeable powder (Entostat, Exosect Ltd., Winchester, United Kingdom). The premise behind this technology is that males are lured to bait stations where they become contaminated with pheromone-laden powder. Laboratory bioassays revealed that forced exposure of males to the powder reduced the responsiveness of contaminated males (Nansen et al., 2007; Huang et al., 2010) and created individuals that were attractive to naïve conspecifics (Huang et al., 2010). However, the extent to which males voluntarily contact the pheromone-laden powder in field-deployed bait stations and subsequent behavioural effects have yet to be examined.

Several lines of evidence intimate competitive attraction as the primary means by which female-equivalent formulations disrupt deciduous fruit pests. Most notably, Oriental fruit moth disruption profiles for wax drops applied at 0–65 drops per tree fit well into the predictions of competitive attraction (Miller et al., 2006b). This outcome is consistent with field observations of Oriental fruit moth and codling moth males often approaching within 20 cm paraffin-wax dispensers (0.1 mL drops containing 5% pheromone) when deployed as 100 point sources per tree (Stelinski et al., 2005b; Epstein et al., 2006). Male visits were brief, generally lasting less than 20 s and none remained near dispensers for more than 2 min. Curves generated from male gypsy moth captures in plots treated with laminate flake pheromone formulations at five densities also are consistent with competitive attraction (Miller et al., 2006b). Mechanisms evoked by wax emulsion formulations may differ depending on the concentration of pheromone in the formulation and the size and density of droplets applied. The low emission rates of female-equivalent formulations are attractive to males and unlikely to provide pheromone concentrations high enough to induce desensitization prior to the onset of males searching for females or synthetic pheromone sources. Wax dispensers that are engineered to emit pheromone at a high rate or release an unattractive blend may operate in a non-competitive manner.

The search for pheromone formulations that could be applied without specialized equipment spurred the development of pheromone encapsulation in microscopic polymer capsules that could be delivered using conventional methods of applying protectants. Microencapsulated (MEC) formulations are sprayed onto a crop at very high densities of a million or more capsules per ha with each releasing an extremely low quantity of pheromone. Although sprayable formulations have been pursued for several deciduous fruit pests, their impact on male orientation has often proved weak compared to other formulations and short-lived in the field (Knight and Larsen, 2004; Stelinski et al., 2007b; Trimble and Appleby, 2004). In small-plot trials, codling moth and obliquebanded leafroller MEC pheromone treatments provided less than 60% catch inhibition (Stelinski et al., 2007b). Furthermore, increasing the rate 20–30-fold did not improve disruption efficacy. On-farm trials demonstrated that a low-volume application method that is approximately ten times more concentrated than a standard high-volume application can significantly improve the efficacy of codling moth MEC treatments (Knight and Larsen, 2004). The most consistent results using sprayable MEC have been obtained for Oriental fruit moth (Il'Ichev et al., 2006; Trimble et al., 2004).

A serious practical limitation to MEC-sprayable pheromone is the short field life. Not only do the minute capsules give up most of their payload quickly, but they are subject to wash-off by rain events (Waldstein and Gut, 2004). Multiple applications are required to achieve season-long control. The addition of latex, pine-resin and other types of stickers to MEC formulations has rarely improved the retention of capsules or increased efficacy (Wins-Purdy et al., 2007, 2008; Stelinski et al., 2007b; Knight et al., 2004). The most economical and efficacious approach for using MEC formulations is to apply low rates at frequent intervals (Polavarapu et al., 2001; Stelinski et al., 2007b). Another option is to only use the sprayable pheromone to supplement insecticide treatments, rather than as the principal season-long control tactic. The approach relies on insecticides coupled with a single pheromone application targeting the peak emergence of a generation to keep the pest population in check.

Evidence gathered from field and flight tunnel experiments suggest that broadcast application of sprayable pheromone disrupts male orientation via a non-competitive mechanism. The ubiquitous presence of high concentrations of pheromone may interfere with the mate-finding process either by camouflage of the female signal or desensitization of the male (Judd et al., 2005b; Stelinski et al., 2005a; Wins-Purdy et al., 2008). In a flight tunnel, wax strips treated with a MEC formulation did not attract codling males to the capsules, but did prevent them from orienting to a female-equivalent lure for 24 h, suggesting that camouflage or desensitization was operating briefly. A concentrated application of the same formulation produced clumps of capsules that were attractive to codling moth males and interfered with their ability to orient to the lure for another 6 days apparently due to competition (Stelinski et al., 2005d). Flight tunnel studies examining the effect of exposing obliquebanded leafroller males to a MEC-treated surface indicated that direct contact with the capsules interfered with orientation behaviour, but the effect diminished quickly upon removal from the treated surface (Wins-Purdy et al., 2008). As argued by Judd et al. (2005a), it seems unlikely that the uniform and highly diluted concentrations of pheromone present on MEC-treated crop and in the atmosphere are sufficient to cause desensitization. Thus, sprayable pheromone formulations probably achieve disruption by jamming or camouflaging the female signal.

2.3.2 Moderate-density dispensers

A majority of the formulations used commercially to disrupt pests of deciduous fruit pests are solid matrix types that are manually applied at moderate densities of 300–1000/ha. The pheromone is enclosed in plastic, wax or other polymers and slowly diffuses from the dispenser at a rate of several milligrams per hour. The challenge is to engineer a dispenser that gradually and consistently disperses pheromone for up to 180 days and provides protection from breakdown prior to releasing it into the atmosphere. Achieving the desired constant rate pheromone emission has proved difficult, with many solid matrix dispensers emitting pheromone faster over the first few weeks followed by a slower release over the remaining field life of the dispenser (McDonough et al., 1992; Tomaszewska et al., 2005; Hebert et al., 2007; Hueskin et al., 2011). The important release characteristics to consider when comparing different solid matrix formulations targeting the same pest is whether the pheromone emission rate is consistently above some established minimum threshold and is maintained at that level throughout the period of adult flight for the targeted pest species.

A major factor impeding adoption of solid matrix formulations is the labour requirement and associated cost of hand-applying many point sources. This can be especially challenging in orchard crops where dispensers may need to be applied high in the canopies of 5–15 m tall trees. A novel disruption formulation that can be readily applied at moderate densities, called the Tangler®, consists of a plastic module containing pheromone and a launcher that enables rapid deployment of numerous dispensers to the upper tree canopy (Gut et al., 2015). Codling moth disruption using the Tangler formulation was equivalent to that provided by a polyethylene rope formulation, when both were deployed at a density of 1000 dispensers/ha. Automated deployment of the Tangler modules using the launcher was five times faster than hand application of the ropes.

The efficacy of solid matrix formulations is often greatly affected by point-source and pest-population densities. For example, the highest levels of disruption are achieved for codling moth where dispenser density is high and pest-population density is low (Gut and Brunner, 1998; Trimble, 1995; Vickers et al., 1998). For a few pest species, solid matrix dispensers can provide near-complete disruption even when initial densities are high. Possible reasons for the differential effect of population density on disruption efficacy are discussed elsewhere in this chapter. Manufacturers typically provide a range of acceptable application rates on the product label to provide the flexibility needed to address differences in pest density encountered by users. The maximum deployment density is largely established based on economic considerations. High-density populations are especially difficult to disrupt and the most cost-effective strategy is to reduce pest density by applying supplemental insecticides along with the pheromone treatment.

A field experiment conducted in Washington State, USA, apple orchards illustrate well the interplay of dispenser density, population density and efficacy of mating disruption using solid matrix dispensers (Jay Brunner, unpublished data). Polyethylene-tube dispensers were applied to commercial apple blocks at four rates of deployment: 125, 250, 500 or 1000/ha. Mating disruption treatments were complemented by a few insecticide sprays. Male captures in pheromone-baited traps and fruit injury at harvest were used to assess disruption efficacy. In blocks with low-population densities, moth captures were fairly low in all pheromone treatments and no fruit injury was found at harvest (Fig. 3). In blocks with high-population densities, inhibition of male captures and fruit injury varied depending on dispenser density; catch and injury were lowest in blocks treated with the highest density of dispensers and both measures of disruption efficacy gradually increased as the number of dispensers deployed decreased.

Because solid matrix formulations release substantially more pheromone than a calling female, considerable attention has been directed towards understanding the neurophysiological effects of male exposure to high doses of pheromone. Confinement in pheromone-permeated air causes long-lasting peripheral adaptation in some key tortricid fruit pests (Stelinski et al., 2005c). Additionally, pre-exposure to polyethylene rope dispensers in a flight tunnel can have an effect on subsequent orientation and flight behaviours 24 h later (Stelinski et al., 2004b). Despite such findings that exposures to high dosages of pheromone in the laboratory or flight tunnel can lead to desensitization, there is substantial evidence in the literature against it occurring in the field. The major inconsistency is that the atmospheric concentrations of pheromone required to desensitize males far exceed that achieved under mating disruption in the field (Schmitz et al., 1997; Judd et al., 2005b). For example, reduction in the numbers of male grapevine moths (*L. botrana*) captured in traps in disrupted

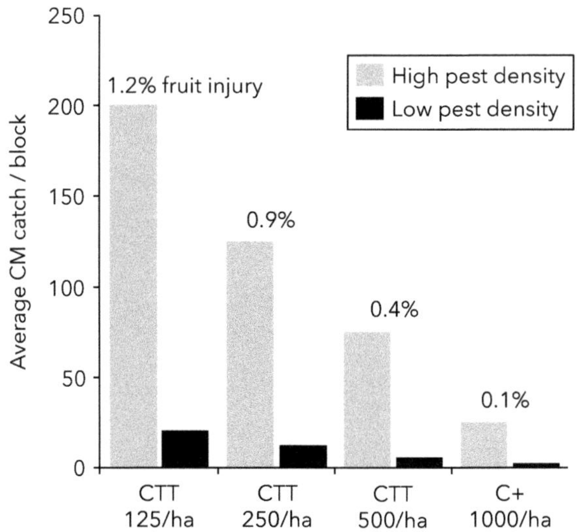

Figure 3 Effectiveness of polyethylene rope dispensers (Isomate ® CTT or C+) applied at four densities for disrupting codling moth at low- and high-population densities. Average captures of males in pheromone-baited traps and fruit injury at harvest (value above each bar) were used to assess efficacy. Fruit injury was not recorded in any of the low-density plots.

vineyards was obtained only when males were pre-exposed in the laboratory to pheromone at concentrations that were orders of magnitude greater than those that have been measured in the field (Schmitz et al., 1997; Koch et al., 1997). Long-lasting peripheral desensitization as measured for obliquebanded leafroller (*Choristoneura rosaceana*) in the laboratory occurred in the field only after a full day's confinement a few centimetres from polyethylene-tube dispensers (Stelinski et al., 2003b).

Disruption profiles reveal that moderately spaced formulations predominantly disrupt via competition. Miller et al. (2006b) found that 11 of 13 data sets allowing graphical analysis of male captures in pheromone-baited traps versus dispenser density generated curves consistent with competitive attraction. The theoretical predictions of dosage-response analysis were experimentally tested in large-cage studies using codling moth and low- or high-releasing solid matrix dispensers (Miller et al., 2010). In all cases, the shapes of disruption profiles were consistent with the predictions of competitive attraction. Among deciduous fruit pests, 14 of 16 dosage-response experiments involving solid matrix dispensers have generated competitive profiles (Table 2). In all cases, plotting male captures on the *y*-axis against dispenser density on the *x*-axis produced a smoothly concave curve and a plot of catch versus dispenser density x catch generated a straight line with a negative slope.

Table 2 Outcome of dosage-response profile analysis for various deciduous fruit pests and dispenser types

Insect	Dispenser type	Dispenser release rate (μg/h)	Mechanism identified by disruption profiles	Reference(s)
Codling moth	Polyethylene rope	5.0	Competitive	Miller et al. (2010)
	0.1 mg lure	0.06–0.1	Competitive	Miller et al. (2010)
Oriental fruit moth	Polyethylene rope	60.0	Non-competitive	Reinke et al. (2014)
	12 mL wax dollop	>40.0	Non-competitive	Miller et al. (2006b)
	1 mL polyethylene tube	15.0	Competitive	Stelinski et al. (2005), Miller et al. (2006b)
	0.1 mL wax drop	0.1–0.4	Competitive	Reinke et al. (2014)
	0.1 mg lure	0.04	Competitive	Reinke et al. (2014)
Obliquebanded leafroller	Polyethylene rope	25.0	Competitive	Lawson et al. (1996), Miller et al. (2006b)
	Polyethylene rope	20.0	Competitive	Deland et al. (1996), Miller et al. (2006b)
European leafroller	Polyethylene rope	20.0	Competitive	Deland et al. (1996), Miller et al. (2006b)
Fruit-tree leafroller	Polyethylene rope	20.0	Competitive	Deland et al. (1996), Miller et al. (2006b)
Light brown apple moth	Polyethylene rope	10.5	Competitive	Suckling and Angerelli (1996), Miller et al. (2006b)
Peachtree borer	Polyethylene rope	6.0	Competitive	Teixeira et al. (2010c)
Lesser peachtree borer	Polyethylene rope	6.0	Competitive	Teixeira et al. (2010c)
American plum borer	1.0 mg lure	<0.05	Competitive	Teixeira et al. (2010c)
Oriental beetle	Wax dollop	0.01–1.0	Competitive	Rodriguez-Saona et al. (2010)

Field observations confirm that searching males are readily attracted to solid matrix dispensers, drawing their attention away from calling females for some period of time. Oriental fruit moth, obliquebanded leafroller and codling moth approached their respective polyethylene rope dispensers in a fully pheromone-treated orchard (Stelinski et al., 2004a, 2005a). Males generally approached to within 1 m of the dispensers before flying away within 20 s. Codling moth males were often observed visiting dispensers even in orchards treated with ten ropes per tree or *ca.* five times the label rate (Stelinski et al., 2005a). Video camera analysis of male peachtree borers (*Synanthedon exitiosa*) showed frequent approaches to dispensers; however, visits were brief, with the majority lasting < 1 s in plots treated with a high density of pheromone dispensers (Teixeira et al., 2010a). Overall, if direct observations of male approach to a single dispenser reflect what is taking place at the majority of dispensers in a treated plot, then the overall visitation rate to this type of source could be high when competitive attraction is the principal mechanism operating.

Observations of male visitations to pheromone dispensers does not itself prove that mating disruption is occurring competitively. If the pheromone treatment quickly elevates their response threshold, males could be attracted to dispensers releasing a high rate of pheromone, but unresponsive to the much lower levels of pheromone emitted by females or lure-baited traps. Then desensitization would be the cause of such mating disruption. This outcome appears to be in play for Oriental fruit moth mating disruption using high-releasing solid matrix dispensers (Reinke et al., 2014). Graphical analysis of dosage-response data revealed that polyethylene ropes (Reinke et al., 2014) or large wax dollops (Miller et al., 2006b) disrupted Oriental fruit moth non-competitively. Disruption achieved in plots treated with these high-releasing, and attractive, devices was attributed to a rise in the male response threshold occurring prior to the onset of sexual activity. Thus, while dispensers releasing pheromone at female-equivalent rates disrupt Oriental fruit moth competitively, exposing males to dispensers releasing much higher amounts of pheromone shifts the mechanism to a non-competitive one.

While moderate-density dispensers primarily operate via competitive attraction, secondary effects that occur after the initial attraction also play an important role in achieving disruption. Most notably, males that approach the source may subsequently incur neurophysiological effects that impede their search for females. For example, codling moth males that approach a polyethylene rope apparently receive a dose of pheromone that deactivates them for the remainder of one night (Miller et al., 2010). The intensity of the complex formed between the male and the disruptant is variable and provides a relative measure of the formulation's effectiveness (Miller et al., 2006a, 2010). The most potent formulations retain the visiting male indefinitely. This is the case for devices that permanently entrap or kill the male (Fig. 4). The strength

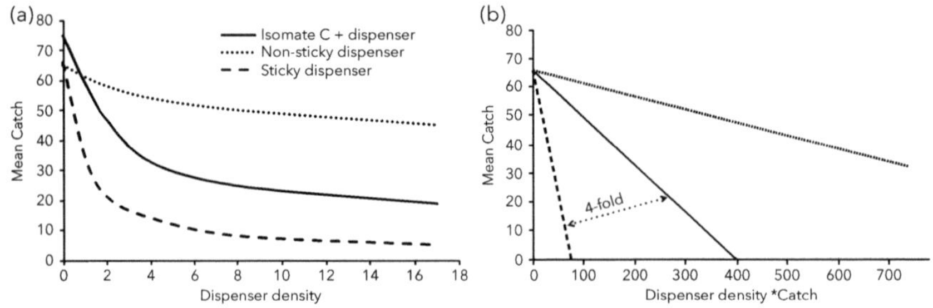

Figure 4 Untransformed (a) and Miller-deLame plots (b) of the disruption profiles for lure-baited sticky traps (dashed line), lure only as the dispenser (dotted line) and Isomate C + dispensers (solid line) disrupting codling moth male captures in monitoring traps. Source: adapted from Miller et al. (2010).

of a temporary complex formed between the male and dispenser is influenced by the amount of pheromone emitted by the dispenser (Miller et al., 2010). Retention is greater for high-releasing compared to low-releasing dispensers (Fig. 4). Female-equivalent formulations that release nanograms of pheromone may only occupy males for a brief period of time, allowing for many searches over the course of an evening. As mentioned previously, a single visit to a more potent reservoir dispenser may preclude males from making subsequent visits for hours or the entire evening.

2.3.3 Low-density dispensers

Early pioneers of communication disruption envisioned that efficacy may not be a function of the spacing between pheromone point sources, but rather may depend upon the amount of sex attractant released per unit area (Farkas et al., 1974). They went on to show that dispensers spaced hundreds of metres apart provided greater than 95% catch suppression when the amount of pheromone released per dispenser was extremely high. Low-density dispensers have now become an important means of dispensing sex pheromones for the purpose of mating disruption. The most highly successful approach formulates and releases insect sex attractants via aerosol-emitting devices, such as puffers (Shorey et al., 1996), misters (Mafra-Neto and Baker, 1996; McGhee et al., 2014) or microsprayers (Isaacs et al., 1999). The widely spaced emitters are deployed at densities of only 2–5/ha, but each unit releases several milligrams of pheromone every 15–30 min over a 6–12-h period. The advantages of aerosol emitters over moderate-density dispensers include reduced cost of application, mechanical dispersion of the pheromone at a constant predetermined rate and a stable environment to protect the active ingredient from degradation prior to release. Additionally, the timing of pheromone emission can be tuned to the period of activity of the target pest, eliminating the waste associated with releasing pheromone during the long periods when males are not searching for females.

Aerosol emitters have been developed and tested for several deciduous fruit pests (Shorey et al., 1996; Fadamiro et al., 1998; Elkins et al., 2005; Knight, 2005; Welter et al., 2005; Suckling et al., 2007a; Steinman et al., 2008; McGhee et al., 2014). The most significant commercial adoption of this low-density approach is for disruption of codling moth in pome fruit and nut crops (Elkins et al., 2005; Steinman et al., 2008; McGhee et al., 2014). The major challenge in using low-density dispensers is that pheromone distribution is dependent on wind flow and areas of little-or-no pheromone coverage can occur. Edges are considered to be at risk and supplemental protection of borders may be required to avoid problems. Aerosol emitters benefit from treatment of large contiguous blocks.

The success achieved by deploying only a few high-releasing dispensers coupled with concern that mechanical failure of a single emitter may leave large areas available for mate finding fostered the development of passive non-mechanical devices designed to release large quantities of pheromone. The aim was to develop low-density dispensers that retain some of the best features of reservoir dispensers, but release pheromone in a manner that impacts males over a large area. The first system developed was a polyethylene 'baggie' consisting of a sealed 8–10 cm plastic sachet that contained a piece of cardboard impregnated with pheromone (Baker et al., 2016). Baggies releasing >0.5 μg/min for more than 80 days significantly reduced captures of male blackheaded fireworm (*Rhopobota naevana*) when deployed in cranberry bogs at a density of 20 bags/ha (Mahr and Baker, 2001). The baggie also proved to be an effective disruptant for Oriental fruit moth and is currently used commercially for mating disruption of several stored product pests (Baker et al., 2016). Manufacturers of standard reservoir formulations have recently developed and tested high-releasing dispensers that are essentially the equivalent of 10–20 standard reservoir dispensers. The approach stems, in part, from research demonstrating the possibility of disrupting codling moth using clusters of standard reservoir dispensers applied at reduced densities (Knight, 2004). Large polyethylene-tube or plastic-matrix dispensers loaded with tenfold more pheromone than the comparable standard dispenser were recently field tested for their potential as passive dispensers for codling moth mating disruption (CMMD) (McGhee, 2014). Deploying the high-releasing dispensers at densities of 10–100/ha generally provided modest reductions in male captures in pheromone-baited traps (<80%) and the low-density treatments were often less effective than the comparable reservoir dispenser applied at a density of 1000 units/ha. Bundling the larger dispensers to further increase the amount of pheromone emitted did not increase disruption efficacy as measured by inhibition of male captures in lure-baited traps. The results are consistent with earlier findings that deploying polyethylene rope dispensers in clusters of 10–100 provided weaker disruption compared to distribution of the same dispensers as individual point sources (Epstein et al., 2006).

Aerosol emitters, thus far, are the most effective low-density approach to dispersing pheromone. The increased impact on mating behaviour is likely related to the highly concentrated pheromone plume released from the dispenser. Baker et al. (2016) compared the structure of pheromone plumes emanating from various types of dispensers and proposed that the best results using widely spaced dispensers will be achieved when strong plume strands reach males far downwind, resulting in moths dosing themselves and becoming desensitized as they move upwind towards the source. They suggest that plume strands emitted from the high-releasing dispenser should be highly concentrated and the best way to achieve this is to release a lot of pheromone from a small surface area. The low-density passive dispensers currently under development release pheromone from very large surfaces, most likely creating highly diffuse plumes.

It seems plausible that the high concentration of pheromone released by widely spaced dispensers and re-emitted by nearby crop surfaces should impact mate finding by impairing the male sensory system or interfering with the female signal. Field experiments reveal that codling moth male captures in lure-baited traps are inhibited for considerable distances downwind of an aerosol dispenser releasing large quantities of pheromone (Gut et al., 2004; Welter et al., 2005). Based on this observation, Welter et al. (2005) concluded that males within the expansive downwind plume of the dispenser become desensitized and unable to respond normally to pheromone. However, they provided no direct evidence that male response to pheromone was impaired. To the contrary, McGhee (2014) found that exposing codling moth males to the high dose of pheromone emitted by aerosol dispensers actually increases their propensity to search for pheromone sources, including lure-baited traps. Others have observed that codling moth males move towards pheromone-treated orchards and search more intensively in disrupted compared to non-disrupted orchards (Witzgal et al., 1999). As demonstrated in flight tunnel assays, exposure to a high pheromone concentration as moths approach dispensers stimulates male responsiveness and elevates their response threshold (Stelinski et al., 2005b). Thus, greatly increasing the pheromone release rate does not switch the principal cause of disruption for codling moth to a non-competitive mechanism, as it did for Oriental fruit moth. Codling moth disruption using aerosol dispensers is achieved competitively through induced allopatry (McGhee et al., 2014; McGhee, 2014), whereby male movement towards and aggregation near the potent emitters diverts males away from females and other sources that are emitting much lower quantities of pheromone. The first few emitters deployed in the orchard provide the greatest reduction in male captures and the impact of each additional one declines as disruption approaches 100% asymptotically. The means by which aerosol emitters disrupt other pest species has not been examined.

2.4 The disruption continuum

In practice, it is easier to interfere with mate finding for some deciduous fruit pests than others. This is best exemplified by the variation in dispenser density and amount of pheromone required for orientational disruption using reservoir devices. Hand application of 250 polyethylene tubes/ha is sufficient for several deciduous fruit pests, while others require deploying as many as 1000 tubes/ha (Table 3). The quantity of pheromone required for disruption is based on field trials comparing efficacy across a range of densities; however, the cost of the active ingredient and the length of time a formulation must release pheromone are taken into consideration as well. Species also differ in the extent to which moth capture in traps is inhibited by the pheromone treatment. The aim is for 100% trap shutdown, but for many pests this is hard to attain. Finally, the impact of initial pest density on efficacy is not the same for all pest species. Trap shutdown and protection of fruit from injury is achievable for some species regardless of pest density, while control for others is highly impacted by population density. In many cases, commercially acceptable fruit

Table 3 Application rates for hand-applied dispensers targeting deciduous tree fruit insect pests

Pest species	Common name	Deployment density[a] (devices/ha)	Amount of pheromone (g AI/ha)
Synanthedon exitosa	Peachtree borer	250-500	3.2-7.9
Vitacea polistiformis	Grape root borer	250-500	8.0-16.0
Synanthedon scitula	Dogwood borer	325-500	13.5-22.4
Synanthedon pictipes	Lesser peachtree borer		
Grapholita molesta	Oriental fruit moth	250-500	25.0-50.0
Lobesia botrana	Grapevine moth	500	38.4
Epiphyas postvittana	Light brown apple moth	500-750	37.1-55.6
Cydia pomonella	Codling moth	500-1000	31.6-63.2
Choristoneura rosaceana	Obliquebanded leafroller	500-1000	45.4-90.7
Pandemis limitata	Three-lined leafroller		
Paralobesia viteana	Grape berry moth	500-1000	48.8-97.5

[a] Rates are for the following Isomate® dispensers (Pacific Biocontrol Corp., https://www.pacificbiocontrol.com/details/): DWB (*S. scitula*), LBAM PLUS (*E. postvittana*), GRB (*V. polistiformis*), PTB Dual (*S. scitula* and *S. pictipes*), OFM MRosso (G. molesta), EGVM (*L. botrana*), C Flex (*C. pomonella*), OBLR/PLR PLUS (*C. rosaceana* and *P. limitata*) and GBM (*E. viteana*).

protection is only realized under low pest pressure and control often requires supplementing the pheromone treatment with insecticide sprays.

Using the above criteria, Gut et al. (2004) proposed a relative ranking of species amenability to a pheromone treatment along a continuum from easy to difficult. Several deciduous fruit pests fall within the easy-to-disrupt end of the spectrum. Clear wing moths, *Synanthedon* spp. and grape root borer (*Vitaea polistiformis*) appear to be especially vulnerable. As few as 250 dispensers per acre and a low pheromone load of only 3.2–13.5 mg of pheromone delivered over the course of the field season is required for communication disruption (Table 3). Inhibition of moth catches in traps usually exceeds 97% and the high level of control appears to be independent of pest pressure (Snow, 1990). Oriental fruit moth is another fruit pest that is fairly easy to disrupt. Applying 250 reservoir dispensers/ha often is sufficient to nearly eliminate male catch in pheromone traps across a range of pest densities and prevent fruit damage (Vickers, 1990; Rice and Kirsch, 1990). Delivery of the pheromone via aerosol emitters (Stelinski et al., 2007a) or sprayable microcapsules Stelinski et al., 2007b; Il'Ichev, 2006) also results in very high-disruption efficacy.

Codling moth and obliquebanded leafroller are appreciably more difficult to disrupt than the aforementioned fruit pests. Practitioners often apply 1000 dispensers/ha to manage codling moth populations (Table 3). The extent to which male catch is suppressed under a pheromone regime is quite variable for codling moth. Trap shutdown may not exceed 70% when pest density is high. Inhibition of male captures in pheromone traps can be reduced by 80–90% at lower pest densities, but only approaches 100% at extremely low densities. The obliquebanded leafroller has proven to be the most difficult deciduous fruit pest to control using pheromones. Deploying 1000 dispensers/ha, with each device loaded with 90 g of pheromone (Table 3), provides levels of trap inhibition in the range of 85–92% (Agnello et al., 1996; Deland et al., 1994; Lawson et al., 1996). Doubling or tripling the application rate does not significantly increase the level of trap shutdown (Lawson et al., 1996). Similarly, up to a 30-fold increase in the quantity of a MEC formulation applied for codling moth or obliquebanded leafroller control does not improve disruption efficacy (Stelinski et al., 2007b).

2.5 Biological factors that influence efficacy

Several biological and ecological factors can play an important role in determining the efficacy of a mating disruption treatment (Rothschild, 1981; Sanders, 1997; Cardé, 2007; Evenden, 2016). The level of control achieved can vary greatly depending upon the ecological setting of the treated area, population density and biological attributes of the target species. Key biological traits that can sway the outcome of a pheromone-based management programme include host specificity, dispersal capacity, number of generations, adult lifespan, fecundity and characteristics of pheromone emission by the female. The male's response to pheromone is an especially important biological characteristic that dictates the success of mating disruption and will be discussed in detail in a later section.

A pheromone treatment provides no protection against immigrating females that have already mated, thus its effectiveness is compromised in settings where gravid females invade from outside the treated area and deposit eggs on the crop. An influx of gravid Oriental fruit moth females into peach and pear orchards is suspected as the key contributor to border infestations in orchards treated with mating disruption for control of this pest (Il'ichev et al., 2002). It follows that high-dispersal capacity and a broad host range are biological attributes that could negatively impact the success of mating disruption. Control using pheromones alone should be more difficult if the pest is highly dispersive or in situations where the area treated with the disruptant is adjacent to untreated areas populated with hosts for the target pest. The enhanced efficacy of mating disruption when deployed on an area-wide basis is partially due to minimizing the opportunities for infestation by invading gravid females.

The density of the target pest population is an especially important variable governing the success of mating disruption. High-density populations in most instances are more difficult to control with this tactic than less-dense populations. For a few pest species, mating disruption can be quite effective even at very high-population densities. The impact of pest density varies depending on the mechanism by which disruption is being achieved. Competitive disruption is a numbers game and is highly density dependent, while non-competitive disruption operates independently of pest density. CMMD formulations act competitively, thus the best control is achieved where pest pressure is low (Gut and Brunner, 1998; Trimble, 1995; Vickers et al., 1998). Attempts to control high-density populations have proven difficult, thus growers typically apply companion insecticide sprays to reduce pest pressure. In contrast, a non-competitive mechanism is in play for Oriental fruit moth disruption using high-releasing solid matrix dispensers (Reinke et al., 2012) and outstanding control is attained even when population densities are high.

The number of generations and adult lifespan vary substantially among different fruit pests and will influence the ease by which their mating behaviour can be interfered with using a disruptant treatment. From a practical standpoint, disruption should be easier to accomplish if the pheromone treatment only needs to be active in the orchard for a short period of time. Thus, univoltine species or those with short adult lifespans in which moths emerge for a restricted time period should be easier to manage using mating disruption than multivoltine species or those with a long adult lifespan. The occurrence of multiple generations also provides an opportunity for pest densities to build, thus potentially increasing the difficulty of achieving mating disruption.

2.6 Environmental factors that influence efficacy

Once the decision is made to implement a mating disruption programme, its commercial success is dictated by environmental and operational factors that influence efficacy. Environmental conditions impact pheromone stability, release rate and movement in the orchard. Exposure to UV radiation and high temperatures can degrade pheromones through oxidative decomposition and other chemical processes (Weatherston, 1990). Temperature directly influences the rate of pheromone emission from all dispenser types except for aerosol emitters. Pheromone release is slower when ambient temperatures are low and faster when temperatures are high. Lower release rates under cool conditions may extend the longevity of the disruption formulation; however, it is also possible that the amount of pheromone emitted may drop below the level required for disruption. Higher release rates under warm conditions may enhance disruption, but it is also likely to reduce longevity of the disruption formulation. From an operational standpoint, disruption formulations may require a higher pheromone-loading rate in hot, compared to cool production regions. Photodegradation and rainfall can substantially reduce the longevity of MEC pheromone formulations (Waldstein and Gut, 2004). Over 75% of Oriental fruit moth pheromone capsules were removed from apple foliage following more than 10 mm of rain (Waldstein and Gut, 2004).

The distribution and retention of pheromone is influenced by several environmental factors, including wind, field or orchard topography and shape and canopy structure. Wind speed and direction can have considerable impact on the distribution of pheromone in disrupted orchards. In apple orchards treated with hand-applied dispensers, pheromone concentrations were highest when wind speeds were low and declined as wind speed increased (Suckling et al., 1999). Wind movement into a disruptant-treated orchard can strongly reduce the pheromone concentration on the perimeter, leaving the border with insufficient levels of pheromone to prevent mating (Milli et al., 1997).

The most favourable circumstances for disruption are when physical conditions promote the uniform distribution of pheromone and facilitate its retention in the crop. Orchard sites that are relatively calm and flat provide a better opportunity for mating disruption than sites that experience frequent high winds or have steep slopes (Gut and Brunner, 1996). Pheromone is also distributed secondarily through a process of adsorption onto plant surfaces and subsequent release back into the orchard or vineyard (Suckling et al., 1996; Gut et al., 2004). Retention may be greater in plantings with full canopies. Seasonal differences in canopy structure also can affect pheromone distribution and retention. In disrupted vineyards, atmospheric pheromone concentration was lower in the spring when the canopy was less developed compared to the summer (Sauer and Karg, 1998). Similarly, it may be harder to maintain pheromone concentrations in orchards with large numbers of missing trees or uneven canopies than in full canopied orchards.

2.7 Operational factors that influence success

In practice, the success of a mating disruption programme is dictated by operational factors that affect pheromone delivery. As discussed earlier, the type of formulation deployed has a major influence on the rate, consistency and longevity of pheromone emission. It also affects the stability of the pheromone prior to release into the environment. The efficacy achieved in the field further depends on application parameters, such as the timing of deployment and distribution of the active ingredient. For tree fruit pests, the general consensus is that the best results are attained when the pheromone treatment is in place prior to any adult activity. For multivoltine species, this means before the start of the first adult flight. The logic behind the early application timing is that the best opportunity for success occurs when pest densities are low. Early implementation of a pheromone-based programme aims to increase the likelihood that pest populations are maintained at densities that are most amenable to mating disruption, especially if a competitive mechanism is in play.

2.7.1 Pheromone coverage

As large-canopied crops, fruit orchards present a unique challenge for attaining an efficacious distribution of pheromone. Pheromone coverage must not only be uniform across the planting, but also extend vertically within the canopy. The standard practice of placing codling moth matrix dispensers in the upper tree canopy is largely based on studies documenting substantially higher captures of males in pheromone-baited or passive traps placed high compared to low in the canopy (Reidl et al., 1979; Weissling and Knight, 1995), as well as a similar increase in mating of tethered virgin females positioned high in the canopy (Weissling and Knight, 1995). However, Epstein et al. (2011a) directly sampled resting females using a vacuum and found that they were evenly distributed in the canopy during periods of female calling and male searching. They went on to show that placing all of the dispensers high in the canopy in 4.5 m tall apple trees was suboptimal for disruption of codling moth (Epstein et al., 2011b). Inhibition of male catch in traps and mating of tethered virgin females were greatest when an equal number of dispensers were distributed in the upper and mid-canopy. Their results are congruent with the predictions of mating disruption operating via competitive attraction. Since efficacy depends on dispensers outcompeting females, placing them in close proximity to calling females enhances mating disruption.

The extent to which vertical dispenser placement influences the efficacy of mating disruption appears to depend on the target species and the canopy size of the treated orchard. In all but a single case, inhibition of Oriental fruit moth male captures in pheromone traps was not influenced by the vertical distribution of dispensers in the tree canopy (DeLame and Gut, 2006; Stelinski et al., 2007b). Only in apple trees taller than 4.5 m was orientational disruption enhanced by deploying dispensers high and low in the canopy. Based on their results, DeLame and Gut (2006) suggested placing dispensers targeting Oriental fruit moth at an easily reachable height in trees <3.5 m tall, in the upper canopy in 3.5–4.5 m tall trees and at both heights in trees taller than 4.5 m.

2.7.2 Monitoring with pheromone-baited traps

Carrying out pheromone-based management programmes that are effective and profitable require reliable means of assessing whether pest densities are being maintained below damaging levels. Growers must have tools for determining whether supplemental insecticide sprays are needed or not. A pheromone-baited trap is the principal tool used to decide whether disruption is being achieved. The reasoning behind this means of measuring mating disruption effectiveness is that if male moths can find a trap baited with a lure-releasing synthetic pheromone, then they also are able to find a calling female and mate with her. Capture of zero (complete shutdown) or very few moths in a pheromone-baited trap is considered to be indicative of males not locating females and an effective disruption programme. Unfortunately, it is not uncommon to record low moth catches in traps and still have inadequate pest control and fruit injury in pheromone-treated plots (refer to several chapters in Ridgway et al. (1990). To guard against the occurrence of these 'false negatives' (Knight, 1995), visual inspection of fruit and foliage for signs of damage should be used in conjunction with trapping.

Considerable effort has been directed towards identifying alternatives to standard pheromone lures that could be used to monitor the density and seasonal activity of pests in pheromone-treated plots. Charmillot (1990) discovered that captures of male codling moth in pheromone-treated orchards increased substantially when traps were baited with 10 or 20 mg rather than the standard 1 mg of pheromone. The utility of a lure loaded with 10 mg of codlemone for assessing codling moth density and seasonal activity in disrupted orchards was validated by researchers in North America (Barrett, 1995; Judd et al., 1996), leading to high-load lures becoming the standard for monitoring codling moth in disrupted orchards.

2.7.3 Monitoring with kairomones

Food baits or kairomones provide a useful alternative to a sex pheromone lure in that host volatiles should not be greatly impeded or suppressed in a background of synthetic sex pheromone. Additionally, kairomone-baited traps capture females as well as males, and thus provide a more direct measure of the potential for larval damage. Traps baited with a brown sugar and terpinyl acetate solution have been used to monitor Oriental fruit moth activity in pheromone-treated pome and stone fruit orchards (Sexton and Il'ichev, 2000; Il'ichev and Sexton, 2002; Il'ichev et al., 2002). They reliably capture females and males of this species and provide a better measure of pest density than pheromone-baited traps. Attractant compounds that can be used to assess male and female activity in disrupted orchards have subsequently been identified for codling moth, Oriental fruit moth and several tortricid pests of deciduous tree fruits (see Chapter 19).

Most notably, USDA-ARS scientists discovered in the late 1990s that an ester present in the odour of ripe Bartlett pears, ethyl (2*E*, 4*Z*)-2, 4-decadienoate (DA), was attractive to both sexes of codling moth (Light et al., 2001). Their initial field trials indicated that DA lure-baited traps provided good resolution of moth flight patterns for both sexes in mating disrupted orchards. However, subsequent evaluation of traps baited with this compound for monitoring codling moth have produced inconsistent results in terms of total moth captures and the ability to attract females in disrupted orchards. Pear ester-baited traps have performed variably depending on the production region (Ioriatti et al., 2003; Thwaite et al., 2004; Trimble and El-Sayed, 2005; Knight and Light, 2005; Kutinokova et al., 2005; Mitchell et al., 2008), as well as the crop or cultivar in which they are placed (Light et al., 2001; Thwaite et al., 2004; Knight and Light, 2005; Knight et al., 2005b). Use of the pear ester in combination with codlemone substantially increased moth captures in pheromone-treated orchards (Knight et al., 2005a, 2010). However, continued inconsistency in capturing females prompted efforts to enhance attraction to females. The addition of acetic acid to the pear ester lure alone or in combination with codlemone significantly increased female captures (Landolt et al., 2007; Knight and Light, 2012). El-Sayed et al. (2013) have identified additional host volatiles that synergize the response of females to pear ester. Further research on cultivar effects, possible synergists and the range of attraction are crucial to the development of a robust pear ester-based monitoring system for codling moth.

2.8 Economic factors that influence adoption

Ultimately, the extent to which fruit growers adopt mating disruption is dictated by cost, ease of application and the availability of a reliable means of assessing the effectiveness of the pheromone treatment. The input costs associated with using mating disruption are high compared to standard control programmes that principally rely on insecticides (Williamson et al., 1996). In addition to the cost of the product, the grower typically incurs higher expenses associated with the application of dispensers, implementation of a more-intensive scouting programme and use of a tactic that in most cases only controls a single pest species. Partial budget analyses of input costs and value of the crop estimated that pest management programmes in apple or peach that used reduced-risk insecticides and mating disruption were on average \$465/ha or \$314/ha, respectively, more expensive compared with those using standard insecticides only (Agnello et al., 2009). In this study conducted across six states, there were no differences in the levels of fruit protection provided by the two programmes. Thus, increased costs resulted in reduced profit in the mating disruption-based programmes. However, in other studies where the pheromone treatment improved pest control and thus fruit quality, mating disruption-based management programmes were equally or more profitable than the conventional programmes (Epstein et al., 2002; Elkins, 2005; McGhee et al., 2011). Apple growers implementing an area-wide CMMD programme using hand-applied dispensers realized a savings of \$55–65/ha compared to those using insecticides only (McGhee et al., 2011). The higher input costs of using disruption were offset by increased yield and fruit quality. Pear growers implementing an area-wide CMMD programme using aerosol emitters were rewarded with an increased 5-year average savings of \$352/ha compared to those using insecticides only (Elkins et al., 2005). Savings principally resulted from a reduction in the use of insecticides, including eliminating some insecticide applications for secondary pest when fewer sprays were used to control primary pests.

2.9 Future trends in mating disruption

2.9.1 Target species selection

If some species are more amenable to disruption than others, what are the factors that allow the more resilient species to operate in environments treated with synthetic pheromone? It has been argued that differences in biological characteristics, such as dispersal capacity, adult lifespan, number of generations, host specificity and fecundity, can influence the ease by which mating behaviour can be interfered with (Evenden, 2016; Cardé, 2007). While we concur that differences among species in these biological capabilities impact the success of a mating disruption treatment, they are not the critical factors that determine the extent to which male orientation to females is impeded by a disruption treatment. Differences in male response to pheromone and the physical-chemical properties of the pheromone, collectively, are the crucial factors that determine how innately tolerant or susceptible a species is to pheromone-based mating disruption.

Laboratory and field studies have demonstrated that there are fundamental differences in the capacities of species to become desensitized when exposed to high doses of synthetic pheromone (Fig. 5). In flight tunnel assays, encounter of males of different species with a high dose of pheromone can induce markedly different subsequent behavioural responses to a female-like lure (Fig. 5; Stelinski et al., 2004b, 2005e, 2006a; Evenden et al., 2000). Some species appear to be readily desensitized by the exposure in that they are unable to orient normally following an encounter with a high-emitting source. Others may experience no change in orientation behaviour, or their search may actually be heightened after exposure to the plume emanating from a high-dose pheromone source. Brief exposure of male redbanded leafroller to a high-releasing rope dispenser reduced their propensity to lock onto the pheromone plume generated by an attractive lure (Fig. 6; Stelinski et al., 2004b). The duration of anemotactic flights was also greatly reduced compared to unexposed males. By contrast, similar exposure of male obliquebanded leafroller males to a high dose of pheromone increased their propensity to fly towards a female-like lure and flights were substantially prolonged (Fig. 5; Stelinski et al., 2004b). Evenden et al. (2000) also showed that constant pheromone exposure did not alter the proportion of obliquebanded leafroller males orienting upwind to pheromone plumes in the flight tunnel. It appears that habituation is a strong possibility as a mechanism of mating disruption for redbanded leafroller but unlikely to be achievable for obliquebanded leafroller.

Species	Proportion to initiate anemotaxis	Intensity of sustained flights	References
Redbanded leafroller *(Argyrotaeniavelutina)*	↓	↓ (in flight tunnel)	Stelinskiet al., 2004
Obliquebanded leafroller *(Chorisoneurarosaceana)*	No Change (15-min) ↑ (24-hr)	↑ (in flight tunnel)	Stelinskiet al., 2004
Oriental fruit moth *(Grapholitamolesta)*	No Change	↓ (in flight tunnel)	Stelinskiet al., 2005
Codling moth *(Cydiapomonella)*	↓ (15-min) No Change (24-hr)	↑ In field ↑ response threshold	Stelinskiet al., 2006 McGhee, 2015

Figure 5 Summary of pheromone rope dispenser pre-exposure effects on male orientation behaviour of four tortricid species.

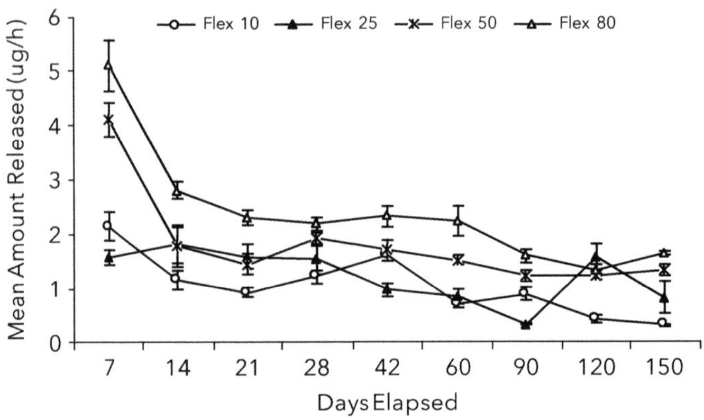

Figure 6 Release profile of codlemone, (*E*, *E*)-8, 10-dodecadienol, from field aged Isomate Flex dispensers, Michigan, USA 2011.

The two principal targets of mating disruption in deciduous tree fruits, Oriental fruit moth and codling moth, also differ markedly in the behavioural changes imparted by a pheromone treatment. Oriental fruit moth males encountering a high-dose pheromone in a flight tunnel did not alter their ability to subsequently initiate anemotaxis to a highly attractive lure 15 min and 24 h later (Fig. 5; Stelinski et al., 2005e). However, encounter with the plume emanating from the dispenser did substantially reduce the duration of sustained flights. The effect declined over time, with longer sustained flights of 24 h compared to 15 min after exposure to the high dose of pheromone. By contrast, brief exposure of codling moth males to a high-releasing pheromone dispenser nearly eliminated moth orientations to a female-equivalent lure 15 min later (Fig. 5; Stelinski et al., 2006a). The effect was likely habituation entailing an elevated response threshold, given that pre-exposed moths often oriented to higher dosages of codlemone (1 and 10 mg), but infrequently to female-equivalent dosages of codlemone (0.1 mg). This is also seen in field trials where more codling moth males are captured in pheromone-treated orchards using high-load lures (10 mg codlemone) compared with standard loadings (1 mg), while the converse is the case in orchards not treated with pheromone (Charmillot, 1990; Barrett, 1995). Some males are still able to orient to a female-equivalent lure in orchards treated with as many as ten pheromone rope dispensers per tree. Similarly, field experiments revealed that forced exposure of male codling moths to a very high dose of pheromone emitted from aerosol devices increased their ability to find a standard lure-baited trap (McGhee et al., 2014). Male codling moths appear to be drawn towards the plume produced by this very high-releasing device. This is consistent with previous observations that codling moth male searching behaviour is enhanced in pheromone-permeated air compared to clean air (Witzgall et al., 1999).

Behavioural responses of males encountering a high dose of pheromone in the flight tunnel assays were normal 24 h following exposure (Fig. 5). This suggests the possibility of codling moth males overcoming the pheromone-induced effects on behaviour by moving to pheromone-free air or to areas of lower pheromone concentrations (Stelinski et al., 2005b). Indeed, codling moth males are readily captured in pheromone-baited traps placed immediately upwind of aerosol-emitting devices (McGhee, 2014).

Sex pheromones that have been synthesized and deployed for disruption are primarily blends of straight chain C_{10}-C_{20} compounds, with acetate, alcohol or aldehyde moieties and a variable number and isomeric configuration of internal double bonds (Mayer and McLaughlin, 1991). All operate as volatiles, invisible as they move through the air, yet conferring essential information about the location of prospective mates. Given that the various pheromones have similar effects on mate finding, it is not surprising that little attention has been given to how differently they may behave chemically as they are dispersed into the environment. Yet, dissimilarity in molecular weight and chemical structure can profoundly affect their rate of evaporation, dispersion in air and adsorption onto foliage and other solid surfaces (Gut et al., 2004).

From a practical standpoint, disparity in physico-chemical properties among pheromones can result in substantial differences in their mobility and persistence in the environment and, accordingly, impact the outcome of mating disruption treatments. Slower evaporation rates and greater ability to adsorb onto solid surfaces result in higher-molecular weight pheromone molecules having longer residence times in the cropping environment compared to lower-molecular weight pheromones that evaporate relatively quickly and are less sticky (Gut et al., 2004). For smaller molecules such as the 12-carbon alcohol pheromone of codling moth, half of a sample exposed to moving air will evaporate in a matter of hours. In contrast, much larger molecules, such as the 18-carbon acetate pheromone of peachtree borer, may take up to 10 days for half the sample to disappear. Thus, the longevity of codling moth pheromone at the sight of release may be quite fleeting compared to that of peachtree borer pheromone.

Following release into the crop as a vapour, pheromone can readily condense onto solid surfaces such as the crop foliage. Adsorbed molecules can subsequently be torn from the surface and re-evaporated into the air. The process of molecules sometimes being in the vapour state while at other times adsorbed onto surfaces is known as *partitioning*. Evidence that pheromone partitions onto crop foliage under field conditions and its buildup can enhance disruption is documented in the literature (Gut et al., 2004; Karg et al., 1994; Suckling et al., 1996). As with evaporation, partitioning is highly influenced by the molecular weight and chemical structure of given pheromone molecules (Gut et al., 2004). Larger molecular weight compounds have a much greater propensity to 'stick' to solid surfaces than do smaller molecules and oxygenation of the hydrocarbons further increases stickiness (Gut et al., 2004). The extent to which pheromone molecules commonly dispensed as disruptants adhere to solid surfaces can vary immensely (Gut et al., 2004). For example, the 18-carbon acetate pheromone of peach tree borer is 100 times stickier than the 12-carbon acetate component of some tortricid moth pheromones. This greater propensity to stick to the foliage can result in a buildup of pheromone in the canopy and contribute to more effective mating disruption (Gut et al., 2004; Table 4).

Developers of mating disruption products have paid little attention to the innate susceptibility or tolerance of a pest species when selecting potential targets. Decisions have largely been based on the identification of potential markets, often focussing on higher-value specialty crops where a $250/ha cost of control can be justified. Not considering how easy or difficult it may be to disrupt a pest when selecting possible targets to pursue commercially also reflects the complexity and cost of laboratory and field research required to make an informed decision. We propose that it may not be necessary to carry out detailed sets of experiments to anticipate whether a pest species is likely to be easy or difficult to disrupt. Three measures that appear to be good indicators of the relative ease of disrupting a pest species are the responsiveness of the male to a synthetic pheromone lure, the female's rate of pheromone emission and the field life of the pheromone (Table 4).

Male orientation to various loading rates of pheromone lures may be the most direct and straightforward means of assessing male responsiveness to pheromone and the ease of impeding male orientation. As pointed out by Sanders (1997), the breadth of the threshold response window varies considerably among species. Males of some species, such as citrus leafminer,

Table 4 Characteristics that can be used to rank the relative ease of disrupting a species using sex pheromones

Insect	Ease of impeding male orientation	Relative lure dose	Relative female pheromone emission	Relative pheromone field life
Citrus leafminer (*Phyllocnistis citrella*)	Easy	Low	Very low	High
Oriental fruit moth (*Grapholita molesta*)	Easy	Low	Low-moderate	Moderate
Three-lined leafroller (*Pandemis limitata*)	Moderately difficult	Moderate	Moderate	Moderate
Codling moth (*Cydia pomonella*)	Difficult	Moderate	Moderate-high	Low
Obliquebanded leafroller (*Choristoneura rosaceana*)	Difficult	High	High	Moderate

Oriental fruit moth, spruce budworm and tomato pinworm, are attracted to a narrow range of concentrations and maximally attracted to low-dose lures (Wyman, 1979; Sanders, 1997; Moreira et al., 2006). For Oriental fruit moth and spruce budworm, high doses arrest upwind progress of males as they approach the lure (Cardé et al., 1975a,b; Baker and Roelofs, 1981; Sanders and Luciak, 1996). Species that have a narrow response threshold and are optimally attracted to low-dose lures appear to have a low capacity to avert the effects of a disruption treatment. In contrast, species that are attracted to higher doses and a wider range of pheromone loadings appear to be more difficult to disrupt. Included in this second group is the obliquebanded leafroller, which is captured in increasingly high numbers as the pheromone dose is increased from 10 to 50 000 μg/lure (Gut et al., 2004). This tortricid species has proven to be unrelenting in its ability to stave off a mating disruption treatment (Lawson et al., 1996).

A less direct, but simple measure of male responsiveness to pheromone, may be the relative pheromone emission rate of calling females. Some species, such as the citrus leafminer and Oriental fruit moth, release relatively low amounts of pheromone. The emissions of as many as 30 female citrus leafminers collected on SPME fibres were often below the GC-MS detection limits (Moreira et al., 2006). A single Oriental fruit moth has been reported to release 0.1–3.2 ng of pheromone per hour (Cardé et al., 1979; Baker et al., 1980). Both species, with females releasing low amounts of pheromone, appear to have little capacity to avoid communication disruption (Table 4). Indeed, nearly complete inhibition of male citrus leafminer captures in pheromone-baited traps and reduced larval infestation was achieved by deploying a remarkably low rate of 1.5 g of pheromone/ha (Stelinski et al., 2008). In contrast, other species, which emit higher amounts of pheromone, appear to be more difficult to disrupt. Included in this group are codling moth and obliquebanded leafroller. Arn et al. (1985) detected an average of 5.4 ng of codlemone, the major component of codling moth sex pheromone in the effluvium of a single female codling moth. An astonishing 14–57 ng of Z11-14: Ac, the major component of obliquebanded leafroller pheromone, was detected in the pheromone gland of individual females from various geographical regions across North America (El-Sayad et al., 2003).

2.9.2 Improving technology based on knowing the mechanism

Although advances in mating disruption technology have continued to progress since the first products were registered 30 years ago, the pheromone industry has largely focussed on improving the ease of application. Mating disruption products of the future should be developed with the aim of enhancing product efficacy and reducing cost to the user. Achieving this requires that they be characterized as to whether disruption is achieved competitively or non-competitively. In a best-case scenario, identifying the principal mode of disruption should be accomplished in the early stage of product development, preferably before commercialization.

Identifying the principal mechanism of disruption and generating disruption profiles readily provides practical guidance in developing better products. Key considerations when disruption is achieved competitively are that (1) the outcome is density dependent, thus high-density populations are more difficult to disrupt; (2) the first few dispensers have the greatest impact with additional dispensers providing less and less benefit; and (3) an optimized dispenser should only release sufficient pheromone to provide the maximal affect possible (Miller and Gut, 2015). In contrast, knowing that the mechanism is non-competitive reveals that (1) the outcome is density independent, mating can be suppressed even at high densities; (2) each dispenser has an equal impact; disruption might fail if pheromone distribution is inadequate due to gaps in coverage; and (3) the dispenser should only release the amount of pheromone necessary to achieve the desired level of disruption; however, the dose should not fall below a level that might shift the mechanism from non-competitive to competitive.

Guided by these considerations, different paths then present themselves with respect to developing cost-effective mating disruption formulations and deployment strategies. The best opportunity for achieving high levels of control will be when mating disruption operates non-competitively. Product development should focus on determining the optimum dispenser release rate and deployment pattern. Those striving to increase adoption of mating disruption should ramp up efforts to identify pests that are amenable to non-competitive disruption (see Section 2.9.1 on target species selection).

Most mating disruption products for pests of deciduous fruits appear to inhibit mate finding via competition. Not surprisingly, a high level of control is only achieved if supplemental insecticides are applied to keep pest densities at levels amenable to disruption. There have been attempts to enhance efficacy by increasing the pheromone emission rate, but for the most part various formulations targeting a particular pest have provided similar levels of disruption and fallen short of 100% inhibition of mating. One practical ramification that can easily be overlooked is that regardless of how much pheromone is released into the orchard, non-competitive disruption and the accompanying near-complete inhibition of mating may not be achievable for some pests. CMMD operates competitively using either low-, moderate- or high-density formulations, although each release increasingly greater amounts of pheromone.

The most meaningful improvements for products that work via competition have centred around the design of formulations that minimize pheromone degradation, last season-long or are easier to deploy. The development of high-density dispensers that can be machine-applied, and low-density dispensers, such as aerosol emitters, was mainly driven by the desire to reduce application costs. Similarly, ease of deployment fuelled the development of novel designs, such as the Tangler. This moderate-density formulation provides up to a fivefold reduction in the time required to apply 1000 dispensers per acre (Gut et al., 2015) Despite these innovations, there has been little progress in reducing the cost of mating disruption. The only way for growers to keep costs low has been to deploy less than the maximum label rate of dispensers in conjunction with companion insecticide sprays. Although economical, this approach has the undesirable feature of reducing the number of point sources and thus the competitive power of the disruption treatment. Reducing the amount of precious active ingredient (AI) used to achieve the desired outcome is the most viable path forward for developing cost-effective disruption products that rely on outcompeting females as the primary means of control. However, this would be better accomplished by reducing the amount of pheromone released per individual dispenser, as opposed to altering dispenser density.

Current reservoir formulations targeting codling moth are loaded with 90–180 mg of codlemone and release 0.5–0.8 mg of the active ingredient per day (Tomaszewska et al., 2005). The experimental basis for this high per dispenser emission rate was essentially a series of field trials conducted in the 1980s that compared trap shutdown and fruit protection in orchards treated with dispensers loaded with various amounts of pheromone (Charmillot, 1990). The outcome was a dispensing system that provided the desired level of disruption for the entire adult codling moth flight of 120–180 days yet kept the pheromone requirement at an economically acceptable level.

Recent studies examining the performance of reservoir dispensers releasing codlemone at different rates suggest that commercial formulations could reduce the amount of AI without compromising efficacy. Dosage-response experiments conducted in large field cages using released moths revealed that CM monitoring lures (Trécé CM L2) placed in unlined traps and Isomate C Plus dispensers both generated competitive profiles and provided 70% and 50% inhibition of moth captures in traps when deployed at the same point-source density (Miller et al., 2010). While 50% disruption would not be commercially viable, it was quite informative considering that the lure released tenfold less pheromone than the dispenser. The authors proposed that a higher rate of pheromone released from the Isomate dispenser prevented further orientations by attracted males for the evening, while low-releasing point sources short-circuited orientations for only a short period allowing multiple visits in the same evening. Further studies examining catch inhibition provided by dispensers releasing codlemone at varying rates indicated that maximum disruptive potency was attained at a release rate of about 0.15 mg/

day, or 20% of the emission rate from current commercial dispensers (McGhee et al., 2010).

Based on these findings, Shin-Etsu Chemical (SEC) company produced twin-tube dispensers with 80%, 50%, 25% or 10% of the standard load of codlemone. The experimental Flex ® dispensers were designed to release pheromone season-long, but at lower hourly release rates (Fig. 6). Disruption efficacy was examined by measuring catch inhibition and fruit injury in commercial apple orchards treated with the different dispensers at a density of 1000/ha (McGhee et al., 2012). Field studies showed that all four dispensers provided a similarly high level of disruption (Fig. 7), despite about a twofold difference in the pheromone emission rate of the Flex 80 and Flex 10 formulations (Fig. 6). The 80% loading rate was commercialized as Isomate CM Flex.

This body of work showing that a substantial reduction in rate of pheromone emission from reservoir dispensers did not reduce CM orientational disruption provided impetus for exploring the possibility of reducing the amount of pheromone emitted by aerosol devices used to disrupt this pest species. The standard amount of pheromone loaded into each aerosol emitter was essentially based on matching that of 500 reservoir dispensers/ha. Additionally, emitters were programmed to release pheromone every 15 min and over a 12-h period. Thus, the expensive AI was emitted over a much longer period than the few hours codling moth females were calling (Geir, 1963), as well as when conditions for moth flight were unfavourable. A suite of field trials revealed that the extremely high amounts released by current technologies were not necessary to achieve satisfactory disruption (McGhee et al., 2016). Three refinements to aerosol release parameters were identified that could conserve

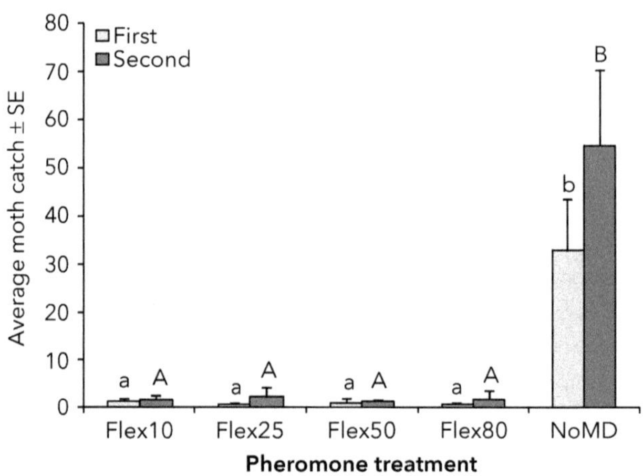

Figure 7 Average first- and second-generation codling moth male captures in pheromone traps baited with Trécé L2 lures in plots treated with Isomate Flex dispensers at four loading rates, or not treated with pheromone (NoMD), Michigan, USA, 2011.

the amount of pheromone used for CM disruption without sacrificing efficacy. The concentration and total amount of pheromone in the canister could be reduced by 50%. The spray interval could be reduced from the standard of once every 15 min or 4/h or to once every 60 min or a single emission/h. Finally, the length of the emission period could be reduced from 12 to 6 h. Combinations of these modifications were also possible without sacrificing efficiency. The innovative approach recently introduced by Semiosbio provides the option of conserving pheromone by shutting off units remotely if traps record low-or-no moth captures indicating a lack of adult activity. The development of lower-releasing, and potentially less-costly, aerosol and reservoir dispensers illustrates how improvements can be achieved through knowledge of the underlying factors that determine the disruption efficacy of a particular technology.

2.9.3 Multi-species disruption

One of drawbacks of using mating disruption is that it is species specific and the accompanying reduction in insecticide use can lead to outbreaks of other pest species. Consequently, there has long been an interest in developing formulations that target more than one pest. A desire to control multiple species with a single formulation first drew the attention of growers and the pheromone industry in the Trentino/South Tyrol area of northern Italy in the early 1990s. The successful adoption of CMMD on an estimated 14 000 ha had resulted in a decrease in the use of cover sprays to control this pest, allowing secondary pests such as the summer fruit tortrix (*Adoxyphyes orana*), the dark fruit-tree tortrix *(Pandemis heparana)* and Oriental fruit moth to increase to levels justifying a control action. In response, several dispenser types were developed and tested that targeted both codling moth and another species. Similar trends occurred simultaneously in Washington State in the United States, where widespread adoption of CMMD led to a reduction in cover sprays and allowed populations of the key leafroller pests, obliquebanded leafroller and three-lined leafroller, *Pandemis limitata,* to reach damaging levels. More recently, *Blastobasis lacticolella* has emerged as a pest in apple and pear orchards treated with mating disruption for other tortricid pests in the UK (M. Fountain, pers. comm.).

Widespread concerns over this resurgence of non-target lepidopterans in disrupted orchards stimulated the research community to explore the possibilities for disrupting multiple species. Simultaneous disruption of codling moth and Oriental fruit moth was achieved using reservoir dispensers (Il'ichev et al., 2007; Stelinski et al., 2009) or aerosol emitters (Stelinski et al., 2007a) loaded with the pheromone of both species. Reservoir dispensers releasing codlemone and (Z)-11-tetradecenyl acetate provided orientational disruption of both codling moth and multiple leafroller species (Judd and Gardiner, 2004). In small-plot trials, Deland et al. (1994) showed that three sympatric leafroller species could be disrupted using a reservoir formulation releasing a 93:7 blend of (*Z*)-11-tetradecenyl acetate and (*E*)-11-tetradecenyl acetate. Similar dispensers loaded with a 98:2 blend of the same acetates provided orientational disruption of obliquebanded and three-lined leafrollers (Evenden et al., 1999). Simultaneous disruption of multiple species of leafrollers that utilize quite different pheromone components proved to be a more complex task. Pfeiffer et al. (1993) were unable to identify a generic blend that could inhibit orientation of variegated leafroller, tufted apple budmoth and redbanded leafroller. Recently, a multi-species formulation containing a blend of six pheromones showed promise for achieving comprehensive control of codling moth and a suite of five leafroller species in Swedish apple orchards (Porcel et al., 2015).

Not surprisingly, the commercial adoption of multi-species disruption products in deciduous fruits has principally centred around combining the pheromones of a key lepidopteran pest with that of a secondary lepidopteran pest. Products targeting codling moth and Oriental fruit moth have been used for many years in fruit production regions where both are major pests of apple, most notably in eastern North America and the Sacramento Delta region in California, USA. Recently, however, Oriental fruit moth has become problematic in Washington State, USA and growers are opting to switch to a multi-species product. The worldwide adoption of a multi-species mating disruption product for codling moth and leafrollers control has increased from approximately 500 ha 5 years ago to 6000 ha in 2018. The necessity of controlling both European grape berry moth and European grapevine moth in Italian vineyards has led to the creation of dual-purpose dispensers that contain the major pheromone component of each (Ioriatti and Lucci, 2016). Growers have the option of using dispensers loaded with 200 mg of each species pheromone when both are abundant, or one containing only 20 mg of Z9–12:Ac, the pheromone for European grape berry moth, if the density of this pest is low. It is estimated that 130 000–140 000 ha are treated in Europe with various products including a multi-species dispenser to control European grape berry and grapevine moths (V. Veronelli, pers. comm.).

There is also a recent interest in developing multi-species mating disruption formulations directed at managing insect species in different orders that have very dissimilar pheromone active ingredients. For example, in response to secondary outbreaks of the vine mealybug (*Planococcus ficus*) Italian researchers are experimenting with dispensers that contain the pheromones of this mealybug and the key lepidopteran pest, European grapevine moth (Ioriatti and Lucchi, 2016).

Given the cost and reduced availability of labour in specialty crops such as pome and stone fruit, there will be increasing emphasis on cost-saving control tactics such as multi-species mating disruption. Moving forward with the approach will require recognizing the benefits and addressing the challenges and technical hurdles of targeting multiple species.

Attributes such as the timing of dispenser application and deployment density can vary considerably for the targeted species. If growers intend to use mating disruption for more than one species, their options are to either apply separate dispensers for each or a single multi-species dispenser. The use of a multi-species dispenser compromises the timing but reduces the cost of application, as only a single trip through the orchard is required. By applying dispensers once, workers are freed up to conduct other tasks rather than make a second trip through the orchard to deploy another dispenser. Additionally, the use of a multi-species dispenser will save the grower money by reducing the number of insecticide sprays required for pest control. Fewer insecticide applications is a good resistance management strategy and also increases the potential for natural enemies to establish and suppress secondary pests such as aphids and mites.

The cost of multi-species disruption is lower if a single dispenser rather than several dispensers is deployed. From a manufacturer's perspective, however, the development of a multi-species dispenser may not be economically viable if the occurrence of the secondary pest being targeted is sporadic. Manufacturers can only justify the development and commercialization of multi-species mating disruption when two or more species are perennial problems that require control measures to prevent crop loss.

A serious challenge with deploying a multi-species dispenser is the differing phenology of the targeted species. For example, Oriental fruit moth flight begins 30–40 days prior to that of codling moth. This impacts the timing of application and potentially the field life of the formulation. The user must decide the best option for deploying a multi-species product. Generally, the earliest flier is targeted. However, this means that the dispenser could be releasing the pheromone of the later flier well before adults are present. The pheromone is wasted and given the cost of pheromone actives and products, this is a poor return on investment. In addition, the dispenser may run out of pheromone needed to disrupt the later flier due to the early season release.

Another constraint to targeting multiple species with a single dispenser is that the optimum application density may differ by species. For example, 500 dispensers/ha may be appropriate for disrupting Oriental fruit moth, while 1000/ha are needed to disrupt codling moth. Opting to apply a multi-species dispenser targeting these two pests would result in over-treating for Oriental fruit moth and wasting a considerable amount of costly pheromone. Despite requiring additional time to make the application, a more cost-effective strategy might be to apply the dual dispenser at the lower rate needed for Oriental fruit moth control, and then boost the codling moth rate by applying a single species dispenser at the reduced rate of 500/ha.

Releasing the correct amount and component ratio for each species with a multi-species dispenser will be difficult, if not impossible. Pheromone differences in molecular chain length, functional groups and the number and location of double bonds impact the rate of emission. Accommodating these differences will entail compromising the release parameters of one or more of the pheromones when producing a multi-species dispenser. Unfortunately, release rate can influence the mechanism and efficacy of mating disruption. A lower emission rate is required to achieve excellent disruption of Oriental fruit moth compared to codling moth. Moreover, as mentioned previously, the mechanism of disruption may shift from non-competitive to competitive at lower release rates, reducing efficacy. Ensuring that pheromone is released from a multi-species dispenser at rates that are near optimal for each targeted species may require innovative changes to dispensing technologies. For example, the Isomate plastic tube dispenser could be designed such that two unique plastics are combined in a manner that would allow for optimizing the release of the pheromone for two insect species. Similarly, the Tangler dispenser provides the option of loading each side of the capsule with a different pheromone.

Producing a multi-species dispenser that contains different pheromones in separate reservoirs also would mitigate the problems associated with blending active ingredients in a single reservoir. Combining actives is not a problem if the pheromones are compatible and one compound does not greatly impact the release of the other. However, some pheromones are not compatible and mixing them can reduce or negate disruption efficacy. Combining pheromones may speed up or slow down the rates of pheromone release. For example, the leafroller pheromone, Z11-14Ac appears to be a good solvent for codlemone and the resulting increase in release rate can reduce the expected longevity of a formulation that blends the two pheromones. Of greater concern is that the pheromone of one species may be antagonistic to another species, thus interfering with the normal response to its pheromone. For example, the presence of Oriental fruit moth pheromone interferes with the response of codling moth to its own sex attractant (Arn et al., 1974; Cardé et al., 1979; Preiss and Priesner, 1988). This negative behavioural effect may be problematic, particularly if the mechanism of disruption is competitive attraction. Companies developing multi-species disruption products must be cognizant of how compatible pheromone active ingredients are as they move forward in developing products targeting more than one species.

3 Mass deployment of attractant-baited traps

The aim of mass trapping is to remove a substantial portion of a pest population before mating, oviposition and crop injury occur. The tactic relies on deploying a number of highly efficient traps baited with species-specific attractants including aggregation and sex pheromones or generalist odours associated with food or ovipositional cues. Studies examining the mechanism of mating disruption indicated that mass trapping has the potential to provide superior control compared with that achieved by mating disruption operating by competitive attraction. Large-cage studies with codling moth revealed that the removal of moths from an apple orchard using pheromone-releasing traps was fourfold more effective at reducing capture in a central monitoring trap than was a mating disruption protocol using hand-applied dispensers (Miller et al., 2010; Fig. 4). The enhanced efficacy of mass trapping was attributed to the removal of individuals from the breeding pool following the first visit to the attractant source, whereas multiple opportunities to find a female or trap were possible in the disruption-treated plots. The potential superiority of mass trapping was confirmed in small-plot trials comparing a high-density trapping scenario to Isomate dispensers for disruption of codling moth and obliquebanded leafroller under high pest pressure (Reinke et al., 2012). Mating disruption using Isomate CM FLEX dispensers reduced codling moth capture in standard monitoring traps by 58%, while sticky microtraps reduced capture by 92%. Mating disruption using Isomate OBLR/PLR Plus dispensers reduced obliquebanded leafroller capture in traps by 69%, while high-density trapping reduced capture by 85%.

Realizing the potential of mass trapping will require that several key practical and biological requirements are met (Cork, 2016; El-Sayed et al., 2006; Miller et al., 2015). The tactic works best if the starting pest density is low, the reproductive capacity of the targeted pest species is not extraordinarily high and there is minimal immigration from outside sources. Favourable attributes of the trapping system include a highly attractive lure or bait and an efficient trap that does not saturate with captured insects. Ideally, attractants should outcompete the target pest's natural attractants such as pheromones or host odours. The prospects for success are enhanced if the lure attracts both males and females to the trap. Ultimately, practical considerations determine the economic viability of a mass-trapping programme. The costs of the traps and lures and the expense associated with maintaining them must be reasonable. Essentially, costs should be competitive with those incurred using other tactics to control the pest. The best mass-trapping systems should operate and require very low maintenance for a substantial portion of the pest life cycle, preferably for at least one generation.

Efforts to develop mass trapping for control of deciduous fruit pests have focussed on moths, fruit flies, midges or beetles (Table 5). Roelofs et al. (1970) demonstrated the potential of limiting fruit damage by moderate populations of redbanded leafroller, *Argyrotaenia velutinana,* by deploying 180–300 pheromone-baited traps per ha. Similarly, reductions in damage in grape by redbanded leafroller and grape berry moth, *Endopiza vitana*, were achieved

Table 5 Applications of mass trapping for management of deciduous tree fruit insect pests

Approach	Pest species	Common name	References
Attract-and-kill stations	*Carpophilus* spp.	Sap beetles	James et al. (2001), Hossain et al. (2006)
Attractant-baited traps	*Rhagoletis pomonella*	Apple maggot	Bostanian et al. (1999), Prokopy (2003)
Sex pheromone-baited traps	*Dasineura mali*	Apple leaf midge	Lo et al. (2015)
Sex pheromone-baited traps	*Cydia pomonella*	Codling moth	Madsen and Cardy (1979), Wilson and Trammel (1980), Reinke et al. (2012)
Sex pheromone-baited traps	*Choristoneura rosaceana*	Obliquebanded leafroller	Wilson and Trammel (1980), Reinke et al. (2012)
Sex pheromone-baited traps	*Argyrotaenia velutinana*	Redbanded leafroller	Roelofs et al. (1970), Taschenberg et al. (1974)
Sex pheromone-baited traps	*Grapholita molesta*	Oriental fruit moth	Wilson and Trammel (1980)
Sex pheromone-baited traps	*Synanthedon scitula*	Dogwood borer	Leskey et al. (2009)
Sex pheromone-baited traps	*Synathedon myopaeformis*	Red-belted clearwing	Bosch et al. (2001)

using 160–300 monitoring traps per ha (Taschenberg et al., 1974). Attempts to use mass trapping for codling moth control have entailed deploying traps baited with standard load pheromone lures at low densities of 5–36 traps/ha (Hagley, 1978; Howell, 1980; Madsen and Carty, 1979; Willson and Trammel, 1980). The technique has been tried alone or in combination with another control tactic. The outcomes have been mixed, with some attempts indicating population suppression and reduced damage compared to controls plots, while others showed no effect. The best results generally were obtained where population densities were low or treated orchards were isolated. Knowing now that the plume reach of these traps was small (Adams et al., 2017), the most likely reason for unsatisfactory levels of suppression were the low trapping densities.

Perimeter trapping has shown potential as an alternative to insecticides for managing apple maggot (*Rhagoletis pomonella*) (Bostanian et al., 1999; Prokopy, 1975; Reissig et al., 1984). The approach is designed to intercept females before they penetrate into the orchard interior and lay eggs on apples. Bostanian et al. (1999) reported a 99–100% reduction in damage by placing sticky traps baited with butyl hexanoate around the perimeter of commercial apple orchards. Traps were spaced at 10–36 m intervals depending on the likelihood of the border serving as an entry point for invading flies. The authors noted that success was dependent on the orchard being free of apple maggot from the onset to ensure that pest pressure resulted from immigrant rather than resident flies.

The apple leaf midge (*Dasineura mali*) has been the target of mass-trapping trials in New Zealand (Suckling et al., 2007b; Lo et al., 2015). Under high pest pressure, deployment of 750 oil traps baited with pheromone lures provided 97% suppression of male captures in pheromone-baited sticky traps compared with untreated plots (Lo et al., 2015). However, there was only a 48% reduction in the number of larvae per shoot in the generation immediately following the deployment of the oil traps. Additionally, there was no difference in shoot damage after generation 3 between the mass trapped and untreated plots. The authors concluded that while mass trapping has potential as a control for apple leaf midge, contending with four generations per year and high-population densities will likely require combining the approach with other tactics to prevent fruit damage.

A few sesiid pests of pome and stone fruits have been candidates for mass trapping. Three consecutive years of mass-trapping red-belted clearwing (*Synathedon myopaeformis*) males in two apple orchards substantially reduced male captures at both locations, but only reduced the number of pupal exuviae in one of the orchards (Bosch et al., 2001). Deployment of pheromone-baited traps in apple at densities of 5 and 20 traps/ha removed many thousands of dogwood borer (*Synanthedon scitula*) males over a 2-year period, yet this did not translate into reduced infestation levels (Leskey et al., 2009).

Combined, the results of these studies make it evident that successfully managing deciduous pests via mass trapping is a tall order. It may be especially difficult to manage pests that cause direct damage to a crop and have tolerances near zero, as a very high proportion of the population will need to be removed. Supplemental use of insecticides will improve the chances of obtaining the desired outcome. Indirect pests should provide a better opportunity for commercially acceptable control using mass trapping. Targeting males only may not be sufficient to limit population growth, since each male can fertilize multiple females. Removing females will have more impact than removing males, as it directly impacts egg production. The best results will be achieved if both sexes are captured. For example, control of *Carpophilus* beetles by mass trapping owes it success in part to the attraction of males and females to the food-based attractants (Hossain et al., 2006). Identifying host odours that are attractive to females provides enhanced opportunities for mass trapping of moth pests such as codling moth (Light et al., 2001).

A high-trapping density will likely be required to achieve a commercially relevant reduction in pest density. Miller et al. (2015) proposed a 'clear and hold' strategy that entails positioning traps throughout the interior of the crop in a large grid evenly and closely spaced to first reduce interior populations

Figure 8 Microtrap consisting of 4 cm^3 cube with entry holes on the sides and top surface, a sticky internal surface and baited with an attractive lure. Source: adapted from Reinke et al. (2012).

and then shield the interior crop from future invasions. Quickly reducing pest numbers is key to preventing crop injury. Many traps are more likely to intercept and head off insurgent pests than a limited number of traps. The optimal-trapping density will vary depending on many factors but should be based on an understanding of plume reach and trapping area (see later section).

A major factor limiting the economic viability of mass trapping is the high cost of deploying a sufficient number of traps. Reinke et al. (2012) developed and tested an innovative approach that relied on inexpensive microtraps with a sticky internal surface that could be deployed at high densities (Fig. 8). Sticky microtraps baited with standard pheromone lures and hung in apple orchards at densities equivalent to those used for mating disruption, 500/ha, provided very high levels of control. A direct comparison with non-sticky microtraps baited with the standard pheromone lure confirmed that sticky microtraps operated via mass trapping rather than just as weak disruptants. Deploying microtraps at high densities has the added benefit of sharing the burden of moth capture among numerous traps, thus reducing the likelihood of trap saturation.

4 Application of attract-and-kill formulations or devices

The terms attract-and-kill, lure-and-kill and male annihilation have been used to describe the tactic of enticing the target pest to a device or location and subsequently removing it from the population using an insecticide or other killing agent. The principles for success and practical applications of the technique have been comprehensively reviewed recently (El-Sayed et al., 2009; Cork, 2016). To a great extent, the success of attract-and-kill depends on many of the same attributes as mass trapping; population density cannot be too high, the lure must be highly attractive and the killing agent must be effective. A major advantage of attract-and-kill over mass trapping is that the problem of trap saturation/maintenance is less severe. This may improve the economics and control, especially when pest density is high. Additionally, it may be possible to develop a formulation that can be mechanically rather than hand-applied. An ambitious goal in developing highly effective and economical formulations for pests of deciduous fruits is for a single application to provide generation- or season-long suppression.

Attract-and-kill formulations for deciduous fruit pests have principally focussed on controlling moth and fruit fly species (Table 6). Four general types have been developed: (1) a mixture of pheromone and insecticide delivered in a paste, gel or wax matrix; (2) a fabric pouch impregnated with insecticide and

Table 6 Applications of attract-and-kill for management of deciduous tree fruit insect pests

Approach	Pest species	Common name	Reference(s)
Pheromone-based toxic pouch	*Grapholita molesta*	Oriental fruit moth	Huang et al. (2015)
Pheromone-based toxic paste	*Grapholita molesta*	Oriental fruit moth	Evenden and McLaughlin (2004a,b)
Pheromone-based toxic paste	*Cydia pomonella*	Codling moth	Charmillot et al. (2000), Ebbinghaus et al. (2001), Lösel et al. (2000), Mansour (2010), Somsai et al. (2010), Stará et al. (2008)
Pheromone-based toxic paste	*Epiphyas postvittana*	Light brown apple moth	Suckling and Brockerhoff (1999)
Pheromone-based toxic paste	*Adoxophyes reticulana*	Summer fruit tortrix	Somsai et al. (2010)
Insecticide bait spray	*Rhagoletis pomonella*	Apple maggot	Pelz et al. (2005)
Insecticide bait spray	*Rhagoletis mendax*	Blueberry maggot	Pelz et al. (2005)
Insecticide bait spray	*Rhagoletis indifferens*	Western cherry fruit fly	Yee and Chapman (2005)
Attractant-based toxic sphere	*Rhagoletis pomonella*	Apple maggot	Prokopy et al. (2000), Wright et al. (2012), Morrison et al. (2016)
Attractant-based toxic sphere	*Rhagoletis mendax*	Blueberry maggot	Ayyappath et al. (2000), Stelinski and Liburd (2001)
Attractant-based toxic sphere	*Drosophila suzukii*	Spotted-wing drosophila	Rice et al. (2017)

baited with a separate attractant lure placed inside the pouch; (3) insecticidal bait sprays; and (4) attracticidal spheres.

4.1 Attracticidal gel, paste or wax droplets

Attract-and-kill using sex pheromones is an approach that has been pursued to gain improved control using reduced release rates as compared to mating disruption formulations (Gut et al., 2004). Attracticides consisting of a combination of low concentrations of pheromone and a pyrethroid insecticide delivered as a paste, gel or wax dollop have been developed for a few key lepidopteran pests of deciduous fruits (Table 6). The approach has been touted as effective for managing codling moth in a number of field trials. Attracticidal formulations applied at densities of 1000–6000 droplets per ha generally provided fruit protection at least equivalent to insecticides (Charmillot et al., 2000; Ebbinghaus et al., 2001; Lösel et al., 2000; Mansour, 2010; Somsai et al., 2010; Stará et al., 2008). Two to three applications were required to achieve control. The best efficacy was achieved if the initial pest density was low and point-source density was high. Krupke (2002) and Lösel et al. (2002) confirmed that a higher density of droplets increased the frequency of male contact with the lethal formulation. In small apple plots, a paste formulation containing Oriental fruit moth pheromone and permethrin significantly reduced male captures in female-baited traps and mating with sentinel females compared to an untreated control (Evenden and McLaughlin, 2004a). Mating behaviour was equally inhibited at application densities of 1500 or 3000 droplets/ha. Similar attracticidal droplets containing light brown apple moth pheromone and permethrin showed promise for control of this pest in New Zealand, suppressing male captures in the centre of 0.3 ha apple plots by 96% compared to untreated plots (Suckling and Brockerhoff, 1999).

Although success using attracticides as an alternative to insecticides has been proclaimed, formulations generally have failed to provide superior control relative to an industry standard mating disruption formulation. They offer the advantage of requiring less active ingredients compared to mating disruption; however, shorter longevity and high application rates seem to negate the economic benefit of this attribute. As with mass trapping and mating disruption, attracticides have proved most effective at low pest-population densities. Additional factors that have impeded adoption include the need for multiple applications due to the short residual of formulations and the deployment of very high droplet densities (Charmillot et al., 2000; Ebbinghaus et al., 2001).

The development of improved attracticidal formulations has been hampered by a limited understanding of how they operate. Formulations containing pheromone, but no toxicant, have rarely been included in field trials and thus, the relative contributions of male kill versus mating disruption to control, cannot be discerned. In the only direct comparison of a codling moth attracticide with or without insecticide, deployment of 5000 droplets/ha containing pheromone alone provided a 50% reduction in male catch in lure-baited traps, while the addition of the toxicant increased trap shutdown to 98% (Charmillot et al., 1996). Using exclusion cages to prohibit light brown apple moth males from touching the insecticide-laced droplets, but allowing pheromone dispersion, Suckling and Brockerhoff (1999) similarly attributed about half of the effect of the treatment to orientation disruption and half to mortality following insecticide exposure. In the case of Oriental fruit moth, mate finding was equally inhibited by formulations with and without insecticide (Evenden and McLaughlin, 2004a). In this case, the measured reduction in mating behaviour was achieved by disruption alone. In order to claim that an insecticide-laden pheromone formulation operates as a potent attracticide rather than primarily as a disruptant, we strongly recommend that future studies include a pheromone-only control in the experimental design.

4.2 Attracticidal pouch

In addition to practical limitations, a major deficiency of current attracticidal formulations is that insects must pick up a lethal dose of insecticide from a very small surface that also contains the pheromone. This presents a behavioural problem because overexposure to the pheromone is likely to result in their vacating the area prior to contacting the toxicant-laden formulation (Baker and Roelofs, 1981; Stelinski et al., 2006a). In behavioural assays conducted in a flight tunnel, separation of the landing substrate from the pheromone source substantially increased contact time by codling moth (Huang et al., 2013). Males spent nearly four times longer on a 7.6 × 12.6 cm card placed immediately upwind of a pheromone lure compared to a lure presented alone. The card and lure were only 1 cm apart, thus very little separation of the two was required to avert overexposure to the pheromone. Placing an insecticide-treated filter paper 1 cm upwind of the lure resulted in 90% knockdown of males 2 h after voluntary exposure.

Future attracticidal formulations that rely on sex pheromones should be designed to minimize direct contact with the attractant, while enticing males to extensively search a nearby insecticide-laced surface. To test this concept, researchers have developed and tested a prototype attract-and-kill device modelled after insecticide-impregnated fabrics successfully used to control mosquitoes and other insects (Curtis et al., 1996; Hogsette et al., 2008). The device consists of a 4″ x 5″ fabric pouch impregnated with deltamethrin and baited with a pheromone lure placed centrally inside the pouch (Fig. 9; Grieshop et al., 2014; Huang et al., 2015). Attracted moths are provided a large insecticide-treated surface to land on and are prevented from

Figure 9 Two types of attract-and-kill devices; an attracticidal pouch consisting of insecticide-impregnated fabric baited with a pheromone lure (left) and an attracticidal sphere with a cap designed to provide controlled release of sucrose and insecticide (right).

contacting the attractant source, thus reducing the potential of overexposure to pheromone.

The attracticidal pouch has been evaluated as a possible control for Oriental fruit moth and Japanese beetle (*Popillia japonica*). In flight tunnel bioassays, duration of contact by Oriental fruit moth males was nearly eight times longer on the pouch compared to a naked lure or SPLAT attract-and-kill wax dollop designed for Oriental fruit moth control (Huang et al., 2015). Contact with the pouch resulted in nearly 100% mortality or knockdown of moths. Surprisingly, deltamethrin in the fabric remained lethal for nearly a year after ageing in the field. In a field experiment, the attracticidal pouch outperformed mating disruption at a low rate of 50 devices/acre. The pouch also showed promise for Japanese beetle control (Grieshop et al., 2014). Forced contact with the deltamethrin-impregnated pouch in a laboratory assay resulted in nearly 100% beetle mortality. Preliminary field studies using the Trécé Japanese Beetle Dual Lure as the attractant source demonstrated that beetles readily contact the device and deployment of 50 attracticidal pouches in half acre plots resulted in a fivefold reduction in beetles captured in a centrally located trap compared to plots treated with attractant-baited pouches without the toxicant or no pouches.

Although the pouch design greatly increases exposure to the insecticide, it retains many of the constraints associated with other attract-and-kill formulations. The required manual application of the devices at a high density presents a formidable economic impediment. However, the impressive lethality of pouches field-deployed for nearly a year partially addresses the desire for season-long control with a single application. The remaining hurdle is the development of a delivery system that releases the attractant at a lure-equivalent rate for an extended period of time.

Further limiting the efficacy of attracticidal formulations is the nearly complete reliance on female sex pheromones that only target males. We anticipate that their performance would be greatly improved if females also were removed from the population. Kairomones, particularly host odours, can be attractive to males and females. The lure used in the Japanese beetle pouch was loaded with an aggregation pheromone and plant volatiles that attracted both sexes. Modelling the combination of sex pheromone and pear ester to attract codling moth to a toxin or sterilizing agent predicted a substantial reduction in mating (Potting and Knight, 2002).

4.3 Bait sprays

Bait sprays have been used for control of *Rhagoletis* species and other tephritid fruit flies for more than a century. Severin (1916) reported early successes in controlling apple maggot using poison bait sprays that mixed sugar molasses with small amounts of arsenical insecticides. A major advance in the composition of baits occurred when Steiner (1952) showed that the inclusion of protein hydrolysate with sugar and insecticide produced highly attractive and effective bait sprays for control of Oriental fruit fly (*Bactrocera dorsalis).* Protein bait sprays have become the principal means of controlling many species of tropical fruit flies. The approach takes advantage of the female flies need for a protein source to maximize egg production. For many years, bait spray formulations for controlling fruit flies were mainly comprised of protein hydrolysate (Nu-Lure) as the bait and malathion as the toxicant (Vargas et al., 2008). Concerns about negative impacts of organophosphorus insecticides on natural enemies and human health prompted the development of improved baits and reliance on reduced-risk insecticides. New formulations based on hydrolysed protein from maize were developed and when combined with the naturally derived insecticide, spinosad, were highly effective in controlling a number of fruit fly species (Moreno and Mangan, 2002; Peck and McQuate, 2000; Vargas et al., 2001). A mixture of spinosad, protein, sugar, ammonium acetate and other ingredients was subsequently produced for control of many fruit fly species as GF-120 or NF Naturalyte (Corteva Agriscience, Indianapolis, IN, USA).

GF-120 has been used commercially for control of blueberry maggot (*Rhagoletis mendax*), western cherry fruit fly (*R. indifferens*), eastern cherry fruit fly (*R. cingulata*) and black cherry fruit fly (*R. fausta*), as well as suppression of apple maggot in deciduous fruit crops. Low-volume applications of this fruit fly bait provided excellent fruit protection from western cherry fruit fly in western US cherry production systems and the approach has been widely adopted as the primary control for this tephritid pest (Yee and Chapman, 2005; Yee and Alston, 2012). Although GF-120 sprays significantly reduced fruit infestation by blueberry maggot and apple maggot compared to untreated controls, they fell short of outperforming the unbaited formulation of spinosad (Pelz et al., 2005). Moreover, bait sprays did not provide the extremely high fruit protection required to meet the stringent quality standards of the blueberry and apple markets. These field tests took place in the Midwestern United States, a region with an abundance of fruit fly hosts outside of the crop and a wet, humid climate. Feeding by female *Rhagoletis* flies is greatest during the week after emergence when they seek nutrients required for egg development. Flies must find and ingest a lethal dose of GF-120 before egg laying for the bait spray to be effective, which may be hard to accomplish in settings where many reproductively mature females immigrate from nearby untreated hosts. Additionally, the components of the bait are highly susceptible to wash-off by rain (Teixeira, 2009). The superior performance of GF-120 in the western United States is likely due to the dry climate and paucity of fruit fly hosts outside of commercial plantings.

In an effort to improve the performance of baited insecticides for control of Rhagoletis species, researchers have examined ways to improve rainfastness and enhance the attractiveness of the bait. The addition of a paraffin-wax matrix to GF-120 increased rainfastness and improved the efficacy of the mixture compared with GF-120 alone when the treatments were exposed to simulated rain (Teixeira et al., 2009). In a field trial, the performance of the wax-enhanced formulation was significantly better than standard GF-120; however, the level of fruit infestation was still well above that required for commercial control. Ammonium acetate is a known attractant for Rhagoletis flies (Dodson, 1948; Frick et al., 1954) and is present at a concentration of 1% in GF-120. Field observations revealed that the bait components of GF-120 failed to attract blueberry maggot flies, rather they acted as arrestants (Pelz et al., 2005). Similar numbers of flies landed on blueberry foliage treated with the bait or water droplets, but flies spent significantly more time in close proximity to bait droplets compared with water droplets. Yee and Chapman (2005) found a similar lack of attraction by western cherry fruit flies to GF-120 in the field. Subsequent studies with eastern cherry fruit fly indicated that the addition of 5–10% ammonium acetate enhanced attraction to droplets, arrestment at droplets and interaction with droplets (Pelz-Stelinski et al., 2006). The attractiveness of GF-120 to apple maggot flies also increased with the addition of 2.5% or 10% ammonia, but this did not improve control. In contrast, aqueous solutions of 1% and 4% ammonium acetate deterred feeding by blueberry maggot flies in laboratory bioassays (Barry and Polavarapu, 2004).

Modifying bait colour is another possible avenue for improving the attraction of fruit fly baits. The location of mates, oviposition sites, food and many other aspects of insect behaviour are guided by visual cues (Giurfa and Menzel, 1997). *Rhagoletis* flies, like many herbivorous insects, are highly responsive to yellow, prompting the speculation that this colour constitutes a supernormal foliage-like stimulus, reflecting peak energy in the same wavelength (Prokopy and Owens, 1983). Yellow traps are highly attractive to Rhagoletis fruit flies (Prokopy, 1972; Prokopy and Boller, 1971). In petri dish arenas, more apple maggot and eastern cherry fruit flies found translucent yellow sucrose deposits, or colours strongly contrasting with the background, than other colours (Teixeira et al., 2010b). This outcome suggests that short-distance interaction between fruit flies and bait droplets may be improved by matching bait colour with their visual preferences.

4.4 Attracticidal spheres

Attracticidal spheres have long shown promise as a behavioural tactic for managing Rhagoletis fruit flies in the Eastern and Midwestern United States (Prokopy et al., 2000; Stelinski and Liburd, 2001; Wright et al., 2012; Morrison et al., 2016a). They rely on a combination of attractants, a feeding stimulant and a toxicant to lure-and-kill the target pest. Adult flies are highly attracted to 8–9 cm diameter coloured spheres, red for apple maggot (Prokopy, 1968) or green for blueberry maggot (Liburd et al., 1998). Originally, these coloured spheres were coated with Tangletrap to ensnare attracted flies for the purpose of monitoring their activity. Deploying them with an appropriate olfactory stimulus enhanced their attractiveness. Subsequent studies revealed that baited and visually attractive sticky spheres had the potential to reduce apple maggot fruit injury when deployed around the perimeter of an apple orchard, in many cases providing control equivalent to insecticide sprays (Prokopy, 2000). Commercial adoption of this control system, however, was severely limited by the weekly maintenance required to remove insects and debris from the sticky surface to retain capturing effectiveness.

Subsequent efforts focussed on switching from a sticky surface to a toxicant as the killing agent. The insecticide, principally imidacloprid, was assimilated in the latex paint used to colour the spheres to aid in maintaining residual activity of the toxicant. Sucrose was added as a feeding stimulant to coerce flies to ingest the toxicant. Spheres laced with sugar and a low dose of insecticide showed promise for control of apple and blueberry maggot (Prokopy et al., 2000; Stelinski and Liburd, 2001). While biodegradable versions that internally incorporated the sugar into the sphere were effective, these starch-based traps quickly degraded under field conditions.

A new generation of attracticidal spheres for control of apple maggot consisted of caps that were designed to provide controlled release of the sucrose and insecticide (Fig. 9; Wright et al., 2012). The red caps were cylindrical or contoured and fixed atop red sphere bases. A perimeter deployment of contoured spheres baited with sucrose and spinosad provided season-long fruit protection equivalent to one to two insecticide applications (Wright et al., 2012). Further refinements of this low-maintenance attract-and-kill system have focussed on the addition of chemical cues to enhance attraction and improve efficacy. Morrison et al. (2016a) found that the specialist olfactory cue, the five-component ripe apple volatile blend, improved attraction and recommended that it be included as part of the attracticidal system in the eastern United States. They also concluded that the generalist olfactory cue, ammonium carbonate, did not increase mortality in laboratory assays or enhance attraction in the field and should be excluded in future formulations.

The development of attracticidal spheres for managing apple maggot well illustrates the long journey from conceptualization to commercial use of a behaviourally based control. It has been 50 years since Prokopy (1968) initiated his investigation of fly response to colour and shape. Generations of scientists have further contributed to our understanding of fruit fly behaviour and used this knowledge to design practical devices for their management. Although commercial viability has been demonstrated, few growers have incorporated attracticidal spheres into their management programme. Economic considerations play a major role in limiting adoption. Ultimately, the density of spheres required to achieve control will determine whether the approach is economical. Thus, we concur with Morrison et al. (2016a) that a crucial piece of missing information is the attractive range of spheres and thus the optimal spacing required for control.

Efforts have been initiated to adapt the lure-and-kill technique using attracticidal spheres to mitigate infestation by spotted-wing drosophila (*Drosophila suzukii*) an invasive fruit fly that has become a devastating pest of cherry and berry crops in North America and elsewhere. Male and female flies are highly responsive to red spherical objects (Kirkpatrick et al., 2016; Rice et al., 2016), suggesting that the insecticide-laden spheres developed for apple maggot could be adapted for managing spotted-wing drosophila. In laboratory trials, attracticidal spheres coated with a neonicotinoid, pyrethroid, organophosphate or spinosyn insecticide killed 100% of exposed flies within 24 h (Rice et al., 2017). Two years of field trials demonstrated promise for the approach, with lower spotted-wing drosophila infestations in raspberry plots treated with spheres compared to untreated plots (Rice et al., 2017). However, in one of the 2 years, infestation in the plots treated with spheres was significantly greater than in insecticide-treated plots. The authors proposed better control was achieved in the year when ripe fruit were removed weekly from the plots. Their logic was that fewer fruit translated into lower pest density and thus better control by attracticidal spheres, as their effectiveness is density dependent. Population densities of spotted-wing drosophila are typically very high in berry fields and cherry orchards, and this may severely hamper the use of lure-and-kill or mass trapping for controlling this pest. Additionally, olfactory cues developed to date do not seem to be highly attractive. Mark release-capture experiments conducted with spotted-wing drosophila in Michigan tart cherry orchards revealed a tiny plume reach (<3 m) for a sticky red panel trap baited with a commercial lure (Kirkpatrick et al., 2018). The inclusion of the same spotted-wing drosophila lure with mass-deployed attracticidal spheres actually increased infestation in raspberry plots compared with use of unbaited spheres alone (Rice et al., 2017). It appears that adults may be attracted to the general vicinity of the lure, but readily visit nearby fruit.

4.5 Future trends in mass trapping and attract-and-kill

4.5.1 Optimal spacing of devices

Having developed an attractive, long-lasting lure and a low cost-efficient trap or attracticidal device, the question remains how many devices need to be deployed to achieve control? Researchers and practitioners have wrestled with this critical question since the earliest attempts to use mass annihilation for insect pest control. A low density of devices is less costly and easier to maintain, but likely will not remove enough individuals from the mating pool to prevent crop damage. A high density of traps would improve the chance of success but is costly and could result in trap interference or worse, trap redundancy. Efforts to determine the optimum density have been made by deploying traps at various densities and collecting catch data and other measures of pest density (Jamieson et al., 2008; Reinke et al., 2012). This trial-and-error approach is quite time-consuming and conclusive results are hard to obtain due to the variation in pest densities across plots and over time. In addition, it is unlikely to provide an understanding of the interaction between traps needed to optimize their deployment.

More efficient methods for determining how many traps or devices to deploy are needed and should focus on providing critical information about plume reach and trapping area. Byers et al. (1989) developed a means of determining the strength and distance of attraction of a lure-baited trap by measuring its effective attraction radius. This novel approach compares catch in a passive trap to catch in a trap baited with an attractive lure and attributes higher catch in the baited trap to the lure's ability to effectively increase the size of a passive trap. While this method does provide a relative measure of the influence of the attractive lure, it does not provide information on plume reach or guidance for spacing traps. In addition, the technique cannot be used for many insect species, especially lepidopterans responding to sex pheromones, where catch in passive traps is often at or near zero.

Others have taken a more direct approach to optimizing deployment density by setting up an array of traps at varying distances apart in an effort to measure plume overlap and trap competition. Gut et al. (2010) compared codling moth catch in pheromone-baited traps placed at a density of 1 trap/4.5 acre, or in 9 trap arrays at densities of 9 traps/9 acre, 9 traps/4.5 acre and 9 traps/0.5 acre. An estimation of trapping area was based on differences in catch when traps were deployed at different densities. Suckling et al. (2015) also proposed using trapping arrays as a means of getting at optimum deployment density, but focussed on the ratio of corner to centre, and edge to centre catch as the critical measure. Although the trapping array methodology is intuitively promising, these attempts to measure trap competition fall short of providing a means of determining optimum trap spacing. In part, this is because the results of catch in a small array are not representative of catch data in the larger array of hundreds of traps that would eventually be deployed in the whole management unit. In addition, catch in a small array of traps does not provide the two essential values needed to optimize trap density, plume reach and trapping area.

Considerable effort has been directed towards measuring plume reach. The earliest studies used direct observation (Elkinton et al., 1987; Baker and Roelofs, 1981) or field EAGs (Baker and Haynes, 1989). Experiments were time-consuming to set up and interpret and, therefore, often were conducted under quite favourable conditions, that is, ideal wind conditions in open fields or aligned between tree rows with no obstacles between the insect and the lure. The resulting measured plume lengths could be quite long. For example, under ideal experimental conditions where plume structures were created by a steady low-speed laminar airflow, male gypsy moths responded to pheromone from as far as 100 m, using the classic zig-zag positive plume following behaviour (Kennedy, 1974). A consequence of this early work is that it established a paradigm of large-plume thinking. Terms such as the 'active space' of the trap, which places the adjective on the trap instead of the insect, is one outcome of this large-plume paradigm. This trap-centric framing of the insect-trap interaction limits our understanding of the insect's dispersive behaviour prior to arriving at and being captured in a trap. Much of the movement that results in arrival at a trap is random dispersal by the searching insect (Miller et al., 2015), rather than long-distance ensnarement by an attractive plume. The large-plume conviction also has affected the way those working with odour sources envisioned the operation of not only pheromone-baited traps, but mass trapping, attract-and-kill and mating disruption. Attractant-baited traps or devices were conceived of as powerful vacuums sucking up insects from considerable distances away. This is not an accurate representation. If large plumes were the rule rather than the exception, one would expect to see more success in using these approaches for managing pests than has thus far been realized.

Determining the trapping area has also proved challenging. Trapping area is comprised of the plume reach plus the dispersive distance of the insects.

The dispersive distance of insects has been investigated in a number of ways including sustained-flight wind tunnels (Miller and Roelofs, 1978), flight mills (Hughes and Dorn, 2002; Schumacher et al., 1997; Pats and Wiktellus, 1989), trapping arrays, (Meats and Edgerton, 2008; Epsky et al., 2010) and capture of released individuals in traps (Howell and Clift, 1974; Mani and Wildbolz, 1977). These methods have provided insights into flight ability, capacity and spatial distribution, but have revealed little about how a population as a whole disperses. Wind tunnels can assess how individuals respond to odour cues, flight mills can measure total flight capacity, but neither can measure an insect's net displacement in the field. Because insects do not move in straight lines or only upwind, but instead meander throughout a crop in search of food, a mate or a place to rest, field measures of movement are essential.

A more robust and field-based method for quantifying the size of the plume, dispersive distance of the insect and trapping area is needed if we are to accurately describe the influence of lure-baited traps or devices and optimize how they are deployed. It requires a technique that takes into account variation in wind speed and direction over time and considers movement by the entire population, not just a few individuals moving under optimal conditions. What is desired is an experimental design that permits researchers to quantify the pheromone plume and movement over a range of normal field conditions, including under less than ideal conditions, across rows, with variable wind directionality and throughout the season.

The single-trap, multiple-release experimental design, along with the graphical interpretations of the catch data detailed in Miller et al. (2015), provides a simple and robust means of quantifying plume reach, dispersive distance of the population and trapping area (Fig. 10). The method consists of a central trap with insects released at regularly spaced intervals in the four cardinal directions away from the trap. The individuals released at each distance are uniquely marked, and the number released is increased with each successive distance to improve the probability of catch where likelihood of catch is very low. To properly plot the data, it is important to have release sites close to the trap where the probability of catch is high and release sites to a distance where very few, if any, individuals are captured. Catch is recorded for a defined period of time (e.g. 1 week) and the proportion caught is calculated for each distance and also calculated for the overall catch. Figure 10b shows the relationship between probability of catch and release distance. The concave line with an asymptotic approach to the *x*-axis illustrates how quickly catch probability falls with distance, and how the catch probability then gradually approaches zero at the furthest distances. This begins to give us some insights into the spacing required between traps in a mass-trapping configuration. Even at the closest distance, the probability of catch may be only 0.5 and falls off sharply at the next closest release point. In fact, when we consider the entire area from which catch

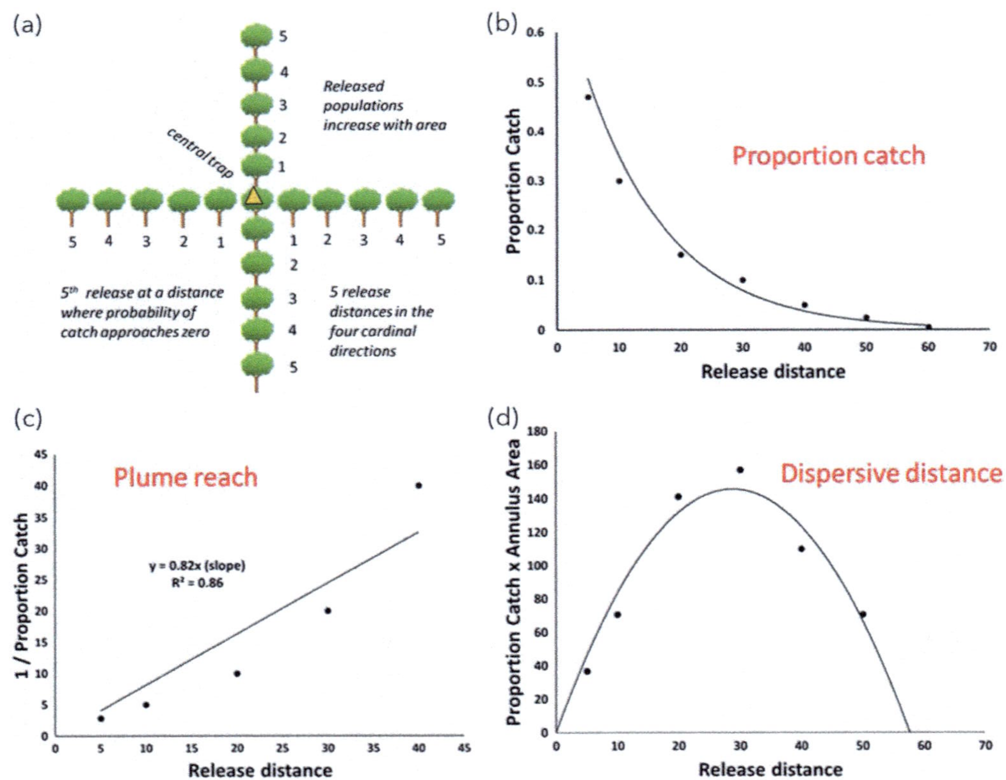

Figure 10 (a) Centrally placed single-trap, multiple-release experimental design. (b) Graph of proportion catch over release distance. (c) Inverse of proportion catch over distance. Slope from this graph can be used to look up the plume reach of the lure. (d) Proportion catch multiplied by the area of the release annulus over distance. Where the line crosses the *x*-axis represents the dispersive distance of 95% of the population.

can occur (trapping area) we find that our trap only samples a small proportion, around 1%, of the total population.

The plot of probability of catch versus release distance also reveals some hurdles in trying to interpret mark recapture results. Because the probability of catch approaches zero very slowly, there is a slight chance of capturing individuals released from extreme distances. Dispersal by these rare individuals is often conflated with the maximal dispersal distance (Schumacher et al., 1997; Mani and Wildbolz, 1977), when more appropriately, the maximum dispersive distance of 95% of the population, or the population that is likely to be in play within the management area, should be considered. Accurately identifying the maximum dispersal distance of 95% of the population has important implications for selecting the optimal intertrap spacing or deployment density for mass-trapping and attract-and-kill strategies.

As mentioned previously, proportion catch decreases quickly with distance and approaches the end of the trapping area asymptotically (Fig. 10b). The ability to measure catch falls sharply at the greatest release distances. This makes it difficult to quantify catch at the furthest distances and to measure maximum dispersal. To address this, catch data can be transformed by multiplying the proportion caught by the annulus area from which the insects originated. A curve can be fitted to the resulting data using a second-order polynomial anchored at the point (0,0). The resulting plot (Fig. 10d) generates a discrete endpoint where it crosses the *x*-axis. This point is the maximum dispersive distance for 95% of the population (Miller et al., 2015). The area under the curve in Fig. 10d reveals the contribution to total catch of each release distance. This figure illustrates that the majority of catch occurs at the central distances.

The most important information required for optimizing the deployment of devices for the purpose of mass annihilation is the reach of the plume. Fortunately, this value also can be derived from the catch data collected in the single-trap multiple-release experiment (Miller et al., 2015). This is accomplished in a two-step process. First, the inverse of proportion catch versus release distance is plotted (Fig. 10c, MAG plot in Miller et al. (2015). The first portion of the plotted data will be linear and the slope from a fitted straight line is used to estimate the plume reach from a standard curve provided as Figure 4.12 in Miller et al. (2015). Consistently up-turning data points of a MAG plot at greater release distances should not be included when estimating plume reach (Miller et al., 2015).

Investigating the plume by this method takes into account the response of the entire population to the lure. Individuals are encountering the plume from all directions and under a range of environmental conditions. Under field conditions, using this method, plume reach appears to be in the range of metres rather than hundreds of metres. Using the single-trap, multiple-release experimental design, Adams et al. (2017) estimated the plume reach from a codling moth sex pheromone-baited trap to be <5 m. A mark recapture study with spotted-wing drosophila revealed that the plume reach of a sticky red panel trap baited with a commercial lure had a very tiny plume reach of <2 m (Kirkpatrick et al., 2018). Indeed, among the published single-trap multiple-release experiments from which a measure of the plume size could be derived using the Miller et al. (2015) method, most generated a plume of only a few metres (Mason et al., 1990; Turchin and Odendall, 1996; Dodds and Ross, 2002; Whamsley et al., 2006; Maki et al., 2011; Sufyan et al., 2011; Adams et al., 2017; Kirkpatrick et al., 2018).

Armed with reliable information on plume reach and dispersal, we can now make decisions about the potential for mass annihilation and optimal deployment strategies for the targeted pest species. Knowing the size of the plume reach has major implications for determining the optimum density for deploying lure-baited traps or devices for controlling a pest. Computer simulations of control by mass trapping conducted by Miller et al. (2015) suggested that the optimal spacing was at about 1.5 times the plum reach of a given trap. Thus, the best chance for commercial success is when the plume reach is large, perhaps 25 m or more. Traps in this case can likely be deployed at a fairly wide and economical spacing. If the plume reach is small, successful mass annihilation will require a close spacing of traps or devices. Opting for a larger spacing will require a compromise between efficacy and cost. If the plume reach is incorrectly considered to be large, the density of traps or devices may be too low to achieve the desired results.

Knowing the probability of catch at various distances away from the trap or kill device also provides valuable insight into how closely traps would need to be spaced to remove most individuals from the population. Figure 10b illustrates how quickly the probability of catch falls with distance. In order to create a trapping grid without gaps in coverage, the distance between adjacent traps must be surprisingly close. Traps spaced further apart than the distance where the probability of success is <0.1 leaves large areas of the crop where individuals can find each other and mate before they are captured or killed. A highly efficacious mass annihilation control programme will likely require that adjacent devices be spaced at a distance at which the probability of successful capture or kill of a searching insect is >0.5 for each trap or the combined probabilities approach 1. Figure 11 illustrates the probabilities of catch over distance for two adjacent traps with small plumes, <5 m, spaced beyond their respective trapping radii. An individual insect located between traps would not be sampled by either trap. As the traps are moved closer together, and their trapping radii begin to overlap, there is a small probability of getting caught in one of the traps but zero probability of being caught in both (i.e. no competition). The reasoning behind this is that for competition to

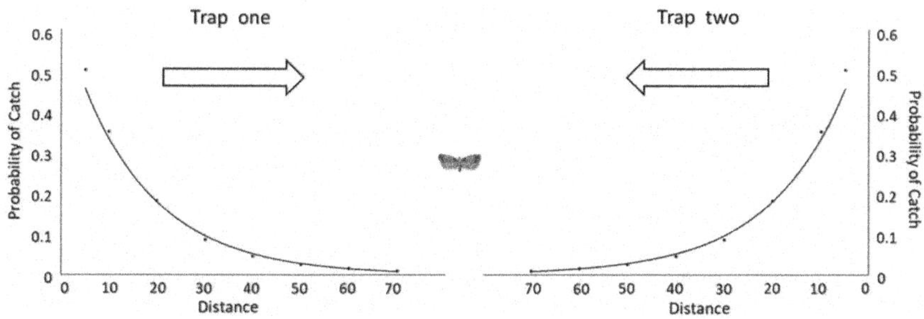

Figure 11 Graphical representation of the catch probabilities of two adjacent traps. As traps are moved closer together, the probabilities begin to overlap. The moth is located centrally, beyond the trapping radii of both traps.

occur the insect would need to travel to, and be caught in, the first trap, and then travel back and be captured in the second trap (assuming the first trap was not there). In the case of the traps in Fig. 11, the probability at a distance of 70 is the product of the two small probabilities (0.001 x 0.001 = 0.000001). This extremely low probability is why competition between adjacent traps does not begin at the edge of the trapping radius. Competition between traps requires a measurable reduction in catch in one trap because of the presence of another, which does not occur until traps are spaced at one half the trapping radii. The implications for mass annihilation are that traps with small plumes will need to be placed very close together to insure a high probability of removing all or most of the individuals in a crop. While competition begins at half the trapping area, the trap spacing needed for an acceptable level of control may be as close as the measured plume reach. Ultimately, an informed decision about how many devices to deploy requires having realistic expectations based on knowing the plume reach, trapping area and cost/benefit ratio of implementing the programme.

4.5.2 Behavioural manipulation using vibrational signals

Mass trapping, attract-and-kill and other behavioural approaches for pest control have largely focussed on the use of semiochemical or visual cues. The tactic has been most successful when specific and potent attractants are used to manipulate behaviour. Achieving control has proved more difficult in cases where attractants only draw insects to the general vicinity of the source. For example, brown marmorated stink bug (*Halyomorpha halys*) and other stink bugs are attracted to aggregation pheromones but are arrested within a few metres of the stimulus source (Morrison et al., 2016b). This makes it difficult to achieve control using mass trapping or attract-and-kill, as only a small portion of the responding individuals contact the trap or kill device.

Many insects, including stink bugs, use additional modes of communication to locate conspecifics, especially at close range. Substrate vibration is a widespread modality in insects that has been an underutilized avenue for developing novel tactics for behavioural manipulation. Sound and substrate vibrations are used by an estimated 195 000 species across many orders of insects (Cocroft and Rodríguez, 2005). Among pests of deciduous tree fruits, this modality appears to be especially prevalent among hemipteran species. A vibration signal produced by brown marmorated stink bug females attracted more than 50% of males to the source, where they tended to remain (Mazzoni et al., 2017). The findings suggest that the efficacy of a trap or kill device may be improved by adding a vibrational stimulus. Playback of vibrational noise also has shown promise in laboratory and small-scale field trials for disrupting the courtship behaviour of American grapevine leafhopper, *Scaphoideus titanus* (Polajnar et al., 2016). While the application of controls based on acoustic signals is still in the development stage, we believe it is a promising approach that warrants further attention. Of course, technological challenges to cost-effective deployment of the signals in the field must be addressed before the practical application of acoustic-based controls can be realized.

5 Case study: CMMD

5.1 Pheromone identification

Roelofs et al. (1971) identified E8, E10 dodecadienol (codlemone) as the sex pheromone of codling moth. In subsequent years, extensive research was conducted to identify secondary components (Einhorn et al., 1984; Arn et al., 1985; Bartell et al., 1988). Despite some evidence for the presence of secondary components in wind tunnel tests, in field experiments the inclusion of the secondary components with codlemone did not increase attraction to pheromone traps versus codlemone alone (Einhorn et al., 1984; Guerin et al., 1985). Preiss and Priesner (1988) were not able to show any specialized receptors for dodecanol, the most abundant secondary component identified in the female sex pheromone gland. Codling moth is one of a few species where a single component and not a blend of major and minor components is sufficient to achieve species-specific communication.

5.2 Early research

Research trials to evaluate the efficacy of CMMD were initiated in the mid-1970s and 1980s in Europe (Charmillot, 1990), Australia (Rothschild and Vickers, 1982) and North America (Cardé et al., 1977; Moffitt and Westigard, 1984; Howell et al., 1992; Judd et al., 1997). Results were generally mixed, and it was not until the early 1990s that the first CMMD products were registered. Isomate C™ was the first CMMD product to be registered in North America (Thomson, 2001). In Europe, the BASF RAK 3™ dispenser was registered for the control of codling moth in the late 1980s (Waldner, 1997).

5.3 Adoption

In the last 25 years, CMMD gained acceptance and is currently registered and sold in countries around the world including the United States, Mexico, Canada, Argentina, Chile, Uruguay, South Africa, Italy, France, Spain, The Netherlands and many other countries in the European Union (EU), Turkey, Australia and New Zealand. CMMD is now used on over 220 000 ha of pome fruit worldwide (Miller and Gut, 2015); primarily in developed countries as growers, in developing countries unfortunately do not have either the financial or technical resources to deploy and support the technology. In developed countries, the largest markets for CMMD include the United States, Mexico, Argentina, South Africa, Italy, Spain and France. In these countries, the market for CMMD is saturated so for the increased use of CMMD to occur, the markets will have to grow in developing countries such as China, eastern EU countries and Russia.

In many areas around the world, including the US states of California and Michigan and countries such as Argentina, South Africa, Italy and France, the adoption of CMMD was driven by a crisis brought on by insecticide resistance and/or the loss of insecticides due to regulations or concerns about food safety (Thomson and Jenkins, 2014). Growers, in an attempt to re-establish control of CM and reduce damage, combined CMMD with insecticides in their pest management programmes. In general, the combination of insecticides and CMMD was very successful in controlling CM populations and decreasing damage to the fruit.

However, not all growers adopted CMMD due to a crisis. In some fruit production regions, such as portions of western North America and the EU, growers adopted CMMD in response to other issues including the cost of implementing worker safety regulations, concerns about the environment, sustaining populations of natural enemies and the minimum residue levels on fruit imposed by domestic and export buyers. In all cases, the adoption of CMMD was slow in the early years following registration but in the mid- to late 1990s, there was a concerted effort by public agencies around the world to demonstrate the efficacy of CMMD and implement area-wide programmes. These government supported programmes greatly accelerated the acceptance and adoption of CMMD by growers in pome fruit production regions around the world.

5.4 Demonstration and area-wide programmes

The success of CMMD in pome fruit systems owed much to publicly funded demonstration and area-wide programmes. The deployment and incorporation of CMMD into existing pest management programmes required a lot of education of agricultural chemical company and other industry representatives and growers to ensure the proper selection of orchards, the optimal timing and placement of dispensers and proper monitoring of orchards (Thomson et al., 2001). Initially, publicly funded entities such as government research and extension agencies and universities were important to the success of CMMD by demonstrating to growers how to deploy the technologies to ensure the efficacy of CMMD (Judd et al., 1997; Waldner, 1997; Barnes and Blomefield, 1997; Gut and Brunner, 1998). Experiments were set up comparing CMMD, CMMD plus insecticides and insecticides alone in side-by-side comparison blocks (Gut and Brunner, 1998). Demonstration trials by publicly funded research scientists gave growers the confidence and expertise needed to incorporate a novel technology into their existing pest management systems. In addition, they provided growers with credible assessments of the comparative efficacy of the commercially available CMMD dispenser technologies.

The idea of area-wide management was most prominently promoted by the USDA scientist E. F. Knipling in the 1950s and 1960s as an effective and coordinated approach to manage insects. Knipling (1978) argued that most insect pests of major importance to agriculture were not properly controlled due to the unorganized approach to control taken by industry and growers. Instead of a farm-to-farm and/or crop-to-crop method of controlling pests, he stressed that the area-wide approach would be much more successful. The eradication of the screw worm (*Cochliomyia hominivorax*) from across the Southern United States was a good example of the success of a coordinated and organized approach to pest management (Baumhover et al., 1955).

In the United States, the Codling Moth Areawide Management Program (CAMP) was initiated in 1993 (Calkins and Faust, 2003). The programme was a joint effort between the University of California, Oregon State University, Washington State University and the US Department of Agriculture/Agricultural Research Service. The objective of the 5-year programme was to implement an area-wide and multi-tactic pest management strategy that would effectively control CM (key pest) in pome fruit while reducing the negative impacts of pesticide use on natural enemies of secondary pests, the environment and farm workers. The major tactic for control of CM was mating disruption using moderate-density dispensers. Five area-wide projects comprising a total of 1294 ha in three western states were initiated in 1995 (Calkins and Faust, 2003). By 2001, the average number of times the most common CM insecticides were applied had declined substantially (Brunner et al., 2007). The success of CAMP elevated the stature of mating disruption as a viable control strategy for CM, especially when used in a coordinated area-wide approach, and set the stage for wide-scale adoption of CMMD not only in US western states but in many other pome fruit production regions around the world. In 2016, approximately 90% of all the pome fruit area (80 000 ha) in Washington State were treated with CMMD (J. Brunner, pers. comm.).

In Michigan, CM resistance to existing insecticides, in particular organophosphates, created a crisis for apple growers by the early 2000s (Mota-Sanchez et al., 2008; McGhee et al., 2011). Following the lead from the CAMP project, a 4-year research and implementation area-wide project began in 2003 (McGhee et al., 2011). The area-wide approach (multiple adjacent blocks treated with CMMD) was compared to blocks where CMMD was applied to individual apple blocks or apple blocks not treated with mating disruption. Captures of male codling moth in pheromone-baited traps were reduced by 93% by year 4 of the project in AW-CMMD orchards. Injury to fruit and the overall number of insecticides targeting codling moth decreased each year in the area-wide blocks. In contrast, fruit injury levels remained about the same each year in individual blocks treated with pheromone and increased significantly in orchards relying on insecticides only. These findings provided the first direct evidence that CMMD was more effective when used on an area-wide basis rather than on individual blocks. Furthermore, growers who implemented AW-CMMD realized a mean savings of $55–65/ha through reduced fruit injury and decreased use of insecticides as compared with orchards using insecticide-only programmes (McGhee et al., 2011). The success of the area-wide project laid the foundation for widespread adoption and continued use of CMMD in Michigan.

Although the South Tyrol project in northern Italy started out as a demonstration trial (Waldner, 1997), it soon evolved into an area-wide project. Mating disruption was originally introduced into this famous European tourist destination due to resistance to the insecticide diflubenzuron and to help mitigate the conflict at the interface of agriculture and tourism. CMMD was first introduced into South Tyrol in 1991 on 110 ha (Waldner, 1997). By 1996, CMMD was successfully applied on 3500 ha. By 2016, all the pome fruit (22 100 ha) was treated with CMMD resulting in better control of CM and a substantial decrease in the use of insecticides. South Tyrol is probably the most successful and continuous area-wide project in the history of CM pest management. In large part, the structure of the industry with cooperatives and private consultants working together with growers and public institutions advocating and supporting CMMD made it feasible.

There were other successful area-wide projects in pome fruit production areas. In Argentina, close to 100% of the apple and pears in the Rio Negro area were treated with CMMD technologies for 3 years. Unfortunately, once the government subsidies were removed at the end of programme, CMMD usage declined due to the perceived high cost of the technology. Still, CMMD is widely used in Argentina today and the government-sponsored area-wide project contributed to its adoption and success. The Republic of South Africa is another success story where close to 80% of the pome fruit ha are treated with CMMD (T. Blomefield, pers. comm.).

5.5 Companion insecticides and CMMD

In 1991, the year CMMD was first registered and sold in the United States, growers were mostly reliant on organophosphate and carbamate insecticides to control CM. The high cost of CMMD (~$330/ha) restricted the number of growers willing to try the new technology. By 1993, the cost of CMMD had decreased by nearly 20% and more growers tried the technology. The combination of CMMD and organophosphates were shown to work extremely well to control CM, and by 1995 and the start of the USDA-sponsored CAMP project, the combination of disruption and insecticides was slowly being adopted by growers. In the mid-1990s, growers in Washington State started deploying CMMD at half of the recommended rate in conjunction with use of organophosphate insecticides. The reduction in cost of CMMD and uncertainty with eliminating insecticides all together made the combination programme very attractive to growers and the programme's success led to increased adoption. The programme was not only attractive to growers due to cost savings but also was widely supported by the agricultural chemical industry as they sold both the CMMD and insecticides. The combination approach also decreased the risk associated with managing CM.

However, change came to the regulation of pesticides with the passage of the Food Quality Protection Act (FQPA) by the US Congress in 1996. The FQPA fundamentally changed EPA's regulation of pesticides by (1) utilizing new safety standards to set residue tolerances, (2) using these new safety standard to reassess, over a 10-year period, all pesticide tolerances that were in place when the FQPA was signed, (3) making special consideration for the susceptibility of children to pesticides, (4) making an aggregate risk assessment from exposure to multiple sources of pesticides with the same mode of action and (5) considering the cumulative exposure to pesticides with common mechanisms of toxicity. By 2006, the EPA had completed its reassessment of 9721 pesticides and the agency revoked or modified ~ 4000 tolerances.

In the mid-1990s, the EPA began the voluntary Reduced Risk Pesticide Program to encourage the development, registration and use of lower-risk pesticides. The EPA expedited and reduced the cost of the registration of reduced-risk pesticides (RRP) in the hope of increasing their use by growers. RRPs had a number of advantages over any of the other conventional insecticides including lower impact on human health, lower toxicity to non-target organisms, lower-use rates and increased compatibility with IPM programmes. The lower costs to register these insecticides and faster approval time encouraged the industry to develop and commercialize RRP products. In conjunction with the FQPA, there was a rapid movement away from conventional chemistries to these products. However, the RRPs were more expensive and required more precise application timing than most of the conventional insecticides (Brunner et al., 2007). Most of the RRPs did not kill the target insect upon contact, as had the organophosphate compounds, but had to be ingested so coverage was critically important. Additionally, they had little effect on adult insects. With the phase-out of azinphosmethyl in 2012 by the EPA, growers were forced to rely primarily on RRPs to supplement CMMD. The lack of a good contact insecticide to reduce adult CM densities when used in conjunction with CMMD made control more difficult but manageable if attention was paid to timing and coverage.

Questions about the perceived benefits and sustainability of some of the RRP products have recently emerged. The spinosyn products are known to be toxic to the hymenopteran parasitoids (Stark et al., 2004), so they can have a negative impact on biological control. Neonicotinoids have generated a lot of controversy because of the potential and perceived impact on pollinators (Chenshen and Warchol, 2014; Lawrence et al., 2016). The EU and Canada are in the process of banning all neonicotinoids and regulatory agencies in other countries are likely to take similar actions. There are increasing signs that CM is becoming resistant to the widely used spinosyn and diamide insecticides. The cost of discovering, developing, registering and commercializing a new pesticide is well over $250 million (Glare et al., 2012). Over the last 30 years, pesticide manufacturers have focussed on registering compounds with low mammalian toxicity, resulting in no new registrations of adulticides with contact activity. In the last 5 years, there have been many mergers of large agriculture chemical companies such as Dow and Dupont, Bayer and Monsanto and ChemChina and Syngenta due to low commodity prices and declining revenues. Unfortunately, these large corporate mergers will likely further reduce research on new chemistries and products. At this point, and in the near future, there appears to be little in the way of new insecticide registrations with unique modes of action. This does not bode well for CM pest management. It is well established that CMMD needs to be integrated into comprehensive pest management strategies in particular with the use of insecticides to successfully control codling moth. With no new chemistries in the pipeline, and the loss of existing chemistries due to food and worker safety issues, regulations and insecticide resistance, it will be increasingly difficult to maintain CM populations at levels where CMMD can be effective. Despite extensive research efforts over the last 10 years to improve the efficacy of CMMD (Miller and Gut, 2015), the dispensing technology is largely the same as when CMMD was first registered. Although our understanding of the mechanisms of CMMD has improved greatly (Miller et al., 2010), this knowledge has not translated into more efficacious dispenser technologies.

5.6 Dispenser development

Over the last 30 years, there have been many dispenser technologies developed and commercialized for CMMD. Only a few have withstood the test of time and are still in use today. The first release device for pheromone-mediated mating disruption was the CONREL (Controlled Release) fibres registered by the US EPA in 1978 for the control of pink bollworm *Pectinophora gossypiella* in cotton (Brooks, 1980). The hollow fibres, filled with pheromone, then chopped and mixed with a sticker, were broadcast by a specially designed ground sprayer or aerially with specialty equipment attached to the wings of the aircraft. The technology was tested and sold for codling moth (Moffitt and Westigard, 1984), but penetration into the market was limited due to reduced efficacy, application issues and the marking of fruit where the fibres adhered to them.

Microencapsulation technology was discovered in 1954 with the production of carbonless paper (Hall and Marrs, 1989). The first application of microencapsulation technology to deliver pheromones was with the gypsy moth (*Lymantria dispar*). In the 1980s and 1990s, the technology was extensively tested by companies such as Suterra, Bend, Oregon and 3M Canada, London and Ontario. 3M Canada gave up its development and commercialization plans for microencapsulation of pheromone for mating disruption based on the less than adequate results for CM, while Suterra still has a commercial formulation registered for CM but acceptance is limited due to concerns about the lack of efficacy and short field life. Growers using the MEC product primarily make a few applications at peak flight to reduce CM density and enhance the efficacy of their insecticidal control programme.

In 1980, SEC, Tokyo, Japan, began research on the large-scale production of insect pheromones. Soon after, SEC developed the hollow-tube dispenser in conjunction with research scientists in Australia (Rothschild, 1975, 1982). Initially, the SEC hollow-tube dispenser, known in the United States as Isomate C™ and Isomate C+™, consisted of two plastic tubes one of which was filled with pheromone and the other with an imbedded aluminium wire. In time, both sides of the dispenser were filled with pheromone in what became known as twin-tube formulations (Isomate CTT™). The SEC dispensers were the first products on the market with a zero-order season-long release rate. No other dispensing technology for CMMD has been so extensively researched and tested by public institutions all over the world. These dispensers from SEC are the most widely used hand-applied technologies in the world. The success of the hand-applied SEC dispensers was largely responsible for the widespread acceptance of mating disruption in fruit-growing regions around the world. Other hand-applied dispensing technologies have been developed and commercialized over the last 25 years. The Hercon Environmental, Emigsville, Pennsylvania, USA, laminate dispenser was an early dispenser commercialized under the name Disrupt™. Other CMMD technologies commercialized and still in use today include the Checkmate CM™, Suterra Inc., Bend, Oregon, USA, Cidetrak™, Trécé Inc., Adair, Oklahoma, USA, NoMate™, Scentry Biologicals Inc., Billings, Montana, USA and RAK®, BASF Crop Protection, Ludwigshafen, Germany.

The hand-applied formulations described above are the most widely used CMMD technologies. However, hand-applied formulations have limitations that are impacting their acceptance today. Hand labour is increasingly expensive and hard to obtain in the most important pome fruit production areas around the world where CMMD is used. Growers are increasingly looking for alternative technologies that are easier and cheaper to apply. Furthermore, the application of hand-applied dispensers in crops such as walnuts and other nut crops with their tall well-developed canopies is very difficult and not cost-effective. Growers have always been interested in MEC-sprayable products such as those from 3M Canada and Suterra, but the efficacy of these formulations is limited for CM and the products expensive relative to the cost of insecticides. In the search for easier-and cheaper-to-apply mating disruption products, growers have started deploying aerosol technologies.

Thomas et al. (1975) successfully developed and formulated the first pheromone aerosol delivery system to disrupt the southern pine beetle (*Dendroctonus frontalis).* The aerosol devices were battery powered and designed to release a 5-s burst of pheromone twice per day. Development of aerosol-type formulations for mating disruption of many species began in earnest in the mid-1990s with the effort of many researchers (Shorey and Gerber, 1996; Mafra-Neto and Baker, 1996; Baker et al., 1997; Fadamiro and Baker, 2002). Similar to the device developed by Thomas, the emitter technologies consisted of battery-operated timers that triggered periodic sprays of insect pheromone. These devices offered several advantages over the traditional hand-applied dispensers including ease of and cheaper application, protection of active ingredient from effects of oxygen and UV light, metered release rate independent of temperature and extended storage. Aerosol emitters have been tested against more than a dozen insects including codling moth. Aerosols are now used on approximately 25% of the ha treated with CMMD in pome fruit in Washington State and on 100% of the walnut ha in California treated with CMMD. In South Tyrol of Italy, aerosol formulations have largely displaced the hand-applied dispensers due to grower concerns about the availability and cost of labour to apply the latter (V. Veronelli, pers. comm.). Aerosols have been extensively tested in the Western United States (Brunner, 2017) and in Italy (Baldessari et al., 2013). The positive results from extensive testing in both areas laid the foundation for grower adoption of aerosol technology.

5.7 Future of CMMD

Almost 30 years have passed since the first CMMD product was registered and sold. In 2012, there were approximately 220000 ha treated with CMMD (Miller and Gut, 2015). Today it is almost impossible to estimate the worldwide-treated area as there are so many suppliers of both active ingredient and formulated product. The increased demand for CMMD over the last 15 years has created a competitive environment and this is good for the pheromone industry. However, it seems that demand for CMMD had flattened out as markets in developed and developing countries have become saturated. There is still huge potential to grow the CMMD market, but the growth will occur in developing countries such as China and Russia where political considerations and the cost and lack of technical resources have limited acceptance. If so, then there will have to be cheaper yet still effective formulations than what is currently being sold. Given the immense size of the pome fruit market in China, it is likely that we will see CMMD products emerge in the near future.

Another interesting trend in CMMD is the introduction of so-called precision agriculture. Semios, Inc., Vancouver, British Columbia, Canada developed and commercialized a wireless mesh network to integrate data from pheromone dispensers, electronic pheromone traps, weather stations and a variety of other sensor technologies including soil moisture and leaf wetness sensors. The backbone of the system has been CMMD pheromone aerosol dispensers which are linked via a wireless system to web-based computer servers. In conjunction with pheromone trap data, the Semios system gives growers easy access to weather information, dispenser performance information and moth catch data via the Web. Despite the higher cost, the Semios system has attracted the interest of some growers and it is now deployed on several 1000 ha of pome fruit in Washington State. It remains to be determined if the information and data generated for CMMD and other features will be attractive enough to justify the added costs and increase use of the technology. There is no doubt that precision agriculture and its associated technologies are coming to pome fruit production. Growers must continually reduce costs and improve efficiencies in a highly competitive global pome fruit market. If precision agriculture becomes widely adopted, it will increase the move of the CMMD market to aerosols and away from hand-applied dispensers.

Serious challenges lie ahead for CMMD. Miller and Gut (2015) identified the mode of action of CMMD as competitive attraction. With competitive attraction, dispensers must compete with females for the attention of males. As such, CM population densities must be low to facilitate dispensers outcompeting females and providing effective disruption. To a large extent, growers have relied on companion application of insecticides in CMMD blocks to keep CM populations low. The lack of insecticides with new modes of action and adulticide activity will make CMMD-based pest management systems very difficult as older technologies are lost due to resistance and regulations. Despite the best efforts of research to refine and improve CMMD, not much has changed in the last 30 years. Increasingly, we are hearing of higher CM populations and damage in orchards where CMMD has been deployed for many years. Increasingly, growers and industry representatives are expressing their frustrations and inquiring about the possibility of CM resistance to its pheromone. In our opinion, CMMD is still showing the same level of efficacy but the lack of and decline in efficacy of companion insecticides are increasingly a cause for serious concern. Additionally, growers are relying on reduced densities of hand-applied dispensers and aerosol emitters, allowing CM populations to build slowly.

If the climate continues to warm up and we get an extra partial or full generation of CM in pome fruit production areas, this will add to the difficulties of controlling CM (Stoecki et al., 2012). We recently summarized data revealing an increase in the total number of degree days accumulated over an average growing season (1 April–30 September). Using data from weather stations near Wenatchee, Washington or Davis, California, the data show an increase of about 400 CM degree days over the past 60 years (Fig. 12). The effect is to add another partial or full generation, thus increasing CM populations and making control more difficult.

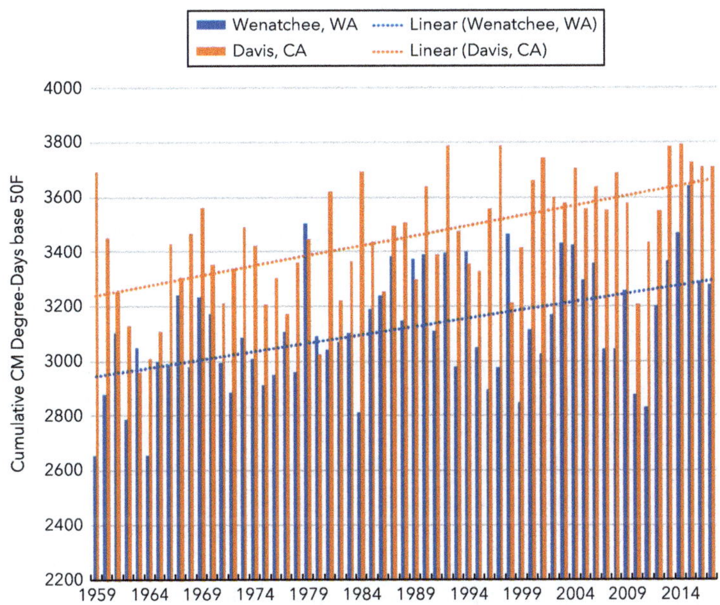

Figure 12 Sixty years of cumulative degree-days base 50°F from 1 April to 31 October for Wenatchee, WA and 1 April to 30 September for Davis, CA. Dotted lines show the linear trend.

6 Summary

Over the past 25 years, semiochemical-based technologies have emerged as essential components of pest management programmes in deciduous fruit production systems. Pheromone-based mating disruption is by far the most widely adopted behavioural control, with nearly 800 000 ha of fruit crops treated worldwide. Our aim in writing this chapter has been to maintain or accelerate this trend of increasing reliance by fruit growers on behaviourally based controls. We have reviewed both the essential progress that researchers and practitioners have made, along with the major hurdles encountered as they conceptualized, developed and implemented behavioural controls. There has been considerable progress in identifying semiochemicals, understanding how insects respond to them and improving formulations and delivery systems. Most recently, methods for identifying the principal means by which mating disruption is achieved have provided a pathway for designing more cost-effective mating disruption formulations. Product improvements if disruption is achieved non-competitively should focus on optimizing dispenser release rate and deployment pattern. If disruption is achieved competitively, the best approaches for reducing cost are to only release sufficient pheromone to provide the maximal effect or deploy fewer dispensers combined with the use of insecticides.

Some major biological and practical limitations continue to impede the use of semiochemicals for controlling pests of deciduous fruit crops. Pest density is often the crucial limiting factor. Mass trapping, attract-and-kill and mating disruption operating non-competitively are only economically feasible if pest densities are low. To counter the problem of controlling high-density populations via mating disruption, deciduous fruit growers often apply companion insecticides to reduce pest-population densities. Those striving to use mating disruption as a stand-alone control should ramp up efforts to identify pests that are amenable to non-competitive disruption, as control is independent of population density.

Perhaps one of the most important uses for mass trapping is protection from invasive species (El-Sayed et al., 2006). Government-coordinated actions can establish quarantine zones where eradication may be possible while pest density is low, and a breeding population has yet to establish. A quick response that combines multiple tactics, including mass trapping, could help mitigate the spread and establishment of detected pests that have slipped past security measures. Mass trapping provides a fast, nontoxic and fairly species-specific tool for rapid deployment even in sensitive urban areas.

Researchers and practitioners continue to struggle with the question of how many dispensers to deploy for successful mass annihilation. While a high density of devices improves the chance of success, the approach is costly and could result in trap interference. Decisions about the potential for mass annihilation and optimal deployment strategies should be based on knowing the plume reach and trapping area of the device, as well as the dispersive distance of the pest. The single-trap multiple-release experimental design, along with the graphical interpretations of the catch data detailed in Miller et al. (2015), provides a simple and robust means of quantifying these critical parameters.

Ultimately, increased reliance by growers on the most successful behaviourally based tactic, mating disruption, will depend on how well technologies can meet the grower needs for enhanced efficacy and lower cost. Understanding how a formulation works will enable the design of more cost-effective technologies. The producers of mating disruption products need to develop long-lasting formulations to address the extended phenology of targeted pests that is accompanying climate warming. Additionally, the pheromone industry needs to address grower concerns about the reduced insecticide sprays that accompany the use of mating leading to outbreaks of non-target pests. This should include ramping up efforts to develop multi-species mating disruption formulations.

There is a heightened need for alternatives to insecticides for managing deciduous fruit pests. As mentioned previously, loss of insecticides due to regulatory action and resistance have fuelled the adoption of behavioural controls. Pest suppression using semiochemicals can provide outstanding pest control leading to better fruit quality and increased profits. Economic benefits are often the result of a substantial reduction in insecticide sprays. The judicious use of insecticides has the added advantages of negligible impact on the environment, pollinators and other beneficials. It is certainly an exhilarating time to be promoting the use of semiochemicals for managing deciduous fruit pests.

7 Where to look for further information

Several book chapters provide an overview on the use of behaviour-modifying chemicals to manage pests. One of the most comprehensive early treatises is the collection of articles in the book *Behavior-Modifying Chemicals for Insect Management* (Ridgway et al., 1990). More recent coverage of the subject includes the following reviews: Foster and Harris (1997), Knight and Weissling (1999), Gut et al. (2004), Baker and Heath (2005) and Witzgall et al. (2010). A forum article by Miller and Gut (2015) and book chapter by Evenden (2016) provide good introductions to the mechanisms of mating disruption and the practical application of the technique. Reviews that broadly treat the use of mass trapping or attract-and-kill for suppressing pest populations include El-Sayed (2006, 2009) and Cork (2016).

Membership in one or more of the societies organized specifically to promote the understanding of chemically mediated interactions between organisms and their environment is a great way to keep abreast of advances in the practical application of semiochemicals. The International Society of Chemical Ecology (ISCE) meets annually to share findings, while the Asia-Pacific Association of Chemical Ecologists (APACE) and the Asociación Latino American de Ecología Quimic (ALAEQ) hold regular meetings every 2 years. The *Journal of Chemical Ecology* is the official journal of ISCE and APACE.

Several companies have websites that provide information on semiochemical-based pest management products:

- ChemTica *Internacional, S. A.* – http://www.chemtica.com/
- Hercon Environmental – https://www.herconenviron.com/
- ISCA Technologies – https://www.iscatech.com/
- Pacific Biocontrol Corporation – https://www.pacificbiocontrol.com/
- Scentry Biologicals Inc. – https://scentry.com/
- Suterra – http://suterra.com/
- Trécé Incorporated – http://trece.com/

8 References

Adams, C. G., Schenker, J. H., McGhee, P. S., Gut, L. J., Brunner, J. F. and Miller, J. R. 2017. Maximizing information yield from pheromone-baited monitoring traps: estimating plume reach, trapping radius, and absolute density of *Cydia pomonella* (Lepidoptera: Tortricidae) in Michigan Apple. *J. Econ. Entomol.* 110(2), 305–18. doi:10.1093/jee/tow258.

Agnello, A. M., Reissig, W. H., Spangler, S. M., Charlton, R. E. and Kain, D. P. 1996. Trap response and fruit damage by obliquebanded leafroller (Lepidoptera: Torticidae) in pheromone-treated apple orchards in New York. *Environ. Entomol.* 25(2), 268–82. doi:10.1093/ee/25.2.268.

Agnello, A. M., Atanassov, A., Bergh, J. C., Biddinger, D. J., Gut, L. J., Haas, M. J., Harper, J. K., Hogmire, H. W., Hull, L. A., Kime, L. F., et al. 2009. Reduced-risk pest management programs for Eastern U.S. apple and peach orchards: a 4-year regional project. *Am. Entomol.* 55(3), 184–97. doi:10.1093/ae/55.3.184.

Arn, H., Schwarz, C., Limacher, H. and Mani, E. 1974. Sex attractant inhibitors of the codling moth *Laspeyresia pomonella* L. *Experientia* 30(10), 1142–4. doi:10.1007/BF01923655.

Arn, H., Guerin, P. M., Buser, H. R., Rauscher, S. and Mani, E. 1985. Sex pheromone blend of the codling moth, *Cydia pomonella*: evidence for a behavioral role of dodecan-1-ol. *Experientia* 41(11), 1482–4. doi:10.1007/BF01950048.

Atterholt, C. A., Delwiche, M. J., Rice, R. E. and Krochta, J. M. 1998. Study of biopolymers and paraffin as potential controlled-release carriers for insect pheromones. *J. Agric. Food Chem.* 46(10), 4429–34. doi:10.1021/jf980642u.

Ayyappath, R., Polavarapa, S. and McGuire, M. R. 2000. Effectiveness of thiamethoxam-coated spheres against blueberry maggot flies (Diptera: Tephritidae). *J. Econ. Entomol.* 93(5), 1473–9. doi:10.1603/0022-0493-93.5.1473.

Baker, T. C. and Haynes, K. F. 1989. Field and laboratory electroantennographic measurements of pheromone plume structure correlated with oriental fruit moth behavior. *Physiol. Entomol.* 14(1), 1–12. doi:10.1111/j.1365-3032.1989.tb00931.x.

Baker, T. C. and Heath, J. J. 2005. Pheromones: function and use in insect control. In: Gilbert, L., Latrou, K. and Gill, S. (Eds), *Comprehensive Molecular Insect Science*. Elsevier, Amsterdam, pp. 406–59.

Baker, T. C. and Roelofs, W. L. 1981. Initiation and termination of oriental fruit moth male response to pheromone concentrations in the field. *Environ. Entomol.* 10(2), 211–8. doi:10.1093/ee/10.2.211.

Baker, T. C., Cardé, R. T. and Miller, J. R. 1980. Oriental fruit moth pheromone component emission rates measured after collection by glass-surface adsorption. *J. Chem. Ecol.* 6(4), 749–58. doi:10.1007/BF00990399.

Baker, T. C., Dittl, T. and Mafra-Neto, A. 1997. Disruption of sex pheromone communication in the blackheaded fireworm in Wisconsin cranberry marshes by using MSTRS(TM) devices. *J. Agric. Entomol.* 14, 449–57.

Baker, T. C., Myrick, A. J. and Park, K. C. 2016. Optimizing the point-source emission rates and geometries of pheromone mating disruption mega-dispensers. *J. Chem. Ecol.* 42(9), 896–907. doi:10.1007/s10886-016-0769-9.

Baldessari, M., Ioriatti, C. and Angeli, G. 2013. Evaluation of Puffer® CM, a release device of pheromone to control codling moth on apple in Italy. *IOBC WPRS Bull.* 91, 199–204.

Barnes, B. and Blomefield, T. 1997. Goading growers towards mating disruption: the South African experience with *Grapholita molesta* and *Cydia pomonella* (Lepidoptera, Tortricidae). *IOBC WPRS Bull.* 20, 1–12.

Barrett, B. A. 1995. Effect of synthetic pheromone permeation on captures of male codling moth (Lepidoptera: Tortricidae) in pheromone and virgin female moth-baited traps at different tree heights in small orchard blocks. *Environ. Entomol.* 24(5), 1201–6. doi:10.1093/ee/24.5.1201.

Barry, J. D. and Polavarapu, S. 2004. Feeding activity and attraction of blueberry maggot (Diptera: Tephritidae) to protein baits, ammonium acetate, and sucrose. *J. Econ. Entomol.* 97(4), 1269–77. doi:10.1093/jee/97.4.1269.

Bartell, R. J. 1982. Mechanisms of communication disruption by pheromone in the control of Lepidoptera: a review. *Physiol. Entomol.* 7, 353–64.

Bartell, R. J., Bellas, T. E. and Whittle, C. P. 1988. Evidence for biological activity of two further alcohols in the sex pheromone of female *Cydia pomonella* (L.) (Lepidoptera: Tortricidae). *Aust. J. Entomol.* 27(1), 11–2. doi:10.1111/j.1440-6055.1988.tb01134.x.

Baumhover, A. H., Graham, A. J., Bitter, B. A., Hopkins, D. E., New, W. D., Dudley, F. H. and Bushland, R. C. 1955. Screw-worm control through release of sterilized flies. *J. Econ. Entomol.* 48(4), 462–6. doi:10.1093/jee/48.4.462.

Bosch, D., Sarasua, M. J. and Avilla, J. 2001. Mass trapping of *Synanthedon myopaeformis* (Borkhausen) in Lleida (Spain) with pheromone traps. *IOBC WPRS Bull.* 24, 167–71.

Bostanian, N. J., Vincent, C., Chouinard, G. and Racette, G. 1999. Managing apple maggot, *Rhagoletis pomonella* [Diptera: Tephritiade] by perimeter trapping. *Phytoprotection* 80(1), 21–33. doi:10.7202/706178ar.

Brooks, T. W., Doane, C. C. and Staten, R. T. 1979. Experience with the first commercial pheromone communication disruptive for suppression of an agricultural pest. In: Ritter, F. D. (Ed.), *Odour communication in Animals*, Elsevier/North Holland, Amsterdam, pp. 375–88.

Brooks, T. W. 1980. Controlled vapor release from hollow fibers: theory and applications with insect pheromones. In: Kydonius, A. E. (Ed.), *Controlled Release Technologies: Methods and Applications Volume II*. CRC Press, Boca Raton, FL, pp. 166–93.

Brunner, J. 2017. Aerosol delivery of pheromones in IFP: a mature technology for plant protection. Integrated protection of fruit crops. *IOBC WPRS Bull.* 123, 1–13.

Brunner, J., Dunley, J., Beers, E. and Jones, V. 2007. Building a multi-tactic biologically intensive pest management system for Washington orchards. In: Felsot, A. and Racke, K. (Eds), *Crop Protection Products for Organic Agriculture. Environment, Health, and Efficacy Assessment*. ACS Symposium Series 947. American Chemical Society, Washington DC, pp. 131–43.

Byers, J. A., Anderbrant, O. and Löfqvist, J. 1989. Effective attraction radius: a method for comparing species attractants and determining densities of flying insects. *J. Chem. Ecol.* 15(2), 749–65. doi:10.1007/BF01014716.

Calkins, C. O. and Faust, R. J. 2003. Overview of areawide programs and the program for suppression of codling moth in the western USA directed by the United States Department of Agriculture – Agriculture Research Service. *Pest Manag. Sci.* 59(6–7), 601–4. doi:10.1002/ps.712.

Cardé, R. T. 1990. Principles of mating disruption. In: Ridgeway, R., Silverstein, R. and Inscoe, M. (Eds), *Behavior-Modifying Chemicals for Insect Management: Applications of Pheromones and Other Attractants*. RCEL Dekker, Inc, New York and Basel, pp. 47–71.

Cardé, R. T. 2007. Using pheromones to disrupt mating of moth pests. In: Kogan, M. (Ed.), *Perspectives in Ecological Theory and Integrated Pest Management*. Cambridge University Press, New York, pp. 122–69.

Cardé, R. T., Baker, T. C. and Roelofs, W. L. 1975a. Ethological function of components of a sex attractant system for oriental fruit moth males, *Grapholita molesta* (Lepidoptera: Tortricidae). *J. Chem. Ecol.* 1, 475–491l.

Cardé, R. T., Trammel, K. and Roelofs, W. L. 1975b. Disruption of sex attraction of the redbanded leafroller (*Argyrotaenia velutinana*) with microencapsulated pheromone components. *Environ. Entomol.* 4(3), 448–50. doi:10.1093/ee/4.3.448.

Cardé, R. T., Baker, T. C. and Castrovillo, P. J. 1977. Disruption of sexual communication in *Laspeyresia pomonella* (codling moth), *Grapholita molesta* (Oriental fruit moth) and *G. prunivora* (lesser appleworm) with hollow fiber attractant sources. *Entomol. Exp. Appl.* 22(3), 280–88. doi:10.1111/j.1570-7458.1977.tb02718.x.

Cardé, A. M., Baker, T. C. and Cardé, R. T. 1979. Identification of a four-component sex pheromone of the female oriental fruit moth, *Grapholitha molesta* (Lepidoptera: Tortricidae). *J. Chem. Ecol.* 5(3), 423–7. doi:10.1007/BF00987927.

Charmillot, P. 1990. Mating disruption technique to control codling moth in western Switzerland. In: Ridgeway, R., Silverstein, R. and Inscoe, M. (Eds), *Behavior-Modifying Chemicals for Insect Management: Applications of Pheromones and Other Attractants*. RCEL Dekker, Inc, New York and Basel, pp. 169–82.

Charmillot, P. J., Pasquier, D., Scalco, A. and Hofer, D. 1996. Essais de lutte contre le carpocapse *Cydia pomonella* L. par un proce´de´ attracticide. *Mitt. Schweiz. Entomol. Gesell* 69, 431–9.

Charmillot, P. J., Hofer, D. and Pasquier, D. 2000. Attract and kill: a new control of the codling moth *Cydia pomonella*. *Entomol. Exp. Appl.* 94(2), 211–6. doi:10.1046/j.1570-7458.2000.00621.x.

Chensheng, L., Warchol, K. M. and Callahan, R. A. 2014. Sub-lethal exposure to neonicotinoids impaired honey bees winterization before proceeding to colony collapse disorder. *Bull. Insectol.* 67, 125–30.

Cocroft, R. B. and Rodríguez, R. L. 2005. The behavioral ecology of insect vibrational communication. *BioScience* 55(4), 323–34. doi:10.1641/0006-3568(2005)055[0323:TBEOIV]2.0.CO;2.

Cork, A. 2016. Pheromone as management tools. Mass-trapping and lure-and-kill. In: Allison, J. D. and Cardé, R. T. (Eds), *Pheromone Communication in Moths: Evolution, Behavior and Application*. University of California Press, Berkeley, CA, pp. 349–63.

Curtis, C. F., Myamba, J. and Wilkes, T. J. 1996. Comparison of different insecticides and fabrics for anti-mosquito bednets and curtains. *Med. Vet. Entomol.* 10(1), 1–11. doi:10.1111/j.1365-2915.1996.tb00075.x.

de Lame, F. M. and Gut, L. J. 2006. Monitoring trap and mating disruption dispenser application heights do not affect captures of male *Grapholita molesta* in pheromone and virgin female-baited traps. *Environ. Entomol.* 35(4), 1058–68. doi:10.1603/0046-225X-35.4.1058.

Deland, J.-P., Judd, G. J. R.. and Roitberg, B. D. 1994. Disruption of pheromone communication in three sympatric leafroller (Lepidoptera: Tortricidae) pests of apple in British Columbia. *Environ. Entomol.* 23(5), 1084–90. doi:10.1093/ee/23.5.1084.

DeLury, N. C., Judd, G. J. R. and Gardiner, M. G. T. 2005. Antennal detection of sex pheromone by female *Pandemis limitata* (Robinson) (Lepidoptera: Tortricidae) and its impact on their calling behaviour. *J. Entomol. Soc. B. C.* 102, 3–11.

den Otter, C. J., Schuil, H. A. and Oosten, A. S. 1978. Reception of host-plant odours and female sex pheromone in *Adoxophyes orana* (Lepidoptera: Tortricidae): electrophysiology and morphology. *Entomol. Exp. Appl* 24(3), 570–8. doi:10.1111/j.1570-7458.1978.tb02818.x.

Doane, C. D. 1999. Controlled-release devices for pheromones. In: Scher, H. B. (Ed.), *Controlled-Release Delivery Systems for Pesticides*. Marcel Dekker, Inc, New York, pp. 295–317.

Dodds, K. J. and Ross, D. W. 2002. Sampling range and range of attraction of *Dendroctonus pseudotsugae* pheromone-baited traps. *Can. Entomol.* 134(3), 343–55. doi:10.4039/Ent134343-3.

Dodson, A. C. 1948. Further studies on lures attractive to the apple maggot. *J. Econ. Entomol.* 41(1), 61–6. doi:10.1093/jee/41.1.61.

Ebbinghaus, D., Losel, P. M., Romeis, J., Cianculli-Teller, M. G., Leusch, H., Olszak, R., Pluciennik, Z. and Scherkenbeck, J. 2001. Appeal: efficacy and mode of action of attract and kill for codling moth control. *IOBC WPRS Bull.* 24, 95–100.

Einhorn, J., Beauvais, F., Gallois, M., Descoins, C. and Causse, R. 1984. Constituants secondaires de la pheromone sexuelle du carpocapse des pommes, *Cydia pomonella* L. (Lepidoptera, Tortricidae). *C. R. Acad. Sci. Paris* 299, 773–8.

Elkins, R. B., DeMoura, R. L. and Klonsky, K. M. 2005. Cost of production of transitioning from conventional codling moth control to aerosol-released mating disruption ('puffers') in pears. *Acta Hort.* 671, 559–63.

Elkinton, J. S., Hal, C., Onot, T. and Cardé, R. T. 1987. Pheromone puff trajectory and upwind flight of male gypsy moths in a forest. *Physiol. Entomol.* 12(4), 399–406. doi:10.1111/j.1365-3032.1987.tb00766.x.

El-Sayed, A. M. 2016. The pherobase: database of pheromones and semiochemicals. Available at: http://www.pherobase.com.

El-Sayed, A. M., Delisle, J., DeLury, N., Gut, L. J., Judd, G. J. R., Legrand, S., Reissig, W. H., Roelofs, W. L., Unelius, C. R. and Trimble, R. M. 2003. Geographic variation in pheromone chemistry, antennal electrophysiology and pheromone-mediated trap catch of North American populations of the obliquebanded leafroller. *Environ. Entomol.* 32(3), 470–6. doi:10.1603/0046-225X-32.3.470.

El-Sayed, A. M., Suckling, D. M., Wearing, C. H. and Byers, J. A. 2006. Potential of mass trapping for long-term pest management and eradication of invasive species. *J. Econ. Entomol.* 99(5), 1550–64. doi:10.1603/0022-0493-99.5.1550.

El-Sayed, A. M., Suckling, D. M., Byers, J. A., Jang, E. B. and Wearing, C. H. 2009. Potential of 'lure and kill' in long-term pest management and eradication of invasive species. *J. Econ. Entomol.* 102(3), 815–35.

El-Sayed, A. M., Cole, L., Revell, J., Manning, L. A., Twidle, A., Knight, A. L., Bus, V. G. and Suckling, D. M. 2013. Apple volatiles synergize the response of codling moth to pear ester. *J. Chem. Ecol.* 39(5), 643–52. doi:10.1007/s10886-013-0277-0.

Epsky, N. D., Espinoza, H. R., Kendra, P. E., Abernathy, R., Midgarden, D. and Heath, R. R. 2010. Effective sampling range of a synthetic protein-based attractant for *Ceratitis capitata* (Diptera: (Diptera: Tephritidae). *J. Econ. Entomol.* 103(5), 1886–95.

Epstein, D., Waldstein, D. and Edson, C. 2002. Michigan apple integrated pest management implementation project. Final Narrative Report, 99pp.

Epstein, D. L., Stelinski, L. L., Reed, T. P., Miller, J. R. and Gut, L. J. 2006. Higher densities of distributed pheromone sources provide disruption of codling moth (Lepidoptera: Tortricidae) superior to that of lower densities of clumped sources. *J. Econ. Entomol.* 99(4), 1327–33. doi:10.1093/jee/99.4.1327.

Epstein, D. L., Miller, J. R., Grieshop, M. J., Stelinski, L. L. and Gut, L. J. 2011a. Direct sampling of resting codling moth (Lepidoptera: Tortricidae) adults in apple tree canopies and surrounding habitats. *Environ. Entomol.* 40(3), 661–8. doi:10.1603/EN10256.

Epstein, D. L., Stelinski, L. L., Miller, J. R., Grieshop, M. J. and Gut, L. J. 2011b. Effects of hand-applied reservoir dispenser height within apple tree canopy on efficacy of pheromone-mediated mating disruption of codling moth (Lepidoptera: Tortricidae). *Pest Manag. Sci.* 67(8), 975–9. doi:10.1002/ps.2141.

Evenden, M. 2016. Mating disruption of moth pests in integrated pest management. A mechanist approach. In: Allison, J. D. and Cardé, R. T. (Eds), *Pheromone Communication in Moths: Evolution, Behavior and Application*. University of California Press, Berkeley, pp. 365–93.

Evenden, M. L. and McLaughlin, J. R. 2004a. Factors influencing the effectiveness of an attracticide formulation against the oriental fruit moth, *Grapholita molesta*. *Entomol. Exp. Appl* 112(2), 89–97. doi:10.1111/j.0013-8703.2004.00181.x.

Evenden, M. L. and McLaughlin, J. R. 2004b. Initial development of an attracticide formulation against the oriental fruit moth *Grapholita molesta* (Lepidoptera: Tortricidae). *Environ. Entomol.* 33, 213–20.

Evenden, M. L., Judd, G. J. R. and Borden, J. H. 1999. Simultaneous disruption of pheromone communication in *Choristoneura rosaceana* and *Pandemis limitata* with pheromone and antagonist blends. *J. Chem. Ecol.* 25(3), 501–17. doi:10.1023/A:1020997619135.

Evenden, M. L., Judd, G. J. R. and Borden, J. H. 2000. Investigations of mechanisms of pheromone communication disruption of Choristoneura rosaceana (Harris) in a wind tunnel. *Journal of Insect Behavior* 13(4), 499–510.

Fadamiro, H. Y. and Baker, T. C. 2002. Pheromone puffs suppress mating by *Plodia interpunctella* and *Sitotroga cerealella* in an infested corn store. *Entomol. Exp. Appl.* 102(3), 239–51. doi:10.1046/j.1570-7458.2002.00945.x.

Fadamiro, H. Y., Cossé, A. A., Dittl, T. and Baker, T. C. 1998. Suppression of mating by blackheaded fireworm (Lepidoptera: Tortricidae) in Wisconsin cranberry marshes by using MSTRS devices. *J. Agric. Entomol.* 15, 377–86.

Farkas, S. R., Shorey, H. H. and Gaston, L. K. 1974. Sex pheromones of Lepidoptera: the use of widely separated evaporators of looplure for the disruption of pheromone communication in *Trichoplusia ni*. *Environ. Entomol.* 3(5), 876–7. doi:10.1093/ee/3.5.876.

Foster, S. P. and Harris, M. O. 1997. Behavioral manipulation methods for insect pest-management. *Annu. Rev. Entomol.* 42, 123–46. doi:10.1146/annurev.ento.42.1.123.

Frick, K. E., Simkover, H. G. and Telford, H. S. 1954. Bionomics of the cherry fruit flies in eastern Washington. *Washington Agri. Expt. Stations Tech. Bull.* 13, 66.

Gaston, L. K., Shorey, H. H. and Sario, S. A. 1967. Insect population control by the use of sex pheromones to inhibit orientation between the sexes. *Nature* 213, 155.

Geir, P. W. 1963. The life history of codling moth, *Cydia pomonella* (L.) (Lepidoptera: Tortricidae), in the Australian Capital Territory. *Aust. J. Zool.* 11(3), 323–67. doi:10.1071/ZO9630323.

Giurfa, M. and Menzel, R. 1997. Insect visual perception: complex abilities of simple nervous systems. *Curr. Opin. Neurobiol.* 7(4), 505–13. doi:10.1016/S0959-4388(97)80030-X.

Glare, T., Caradus, J., Gelernter, W., Jackson, T., Keyhani, N., Köhl, J., Marrone, P., Morin, L. and Stewart, A. 2012. Have biopesticides come of age? *Trends Biotechnol.* 30(5), 250–8. doi:10.1016/j.tibtech.2012.01.003.

Gökçe, A., Stelinski, L. L., Gut, L. J. and Whalon, M. E. 2007. Comparative behavioral and EAG responses of female obliquebanded and redbanded leafroller moths (Lepidoptera: Tortricidae) to their sex pheromone components. *Eur. J. Entomol.* 104(2), 187–94. doi:10.14411/eje.2007.029.

Grieshop, M., Huang, J. and Gut., L. 2014. Development of an attract and kill device for orchard pests. *N. Y. Fruit Q.* 22, 11–4.

Guerin, P., Arn, H. and Buser, H. 1985. Characterization of some tortricid sex pheromones using the insect antenna as a gas chromatographic detector. In: Acree, T. and Soderlund, D. (Eds), *Semiochemistry: Flavors and Pheromones*. Walter de Gruyter & Co., Berlin, pp. 239–50.

Gut, L. and Brunner, J. 1996. Implementing codling moth mating disruption in Washington pome fruit orchards. Tree Fruit Research Extension Center Information Series, No. 1. Washington State University, Wenatchee, WA.

Gut, L. and Brunner, J. 1998. Pheromone-based management of codling moth (Lepidoptera: Tortricidae) in Washington apple orchards. *J. Agric. Entomol.* 15, 387–405.

Gut, L. J., Stelinski, L. L., Thomson, D. R. and Miller, J. R. 2004. Behaviour-modifying chemicals: prospects and constraints in IPM. In: Koul, O., Dhaliwal, G. and Cuperus, G. (Eds), *Integrated Pest Management: Potential, Constraints, and Challenges*. CABI Publishing, New York, pp. 73-120.

Gut, L., McGhee, P., Grieshop, M. and Miller, J. 2010. Unequal trapping area challenges trap shutdown as an intuitive measure of mating disruption. Abstracts of the *84th Annual Orchard Pest and Disease Management Conference*, p. 51.

Gut, L., Huang, J. and McGhee, P. 2015. A novel mating disruption system designed for rapid deployment of reservoir pheromone dispensers. *N. Y. Fruit Q.* 23(3), 10–3.

Hagley, E. A. C. 1978. Sex pheromones and suppression of the codling moth (Lepidoptera: Olethreutidae). *Can. Entomol.* 110(7), 781–3. doi:10.4039/Ent110781-7.

Hall, D. and Marrs, G. 1989. Microcapsules. In: Jutsum, A. and Gordon, R. (Eds), *Insect Pheromones in Plant Protection*. John Wiley and Sons, New York, pp. 199–248.

Harari, A. R., Zahavi, T. and Steinitz, H. 2015. Female detection of the synthetic sex pheromone contributes to the efficacy of mating disruption of the European grapevine moth, *Lobesia botrana*. *Pest Manag. Sci.* 71(2), 316–22. doi:10.1002/ps.3830.

Hardie, J. and Minks, A. K. 1999. *Pheromones of Non-Lepidopteran Insects Associated with Agricultural Plants*. CAB International, Wallingford, UK, 466pp.

Hebert, V. R., Tomaszewska, E., Brunner, J. F., Jones, V. P. and Doerr, M. 2007. Evaluating the pheromone release rate characteristics of commercial mating disruption devices. In: Felsot, A. S. and Racke, K. D. (Eds), *Crop Protection Products for Organic Agriculture*. ACS Symposium Series (vol. 947). American Chemical Society, Washington DC, pp. 144–57.

Heuskin, S., Verheggen, F. J., Haubruge, E. and Wathelet, J. P. 2011. The use of semiochemical slow-release devices in integrated pest management strategies. *Biotech. Agro. Soc. Environ.* 15, 459–70.

Hogsette, J. A., Nalli, A. and Foil, L. D. 2008. Evaluation of different insecticides and fabric types for development of treated targets for stable fly (Diptera: Muscidae) control. *J. Econ. Entomol.* 101(3), 1034–8. doi:10.1093/jee/101.3.1034.

Hossain, M. S., Williams, D. G., Mansfield, C., Bartelt, R. J., Callinan, L. and Il'ichev, A. L. 2006. An attract-and-kill station to control *Carpophilus* spp. in Australian stone fruit orchards. *Entomol. Exp. Appl* 118, 11–9.

Howell, J. F. 1980. Codling moth: measuring removal of males by sex pheromone trapping. U.S. Dep. Agric. SEA/ARS Western Ser. 14, 1–6.

Howell, J. F. and Clift, A. E. 1974. The dispersal of sterilized codling moths released in the Wenas Valley, Washington. *Environ. Entomol.* 3(1), 75–81. doi:10.1093/ee/3.1.75.

Howell, J. F., Knight, A. L., Unruh, T. R., Brown, D. F., Krysan, J. L., Sell, C. R. and Kirsch, P. A. 1992. Control of codling moth in apple and pear with sex pheromone-mediated mating disruption. *J. Econ. Entomol.* 85(3), 918–25. doi:10.1093/jee/85.3.918.

Hughes, J. and Dorn, S. 2002. Sexual differences in the flight performance of the oriental fruit moth *Cydia molesta*. *Entomol. Exp. Appl.* 103(2), 171–82. doi:10.1046/j.1570-7458.2002.00967.x.

Huang, J., Stelinski, L. L. and Gut, L. J. 2010. Mating behaviors of *Cydia pomonella* (Lepidoptera: Tortricidae) as influenced by sex pheromone in electrostatic powder. *J. Econ. Entomol.* 103(6), 2100–6. doi:10.1603/EC10063.

Huang, J., Gut, L. J. and Miller, J. R. 2013. Separating the attractant from the toxicant improves attract-and-kill of codling moth (Lepidoptera: Tortricidae). *J. Econ. Entomol.* 106(5), 2144–50. doi:10.1603/EC13152.

Huang, J., Gut., L. J. and Grieshop, M. 2015. Development of a new attract-and-kill technology for oriental fruit moth control using insecticide-impregnated fabric. *Entomol. Exp. Appl* 154(2), 102–9. doi:10.1111/eea.12259.

Il'ichev, A. and Sexton, S. B. 2002. Reduced application rates of mating disruption for effective control of oriental fruit moth *Grapholita molesta* (Busck) (Lepidoptera: Tortricidae) on pears. *Gen. Appl. Entomol.* 31, 47–51.

Il'ichev, A. L., Gut, L. J., Hossain, M. S., Williams, D. G. and Jerie, P. H. 2002. Area-wide approach for improved control of oriental fruit moth, *Grapholita molesta* (Busck) (Lepidoptera: Tortricidae) by mating disruption. *Gen. Appl. Entomol.* 31, 7–15.

Il'ichev, A. L., Stelinski, L. L., Williams, D. G. and Gut, L. J. 2006. Sprayable microencapsulated sex pheromone formulation for mating disruption of oriental fruit moth (Lepidoptera: Tortricidae) in Australian peach and pear orchards. *J. Econ. Entomol.* 99(6), 2048–54.

Il'ichev, A. L., Stelinski, L. L., Williams, D. G. and Gut, L. J. 2007. Dual pheromone dispenser for combined control of oriental fruit moth, *Grapholita molesta* Busck, and codling moth, *Cydia pomonella* L. (Lep., Tortricidae) in pears. *J. Appl. Entomol.* 131, 368–76.

Ioriatti, C. and Lucchi, A. 2016. Semiochemical strategies for tortricid moth control in apple orchards and vineyards in Italy. *J. Chem. Ecol.* 42(7), 571–83. doi:10.1007/s10886-016-0722-y.

Ioriatti, C., Molinari, F., Pasqualini, E., DeCristofaro, A., Schmidt, S. and Espinha, I. 2003. The plant volatile attractant (E, Z)-2,4-ethyldecadienoate (DA2313) for codling moth monitoring. *Boll. Zool. Agrar. Bachic* 35, 127–37.

Ioriatti, C., Lucchi, A. and Bagnoli, B. 2008. Grape areawide pest management in Italy. In: Koul, O., Cuperus, G. and Elliott, N. (Eds), *Areawide Pest Management: Theory and Implementation*. CAB International, Wallingford, UK, pp. 208–25.

Isaacs, R., Ulczynski, M., Wright, B., Gut, L. J. and Miller, J. R. 1999. Performance of a microsprayer specifically designed for pheromone-mediated control of insect pests. *J. Econ. Entomol.* 92, 1157–64.

James, D. G., Vogel, B., Faulder, R. J., Bartelt, R. J. and Moore, C. J. 2001. Pheromone-mediated mass trapping and population diversion as strategies for suppressing *Carpophilus* spp. (Coleoptera: Nitidulidae) in Australian stone fruit orchards. *Agric. Forest Entomol.* 3(1), 41–7. doi:10.1046/j.1461-9563.2001.00086.x.

Jamieson, L. E., Suckling, D. M. and Ramankutty, P. 2008. Mass trapping of *Prays nephelomima* (Lepidoptera: Yponomeutidae) in citrus orchards: optimizing trap design and density. *J. Econ. Entomol.* 101(4), 1295–301. doi:10.1093/jee/101.4.1295.

Judd, G. J. R. and Gardiner, M. G. T. 2004. Simultaneous disruption of pheromone communication and mating in *Cydia pomonella*, *Choristoneura rosaceana*, and *Pandemis limitata* (Lepidoptera: Tortricidae) using Isomate-CM/LR in apple orchards. *J. Entomol. Soc. B. C.* 101, 3–13.

Judd, G. J. R., Gardiner, M. G. T. and Thomson, D. R. 1996. Commercial trials of pheromone-mediated mating disruption with Isomate-C® to control codling moth in British Columbia apple and pear orchards. *J. Entomol. Soc. B. C.* 92, 23–34.

Judd, G. J. R., Gardiner, M. G. T. and Thomson, D. R. 1997. Control of codling moth in organically-managed apple orchards by combining pheromone-mediated mating disruption, post-harvest fruit removal and tree banding. *Entomol. Exp. Appl.* 83(2), 137–46. doi:10.1046/j.1570-7458.1997.00166.x.

Judd, G. J. R., DeLury, N. C. and Gardiner, M. G. T. 2005a. Examining disruption of pheromone communication in *Choristoneura rosaceana* and *Pandemis limitata* using microencapsulated (Z)-11-tetradecenyl acetate applied in a laboratory flight tunnel. *Entomol. Exp. Appl* 114(1), 35–45. doi:10.1111/j.0013-8703.2005.00230.x.

Judd, G. J. R., Gardiner, M. G. T., DeLury, N. C. and Karg, G. 2005b. Reduced antennal sensitivity, behavioural response, and attraction of male codling moths, *Cydia pomonella*, to their pheromone (E,E)-8-10-dodecadien-1-ol following various pre-exposure regimes. *Entomol. Exp. Appl.* 114(1), 65–78. doi:10.1111/j.0013-8703.2005.00231.x.

Karg, G., Suckling, D. M. and Bradley, S. J. 1994. Adsorption and release of pheromone of *Epiphyas postvittana* (Lepidoptera: Tortricidae) by apple leaves. *J. Chem. Ecol.* 20(8), 1825–41. doi:10.1007/BF02066225.

Kennedy, J. S. and Marsh, D. 1974. Pheromone-regulated anemotaxis in flying moths. *Science* 184(4140), 999–1001. doi:10.1126/science.184.4140.999.

Kirkpatrick, D. M., McGhee, P. S., Hermann, S. L., Gut, L. J. and Miller, J. R. 2016. Alignment of spotted wing drosophila (Diptera: Drosophilidae) on odorless disks varying in color (Diptera: Drosophilidae). *Environ. Entomol.* 45(1), 185–91. doi:10.1093/ee/nvv155.

Kirkpatrick, D. M., Gut, L. J. and Miller, J. R. 2018. Estimating monitoring trap plume reach and trapping area of *Drosophila suzukii* (Diptera: Drosophilidae) in Michigan tart cherry. *J. Econ. Entomol.* 111(3), 1285–9. doi:10.1093/jee/toy062.

Knight, A. L. 1995. The impact of codling moth (Lepidoptera: Tortricidae) mating disruption on apple pest management in Yakima Valley, Washington. *J. Entomol. Soc. B. C.* 92, 29–38.

Knight, A. L. 2005. Managing codling moth (Lepidoptera: Tortricidae) with an internal grid of either aerosol puffers or dispenser clusters plus border applications of individual dispensers. *J. Entomol. Soc. B. C.* 101, 69–78.

Knight, A. L. and Larsen, T. E. 2004. Improved deposition and performance of a microencapsulated sex pheromone formulation for codling moth (Lepidoptera: Tortricidae with a low volume application. *J. Entomol. Soc. B. C.* 101, 79–86.

Knight, A. L. and Light, D. M. 2005. Factors affecting the differential capture of male and female codling moth (Lepidoptera: Tortricidae) in traps baited with ethyl (E, Z)-2,4-decadienoate. *Environ. Entomol.* 34(5), 1161–9. doi:10.1093/ee/34.5.1161.

Knight, A. L. and Light, D. M. 2012. Monitoring codling moth (Lepidoptera: Tortricidae) in sex pheromone-treated orchards with (E)-4,8-dimethyl-1,3,7-nonatriene or pear ester in combination with codlemone and acetic acid. *Environ. Entomol.* 41(2), 407–14. doi:10.1603/EN11310.

Knight, A. L. and Weissling, T. J. 1999. Behavior-modifying chemicals in management of arthropod pests. In: Ruberson, J. R. (Ed.), *Handbook of Pest Management*. Marcel Dekker, Inc, New York, pp. 521–46.

Knight, A. L., Larsen, T. E. and Ketner, K. C. 2004. Rainfastness of microencapsulated sex pheromone formulation for codling moth (Lepidoptera: Tortricidae). *J. Econ. Entomol.* 97(6), 1987–92. doi:10.1093/jee/97.6.1987.

Knight, A. L., Hilton, R. and Light, D. M. 2005a. Monitoring codling moth (Lepidoptera: Tortricidae) in apple with blends of ethyl (E, Z)-2, 4-decadienoate and codlemone. *Environ. Entomol.* 34(3), 598–603. doi:10.1603/0046-225X-34.3.598.

Knight, A. L., VanBuskirk, P., Hilton, R., Zoller, B. and Light, D. M. 2005b. Monitoring codling moth (Lepidoptera: Tortricidae) in four cultivars of pear. *Acta Hort.* 671, 565–9.

Knight, A. L., Haworth, J., Lingren, B. and Hebert, V. 2010. Combining pear ester with codlemone improves management of codling moth. *IOBC WPRS Bull.* 72, 145–9.

Knipling, E. F. 1978. Eradication of plant pests-pro-advances in technology for insect population eradication and suppression. *Bull. Entomol. Soc. Am.* 24(1), 44–52. doi:10.1093/besa/24.1.44.

Knipling, E. F. 1979. The Basic Principles of Insect Population Suppression and Management. Agric. Handbook No. 512. United States Department of Agriculture, Washington DC.

Koch, U. T., Lüder, W., Clemenz, S. and Cichon, L. I. 1997. Pheromone measurement by field EAG in apple orchards. *IOBC WPRS Bull.* 20, 181–90.

Krupke, C. H., Roitberg, B. D. and Judd, G. J. R. 2002. Field and laboratory responses of male codling moth (Lepidoptera: Tortricidae) to a pheromone-based attract-and-kill strategy. *Environ. Entomol.* 31(2), 189–97. doi:10.1603/0046-225X-31.2.189.

Kuhns, E. H., Pelz-Stelinski, K. and Stelinski, L. L. 2012. Reduced mating success of female tortricid moths following intense pheromone auto-exposure varies with sophistication of mating system. *J. Chem. Ecol.* 38(2), 168–75. doi:10.1007/s10886-012-0076-z.

Kutinkova, H., Subchev, M., Light, D. M. and Lingren, B. 2005. Interactive effects of ethyl (2E, 4Z)-2,4-decadienoate and sex pheromone lures to codling moth: apple orchard investigations in Bulgaria. *J. Plant Prot. Res.* 45, 49–53.

Kydonius, A. E. and Quisumbing, A. R. 1980. Multilayered laminated structures. In: Kydonius, A. E. (Ed.), *Controlled Release Technologies: Methods and Applications Vol. I*. CRC Press, Boca Raton, FL, pp. 183–238.

Landolt, P. J., Suckling, D. M. and Judd, G. J. R. 2007. Positive interaction of a feeding attractant and a host kairomone for trapping the codling moth, *Cydia pomonella* (L.). *J. Chem. Ecol.* 33(12), 2236–44. doi:10.1007/s10886-007-9391-1.

Lawrence, T. J., Culbert, E. M., Felsot, A. S., Hebert, V. R. and Sheppard, W. S. 2016. Survey and risk assessment of *Apis mellifera* (Hymenoptera: Apidae) exposure to neonicotinoid pesticides in urban, rural, and agricultural settings. *J. Econ. Entomol.* 109(2), 520–8. doi:10.1093/jee/tov397.

Lawson, D. S., Reissig, W. H., Agnello, A. M., Nyrop, J. P. and Roelofs, W. L. 1996. Interference with the mate-finding communication system of the obliquebanded leafroller (Lepidoptera: Tortricidae) using synthetic sex pheromones. *Environ. Entomol.* 25(5), 895–905. doi:10.1093/ee/25.5.895.

Leskey, T. C., Bergh, J. C., Walgenbach, J. F. and Zhang, A. 2009. Evaluation of pheromone-based management strategies for dogwood borer (Lepidoptera: Sesiidae) in commercial apple orchards. *J. Econ. Entomol.* 102(3), 1085–93. doi:10.1603/029.102.0329.

Liburd, O. E., Alm, S. R., Casagrande, R. A. and Polavarapu, S. 1998. Effect of trap color, bait, shape and orientation in attraction of blueberry maggot (Diptera: Tephritidae). *J. Econ. Entomol.* 91(1), 243–9. doi:10.1093/jee/91.1.243.

Light, D. M., Knight, A. L., Henrick, C. A., Rajapaska, D., Lingren, B., Dickens, J. C., Reynolds, K. M., Buttery, R. G., Merrill, G., Roitman, J., et al. 2001. A pear-derived kairomone with pheromonal potency that attracts male and female codling moth, *Cydia pomonella* (L.). *Naturwissenschaften* 88(8), 333–8. doi:10.1007/s001140100243.

Lo, P. L., Walker, J. T. and Suckling, D. M. 2015. Prospects for the control of apple leaf midge *Dasineura mali* (Diptera: Cecidomyiidae) by mass trapping with pheromone lures. *Pest Manag. Sci.* 71(7), 907–13. doi:10.1002/ps.3857.

Lösel, P. M., Penners, G., Potting, R. P. J., Ebbinghaus, D., Elbert, A. and Scherkenbeck, J. 2000. Laboratory and field experiments towards the development of an attract and kill strategy for the control of the codling moth, *Cydia pomonella*. *Entomol. Exp. Appl* 95(1), 39–46. doi:10.1046/j.1570-7458.2000.00639.x.

Lösel, P. M., Potting, R. P. J., Ebbinghaus, D. and Scherkenbeck, J. 2002. Factors affecting the field performance of an attracticide against the codling moth *Cydia pomonella*. *Pest Manag. Sci.* 58(10), 1029–37. doi:10.1002/ps.577.

Madsen, H. F. and Carty, B. E. 1979. Codling moth (Lepidoptera: Olethreutidae): suppression by male removal with sex pheromone traps in three British-Columbia Canada orchards. *Can. Entomol.* 111(5), 627–30. doi:10.4039/Ent111627-5.

Mafra-Neto, A. and Baker, T. C. 1996. Timed, metered sprays of pheromone disrupt mating of *Cauda cautella* (Lepidoptera: Pyralidae). *J. Agric. Entomol.* 13, 149–68.

Mahr, D. and Baker, T. C. 2001. Mating disruption for insect control: where are we? In: *Proceedings of the 2001 Wisconsin Cranberry School*. University Wisconsin, Madison, pp. 1–4.

Maki, E. C., Millar, J. G., Rodsstein, J., Hanks, L. M. and Barbour, J. D. 2011. Evaluation of mass trapping and mating disruption for managing *Prionus californicus* (Coleoptera: Cerambycidae) in hop production yards. *J. Econ. Entomol.* 104(3), 933–8. doi:10.1603/EC10454.

Mani, E. and Wildbolz, T. 1977. The dispersal of male codling moths (*Laspeyresia pomonella* L.). in The Upper Rhine Valley. *J. Appl. Entomol.* 83, 161–8.

Mansour, M. 2010. Attract and kill for codling moth *Cydia pomonella* (Linnaeus) (Lepidoptera: Tortricidae) control in Syria. *J. Appl. Entomol.* 134(3), 234–42. doi:10.1111/j.1439-0418.2009.01393.x.

Mason, L. J., Jansson, R. K. and Heath, R. R. 1990. Sampling range of male sweetpotato weevils (*Cylas Formicarius elegantulus*) (summers) (Coleoptera: Curculionidae) to pheromone traps: influence of pheromone dosage and lure age. *J. Chem. Ecol.* 16(8), 2493–502. doi:10.1007/BF01017472.

Mayer, M. S. and McLaughlin, J. R. 1991. *Handbook of Insect Pheromones and Sex Attractants*. CRC Press, Boca Raton, FL, 1083pp.

Mazzoni, V., Polajnar, J., Baldini, M., Stacconi, M. V. R., Anfora, G., Guidetti, R. and Maistrello, L. 2017. Use of substrate-borne vibrational signals to attract the Brown Marmorated Stink Bug, *Halyomorpha halys*. *J. Pest. Sci.* 90(4), 1219–29. doi:10.1007/s10340-017-0862-z.

McDonough, L. M., Aller, W. C. and Knight, A. L. 1992. Performance characteristics of a commercial controlled-release dispenser of sex pheromone for control of codling moth (*Cydia pomonella*) by mating disruption. *J. Chem. Ecol.* 18(12), 2177–89. doi:10.1007/BF00984945.

McGhee, P. S. 2014. Impact of high releasing mating disruption formulations on (male) codling moth, *Cydia pomonella* L., behavior. Ph.D. Dissertation. Michigan State University, East Lansing, MI, 122pp.

McGhee, P. S., Gut, L., Siegert, P., Adams, C., Reinke, M., Miller, J., Doerr, M. and Brunner, J. 2010. Improve hand-applied dispenser mating disruption systems for codling moth by determining an optimized dispenser release rate. Abstracts of the *84th Annual Orchard Pest and Disease Management Conference*, p. 36.

McGhee, P. S., Epstein, D. L. and Gut, L. J. 2011. Quantifying the benefits of areawide pheromone mating disruption programs that target codling moth (Lepidoptera: Tortricidae). *Amer. Entomol.* 57(2), 94–100. doi:10.1093/ae/57.2.94.

McGhee, P. S., Gut, L., Haas, M., Miller, J. and Huang, J. 2012. Field evaluations of Isomate Flex CM and Cidetrak CM 'Meso' hand-applied mating disruption. Abstracts of the *86th Annual Orchard Pest and Disease Management Conference*, p. 21.

McGhee, P. S., Gut, L. J. and Miller, J. R. 2014. Aerosol emitters disrupt codling moth, *Cydia pomonella*, competitively. *Pest Manag. Sci.* 70(12), 1859–62. doi:10.1002/ps.3732.

McGhee, P. S., Miller, J. R., Thomson, D. R. and Gut, L. J. 2016. Optimizing aerosol emitters for mating disruption of codling moth, *Cydia pomonella* L. *J. Chem. Ecol.* 42(7), 612–6. doi:10.1007/s10886-016-0724-9.

Meats, A. and Edgerton, J. E. 2008. Short- and long-range dispersal of the Queensland fruit fly, *Bactrocera tryoni* and its relevance to invasive potential, sterile insect technique and surveillance trapping. *Aust. J. Exp. Agric.* 48(9), 1237–45. doi:10.1071/EA07291.

Miller, J. R. and Cowles, R. S. 1990. Stimulo-deterrent diversion: A concept and its possible application to onion maggot control. *J. Chem. Ecol.* 16(11), 3197–212. doi:10.1007/BF00979619.

Miller, J. R. and Gut, L. J. 2015. Mating disruption for the 21st century: matching technology with mechanism. *Environ. Entomol.* 44(3), 427–53. doi:10.1093/ee/nvv052.

Miller, J. R. and Roelofs, W. L. 1978. Sustained-flight tunnel for measuring insect responses to wind-borne sex pheromones. *J. Chem. Ecol.* 4(2), 187–98. doi:10.1007/BF00988054.

Miller, J. R., Gut, L. J., de Lamé, F. M. and Stelinski, L. L. 2006a. Differentiation of competitive vs. non-competitive mechanisms mediating disruption of moth sexual communication by point sources of sex pheromone: (part 1) Theory. *J. Chem. Ecol.* 32(10), 2089–114. doi:10.1007/s10886-006-9134-8.

Miller, J. R., Gut, L. J., de Lame, F. M. and Stelinski, L. L. 2006b. Differentiation of competitive vs. non-competitive mechanisms mediating disruption of moth sexual communication by point sources of sex pheromone: (Part 2) case studies. *J. Chem. Ecol.* 32(10), 2115–43. doi:10.1007/s10886-006-9136-6.

Miller, J. R., McGhee, P. S., Siegert, P. Y., Adams, C. G., Huang, J., Grieshop, M. J. and Gut, L. J. 2010. General principles of attraction and competitive attraction as revealed by large-cage studies of moths responding to sex pheromone. *Proc. Natl. Acad. Sci. U.S.A.* 107(1), 22–7. doi:10.1073/pnas.0908453107.

Miller, J. R., Adams, C. G., Weston, P. A. and Schenker, J. H. 2015. *Trapping of Small Organisms Moving Randomly; Principles and Applications to Pest Monitoring and Management*. Springer Briefs in Ecology, New York, 114pp.

Milli, R., Koch, U. T. and de Kramer, J. J. 1997. EAG measurement of pheromone distribution in apple orchards treated for mating disruption of *Cydia pomonella*. *Entomol. Exp. Appl.* 82(3), 289–97. doi:10.1046/j.1570-7458.1997.00143.x.

Mitchell, V. J., Manning, L. A., Cole, L., Suckling, D. M. and El-Sayed, A. M. 2008. Efficacy of the pear ester as a monitoring tool for codling moth *Cydia pomonella* (Lepidoptera: Tortricidae) in New Zealand apple orchards. *Pest Manag. Sci.* 64(3), 209–14. doi:10.1002/ps.1479.

Moffitt, H. R. and Westigard, P. H. 1984. Suppression of the codling moth (Lepidoptera: Tortricidae) population on pears in Southern Oregon through mating disruption with sex pheromone. *J. Econ. Entomol.* 77(6), 1513–9. doi:10.1093/jee/77.6.1513.

Moreira, J. A., McElfresh, J. S. and Millar, J. G. 2006. Identification, synthesis, and field testing of the sex pheromone of the citrus leafminer, *Phyllocnistis citrella*. *J. Chem. Ecol.* 32(1), 169–94. doi:10.1007/s10886-006-9359-6.

Moreno, D. S. and Mangan, R. L. 2002. A bait matrix for reduced-risk insecticides used against fruit flies (Diptera: Tephritidae). In: Hallman, G. and Schwalbe, C. P. (Eds), *Invasive Arthropods in Agriculture*. Science Publishers, Enfield, NH, pp. 333–62.

Morrison, W. R., Lee, D. H., Reissig, W. H., Combs, D., Leahy, K., Tuttle, A., Cooley, D. and Leskey, T. C. 2016a. Inclusion of specialist and generalist stimuli in attract-and-kill programs: their relative efficacy in apple maggot fly (Diptera: Tephritidae) pest management. *Environ. Entomol.* 45(4), 974–82. doi:10.1093/ee/nvw043.

Morrison, W. R., Lee, D. H., Short, B. D., Khrimian, A. and Leskey, T. C. 2016b. Establishing the behavioral basis for an attract-and-kill strategy to manage the invasive *Halyomorpha halys* in apple orchards. *J. Pest. Sci.* 89(1), 81–96. doi:10.1007/s10340-015-0679-6.

Mota-Sanchez, D., Wise, J. C., Vander Poppen, R., Gut, L. and Hollingsworth, R. M. 2008. Resistance of codling moth, *Cydia pomonella* (L.) (Lepidoptera: Tortricidae) larvae in Michigan to different modes of action causes reduced residual activity. *Pest Manag. Sci.* 64, 881–90.

Nansen, C., MacDonald, K. M., Rogers, C. D., Thomas, M., Poppy, G. M. and Baxter, I. H. 2007. Effects of sex pheromone in electrostatic powder on mating behavior by *Lobesia botrana* males. *J. Appl. Entomol.* 131(5), 303–10. doi:10.1111/j.1439-0418.2007.01167.x.

Pats, P. and Wiktellus, S. 1989. Tethered flight of *Chilo partellus* (Swinhoe) (Lepidoptera: Pyralidae). *Bull. Entomol. Res.* 79(1), 109–14. doi:10.1017/S0007485300018617.

Peck, S. L. and McQuate, G. T. 2000. Field tests of environmentally friendly Malathion replacements to suppress wild Mediterranean fruit fly populations. *J. Econ. Entomol.* 93(2), 280–9. doi:10.1603/0022-0493-93.2.280.

Pelz, K. S., Isaacs, R., Wise, J. C. and Gut, L. J. 2005. Protection of fruit against infestation by apple maggot and blueberry maggot flies (Diptera: Tephritidae) using GF-120 fruit fly bait. *J. Econ. Entomol.* 98(2), 432–7. doi:10.1093/jee/98.2.432.

Pelz-Stelinski, K. S., Gut, L. J. and Isaacs, R. 2006. Behavioral responses of *Rhagoletis cingulata* (Diptera: Tephritidae) to GF-120 insecticidal bait enhanced with ammonium acetate. *J. Econ. Entomol.* 99(4), 1316–20. doi:10.1093/jee/99.4.1316.

Pfeiffer, D. G., Kaakeh, W., Killian, J. C., Lachance, M. W. and Kirsch, P. 1993. Mating disruption to control damage by leafrollers in Virginia apple orchards. *Entomol. Exp. Appl.* 67(1), 47-56. doi:10.1111/j.1570-7458.1993.tb01650.x.

Polajnar, J., Eriksson, A., Virant-Doberlet, M. and Mazzoni, V. 2016. Mating disruption of a grapevine pest using mechanical vibrations: from laboratory to the field. *J. Pest. Sci.* 89(4), 909–21. doi:10.1007/s10340-015-0726-3.

Polavarapu, S., Lonergan, G., Peng, H. and Neilsen, K. 2001. Potential for mating disruption of *Sparganothis sulfureana* (Lepidoptera: Tortricidae) in cranberries. *J. Econ. Entomol.* 94(3), 658–65.

Porcel, M., Sjöberg, P., Swiergiel, W., Dinwiddie, R., Rämert, B.. and Tasin, M. 2015. Mating disruption of *Spilonota ocellana* and other apple orchard tortricids using a multispecies reservoir dispenser. *Pest Manag. Sci.* 71(4), 562–70. doi:10.1002/ps.3844.

Potting, R. P. J. and Knight, A. L. 2002. Predicting the efficacy of modified modes of action of a pheromone-based attracticide: a bisexual attractant and autosterilisation. *IOBC WPRS Bull.* 25(9), 187–94.

Preiss, R. and Priesner, E. 1988. Responses of male codling moths (*Laspeyresia pomonella*) to codlemone and other alcohols in a wind tunnel. *J. Chem. Ecol.* 14(3), 797–813. doi:10.1007/BF01018774.

Prokopy, R. J. 1968. Visual responses of apple maggot flies, *Rhagoletis pomonella* (Diptera: Tephritidae): orchard studies. *Entomol. Exp. Appl.* 11(4), 403–22. doi:10.1111/j.1570-7458.1968.tb02070.x.

Prokopy, R. J. 1972. Responses of apple maggot flies to rectangles of different colors and shades. *Environ. Entomol.* 1(6), 720–6. doi:10.1093/ee/1.6.720.

Prokopy, R. J. 1975. Apple maggot control by sticky spheres. *J. Econ. Entomol.* 68(2), 197–8. doi:10.1093/jee/68.2.197.

Prokopy, R. J. 2003. Two decades of bottom-up, ecologically based pest management in a small commercial apple orchard in Massachusetts. *Agric. Ecosyst. Environ.* 94(3), 299–309. doi:10.1016/S0167-8809(02)00036-1.

Prokopy, R. J. and Boller, E. F. 1971. Response of European cherry fruit flies to colored rectangles. *J. Econ. Entomol.* 64(6), 1444–7. doi:10.1093/jee/64.6.1444.

Prokopy, R. J. and Owens, E. D. 1983. Visual detection of plants by herbivorous insects. *Annu. Rev. Entomol.* 28(1), 337–64. doi:10.1146/annurev.en.28.010183.002005.

Prokopy, R. J., Wright, S. E., Black, J. L., Hu, X. P. and McGuire, M. R. 2000. Attracticidal spheres for controlling apple maggot flies: commercial-orchard trials. *Entomol., Exp. Appl* 97(3), 293–9. doi:10.1046/j.1570-7458.2000.00742.x.

Reinke, M. D., Miller, J. R. and Gut, L. J. 2012. Potential of high-density pheromone-releasing microtraps for control of codling moth, *Cydia pomonella* and obliquebanded leafroller *Choristoneura rosaceana*. *Physiol. Entomol.* 37(1), 53–9. doi:10.1111/j.1365-3032.2011.00816.x.

Reinke, M. D., Siegert, P. Y., McGhee, P. S., Gut, L. J. and Miller, J. R. 2014. Pheromone release rate determines whether sexual communication of Oriental fruit moth is disrupted by competitive or non-competitive mechanisms. *Entomol. Exp. Appl.* 150(1), 1–6. doi:10.1111/eea.12137.

Reissig, W. H., Weires, R. W., Forshey, C. G., Roelofs, W. L., Lamb, R. C., Aldwinckle, H. S. and Alm, S. R. 1984. Management of the apple maggot, *Rhagoletis pomonella* (Walsh) (Diptera: Tephritidae), in disease resistant dwarf and semi-dwarf apple trees. *Environ. Entomol.* 13(3), 684–90. doi:10.1093/ee/13.3.684.

Rice, R. E. and Kirsch, P. 1990. Mating disruption of Oriental fruit moth in the United States. In: Ridgway, R. L., Silverstein, R. M. and Inscoe, M. N. (Eds), *Behavior-Modifying Chemicals for Insect Management*. Marcel Dekker, New York, pp. 193–211.

Rice, K. B., Short, B. D., Jones, S. K. and Leskey, T. C. 2016. Behavioral responses of *Drosophila suzukii* (Diptera: Drosophilidae) to visual stimuli under laboratory, semifield, and field conditions. *Environ. Entomol.* 45(6), 1480–8. doi:10.1093/ee/nvw123.

Rice, K. B., Short, B. D. and Leskey, T. C. 2017. Development of an attract-and-kill strategy for *Drosophila suzukii* (Diptera: Drosophilidae): evaluation of attracticidal spheres under laboratory and field conditions. *J. Econ. Entomol.* 110(2), 535–42. doi:10.1093/jee/tow319.

Ridgway, R. L., Silverstein, R. M. and Inscoe, M. N. 1990. *Behavior-Modifying Chemicals for Insect Management*. Marcel Dekker, New York, 761pp.

Riedl, H., Hoying, S. A., Barnett, W. W. and Detar, J. E. 1979. Relationship of within-tree placement of the pheromone trap to codling moth catches. *Environ. Entomol.* 8(4), 765–9. doi:10.1093/ee/8.4.765.

Rodriguez-Saona, C., Polk, D., Holdcraft, R., Chinnasmy, D. J. and Mafra-Neto, A. 2010. SPLAT-OrB reveals competitive attraction as a mechanism of mating disruption in oriental beetle (Coleoptera: Scarabaeidae). *Environ. Entomol.* 39(6), 1980–9. doi:10.1603/EN10062.

Roelofs, W. L., Glass, E. H., Tette, J. and Comeau, A. 1970. Sex pheromone trapping for red-banded leaf roller control theoretical and actual. *J. Econ. Entomol.* 63(4), 1162–7. doi:10.1093/jee/63.4.1162.

Roelofs, W. L., Comeau, A., Hill, A. and Milicevic, G. 1971. Sex Attractant of the codling moth: characterization with electroantennogram technique. *Science* 174(4006), 297–9. doi:10.1126/science.174.4006.297.

Rothschild, G. H. L. 1975. Control of oriental fruit moth (*Cydia molesta* [Busck][Lepidoptera, Tortricidae]) with synthetic female pheromone. *Bull. Entomol. Res.* 65(3), 473–90. doi:10.1017/S0007485300006143.

Rothschild, G. H. L. 1981. Mating disruption of lepidopterous pests: current status and future prospects. In: Mitchell, E. R. (Ed.), *Management of Insect Pests with Semiochemicals: Concepts and Practice*. Plenum Press, New York, pp. 207–28.

Rothschild, G. H. L. 1982. Suppression of mating in codling moths with synthetic sex pheromone and other compounds. In: Kydonius, A. and Beroza, M. (Eds), *Insect Suppression with Controlled Release Pheromone Systems*. CRC Press, Boca Raton, pp. 117–34.

Rothschild, G. H. L. and Vickers, R. 1982. Preliminary trials on mating disruption of codling moth in S.E. Australia. *DSIR Ent. Div. Rept.* 2, 103–7.

Sanders, C. J. 1997. Mechanisms of mating disruption in moths. In: Cardé, R. T. and Minks, A. K. (Eds), *Insect Pheromone Research, New Directions*. Chapman and Hall, New York, pp. 333–46.

Sanders, C. J. and Lucuik, G. S. 1996. Disruption of male oriental fruit moth to calling females in a wind tunnel by different concentrations of synthetic pheromone. *J. Chem. Ecol.* 22(11), 1971–86. doi:10.1007/BF02040089.

Sauer, A. E. and Karg, G. 1998. Variables affecting the pheromone concentration in vineyards treated for mating disruption of grape vine moth. *J. Chem. Ecol.* 24(2), 289–302. doi:10.1023/A:1022532425253.

Schmitz, V., Renou, M., Roehrich, R., Stockel, J. and Lecharpentier, P. 1997. Disruption mechanisms of pheromone communication in the European grape moth *Lobesia botrana* Den & Schiff. III. Sensory adaptation and habituation. *J. Chem. Ecol.* 23(1), 83–95. doi:10.1023/B:JOEC.0000006347.86822.17.

Schumacher, P., WeyEneth, A., Weber, D. C. and Dorn, S. 1997. Long flights in *Cydia pomonella* L. (Lepidoptera: Tortricidae) measured by a flight mill: influence of sex, mated status and age. *Physiol. Ent.* 22(2), 149–60. doi:10.1111/j.1365-3032.1997.tb01152.x.

Severin, H. H. P. 1916. Soluble poisons in the poisoned bait spray to control the adult of the apple maggot (*Rhagoletis pomonella* Walsh). *Maine Agric. Exp. Stn Bull.* 251, 168p.

Sexton, S. B. and Il'ichev, A. 2000. Pheromone mating disruption with reference to oriental fruit moth *Grapholita molesta* (Busck) (Lepidoptera: Tortricidae) Literature review. *Gen. Appl. Entomol.* 29, 63–8.

Shorey, H. H. 1977. Manipulation of insect pests of agricultural crops. In: Shorey, H. H. and McKelvey, Jr., J. J. (Eds), *Chemical Control of Insect Behavior: Theory and Application*. John Wiley and Sons, Toronto, pp. 353–67.

Shorey, H. H. and Gerber, R. G. 1996. Use of puffers for disruption of sex pheromone communication of codling moths (Lepidoptera: Tortricidae) in walnut orchards. *Environ. Entomol.* 25(6), 1398–400. doi:10.1093/ee/25.6.1398.

Shorey, H. H., Sisk, C. B. and Gerber, R. G. 1996. Widely separated pheromone release sites for disruption of sex pheromone communication in two species of Lepidoptera. *Environ. Entomol.* 25(2), 446–51. doi:10.1093/ee/25.2.446.

Snow, J. W. 1990. Peachtree borer and lesser peachtree borer control in the United States. In: Ridgway, R. L., Silverstein, R. M. and Inscoe, M. N. (Eds), *Behavior-Modifying Chemicals for Insect Management*. Marcel Dekker, New York, pp. 241–53.

Somsai, A. P., Oltean, I., Gansca, L., Oprean, I., Raica, P. and Harsan, E. 2010. Control of two sympatric tortricids, *Cydia pomonella* and *Adoxophyes reticulana* (Lepidoptera: Tortricidae) by an experimental 'attract and kill' formulation. *Bull. UASVM Hortic.* 67, 453–7.

Stará, J., Kocourek, F. and Falta, V. 2008. Control of codling moth *Cydia pomonella* L., (Lepidoptera: Tortricidae) by the 'attract and kill' strategy. *J. Plant Dis. Prot.* 115(2), 75–9. doi:10.1007/BF03356242.

Stark, J. D., Vargas, R. and Miller, N. 2004. Toxicity of spinosad in protein bait to three economically important tephritid fruit fly species (Diptera: Tephritidae) and their parasitoids (Hymenoptera: Braconidae). *J. Econ. Entomol.* 97(3), 911–5. doi:10.1093/jee/97.3.911.

Steiner, L. F. 1952. Fruit fly control in Hawaii with poisoned sprays containing protein hydrolysate. *J. Econ. Entomol.* 45(5), 838–43. doi:10.1093/jee/45.5.838.

Steinmann, K. P., Zhang, M. H., Grant, J. A., Pickel, C. and Goodhue, R. E. 2008. Pheromone-based pest management can be cost-effective for walnut growers. *Calif. Agric.* 62(3), 105–10. doi:10.3733/ca.v062n03p105.

Stelinski, L. L. and Liburd, O. E. 2001. Evaluation of various deployment strategies of imidacloprid-treated spheres in highland blueberries for control of *Rhagoletis mendax* (Diptera: Tephritidae). *J. Econ. Entomol.* 94(4), 905–10.

Stelinski, L. L., Gut, L. J. and Miller, J. R. 2003a. Concentration of air-borne pheromone required for long-lasting adaptation in the obliquebanded leafroller, *Choristoneura rosaceana*. *Physiol. Entomol.* 28(2), 97–107. doi:10.1046/j.1365-3032.2003.00319.x.

Stelinski, L. L., Miller, J. R. and Gut, L. J. 2003b. Presence of long-lasting peripheral adaptation in oblique-banded leafroller, *Choristoneura rosaceana* and absence of such adaptation in redbanded leafroller, *Argyrotaenia velutinana*. *J. Chem. Ecol.* 29(2), 405–23.

Stelinski, L. L., Gut, L. J., Pierzchala, A. V. and Miller, J. R. 2004a. Field observations quantifying attraction of four tortricid moth species to high-dosage pheromone rope dispensers in untreated and pheromone-treated apple orchards. *Entomol. Exp. Appl.* 113, 187–96.

Stelinski, L. L., Gut, L. J., Vogel, K. J. and Miller, J. R. 2004b. Behaviors of naive vs. pheromone-exposed leafroller moths in plumes from high-dosage pheromone dispensers in a sustained-flight wind tunnel: implications for mating disruption of these species. *J. Insect Behav.* 17(4), 533–54. doi:10.1023/B:JOIR.0000042540.09188.eb.

Stelinski, L. L., Gut, L. J., Epstein, D. and Miller, J. R. 2005a. Attraction of four tortricid moth species to high dosage pheromone rope dispensers: observations implicating false plume following as an important factor in mating disruption. *IOBC WPRS Bull.* 28, 313-7.

Stelinski, L. L., Gut, L. J. and Miller, J. R. 2005b. Occurrence and duration of long-lasting peripheral adaptation among males of three species of economically important tortricid moths. *Ann. Entomol. Soc. Amer.* 98(4), 580–6. doi:10.1603/0013-8746(2005)098[0580:OADOLP]2.0.CO;2.

Stelinski, L. L., Gut, L. J., Mallinger, R. E., Epstein, D., Reed, T. P. and Miller, J. R. 2005c. Small plot trials documenting effective mating disruption of oriental fruit moth, *Grapholita molesta* (Busck), using high densities of wax-drop pheromone dispensers. *J. Econ. Entomol.* 98(4), 1267–74.

Stelinski, L. L., Gut, L. J., Ketner, K. C. and Miller, J. R. 2005d. Orientational disruption of codling moth, Cydia pomonella (L.) (Lep. Tortricidae), by concentrated formulations of microencapsulated pheromone in flight tunnel assays. *J. Appl. Entomol.* 129, 481–8.

Stelinski, L. L., Vogel, K. J., Gut, L. J. and Miller, J. R. 2005e. Seconds-long pre-exposures to pheromone from rubber septum or polyethylene-tube dispensers alter subsequent behavioral responses of male *Grapholita molesta* (Lepidoptera: Tortricidae) in a sustained-flight tunnel. *Environ. Entomol.* 34(3), 696–704. doi:10.1603/0046-225X-34.3.696.

Stelinski, L. L., Gut, L. J. and Miller, J. R. 2006a. Orientational behaviors and EAG responses of male codling moth after exposure to synthetic sex pheromone from various dispensers. *J. Chem. Ecol.* 32(7), 1527–38. doi:10.1007/s10886-006-9067-2.

Stelinski, L. L., Il'ichev, A. L. and Gut, L. J. 2006b. Antennal and behavioral responses of virgin and mated oriental fruit moth (Lepidoptera: Tortricidae) females to their sex pheromone. *Ann. Entomol. Soc. Amer.* 99(5), 898–904. doi:10.1603/0013-8746(2006)99[898:AABROV]2.0.CO;2.

Stelinski, L. L., Miller, J. R., Ledebuhr, R. and Gut, L. J. 2006c. Mechanized applicator for large-scale field deployment of paraffin-wax dispensers of pheromone for mating disruption in tree fruit. *J. Econ. Entomol.* 99(5), 1705–10. doi:10.1093/jee/99.5.1705.

Stelinski, L. L., Gut, L. J., Haas, M., McGhee, P. and Epstein, D. 2007a. Evaluation of aerosol devices for simultaneous disruption of sex pheromone communication in *Cydia pomonella* and *Grapholita molesta* (Lepidoptera: Tortricidae). *J. Pest. Sci.* 80(4), 225–33. doi:10.1007/s10340-007-0176-7.

Stelinski, L. L., McGhee, P., Haas, M., Il'ichev, A. L. and Gut, L. J. 2007b. Sprayable microencapsulated sex pheromone formulations for mating disruption of four tortricid species: effects of application height, rate, frequency, and sticker adjuvant. *J. Econ. Entomol.* 100(4), 1360–9.

Stelinski, L. L., Miller, J. R., Ledebuhr, R., Siegert, P. and Gut, L. J. 2007c. Season-long mating disruption of *Grapholita molesta* (Lepidoptera: Tortricidae) by one machine application of pheromone in wax drops (SPLAT-OFM). *J. Pest. Sci.* 80(2), 109–17. doi:10.1007/s10340-007-0162-0.

Stelinski, L. L., McGhee, P., Grieshop, M., Brunner, J. and Gut, L. J. 2008a. Efficacy and mode of action of female equivalent dispensers of pheromone for mating disruption of codling moth, *Cydia pomonella* (L.). *Agric. For. Entomol.* 10, 389–97.

Stelinski, L. L., Miller, J. R. and Rogers, M. E. 2008b. Mating disruption of citrus leafminer mediated by a noncompetitive mechanism at a remarkably low pheromone release rate. *J. Chem. Ecol.* 34(8), 1107–13. doi:10.1007/s10886-008-9501-8.

Stelinski, L. L., Il'ichev, A. L. and Gut, L. J. 2009. Efficacy and release rate of reservoir pheromone dispensers for simultaneous disruption of codling moth and Oriental fruit moth (Lepidoptera: Tortricidae). *J. Econ. Entomol.* 102(1), 315–23.

Stoeckli, S., Hirschi, M., Spirig, C., Calanca, P., Rotach, M. W. and Samietz, J. 2012. Impact of climate change on voltinism and prospective diapause induction of a global pest insect-*Cydia pomonella* (L.). *PLoS ONE* 7(4), e35723. doi:10.1371/journal.pone.0035723.

Suckling, D. M. 2016. Monitoring for surveillance and management. In: Allison, J. D. and Cardé, R. T. (Eds), *Pheromone Communication in Moths: Evolution, Behavior and Application*. University of California Press, Berkeley, CA, pp. 337–48.

Suckling, D. M. and Brockerhoff, E. G. 1999. Control of lightbrown apple moth (Lepidoptera: Tortricidae) by using an attracticide. *J. Econ. Entomol.* 92(2), 367–72. doi:10.1093/jee/92.2.367.

Suckling, D. M. and Karg, G. 2000. Pheromones and other semiochemicals. In: Rechcigl, J. and Rechcigl, N. (Eds), *Biological and Biotechnological Control of Insect Pests*. CRC Press LLC, Boca Raton, pp. 63–99.

Suckling, D. M., Karg, G. and Bradley, S. J. 1996. Apple foliage enhances mating disruption of lightbrown apple moth. *J. Chem. Ecol.* 22(2), 325–41. doi:10.1007/BF02055102.

Suckling, D. M., Green, S. R., Gibb, A. R. and Karg, G. 1999. Predicting atmospheric concentration of pheromone in treated apple orchards. *J. Chem. Ecol.* 25(1), 117–39. doi:10.1023/A:1020893201446.

Suckling, D. M., Daly, J. M., Chen, X. and Karg, G. 2007a. Field electroantennogram and trap assessments of aerosol pheromone dispensers for disrupting mating in *Epiphyas postvittana*. *Pest Manag. Sci.* 63(2), 202–9. doi:10.1002/ps.1312.

Suckling, D. M., Walker, J. T. S., Shaw, P. W., Manning, L. A., Lo, P., Wallis, R., Bell, V., Sandanayaka, W. R., Hall, D. R., Cross, J. V., et al. 2007b. Trapping *Dasineura mali* (Diptera: Cecidomyiidae) in apples. *J. Econ. Entomol.* 100, 745–51.

Suckling, D. M., Stringert, L. D., Kean, J. M., Lo, P. L., Bell, V., Walker, J. Ts, Twidle, A. M., Jimenez-Perez, A. and El-Sayed, A. M. 2015. Spatial analysis of mass trapping: how close is close enough? *Pest Manag. Sci.* 71(10), 1452–61. doi:10.1002/ps.3950.

Sufyan, M., Neuhoff, D. and Furlan, L. 2011. Assessment of the range of attraction of pheromone traps to *Agriotes lineatus* and *Agriotes obscurus*. *Agric. Forest Entomol.* 13(3), 313–9. doi:10.1111/j.1461-9563.2011.00529.x.

Swenson, D. W. and Weatherston, I. 1989. Hollow-fiber controlled-release systems. In: Jutsum, A. R. and Gordon, R. F. S. (Eds), *Insect Pheromones in Plant Protection*. Wiley, Chichester, UK, pp. 173–97.

Taschenberg, E. F., Cardé, R. T. and Roelofs, W. L. 1974. Sex pheromone mass trapping and mating disruption for control of red-banded leafroller and grape berry moths in vineyards. *Environ. Entomol.* 3(2), 239–42. doi:10.1093/ee/3.2.239.

Teixeira, L. A. F., Wise, J. C., Gut, L. J. and Isaacs, R. 2009. Paraffin wax emulsion for increased rainfastness of insecticidal bait to control apple maggot *Rhagoletis pomonella* (Walsh) (Diptera: Tephritidae). *J. Econ. Entomol.* 102(3), 1108–15.

Teixeira, L. A. F., Grieshop, M. J. and Gut, L. J. 2010a. Effect of pheromone dispenser density on timing and duration of approaches by peachtree borer. *J. Chem. Ecol.* 36(10), 1148–54. doi:10.1007/s10886-010-9853-8.

Teixeira, L. A. F., Gut, L. J. and Isaacs, R. 2010b. Response of apple maggot and cherry fruit fly (Diptera: Tephritidae) to color and contrast cues from small deposits. *J. Entomol. Sci.* 45(2), 65–74. doi:10.18474/0749-8004-45.2.65.

Teixeira, L. A. F., Miller, J. R., Epstein, D. L. and Gut, L. J. 2010c. Comparison of mating disruption and mass trapping with Pyralidae and Sesiidae moths. *Entomol. Exp. Appl.* 137, 176–83. DOI: 10.1111/j.1570-7458.2010.01047.x.

Thomas, H., White, J. and Speers, C. 1975. Dispensing pressurized aerosols of southern pine beetle *Dendroctonus frontalis* pheromone under field conditions. *J. Georgia Entomol. Soc.* 10, 265–71.

Thomson, D. and Jenkins, J. 2014. Successes with area-wide mating disruption: moving from crisis management to sustainable pheromone-based pest management. *IOBC WPRS Bull.* 99, 9–11.

Thomson, D. R., Gut, L. J. and Jenkins, J. W. 1998. Pheromones for insect control. In: Hall, F. R. and Menn, J. J. (Eds), *Methods in Biotechnology, vol 5: Biopesticides: Use and Delivery*. Humana Press Inc, Totowa, NJ, pp. 385–412.

Thomson, D., Brunner, J., Gut, L., Judd, G. and Knight, A. 2001. Ten years implementing codling moth mating disruption in the orchards of Washington and British Columbia: starting right and managing for success! *IOBC WPRS Bull.* 24, 23-30.

Thwaite, W. G., Mooney, A. M., Eslick, M. A. and Nichol, H. L. 2004. Evaluating pear-derived kairomone lures for monitoring *Cydia pomonella* (L.) (Lepidoptera: Tortricidae) in Granny Smith apples under mating disruption. *Gen. Appl. Entomol.* 33, 56–60.

Tomaszewska, E., Hebert, V. R., Brunner, J. F., Jones, V. P., Doerr, M. and Hilton, R. 2005. Evaluation of pheromone release from commercial mating disruption dispensers. *J. Agric. Food Chem.* 53(7), 2399–405. doi:10.1021/jf048163k.

Trimble, R. M. 1995. Mating disruption for control of codling moth *Cydia pomonella* (L.) (Lepidoptera: Tortricidae), in organic apple production in Southwestern Ontario. *Can. Entomol.* 127, 493–505.

Trimble, R. M. and Appleby, M. E. 2004. Comparison of efficacy of programs using insecticide and insecticide plus mating disruption for controlling the obliquebanded leafroller in apple (Lepidoptera: Tortricidae). *J. Econ. Entomol.* 97(2), 518–24. doi:10.1093/jee/97.2.518.

Trimble, R. M. and El-Sayed, A. M. 2005. Potential of pear kairomone for monitoring activity of codling moth in Ontario apple orchards. *Can. Entomol.* 137(1), 110–6. doi:10.4039/n04-059.

Trimble, R. M., Pree, D. J., Barszcz, E. S. and Carter, N. J. 2004. Comparison of a sprayable pheromone formulation and two hand-applied pheromone dispensers for use in the integrated control of Oriental fruit moth (Lepidoptera: Tortricidae). *J. Econ. Entomol.* 97(2), 482–9. doi:10.1093/jee/97.2.482.

Turchin, P. and Odendaal, F. J. 1996. Measuring the effective sampling area of a pheromone trap for monitoring population density of southern pine beetle (Coleoptera: Scolytidae). *Environ. Entomol.* 25(3), 582–8. doi:10.1093/ee/25.3.582.

Vargas, R. I., Peck, S. L., McQuate, G. T., Jackson, C. G., Stark, J. D. and Armstrong, J. W. 2001. Potential for areawide integrated management of Mediterranean fruit fly (Diptera: Tephritidae) with a braconid parasitoid and a novel bait spray. *J. Econ. Entomol.* 94(4), 817–25. doi:10.1603/0022-0493-94.4.817.

Vargas, R. I., Mau, R. F. L., Jang, E. B., Faust, R. M. and Wong, L. 2008. The Hawaii fruit fly areawide pest management programme. In: Koul, O., Cuperus, G. and Elliott, N. (Eds), *Areawide Pest Management Theory and Implementation*. CAB International, Oxfordshire, UK, pp. 300–25.

Vickers, R. A. 1990. Oriental fruit moth in Australia and Canada. In: Ridgway, R. L., Silverstein, R. M. and Inscoe, M. N. (Eds), *Behavior-Modifying Chemicals for Insect Management*. Marcel Dekker, New York, pp. 183–92.

Vickers, R. A., Thwaite, W. G., Williams, D. G. and Nicholas, A. H. 1998. Control of codling moth in small plots by mating disruption: alone and with limited insecticide. *Entomol. Exp. Appl.* 86(3), 229–39. doi:10.1046/j.1570-7458.1998.00285.x.

Waldner, W. 1997. Three years of large-scale control of codling moth by mating disruption in the South Tyrol, Italy. *IOBC WPRS Bull.* 20, 35–44.

Waldstein, D. W. and Gut, L. J. 2004. Effects of rain and sunlight on oriental fruit moth (Lepidoptera: Tortricidae) microcapsules applied to apple foliage. *J. Agric. Urban Entomol.* 21, 117–28.

Wamsley, C., Wilde, G. C. and Higgins, R. 2006. Preliminary results of use of a mark-release-recapture technique for determining the sphere of influence of a kairomone-baited lure trap attractive to adult western corn rootworm (Coleoptera: Chrysomelidae). *J. Kansas Entomol. Soc.* 79(1), 23–7. doi:10.2317/0404.06.1.

Weatherston, I. 1990. Principles of design of controlled-release formulations. In: Ridgway, R. L., Silverstein, R. M. and Inscoe, M. N. (Eds), *Behavior-Modifying Chemicals for Insect Management*. Marcel Dekker, New York, pp. 93–112.

Weissling, T. J. and Knight, A. L. 1995. Vertical distribution of codling moth adults in pheromone-treated and untreated plots. *Entomol. Exp. Appl* 77(3), 271–5. doi:10.1111/j.1570-7458.1995.tb02324.x.

Weissling, T. J. and Knight, A. L. 1996. Oviposition and calling behavior of codling moth (Lepidoptera: Tortricidae) in the presence of codlemone. *Ann. Entomol. Soc. Amer.* 89(1), 142–7. doi:10.1093/aesa/89.1.142.

Welter, S. C., Pickel, C., Millar, J., Cave, F., Van Steenwyk, R. A. and Dunley, J. 2005. Pheromone mating disruption offers selective management options for key pests. *Calif. Agric.* 59(1), 16–22. doi:10.3733/ca.v059n01p16.

Williamson, E. R., Folwell, R. J., Knight, A. and Howell, J. F. 1996. Economics of employing pheromones for mating disruption of the codling moth, *Carpocapsa pomonella*. *Crop Prot.* 15(5), 473–7. doi:10.1016/0261-2194(96)00013-0.

Wilson, H. R. and Trammel., K. 1980. Sex-pheromone trapping for control of codling moth (Lepidoptera, Olethreutidae), oriental fruit moth (Lepidoptera, Olethreutidae), lesser appleworm (Lepidoptera, Olethreutidae), and 3 tortricid leafrollers in a New-York apple orchard. *J. Chem. Ecol.* 73, 291–5.

Wins-Purdy, A. H., Judd, G. J. R. and Evenden, M. L. 2007. Disruption of pheromone communication of *Choristoneura rosaceana* (Lepidoptera: Tortricidae) using microencapsulated sex pheromones formulated with horticultural oil. *Environ. Entomol.* 36(5), 1189–98.

Wins-Purdy, A. H., Judd, G. J. R. and Evenden, M. L. 2008. Mechanisms of pheromone communication disruption in *Choristoneura rosaceana* exposed to microencapsulated (Z)-11 tetradecenyl acetate formulated with and without horticultural oil. *J. Chem. Ecol.* 34(8), 1096–106. doi:10.1007/s10886-008-9500-9.

Witzgall, P., Bäckman, A., Svensson, M., Koch, U., Rama, F., El-Sayed, A., Brauchli, J., Arn, H., Bengtsson, M. and Löfqvist, J. 1999. Behavioral observations of codling moth, *Cydia pomonella*, in orchards permeated with synthetic pheromone. *BioControl* 44(2), 211–37. doi:10.1023/A:1009976600272.

Witzgall, P., Stelinski, L., Gut, L. and Thomson, D. 2008. Codling moth management and chemical ecology. *Annu. Rev. Entomol.* 53, 503–22. doi:10.1146/annurev.ento.53.103106.093323.

Witzgall, P., Kirsch, P. A. and Cork, A. 2010. Sex pheromones and their impact on pest management. *J. Chem. Ecol.* 36(1), 80–100. doi:10.1007/s10886-009-9737-y.

Wright, S. E., Leskey, T. C., Jacome, I., Pinero, J. C. and Prokopy, R. J. 2012. Integration of insecticidal, phagostimulatory, and visual elements of an attract and kill system for apple maggot fly (Diptera: Tephritidae). *J. Econ. Entomol.* 105(5), 1548–56. doi:10.1603/EC10104.

Wyman, J. A. 1979. Effect of trap design and sex attractant release on tomato pinworm catches. *J. Econ. Entomol.* 72(6), 865–8. doi:10.1093/jee/72.6.865.

Yee, W. L. and Alston, D. G. 2012. Control of *Rhagoletis indifferens* using thiamethoxam and Spinosad baits under external fly pressure and its relation to rapidity of kill and residual bait activity. *Crop Prot.* 41, 17–23. doi:10.1016/j.cropro.2012.05.013.

Yee, W. L. and Chapman, P. S. 2005. Effects of GF-120 fruit fly bait concentrations on attraction, feeding, mortality, and control of *Rhagoletis indifferens* (Diptera: Tephritidae). *J. Econ. Entomol.* 98(5), 1654–63. doi:10.1093/jee/98.5.1654.

Chapter 22

Optimizing insecticide use in integrated management of fruit insect pests

Claudio Ioriatti and Gino Angeli, Fondazione Edmund Mach, Italy; Greg Krawczyk, The Pennsylvania State University, USA; and Carlo Duso, University of Padova, Italy

1 Introduction

Since its inception, integrated pest management (IPM) was conceived to reduce grower reliance on pesticides by combining and integrating biological, physical and chemical control tools (Kogan, 1998). The concept, reported by Stern et al. (1959) in the United States and by the pioneering activity of fruit entomologists operating within the International Organization for Biological Control (IOBC) in Europe (Boller et al., 2006), provided the foundations for understanding the relationship between damaging pests and beneficial insects in agricultural systems.

Many definitions of IPM have been suggested, but the standard definition adopted by the FAO Panel of Experts on Integrated Control is:

> Integrated Pest Control is a pest management system that, in the context of the associated environment and population dynamics of the pest species, utilizes all suitable techniques and methods in as compatible manner as possible and maintains the pest population at the levels below those causing economic injury (FAO, 1968).

Some years later, the term 'integrated pest management' gained more consensus because 'management' is considered as having a broader meaning and more clearly implies human interference (Minks et al., 1998).

http://dx.doi.org/10.19103/AS.2019.0046.28

IPM theory recognizes that each insect population density fluctuates around the average population density (GEP = general equilibrium position) over a period of time in the absence of permanent environmental change. Other keystone concepts included in the IPM theory are the 'economic injury level' (EIL), the lowest population density that will cause economic damage and the 'economic threshold' (ET), the density at which control measures should be determined to prevent an increasing pest population from reaching the EIL. The ET is an operational criterion instructing when to apply the control measures to reduce pest status (Stejskal, 2003). For a number of key fruit pests, the ET was established during the late 1960s and early 1970s (Boller et al., 2010; Waldner, 1996; Baggiolini et al., 1984; Brader, 1974) and is still used as a reference despite the changes in the cropping systems, the crop value and the input costs.

The combination of regular monitoring, use of injury threshold and timely application of pesticides contributed to the reduction of insecticide input, but was not an exhaustive definition of an IPM programme. The adoption of pesticides having fewer side effects on natural enemies should also be included in IPM programmes (Minks et al., 1998). In fact, natural control may prevent a pest species from reaching an EIL or it may allow economic outbreaks depending on the pesticide programme implemented (Stern et al., 1959).

In particular, fruit orchards (Fig. 1), as most other perennial crops, harbour many species of herbivorous arthropods whose population levels fluctuate around their respective general equilibrium position (Gardiner et al., 2009; Szentkiràyi and Kozàr, 1991). For the majority of arthropods the average population density is generally below the EIL as a result of the combined actions of abiotic and biotic environmental factors. Consequently, most herbivorous, potentially harmful, species are classified as 'sub-economic pests' because their

Figure 1 Apple orchard agroecosystem in Northern Italy.

population density is limited by natural controlling forces (Stern et al., 1959). In particular, within a community of organisms, in an agroecosystem, biological control could be exerted by the so-called natural enemy specialists, which become abundant following a rise in pest populations, and the generalists which may already be present in an orchard feeding on non-pest prey and/or alternative foods and promptly becoming active when pest numbers increase (Settle et al., 1996; DeBach and Rosen, 1991).

Conservation and enhancement of the biological control agents are the primary objectives of IPM programmes, and can be pursued by establishing and maintaining ecological infrastructures. An increase in landscape complexity generally improves insect community diversity and effectiveness of biological control agents by providing natural enemies with overwintering sites, shelters and alternative hosts and food (including nectar and pollen) (Szentkiràyi and Kozàr, 1991; Cross et al., 2010, 2015). It has been reported that diverse landscape structure around orchards significantly influences the abundance of predators and masks detrimental effects of pesticides (Markó et al., 2017). For example, Cross et al. (2010) identified tree and herbaceous plant species that are a good source of *Anthocoris nemoralis* (F.), predator of pear psylla, when planted in hedgerows and/or windbreaks around pear orchards. Wyss et al. (1995) demonstrated that wildflower strips augmented the density of aphidophagous insects and spiders which reduced numbers of *Dysaphis plantaginea* Pass., able to return on the apple trees from their summer host plants, *Plantago* spp.

In contrast, the simplification of agricultural landscapes is recognized as the first factor responsible for reducing the abundance and diversity of natural enemies and jeopardizing their natural ability to control potential insect pests. Under these circumstances, a population outbreak of some insect pest species may occur, and the application of control measures to protect the crop and avoid economic damage may be required (Kogan, 1998).

The frequency of pest outbreaks varies from occasional to regular according to a mutual position in GEP and EIL and the type of the fluctuations so that pests are classified as either 'occasional/sporadic', 'perennial' or 'chronic/severe' (Gray et al., 2009; Pringle, 2006; Amaro and Baggiolini, 1982; Stern et al., 1959).

To prevent fruit injury, modern agriculture largely relies on the application of synthetic pesticides whose injudicious use causes serious public concern about potential impacts on human health and the environment. The most frequently claimed adverse consequences of pesticides include pesticide residues on food, chronic health effects, contamination of water and soil, pest resistance, upsurge of secondary pests and negative impacts on non-target organisms.

Adoption of IPM strategies requires pest population assessments and prediction, an economic analysis based on an appreciation of crop responses to the pest damage, the application of control measures only when necessary and the integrated use of several selective and compatible tactics (Pedigo, 1995).

The goal of this chapter is to demonstrate how to optimize the use of pesticides when they are necessary for controlling fruit insect pests in order to enhance the crucial contribution of biological control agents and to mitigate the detrimental impacts of chemical control on the environment and ecosystem services.

2 Optimizing insecticide use in IPM

Fruit orchards were one of the initial agroecosystems selected for testing the implementation of IPM to successfully combine biological control and selected pesticide applications (Kogan, 1998). For a successful IPM programme it is essential to recognize that there is an interaction between pest populations and their natural enemies at the ecosystem-level, and that fruit orchards are in close contact and interaction with the surrounding environment. In this context, judicious and harmonious use of selective pesticides in combination with non-chemical tools and control tactics are expected to maintain the pest populations below EILs and to minimize the potential deleterious side effects of pesticides. Implementation of these ecological principles in an effective pest control strategy is often influenced by conflicting social interests. Increased public fears about the impact of pesticides such as the potential adverse effects on human health and overall environmental quality, promote the development of tools with lower risk factors. However, at the same time, high cosmetic standards for fresh fruit required by consumers represent formidable obstacles in the adoption of alternative pest control tools (Damos et al., 2015; Altieri and Nicholls, 2005). While considering all these aspects, the resultant insect pest management strategy 'should be based on cost/benefit analyses that take into account the interests of and impacts on producers, society and the environment' (Kogan, 1998). This goal is, at least partially, achievable by avoiding unnecessary application of harmful products and by optimizing the unavoidable use of insecticides which should always be based on accurate information about the actual pest-beneficial insect equilibrium in fruit orchards.

2.1 Insect pest monitoring

The assessment of the real need for the use of an insecticide and determination of the best timing of the application are the first crucial steps to reduce and optimize insecticide usage under IPM principles.

As commercial fruit orchards are usually too large to be scrutinized by individual trees for pest presence, decisions about the need for intervention are reached by sampling potentially infested plant organs (e.g. leaves, fruits, shoots etc.) and inferring the general pest abundance from the obtained data. Sampling plans in IPM can have different goals: detect the presence, quantify the insect population size or make decisions about the need for intervention (Moon and Wilson, 2009).

Insect populations can be monitored using a variety of tools such as 'active' traps luring insect pests by means of attractants (e.g. semiochemicals, light, colour, feeding bait), or 'passive' tools like suction devices, tap sampling, artificial shelters or sticky traps.

Because of their high selectivity and cost-effectiveness, traps baited with sex pheromones are widely used in many IPM strategies. They can either support or replace the frequent field scouting of target pests to enable decisions on whether an intervention is needed and what timing is required (Witzgall et al., 2010). Hundreds of semiochemical compounds have been identified that are regularly used by IPM practitioners to monitor the presence and abundance of insect pests (El-Sayed, 2018). Traps baited with pheromones have also allowed the development of phenology models for pests, while the numbers of trapped insects have also been used to set action thresholds for several fruit pests (Ioriatti and Lucchi, 2016; Faccioli et al., 1993; Alford et al., 1979).

Proper identification of pests and naturally occurring biological control agents is an important prerequisite to effectively handle pest problems. Another critical aspect for a successful IPM programme is the availability of trained personnel who have access to modern tools for rapid identification of insect species (Daane et al., 2018). Use of appropriate tools and the implementation of efficient field sampling plans provide growers and consultants with robust information about the current status of dynamic pest-natural enemy ratios which can be used to determine the best pest management option (Binns et al., 2000; Jones, 1991).

2.2 Predicting pest occurrence

Accurate prediction of pest occurrence and precise assessment of observed developmental stages are the key elements used to decide if an insecticide treatment is needed. Predictions of occurrence also enable optimal treatment timing and avoidance of unnecessary insecticide applications. Insects are ectothermic organisms and their growth rate is strongly affected by temperature. Relationships between heat accumulation and development rates for the relevant life stages of the most important fruit tree pests were established under controlled laboratory conditions (Nietschke et al., 2007). By using a heat-unit accumulation (e.g. degree-days) it is possible to forecast the phenology of pests and correlate values of heat-unit accumulation to key physiological life events, such as egg deposition, egg hatching and adult flight, or to model the age structure of pest populations. Phenology models can be used to explain past biological development in addition to forecasting crucial future events important for management activities. Insect pest forecasting tools are now available to growers and consultants to optimize timing of monitoring, management and control measures for a number of pest species in fruit orchards. For example, in Europe, phenology models in SOPRA (Schadorganismen-Prognose auf Apfel), a forecasting tool for insect pests on apple, are established for rosy apple aphid, *D. plantaginea*, apple sawfly, *Hoplocampa testudinea* (Klug), codling moth, *Cydia pomonella* (L.), smaller fruit tortrix, *Grapholita lobarzewskii* (Nowicki, 1860), apple blossom weevil, *Anthonomus pomorum* (L.), summer fruit tortrix, *Adoxophyes orana* (Fischer von Röslerstamm), European pear psylla, *Cacopsylla pyri* (L.), European cherry fruit fly, *Rhagoletis cerasi* (L.), European red spider mite, *Panonychus ulmi* (Koch), and plum tortrix, *Grapholita funebrana* (Tr.) (Samietz et al., 2015). In the United States, the WSU-Tree Fruit Decision Aid System (DAS) is the premier source of time-sensitive information for management in Washington State tree fruits and according to surveys carried out in 2008, 2010 and 2013, it is used by the majority of IPM decision makers in the WA state fruit industry. DAS collects daily weather data along with forecast data to predict phenology information of ten insects: apple maggot, *Rhagoletis pomonella* (Walsh), campylomma bug, *Campylomma verbasci* (Meyer), *C. pomonella*, lacanobia fruit worm, *Lacanobia subjuncta* (Grote & Robinson), pandemis leaf roller, *Pandemis pyrusana* (Kearfott), peach twig borer, *Anarsia lineatella* (Zeller), obliquebanded leaf roller, *Choristoneura rosaceana* (Harris), oriental fruit moth, *G. molesta* (Busck), San Jose scale, *Quadraspidiotus perniciosus* (Comstock), and western cherry fruit fly, *Rhagoletis indifferens* (Curran) (Chambers et al., 2015; Jones et al., 2010).

2.3 Monitoring natural enemies

Monitoring of insects is important, not only to evaluate the presence and abundance of the pest and key natural enemies, but it is equally essential to assess seasonal phenology and inter-annual variation and to identify periods of high vulnerability to pesticides (Jones et al., 2009). The diversity of natural enemy groups in fruit orchards can be quite high (Mills et al., 2016b; Brader, 1974), but to implement IPM programmes with high ecological benefit, we need to know the most effective natural enemies and when they are active during the growing season (Jones et al., 2016a). Monitoring techniques for natural enemies are similar to those mentioned for insect pests and include such activities as tap sampling, sticky traps, Malaise traps, vacuum samples, rearing infested hosts, use of sentinel hosts, leaf samples, pitfall traps and visual counts.

Some of these tools are used mostly by professional staff and some employed only for research purpose. Malaise traps and vacuum samples provide a reliable description of the insect community structure in certain habitats through time and space (Lucchi et al., 2017; Mates et al., 2012; Pfannenstiel et al., 2012), while the use of sentinel hosts or rearing infested insect stages provides information about indigenous parasitoids that develop on native and invasive pest species (Roversi et al., 2016; Miller et al., 2015). Malaise and vacuum samples also help to verify the establishment and the efficacy of inundative releases of imported natural enemies (Argov and Gazit, 2008).

Growers and consultants generally use more user-friendly and simpler methods for the assessment of insect pests and natural enemy abundance. Direct inspection of foliage with a hand lens is common practice by growers and consultants to estimate the abundance of predatory mites (Baillod et al., 1989), while a more precise estimation of populations is achieved using washing techniques and subsequent identification using a microscope (Boller, 1984). Tap sampling (beating method) over either a tray or an umbrella is the main monitoring method for estimating the abundance of anthocorids, the key natural enemies of pear psylla (Cross et al., 2010; Stäubli et al., 1992), whereas artificial refugia made by corrugated cardboard tied into trees can be deployed to monitor the presence and seasonal dynamics of earwig populations (Fountain and Harris, 2015; Beliën et al., 2012).

Similarly, as with insect pest monitoring, there is a potential to use synthetic parasitoid sex pheromones to detect parasitoid presence, to evaluate their synchronization with pest hosts and to quantify their relative abundance. It may also be possible to determine the field effectiveness of conservation biocontrol measures and selectiveness of applied pest management practices. To date only a few parasitoid sex pheromones have been identified (Ayasse et al., 2001). A field application of a parasitoid sex pheromone to survey *Ascogaster quadridentata* Wesmael (Hymenoptera: Braconidae) populations over time was conducted in four regions of New Zealand (Suckling et al., 2002). Regrettably, the authors concluded that trap efficiency was too low and that numbers of male parasitoids captured during the season were inadequate for practical use in pest management decisions in this instance.

Pheromones released by herbivores may also serve as semiochemical attractants in upper trophic levels by attracting natural enemies. The aggregation pheromone of *Frankliniella occidentalis* (Pergande) is reported as a potential kairomone for its natural enemy, the predatory bug *Orius laevigatus* (Fieber) (Vaello et al., 2017). Similarly, the sex pheromone of *Planococcus ficus* (Signoret) was attractive to its parasitoid, *Anagyrus* near *pseudococci* (Franco et al., 2008, 2011). From a practical perspective the response of natural enemies to the pheromones of their host species could also be exploited to monitor the presence of specific natural enemies.

The interaction between pests and their natural enemies is frequently affected, either positively or negatively, by host plants, which influence interactions by producing allelochemicals in response to herbivorous feeding (Dicke et al., 1990). These volatile compounds are defined as the herbivore-induced plant volatiles (HIPVs) and can manipulate the foraging behaviour of natural enemies (James, 2003, 2005). Synthetic HIPVs have been demonstrated to be effective in attracting beneficial insects and can be used as lures in traps to monitor specific natural enemies in the field (Rodriguez-Saona et al., 2011 and references herein). Several HIPVs, both alone and in combination, have been evaluated for monitoring natural enemies in fruit orchards (Jones et al., 2016b). For example, traps baited with the herbivore-induced plant volatile squalene have been demonstrated to be effective in monitoring green lacewings *Chrysopa nigricornis* Burmeister (Neuroptera: Chrysopidae), a common predator of aphids and other soft-bodied arthropods in the North American fruit orchards (Jones et al., 2011).

Methyl salicylate (MeSA) is a common HIPV which when applied in the field is reported to be attractive to several taxa of predaceous arthropods and potentially able to enhance natural enemy abundance. A commercial formulation of MeSA is available to growers as a slow-release dispenser (PredaLure; AgBio Inc., Westminster, Colorado, USA). According to Rodriguez-Saona et al. (2011), MeSA could be recognized as a universal lure for beneficial insects, particularly hoverflies (Syrphidae), lady beetles (Coccinellidae) and green lacewings (Chrysopidae). For other taxa such as Anthocoridae, MeSA lures gave a context-dependent response that would need further investigation before it could be considered as a reliable pest management tool (Lee, 2010; Rodriguez-Saona et al., 2011). The attraction efficacy was improved by combining plant volatiles and it was possible to fine-tune traps for specific natural enemy groups by selecting a trap design colour, shape and lure composition (Jones et al., 2011, 2016c).

The release rate and longevity of different lure formulations, the range of attraction and the effect of the crop context on the attraction efficacy are key factors that need further investigation. Nevertheless, the use of HIPVs represents a potential tool for enhancing pest management, by providing practitioners with information on natural enemy population abundance, diversity and phenology as well as an assessment of the effects of pesticide applications (Jones et al., 2009, 2016c).

Information about the actual status of natural enemy populations could also be obtained by utilization of phenology models. Some models are already available, and can be used to provide recommendations for various orchard management actions or the optimal time window for pesticide applications (Jones et al., 2016a; Moerkens et al., 2011). Jones et al. (2016a) used HIPV-based natural enemy trap data in combination with the developmental rate data to develop a phenology model for *C. nigricornis*, which provides information over a period of time when different pesticides can be applied in fruit orchards without disrupting natural enemy populations.

The European earwig *Forficula auricularia* L. (Dermaptera: Forficulidae) is a worldwide distributed natural enemy of pest species in pipfruit orchards. It is recognized as capable of suppressing several pest species of fruit orchard including scale insects, aphids, codling moth, midges and pear psylla. A day degree model capable of predicting the phenology of local earwigs was developed and integrated in a software tool (Moerkens et al., 2011). The programme provides growers with management recommendations that take into consideration the side effects of pesticides and soil tillage in relation to actual earwig phenology (Beliën et al., 2012).

2.4 Selection of pest control measures

According to the European Union Directive 2009/128/EC on sustainable use of pesticides, 'the member states shall establish or support the establishment of necessary conditions for the implementation of integrated pest management'. In order to fulfil recommendations provided by this Directive, national guidelines for IPM (e.g. https://www.reterurale.it/flex/cm/pages/ServeBLOB.php/L/IT/IDPagina/17765) should be developed to provide professional users with information and tools that are needed for implementing pest monitoring and making decisions about pesticide application. Guidelines provide action thresholds, when available, and indicate the control measures needed to promote low pesticide-input pest management. Wherever possible they give priority to non-chemical methods. Pesticides with the lowest risk to human health and the environment, including side effects on beneficial organisms, are selected among those available for each pest. In an advanced IPM programme, control actions should be taken after evaluation of an increase of pest status and the status of their biological control agents (Janssen and Sabelis, 1992). The availability of tools that predict both pest and natural enemy phenology enable implementation of the best pest management strategies with minimal detrimental effects on the agroecosystem equilibrium. Following those recommendations growers can select and implement those pest control measures that maximize the pest control and minimize the disruptive effect on natural enemies.

A reduction in harmful effects of pest control strategies on beneficial insects and the environment can be achieved by giving priority to the implementation of preventive tactics and non-chemical control measures. These approaches include using pest-resistant plants, semiochemical-based disruption tactics and cultural practices including mechanical or physical control. Insect-resistant rootstocks are already available to reduce the impact of various fruit pests such as *G. molesta*, *Cacopsylla bidens* (Šulc), *D. plantaginea* and *E. lanigerum* and additional, potential new genetic sources of resistance are being exploited in fruit tree breeding programmes (Shaltiel-Harpaz, et al., 2018; Bernardi et al., 2017; Fazio et al., 2013; Angeli and Forti, 1999).

Pheromone mating disruption is widely implemented around the world to control a number of pests and is recognized as one of the most selective control tactics (Witzgall et al., 2010; Gut and Brunner, 1998). Replacement of broad-spectrum insecticides with mating disruption for codling moth resulted in a decrease of pear psylla (*Cacopsylla pyricola*) and two-spotted spider mite (*Tetranychus urticae*) populations in Californian pear orchards (Varela and Elkins, 2008), most likely because of an increase in natural enemies.

Exclusion netting has also been used successfully against various fruit pests (Chouinard et al., 2017; Del Fava et al., 2017; Lloyd et al., 2005), but may also create favourable environmental conditions for other pests (e.g. aphids) and disease outbreaks depending on the timing of net installation and its shading effect on the canopy. The effect of exclusion netting on natural enemy abundance is not always clear-cut and depends on the taxa considered (Lefebvre et al., 2016; Marliac et al., 2015; Dib et al., 2010). Nets may also offer optimal conditions for natural enemy releases to control pests; for this purpose, the predators, *Episyrphus balteatus* De Geer and *F. auricularia*, were identified as the best candidates for controlling *E. lanigerum* with early spring augmentative release (Dib et al., 2010).

Adhesive bands are applied as physical barriers around tree trunks to exclude ants tending aphids enabling predation by natural enemies and fungal infection in aphid colonies. However, this prevents the generalist predator *F. auricularia* from climbing the trees (Nagy et al., 2013; Miñarro et al., 2010; Fontanari et al., 1993). Providing the black garden ant, *Lasius niger* (L.), with alternative sugar sources in early spring can reduce *D. plantaginea* infestations on apple trees and does not disturb the activity of *F. auricularia* (Nagy et al., 2013). Cultural practices like pruning, irrigation and fertilization can also be used to reduce aphid colonization in trees since these techniques contribute to the control of shoot development (Jordan et al., 2017).

When proposed ETs are reached and corrective pesticide application is necessary, careful choice of available active ingredients, pesticide dose, application timing and location of application is needed. Effective conservation of natural enemies relies on understanding the impact of pesticides and taking into consideration the ecological and physiological compatibility of the pesticide. Limited and/or selective placement of pesticide (e.g. baits, local application) or specific timing of an application can minimize natural enemy exposure to potentially harmful insecticides.

Since the first implementation of IPM in tree fruit orchards, the conservation biological control of phytophagous mites was of paramount importance due to their frequent outbreaks caused by the use of broad-spectrum insecticides (Solomon et al., 1993, 2000; Hoyt, 1969). Natural enemies of phytophagous mites such as predatory mites, particularly of the family Phytoseiidae, were used as initial indicators of the selectivity of pesticides. Evaluation methods have been standardized for the side effects of pesticides on beneficial insects and a specific database has been compiled in an IOBC document on the pesticide effects on at least some beneficial insects (Jansen, 2018).

To improve the overall understanding of the selectivity of pesticides on natural enemies it is necessary to perform bioassays on the various life stages and where possible, not only in the laboratory, but also under more realistic field conditions. Estimates of the lethal and sublethal effects are more realistic when tested directly on the crop. Traditionally, only lethal effects were estimated, but more recently sublethal effect data are integrated and can include those on development time, feeding activity and fecundity (Mills et al., 2016a; Fountain and Harris, 2015).

3 Case study 1: woolly apple aphid

The woolly apple aphid, *Eriosoma lanigerum* (Hausmann), is native to eastern North America and from there has spread to all fruit growing regions of the world through infested nursery stock. Commonly this aphid colonizes and attacks both the aerial and root system of the tree.

In Europe, this aphid has ten to twelve generations per year (Cross et al., 1999) and can overwinter on the aerial portions of the canopy (pruning wounds and cracks). Winter mortality is relatively high especially under severe winter conditions. More frequently woolly apple aphid overwinters on root suckers and in colonies on the roots, where it is protected from the harsh winter weather. Throughout the spring and early summer aphid crawlers disperse and infest the aerial portions of the tree, colonizing new shoots, where most of the woolly aphid population develops (Baldessari and Angeli, 2018a; Lordan et al., 2015; Heunis and Pringle, 2006). Apple trunks, large branches, new growth shoots and roots colonized by *E. lanigerum* form densely packed colonies covered with a white, waxy, filamentous secretion (Baldessari and Angeli, 2018b; Baldessari et al., 2015; Beers, 2014; Mueller et al., 1992). On old wood, trunks and large branches, most colonies can decline but do not disappear during the summer (Mueller et al., 1992; Hoyt and Madsen, 1960). High aphid populations can infest the fruit directly (stem, calyx) and produce wool and honeydew on fruits making them unmarketable (Beers et al., 2010).

On apple trees, the pest status of woolly apple aphid fluctuates between years depending on biotic and abiotic factors (Lordan et al., 2015). The increased susceptibility of some apple varieties and dwarfing rootstocks, changes in pesticide programmes, the disruption of biological control agents and some recently introduced agronomic techniques (e.g. single row exclusion netting) contribute to more frequent outbreaks (Gontijo et al., 2012; Beers et al., 2010; Angeli and Simoni, 2006).

Earwigs (Dermaptera), ladybird beetles (Coleoptera: Coccinellidae), hoverflies (Diptera: Syrphidae), lacewings (Chrysopoidea) and spiders (Araneae) are reported to be predators of woolly apple aphids (Lordan et al., 2015; Wyss et al., 1999). Research on the natural control of *E. lanigerum* by predators has generated a wealth of literature, but often with contradictory results (Blommers, 1994). One of the potential reasons for the discrepancies among observed results is the great complexity and geographic variability of the apple orchard ecosystem. Nevertheless, earwigs are frequently cited as the most important (Lordan et al., 2015; Gontijo et al., 2012; Nicholas et al., 2005; Asante, 1995; Mueller et al., 1988).

The control of woolly apple aphid by the solitary, host-specific parasitoid, *Aphelinus mali* (Haldeman) (Hymenoptera: Aphelinidae), is cited as one of the most successful examples of classical biological control of insect pests (Goossens et al., 2011; Mueller et al., 1992; Asante et al., 1991; Lundie, 1924) (Fig. 2). DeBach (1964) reported *A. mali* as a successful biological control agent in at least 40 countries. Therefore, this host-parasitoid system has been extensively studied worldwide (Mols and Boers, 2001; Bonnemaison, 1965; Dumbleton and Jeffreys, 1938).

For the most effective implementation of IPM principles, monitoring of both pest and its natural enemy abundance are equally important. Knowledge about the presence and seasonal phenology of *E. lanigerum* and of its two key

Figure 2 *Eriosoma lanigerum*, woolly apple aphid colony on apple shoot; the proportion of black, mummified aphids is used as a relative measure of parasitism. Courtesy of Mario Baldessari.

natural enemies, *A. mali* and earwigs, are essential to dictate rational control actions. Aphid movement from roots to the aerial parts of the tree and *vice versa* could be monitored by installing two 5-cm wide adhesive tapes around a sample of selected trunks (Fig. 3) (Baldessari and Angeli, 2018b; Lordan et al., 2015; Beers et al., 2010). Tapes should be replaced weekly and aphid numbers on each tape should be visually estimated by a qualitative index of categories. During the season, woolly apple aphid aerial colonies are monitored visually as percentage of infested shoots. The infestation level is rated using a qualitative scale where: 0 = no infestation; 1 = trace infestation, 2 = up to 10% of the tree branches with severe infestation; 3 = 11–25% of the tree branches with severe infestation; 4 = more than 25% of the tree branches with severe infestation (Nicholas et al., 2005).

Earwigs can be monitored using rolled strips of corrugated cardboard as artificial shelters (Fig. 4), which are attached to the trunk and checked every other week (Lordan et al., 2015).

Monitoring of the *A. mali* adult population during the season should be done using yellow sticky traps placed on the tree at a height of 1.5–2 m (Fig. 5) (Beers, 2012). Two traps per orchard should be hung before bloom and

Figure 3 Aphid migration monitored by adhesive tapes applied around apple trunk. Courtesy of Mario Baldessari.

Figure 4 Rolled strip of corrugated cardboard (refuge) hung on an apple tree and used as shelter for earwig. Courtesy of Tim Beliën.

changed weekly in spring and every 2 weeks during the summer. The number of *A. mali* adults is determined under a stereoscope on both sides of the trap.

Predictive models were developed for these two species based on temperature thresholds and accumulated degree-days temperatures. Developmental times of *A. mali* and *E. lanigerum* were studied at different temperature regimes and a good linear model fit between developmental rate and temperature was observed in the range 13–30°C (Asante and Danthanarayana, 1992). *E. lanigerum* becomes active from 267 degree-days

Figure 5 Yellow sticky traps used to monitor *A. mali* flight. Courtesy of Mario Baldessari.

above the temperature threshold of 5.2°C (Asante et al., 1991), which generally corresponds to crop stage green bud (BBCH 56), whereas for *A. mali*, the lowest developmental threshold is 8.3°C and it needs 255 degree-days to begin its activity (Asante and Danthanarayana, 1992). Consequently, developmental time of *A. mali* generally lags behind that of its aphid host throughout the year and a model integration of an early season predation activity of *E. lanigerum* is needed (Lordan et al., 2015).

The phenology and abundance of *F. auricularia* in relation to the temperature sum in day-degrees above a thermal threshold of 6°C accumulated from 1st of January was also studied by Helsen et al. (1998). Maximum numbers of earwigs were trapped in the trees when the first brood nymphs entered the fourth instar, generally between 600 and 750 day-degrees. A minimum of seven earwigs per tree was reported to control *E. lanigerum* in apple orchards from Northern Europe (Phillips, 1981). However, this likely varies with tree canopy density, crop husbandry and actual pest pressure.

According to Nicholas et al. (2005), earwigs in combination with *A. mali* reduced woolly apple aphid infestations below the action threshold set for commercial orchards. A beneficial interaction between *A. mali* and *F. auricularia* was recently confirmed by Quarrell et al. (2017) and the optimal numbers for each species (>14 earwigs/trunk trap/week within the first seven weeks post-blossom or >0.5 wasps/tree during the first generation of *A. mali*) were suggested as the thresholds to avoid severe outbreaks of *E. lanigerum*.

Whenever woolly apple aphid populations become too high, the pest is typically controlled by canopy sprays. No specific economic injury threshold has been developed because the nature of the damage is difficult to estimate (Baldessari et al., 2015; Beers, 2014). Infestations in the orchard tend to be chronic because of root colonies. Starting from flowering, crawlers migrate up the trunk towards the canopy where they establish colonies. Close to the flowering time, apple orchards should be inspected to determine the need for insecticide application by checking for the presence of aphid-infested flower clusters. Applications before flowering targeting rosy apple aphid (*D. plantaginea*) might also be effective on woolly apple aphid crawlers during their migration to the upper part of the tree. Historically, organophosphate insecticides were commonly applied in pre- and post-flowering and they were generally effective in controlling woolly apple aphid. In recent years, neonicotinoid insecticides and spirotetramat, alone or in combination with mineral oil, frequently replace organophosphates. Well-timed combined control of woolly apple aphid and rosy apple aphid has been reported when spirotetramat in combination with mineral oil was applied on unprotected crawlers at the phenological growth stages BBCH 71-73 (Schoevaerts et al., 2010). It is important that there is active growth of the tree for spirotetramat to be effective via take up through phloem feeding. Management programmes of primary pests are also reported as being potentially responsible for disruption of the biological control of *E. lanigerum* (Fountain and Harris, 2015; Angeli, 2015; Beers, 2014; Gontijo et al., 2012; Goossens et al., 2011; Baldessari et al., 2007).

Even though withdrawals of registration and restriction on the use of broad-spectrum insecticides have been implemented in many countries, a replacement by newer compounds does not always help to avoid outbreaks of *E. lanigerum*. In Washington State, for example, increased levels of *E. lanigerum* were consistently associated with the use of the new generation insecticides, spinetoram and novaluron, which showed a negative impact on earwigs and *A. mali* (Beers et al., 2016). The selectivity towards *A. mali* of insecticides commonly used in apple orchards has been widely investigated (Ateyyat, 2012; Rogers et al., 2011; Vogt and Ternes, 2006; Cohen et al., 1996). Most broad-spectrum insecticides (e.g. organophosphates, pyretroids, chlorinates) were highly toxic to *A. mali* adults and generalist predators and the numbers of parasitoids and predators increased significantly in orchards when organophosphates, pyrethroids and chlorinates were limited or banned. Imidacloprid and spinosad were more toxic to adult parasitoids than pirimicarb and triazamate. Neither cyhexatin (acaricide), indoxacarb, methoxyfenozide, thiacloprid, emamectin benzoate and chorantraniliprole nor the fungicides penconazole, cyprodinil and trifloxystrobin had a considerable toxic effect upon *A. mali*. Sulphur was found to be moderately toxic to *A. mali* (Rogers et al., 2011). Other studies reported diflubenzuron, fenoxycarb, lufenuron, *B. thuringiensis*, sulphur and oil as selective or slightly harmful, while the insect growth regulators, tebufenozide and pyripoxyfen, were classified as harmless. Similarly acetamiprid, spirotetramat and abamectin were classified as partially selective (Brück et al., 2009).

In Table 1 we report the toxicity rating of pesticide to *A. mali*, adapted from Croft and Whalon (1982). Tests were conducted on adult insects exposed to residues either on leaves or on filter paper.

When the application of a pesticide with harmful effects on natural enemies is required, selection of a less disruptive timing could also help to preserve beneficial insect species. All non-selective pesticides have to be avoided during the flight of the first-generation *A. mali* adults. Some potentially harmful compounds such as chlorpyrifos can still be selective if used during the endoparasitic phase of *A. mali*, between the first and summer generations of the parasitoid, although in the United States the use of this product as a foliar application during the season is banned. Some harmful pesticides sprayed on mummies do not affect the survival of the parasitoid inside and the life cycle of *A. mali* is not disrupted, allowing the next *A. mali* generation to develop (Goossens et al., 2011; Vogt and Ternes, 2006).

A decision support system combining a phenological degree-day model predicting earwig development and a database with known side effects of

Table 1 Relative toxicity of the most common insecticides to *Aphelinus mali*

Chemical	Toxicity	References
Acetamiprid	++	Gonjito (2011)
Chlorpyrifos-ethyl	+++++	Bradley et al. (1997)
Carbaryl	+++++	Bradley et al. (1997), Gontijo (2011)
Cyantraniliprole	+	Gontijo (2011)
Chlorantraniliprole	+	Gontijo (2011), Brugger (2010), Rogers et al. (2011)
Diazinon	++++	Bradley et al. (1997), Rogers et al. (2011)
Emamectin benzoate	+	Rogers et al. (2011)
Fenoxycarb	+	Bradley et al. (1997)
Flupyradifurone	++	Jeschke et al. (2015)
Imidacloprid	++++	Bradley et al. (1997), Cohen et al. (1996), Vogt and Ternes (2006), Ateyyat (2012)
Indoxacarb	+	Rogers et al. (2011)
Lambda-cyhalothrin	++	Gontijo (2011)
Lime sulphur	++/++++	Rogers et al. (2011), Vogt and Ternes (2006)
Lufenuron		Bradley et al. (1997)
Methoxyfenozide	+	Rogers et al. (2011)
Novalum	++	Gontijo (2011)
Pyrethrum/pip. but	+++++	Bradley et al. (1997)
Pirimicarb	+	Bradley et al. (1997), Ateyyat (2012)
Spinetoram	++++	Gontijo (2011), Rogers et al. (2015)
Spinosad	+++/++++	Gontijo (2011), Rogers et al. (2015)
Spirotetramat	+	Rogers et al. (2015)
Tebufenozide	+	Shaw and Walker (1996), Bradley et al. (1997)
Thiacloprid	+++/++++	Gontijo (2011), Rogers et al. (2015), Vogt and Ternes (2006)

Toxicity rating (+) adapted from Croft and Whalon (1982): + (<10%) no or little toxicity; ++ (10%–40%) low toxicity; +++ (50%–75%) moderate toxicity; ++++ (75%–99.99%) high toxicity; +++++ (100%) completely toxic.

pesticides has also been developed. The output gives the current status of earwig population and management recommendations to avoid disruptive effects of pesticides during a critical period for predator development (Beliën et al., 2012).

Some cultural practices could add a valuable component to the woolly apple aphid management programme. Biological control could be enhanced by integrating flowering plants with apple trees. *A. mali* and generalist predators are attracted to the nectar and pollen of flowers which constitutes an additional food source (Gontijo, 2011). Exploitation of genetic tolerance or resistance of plant material could be another approach for reducing pesticide pressure and improving biological control. Some *E. lanigerum*-resistant apple rootstock

(MM 106, MM 111 from East Malling Research Station, England) have been used in the past, but at present they are almost completely replaced by less vigorous M9 rootstock, which is not resistant to *E. lanigerum*. New rootstock selections combining reduced vigour and resistance to *E. lanigerum* are under development and are expected to be commercially available (Anon., 2015; Fazio et al., 2013).

4 Case study 2: pear psylla

In pear orchards *Psylla* spp. were considered of secondary importance worldwide until the 1960s, when frequent outbreaks began to occur in parallel with an increase in the use of pesticides and intensification of commercial fruit production (Forti et al., 1993; Riedl, 1991; Geoffrion, 1984). At present, *Psylla* spp. represent, together with codling moth, the two major pests of pear orchards (Agnello, 2014; Pasqualini et al., 1999).

Several species of psyllids colonize commercial pear orchards and although they can be present as mixed populations, *Cacopsylla pyri* (L.) is the most important species in Europe and *Cacopsylla* (=*Psylla*) *pyricola* (Foerster) in North and South Americas (Jenser et al., 2010). In the United Kingdom, psyllid populations were originally identified as *C. pyricola*, but subsequent identification by Nagy et al. (2010) revealed *C. pyri* as the most dominant psyllid species. These two species are almost equal in size and have very similar life cycles.

In southern Europe, *C. pyri* has four to six generations a year and adults are active in the pear orchard for 9–10 months (Briolini et al., 1989). In North America and Northern Europe, *C. pyricola* has two to six generations per year, and adults are active for 7–8 months (Scutareanu et al., 1999; Horton et al., 1994). Both species overwinter as adults in a state of reproductive diapause in the cracks of tree bark or in litter on the ground (Agnello, 2014).

Pear psyllids become active at temperatures between 4.4°C and 10°C and can be found on the trunk and branches even prior to the trees breaking dormancy. In the spring, mating occurs and the first eggs are laid on the spurs of developing buds between 10°C and 15.6°C, even before bud burst (Agnello, 2014; Trapman and Blommers, 1992; Faudrin, 1984).

C. pyri and *C. pyricola* damage pears in several ways: the honeydew excreted by the juvenile stages, falls onto leaves and fruits and may kill leaf tissue and russet the fruit. Honeydew serves as a growth medium for black sooty mould fungi, whose presence on fruit reduces market value. Excessive feeding and injection of toxic saliva by psylla weaken the tree and cause premature defoliation; this affects tree vigour and, if present for several years, it can cause injury called psylla shock (Agnello, 2014). In some areas, the adults are also known to be vectors of the mycoplasma-like organism causing pear decline (PD) syndrome, especially during spring growth when new phloem forms (Carraro et al., 1998; Davies et al., 1998).

In pear orchards, under IPM, pear psylla pest status fluctuates between years depending on biotic and abiotic factors. The high susceptibility of some pear varieties, changes in pesticide programmes, disruption of biological control, insecticide resistance and some recently introduced agronomic techniques (such as high-density plantings, high doses of fertilizers, intense irrigation and net covering) represent the key factors increasing the frequencies of outbreaks (Civolani, 2012; Riedl, 1991).

The combined effect of a complex of predatory species and few parasitoids is, theoretically, capable of effectively controlling pear psylla (Agnello, 2014). In Europe, 56 beneficial organisms have been associated with pear psylla, while 35 natural enemies have been documented in Oregon (Herard and Chen, 1985).

The generalist predators include various species including Anthocoridae (Hemiptera), recognized worldwide as the most effective example of classical biological control of pear psylla (Souliotis and Moschos, 2008; Solomon et al., 1989; Herard and Chen, 1985). This group is characterized by a high searching efficiency, strong aggregative behaviour and rapid population response to changes in prey density (Shaltiel and Coll, 2004).

To achieve successful biological control of psylla it is important that interactions between the pest and natural enemies begin early in the season and continue throughout the season. Successful biological control of pear psylla is enhanced if beneficial organisms reproduce and overwinter in or nearby the orchard (Riedl, 1991). The seasonal phenology of anthocorids is well synchronized with pear psylla and the positive outcome of prey-predator interaction is due to their searching ability, consumption potential and rate of reproduction (Scutareanu et al., 1999).

Knowledge of pest presence, its seasonal phenology and abundance of key natural enemies are essential for taking rational pest management actions. The main technique for monitoring psyllid and anthocorid are the beating method (see above) and coloured sticky traps. The beating tray method requires, at each sampling date, 100 branches (1–2 per tree) to be sampled. A branch is tapped with a stick and the dislodged insects fall into a funnel-like structure (umbrella) or on a flat sheet (tray). This approach allows monitoring of population dynamics of both pear psylla adults and mobile stages of anthocorids (Figs. 6 and 7). It also permits the monitoring of mobile stages of secondary predators like pirate bugs (*Orius* spp.), Coccinellidae, spiders, Chrysopidae and earwigs (*Forficula* spp.) (Stäubli et al., 1992).

When PD syndrome symptoms are present, specific treatment thresholds are suggested against overwintering adults, based on beating tray counts. In

Figure 6 Larva of *Anthocoris nemoralis*. Courtesy of Massimiliano Melandri.

the United States, a threshold of 100–400 of *C. pyricola* adults per 100 beats at leaf fall is proposed (Roger and Lorraine, 1989), while in Switzerland, at the same timing, a threshold of 50–100 adults of *C. pyri* per 100 beats is suggested (Stäubli and Antonin, 1984).

Figure 7 Adult of *Anthocoris nemoralis*. Courtesy of Massimiliano Melandri.

Alternatively five to ten yellow panel traps (about 20 × 15 cm) can be hung at 2 m height in the canopy and checked weekly. To decide the timing for treatment, the sticky traps provide less precise information on the real population density than the beating tray and they are also not as effective for monitoring anthocorids and other beneficial organisms (Civolani and Pasqualini, 2003). Neither of these methods provides a good estimation of the abundance of psyllid nymphs causing the most harm to the tree and fruit.

Hence, a more precise estimation of pear psylla and natural enemy population density is possible by visual inspection of shoots for eggs and/or young stage of psylla present on leaf petioles or leaf veins (both psylla and anthocorid) using a hand lens or dissecting microscope.

Corrugated cardboard paper wrapped around the base of tree trunks are also used to observe overwintering predators and can give an early indication of the potential biological control of pear psylla populations (Jenser et al., 2010).

A phenological model for *C. pyri* was developed and validated with field observations, both in Italy and Switzerland (Schaub et al., 2005). The model provided a good simulation of the insect phenology from the beginning of the year to the second generation and can be used as a forecasting tool to improve timing of monitoring and control measures (Samietz et al., 2015).

In Europe, chemical control of the overwintering *C. pyri* generation is generally not recommended in IPM programmes (Civolani and Pasqualini, 2003), while early juvenile stages of spring (second generation) are the primary target for control strategies (Pasqualini et al., 1997).

In modern IPM programmes, the organophosphate insecticides, widely used in the past to control mobile stages, were replaced, firstly, by the products from the group of chitin inhibitors (e.g. diflubenzuron and teflubenzuron) applied at egg stage or newborn larvae (Stäubli et al., 1992) and more recently, by products such as abamectin, spirodiclofen and spirotetramat.

Abamectin is primarily used to target young nymphs; optimal timing for its application is when yellow eggs are present and the egg hatching peak has not yet been achieved. Moreover, a significant improvement in its efficacy is obtained by the addition of mineral oil (Civolani, 2012).

Abamectin can be used in rotation with the tetronic acid derivate, spirodiclofen, as part of the pesticide resistance management because of the different modes of action. Similar to abamectin, the optimal timing for spirodiclofen is the yellow-coloured eggs stage, and the product activity is improved when mixed with mineral oil. If high and prolonged infestations occur, a spirodiclofen application followed by treatment with mineral oil may be a good alternative (Civolani, 2012).

Spirotetramat, a lipid biosynthesis inhibitor similar to the tetronic acid derivate spirodiclofen, is considered to be safe to most beneficial insects

(Brück et al., 2009) and may represent a valuable alternative to abamectin, when included in an integrated resistance management (Nin et al., 2012). Spirotetramat is particularly effective against juvenile stages of sucking pests, including psyllids (Civolani et al., 2015; Jaworska et al., 2012; Nauen et al., 2008) and because of its translocation properties within the plant after foliar uptake, the insecticidal activity is translocated within the entire vascular system allowing the protection of new shoots or leaves appearing after foliar application (Nauen et al., 2008). It should be noted that spirotetramat and spirodiclofen are from the same class of active ingredients and hence there are restrictions on use within the same cropping season in order to prevent insecticide resistance.

Sprayable kaolin particle film can be applied to deter the oviposition of winter-form pear psylla and is useful both in IPM and organic pear production (Saour et al., 2010; Erler and Cetin, 2007). Applications are recommended for use early in the season before the peak foliage as the film may inhibit photosynthesis in areas with low light levels.

Whatever insecticide strategy is applied to control pear psylla, control can only be achieved with the integration of balanced tree growth and simultaneous preservation of natural antagonists of pear psylla. For this purpose, the key factor for biological control of psyllids is to protect day active anthocorids and night active earwigs throughout the season.

To promote the biological control of pear psylla, all non-selective pesticides should be avoided. During the last two decades new pesticides have been developed which pose, generally, lower toxicity towards beneficial species in comparison and are more suitable for the integrated pest control programmes against psyllids. Several studies have been performed to evaluate the selectivity of the pesticides towards anthocorids and other natural enemies of pear psylla (Fountain and Harris, 2015; Angeli et al., 2005; Stäubli and Pasquier, 1988). The investigated active ingredients include those directly employed against psyllids and also those generally applied against other key pests present in pear orchards (Table 2).

Conservation biocontrol also plays a role in IPM programmes and includes habitat manipulation creating a favourable environment for natural enemies by providing them with alternative food sources and shelter. Hedgerows along pear borders provide food and refugia where natural enemies can develop and disperse to nearby orchards (Shaltiel and Coll, 2004). Hedgerows/windbreaks rich in goat and grey willow (*Salix caprea* and *S. cinerea*) and hazel (*Corylus avellana*) with an abundant understorey of nettle (*Urtica dioca*) are a good source of anthocorids from the early spring (*Anthocoris nemoralis*) through the season (*A. nemorum*) and are an example of the surrounding landscape supporting the conservation biocontrol of pear psyllids (Cross et al., 2010).

Table 2 Relative toxicity of common orchard insecticides to *Anthocoris* spp. (J-young; A-adult)

Chemical	Toxicity Juvenile	Toxicity Adult	References
Abamectin	++++	++++	Schaub et al. (1999, 2002), Biobest (2018)
Acequinocyl	++	++	Biobest (2018)
Acetamiprid	+++	++	Biobest (2018)
Acrinathrin	++++	++++	Biobest (2018), Civolani and Pasqualini (1999)
Amitraz	++++	+++	Schaub et al. (1999, 2002)
Azadiracthin	+	+	Biobest (2018)
Buprofezin	+	+	Biobest (2018)
Chlorpyrifos-ethyl	++++	+++	Pasqualini et al. (1999), Biobest (2018)
Deltamethrin			
Diflubenzuron	++	+	Stäubli and Pasquier (1988)
Dimethoate	++++	++++	Bengochea et al. (2010), Biobest (2018)
Emamectin benzoate	++	++	Fusarini et al. (2010)
Fenoxycarb	++/+++	+	Solomon and Fitzgerald (2015), Biobest (2018)
Flufenoxuron	++++	+++	Caroli and Pasqualini (2004), Girolami et al. (2001)
Kaolin	++	++	Bengochea et al. (2010)
Imidacloprid	++	+	Biobest (2018)
Indoxacarb	+	+	Civolani et al. (2007), Biobest (2018)
Insecticide soaps	+	+	Pasqualini et al. (1999)
Methoxyfenozide	++	+	Pasqualini and Civolani (2007);
Mineral oil	+	+	Pasqualini et al. (1999)
Pirimicarb	+	+	Biobest (2018)
Rynaxypyr	+	+	Marchesini et al. (2008)
Spinosad	++	++	Civolani et al. (2007), Biobest (2018)
Spirodiclofen	++	+	Pasqualini and Civolani (2007), Biobest (2018)
Spirotetramat	++	+	Biobest (2018), Brück et al. (2009)
Tebufenozide	++	+	Caroli and Pasqualini (2004)
Thiacloprid	+++	+++	Biobest (2018)
Thiamethoxam	++++	++++	Biobest (2018)

Toxicity rating (+) adapted from IOBC (Hassan et al., 1985; Sterk et al., 1999). + Harmless <25% mortality; ++ Slightly harmful 26%–50% mortality; +++ Moderately harmful 51%–75% mortality; ++++ Harmful >75% mortality.

5 Case study 3: European red spider mite

Outbreaks of spider mites (Tetranychidae), in particular of the European red spider mite, *Panonychus ulmi*, and the two-spotted spider mite, *Tetranychus urticae*, after the Second World War attracted the interest of researchers. Early studies observed that a number of factors caused population increases including the use of insecticides (e.g. chlorinates and organophosphates) to control codling moth and fungicides (e.g. dithiocarbamates) to control apple scab (van de Vrie et al., 1972). Cultural practices, in particular fertilization, were also considered to promote spider mite outbreaks. The use of acaricides became popular but resistance soon emerged and control of spider mites became problematic (Cranham and Helle, 1985).

Phytophagous mite management followed three main lines within an IPM framework: the evaluation of mite pest damage, the conservation/augmentation of the most promising biocontrol agents and the evaluation of pesticide side effects on the latter. Nevertheless, a number of studies were devoted to the crucial aspects of the biology and the ecology of the most important phytophagous mites. In this case study we focus on the European red spider mite *P. ulmi*.

P. ulmi has a worldwide distribution and is recorded on 147 host plants (Migeon and Dorkeld, 2006–17). Overwintered eggs are laid at the base of buds, spurs and in bark crevices. Eggs hatch in the spring and the first generation completes development in 20–40 days, depending on temperature. During the growing season three to four generations can be completed in Northern Europe and six to eight generations in southern Europe. Population growth is positively affected by relatively high temperatures, thus mite populations peak from early summer onwards.

Demographic parameters can be affected by the fruit variety as shown comparing 'Fuji' with 'Golden Delicious' (Yin et al., 2013). *P. ulmi* feeds on the spongy mesophyll and palisade cells causing leaf bronzing, followed by defoliation and yield reduction; when serious infestations occur the number of flowers can be reduced in the subsequent season (e.g. van de Vrie, 1985). Serious damage on pear (leaf burn and defoliation) can occur on cv. 'William', 'Conference' and 'Kaiser' when the temperature exceeds 30°C and relative humidity decreases to 30%–40% (Barbieri et al., 1993).

Damage caused to apple production by *P. ulmi* has been widely investigated and includes reduction in leaf net CO_2 exchange rates (NCER), reduction in fruit size and number, alteration in fruit quality colour and sugar content, and poor return bloom (Lakso et al., 1996). Studies aimed at investigating the impact of mites on fruit crops led to different results because there are a number of environmental and cultural factors influencing mite–plants interactions (van de Vrie, 1985; Hoyt et al., 1979; Hoyt, 1969). This variability influenced the definition of ETs. Monitoring is essential in spider mite management and, in the case of *P. ulmi,* sampling should involve a relatively high number of plant organs.

The prediction of the infestations has also been investigated. The relationship between *P. ulmi* winter eggs and cumulate mite densities in early to mid-summer was studied in Quebec, Canada (Bostanian et al., 2007). Stepwise regression was used to compute mite-days over a 4-year period using winter egg density, cumulative rainfall and degree-days (their squared values and products) as predictors. The best fit included terms for winter eggs, rainfall and the product of rainfall per degree-days. Moreover, the number of winter eggs, adjusted for current year rainfall and degree-days, were significant predictors of mite-days in the summer.

More recently, phenology models (based on time-varying distributed delay approach) were developed considering data on radiation and temperature collected an hourly basis. They were incorporated in a forecasting tool aimed at optimizing surveillance and control of mite and insect pests (Samietz et al., 2015).

Optimizing chemical control of mites needs to take into consideration the hazard of the development of resistance due to their reproductive potential and their short life cycle. Early acaricides such as dicofol, tetradifon and propargite were replaced successfully by cyhexatin in the 1980s. Clofentezine and hexythiazox became popular in the 1990s, but resistance developed after their intensive use (Pree et al., 2005; Nauen et al., 2001). The next in line were the mitochondrial complex electron transport inhibitors; these acaricides were effective when compared to previous compounds but their efficacy also did not persist (Auger et al., 2003; Ioriatti et al., 2000). Acaricide resistance monitoring is an essential component of resistance management strategies and in this context mite baseline susceptibility studies are crucial. Resistance to spirodiclofen, a relatively new acaricide, has been monitored since its early use, and a shifting in susceptibility of *P. ulmi* has been reported (Kramer and Nauen, 2011).

Mechanisms of resistance in *P. ulmi* were not investigated in depth for each group of acaricides. Strand-specific RNA-seq-based transcriptome resource for *P. ulmi* strains showing susceptible and resistant strains allowed the detection of detoxification enzymes, target sites of acaricides and horizontally transferred genes implicated in pesticide resistance (Bajda et al., 2015).

Failure of chemical control strategies against spider mites has promoted *ad hoc* studies on alternative mite management methods. In fact, steps in development of an effective mite control strategy include the utilization of an ET, which is a function of the number of cumulative mite-days and the relative densities of the populations of the predator and the prey (Avilla et al., 1993).

Effective biocontrol agents were identified and factors affecting their performance studied in detail. In Europe the most effective predatory insects in keeping mite populations under acceptable levels belong to the Hemiptera (e.g. *Orius vicinus* Ribaut), Neuroptera (e.g. *Chrysoperla carnea* Stephens) and Coleoptera (e.g. *Stetorus punctillum* Weisse and *Oligota flavicornis* Boisduval and Lacordaire) (Fauvel, 1971; Collyer, 1953; Solomon, 1975; Pasqualini and Malavolta, 1986). This complex comprises other species belonging to the same or other families in North America (McMurtry et al., 1970). Macropredators are voracious feeders and sometimes exhibit a high fecundity. However, their developmental times are longer than those of their mite prey. They are attracted by dense populations of spider mites and can contribute significantly to the reduction of mite numbers and related plant damage. However, they do not persist when prey densities decline precluding any preventive action towards increases in new mite population.

The most common predatory mites occurring in apple orchards belong to the family Phytoseiidae. Generalist phytoseiid mites dominate European orchards, while specialist phytoseiid mites are also occurring in North America (McMurtry and Croft, 1997). The life history traits and life table parameters of generalist predatory mites have been studied in depth to delineate their role under realistic conditions. Among the most common phytoseiid species in European orchards are *Typhlodromus pyri* Scheuten, *Amblyseius andersoni* (Chant) and *Kampimodromus aberrans* (Oudemans).

T. pyri has been considered, for a long time, the most important predatory mite in European apple orchards (Collyer, 1964; Solomon, 1975). The ability of *T. pyri* to persist for long periods in orchards can be explained by its supplemental feeding on pollen, fungi and other non-pest foods (Pozzebon et al., 2009; Pozzebon and Duso, 2008; Addison et al., 2000; Solomon et al., 2000; Zemek and Prenerivà, 1997). Additional factors explaining its occurrence in fruit orchards is the competitiveness towards other predatory mite species (Schausberger, 1997, 1998, 1999a,b; Croft and McRae, 1992a,b) and pesticide resistance (Bonafos et al., 2008; Solomon et al., 1993; Duso et al., 1992).

The predatory mite, *A. andersoni,* has been reported as a dominant species in northern Italy and other areas. Resistance to insecticides and fungicides and interspecific predation probably favoured this situation (Angeli and Ioriatti, 1994; Croft, 1994; Duso et al., 1992; Ioriatti et al., 1992; Baillod et al., 1985; Ivancich Gambaro, 1975). Studies on the biology of *A. andersoni* focused on its voracity, but this factor can affect its persistence in field conditions (Lorenzon et al., 2012).

K. aberrans has been reported as more common in neglected orchards rather than in commercial orchards probably due to its susceptibility to pesticides. This situation changed in the 1990s when *K. aberrans* resistance to insecticides was first reported in vineyards (Duso et al., 2009). Demographic studies reveal a preference of *K. aberrans* for pollen and eriophyoid mites compared to spider mites. However, its persistence on alternative foods (mainly pollen) in conditions of prey scarcity can play a role in preventing outbreaks of phytophagous mites (Lorenzon et al., 2012). Moreover, populations can recover easily from pesticide applications when pollen grain density is abundant on leaves (Pozzebon et al., 2014).

The compatibility of pesticides with the survival and ability to colonize orchards by beneficial arthropods are the key aspects of IPM strategies. Recent studies have demonstrated the importance of evaluating the direct and indirect toxicity, as well as the lethal and sublethal effects of pesticides to enhance the conservation of biocontrol agents (Duso et al., 2014; Pozzebon et al., 2014). The release of pesticide-resistant strains of predatory mites has also been evaluated. The most recent trials involving *K. aberrans* suggested this species is an excellent candidate for the control of both spider mites and rust mites in light of climate change (Ahmad et al., 2013; Duso et al., 2009).

6 Future trends and conclusion

Recent technological advances provide new opportunities to optimize the use of insecticides in integrated management of fruit insect pests. The proper implementation of the ETs represents a vital component of IPM programmes in order to avoid unnecessary insecticide applications. Since most of the ETs for key fruit pests may be based on outdated or even unsubstantiated research, then more robust and scientifically based ETs for orchard pests need to be established to be considered reliable and practical by farmers and advisers (Ramsden et al., 2017).

Research is expected to advance new technologies, for example remote sensing of plant injury, for deepening knowledge on temporal and spatial dynamics of pest–natural enemy relationships, to identify less labour-intense field scouting and monitoring activities, to improve forecasting tools and decision support systems. Increased adoption of innovative application technologies and selection and release of more pest-resistant fruit tree cultivars is needed.

Small, sensitive and spectral digital cameras are now available for real-time monitoring of behaviours and interactions between pest and natural enemies to improve our understanding of round-the-clock relationship between pests and their antagonists (Merfield et al., 2004). Learning to recognize and identify naturally occurring natural enemies in a managed system is essential for taking fuller advantage of biological control agents present in the orchard system (Cuthbertson and Murchie, 2014). Molecular analyses including specific primer sets, restriction enzymes and DNA barcoding provide a useful and reliable tool for improved insect identification (Daane et al., 2018). Predator gut content analysis performed with serological methods and new molecular approaches provides useful information about diet diversity and consequently helps to explain the relative contribution of different beneficial species to pest control (Unruh et al., 2008; Symondson, 2002).

Improvement of IPM incorporating rational and judicious pesticide application could be enhanced with the implementation of electronic traps for sequential remote monitoring of fruit tree pests that are expected to discriminate several pests and to provide continuous monitoring of population fluctuations (Khot et al., 2018; Goldshtein et al., 2017; Krawczyk et al., 2015; Guarnieri et al., 2011).

Geo-statistical methods applied to estimate pest population spatial and temporal dynamics at the orchard level and area-wide scale could provide useful insight on landscape drivers of the spatial patterns of pests and their natural enemies (Sciaretta et al., 2018; Garcia et al., 2017). Furthermore, precise mapping of pest distribution could be achieved with geo-referenced crop scouting by combining automatic monitoring devices with geographic information systems or global positioning systems.

Development of demographic models in which the phenological model of the pest is integrated with the pesticide lethal and sublethal effects on the actual population structure can be used to simulate the impact of control actions applied at key times during the season (Jones et al., 2009; Wiman et al., 2016) and to compare the consequences of alternative pesticide applications. This demographic model can be improved by including the side effects of the pesticide on beneficial organism population structure in order to assess an overall impact of the control measure on the pest population growth (Stavrinides and Mills, 2009).

Since the first identification of an insect pheromone, hundreds of others have been discovered and used in monitoring and direct control of number of fruit pests. The new frontier of the research on chemical ecology is now the identification of countless molecules involved in insect olfaction in the natural environment. Host plant volatiles, in addition to sex and aggregation pheromones, shape the reproductive behaviour of herbivorous insects and have application not only for monitoring, but also for disrupting the normal behaviours. Synthetic HIPVs could be exploited for enhancing populations of beneficial insects within fruit orchards to increase biological control of pests and, potentially, make the pesticide application unnecessary (Witzgall et al., 2010; Franco et al., 2008; Powell and Pickett, 2003).

Sustainable fruit production can also benefit from the innovation of precision application technologies that improve the efficacy of the pesticide application by optimizing canopy coverage and reducing spray drift (Cross et al., 2013; Zijlstra et al., 2011). Variable spray rate technology utilizes information from real-time sensor data to guide spray treatment by selecting the target area and adjusting the amount of applied pesticide, taking into account the variability of the crop growth and the pest presence in the orchard (Khot et al., 2018). Fixed spraying system and surgical spraying are other innovative application technologies under development that, if applied on modern vertical tree architecture, could improve timely treatment and reduce chemical use (Khot et al., 2018; Bondesan et al., 2016; Dorigoni, 2016).

Host plant resistance represents another key component of advanced IPM programmes. Commercial fruit varieties resistant to various pathogens are quite common in fruit trees, while there are very few examples of commercial material recognized as resistant to arthropod pests. Investigations on insect susceptibility have been reported for different fruit tree pests (Fotiric Aksic et al., 2015; Mody et al., 2015; Hogmire and Miller, 2005). Mechanisms involved in resistance to woolly apple aphid have been studied (Costa et al., 2014) and genes conferring resistance have been identified including the molecular markers for such genes. The availability of molecular markers for this as for other genes responsible for arthropods resistance will accelerate breeding selection of durable resistance (Bus et al., 2008; Sandanayaka et al., 2003, 2005). New breeding programmes for arthropod resistance, affecting plant pest defensive traits (physical and chemical), should be considered including their positive or negative effects on biological control (Peterson et al., 2016). For example, physically mediated traits as well as production of HIPV are mechanisms that involve the interactions between plants, herbivores and their natural enemies. Hence, improvement or enhancement of plant traits that could increase indirect defences by promoting the activity of natural enemies, should be included in future plant breeding programmes.

7 Where to look for further information

References listed at the end of the chapter allow the readers to follow up some topics by accessing the appropriate source of information. In order to get continuously updated information on the topic, we suggest accessing the website of the IOBC (https://www.iobc-wprs.org/), which promotes environmentally safe methods of pest and disease control in plant protection since its establishment in 1955.

The organization is divided into six world regional sections. The West Palearctic Regional Section (WPRS) is particularly active in fostering research on new control methods within an IPM context. Integration and optimization of pesticides in IPM strategies in orchards is the focus of the fruit crops working group.

To enhance biological control, and reduce impact of pesticides on non-target organisms, a dedicated working group operating within the IOBC-WPRS develops standardized testing methods for effects of pesticides on beneficial species and promotes the use of the most selective products.

IOBC-WPRS organizes meetings, symposia and conferences, which constitute valuable opportunities for acquiring updated results, especially of biological control methods, but also of all pest control methods, including chemicals. The events stimulate contacts and cooperation between scientists, regulatory authorities and industry, and they are an attractive forum for exchange of information on damage assessment, pest and disease modelling, practical implementation of biological and integrated controls, but also for inspiring policy makers involved in IPM. In this context, IOBC-WPRS produces and publishes technical guidelines for integrated fruit production, which are regularly updated in relation to the new knowledge and novel tools available from the research in IPM. Proceedings of the IOBC meetings and results of the specialized workshops are published in the IOBC-WPRS bulletins.

8 References

Addison, J. A., Hardman, J. M. and Walde, S. J. (2000), 'Pollen availability for predaceous mites on apple: Spatial and temporal heterogeneity', *Exp. Appl. Acarol.*, 24, 1-18.

Agnello, M. A. (2014), 'Pear psylla', *in* T. B. Sutton, H. S. Aldwinckle, M. A. Agnello, and J. F. Walgenbach (Eds), *Compendium of Apple and Pear Diseases and Pest*, pp. 175-7. APS Press.

Ahmad, S., Pozzebon, A. and Duso, C. (2013), 'Augmentative releases of the predatory mite *Kampimodromus aberrans* in organic and conventional apple orchards', *Crop Prot.*, 52, 47-56.

Alford, D. V., Carden, P. W., Dennis, E. B., Gould, H. J. and Vernon, J. D. R. (1979), 'Monitoring codling and tortrix moths in United Kingdom apple orchards using pheromone traps', *Ann. Appl. Biol.*, 91(2), 165-78.

Altieri, M. A. and Nicholls, C. I. (2005), *Agroecology and the Search for a Truly Sustainable Agriculture*, United Nations Environmental Programme, Environmental Training Network for Latin America and the Caribbean.

Amaro, P. and Baggiolini, M. (1982), *Introdução à protecção integrada*, FAO/DGPPA, Lisboa.

Angeli, G. (2015), 'Nuove prospettive di difesa dagli afidi del melo', *Inf. Agrar.*, 13, 50-1.

Angeli, G. and Forti, D. (1999), 'Valutazione di alcuni parametri biologici e demografici di *Dysaphis plantaginea* Passerini (Homoptera: Aphididae) allevato su sei differenti varietà di melo', *Boll. Ent. 'Guido Grandi' Univ. Bologna*, 53, 121-31.

Angeli, G. and Ioriatti, C. (1994), 'Susceptibility of two strains of *Amblyseius andersoni* Chant. (Acari: Phytoseiidae) to dithiocarbamate fungicides', *Exp. Appl. Acarol.*, 18, 669-79.

Angeli, G. and Simoni, S. (2006), 'Apple cultivars acceptance by *Dysaphis plantaginea* Passerini (Homoptera: Aphididae)', *J. Pest Sci.*, 79, 175-9.

Angeli, G., Baldessari, M., Maines, R. and Duso, C. (2005), 'Side-effects of pesticides on the predatory bug *Orius laevigatus* Fieber (Heteroptera: Anthocoridae) in the laboratory', *Biocontrol Sci. Technol.*, 15(7), 745-54.

Anon. (2015), '*East Malling Rootstock Club Annual Report 2014-15*', http://www.emr.ac.uk/wp-content/uploads/2015/10/EMRC_Annual-Report-14-15_INN.pdf (accessed on 25 October 2018).

Argov, Y. and Gazit, Y. (2008), 'Biological control of the Mediterranean fruit fly in Israel: Introduction and establishment of natural enemies', *Biol. Control*, 46(3), 502-7.

Asante, S. K. (1995), 'Functional responses of the European earwig and two species of coccinellids to densities of *Eriosoma lanigerum* Hausmann (Hemiptera: Aphididae)', *J. Aust. Ent. Soc.*, 34, 105–9.

Asante, S. K. and Danthanarayana, W. (1992), 'Development of *Aphelinus mali* an endoparasitoid of woolly apple aphid, *Eriosoma lanigerum* at different temperatures', *Entomol. Exp. Appl.*, 65(1), 31–7.

Asante, S. K., Danthanaryana, W. and Heatwole, H. (1991), 'Bionomics and population growth statistics of apterous virginoparae of woolly apple aphid, *Eriosoma lanigerum*, at constant temperatures', *Entomol. Exp. Appl.*, 60, 261–70.

Ateyyat, M. (2012), 'Selectivity of four insecticides to woolly apple aphid, *Eriosoma lanigerum* (Haus.) and its sole parasitoid, *Aphelinus mali* (Hald.)', *World Appl. Sci. J.*, 16(8), 1060–4.

Auger, P., Bonafos, R., Guichou, S. and Kreiter, S. (2003), 'Resistance to fenazaquin and tebufenpyrad in *Panonychus ulmi* Koch (Acari: Tetranychidae) populations from South of France apple orchards', *Crop Prot.*, 22, 1039–44.

Avilla, J., Bosch, D., Sarasua, M. J. and Costa-Comelles, J. (1993), 'Biological control of *Panonychus ulmi* in apple orchards in Lleida (NE of Spain)', *Acta Hortic.*, 347, 267–72.

Ayasse, M., Paxton, R. J. and Tengo, J. (2001), 'Mating behavior and chemical communication in the order Hymenoptera', *Annu. Rev. Entomol.*, 46, 31–78.

Baggiolini, M., Keller, E., Milaire, H. G. and Steiner, H. (1984), *Guida per i rilievi periodici dei nemici del melo*, IOBC-WPRS, Verona.

Baillod, M., Guignard, E. and Kreiter, S. (1985), 'Toxicité de quelques insecticides pour une souche de *Typhlodromus pyri* Scheuten (Fully, Valais, Suisse) et une souche de *Amblyseius andersoni* Chant (Cugnasco, Tessin, Suisse) determination de leur résistance en laboratoire (Acari, Phytoseiidae)', *Int. Symp.: Influenza degli antiparassitari sulla fauna utile in frutticoltura*, pp. 25–30.

Baillod, M., Antonin, P., Guignard, E. and Jermini, M. (1989), 'Vers une generalisation de la lutte biologique contre les acariens phytophages en vergers de pommiers', *Rev. Suisse Viticul. Arboric. Hortic.*, 21(5), 279–84.

Bajda, S., Dermauw, W., Greenhalgh, R. and Nauen, R. (2015), 'Transcriptome profiling of a spirodiclofen susceptible and resistant strain of the European red mite *Panonychus ulmi* using strand-specific RNA-seq', *BMC Genom.*, 16, 974.

Baldessari, M. and Angeli, G. (2018a), 'Efficacia di spirotetramat verso gli afidi del melo', *Inf. Agrar.*, 13, 33–8.

Baldessari, M. and Angeli, G. (2018b), 'Efficacy of spirotetramat (Movento®) on apple aphids', *Atti Gior. Fitopatolog.*, 1, 141–50.

Baldessari, M., Rizzi, C. and Angeli, G. (2007), 'La difesa dell'afide lanigero sul melo in Trentino', *Inf. Agrar.*, 63, 20–2.

Baldessari, M., Angeli, G. and Tomasi, C. (2015), 'Gestione integrata degli afidi cenerognolo e lanigero del melo', *Inf. Agrar.*, 13, 48–52.

Barbieri, R., Scudellari, D., Zinoni, F., Marangoni, B., Tassinari, V. and Losi, G. (1993), 'Influenza dell'irrigazione e dei trattamenti acaricidi sul "brusone" e sulla produzione del pero "Conference"', *Rivista di Frutticoltura*, 4, 79–84.

Beers, E. H. (2012), 'Effect of trap color and orientation on the capture of *Aphelinus mali* (Hymenoptera: Aphelinidae), a parasitoid of woolly apple aphid (Hemiptera: Aphididae)', *J. Econ. Entomol.*, 105(4), 1342–9.

Beers, E. H. (2014), 'Woolly apple aphid', *in* T. B. Sutton, H. S. Aldwinckle, A. M. Agnello, and J. F. Walgenbach (Eds), *Compendium of Apple and Pear Diseases and Pests*, 2nd Edition, pp. 174–5. APS Press.

Beers, E. H., Cockfield, S. D. and Gontijo, L. M. (2010), 'Seasonal phenology of woolly apple aphid (Hemiptera: Aphididae) in Central Washington', *Environ. Entomol.*, 39(2), 286–94.

Beers, E. H., Horton, D. R. and Miliczky, E. (2016), 'Pesticides used against *Cydia pomonella* disrupt biological control of secondary pests of apple', *Biol. Control*, 102, 35–43.

Beliën, T., Moerkens, R., Leirs, H. and Peusens, G. (2012), 'Earwig management tool: An IPM decision aid system for augmentation of European earwig populations (Dermaptera: Forficulidae) in pip fruit orchards', *Commun. Agric. Appl. Biol. Sci.*, 77(4), 657–62.

Bengochea, P., Hernando, S., Saelices, R., Adán, Á., Budia, F., González-Núñez, M., et al. (2010), 'Side effects of kaolin on natural enemies found on olive crops', *IOBC-WPRS Bull.*, 55, 61–7.

Bernardi, D., Lazzari, J. C., Andreazza, F., Mayer, N. A., Botton, M. and Nava, D. E. (2017), 'Susceptibility, oviposition preference, and biology of *Grapholita molesta* (Lepidoptera: Tortricidae) in *Prunus* spp. rootstock genotypes', *Environ. Entomol.*, 46(4), 871–7.

Biobest, (2018), *'Side Effect Manual'*, https://www.biobestgroup.com/en/side-effect-manual (accessed on 30 July 2018).

Binns, M. R., Nyrop, J. P. and van der Werf, W. (2000), *Sampling and Monitoring in Crop Protection: The Theoretical Basis for Developing Practical Decision Guides*, CABI Publishing, Wallingford, UK.

Blommers, L. H. M. (1994), 'Integrated pest management in European apple orchards', *Annu. Rev. Entomol.*, 39, 213–41.

Boller, M. (1984), 'Eine einfache Ausschwemm-methode zür schnellen Erfassung von Raubmilben, Thrips, und anderen Kleinarthropodes im Weinbau', *Schweiz. Zeitschrift für Obst und Weinbau*, 120, 16–17.

Boller, E. F., van Lenteren, J. C. and Delucchi, V. (2006), *International Organization for Biological Control of Noxious Animals and Plants: History of the First 50 Years (1956–2006)*, IOBC, Zurich, Switzerland.

Boller, E. F., Minks, A. K., Cross, J. V., van Lenteren, J. C. and Wildbolz, T. (2010), 'The working group "Integrated Protection of Fruit Crops" is celebrating its 50th anniversary', *IOBC-WPRS Bull.*, 54, 193–217.

Bonafos, R., Vigues, V., Serrano, E. and Auger, P. (2008), 'Resistance monitoring to deltamethrin and chlorpyriphos-ethyl in 13 populations of *Typhlodromus pyri* Scheuten (Acari: Phytoseiidae) from vineyards in the southwest of France', *Crop Prot.*, 27, 855–8.

Bondesan, D., Rizzi, C., Dorigoni, A., Pasqualini, J. and Micheli, F. (2016), 'First assessments of fixed spray application systems in narrow-wall apple orchards', *in* P. Balsari, S. E. Cooper, E. Gil, C. R. Glass, W. Jones, B. Magri, et al. (Eds), *International Advances in Pesticide Application*, pp. 411–14. The Association of Applied Biologists, Wellesbourne, UK.

Bonnemaison, L. (1965), 'Observations ecologiques sur *Aphelinus mali* Haldeman parasite du puceron lanigere *Eriosoma lanigerum* Hausmann', *Ann. Soc. entomol. Fr.*, 1, 143–76.

Bostanian, N. J., Hardman, J. M., Racette, G. and Franklin, J. L. (2007), 'The relationship between winter egg counts of the European red mite *Panonychus ulmi* (Acari: Tetranychidae) and its summer abundance in a reduced spray orchard', *Exp. Appl. Acarol.*, 42, 185–95.

Brader, L. (1974), *Les organismes auxiliaires en verger de pommiers*, Brochure n.3, Centrale Offsetdrukkerij Pudoc, Wageningen (NL).

Bradley, S. J., Murrell, V. C., Shaw, P. W. and Walker, J. T. S. (1997), 'Effect of orchard pesticides on Aphelinus mali, the woolly apple aphid parasitoid', *Proceedings of the New Zealand Plant Protection Conference*, pp. 218–22. New Zealand Plant Protection Society Inc.

Briolini, G., Faccioli, G. and Pasqualini, E. (1989), 'A seven-year research on alternative methods to control Pear Psylla', *IOBC-WPRS Bull.*, 13(2), 89–92.

Brück, E., Elbert, A., Fischer, R., Krueger, S., Kühnhold, J., Klueken, A. M., et al. (2009), 'Movento®, an innovative ambimobile insecticide for sucking insect pest control in agriculture: Biological profile and field performance', *Crop Prot.*, 28(10), 838–44.

Brugger, K. E., Cole, P. G., Newman, I. C., Parker, N., Scholz, B., Suvagia, P., et al. (2010), 'Selectivity of clorantraniliprole to parasitoid wasp', *Pest Manag. Sci.*, 66, 1075-81.

Bus, V. G. M., Chagné, D., Bassett, H. C. M., Bowatte, D., Calenge, F., Celton, J.-M., et al. (2008), 'Genome mapping of three major resistance genes to woolly apple aphid (*Eriosoma lanigerum* Hausm.)', *Tree Genet. Genom.*, 4, 233–6.

Caroli, F. L. and Pasqualini, E. (2004), 'Laboratory tests of the impact of insect growth regulators on *Anthocoris nemoralis*', *IOBC-WRPS Bull.*, 27(6), 81–6.

Carraro, L., Loi, N., Ermacora, N., Gregoris, A., Osler, R. and Hadidi, A. (1998), 'Transmission of pear decline by using naturally infected *Cacopsylla pyri* L.', *Acta Hortic.*, 472, 665–8.

Chambers, U., Petit, B. and Jones, V. P. (2015), 'WSU-DAS – The online Pest Management support system for tree fruits in Washington State', *Acta Hortic.*, 1068, 27–34.

Chouinard, G., Veilleux, J., Pelletier, F., Larose, M., Philion, V. and Cormier, D. (2017), 'Impact of exclusion netting row covers on arthropod presence and crop damage to "Honeycrisp" apple trees in North America: A five-year study', *Crop Prot.*, 98, 248–54.

Civolani, S. (2012), 'The past and present of pear protection against the pear psylla, *Cacopsylla pyri* L.', *in* F. Perveen (Ed.), *Insecticides-Pest Engineering*. InTech, Rijeka, Croatia.

Civolani, S. and Pasqualini, E. (1999), 'Tossicità nel breve periodo di diverse formulazioni di insetticidi su alcuni gruppi di entomofagi', *Inf. Agrar.*, 20, 87–90.

Civolani, S. and Pasqualini, E. (2003), '*Cacopsylla pyri* L. (Hom. Psyllidae) and its predators relationship in Italy's Emilia-Romagna region', *J. Appl. Entomol.*, 127, 214–20.

Civolani, S., Peretto, R., Caroli, L., Pasqualini, E., Chicca, M. and Leis, M. (2007), 'Preliminary resistance screening on abamectin in pear psylla (Hemiptera: Psyllidae) in Northern Italy', *J. Econ. Entomol.*, 100(5), 1637–41.

Civolani, S., Boselli, M., Butturini, A., Chicca, M., Cassanelli, S., Tommasini, M. G., et al. (2015), 'Testing spirotetramat as an alternative solution to abamectin for *Cacopsylla pyri* (Hemiptera: Psyllidae) control: Laboratory and field tests', *J. Econ. Entomol.*, 108(6), 2737–42.

Cohen, H., Horowitz, A. R., Nestel, D. and Rosen, D. (1996), 'Susceptibility of the woolly apple aphid parasitoid, *Aphelinus mali* (Hym: Aphelinidae), to common pesticides used in apple orchards in Israel', *Entomophaga*, 41, 225–33.

Collyer, E. (1953), 'Biology of some predatory insects and mites associated with the Fruit tree red spider mite (*Metatetranychus ulmi* Koch) in South-East England. II. Some important predators of the mite', *J. Hortic. Sci.*, 28, 85–97.

Collyer, E. (1964), 'A summary of experiments to demonstrate the role of *Typhlodromus pyri* Scheuten in the control of *Panonychus ulmi* (Koch) in England', *Acarologia*, 363–71.

Costa, A., Williams, D. G. and Powell, K. S. (2014), 'Discovery of three woolly apple aphid *Eriosoma lanigerum* (Hemiptera: Aphididae) biotypes in Australia: The role of antixenosis and antibiosis in apple tree resistance', *Austral. Entomol.*, 53(3), 280–7.

Cranham, J. E. and Helle, W. (1985), 'Pesticide resistance in Tetranychidae', *in* W. Helle, and M. W. Sabelis (Eds), *Spider Mites: Their Biology, Natural Enemies and Control*, pp. 405–21. Elsevier, Amsterdam.

Croft, B. A. (1994), 'Biological control of apple mites by a phytoseiid mite complex and *Zetzellia mali*: Long-term effects and impact of azinphosmethyl on colonization by *Amblyseius andersoni* (Acari: Phytoseiidae)', *Environ. Entomol.*, 23, 1317–25.

Croft, B. A. and MacRae, I. V. (1992a), 'Biological control of apple mites by mixed populations of *Metaseiulus occidentalis* (Nesbitt) and *Typhlodromus pyri* Scheuten (Acari: Phytoseiidae)', *Environ. Entomol.*, 21, 202–9.

Croft, B. A. and MacRae, I. V. (1992b), 'Persistence of *Typhlodromus pyri* and *Metaseiulus occidentalis* on apple after inoculative release and competition with *Zetzellia mali*', *Environ. Entomol.*, 21, 1168–77.

Croft, B. A. and Whalon, M. E. (1982), 'Selective toxicity of pyrethroid insecticides to arthropod natural enemies and pests of agricultural crops', *Entomophaga*, 27, 3–21.

Cross, J. V., Solomon, M. G., Babandreier, I. D., Blommers, L., Easterbrook, M. A., Jay, C. N., et al. (1999), 'Biocontrol of pests of apples and pears in Northern and Central Europe: 2. parasitoids', *Biocontrol Sci. Technol.*, 9, 277–314.

Cross, J., Nagy, C. and Batki, M. (2010), 'Conservation biocontrol of pear psyllids', *Mitt. Klosterneuburg*, 60(4), 403–12.

Cross, J., Balsari, P., Doruchowski, G., Douzals, J. P., Herbst, A., Marucco, P., et al. (2013), 'Orchard spray application in Europe – State of the art and research challenges', *IOBC-WPRS Bull.*, 91, 465–75.

Cross, J., Fountain, M., Marko, V. and Nagy, C. (2015), 'Arthropod ecosystem services in apple orchards and their economic benefits', *Ecol. Entomol.*, 40(1), 82–96.

Cuthbertson, A. G. S. and Murchie, A. K. (2014), 'Unnecessary pesticide applications: The need for correct identification of a beneficial mite species in Northern Irish apple orchards', *IOBC-WPRS Bull.*, 103, 61–5.

Daane, K. M., Vincent, C., Isaacs, R. and Ioriatti, C. (2018), 'Entomological opportunities and challenges of sustainable viticulture in a global market', *Annu. Rev. Entomol.*, 63, 193–214.

Damos, P., Escudero Colomar, L. A. and Ioriatti, C. (2015), 'Integrated fruit production and pest management in Europe: The apple case study and how far we are from the original concept?', *Insects*, 6, 626–57.

Davies, D. L., Clark, M. F., Adams, A. N. and Hadidi, A. (1998), 'The epidemiology of pear decline in the UK', *Acta Hortic.*, 472, 669–72.

DeBach, P. (1964), *Biological Control of Insect Pests and Weeds*, Reinhold Publishing Corporation, New York.

DeBach, P. and Rosen, D. (1991), *Biological Control by Natural Enemies*, Cambridge University Press, Cambridge, UK.

Del Fava, E., Ioriatti, C. and Melegaro, A. (2017), 'Cost-benefit analysis of controlling the spotted wing drosophila (*Drosophila suzukii* (Matsumura)) spread and infestation of soft fruits in Trentino, Northern Italy', *Pest Manag. Sci.*, 73(11), 2318–27.

Dib, H., Sauphanor, B. and Capowiez, Y. (2010), 'Effect of codling moth exclusion nets on the rosy apple aphid, *Dysaphis plantaginea*, and its control by natural enemies', *Crop Prot.*, 29(12), 1502–13.

Dicke, M., Sabelis, M. W., Takabayashi, J., Bruin, J. and Posthumus, M. A. (1990), 'Plant strategies of manipulating predator-prey interactions through allelochemicals: Prospects for application in pest control', *J. Chem. Ecol.*, 16, 3091–118.

Dorigoni, A. (2016), 'Innovative fruit tree architecture as a nexus to improve sustainability in orchards', *Acta Hortic.*, 1137, 1–10.

Dumbleton, L. J. and Jeffreys, F. J. (1938), 'The control of the woolly aphis by *Aphelinus mali*', *N. Z. J. Sci. Technol.*, 20(3), 183–90.

Duso, C., Camporese, P. and van de Geest, L. P. S. (1992), 'Toxicity of a number of pesticides to strain of *Typhlodromus pyri* and *Amblyseius andersoni* (Acari: Phytoseiidae)', *Entomophaga*, 37, 363–72.

Duso, C., Fanti, M., Pozzebon, A. and Angeli, G. (2009), 'Is the predatory mite *Kampimodromus aberrans* a candidate for the control of phytophagous mites in European apple orchards?', *Biocontrol*, 54, 369–82.

Duso, C., Ahmad, S., Tirello, P., Pozzebon, A., Klaric, V., Baldessari, M., et al. (2014), 'The impact of insecticides applied in apple orchards on the predatory mite *Kampimodromus aberrans* (Acari Phytoseiidae)', *Exp. Appl. Acarol.*, 62, 391–414.

El-Sayed, A. M. (2018), '*The Pherobase: Database of Pheromones and Semiochemicals*', http://www.pherobase.com (accessed 22 July 2018).

Erler, F. and Cetin, H. (2007), 'Effect of kaolin particle film treatment on winterform oviposition of the pear psylla *Cacopsylla pyri*', *Phytoparasitica*, 35(5), 466–73.

Faccioli, G., Antropoli, A. and Pasqualini, E. (1993), 'Relationship between males caught with low pheromone doses and larval infestation of *Argyrotaenia pulchellana*', *Entomol. Exp. Appl.*, 68(2), 165–70.

FAO (1968), '*Report of FAO Panel of Experts on Integrated Pest Control*', FAO-UN Meeting report PL/1967/M/7, Rome, Italy.

Faudrin, J. C. (1984), 'Avertissement agricoles et Psylle du poireir', *IOBC-WPRS Bull.*, 17(2), 358–67.

Fauvel, G. (1971), 'Influence de l'alimentation sur la biologie de l'*Orius vicinus* (Ribaut) (Heteroptera, Anthocoridae)', *Ann. Zool. Ecol. Anim.*, 3, 31–42.

Fazio, G., Aldwinckle, H. and Robinson, T. (2013), 'Unique characteristics of Geneva® apple rootstocks', *N. Y. Fruit Quart.*, 21(2), 25–8.

Fontanari, M., Sacco, M. and Girolami, V. (1993), 'Influenza delle formiche sugli afidi e loro predatori nei fruttiferi', *Inf. Fitopatol.*, 43(4), 47–55.

Forti, D., Angeli, G. and Ioriatti, C. (1993), 'Effets secondaires sur l'equilibre *Cacopsilla pyri-Anthocoris nemoralis* de deux ICI utilises pour la lutte contre *Cydia pomonella* L., dans un verger de poirier du Trentino (Italie)', *Med. Fac. Landbouww. Univ. Gent.*, 58(2b), 533–42.

Fotiric Aksic, M. M., Dabić, D. C., Gašić, U. M., Zec, G. N., Vulić, T. B., Tešić, Z. L., et al. (2015), 'Polyphenolic profile of pear leaves with different resistance to pear psylla (*Cacopsylla pyri*)', *J. Agric. Food Chem.*, 63(34), 7476–86.

Fountain, M. T and Harris, A. L. (2015), 'Non-target consequences of insecticides used in apple and pear orchards on *Forficula auricularia* L. (Dermaptera: Forficulidae),' *Biol. Control*, 91, 27–33.

Franco, J. C., da Silva, E. B., Cortegano, E., Branco, M., Zada, A. and Mendel, Z. (2008), 'Kairomonal response of the parasitoid *Anagyrus* spec. nov. near *pseudococci* to the sex pheromone of the vine mealybug', *Entomol. Exp. Appl.*, 126(2), 122-30.

Franco, J. C., da Silva, E. B., Fortuna, T., Cortegano, E., Branco, M., Suma, P., et al. (2011), 'Vine mealybug sex pheromone increases citrus mealybug parasitism by *Anagyrus* sp. near *pseudococci* (Girault)', *Biol. Control*, 58(3), 230-8.

Fusarini, L., Valente, M., Bianchi, P., Mori, N., Civolani, S. and Pasqualini, E. (2010), 'Selettività di emamectina benzoato (Affirm®) su popolazioni naturali di *Anthocoris nemoralis*', *Atti Gior. Fitopatolog.*, 1, 79-86.

Garcia, A. G., Araujo, M. R., Uramoto, K., Walder, J. M. M. and Zucchi, R. A. (2017), 'Geostatistics and geographic information system to analyze the spatial distribution of the diversity of *Anastrepha* species (Diptera: Tephritidae): The effect of forest fragments in an urban area', *Environ. Entomol.*, 46(6), 1189-94.

Gardiner, M. M., Fiedler, A. K., Costamagna, A. C. and Landis, D. A. (2009), 'Integrating conservation biological control into IPM system', *in* E. B. Radcliffe, W. D. Hutchison, R. E. Cancelado (Eds), *Integrated Pest Management: Concept, Tactics, Strategies and Case Studies*, pp. 151-62. Cambridge University Press.

Geoffrion, R. (1984), 'Les psylles du Poirier - historique, importance economique', *IOBC-WPRS Bull.*, 7, 13-15.

Girolami, V., Berti, M. and Coiutti, C. (2001), 'Tossicità di nuovi insetticidi IGR su antocoridi predatori della psilla', *Inf. Agrar.*, 30, 63-6.

Goldshtein, E., Cohen, Y., Hetzroni, A., Gazit, Y., Timar, D., Rosenfeld, L. et al. (2017), 'Development of an automatic monitoring trap for Mediterranean fruit fly (*Ceratitis capitata*) to optimize control applications frequency', *Comput. Electron. Agric.*, 139, 115-25.

Gontijo, L. M. (2011), '*Integrated Biological Control of Woolly Apple Aphid in Washington State*', Doctor of Philosophy, Washington State University, Pullman, WA.

Gontijo, L. M., Cockfield, S. D. and Beers, E. H. (2012), 'Natural enemies of woolly apple aphid (Hemiptera: Aphididae) in Washington State', *Environ. Entomol.*, 41, 1364-71.

Goossens, D., Bangels, E., Belien, T., Schoevaerts, C. and de Maeyer, L. (2011), 'Optimal profit of the parasitation by *Aphelinus mali* in an IPM complementary strategy for the control of *Eriosoma lanigerum*', *Commun. Agric. Appl. Biol. Sci.*, 76(3), 457-65.

Gray, M. E., Ratcliffe, S. T., Rice, M. E. (2009), 'The IPM paradigm: Concept, strategies and tactics', *in* E. B. Radcliffe, W. D. Hutchison, and R. E. Cancelado (Eds), *Integrated Pest Management: Concept, Tactics, Strategies and Case Studies*, pp. 1-13, Cambridge University Press.

Guarnieri, A., Maini, S., Molari, G. and Rondelli, V. (2011), 'Automatic trap for moth detection in integrated pest management', *Bull. Insectol.*, 64(2), 247-51.

Gut, L. J. and Brunner, J. F. (1998), 'Pheromone-based management of codling moth (Lepidoptera: Tortricidae) in Washington apple orchards', *J. Agric. Urban Entomol.*, 15(4), 387-406.

Hassan, S. A., Bigler, F., Blaisinger, P., Bogenschütz, H., Brun, J., Chiverton, P., et al. (1985), 'Standard methods to test the side-effects of pesticides on natural enemies of insects and mites developed by the IOBC/WPRS Working Group Pesticides and Beneficial Organisms', *EPPO Bull.*, 5(2), 214-55.

Helsen, H., Vaal, F. and Blommers, L. (1998), 'Phenology of the common earwig *Forficula auricularia* L. (Dermaptera: Forficulidae) in an apple orchard', *Int. J. Pest Manag.*, 44(2), 75-9.

Herard, F. and Chen, K. (1985), 'Ecology of *Anthocoris nemorum* (L.) (Het.: Anthocoridae) and evaluation of its potential effectiveness for biological control of pear psylla', *Agronomie*, 5, 855–63.

Heunis, J. M. and Pringle, K. L. (2006), 'Field biology of woolly apple aphid, *Eriosoma lanigerum* (Hausmann), and its natural enemy, *Aphelinus mali* (Haldeman), in apple orchards in the Western Cape Province', *Afr. Entomol.*, 14, 77–86.

Hogmire, H. W. and Miller, S. S. (2005), 'Relative susceptibility of new apple cultivars to arthropod pests', *Hortscience*, 40(7), 2071–5.

Horton, D. R., Higbee, B. S. and Krysan, J. L. (1994), 'Postdiapause development and mating status of pear psylla (Homoptera: Psyllidae) affected by pear and nonhost species', *Ann. Entomol. Soc. Am.*, 87, 241–9.

Hoyt, S. C. (1969), 'Integrated chemical control of insects and biological control of mites on apple in Washington', *J. Econ. Entomol.*, 62, 74–86.

Hoyt, S. C. and Madsen, H. F. (1960), 'Dispersal behaviour of the first instar nymphs of the woolly apple aphid', *Hilgardia*, 30, 267–99.

Hoyt, S. C., Tanigoshi, L. K. and Browne, R. W. (1979), 'Economic injury level studies in relation to mites on apple', *Recent Adv. Acarol.*, 1, 3–12.

Ioriatti, C. and Lucchi, A. (2016), 'Semiochemical strategies for tortricid moth control in apple orchards and vineyards in Italy', *J. Chem. Ecol.*, 42(7), 571–83.

Ioriatti, C., Pasqualini, E. and Toniolli A. (1992), 'Effects of the fungicides mancozeb and dithianon on mortality and reproduction of the predatory mite *Amblyseius andersoni*', *Exp. Appl. Acarol.* 15, 109–16.

Ioriatti, C., Sofia, M. and Menapace, L. (2000), 'Studio della suscettibilità delle popolazioni trentine di *Panonychus ulmi* a tre acaricidi del gruppo M. E.T. I. (fenazaquin, fenpyroximate, tebufenpyrad) e a propargite', *Atti Gior. Fitopatolog.*, 319–26.

Ivancich Gambaro, P. (1975), 'Selezione di popolazioni di Acari predatori resistenti ad alcuni insetticidi fosforati-organici', *Inf. Fitopatol.*, 7, 21–5.

James, D. G. (2003), 'Synthetic herbivore-induced plant volatiles as field attractants for beneficial insects', *Environ. Entomol.*, 32, 977–82.

James, D. G. (2005), 'Further field evaluation of synthetic herbivore-induced plant volatiles as attractants for beneficial insects', *J. Chem. Ecol.*, 31, 481–95.

Jansen, J. P. (2018), '*IOBC Pesticide Side Effect Database*', https://www.iobc-wprs.org/ip_ipm/IOBC_Pesticide_Side_Effect_Database.html (accessed 28 July 2018).

Janssen, A. R. M. and Sabelis, M. W. (1992), 'Phytoseiid life-histories, local predator-prey dynamics, and strategies for control of tetranychid mites', *Exp. Appl. Acarol.*, 14(3–4), 233–50.

Jaworska, K., Olszak, R. W., Łabanowska, B. H. and Korzeniowski, M. (2012), 'Efficacy of spirotetramat in the control of pear psylla (*Cacopsylla pyri* L.) on pear trees in Poland', *J. Fruit Ornam. Plant Res.*, 20(2), 91–106.

Jenser, G., Szita, E. and Balint, J. (2010), 'Measuring pear psylla population density (*Cacopsylla pyri* L. and *C. pyricola* Förster): Review of previous methods and evaluation of a new technique', *North West J. Zool.*, 6(1).

Jeschke, P., Haas, M., Nauen, R., Gutbrod, O., Beck, M. E., Matthiesen, S., et al. (2015), 'Sivanto®-a novel insecticide with a sustainable profile', *Discovery and Synthesis of Crop Protection Products, ACS Symposium Series* 1204, 331–44.

Jones, V. P. (1991), 'Sampling for integrated pest management', *in New Directions in Tree Fruit Pest Management*, pp. 61–73, Good Fruit Grower, Yakima, WA.

Jones, V. P., Unruh, T. R., Horton, D. R., Mills, N. J., Brunner, J. F., Beers, E. H., et al. (2009), 'Tree fruit IPM programs in the western United States: The challenge of enhancing biological control through intensive management', *Pest Manag. Sci.*, 65(12), 1305–10.

Jones, V. P., Brunner, J. F., Grove, G. G., Petit, B., Tangren, G. V. and Jones, W. E. (2010), 'A web-based decision support system to enhance IPM programs in Washington tree fruit', *Pest Manag. Sci.*, 66(6), 587–95.

Jones, V. P., Steffan, S. A., Wiman, N. G., Horton, D. R., Miliczky, E. R., Zhang, Q.-H., et al. (2011), 'Evaluation of herbivore-induced plant volatiles for monitoring green lacewings in Washington apple orchards', *Biol. Control*, 56, 98–105.

Jones, V. P., Horton, D. R. Mills, N. J., Unruh, T. R., Miliczky, E., Shearer, P. W., et al. (2016a), 'Using plant volatile traps to develop phenology models for natural enemies: An example using *Chrysopa nigricornis* (Burmeister) (Neuroptera: Chrysopidae)', *Biol. Control*, 102, 77–84.

Jones, V. P., Horton, D. R., Mills, N. J., Unruh, T. R., Baker, C. C., Melton, T. D., et al. (2016b), 'Evaluating plant volatiles for monitoring natural enemies in apple, pear and walnut orchards', *Biol. Control*, 102, 53–65.

Jones, V. P., Mills, N. J., Brunner, J. F, Horton, D. R., Beers, E. H., Unruh, T. R., et al. (2016c), 'From planning to execution to the future: An overview of a concerted effort to enhance biological control in apple, pear, and walnut orchards in the western U.S.', *Biol. Control*, 102, 1–6.

Jordan, M. O., Rousselin, A., Vercambre, G. and Sauge, M.-H. (2017), 'Combining irrigation, fertilisation and pruning techniques helps control aphid populations in apple and peach orchards', *IOBC-WPRS Bull.*, 123, 79–85.

Khot, L., Hoheisel, G.-A., Osroosh, Y. and Ehsani, R. (2018), 'Precision technology for pest and disease management', *in* Q. Zhang (Ed.), *Automation in Tree Fruit Production: Principles and Practice*, pp. 112–33, CAB International, Wallingford, UK.

Kogan, M. (1998), 'Integrated Pest Management: Historical perspectives and contemporary developments', *Annu. Rev. Entomol.*, 43, 243–70.

Kramer, T. and Nauen, R. (2011), 'Monitoring of spirodiclofen susceptibility in field populations of European red mites, *Panonychus ulmi* (Koch) (Acari: Tetranychidae), and the cross-resistance pattern of a laboratory-selected strain', *Pest Manag. Sci.*, 67, 1285–93.

Krawczyk, G., Lehman, B. L., Park, J. and Hull, L. A. (2015), 'Field validation of digital insect traps for monitoring lepidopteran pests in orchards', *IOBC-WPRS Bull.*, 109, 47–51.

Lakso, A. N., Mattii, G. B., Nyrop, J. P. and Denning, S. S. (1996), 'Influence of European red mite on leaf and whole-canopy carbon dioxide exchange, yield, fruit size, quality, and return cropping in "Starkrimson Delicious" apple trees', *J. Am. Soc. Hortic. Sci.*, 121(5), 954–8.

Lee, J. C. (2010), 'Effect of methyl salicylate-based lures on beneficial and pest arthropods in strawberry', *Environ. Entomol.*, 39(2), 653–60.

Lefebvre, M., Franck, P., Toubon, J.-F., Bouvier, J.-C. and Lavigne, C. (2016), 'The impact of landscape composition on the occurrence of a canopy dwelling spider depends on orchard management', *Agric. Ecosyst. Environ.*, 215, 20–9.

Lloyd, A., Hamacek, E., George, A., Nissen, R. and Waite, G. (2005), 'Evaluation of exclusion netting for insect pest control and fruit quality enhancement in tree crops', *Acta Hortic.*, 694, 253–8.

Lordan, J., Alegre, S., Gatius, F., Sarasúa, M. J. and Alins, G. (2015), 'Woolly apple aphid *Eriosoma lanigerum* Hausmann ecology and its relationship with climatic variables and natural enemies in Mediterranean areas', *Bull. Entomol. Res.*, 105(1), 60–9.

Lorenzon, M., Pozzebon, A. and Duso, C. (2012), 'Effects of potential food sources on biological and demographic parameters of the predatory mites *Kampimodromus aberrans*, *Typhlodromus pyri* and *Amblyseius andersoni*', *Exp. Appl. Acarol.*, 58, 259–78.

Lucchi, A., Loni, A., Gandini, L. M., Scaramozzino, P., Ioriatti, C., Ricciardi, R., et al. (2017), 'Using herbivore-induced plant volatiles to attract lacewings, hoverflies and parasitoid wasps in vineyards: Achievements and constraints', *Bull. Insectol.*, 70(2), 273–82.

Lundie, A. E. (1924), *A Biological Study of Aphelinus mali Hald., A Parasite of the Woolly Apple Eriosoma lanigera Hausman*, Cornell University, Ithaca, New York.

Marchesini, E., Mori, N., Pasini, M. and Bassi, A. (2008), 'Selettività di Rynaxypyr su artropodofauna utile in agroecosistemi diversi', *Atti Gior. Fitopatolog.*, 1, 71–6.

Marliac, G., Penvern, S., Barbier, J.-M., Lescourret, F. and Capowiez, Y. (2015), 'Impact of crop protection strategies on natural enemies in organic apple production', *Agron. Sustain. Dev.*, 35(2), 803–13.

Markó, V., Elek, Z., Kovács-Hostyánszki, A., Kőrösi, A., Somay, L., Földesi, R., et al. (2017), 'Landscapes, orchards, pesticides – Abundance of beetles (Coleoptera) in apple orchards along pesticide toxicity and landscape complexity gradients', *Agric. Ecosyst. Environ.*, 247, 246–54.

Mates, S. G., Perfecto, I. and Badgley, C. (2012), 'Parasitoid wasp diversity in apple orchards along a pest-management gradient', *Agric. Ecosyst. Environ.*, 156, 82–8.

McMurtry, J. A. and Croft, B. A. (1997), 'Life-styles of phytoseiid mite and their roles in biological control', *Ann. Rev. Entomol.*, 42, 291–321.

McMurtry, J., Huffaker, C. and van de Vrie, M. (1970), 'Ecology of tetranychid mites and their natural enemies: A review: I. Tetranychid enemies: Their biological characters and the impact of spray practices', *Hilgardia*, 40(11), 331–90.

Merfield, C. N, Wratten, S. D. and Navntoft, S. (2004), 'Video analysis of predation by polyphagous invertebrate predators in the laboratory and field', *Biol. Control*, 29, 5–13.

Migeon, A. and Dorkeld, F. (2006–17), *'Spider Mites Web: A Comprehensive Database for the Tetranychidae'*, http://www.montpellier.inra.fr/CBGP/spmweb (accessed 23 July 2018).

Miller, B., Anfora, G., Buffington, M., Daane, K. M., Dalton, D. T., Hoelmer, K. M., et al. (2015), 'Seasonal occurrence of resident parasitoids associated with *Drosophila suzukii* in two small fruit production regions of Italy and the USA', *Bull. Insectol.*, 68(2), 255–63.

Mills, N. J., Beers, E. H., Shearer, P. W., Unruh, T. R. and Amarasekare, K. G. (2016a), 'Comparative analysis of pesticide effects on natural enemies in western orchards: A synthesis of laboratory bioassay data', *Biol. Control*, 102, 17–25.

Mills, N. J., Jones, V. P., Baker, C. C., Melton, T. D., Steffan, S. A., Unruh, T. R., et al. (2016b), 'Using herbivore-induced plant volatiles and floral volatiles to attract natural enemies for studies of ecosystem structure and function', *Biol. Control*, 102, 66–76.

Miñarro, M., Fernández-Mata, G. and Medina, P. (2010), 'Role of ants in structuring the aphid community on apple', *Ecol. Entomol.*, 35, 206–15.

Minks, A. K., Blommers, L. H. M., Ramakers, P. M. J. and Theunissen, J. (1998), 'Fifty years of biological and integrated control in western Europe: Accomplishments and future prospects', *Med. Fac. Landbouww. Univ. Gent.*, 62(2a), 165-81.

Mody, K., Collatz, J. and Dorn, S. (2015), 'Plant genotype and the preference and performance of herbivores: Cultivar affects apple resistance to the florivorous weevil *Anthonomus pomorum*', *Agric. For. Entomol.*, 17(4), 337-46.

Moerkens, R., Gobin, B., Peusens, G., Helsen, H., Hilton, R., Dib, H., et al. (2011), 'Optimizing biocontrol using phenological day degree models: The European earwig in pipfruit orchards', *Agric. For. Entomol.*, 13(3), 301-12.

Mols, P. J. M. and Boers, J. M. (2001), 'Comparison of a Canadian and a Dutch strain of the parasitoid *Aphelinus mali* (Hald) (Hym., Aphelinidae) for control of woolly apple aphid *Eriosoma lanigerum* (Haussmann) (Hom., Aphididae) in the Netherlands: A simulation approach', *J. Appl. Entomol.*, 125(5), 255-62.

Moon, R. D. and Wilson, L. T (2009), 'Sampling for detection, estimation and IPM decision making', *in* E. B. Radcliffe, W. D. Hutchison, and R. E. Cancelado (Eds), *Integrated Pest Management: Concept, Tactics, Strategies and Case Studies*, pp. 75-89. Cambridge University Press.

Mueller, T. F., Blommers, L. H. M. and Mols, P. J. M. (1988), 'Earwig predation on the woolly apple aphid, *Eriosoma lanigerum*, in the Netherlands', *Entomol. Exp. Appl.*, 47, 145-52.

Mueller, T. F., Blommers, L. H. M. and Mols, P. J. M. (1992), 'Woolly apple aphid (*Eriosoma lanigerum* Hausm., Hom., Aphidae) parasitism by *Aphelinus mali* Hal. (Hym., Aphelinidae) in relation to host stage and host colony size, shape and location', *J. Appl. Entomol.*, 114, 143-54.

Nagy, C., Cross, J., Luton, M. and Ashdown, C. (2010), 'Mixed deciduous hedgerows as sources of anthocorids and other predators of pear psyllids in the UK', *IOBC-WPRS Bull.*, 54(2), 471-7.

Nagy, C., Cross, J. and Marko, V. (2013), 'Sugar feeding of the common black ant, *Lasius niger* (L.), as a possible indirect method for reducing aphid populations on apple by disturbing ant-aphid mutualism', *Biol. Control*, 65(1), 24-36.

Nauen, R., Stumpf, N., Elbert, A., Zebitz, C. P. W. and Kraus, W. (2001), 'Acaricide toxicity and resistance in larvae of different strains of *Tetranychus urticae* and *Panonychus ulmi* (Acari: Tetranychidae)', *Pest Manag. Sci.*, 57, 253-61.

Nauen, R., Reckmann, U., Thomzik, J. and Thielert, W. (2008), 'Biological profile of spirotetramat (Movento®)- a new two-way systemic (ambimobile) insecticide against sucking pest species', *Bayer CropSci. J.*, 61(2): 245-78.

Nicholas, A. H., Spooner-Hart, R. N. and Vickers, R. A. (2005), 'Abundance and natural control of the woolly aphid *Eriosoma lanigerum* in an Australian apple orchard IPM program', *BioControl*, 50(2), 271-91.

Nietschke, B. S., Magareya, R. D., Borchert, D. M., Calvin, D. D. and Jones, E. (2007), 'A developmental database to support insect phenology models', *Crop Prot.*, 26, 1444-8.

Nin, S., Ferri, A., Sacchetti, P. and Giordani, E. (2012), 'Pear resistance to Psilla (*Cacopsylla pyri* L.): A review', *Adv. Hortic. Sci.*, 26, 59-74.

Pasqualini, E. and Civolani, S. (2007), 'Spirodiclofen, nuovo insetticida efficace contro la psilladel pero', *Inf. Agrar.*, 11, 89-93.

Pasqualini, E. and Malavolta, C. (1986), 'Natural control of P. ulmi in apple orchards of Emilia-Romagna', *Bull. IOBC-WPRS*, 7, 29-33.

Pasqualini, E., Natale, D., Civolani, S. and Vergnani, S. (1997), 'Indagini sull'attività di alcuni prodotti non convenzionali per il contenimento di *Cacopsylla pyri* L. (Homoptera Psyllidae)', *Inf. Agrar.*, 21, 65–8.

Pasqualini, E., Civolani, S., Vergnani, S., Cavazza, C. and Ardizzoni, M. (1999), 'Selettività di alcuni insetticidi su *Anthocoris nemoralis* F. (Heteroptera Anthocoridae)', *Inf. Agrar.*, 46, 71–4.

Pedigo, L. P. (1995), 'Closing the gap between IPM theory and practice', *J. Agric. Entomol.*, 12(4), 171–81.

Peterson, J. A., Ode, P. J., Oliveira-Hofman, C. and Harwood, J. D. (2016), 'Integration of plant defense traits with biological control of arthropod pests: Challenges and opportunities', *Front. Plant Sci.*, 7, 1794.

Pfannenstiel, R. S., Mackey, B. E. and Unruh, T. R. (2012), 'Leafroller parasitism across an orchard landscape in central Washington and effect of neighboring rose habitats on parasitism', *Biol. Control*, 62(3), 152–61.

Phillips, M. L. (1981), *'The Ecology of the Common Earwig Forficula auricularia in Apple Orchards'*, PhD Thesis, Department of Agriculture and Horticulture, Long Ashton Research Center, Bristol, UK.

Pozzebon, A. and Duso, C. (2008), 'Grape downy mildew *Plasmopara viticola*, an alternative food for generalist predatory mites occurring in vineyards', *Biol. Control*, 45, 441–9.

Pozzebon, A., Loeb, G. M. and Duso C. (2009), 'Grape powdery mildew as a food source for generalist predatory mites occurring in vineyard: Effects on life history traits', *Ann. Appl. Biol.* 115, 81–9.

Pozzebon, A., Ahmad, S., Tirello, P., Lorenzon, M. and Duso, C. (2014), 'Does pollen availability mitigate the impact of pesticides on generalist predatory mites?', *BioControl*, 59, 585–59.

Powell, W. and Pickett, J. A. (2003), 'Manipulation of parasitoids for aphid pest management: Progress and prospects', *Pest Manag. Sci.*, 59(2), 149–55.

Pree, D. J., Whitty, K. J. and van Driel, L. (2005), 'Baseline susceptibility and cross resistances of some new acaricides in the European red mite, *Panonychus ulmi*', *Exp. Appl. Acarol.*, 37, 165–71.

Pringle, K. L. (2006), 'The use of economic thresholds in pest management: Apples in South Africa: Research in action', *S. Afr. J. Sci.*, 102(5–6), 201–4.

Quarrell, S. R., Corkrey, R. and Allen, G. R. (2017), 'Predictive thresholds for forecasting the compatibility of *Forficula auricularia* and *Aphelinus mali* as biological control agents against woolly apple aphid in apple orchards', *BioControl*, 62(2), 243–56.

Ramsden, M. W, Kendall, S. L., Ellis, S. A. and Berry, P. M. (2017), 'A review of economic thresholds for invertebrate pests in UK arable crops', *Crop Prot.*, 96, 30–43.

Riedl, H. (1991), 'Beneficial arthropods for pear pest management', *in* K. M. Williams (Ed.), *New Directions in Tree Fruit Pest Management*, pp. 101–18. Good Fruit Grower.

Rodriguez-Saona, C, Kaplan, I., Braasch, J., Chinnasamy, D. and Williams, L. (2011), 'Field responses of predaceous arthropods to methyl salicylate: A meta-analysis and case study in cranberries', *Biol. Control*, 59(2), 294–303.

Roger, G. A. and Lorraine, M. L. (1989), 'Use of sticky traps and limb jarring to aid in pest management decisions for summer populations of the pear psylla (Homoptera: Psyllidae) in Connecticut', *J. Econ. Entomol.*, 82(5), 1448–54.

Rogers, D. J., Sharma, N., Stretton, D. C. and Walker, J. T. S. (2011), 'Toxicity of pesticides to *Aphelinus mali*, the parasitoid of woolly apple aphid', *N. Z. Plant Prot.*, 64, 235–40.

Rogers, D. J., Lo, P. L. and Walker, J. T. S. (2015), 'Vulnerability of pest management in New Zealand apples: Pesticide toxicity to *Aphelinus mali* (Hymenoptera: Encyrtidae)', *N. Z. Plant Prot.*, 68, 299–305.

Roversi, P. F., Binazzi, F., Marianelli, L., Costi, E., Maistrello, L. and Sabbatini Peverieri, G. (2016), 'Searching for native egg-parasitoids of the invasive alien species *Halyomorpha halys* Stål (Heteroptera Pentatomidae) in Southern Europe', *Redia*, 99, 63–70.

Samietz, J, Höhn, H., Razavi, E., Schaub, L. and Graf, B. (2015), 'Decision support for sustainable orchard pest management with the Swiss forecasting system SOPRA', *Acta Hortic.*, 1099, 383–90.

Sandanayaka, W. R. M., Bus, V. G. M., Connolly, P. and Newcomb, R. (2003), 'Characteristics associated with Woolly Apple Aphid *Eriosoma lanigerum*, resistance of three apple rootstocks', *Entomol. Exp. Appl.*, 109, 63–72.

Sandanayaka, W. R. M., Bus, V. G. M. and Connolly, P. (2005), 'Mechanisms of woolly aphid [*Eriosoma lanigerum* (Hausm.)] resistance in apple', *J. Appl. Entomol.*, 129(9/10), 534–41.

Saour, G., Ismail, H. and Hashem, A. (2010), 'Impact of kaolin particle film, spirodiclofen acaricide, harpin protein, and an organic biostimulant on pear psylla *Cacopsylla pyri* (Hemiptera: Psyllidae)', *Int. J. Pest Manag.*, 56(1), 75–9.

Schaub, L., Bloesch, B. and Garnier, G. (1999), 'Effets secondaires de l'amitraze sur *Anthocoris nemoralis*, punaise predatrise du psylle commun du poirier', *Rev. Suisse Vitic. Arboric. Hortic.*, 31(3), 146–8.

Schaub, L., Garnier, G. and Bloesch, B. (2002), 'An extended laboratory method maximizing plant contact time to study the side-effects of pesticides on plant-dwelling predators', *J. Appl. Entomol.*, 126(4), 155–8.

Schaub, L., Graf, B. and Butturini, A. (2005), 'Phenological model of pear psylla *Cacopsylla pyri*', *Entomol. Exp. Appl.*, 117(2), 105–11.

Schausberger, P. (1997), 'Inter- and intraspecific predation on immatures by adult females in *Eiseius finlandicus*, *Typhlodromus pyri* and *Kampimodromus aberrans* (Acari: Phytoseiidae),' *Exp. Appl. Acarol.*, 21, 131–50.

Schausberger, P. (1998), 'Population growth and persistence when prey is diminishing in single-species and two-species systems of the predatory mites *Eiseius finlandicus*, *Typhlodromus pyri* and *Kampimodromus aberrans*', *Entomol. Exp. Appl.*, 88, 275–86.

Schausberger, P. (1999a), 'Predation preference of *Typhlodromus pyri* and *Kampimodromus aberrans* (Acari, Phytoseiidae) when offered con- and heterospecific immature life stages', *Exp. Appl. Acarol.*, 23, 389–98.

Schausberger, P. (1999b), 'Juvenile survival and development in *Euseius finlandicus*, *Typhlodromus pyri* and *Kampimodromus aberrans* (Acari, Phytoseiidae) feeding on con and heterospecific immatures', *Exp. Appl. Acarol.*, 23, 297–307.

Schoevaerts, C., Goossens, D., D'Haemer, K., van Dyck, H. and de Maeyer, L. (2010), 'The multitarget use of spirotetramat (Movento® 100 SC): Simultaneous control of key pests in apples', *Acta Hortic.*, 917, 69–76.

Sciarretta, A., Tabilio, M. R., Lampazzi, E., Ceccaroli, C., Colacci, M. and Trematerra, P. (2018), 'Analysis of the Mediterranean fruit fly [*Ceratitis capitata* (Wiedemann)] spatiotemporal distribution in relation to sex and female mating status for precision IPM', *PLoS ONE*, 13(4), e0195097.

Scutareanu, P., Lingeman, R., Drukker, B. and Sabelis, M. W. (1999), 'Cross-correlation analysis of fluctuations of pear psyllids and anthocorid bugs', *Ecol. Entomol.*, 24, 354–62.

Settle, W. H., Ariawan, H., Astuti, E. T., Cahyana, W., Hakim, A. L., Hindayana, D., et al. (1996), 'Managing tropical rice pests through conservation of generalist natural enemies and alternative prey', *Ecology*, 77(7), 1975–88.

Shaltiel, L. and Coll, M. (2004), 'Reduction of pear psylla damage by the predatory bug *Anthocoris nemoralis* (Heteroptera: Anthocoridae): The importance of orchard colonization time and neighboring vegetation', *Biocontrol Sci. Technol.*, 14(8), 811–21.

Shaltiel-Harpaz, L., Gerchman, Y., Ibdah, M., Kedoshim, R., Rachmany, D., Hatib, K., et al. (2018), 'Grafting on resistant interstocks reduces scion susceptibility to pear psylla, *Cacopsylla bidens*', *Pest Manag. Sci.* 74(3), 617–26.

Shaw, P. W. and Walker, J. T. S. (1996), 'Biological control of woolly apple aphid by *Aphelinus mali* in an integrated fruit production programme in Nelson', *in Proceedings 49th New Zealand Plant Protection Conference*, pp. 59–63. New Zealand Plant Protection Society.

Solomon, M. (1975), 'The colonization on an apple orchard of predators of the tree red spider mite, '*Ann. Appl. Biol.*, 80, 119–22.

Solomon, M. G. and Fitzgerald, J. D. (2015), 'Fenoxycarb, a selective insecticide for inclusion in integrated pest management systems for pear in the UK', *J. Hortic. Sci.*, 65(5), 535–7.

Solomon, M. G., Granham, J. E., Easterbrook, M. A. and Fitzgerald, J. D. (1989), 'Control of the pear psyllid, *Cacopsylla pyricola* in South East England by predators and pesticides. *Crop Prot.*, 8, 197–205.

Solomon, M. G., Easterbrook, M. A. and Fitzgerald, J. D. (1993), 'Mite-management programmes based on organophosphate-resistant *Typhlodromus pyri* in UK apple orchards', *Crop Prot.*, 12(4), 249–54.

Solomon, M. G., Cross, J. V., Fitz Gerald, J. D., Campbell, C. A. M., Jolly, R. L., Olszak, R. W., et al. (2000), 'Biocontrol of pests of apples and pears in northern and central Europe-3 Predators', *Biocontrol Sci. Technol.*, 10, 91–128.

Souliotis, C. and Moschos, T. (2008), 'Effectiveness of some pesticides against *Cacopsylla pyri and* impact on its predator *Anthocoris nemoralis* in pear-orchards', *Bull. Insectol.*, 1, 25–30.

Stäubli, A. and Antonin, P. (1984), 'Aménagement d'une lutte raisonnée contre les psylles du poirier (*Psylla pyri* L.) en Suisse particulièrement en Valais et dans le Bassin lémanique', *IOBC-WPRS Bull.*, 6, 258–64.

Stäubli, A. and Pasquier, D. (1988),'Méthode de laboratoire pour tester l´action secondaire des pesticides sur *Anthocoris nemoralis* F. (Anthocoridae, Heteroptera) laboratory test', *IOBC-WPRS Bull.*, 11(4), 91–8.

Stäubli, A., Hächler, M., Pasquer, D., Antonin, P. and Mittaz, C. (1992), 'Dix années d'experiences et d'observations sur le psylle commun du poirier', *Revue Suisse Vitic. Arboric. Hortic.*, 24(2), 89–104.

Stavrinides, M. C. and Mills, N. J. (2009), 'Demographic effects of pesticides on biological control of Pacific spider mite (*Tetranychus pacificus*) by the western predatory mite (*Galendromus occidentalis*)', *Biol Control*, 48, 267–73.

Stejskal, V. (2003), '"Economic Injury Level" and preventive pest control', *J. Pest Sci.*, 76(6), 170–2.

Sterk, G., Hassan, S. A., Baillod, M., Bakker, F., Bigler, F., Blümel, S., et al. (1999), 'Results of the seventh joint testing programme carried out by the IOBC/WPRS–Working Group "Pesticides and Beneficial Organisms"', *BioControl*, 44, 99–117.

Stern, V. M., Smith, R. F., van den Bosch, R. and Hagen, K. S. (1959), 'The integration of chemical and biological control of the spotted alfalfa aphid: The integrated control concept', *Hilgardia*, 29, 81–101.

Suckling, D. M., Gibb, A. R., Burnip, G. M. and Delury, N. C. (2002), 'Can parasitoid sex pheromones help in insect biocontrol? A case study of codling moth (Lepidoptera: Tortricidae) and its parasitoid *Ascogaster quadridentata* (Hymenoptera: Braconidae)', *Environ. Entomol.*, 31(6), 947–52.

Symondson, W. O. (2002), 'Molecular identification of prey in predator diets', *Mol. Ecol.*, 11, 627–41.

Szentkiràyi, F. and Kozàr, F. (1991), 'How many species are there in apple insect communities? Testing the resource diversity and intermediate disturbance hypotheses', *Ecol Entomol.*, 16, 491–503.

Trapman, M. and Blommers, L. (1992), 'An attempt to pear sucker management in the Netherlands', *J. Appl. Entomol.*, 114, 38–51.

Unruh, T. R., Yu, T., Willett, L. S., Garczynski, S. F. and Horton, D. R. (2008), 'Development of monoclonal antibodies to pear psylla (Hemiptera: Psyllidae) and evaluation of field predation by two key predators', *Ann. Entomol. Soc. Am.*, 101, 887–98.

Vaello, T., Casas, J. L., Pineda, A., de Alfonso, I. and Marcos-García, M. A. (2017), 'Olfactory Response of the Predatory Bug *Orius laevigatus* (Hemiptera:Anthocoridae) to the Aggregation Pheromone of Its Prey, *Frankliniella occidentalis* (Thysanoptera: Thripidae)', *Environ. Entomol.*, 46(5), 1115–19.

van de Vrie, M. (1985), 'Apple', *in* W. Helle, and M. W. Sabelis (Eds), *Spider Mites: Their Biology, Natural Enemies and Control*, volume 1B, pp. 311–25. Elsevier, Amsterdam.

van de Vrie, M., McMurtry, J. A. and Huffaker, C. B. (1972), 'Ecology of tetranychid mites and their natural enemies: A review III. Biology, ecology and pest status and host plant relations of tetranychids', *Hilgardia*, 4, 343–432.

Varela, L. G. and Elkins, R. B. (2008), 'Conversion from use of organophosphate insecticides to codling moth mating disruption in California pear orchards', *Acta Hortic.*, 800, 955–60.

Vogt, H. and Ternes, P. (2006), 'Side effects of pesticides on *Aphelinus mali* and other antagonists of the woolly apple aphid', *IOBC-WPRS Bull.*, 29(10), 115.

Waldner, W. (1996), 'IFP – More than preservation of predatory mites', *Acta Hortic.*, 422, 85–9.

Wiman, N. G., Dalton, D. T., Anfora, G., Biondi, A., Chiu, J., Daane, K. M., et al. (2016), '*Drosophila suzukii* population response to environment and management strategies', *J. Pest Sci.*, 89(3), 653–65.

Witzgall, P., Kirsch, P. and Cork, A. (2010), 'Sex pheromones and their impact on pest management', *J. Chem. Ecol.*, 36, 80–100.

Wyss, E., Niggli, U. and Nentwig, W. (1995), 'The impact of spiders on aphid populations in a strip-managed apple orchard', *J. Appl. Entomol.*, 119(1–5), 473–8.

Wyss, E., Villiger, M. and Müller-Schärer, H. (1999), 'The potential of three native insect predators to control the rosy apple aphid, *Dysaphis plantaginea*', *BioControl*, 44(2), 171–82.

Yin, W. D., Qiu, G. S., Yan, W. T., Sun, L. N., Zhang, H. J., Ma, C. S., et al. (2013), 'Age-stage two-sex life tables of *Panonychus ulmi* (Acari: Tetranychidae), on different apple varieties', *J. Econ. Entomol.*, 106, 2118–25.

Zemek, R. and Prenerivà, E. (1997), 'Powdery mildew (Ascomycotina: Erysiphales) - an alternative food for the predatory mite *Typhlodromus pyri* Scheuten (Acari: Phytoseiidae)', *Exp. Appl. Acarol.*, 21, 405–14.
Zijlstra, C., Lund, I., Justesen, A. F., Nicolaisen, M., Kryger Jensen, P., Bianciotto, V., Posta, K., et al. (2011), 'Combining novel monitoring tools and precision application technologies for integrated high-tech crop protection in the future (a discussion document)', *Pest Manag. Sci.*, 67, 616–25.

Index